Molecular Biology Intelligence Unit

Zinc Finger Proteins:

From Atomic Contact to Cellular Function

Molecular Biology Intelligence Unit

Zinc Finger Proteins:
From Atomic Contact to Cellular Function

Shiro Iuchi
Department of Cell Biology
Harvard Medical School
Boston, Massachusetts, U.S.A.

Natalie Kuldell
Biological Engineering Division
Massachusetts Institute of Technology
Cambridge, Massachusetts, U.S.A.

Landes Bioscience / Eurekah.com
Georgetown, Texas
U.S.A.

Kluwer Academic / Plenum Publishers
New york, New York
U.S.A.

Zinc Finger Proteins:

From Atomic Contact to Cellular Function

Molecular Biology Intelligence Unit

LANDES BIOSCIENCE / EUREKAH.COM
KLUWER ACADEMIC / PLENUM PUBLISHERS

Printed in the U.S.A.

Kluwer Academic / Plenum Publishers, 233 Spring Street, New York, New York, U.S.A. 10013
http://www.wkap.nl

Please address all inquiries to the Publishers:
Landes Bioscience / Eurekah.com, 810 South Church Street, Georgetown, Texas, U.S.A. 78626
Phone: 512/ 863 7762; FAX: 512/ 863 0081
http://www.landesbioscience.com
http://www.eurekah.com

Zinc Finger Proteins: From Atomic Contact to Cellular Function, edited by Shiro Iuchi and Natalie Kuldell, Landes / Kluwer dual imprint / Landes series: Molecular Biology Intelligence Unit

ISBN: 0-306-48229-0

Library of Congress Cataloging-in-Publication Data

Zinc finger proteins : from atomic contact to cellular function / [edited by] Shiro Iuchi, Natalie Kuldell.
p. ; cm. -- (Molecular biology intelligence unit)
Includes bibliographical references and index.
ISBN 0-306-48229-0
1. Zinc-finger proteins. I. Iuchi, Shiro. II. Kuldell, Natalie. III. Series: Molecular biology intelligence unit (Unnumbered)
[DNLM: 1. Zinc Fingers--physiology. QU 55 Z767 2005]
QP552.Z55Z56 2005
572.8'645--dc22

2004023431

CONTENTS

BIOLOGY OF ZINC FINGER PROTEINS

BIOLOGY OF ZINC ION

EDITORS

Shiro Iuchi
Department of Cell Biology
Harvard Medical School
Boston, Massachusetts, U.S.A.
E-mail: shiro_iuchi@hms.harvard.edu
Chapters 2, 21

Natalie Kuldell
Biological Engineering Division
Massachusetts Institute of Technology
Cambridge, Massachusetts, U.S.A.
E-mail: nkuldell@mit.edu
Chapter 26

CONTRIBUTORS

K. Farid Ahmad
Department of Medical Biophysics
University of Toronto & Division of Molecular and Structural Biology
Ontario Cancer Institute
Toronto, Ontario, Canada
Chapter 20

Asifa Akhtar
Gene Expression Programme
European Molecular Biology Laboratory
Heidelberg, Germany
Chapter 33

Carl E. Allen
Department of Pediatrics
Columbus Children's Hospital Research Institute
Columbus, Ohio, U.S.A.
Chapter 29

Peter B. Becker
Adolf-Butenandt-Institut, Molekularbiologie
Ludwig-Maximilians-Universität München
München, Germany
E-mail: pbecker@mol-bio.med.uni-muenchen.de
Chapter 33

Marlene Belfort
Wadsworth Center
New York State Department of Health
Department of Biomedical Sciences
School of Public Health
University at Albany
Albany, New York, U.S.A.
Chapter 7

Maria José Benitez
Departamento de Quimica Fisica Aplicada
Universidad Autonoma de Madrid
Cantoblanco
Madrid, Spain
E-mail: juans.jimenez@uam.es
Chapter 18

Perry J. Blackshear
Office of Clinical Research and the Laboratory of Signal Transduction
National Institute of Environmental Health Sciences
National Institutes of Health
Research Triangle Park, North Carolina, U.S.A.
and
Departments of Medicine and Biochemistry
Duke University Medical Center
Durham, North Carolina, U.S.A.
E-mail: black009@niehs.nih.gov
Chapter 13

Raymond S. Brown
Division of Experimental Medicine
Beth Israel Deaconess Medical Center
Harvard Institutes of Medicine
Boston, Massachusetts, U.S.A.
Chapter 3

Yuen-Ling Chan
Department of Biochemistry and Molecular Biology
The University of Chicago
Chicago, Illinois, U.S.A.
Chapter 14

G. Marius Clore
Laboratory of Chemical Physics
National Institute of Diabetes and Digestive and Kidney Diseases
National Institutes of Health
Bethesda, Maryland, U.S.A.
Chapter 4

Claude Cochet
INSERM EMI 104
Département Réponse et Dynamique Cellulaires
CEA
Grenoble, France
E-mail: ccochet1@cea.fr
Chapter 18

Tucker Collins
Department of Pathology
Children's Hospital Boston
and Harvard Medical School
Boston, Massachusetts, U.S.A.
E-mail: tcollins@rics.bwh.harvard.edu
Chapter 22

Nicoletta Corbi
CNR
Istituto Biologia e Patologia Molecolari
Sezione di Patologia Molecolare
Rome, Italy
E-mail: nicoletta.corbi@ibpm.cnr.it
Chapter 9

Martyn K. Darby
National Institute of Environmental Health Sciences
National Institutes of Health
Research Triangle Park, North Carolina, U.S.A.
E-mail: darby1@niehs.nih.gov
Chapter 11

Roberto N. De Guzman
Department of Molecular Biology
The Scripps Research Institute
La Jolla, California, U.S.A.
Chapter 17

Victoria Derbyshire
Wadsworth Center
New York State Department of Health
and
Department of Biomedical Sciences
School of Public Health
University at Albany
Albany, New York, U.S.A.
Chapter 7

John Dresios
Department of Biochemistry and Molecular Biology
The University of Chicago
Chicago, Illinois, U.S.A.
Chapter 14

H. Jane Dyson
Department of Molecular Biology
The Scripps Research Institute
La Jolla, California, U.S.A.
Chapter 17

David J. Eide
Departments of Nutritional Sciences and Biochemistry
University of Missouri
Columbia, Missouri, U.S.A.
E-mail: eided@missouri.edu
Chapter 35

Odile Filhol
INSERM EMI 104
Département Réponse et Dynamique Cellulaires
CEA
Grenoble, France
E-mail: ofilholcochet@cea.fr
Chapter 18

Kenneth H. Fischbeck
Neurogenetics Branch
National Institute of Neurological Disorders
and Stroke
National Institutes of Health
Bethesda, Maryland, U.S.A.
E-mail: fischbek@ninds.nih.gov
Chapter 31

Jane Flint
Department of Molecular Biology
Lewis Thomas Laboratory
Princeton University
Princeton, New Jersey, U.S.A.
Chapter 3

Katia Georgopoulos
Cutaneous Biology Research Center
Massachusetts General Hospital
and Harvard Medical School
Charlestown, Massachusetts, U.S.A.
E-mail: katia.georgopoulos@cbrc2.mgh.harvard.edu
Chapter 27

Pablo Gómez-del Arco
Cutaneous Biology Research Center
Massachusetts General Hospital
and Harvard Medical School
Charlestown, Massachusetts, U.S.A.
Chapter 27

Howard Green
Department of Cell Biology
Harvard Medical School
Boston, Massachusetts, U.S.A.
E-mail: hgreen@hms.harvard.edu
Chapter 28

Angela M. Gronenborn
Laboratory of Chemical Physics
National Institute of Diabetes
and Digestive and Kidney Diseases
National Institutes of Health
Bethesda, Maryland, U.S.A.
E-mail: gronenborn@nih.gov
Chapter 5

Marc Haenlin
Centre de Biologie du Développement
CNRS UMR
Toulouse, France
E-mail: haenlin@cict.fr
Chapter 30

Jay S. Hanas
University of Oklahoma College of Medicine
Oklahoma City, Oklahoma, U.S.A.
E-mail: jay-hanas@ouhsc.edu
Chapter 8

Fredrik Hellborg
Department of Oncology-Pathology
Karolinska Institute
Cancer Center Karolinska (CCK)
Stockholm, Sweden
Chapter 12

James R. Hocker
University of Oklahoma College of Medicine
Oklahoma City, Oklahoma, U.S.A.
Chapter 8

Kazuhiro Ikeda
Research Center for Genomic Medicine
Saitama Medical School
Saitama, Japan
Chapter 16

Takahisa Ikegami
Laboratory of Structural Proteomics
Institute for Protein Research
Osaka University
Chapter 32

Satoshi Inoue
Research Center for Genomic Medicine
Saitama Medical School
Saitama, Japan
Chapter 16

Ho-Shik Kim
Genetics of Development and Disease Branch
National Institute of Diabetes and Digestive and Kidney Diseases
National Institutes of Health
Bethesda, Maryland, U.S.A.
Chapter 24

Aaron Klug
MRC Laboratory of Molecular Biology
Cambridge, United Kingdom
Chapter 1

Dimitri Krainc
MassGeneral Institute for Neurodegenerative Disease
Harvard Medical School
Massachusetts General Hospital
Charlestown, Massachusetts, U.S.A.
E-mail: dkrainc@partners.org
Chapter 23

Seiki Kuramitsu
Department of Biology
Osaka University
Osaka, Japan
E-mail: kuramitsu@bio.sci.osaka-u.ac.jp
Chapter 6

Wi S. Lai
Office of Clinical Research and the Laboratory of Signal Transduction
National Institute of Environmental Health Sciences
National Institutes of Health
Research Triangle Park, North Carolina, U.S.A.
Chapter 13

Jason L. Larabee
University of Oklahoma College of Medicine
Oklahoma City, Oklahoma, U.S.A.
Chapter 8

Sean Bong Lee
Genetics of Development and Disease Branch
National Institute of Diabetes and Digestive and Kidney Diseases
National Institutes of Health
Bethesda, Maryland, U.S.A.
E-mail: seanl@intra.niddk.nih.gov
Chapter 24

Hongjie Li
Genetics of Development and Disease Branch
National Institute of Diabetes and Digestive and Kidney Diseases
National Institutes of Health
Bethesda, Maryland, U.S.A.
Chapter 24

Valentina Libri
CNR
Istituto Biologia e Patologia Molecolari
Sezione di Patologia Molecolare
Rome, Italy
E-mail: valentina.libri@ibpm.cnr.it
Chapter 9

Jonathan D. Licht
Division of Hematology and Oncology
Department of Medicine
Mount Sinai School of Medicine
New York, New York, U.S.A.
E-mail: Jonathan.licht@mssm.edu
Chapter 20

Huifei Liu
Department of Pathology
Harvard Medical School
Boston, Massachusetts, U.S.A.
Chapter 25

Kirsteen H. Maclean
Department of Biochemistry
St. Jude Children's Research Hospital
Memphis, Tennessee, U.S.A.
E-mail: kirsteen.maclean@stjude.org
Chapter 36

Carl G. Maki
The University of Chicago
Department of Radiation and Cellular Oncology
Chicago, Illinois, U.S.A.
E-mail:cmaki@rover.uchicago.edu
Chapter 34

Maria A. Martinez-Yamout
Department of Molecular Biology
The Scripps Research Institute
La Jolla, California, U.S.A.
Chapter 17

Ryoji Masui
Department of Biology
Osaka University
Osaka, Japan
Chapter 6

Ari Melnick
Department of Developmental and Molecular Biology
Albert Einstein College of Medicine
Bronx, New York, U.S.A.
Chapter 20

Cristina Mendez-Vidal
Department of Oncology and Pathology
Karolinska Institute
Cancer Center Karolinska (CCK)
Stockholm, Sweden
Chapter 12

Masami Muramatsu
Research Center for Genomic Medicine
Saitama Medical School
Saitama, Japan
E-mail: muramasa@saitama-med.ac.jp
Chapter 16

Taku Naito
Cutaneous Biology Research Center
Massachusetts General Hospital
and Harvard Medical School
Charlestown, Massachusetts, U.S.A.
Chapter 27

Noriko Nakagawa
RIKEN Harima Institute
Hyogo, Japan
Chapter 6

James G. Omichinski
Département de Biochimie
Université de Montréal
Montréal, Quebec, Canada
Chapter 4

Claudio Passananti
CNR
Istituto Biologia e Patologia Molecolari
Sezione di Patologia Molecolare
Rome, Italy
E-mail: claudio.passananti@ibpm.cnr.it
Chapter 9

Ruth S. Phillips
Office of Clinical Research and the Laboratory
of Signal Transduction
National Institute of Environmental Health Sciences
National Institutes of Health
Research Triangle Park, North Carolina, U.S.A.
Chapter 13

Federica Piccioni
Neurogenetics Branch
National Institute of Neurological Disorders
and Stroke
National Institutes of Health
Bethesda, Maryland, U.S.A.
Chapter 31

Gilbert G. Privé
Department of Medical Biophysics
University of Toronto and Division of Molecular
and Structural Biology
Ontario Cancer Institute
Toronto, Ontario, Canada
Chapter 20

Jun Qin
Structural Biology Program
The Lerner Research Institute
The Cleveland Clinic Foundation
Cleveland, Ohio, U.S.A.
E-mail: qinj@ccf.org
Chapter 15

Paul J. Romaniuk
Department of Biochemistry and Microbiology
University of Victoria
Victoria, British Columbia, Canada
E-mail: pjr@uvic.ca
Chapter 10

Tara L. Sander
Division of Pediatric Surgery
Medical College of Wisconsin
Cardiovascular Research Center
Milwaukee, Wisconsin, U.S.A.
Chapter 22

John Seavitt
Cutaneous Biology Research Center
Massachusetts General Hospital
and Harvard Medical School
Charlestown, Massachusetts, U.S.A.
Chapter 27

Yang Shi
Department of Pathology
Harvard Medical School
Boston, Massachusetts, U.S.A.
E-mail: yang_shi@hms.harvard.edu
Chapter 25

Masahiro Shirakawa
Graduate School of Integrated Science
Yokohama City University
Yokohama, Kanagawa, Japan
E-mail: shirakawa@tsurumi.yokohama-cu.ac.jp
Chapter 32

Harald Stenmark
Department of Biochemistry
Institute for Cancer Research
The Norwegian Radium Hospital
Oslo, Norway
E-mail: stenmark@ulrik.uio.no
Chapter 19

Charlotte J. Sumner
Neurogenetics Branch
National Institute of Neurological Disorders
and Stroke
National Institutes of Health
Bethesda, Maryland, U.S.A.
Chapter 31

Magdalena Tarkowska
Department of Oncology and Pathology
Karolinska Institute
Cancer Center Karolinska (CCK)
Stockholm, Sweden
Chapter 12

Hung Tseng
Department of Dermatology
University of Pennsylvania School of Medicine
Cell and Developmental Biology
and Center for Research on Reproduction
and Women's Health
Philadelphia, Pennsylvania, U.S.A.
Chapter 28

Patrick Van Roey
Wadsworth Center
New York State Department of Health
Albany, New York, U.S.A.
E-mail: vanroey@wadsworth.org.
Chapter 7

Algirdas Velyvis
Department of Pharmacology
School of Medicine
Case Western Reserve University
Cleveland, Ohio, U.S.A.
Chapter 15

Lucas Waltzer
Centre de Biologie du Développement
CNRS UMR
Toulouse, France
Chapter 30

Margareta T. Wilhelm
Department of Oncology and Pathology
Karolinska Institute
Cancer Center Karolinska (CCK)
Stockholm, Sweden
Chapter 12

Christine Williams
Cutaneous Biology Research Center
Massachusetts General Hospital
and Harvard Medical School
Charlestown, Massachusetts, U.S.A.
Chapter 27

Klas G. Wiman
Department of Oncology-Pathology
Karolinska Institute
Cancer Center Karolinska (CCK)
Stockholm, Sweden
E-mail: Klas.Wiman@mtc.ki.se
Chapter 12

Ira G. Wool
Department of Biochemistry and Molecular Biology
The University of Chicago
Chicago, Illinois, U.S.A.
E-mail: irawool@midway.uchicago.edu
Chapter 14

Peter E. Wright
Department of Molecular Biology
The Scripps Research Institute
La Jolla, California, U.S.A.
E-mail: wright@scripps.edu
Chapter 17

Lai-Chu Wu
Departments of Molecular
and Cellular Biochemistry and Internal Medicine
Division of Immunology, Molecular Virology
and Medical Genetics
The Ohio State University
College of Medicine and Public Health
Columbus, Ohio, U.S.A.
E-mail: wu.39@osu.edu
Chapter 29

Liqing Wu
Department of Radiation and Cellular Oncology
The University of Chicago
Chicago, Illinois, U.S.A.
Chapter 34

Toshimi Yoshida
Cutaneous Biology Research Center
Massachusetts General Hospital
and Harvard Medical School
Charlestown, Massachusetts, U.S.A.
Chapter 27

PREFACE

In the early 1980s, a few scientists started working on a *Xenopus* transcription factor, TFIIIA. They soon discovered a novel domain associated with zinc, and named this domain "zinc finger." The number of proteins with similar zinc fingers grew quickly and these proteins are now called C_2H_2, Cys_2His_2 or classical zinc finger proteins. To date, about 24,000 C_2H_2 zinc finger proteins have been recognized. Approximately 700 human genes, or more than 2% of the genome, have been estimated to encode C_2H_2 finger proteins. From the beginning these proteins were thought to be numerous, but no one could have predicted such a huge number. Perhaps thousands of scientists are now working on C_2H_2 zinc finger proteins from various viewpoints. This field is a good example of how a new science begins with the insight of a few scientists and how it develops by efforts of numerous independent scientists, in contrast to a policy-driven scientific project, such as the Human Genome Project, with goals clearly set at its inception and with work performed by a huge collaboration throughout the world.

As more zinc finger proteins were discovered, several subfamilies, such as C_2C_2, CCHC, CCCH, LIM, RING, TAZ, and FYVE emerged, increasing our understanding of zinc fingers. The knowledge was overwhelming. Moreover, scientists began defining the term "zinc finger" differently and using various names for identical zinc fingers. These complications may explain why no single comprehensive resource of zinc finger proteins was available before this publication.

This book adopts a broad definition of zinc finger as a peptide domain with a special tertiary structure stabilized by Zn^{2+} coordination. These tertiary structured fingers confer specific binding activities to various molecules such as DNA, RNA, proteins or small molecules. Some groups of scientists exploited the binding specificity to develop tools that target zinc finger proteins to any DNA and RNA segments at will. Other groups studied additional domains that are required for executing functions of zinc finger protein molecules. Some of these domains, such as BTB, KRAB and SCAN, are often present in C_2H_2 zinc finger proteins, while other associated domains are less common but may also play important roles. Functions of zinc finger proteins include gene expression, signal transduction, cell growth, differentiation and development. Given the broad role zinc finger proteins play in the cell, it is logical that mutations in some zinc finger proteins have been found to cause diseases such as cancer and neurological disorders. Likewise, zinc ion deficiency leads to cell death, probably because of the essential nature of zinc finger proteins.

This book systematically describes features of various zinc finger proteins, presenting specific activities of their domains in the earlier chapters and cellular activities of the protein molecules in the later chapters. The chapters are written by authors outstanding in the field. These chapters are grouped according to their binding function and typically include several illustrative examples of zinc finger proteins for each function. The arrangement reflects this book's intention to encompass the principles and the paradigms of zinc finger proteins at levels from atomic through cellular to organismal. This book should be a useful resource for experts in the field, but also valuable for scientists and graduate students of biological science.

We thank Drs. J.M. Baraban, D.A. Haber, A.G. Jochemsen, F.L. Rauscher III and B. Schulz for their suggestions regarding experts of research fields, Ms. K. Easley for her assistance, and Dr. H. Green for his support.

Shiro Iuchi
Natalie Kuldell

ABBREVIATIONS

aACRYBP1	aA-crystallin binding protein
AAV	adeno associated virus
ADPRT	poly(ADP-ribosyl) transferase
AF-1	activation function 1
AF-2	activation function 2
AFC	antibody forming cell
AGM	aorta gonad mesonephros
AIF	apoptosis-inducing factor
Aio	Aiolos
AR	androgen receptor
ARE	androgen responsive element
AT-BP2	a1-antitrypsin promoter binding protein
ATM	ataxia telangiectasia mutated
ATR	ATM and Rad3-related
BAC	bacterial artificial chromosome
BAF	brg1 associated factors
BLAST	basic local alignment search tool
BM	bone marrow
BMI-1	B-cell-specific Moloney murine leukemia virus integration site 1
BMP	bone morphogenetic protein
Bn1	Basonuclin 1
Bn2	Basonuclin 2
bp	base pairs
BRCA1	breast cancer predisposition gene 1
BTB	broad-complex, tramtrack, and bric-a-brac
CARD	caspase recruitment domain
Cbl	casitas B-lineage lymphoma protein
CBP	cAMP response element binding protein
CD	circular dichroism; chromo domain; chromosome modifier domain
Cdk	cyclin-dependent kinase
CFTR	cystic fibrosis trans-membrane conductance regulator
CHD	congenital heart defect
ChIP	chromatin immunoprecipitation
CHX	cyclohexamide
CK2	protein kinase CK2 (formerly known as casein kinase II)
Cys	cysteine
CKII	casein kinase II
CLP	common lymphoid progenitor
CoA	coenzyme A
Col2a1	type II collagen gene
CpG	cytidine-guanidine dinucleotide pairs
CREB	cAMP response element binding
CRIP	cysteine-rich intestinal protein
CRP	cysteine-rich protein
CS	cleavage site
CSD	chromo shadow domain
CtBP	c-terminal binding protein
5S DNA	5S ribosomal RNA gene
DBD	DNA binding domain
DCC	dosis compensation complex
DDS	Denys-Drash syndrome
DHFR	dihydrofolate reductase
DHS	Dnase I hypersensitivity site
DHT	5a-dihydrotestosterone
DMD	Duchenne muscular dystrophy
dn	dominant negative
DN	double negative
DNase	deoxyribonuclease
DP	double positive
Dpp	decapentaplegic
ds	double-stranded
DTT	dithiothreitol
E6AP	viral E6-associated protein
ED	effector domain
Edn	endothelin
EED	embryonic ectoderm development
Efp	estrogen-responsive finger protein
EGR1	early growth response gene 1
ERCC	excision repair cross-complementing rodent repair deficiency
ES cells	embryonic stem cells
EWS	Ewings sarcoma gene
EXAFS	extended X-ray absorption fine structure
EZF-2	endothelial zinc finger protein-2
EZH	human homolog of enhancer of Zeste
FA	focal adhesion
FKBP25	FK506-binding protein
FL(2)D	female-lethal(2)-d
FS	Frasier syndrome
FSHD	facioscapulo-humeral muscular dystrophy
FYVE	conserved in Fab1, YOTB Vac1 and EEA1
GATA family	GATA sequence-binding zinc finger protein
GEF	GDP/GTP exchange factor
GlcNAc	N-acetylglucosamine
Gly	glycine
GO	8-oxoguanine
GST	glutathione transferase
GTF	general transcription factor
H2TH	helix-two turns-helix
HAT	histone acetyl transferase
HD	Huntington's disease
HDAC	histone deacetylase
HIF-1a	hypoxia-inducible factor-1a
HIV	human immunodeficiency virus
HIV-1	human immunodeficiency virus type I
HIV-EP	human immunodeficiency virus type 1 enhancer binding protein
HMG	high-mobility group
HmL37ae	Haloarcula marismortui ribosomal protein L37ae
HMQC	heteronuclear multiple quantum coherence
HMTs	histone methytransferases
HP1a	heterochromatin protein 1 alpha
HPC	human homolog of Polycomb
HPH	human homolog of Polyhomeotic
HPV	human papillomavirus
HSC	hemopoietic stem cell
HSED	Heteronuclear Spin-Echo Difference
HSP	heat shock protein
HSQC	heteronuclear single quantum correlation
ICAD	inhibitor of calcium-activated DNAase
ICR	internal control region
ICR	imprinted-control region
Ig	immunoglobulin
IGF-1	insulin growth factor-1

IL-2	interleukin-2
Ileu	isoleucine
ILF3	interleukin enhancer binding factor 3
ILK	integrin-linked kinase
IS	insertion site
KAP-1	KRAB-associated protein 1
kb	kilobase
kD (KDa)	kilodalton
KLF	Krüppel-like factor
KRAB	Krüppel-associated box
LBD	ligand binding domain
LBP-1	leader binding protein-1
Ldb	LIM domain binding
LDL	low density lipoprotein
LeR	leucine rich region
Leu	leucine
LFSE	ligand field stabilization energy
Lhx	LIM homeobox
LID	LIM interaction domain
LIM-HD	LIM homeodomain
LMO	LIM only protein
LSF	late SV40 factor
LTR	long terminal repeat
Lys	lysine
MBP	major histocompatability complex enhancer binding protein; maltose binding protein
MBT	midblastula transition
MDM	murine double minute
Mdm2	murine double minute 2
MEFs	mouse embryo fibroblasts
MGI	mouse genome informatics
MIS	Müllerian inhibiting substance
MMLV	Moloney murine leukemia virus
MMPs	matrix metalloproteinases
MOF	males-absent-on-the-first
MOZ	monocytic leukemia zinc finger protein
MTA2	metastasis-associated protein 2
MTF	metalloregulatory transcription factor
MVE	multivesicular endosome
MZF1	myeloid zinc finger gene 1
NB	nuclear body
NCBI	National Center for Biotechnology Information
NER	nucleotide excision repair
NES	nuclear export signal
NLS	nuclear localization signal
NMR	nuclear magnetic resonance
NO	nitric oxide
NOE	nuclear Overhauser effect
nt	nucleotide
NuRD	nucleosome remodeling deacetylase
OS	Opitz syndrome
oTFIIIA	oocyte form of TFIIIA
PAR4	prostate apoptosis response gene 4
Pc	polycomb
PCAF	p300-CBP associated factor
PcG	polycomb group
Pcl	polycomb-like
PCNA	proliferating cell nuclear antigen
Peg3	paternally expressed gene 3
PGC-2	PPARg coactivator-2
Ph	polyhomeotic
PHA	phytohemagglutinin
PHAX	phosphorylated adaptator for RNA export
PHD	plant homeobox domain protein; plant homeodomain
Pho	pleiohomeotic
PI 3-kinase	phosphatidylinositol 3-kinase
$PI(3,5)P_2$	phosphatidylinositol 3,5-bisphosphate
PI3P	phosphatidylinositol 3-phosphate
PMA	phorbol myristate acetate
PML	promyelocytic leukemia protein
PML-Nbs	promyelocytic leukemia protein-nuclear bodies
pol III	RNA polymerase III
POZ	poxvirus and zinc finger
PPARγ	peroxisome proliferator-activated receptor gamma
ppm	parts per million
PRDII-BF1	positive regulatory domain II binding factor
PRE	polycomb response element
Pro	proline
R	DNA degenerate alphabet A or G
5S RNP	5S ribonucloeprotein particle
5S rRNA	5S ribosomal RNA
7S RNP	7S storage particle; 7S ribonucleoprotein particle
RAG	recombination activating gene
RAR	retinoic acid receptor
RAZ1	SCAN-related protein associated with MZF1B
Rb	retinoblastoma protein
RBCC	RING finger-B box-coiled coil
RE	response element
Rex-1	reduced expression 1
RFC	replication factor C
RFP	ret finger protein
RING	really interesting new gene
RMSD	root-mean-square-deviation
RNA pol II	RNA polymerase II
RNAi	RNA interference
RNP	ribonucleoprotein particle
rp	ribosomal protein
RPA	replication protein A
RPB9	RNA polymerase II subunit 9
RRM	RNA recognition motif
rRNA	ribosomal RNA
RSS	recombination signal sequences of V(D)J recombination
RYBP	ring1- and YY1- binding protein
SAA1	serum amyloid A1
SAS	something about silencing
SBMA	spinal and bulbar muscular atrophy
SCAN	SRE-ZBP, Ctfin51, AW-1 (ZNF 174), and Number 18
SCAN-ZFP	SCAN domain-containing C_2H_2 zinc finger protein
Scm	sex combs on midleg
SDP1	SCAN-domain containing protein 1
SELEX	systematic evolution of ligands by exponential enrichment
SET	SuVar3-9, Enhancer of Zeste, Trithorax
SF1	steroidogenic factor 1
SKIP	ski-interacting protein
SLC39	solute carrier family 39
SNOC	S-nitrosocysteine
SOX9	SRY-related HMG box
SP	single positive
SPRY	SPla and the RYanodine Receptor
SRC-1	steroid receptor coactivator-1
SRE	serum response element
SREBP	serum response element binding protein
SRF	serum response factor
SRY	sex-determining region Y gene
ss	single-stranded
SSTR2	somatostatin receptor type II
sTFIIIA	somatic form of TFIIIA
SUMO	small ubiquitin-related modifier
TBP	TATA-box binding protein; TATA-binding protein
td	thymidylate synthase gene
TF	transcription factor

TFIIA	transcription factor IIA
TFIIH	transcription factor IIH
TFIIIA	transcription factor IIIA
RRE	Rev response element
RRE-IIB	Rev response element stem loop IIB
TTP	tristetraprolin
TGF-β	transforming growth factor beta
TIF1	transcriptional intermediary factor 1
TIF1β	transcriptional intermediary factor 1 beta
Tip	TAT-interacting protein
tkv	thick veins
TM	transmembrane domain
TNF	tumor necrosis factor
TPEN	N,N,N',N'-tetrakis-(2-pyridylmethyl)ethylenediamine
TRAF	tumor necrosis factor receptor-associated factor
TRIM	tripartite motif
ts	temperature sensitive
TSA	trichostatin A
Tyr	tyrosine
URR	upstream regulatory region
UTR	untranslated region
V(D)J	variable, diversity, and joining
Val	valine
VEGF	vascular endotherial growth factor
VHL	von Hipple Lindau
W	DNA degenerate alphabet A or T
WAGR	Wilms tumor, aniridia, genitourinary malformations, and mental retardation
wt	wild type
WT1	Wilms tumor suppressor gene 1
WTAP	WT1-associating protein
X. laevis	*Xenopus laevis*
Xi	X chromosome inactivation
Xlo	Xenopus oocyte type
Xls	Xenopus somatic type
XP	xeroderma pigmentosum
XPA	Xeroderma pigmentosum A
YAF2	YY1-associated factor 2
YL37a	yeast ribosomal protein L37a
YY1	Yin Yang 1
ZBRK1	zinc finger and BRCA1-interacting proteins with KRAB domain 1
ZF	zinc finger
ZFP	zinc finger protein
ZIP	Zrt-, Irt-like protein
ZNF	zinc finger

CHAPTER 1

The Discovery of Zinc Fingers and Their Practical Applications in Gene Regulation: A Personal Account

Aaron Klug

Abstract

An account is given of the discovery of the classical Cys_2His_2 (C_2H_2) zinc finger, arising from biochemical studies on the protein transcription factor IIIA found in *Xenopus ooctyes*, and of subsequent structural studies on its 3D structure and its interaction with DNA. Each finger is a self-contained domain stabilized by a zinc ion ligated to a pair of cysteines and a pair of histidines, and by an inner structural hydrophobic core. This work showed not only a novel protein fold but also a novel principle of DNA recognition. Whereas other DNA binding proteins generally make use of the symmetry of the double helix, zinc fingers can be linked linearly in tandem to recognize nucleic acid sequences of different lengths. This modular design offers a large number of combinatorial possibilities for the specific recognition of DNA (or RNA). It is therefore not surprising that this zinc finger is found widespread in nature, in 3% of the genes of the human genome.

It had long been the goal of molecular biologists to design DNA binding proteins for specific control of gene expression. It has been demonstrated that the zinc finger design is ideally suited for such purposes, discriminating between closely related sequences both in vitro and in vivo. The first example of the potential of the method was in 1994 when a three-finger protein was constructed to block the expression of an oncogene transformed into a mouse cell line. By fusing zinc finger peptides to repression or activation domains, genes can be selectively switched off and on. Several recent applications are described.

After the initial discovery, other types of zinc-binding domains which fold and interact with DNA or RNA in a different way were found, and these have become loosely grouped under the name of zinc finger proteins.

Introduction

After ten years of research on the structure of chromatin which had led to the discovery of the nucleosome and an outline of its structure, as well as the next level of folding of DNA in the 300Å chromatin fiber,[1,2] I became interested in the then so-called "active chromatin," the chromatin which is involved in transcription or poised to do so. I looked for a system which was tractable, that is, offered the possibility of extracting relatively large amounts of material for biochemical and structural studies.

I became intrigued by the work of Robert Roeder, then at Washington University, and Donald Brown, of the Carnegie Institute of Washington in Baltimore on the 5S RNA genes of *Xenopus laevis*, which are transcribed by RNA polymerase III (reviewed in Brown).[3] They discovered that the correct initiation of transcription requires the binding of a 40 kD protein factor, variously called factor A or transcription faction IIIA (TFIIIA), which had been purified from oocyte extracts. By deletion mapping it was found that this factor interacts with a region about 50 nucleotides long within the gene, called the internal control region.

Immature oocytes store 5S RNA molecules in the form of 7S ribonucleoprotein particles,[4] each containing a single 40 kD protein which was later shown[5] to be identical with transcription factor IIIA. TFIIIA therefore binds both 5S RNA and its cognate DNA and it was therefore suggested that it may mediate autoregulation of 5S gene transcription.[5] Whether this autoregulation occurred in vivo or not, the dual interaction provided an interesting structural problem with could be approached because of the presence of large quantities of the protein TFIIIA in immature *Xenopus* oocytes.

In the autumn of 1982, I therefore proposed to a new graduate student, Jonathan Miller, that he begin studies on TFIIIA. This led to the discovery of a remarkable repeating motif within the protein, which we later, in laboratory jargon, called zinc fingers, because they contained zinc and gripped or grasped the DNA. The full story of the experiments is told in our first paper[6] and I will only summarize it here. I should however emphasize that the repeating structure was discovered through biochemistry, not as some reviews have stated, by computer sequence analysis. When the sequence was published I looked for, and found by eye, a repeating pattern, which was then confirmed and aligned as a motif of 30 amino acids by Andrew McLachlan's computer analysis.[6]

The 7S RNP Particle

When Miller repeated the published protocols for purifying the 7S particle, he obtained very low yields, which we attributed to dissociation. Brown and Roeder had used buffers which contained variously dithiothreitol (DTT), used because the protein had a high cysteine content, and/or EDTA to remove any contamination by metals which hydrolyse nucleic acids. We observed that gel filtraton of the complex in 0.1mM DTT resulted in separate elution of protein and 5S RNA. However, when we found that the strong reducing agent sodium borohydride did not disrupt the complex, we realized that the protein was not being held

Zinc Finger Proteins: From Atomic Contact to Cellular Function, edited by Shiro Iuchi and Natalie Kuldell.

together by disulfide bridges, and that a metal might be involved. When the particle was incubated with a variety of chelating agents, particle dissociation could be prevented only by prior addition of Zn^{2+}, and not by a variety of other metals. Analysis of a partially purified 7S preparation by atomic absorption spectroscopy also revealed a significant concentration of Zn, with at least 5 mol Zn per mol particle.

While these experiments were in progress, Hanas et al[7] reported the presence of Zn in the 7S particle, at a ratio of two per particle. We thought this must be an underestimate since their buffers contained 0.5mM or 1mM DTT, which has a high binding constant for Zn of about 10^{10}. We therefore repeated the analysis with pure and undissociated particle preparations, taking great care to ensure no contamination. We concluded that the native 7S particle contains between 7 and 11 zinc ions. This result was consistent with the fact that the protein contains large numbers of histidine and cysteine residues, the commonest ligands for zinc in enzymes and other proteins. This hinted at some kind of internal substructure.

A natural step was therefore to see if any such substructure could be revealed by proteolytic digestion, and Miller had already begun such studies. He found two products, an intermediate 33 kD fragment, and a limit 23 kD. At about that time Brown's group[8] also showed that, on treatment with proteolytic enzymes, the 40 kD TFIIIA protein breaks down to a 30 kD product, which is then converted to a 20 kD product. They proposed that TFIIIA consists of three structure domains which they identified as binding to different parts of the 50 base-pair internal control region of the 5S RNA gene.

Carrying on these proteolytic studies, we found that on prolonged proteolysis the TFIIIA protein breaks down further, finally to a limit digest of about 3 kD. In the course of this breakdown, periodic intermediates differing in size by about 3 kD could be seen. The correspondence in size between these last two values suggested that the 30 3kD domain of TFIIIA might contain a periodic arrangement of small, compact domains of kD. If each such domain contained one Zn atom, then the observed high Zn content would be accounted for.

This novel idea of small Zn-stabilized domains was strengthened by the timely publication by Roeder's group[9] of the sequence of TFIIIA derived from a cDNA clone. By inspection, it could be seen that the large number of cysteines and histidines present in the protein appeared to occur in more or less regular patterns. A rigorous computer analysis showed that, of the 344 amino acids of the TFIIIA sequence residues, numbers 13-276 form a continuous run of nine tandemly repeated, similar units of about 30 amino acids, each containing two invariant pairs of histidines and cysteines.[6] Repeating patterns in the sequence were also noticed by R S Brown, Sander and Argos[10] who concluded, however, that the whole protein was divided into twelve repeats, indexed on a 39 amino acid unit (although their abstract states "about 30").

A Repeating Structure for TFIIIA

From the three different lines of evidence described above, namely (1) a 30 amino acid repeat in the sequence, which (2) corresponds in size to the observed periodic intermediates and the limit-digest product of 3 kD, and (3) the measured Zn content of 7-11 atoms, we proposed[6] that most of the TFIIIA protein has a repeating structure in which each of the nine 30 amino acid units folds around a Zn ion to form a small independent structural domain, the "finger." 25 of the 30 amino acid residues form a loop around the central Zn ion and the five intervening amino acids provide the linkers between consecutive fingers. The Zn ion forms the basis of the folding by being tetrahedrally coordinated to the two invariant pairs of cysteines and histidines. Each repeat also contained besides this unique conserved pattern of Cys-Cys .... His-His, three other conserved amino acids, namely Tyr6 (or Phe6), Phe17 and Leu23, all of which are large hydrophobic residues. The whole of the 30 amino acid repeat is rich in basic and polar residues but the largest number are found concentrated in the region between the second cysteine and the first histidine, implicating this region in particular in nucleic acid binding.

Formally, when indexed on a 30 amino acid repeat, the repeating structure could be written at

1 – 5	6	8	13	17	23	26	30
linker	$h X_1$	$C X_{2,4}$	$C X_3$	$h X_5$	$h X_{2,3}$	$H X_{3,4} H$	

where h represents a conserved (large) hydrophobic residue. The proposal that each 30 amino acid unit formed an independently folded, Zn-stabilized domain soon gained support from two lines of research. First, we carried out a study using EXAFS (extended X-ray absorption fine structure) confirming that the Zn ligands are two cysteines and two histidines. Secondly it was found by Tso et al[12] that, in the DNA sequence of the gene for TFIIIA, the position of the intron-exon boundaries mark most of the 9 proposed finger domains.

In evolutionary terms, the multi-fingered TFIIIA may have arisen by gene duplication of an ancestral domain comprising about 30 amino acids. Because one such self-contained small domain would have had the ability to bind to nucleic acids, and could be passed on by exon shuffling, we suggested that these domains might occur more widely in gene control proteins than in just this case of TFIIIA. The extent to which this prediction has been borne out (3% of the genes of the human genome, at the latest count) still, on occasion, astonishes me. Indeed, within months of the publication of our paper, I received word of sequences homologous to the zinc finger motif of TFIIIA. The first two were from *Drosophila*, the serendipity gene from Roshbash's group[13] and Kruppel from Jackle's group.[14]

The key point that emerged from our first paper was that, not only had there emerged a novel protein fold for nucleic acid binding, but also a novel principle of DNA recognition. The overall design for specific DNA recognition was distinctly different from that of the helix-turn-helix motif found in the first DNA binding proteins to be described. The latter binds to DNA as a symmetric dimer to a palindromic sequence on the DNA, thus making use of both the 2-fold symmetry of the DNA helix backbones and also the nucleotide sequence. (Later heterodimeric variations of this and related designs were found, but they still make use of the helix symmetry).

In contrast, the zinc finger is a module that can be used singly or linked tandemly in a linear fashion to recognize DNA (or RNA) sequences of different lengths. Each finger domain has a similar structural framework but can achieve chemical distinctiveness through variations in a number of key amino acid residues. This modular design offers a large number of combinational possibilities for the specific recognition of DNA (or RNA). It was not surprising that it is found widespread throughout so many different types of organisms.

The Structure of the Zinc Finger and Its Interaction with DNA

We had noted,[6] that in addition to the characteristic arrangements of conserved cysteines and histidines which are fundamental in the folding of the finger by the coordinating Zn, there are several other conserved amino acids, notably Tyr6, Phe17 and Leu23, and that were likely to form a hydrophobic structural core of the folded structure. In other words, the seven conserved amino acids in each unit provide the framework of tertiary folding, whereas some of the variable residues determine the specificity of each domain. Jeremy Berg[15] built on these original observations by fitting known structural motifs from other metallo-proteins to the consensus sequence of the TFIIIA finger motifs. His proposed model consisted of an antiparallel β-sheet, which contains the loop formed by the two cysteines and an α-helix containing the His-His loop. The two structural units are held together by the Zn atom. In analogy with the way in which the bacterial helix-turn-helix motif binds DNA, DNA recognition was postulated to reside mostly in the helical region of the protein structure.

Berg's model was soon confirmed by the NMR studies of Peter Wright's group[16] on a single zinc finger in solution, and subsequently by David Neuhaus in our laboratory[17,18] on a two-finger peptide. Our work took longer to solve but had the merit of showing that adjacent zinc fingers are structurally independent in solution, being joined by flexible linkers.

The question remained of the precise pattern of amino acid interactions of zinc fingers with DNA. The breakthrough came in 1991 when Nikola Pavletich and Carl Pabo,[19] both then at Johns Hopkins, solved the crystal structure of a complex of a DNA oligonucleotide specifically bound to the three-finger DNA binding domain of the mouse transcription factor Zif268, an early response gene. The primary contacts are made by the α-helix which binds in the DNA major groove through primary hydrogen-bond interactions from helical positions –1, 3 and 6 to three successive bases (a triplet) on one strand of the DNA, and through a secondary interaction from helical position 2 on the other strand. This is the canonical docking arrangement, but there are however some wide variations from this arrangement in the family of zinc finger – DNA complexes now known. There are also, of course, other minor secondary interactions.

Zinc Finger Peptides for the Regulation of Gene Expression

The mode of DNA recognition by a finger is principally a one-to-one interaction between amino acids from the recognition helix and the DNA bases. Moreover, because the fingers function as independent modules, fingers with different triplet specificities are combined to give specific recognition of longer DNA sequences. For this reason, the zinc finger motifs are ideal natural building blocks for the de novo design of proteins for recognizing any given sequence of DNA. Indeed the first protein engineering experiments by Berg and others, using site-directed mutagenesis, showed that it is possible to alter rationally the DNA-binding characteristics of individual zinc fingers when one or more of the α-helical positions are varied in a number of proteins. As a collection of these mutants accumulated, it became possible to propose some rules relating amino acids on the recognition α-helix to corresponding bases in the bound DNA sequence.

In our laboratory, my colleagues and I adopted a different approach. The reason was that the "rules" did not take into account the fact that real DNA structures are not fixed in the canonical B form but there are wide departures, depending on the DNA sequence.[20,21] This was further brought home to us by the structure of the second zinc finger DNA complex to be solved, that with the *Drosophila* tramtrack protein, by Fairall et al[22] in Daniela Rhodes's group our laboratory. Here the helical position used for the primary contact with the 3'-most base of one of the triplets (thymine) is not -1, but 2. The reason was that the DNA structure was much distorted from the B form, with the thymine followed by the adenine at a helical rotation angle of 39° rather than the canonical 36°. The reason is that the T-A step in DNA is unstable as I had noted long ago.[20]

Affinity Selection from a Library of Zinc Fingers by Phage Display

The alternative to this rational but biased design of proteins with new specificities is the isolation of desirable variants from a large pool or library. A powerful method of selecting such proteins is the cloning of peptides[23] or protein domains[24] as fusions to the minor coat protein (pIII) of bacteriophage fd, which leads to their expression on the tip of the capsid. Phage displaying the peptides of interest can then be affinity purified by binding to the target and then amplified for use in further rounds of selection and for DNA sequencing of the cloned gene. We applied this technology to the study of zinc finger-DNA interactions, after my colleague, Yen Choo, demonstrated that functional zinc finger proteins could be displayed on the surface of fd phage, and that such engineered phage could be captured on a solid support coated with the specific DNA.[25,26] The phage display method was also adopted by other groups working on zinc fingers, including those of Carl Pabo and Carlos Barbas.

We created phage display libraries comprising about 10^7 variants of the middle finger from the DNA-binding domain of Zif268. A DNA oligonucleotide of fixed sequence was used to bind and hence purify phage from this library over several rounds of selection, returning a number of different but related zinc fingers which bind the given DNA. By comparing similarities in the amino acid sequence of functionally equivalent fingers, we could deduce the likely mode of interaction of these fingers with DNA. Remarkably we found that many base contacts occur from three primary positions on the α-helix of the zinc finger, correlating with the implications of the crystal structure of Zif268 bound to DNA. This demonstrated ability to select zinc fingers with desired specificity meant that DNA-binding proteins containing zinc fingers could be made to measure, and that some general rules could be devised for a crude recognition code.

Use of Engineered Zinc Fingers to Repress Gene Expression in a Mouse Cell Line

In the course of the above work, we showed that a zinc finger mini-domain could discriminate between closely related DNA triplets, and proposed that they could be linked together to form longer peptides for the specific recognition of longer DNA sequences. One interesting possibility for the use of such protein peptides is to target selectively genetic differences in pathogens or transformed cells. In December 1994 we reported the first such application,[27] in which we built a protein which recognized a specific DNA sequence both in vitro and in vivo. This was a crucial test of our understanding of the mechanism of zinc-finger DNA recognition. The proof of principle stimulated ourselves, and later others, to devote our future studies to potential applications in gene regulation for research purposes or for therapeutic correction.

In summary, we created a three finger peptide able to bind site-specifically to a unique nine base-pair region of the p190 *bcr-abl* cDNA: this is a transforming oncogene which arises by translocation between the tips of chromosomes 9 and 22, of which one product is the Philadelphia chromosome. The latter contains a novel DNA sequence at the junction of two exons, one each from the two genomic parent *bcr* and *abl* genes. Our engineered peptide discriminated in vitro against like regions of the parent *bcr* and *c-abl* genes, differing in only a single base, by factors greater than one order of magnitude.

Our peptide also contained a nuclear localization signal fused to the zinc finger domain so that the peptide could accumulate in the nucleus. Consequently, stably transformed mouse cells, made interleukin-3 independent by the action of the oncogene, were found to revert to IL-3 dependence on transient transfection with a vector expressing the peptide. Our construct was also engineered to contain a c-myc epitope, which enabled us to follow by immunofluorescence the localization of the peptide to the nuclei of the transfected cells. When IL-3 is subsequently withdrawn from cell culture, over 90% of the transfected p190 cells become apoptotic (that is, showing chromosome degradation) within 24 hours. Our experiments were repeated on cells transformed by another related oncogene p210 *bcr-abl*, which served as a control. All transfected p210 cells maintained their IL-3 dependence, and remained intact on entry of the engineered peptide.

Measurements of the levels of p190 *bcr-abl* mRNA extracted from cells treated with the peptide showed that the repression of oncogenic expression by the zinc finger peptide was due to a transcriptional block imposed by the sequence-specific binding of the peptide, which presumably obstructed the path of the RNA polymerase.

Promoter-Specific Activation by Zinc Finger

These experiments showed that a zinc finger peptide could be engineered to switch off gene expression in vivo. In the same paper[27] we described other experiments on a different cell system (cultured mouse fibroblasts) to show that a gene could also be switched on in a similar way. We used the same nine base pair sequence, but this time as a promoter for a CAT reporter gene contained in a plasmid. The peptide, which recognized the promoter, was fused to a VP16 activation domain and, on transient transfection, stimulated expression of the reporter gene by a factor of 30-fold above controls.

Improving Zinc Finger Specificity

(i) Our more recent work has focused on improving the specificity of recognition by zinc fingers of the DNA target. While the main source of specificity lies in the amino acids at positions –1, 3 and 6 of the recognition α-helix of a zinc finger for successive bases lying on one strand of a DNA triplet, we found that the "cross-stand" interaction described above from helical position 2 to the neighboring base pair on the adjacent triplet can significantly influence the specificity.[28] Therefore it has been necessary to revise the simple model that zinc fingers are essentially independent modules that bind three base-pair subsites to a model that considers functional synergy between adjacent independently folded zinc fingers. In this revised model, Zif268-like zinc fingers potentially bind four base-pair overlapping subsites. We therefore redesigned our method of phage library construction to take account of this refinement,[29] which also has the merit of being more widely applicable and rapid.

(ii) An important step forward has been to increase the length of the DNA sequence targeted and hence its degree of rarity. Three zinc fingers recognize nine base pairs, a sequence which would occur randomly several times in a large genome. However six fingers linked together would recognize a DNA sequence 18 base pairs in length, sufficiently long to constitute a rare address in the human genome. One cannot simply go on adding fingers, because the periodicity of packed fingers does not quite match the DNA periodicity, so that they get out of register. We have learned how to engineer longer runs of zinc fingers which can target longer DNA sequences.[30,31] By fusing functional groups to the engineered DNA binding domains, for example silencing or activation domains, highly specific transcription factors can be generated to up- or down-regulate expression of a target gene.

Some Applications of Engineered Zinc Fingers

Examples of some recent applications by ourselves and others, using either three finger or six finger peptides are:

(i) inhibition of HIV-1 expression;[32]

(ii) the disruption of the infective cycle of infection by herpes simplex virus;[33]

(iii) activating the expression of VEGF-A in a monkey kidney cell line;[34]

(iv) activating the expression of vascular endothelial growth factor (VEGF) in a human cell line, and in an animal model; and[35]

(v) regulation of zinc finger expression by small molecules.[36]

Other Classes of Zinc Fingers and Zinc-Binding Domains

Shortly after the classical C_2H_2 zinc finger was discovered in TFIIIA, sequence motifs that appeared to be related were found in several other protein or cDNA sequences of molecules which bound DNA. It was therefore at first thought that these might have a rather similar structure to the TFIIIA type finger domains.

The most important and widespread examples are those from members of the superfamily of hormone-activated nuclear receptors which play a central role in the control of eukaryotic gene expression, and are indeed transcription factors.[37,38] The DNA-binding domains (DBDs) of such receptors all include two motifs in tandem, each about 30 amino acids long, but each motif contains two pairs of cysteines rather than a pair of cysteines and a pair of histidines as in the first class. They do indeed bind Zn^{2+}, but the three-dimensional structure of two such DBDs, determined in solution using 2-D NMR spectroscopy, showed that the receptor DBD is structurally distinct from the TFIIIA type of zinc finger.[39,40] The two motifs in each domain each fold up into an irregular loop followed by an α-helix, but the two together form a single structural unit with their helices crossing at right angles, so that the DNA recognition helix (from the first motif) is supported by the helix from the second motif.

Hormone receptors bind to palindromic sites (response elements, *RE*) on the DNA as dimers, and the DBDs alone also form dimers, the dimer interface arising from a region of the loop of the second motif of each receptor. These different roles for the two motifs within on structural unit, namely helix recognition and dimerization, were deduced by mapping on to the three-dimensional structures the site-directed mutagenesis data from a number of laboratories, particularly those of P. Chambon, R. Evans and G. Ringold. This combination of structural analysis and biochemical and genetic experiments pointed toward a mechanism of interaction with DNA and a general model was proposed by Härd et al[39] and Schwabe et al.[40]

The detailed chemistry of the interactions at the protein-DNA interface, however, had to await a crystal structure of a complex. The first to be determined was that at 2.9Å resolution by Luisi and Sigler[41] of the DBD of the glucocorticoid receptor (GR) complexed with a DNA segment 18 bp long. However, the DNA segment used was composed of two half sites (each of 6 bp) separated by a nonnative spacing, with four (rather than three) intervening base pairs. As a consequence of this, the two DBDs did not bind equivalently to DNA.

The crystal structure of a second hormone-receptor-DNA complex with a correct cognate binding site was solved (at 2.4Å resolution) by Schwabe et al[42] in our laboratory. This was of the DBD of the estrogen receptor which recognizes a different DNA half-site from GR but with the same native separation of 3 base pairs between half sites ($ERE_{s}3$). The protein binds as a symmetrical dimer in an equivalent manner to both half-sites. The interactions seen in this complex are characteristic in number and type of those later found in specific interactions in other families of protein-DNA complexes.

Clearly the second class of zinc finger DNA-binding domains is not a simple variant of the first TFIIIA class. They differ both in their structure and the way in which they interact with DNA. Above all the GR and ER receptors operate as dimers which bind to palindromic DNA sites, whereas the binding of zinc fingers of the first C_2H_2 class makes no use of the symmetry of the DNA structure nor of base sequence. The latter function as independent modules which can be strung together in a directly repeating (tandem) fashion with no restriction on their number.

It should however be added that some members of the nuclear receptor family (for example thyroid, vitamin D, retinoic acid) also bind as dimers, but as nonsymmetrical dimers to a DNA binding site made of two directly repeated identical "half-sites." Here clearly another interface is brought into play, but again the discrimination depends entirely on the separation between the repeats in the DNA sequence.[43]

The structure of a member of a third class of zinc binding domain of a distinct structural type emerged soon afterwards.[44] This was the GAL4 transcriptional activator, representative of a small family found only in yeast. Here two zinc ions and six cysteines in the DBD form a binuclear cluster with each Zn^{2+} coordinated by four cysteines, so that two of the cysteines are shared. This cluster holds together two short helices, related by a quasi dyad, one of which inserts into the major groove of the DNA as in the hormone receptors; we thus see the recognition helix supported by a second helix. Again, the molecule binds as a dimer to a palindromic DNA binding site, with two short half-sites separated by approximately 1.5 turns of the DNA helix, so that the DBDs bind on opposite faces of the DNA.

The fourth class of zinc fingers discovered was constituted by the Cys-X_2-Cys-X_4-His-X_4-Cys sequences found in the nucleocapsid proteins of retroviruses, which form 'stubby' fingers. Thereafter several more types were found, which are described in later chapters in this book.

This general use of zinc became increasingly clear, even though the structural information was often still limited. A diverse set of families of proteins which bind Zn^{2+} and interact with nucleic acids were uncovered and became loosely grouped under the name of zinc finger proteins. The term was not inappropriate since the zinc-binding domains in all these cases do grip or grasp the double helix. Nor would it be inappropriate if the domain does not bind DNA or RNA but another protein, as was foreshadowed,[46] and as has been increasingly found.

However in more recent years the term zinc finger has begun to be used even when the zinc-binding motif or amino acid sequence does not form an independent, self-folding domain or minidomain. I personally would prefer the term to be restricted to the latter cases. The cases where a Zn ion, say, is merely used to link together two regions of a protein, belong to the field of the bioinorganic chemistry of zinc. Its widespread use in proteins for such structural purposes must be attributed to the fact that zinc is what I would call a "safe" metal, in that it has only one main oxidation state and hence has no redox chemistry, as for instance copper and iron *do*. As a structural element in a protein it has the advantage over a disulfide bridge in bringing together two parts of a protein, because it cannot be reduced in the reducing atmosphere inside a cell.

Postscript

While, as described in the Introduction, it has long been known that TFIIIA binds RNA as well as DNA, there is now increasing evidence that zinc fingers are widely used to recognize RNA. However the molecular basis of the recognition has remained elusive. In this laboratory, we had not forgotten that TFIIIA binds 5S RNA. After our failure to crystallize the native 7S RNP particle itself (or a reconstituted version) we set out[47] to find relevant subcomplexes which could show the main interactions between TFIIIA and the RNA. We succeeded and have recently determined the X-ray structure of a zinc finger RNA complex which reveals two modes of zinc finger binding, both different from that for DNA.[48]

References

1. Kornberg RD. Structure of chromatin. Ann Rev Biochem 1977; 46:931-954.
2. Klug A. From macromolecules to biological assemblies. In: The Nobel Foundation 1983; 93-125 Les Prix Nobel en 1982.
3. Brown DD. The role of stable complexes that repress and activate eukaryotic genes. Cell 1984; 37:359-365.
4. Picard B, Wegnez M. Isolation of a 7S particle from Xenopus laevis oocytes: A 5S RNA-protein complex. Proc Nat Acad Sci USA 1979; 76:241-245.
5. Pelham HRB, Brown DD. A specific transcription factor that can bind either the 5S RNA gene or 5S RNA. Proc Nat Acad Sci USA 1980; 77:4170-4174.
6. Miller J, McLachlan AD, Klug A. Repetitive zinc-binding domains in the protein transcription factor III A from Xenopus oocytes. EMBO J 1985; 4:1609-1614.
7. Hanas JS, Hazuda DJ, Bogenhagen DF et al. Xenopus transcription factor A requires zinc for binding to the 5S RNA gene. J Biol Chem 1983; 258(14):120-125.
8. Smith DR, Jackson IJ, Brown DD. Domains of the positive transcription factor specific for the Xenopus 5S RNA gene. Cell 1984; 37:645-652.

9. Ginsberg AM, King BO, Roeder RG. Xenopus 5S gene transcription factor, TFIIIA: characterisation of a cDNA clone and measurement of RNA levels throughout development. Cell 1984; 39:479-489.
10. Brown RS, Sander C, Argos P. The primary structure of transcription factor TFIIIA has 12 consecutive repeats. FEBS Lett 1985; 186:271-274.
11. Diakun GP, Fairall L, Klug A. EXAFS study of the zinc-binding sites in the protein transcription factor IIIA. Nature 1986; 324:698-699.
12. Tso JY, van den Berg DJ, Korn LT. Structure of the gene for Xenopus transcription factor TFIIIA. Nucl Acids Res 1986; 14:2187-2200.
13. Vincent A, Colot HV, Rosbash M. Sequence and structure of the serendipity locus of Drosophila melanogaster: A densely transcribed region including a blastoderm-specific gene. J Mol Biol 1985; 185:146-166.
14. Rosenberg UB, Schröder C, Preiss A et al. Structural homology of the product of the Drosophila Krüppel gene with Xenopus transcription factor IIIA. Nature 1986; 319:336-339.
15. Berg JM. Proposed structure for the zinc binding domains from transcription factor IIIA and related proteins. Proc Nat Acad Sci USA 1988; 85:99-102.
16. Lee MS, Gippert GP, Soman KV et al. Three-dimensional solution structure of a single zinc finger DNA-binding domain. Science 1989; 245:635-637.
17. Nakaseko Y, Neuhaus D, Klug A et al. Adjacent zinc finger motifs in multiple zinc finger peptides from SW15 form structurally independent flexibly linked domains. J Mol Biol 1992; 228:619-636.
18. Neuhaus D, Nakaseko Y, Schwabe JW et al. Solution structures of two zinc-finger domains from SWI5 obtained using two-dimensional 1H nuclear magnetic resonance spectroscopy. A zinc-finger structure with a third strand of beta-sheet. J Mol Biol 1992; 228(2):637-651.
19. Pavletich NP, Pabo CO. Zinc finger-DNA recognition: Crystal structure of a Zif268-DNA complex at 2.1Å. Science 1991; 252:809-817.
20. Klug A, Jack A, Viswamitra MA et al. A hypothesis on a specific sequence-dependent conformation of DNA and its relation to the binding of the lac-repressor protein. J Mol Biol 1979; 131(4):669-680.
21. Rhodes D, Klug A. Sequence dependent helical periodicity of DNA. Nature 1981; 292:378-380.
22. Fairall L, Schwabe JW, Chapman L et al. The crystal structure of a two zinc-finger peptide reveals an extension to the rules for zinc-finger/DNA recognition. Nature 1993; 366(6454):483-487.
23. Smith GP. Filamentous fusion phage: Novel expression vectors that display cloned antigens on the virion surface. Science 1985; 228:1315-1317.
24. McCafferty J, Griffiths AD, Winter G et al. Phage antibodies: Filamentous phage displaying antibody variable domains. Nature 1990; 348(6301):552-554.
25. Choo Y, Klug A. Towards a code for the interactions of zinc fingers with DNA: Selection of randomised fingers displayed on phage. Proc Natl Acad Sci USA 1994; 91:11163-11167.
26. Choo Y, Klug A. Selection of DNA binding sites for zinc fingers using rationally randomised DNA reveals coded interactions. Proc Natl Acad Sci USA 1994; 91:11168-11172.
27. Choo Y, Sánchez-García I, Klug A. In vivo repression by a site-specific DNA-binding protein designed against an oncogenic sequence. Nature 1994; 372:642-645.
28. Isalan M, Choo Y, Klug A. Synergy between adjacent zinc fingers in sequence-specific DNA recognition. Proc Natl Acad Sci USA 1997; 94:5617-5621.
29. Isalan M, Klug A, Choo Y. A rapid, generally applicable method to engineer zinc fingers illustrated by targeting the HIV-1 promoter. Nat Biotechnol 2001; 19:656-660.
30. Moore M, Choo Y, Klug A. Design of polyzinc finger peptides with structured linkers. PNAS 2001; 98:1432-1436.
31. Moore M, Klug A, Choo Y. Improved DNA binding specificity from polyzinc finger peptides by using strings of two-finger units. PNAS 2001; 98:1437-1441.
32. Reynolds L, Ullman C, Moore M et al. Repression of the HIV-1 5' LTR promoter and inhibition of HIV-1 replication by using engineered zinc-finger transcription factors. Proc Natl Acad Sci USA 2003; 100(4):1615-1620.
33. Papworth M, Moore M, Isalan M et al. Inhibition of herpes simplex virus 1 gene expression by designer zinc-finger transcription factors. Proc Natl Acad Sci USA 2003; 100(4):1621-1626.
34. Liu PQ, Rebar EJ, Zhang L et al. Regulation of an endogenous locus using a panel of designed zinc finger proteins targeted to accessible chromatin regions. Activation of vascular endothelial growth factor A. J Biol Chem 2001; 276(14):11323-11334.
35. Rebar E, Hung Y, Hickey R et al. Induction of angiogenesis in a mouse model using engineered transcription factors. Nature Med 2002; 8:1427-1432.
36. Beerli RR, Schopfer U, Dreier B et al. Chemically regulated zinc finger transcription factors. J Biol Chem 2000; 275(42):32617-32627.
37. Evans RM. The steroid and thyroid hormone receptor superfamily. Science 1988; 240:889-898.
38. Green S, Chambon P. Nuclear receptors enhance our understanding of transcription regulation. Trends Genet 1988; 4:309-315.
39. Hard T, Kellenbach E, Boelens R et al. Solution structure of the glucocorticoid receptor DNA-binding domain. Science 1990; 249(4965):157-160.
40. Schwabe JWR, Neuhaus D, Rhodes D. Solution structure of the DNA-binding domain of the oestrogen receptor. Nature 1990; 348:458-461.
41. Luisi BF, Xu WX, Otwinowski Z et al. Crystallographic analysis of the interaction of the glucocorticoid receptor with DNA. Nature 1991; 352(6335):497-505.
42. Schwabe JW, Chapman L, Finch JT et al. The crystal structure of the estrogen receptor DNA-binding domain bound to DNA: How receptors discriminate between their response elements. Cell 1993; 75(3):567-578.
43. Umesono K, Murakami KK, Thompson CC et al. Direct repeats as selective response elements for the thyroid hormone, retinoic acid, and vitamin D3 receptors. Cell 1991; 65(7):1255-1266.
44. Marmorstein R, Carey M, Ptashne M et al. DNA recognition by GAL4: Structure of a protein-DNA complex. Nature 1992; 356(6368):408-414.
45. Summers MF, Henderson LE, Chance MR et al. Nucleocapsid zinc fingers detected in retroviruses: EXAFS studies of intact viruses and the solution-state structure of the nucleocapsid protein from HIV-1. Protein Sci 1992; 1(5):563-574.
46. Klug A, Rhodes D. 'Zinc fingers': A novel protein motif for nucleic acid recognition. Trends Biochem 1987; 12:464-469.
47. Searles MA, Lu D, Klug A. The role of the central zinc fingers transcription factor IIIA in binding to 5S RNA. J Mol Biol 2000; 301:47-60.
48. Lu D, Searles MA, Klug A. Crystal structure of a zinc finger RNA complex reveals two modes of molecular recognition. Nature 2003; 426:96-100.

CHAPTER 2

C_2H_2 Zinc Fingers As DNA Binding Domains

Shiro Iuchi

Abstract

A great number of C_2H_2 zinc finger proteins selectively bind to specific DNA sequences and play a critical role in controlling transcription of genes. The specific binding is achieved by zinc finger domains with ββα structure that is formed by tetrahedral binding of Zn^{2+} ion to the canonical cysteine and histidine residues. Two to three tandem zinc fingers are necessary and sufficient for the specific binding without participation of any other domains. Zinc fingers bind in the major groove of the DNA, wrapping around the strands, with specificity conferred by side chains of several amino acid on the α helices. Some zinc finger proteins undergo homodimerization by hydrophobic interactions or by finger-finger binding and reinforce the specific binding to DNA. Conserved linkers between tandem fingers are necessary for stabilizing the DNA complex. Regulatory mechanisms of zinc finger binding to DNA are emerging. Some cellular factors are found to acetylate and phosphorylate zinc fingers and the linkers of a few proteins. These modifications alter the binding activity of the zinc finger proteins and hence control expression of their target genes. Other factors can methylate promoter regions of genes. This modification alters affinity of zinc finger proteins for the DNA segments and hence controls expression of their target genes.

Introduction

The C_2H_2 zinc finger consists of twenty to thirty amino acid residues that have a special secondary structure stabilized by zinc tetrahedral binding to two cysteine and two histidine residues.[1-6] Proteins with these zinc fingers are called C_2H_2 zinc finger proteins. The C_2H_2 zinc finger protein family is the largest group of all zinc finger protein families (Table 1) and the second largest group of all protein classes after the envelope glycoprotein GP120 family. The proteins are present in prokaryotes as well as eukaryotes and are abundant in mammals. More than 700 human genes, or greater than 2% of the total human genes, encode C_2H_2 zinc finger proteins.[7] Not surprisingly, C_2H_2 zinc finger proteins participate in a variety of cellular activities including development, differentiation, and tumor suppression. Among C_2H_2 zinc finger proteins, many bind to DNA duplexes in a finger-mediated specific manner and participate in controlling expression of the target genes. C_2H_2 zinc fingers, which are often described as $X_2CX_{2-4}CX_{12}HX_{2-8}H$ to show the intervals between the zinc binding residues, contain two β strands and one α helix. All the primary, secondary and tertiary structures are important for binding to DNA duplexes. In this chapter the general features of C_2H_2 zinc fingers-DNA binding are described.

Folding of C_2H_2 Zinc Fingers

Requirement of zinc ion for transcription factors to bind their cognate DNA was found in *Xenopus* TFIIIA (Transcription Factor IIIA) first.[8] The requirement was due to the ion coordination with a small peptide domain, named zinc finger, that contains two canonical cysteine and histidine residues.[1,4-6] Now, thousands of proteins are known to have C_2H_2 zinc fingers and the majority of the fingers are thought to bind to DNA. Zinc finger proteins can take three states: unfolded, folded and DNA-bound forms (Fig. 1 and for details of folded zinc finger domain see ref. 1 and 3 as well as Fig. 1 of Chapter 8). Unfolded zinc fingers do not bind to target DNA, but folded fingers bind to the cognate DNA duplexes. The protein molecules of the DNA complex are usually associated with other transcription factors that bind to different domains on the zinc finger protein, or to different parts of the zinc fingers. The C_2H_2 zinc finger motif contains all the information necessary for its folding but folds properly only when Zn^{2+} binds to the canonical residues, two cysteines and two histidines.[9,10] The change in Gibbs free energy (ΔG) of the folding is enthalpy-driven. The amount, about -8.8 kcal per mole (Table 2),[11] indicates that the folded fingers are very stable. C_2H_2 zinc fingers contain three conserved hydrophobic amino acids at position -12, -3 and +4 in addition to the two canonical cysteines and histidines (see bolded amino acids in the sequence below where the first amino acid residue of the α helix is designated as position 1). It has been shown that these seven amino acid residues are necessary and sufficient to fold peptides properly by using a designed-synthetic peptide, K(-13)-**Y**ACAACAAA**F**AAKAA**L**AAHAAAHA-K13.[9] This peptide binds Co^{2+}, a substitute for Zn^{2+}, to the cysteines with a higher affinity than to the histidines, and the proper folding occurs only when the ratio of the ion to the peptide is one or higher. Similar binding of Zn^{2+} to the zinc finger motif has also been observed in the synthetic peptide of the zif268 third finger [**F**(-12)-ACDICGRK**F**ARSDERKRHTKIHLRQ-K15].[10,12] The authors of this work have proposed that the zinc finger folding begins with binding of Zn^{2+} to the canonical cysteines and then establishes the tetrahedral structure involving the histidines. They have also suggested that the α helix emerges at the S(1)-DERKRHTKI-H11 sequence as the metal ion binding proceeds.

Tandem C_2H_2 Zinc Fingers

C_2H_2 zinc finger proteins often contain the fingers as tandem repeats connected by short oligopeptides, called linkers. Based on the number and repeat pattern of the fingers, C_2H_2 zinc

Zinc Finger Proteins: From Atomic Contact to Cellular Function, edited by Shiro Iuchi and Natalie Kuldell.

Table 1. Number of zinc finger proteins reported as of May 2003

Type	Number of Proteins	Average Domain's Length (AA)
C_2H_2	23,989	23
CCHC (C_2HC)	5,215	17
RING	2,010	41
LIM	1,246	58
C_2C_2 (2x)*	1,243	65
CCCH (C_2CH)	1,019	26
TAZ	51	81

Data are taken from Pfam 9.0 at Washington Uninversity in St. Louis. *The number is for proteins containing two C_2C_2 fingers.

finger proteins can be divided into four classes (Fig. 2), (A) single C_2H_2, (B) triple C_2H_2, (C) multiple-adjacent C_2H_2, and (D) separated-paired C_2H_2 zinc finger proteins.[13] This classification is useful to predict how the zinc fingers of the proteins exert their binding activity. The single zinc finger differs from the other zinc fingers in that it requires an additional, non-zinc finger domain to establish the binding to the target DNA.[14] The other classes, that is the triple, multiple-adjacent and separated-paired C_2H_2 fingers, bind to the specific DNA sequence without the aid of other domains. Both the triple C_2H_2 and the multiple-adjacent C_2H_2 zinc fingers bind to the cognate DNA at the three consecutive fingers.[13] Another experiment has shown that a finger peptide with four-tandem identical repeats binds to the target DNA sequence at the three consecutive fingers only.[15] Furthermore, the separated-paired C_2H_2 finger can also specifically bind to the cognate DNA, often at the one pair finger. Taking all these results into account, it may be concluded that two to three successive C_2H_2 zinc fingers are the most suitable unit to specifically bind to the cognate DNA.

Multiple-adjacent C_2H_2 zinc fingers bind to the DNA, based on the rule that two to three successive fingers are responsible for the specific DNA binding, but the fingers have additional DNA contacts. For example, TFIIIA with nine zinc fingers establishes the DNA binding at fingers 1-3, but briefly touches the DNA at finger 5 and weakly binds to the DNA at fingers 7-9.[16-20] Another example is Zac. This has seven zinc fingers and binds to a GC rich DNA duplex. Biochemical and genetic analysis have shown that Zac binds to the DNA at the two consecutive fingers (finger 2-3) and also at the three consecutive fingers (finger 5-7), without involving finger 4.[21] Accordingly, Zac contacts the DNA at more than three fingers, keeping the two to three finger-DNA binding rule. It is curious that extra fingers are present in multiple-adjacent zinc finger and separated-paired zinc finger proteins, but it is becoming increasingly clear that the extra fingers are engaged in other interactions not only with a secondary locus of DNA but also with distinct molecules such as RNA and proteins.[13,22-24]

It is interesting to speculate how these tandem zinc fingers have evolved from a single finger. The *Escherichia coli* gene, *arcA*, encodes a helix-turn-helix DNA binding repressor protein for the genes directing aerobic respiration enzymes.[25] This gene has an identical eleven-nucleotide sequence flanking a short region, and this genetic organization prompts the region to duplicate spontaneously (Iuchi and Lin, unpublished). Similarly, the identical

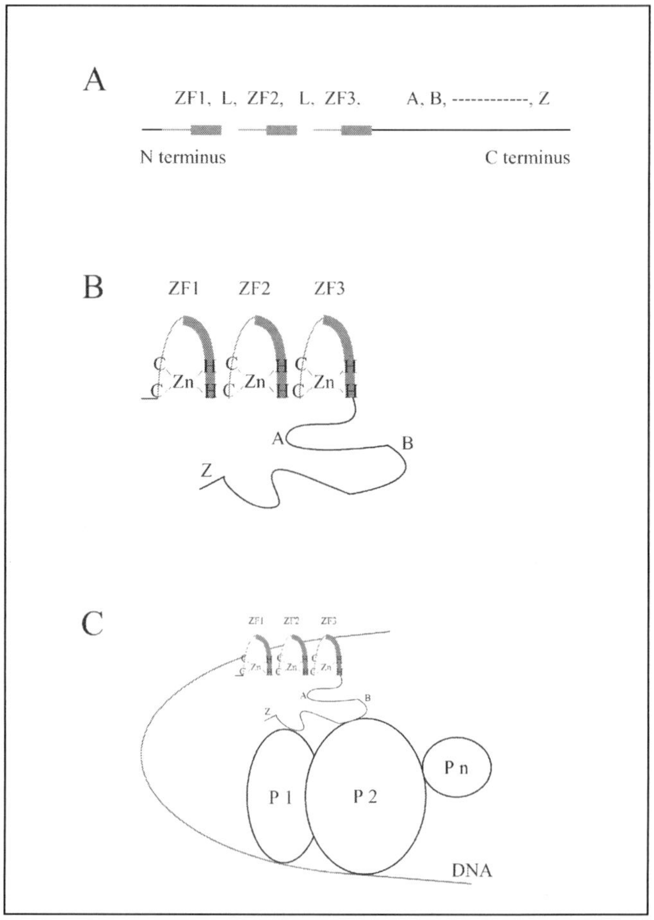

Figure 1. Schematic representation of three states of C_2H_2 zinc finger proteins. A) Primary structure of proteins. Polypeptide has three tandem fingers (ZF1, ZF2 and ZF3), each of which has α helix region (thick line) and is connected by linker (L). Other domains of the polypeptide are presented by A, B, -----, Z. B) Folded zinc finger proteins. Zn^{2+} tetrahedrally binds to the canonical cysteine and histidine residues and stabilizes the zinc finger's tertiary structure. (For details of folded C_2H_2 zinc fingers, see ref. 1 and 3 as well as Fig. 1 of Chapter 8). C) Zinc finger protein-DNA complex. The α helices of three zinc fingers bind to the DNA duplex in the major groove. Various domains of zinc finger proteins, including a part of zinc fingers, interact with other transcription factors.

linker sequences flanking the C_2H_2 zinc finger DNA may allow a zinc finger sequence to duplicate together with the linker itself (Fig. 3). The duplication would in turn stimulate the gene duplication further by using the linkers again or the zinc finger itself. The multiple zinc fingers can gain a better affinity for a DNA sequence and give the zinc finger protein molecules a selective advantage over the original finger proteins to function as transcription factors. This speculation is consistent with the fact that many of the separated-paired C_2H_2 and triple C_2H_2 zinc fingers conserve amino acid residues between tandem zinc fingers.[13] After the duplication, point mutations would further improve its affinity for the target. Alternative splicing within the gene[26,27] or recombination with another zinc finger gene[28] would allow the finger peptide to acquire tandem hybrid zinc fingers with a very different specificity. It should not be difficult to show that gene duplication indeed causes the evolution in living cells using well-established bacterial and yeast genetic systems.

Table 2. Thermodynamic parameters for zinc finger reactions

Reaction	ZF	K_a (M^{-1})	ΔG	ΔH	TΔS	Reference
$ZFP + Zn^{2+} \leftrightarrow ZFP \bullet Zn^{2+}$						
	1	3.5×10^6	-8.8	-9.3	-0.5	11
$ZFP \bullet Zn^{2+} + DNA \leftrightarrow ZFP \bullet Zn^{2+} \bullet DNA$						
Zif268	3	2.2×10^{10}	NA	NA	NA	33
Zif268	3	2.8×10^8	-11	-6.9	4.5	29
TFIIIA	3	1.3×10^7	-9	-6.9	2.6	30
SP1	3	2.6×10^7	-10	NA	NA	32
WT1	4	8.8×10^8	-12	7	19	29
YY1	4	1.8×10^6	- 8	-11	-2.8	31

ZFP denotes zinc finger peptide. ZF shows number of zinc fingers per finger peptide. The relation between Gibbs free energy, enthalpy and entropy is given by ΔG = ΔH -TΔS, where T is absolute temperature. Negative ΔG value suggests that each above reaction favors the association- over the dissociation- reaction. The unit of energy is kcal/ mol.

Overall Features of the Triple-C_2H_2 Zinc Fingers Binding

The change in Gibbs free energy of zinc finger-DNA binding is similar to or higher than that of the zinc finger folding (Table 2)[11,29-33] and the binding requires no enzyme action. ΔG is constant over a biologically-meaningful temperature range, 5 to 45 °C.[31] The reactions are mostly enthalpy- and entropy-driven reactions, but some are only enthalpy- or only entropy-driven. The ΔG values suggest that the C_2H_2 zinc fingers bind to DNA as strongly as some antibodies, whose K_a for their antigens is 10^5 to 10^{12} M^{-1}.[34] These zinc finger-DNA complexes are of such high affinity that they routinely display an electrophoretic mobility shift on native polyacrylamide gels.

When an amino acid and guanine are mixed in water, the amino acid starts to associate with the nucleotide base and the reaction soon reaches equilibrium. Tendency of the reaction is described by the affinity constant, $K_a = 1/K_d$ = [aa-guanine] / [aa] [guanine]. The K_a of amino acids for guanine is in the order arginine > lysine > glutamine > glutamate > glycine.[35] Amino acids also have an inherent K_a for the three other bases. Accordingly, it is predictable that the amino acid-base associations are key for zinc finger peptides to specifically bind to the DNA duplexes. Indeed, the arginine-guanine contact, whose association is the greatest of all the combinations, is quite often present in zinc finger-DNA complexes (Fig. 4B and C). However, each amino acid residue of the zinc finger peptides does not have free mobility to access to the favorite bases due to the rigid zinc finger structure. How the zinc finger peptides recognize and bind to the cognate DNA duplex is a big question for the zinc finger-DNA interactions and the answer has come mostly from structure-oriented and genetic oriented analyses of zinc finger-DNA complexes. In particular, the structure of Zif268-DNA and TFIIIA-DNA complexes contributed enormously to the understanding of how the triple fingers align with the DNA duplex and how each finger interacts with the nucleotides of the DNA.[17,18,36] Three α helices of the Zif268 triple zinc fingers bind in the major groove of the target DNA duplex antiparallel to the primary strand (defined as the strand to which zinc finger contacts most, Fig. 4A and for details of the binding see ref. 2), making hydrogen bonds and forming hydrophobic interactions with nucleotide bases and wrapping around the DNA for almost one turn. In addition to hydrogen bonds and hydrophobic interactions, phosphate contacts also participate in the zinc finger-DNA complex formation. Phosphate contacts, linking to the DNA backbone, may not be significant in determining the specificity but appear to be important for strengthening the binding.[37] Through further analysis of various zinc finger–DNA complexes, it was found that the overall DNA binding mode of Zif268 is shared with other tandem C_2H_2 zinc fingers. These fingers include TFIIIA, SP1, GL1, WT1 and Tramtrack[17,18,38-41] as well as the single finger GAGA.[14] In addition to these GC-rich DNA binding transcription factors, an AT-rich DNA recognizing transcription factor, CF2II, is also thought to take the same binding mode.[27] However, one AT-rich DNA binding transcription factor, Nmp4, is proposed to associate with minor groove of the DNA.[42]

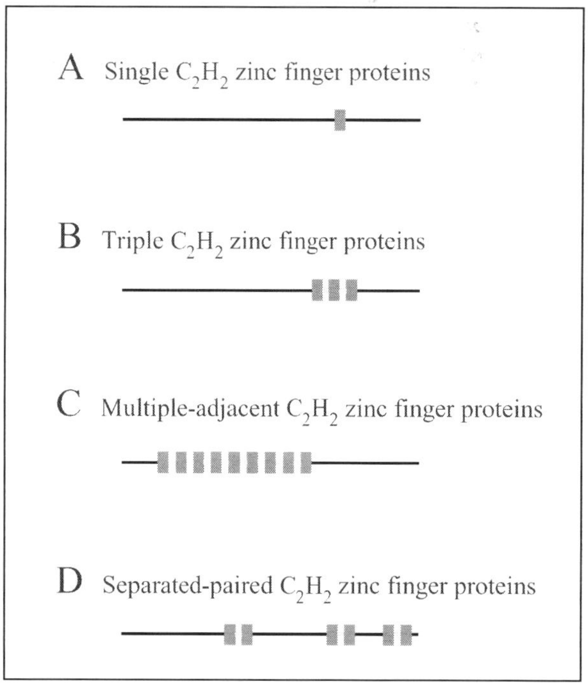

Figure 2. Schematic representation of four classes of C_2H_2 zinc finger proteins. Only one example is shown for each class. Some multiple-adjacent C_2H_2 zinc finger proteins contain more than thirty zinc fingers. The number and the pattern of C_2H_2 zinc fingers indicate how the zinc fingers are involved in the DNA binding (see text).

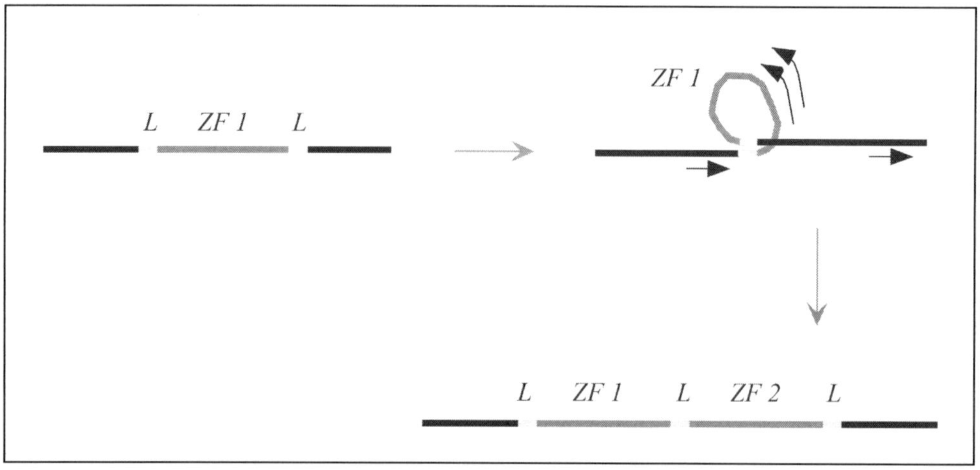

Figure 3. A model of intramolecular duplication of the *zinc finger* region. *ZF1* flanked by two identical linker sequences for (*L*) would form a loop overlapping at the identical sites, or form a hairpin structure complementing within the single strand at the identical sites. When DNA polymerase reads the loop or the hairpin structure twice, then the *zinc finger* region is duplicated together with the linker sequence.

DNA of the Zif268-DNA complex takes a slightly unwound B form so that the major groove is still wide and can be deep.[2] This form contains 11.3 bp per turn, which is slightly more bases than the B-form itself (10.5 bp per turn).

Side Chain-Base Contacts in C_2H_2 Zinc Finger-DNA Complexes

Amino acid side chains on the surface of α helices in Zif268 are exposed to the cognate DNA duplex in the major groove, and the side chains at position 6, 3, 2 and -1 contact selectively with four successive bases (subsite) (Fig. 4B). Residues at position 6, 3 and -1 bind to three successive bases of the primary strand, the strand contacted most by the side chains, and the residue at position 2 binds to the fourth base present in the complementary strand. In this way, Zif268 finger 3, 2 and 1 recognize the primary strand's subsite, 5'-GCGT-3', 5'-TGGG-3' and 5'-GCGT-3', respectively. Of the four base pairs in the subsites, the base pair(s) at the end is shared by adjacent fingers. Consequently, the Zif268 triple zinc fingers bind to the ten nucleotide base pair, 5'-GCGTGGGCGT-3'. Based on statistics, one can predict with a high accuracy which side chain-base contacts can happen at the key positions of the α helices in the Zif268 finger context.[43] Stereochemistry between amino acids and bases can also predict the contacts with a similar accuracy or somewhat less accuracy.[44]

The rule that side chains positioned at 6, 3, 2 and -1 contact bases is well preserved throughout the family of C_2H_2 zinc fingers, but additional contacts can occur in other fingers. Such an example has been observed in the TFIIIA-DNA complex (Fig. 4C).[17,18] There is an additional contact of the side chain at position 10. Furthermore, some side chains at the regular positions 6, 3, 2 and -1 reach bases out of the subsites. When Zif268 was engineered to bind to AT-rich DNA duplexes, many irregular contacts occurred.[2,45] In fact, many of these irregularities occurred in the complementary strand but not in the primary strand.[46] These facts, together with results obtained by mutational analyses, have led to the conclusion that there is no simple code for side chain-base contacts.[2,47] Superimposition of several finger-DNA complex images showed that the irregular contacts coincide with the slight difference in docking angle of the α helix to DNA. The difference may reveal the influences of all of the factors involved in specific zinc finger–DNA binding, including amino acid residues within the subsite,[47] linkers, and adjacent fingers.[2,37]

Absence of a strict rule in the side chain-base contacts makes it impossible to predict the side chain-base contact with 100%

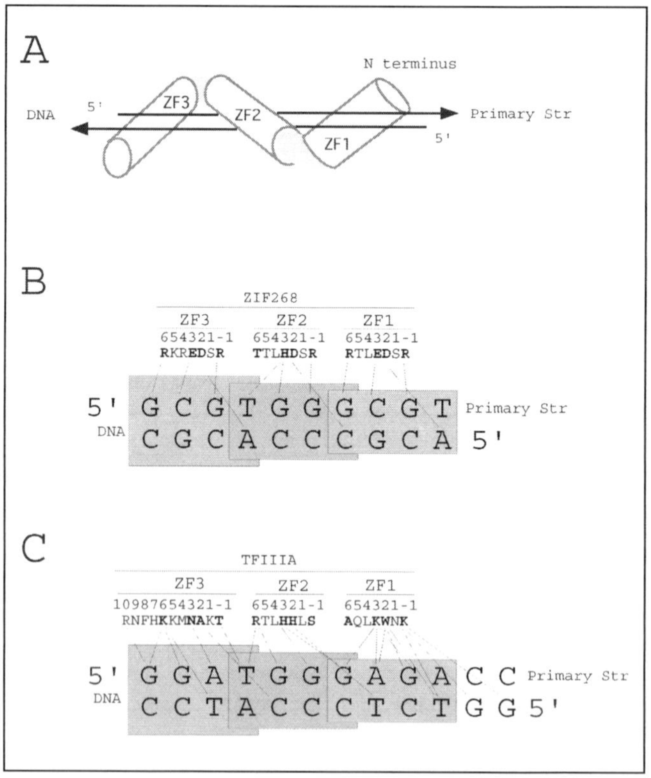

Figure 4. A) Antiparallel binding of zinc fingers to DNA. The orientation of the zinc fingers and the primary strand are antiparallel. B) and C) show amino acid side chain-nucleotide base contacts. Boxes show subsites. The side chain-DNA contacts are drawn, based on the results obtained with the crystallized Zif268-DNA complex B)[2,36] and with the TFIIIA-DNA complex in solution C).[17] Similar results of the side chain-DNA contact for the TFIIIA-DNA complex has been obtained with the crystallized complex.[18]

accuracy, but it rather indicates that C_2H_2 zinc fingers are able to bind to almost any DNA duplexes. The versatility of the zinc finger binding relies on hydrogen bonds that can (i) make long distance contact to bases with and without participation of a water molecule (>2.75 angstroms),[35,48] (ii) make contacts to more than one base, and (iii) make contacts to bases with flexible angles.[35] Zinc fingers' binding also relies on nonspecific hydrophobic interactions. Although it is difficult to predict exactly which zinc finger peptide sequence specifically binds to a DNA sequence, one can obtain desired zinc fingers by the phage display methods and manipulate expression of the target genes with the obtained finger proteins.[2,49-53]

Linkers

About half of zinc finger proteins have a well-conserved linker, TGEKP, between adjacent fingers.[37] The importance of the linker in zinc finger-DNA binding has been revealed by analyzing the effect of mutation on DNA affinity after making substitutions for the conserved residues[54,55] and by analyzing structures of finger-DNA complexes.[2,3,56,57] The linker is flexible in solution without DNA. Upon forming the finger-DNA complex, however, the conserved lysine residue of the TGEKP sequence (Fig. 1) interacts with the phosphate backbone. Moreover, linkers contact the C terminus of the preceding α helix by involving threonine and glycine residues so that the zinc finger-DNA complex becomes more stable (C capping).[17,57,58] Alternative splicing of the *WT 1* message disrupts the conserved linker between finger 3 and 4 by inserting the sequence KTS. The change from TGEKP to sequence TGKTSEP is accompanied by a severe decrease in DNA binding. An NMR study of the finger peptide-DNA complex has shown that the insertion increases the flexibility between finger 3 and 4 and abrogates binding of finger 4 to its cognate site.[41] These findings have demonstrated that the conserved linker is not only necessary to promote fingers to fit completely into the DNA major groove, but also necessary to strengthen the DNA binding. A separated-paired zinc finger class, ZAS family, also contains the TGEKP linker,[59] but fingers belonging to a subtype of the same class, basonuclin-type fingers, have a distinct linker, LR(K)MHK.[13,60] Tramtrack also has a distinct linker sequence, KRNVKVYP.[39]

Dimerization of Zinc Finger Protein

DNA binding proteins often bind to the target DNA duplex as dimers in order to increase their binding affinity and modulate their regulatory activities.[61] This is true for C_2H_2 zinc finger proteins. Many multiple-adjacent C_2H_2 zinc finger proteins, such as Ikaros, Roaz, GL1, SW15, TRPS-1 and Zac, form homodimers on the target DNA duplex using their zinc fingers.[23,24,62-64] GL1 forms a homodimer through the hydrophobic surface of zinc finger 1 that is not involved in the DNA binding. Similarly, SWI5 dimerizes through finger 1 at the hydrophobic surface of both the β strand and the α helix's C-terminal half. This hydrophobic dimerization was applied to make the homodimer of an engineered two-finger peptide on the DNA.[63] However, the hydrophobic binding is not the only dimerization mechanism for zinc fingers. Ikaros contains six zinc fingers. The N-terminal four fingers participate in the specific binding to the DNA, and the C-terminal two fingers, separated from the four and making a pair of fingers, are responsible for the homodimerization.[23] Recently, a mutational analysis of zinc fingers 5 and 6 demonstrated that amino acid residues on the α helix are responsible for the specific finger-finger interaction and hence for the Ikaros homodimerization.[62] These amino acids correspond to those of C_2H_2 zinc fingers with DNA binding activity, and differ from those needed for the hydrophobic dimerization.

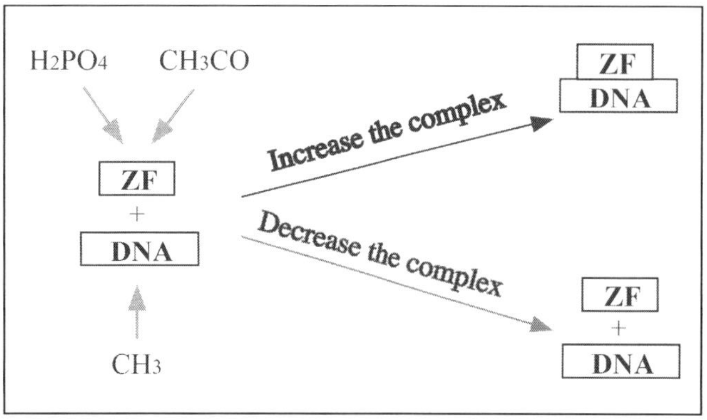

Figure 5. Modification of zinc finger proteins and the target DNA.

Regulation of Zinc Finger Binding to DNA

In vitro zinc finger-DNA binding proceeds as long as both substances are present in a suitable buffer, but the in vivo binding does not happen automatically and proceeds only under certain conditions. The regulation is mediated by modification of either the zinc finger protein itself or the target DNA (Fig. 5). The former can be executed by acetylation and also by phosphorylation, whereas the latter is mediated by methylation. All these modifications increase or decrease the zinc finger protein-DNA complex formation. The direction in which the regulation proceeds, repression or activation, depends on the fingers and the DNA, as well as the regulatory signal involved.

Acetylation of Zinc Fingers

YY1 (Yin Yang 1), with four tandem C_2H_2 zinc fingers, binds to various genes and represses or activates their target gene expression.[65,66] PCAF (p300/CBP associated factor) acetylates YY1 at the zinc finger domain and inhibits DNA binding. PCAF also acetylates members of the KLF (Krüppel-like factor) family belonging to the triple C_2H_2 zinc finger protein class.[67,68] It acetylates KLF13 at two lysines of the TGEKK linker between finger 2 and 3 and results in stimulating the fingers to bind to DNA. Another coactivation factor CBP/p300 acetylates a total of six lysines including the linker's two lysines, but it prevents the fingers from binding to DNA. Of the six residues, the lysine of finger 1 directly contacts bases. However, acetylation of this lysine is not enough to interrupt the DNA binding and acetylation of all six lysines is required. Moreover in vitro experiments showed that KLF2 zinc finger cotransfected with PCAF increases the target γ globulin gene expression in vivo while the zinc finger cotransfected with p300 decreases gene expression.[68] Therefore, it can be concluded that acetylation of KLF2 regulates the target gene expression both positively and negatively. Because lysine residues are abundant in a variety of zinc fingers, acetylation of C_2H_2 zinc fingers is likely to be a common mechanism to regulate gene expression by modulating the zinc fingers' DNA binding activity.

Another domain of YY1, the central glycine-lysine-rich domain, is acetylated by p300 and PCAF. This acetylation has nothing to do with the DNA binding but is required for fully repressing the target gene transcription.[69]

Phosphorylation of Zinc Fingers

During mitosis, Ikaros is phosphorylated on threoine/serine residues of the three linkers and interrupted for its DNA binding.[70] Phosphoryaltion of linkers also occurs to Sp1. This suggests that phosphorylation is a global inhibitory mechanism to keep C_2H_2 zinc fingers out of DNA during mitosis. Interestingly, phosphorylation modifies the linkers as the favored sites. A few interpretations can be given. First, modification of the conserved linker is the most effective way to modulate zinc finger binding activity because the linker is essential for the high affinity zinc finger-DNA binding. Second, the modification sites, theonines/ serines, are always available in the conserved linkers. Finally, linkers may be accessible even in the finger-DNA complexes so that zinc finger peptides can be separated from the DNA as soon as the regulatory signal is delivered to the fingers. The same arguments can be applied to regulation by acetylation since it also modifies linkers.

A serine residue of the C terminus immediately after a zinc finger can be phosphorylated in the case of Cre1 (catabolite repression) of *Hypocrea jecorina*, perhaps by caseinkinasc-II-likc enzyme.[71] Surprisingly, this phosphorylation increases the finger's DNA biding activity, suggesting that the unphosphorylated serine residue may mask the finger's DNA binding activity directly by interacting with the finger or indirectly by interacting with other intramolecular domains.

Methylation of DNA

Engineered zinc fingers have been shown to distinguish methylated DNA duplex from unmethylated.[72] It is becoming increasingly clear that natural zinc fingers differentially recognize methylated and unmethylated CpG (cytidine-guanidine dinucleotide pairs) and play important regulatory roles in the expression of target genes in vivo. CTCF (CCCTC-binding factor) has eleven C_2H_2 zinc fingers and binds to an element present between a promoter and an enhancer in order to block the enhancer activity.[73] The CTCF binding sites are widely distributed in vertebrates and several similar CpG-rich sequences are present in ICR (imprinted-control region) of the *ifg2/H19* as well. Electrophoretic mobility shift experiments have demonstrated that CTCF binds to the CpG-rich elements of ICR but only when these are unmethylated.[74] Although more experiments have to be done to obtain the CTCF function in vivo, the zinc finger proteins are likely to participate in the genomic imprinting process by differentially binding to the methylated- and unmethylated CpG-rich elements.

Kaiso, with triple C_2H_2 zinc fingers, is another example of this type of DNA binding. However, in this case, the zinc fingers recognize the methylated CpG-rich sequence but not the unmethylated sequence.[75,76] In vivo transfection experiments have shown that kaiso represses expression of genes with methylated binding sites in the promoter.[75]

Acknowledgements

I thank Dr. R. Freeman for information on the Pfam database.

References

1. Klug A, Schwabe JW. Protein motifs 5. Zinc fingers. FASEB J 1995; 9(8):597-604.
2. Pabo CO, Peisach E, Grant RA. Design and selection of novel Cys2His2 zinc finger proteins. Annu Rev Biochem 2001; 70:313-340.
3. Laity JH, Lee BM, Wright PE. Zinc finger proteins: new insights into structural and functional diversity. Curr Opin Struct Biol 2001; 11(1):39-46.
4. Miller J, McLachlan AD, Klug A. Repetitive zinc-binding domains in the protein transcription factor IIIA from Xenopus oocytes. EMBO J 1985; 4(6):1609-1614.
5. Brown RS, Sander C, Argos P. The primary structure of transcription factor TFIIIA has 12 consecutive repeats. FEBS Lett 1985; 186(2):271-274.
6. Diakun GP, Fairall L, Klug A. EXAFS study of the zinc-binding sites in the protein transcription factor IIIA. Nature 1986; 324(6098):698-699.
7. Lander ES, Linton LM, Birren B et al. Initial sequencing and analysis of the human genome. Nature 2001; 409(6822):860-921.
8. Hanas JS, Hazuda DJ, Bogenhagen DF et al. Xenopus transcription factor A requires zinc for binding to the 5 S RNA gene. J Biol Chem 1983; 258(23):14120-14125.
9. Michael SF, Kilfoil VJ, Schmidt MH et al. Metal binding and folding properties of a minimalist Cys2His2 zinc finger peptide. Proc Natl Acad Sci USA 1992; 89(11):4796-4800.
10. Cox EH, McLendon GL. Zinc-dependent protein folding. Curr Opin Chem Biol 2000; 4(2):162-165.
11. Lachenmann MJ, Ladbury JE, Phillips NB et al. The hidden thermodynamics of a zinc finger. J Mol Biol 2002; 316(4):969-989.
12. Miura T, Satoh T, Takeuchi H. Role of metal-ligand coordination in the folding pathway of zinc finger peptides. Biochim Biophys Acta 1998; 1384(1):171-179.
13. Iuchi S. Three classes of C2H2 zinc finger proteins. Cell Mol Life Sci 2001; 58(4):625-635.
14. Omichinski JG, Pedone PV, Felsenfeld G et al. The solution structure of a specific GAGA factor-DNA complex reveals a modular binding mode [see comments]. Nat Struct Biol 1997; 4(2):122-132.
15. Nagaoka M, Kaji T, Imanishi M et al. Multiconnection of identical zinc finger: implication for DNA binding affinity and unit modulation of the three zinc finger domain. Biochemistry 2001; 40(9):2932-2941.
16. Liao XB, Clemens KR, Tennant L et al. Specific interaction of the first three zinc fingers of TFIIIA with the internal control region of the Xenopus 5 S RNA gene. J Mol Biol 1992; 223(4):857-871.
17. Wuttke DS, Foster MP, Case DA et al. Solution structure of the first three zinc fingers of TFIIIA bound to the cognate DNA sequence: determinants of affinity and sequence specificity. J Mol Biol 1997; 273(1):183-206.
18. Nolte RT, Conlin RM, Harrison SC et al. Differing roles for zinc fingers in DNA recognition: structure of a six-finger transcription factor IIIA complex. Proc Natl Acad Sci USA 1998; 95(6):2938-2943.
19. Clemens KR, Zhang P, Liao X et al. Relative contributions of the zinc fingers of transcription factor IIIA to the energetics of DNA binding. J Mol Biol 1994; 244(1):23-35.
20. Fairall L, Rhodes D, Klug A. Mapping of the sites of protection on a 5 S RNA gene by the Xenopus transcription factor IIIA. A model for the interaction. J Mol Biol 1986; 192(3):577-591.
21. Hoffmann A, Ciani E, Boeckardt J et al. Transcriptional activities of the zinc finger protein Zac are differentially controlled by DNA binding. Mol Cell Biol 2003; 23(3):988-1003.
22. Fan CM, Maniatis T. A DNA-binding protein containing two widely separated zinc finger motifs that recognize the same DNA sequence. Genes Dev 1990; 4(1):29-42.
23. Sun L, Liu A, Georgopoulos K. Zinc finger-mediated protein interactions modulate Ikaros activity, a molecular control of lymphocyte development. EMBO J 1996; 15(19):5358-5369.
24. Mackay JP, Crossley M. Zinc fingers are sticking together. Trends Biochem Sci 1998; 23(1):1-4.
25. Iuchi S, Lin EC. Adaptation of Escherichia coli to respiratory conditions: regulation of gene expression. Cell 1991; 66(1):5-7.
26. Hsu T, Gogos JA, Kirsh SA et al. Multiple zinc finger forms resulting from developmentally regulated alternative splicing of a transcription factor gene. Science 1992; 257(5078):1946-1950.
27. Gogos JA, Hsu T, Bolton J et al. Sequence discrimination by alternatively spliced isoforms of a DNA binding zinc finger domain. Science 1992; 257(5078):1951-1955.
28. Greisman HA, Pabo CO. A general strategy for selecting high-affinity zinc finger proteins for diverse DNA target sites. Science 1997; 275(5300):657-661.

29. Hamilton TB, Borel F, Romaniuk PJ. Comparison of the DNA binding characteristics of the related zinc finger proteins WT1 and EGR1. Biochemistry 1998; 37(7):2051-2058.
30. Liggins JR, Privalov PL. Energetics of the specific binding interaction of the first three zinc fingers of the transcription factor TFIIIA with its cognate DNA sequence. Proteins 2000; Suppl 4:50-62.
31. Houbaviy HB, Burley SK. Thermodynamic analysis of the interaction between YY1 and the AAV P5 promoter initiator element. Chem Biol 2001; 8(2):179-187.
32. Matsushita K, Sugiura Y. Effect of arginine mutation of alanine-556 on DNA recognition of zinc finger protein Sp1. Bioorg Med Chem 2001; 9(9):2259-2267.
33. Miller JC, Pabo CO. Rearrangement of side-chains in a Zif268 mutant highlights the complexities of zinc finger-DNA recognition. J Mol Biol 2001; 313(2):309-315.
34. Harlow E, Lane D. Antibodies. New York: Cold Spring Harbor Labortory; 1988.
35. Saenger W. Principles of nucleic acid structure. New York: Springer-Verlag; 1983.
36. Pavletich NP, Pabo CO. Zinc finger-DNA recognition: crystal structure of a Zif268-DNA complex at 2.1 A. Science 1991; 252(5007):809-817.
37. Wolfe SA, Nekludova L, Pabo CO. DNA recognition by Cys2His2 zinc finger proteins. Annu Rev Biophys Biomol Struct 2000; 29:183-212.
38. Pavletich NP, Pabo CO. Crystal structure of a five-finger GLI-DNA complex: new perspectives on zinc fingers. Science 1993; 261(5129):1701-1707.
39. Fairall L, Schwabe JW, Chapman L et al. The crystal structure of a two zinc-finger peptide reveals an extension to the rules for zinc-finger/DNA recognition. Nature 1993; 366(6454):483-487.
40. Yokono M, Saegusa N, Matsushita K et al. Unique DNA binding mode of the N-terminal zinc finger of transcription factor Sp1. Biochemistry 1998; 37(19):6824-6832.
41. Laity JH, Dyson HJ, Wright PE. Molecular basis for modulation of biological function by alternate splicing of the Wilms' tumor suppressor protein. Proc Natl Acad Sci USA 2000; 97(22):11932-11935.
42. Torrungruang K, Alvarez M, Shah R et al. DNA binding and gene activation properties of the Nmp4 nuclear matrix transcription factors. J Biol Chem 2002; 277(18):16153-16159.
43. Benos PV, Lapedes AS, Stormo GD. Probabilistic code for DNA recognition by proteins of the EGR family. J Mol Biol 2002; 323(4):701-727.
44. Suzuki M, Brenner SE, Gerstein M et al. DNA recognition code of transcription factors. Protein Eng 1995; 8(4):319-328.
45. Choo Y, Klug A. Designing DNA-binding proteins on the surface of filamentous phage. Curr. Opin Biotechnol 1995; 6(4):431-436.
46. Wolfe SA, Grant RA, Elrod-Erickson M et al. Beyond the "recognition code": structures of two Cys2His2 zinc finger/TATA box complexes. Structure (Camb) 2001; 9(8):717-723.
47. Gogos JA, Jin J, Wan H et al. Recognition of diverse sequences by class I zinc fingers: asymmetries and indirect effects on specificity in the interaction between CF2II and A+T-rich elements. Proc Natl Acad Sci USA 1996; 93(5):2159-2164.
48. Tsui V, Radhakrishnan I, Wright PE et al. NMR and molecular dynamics studies of the hydration of a zinc finger-DNA complex. J Mol Biol 2000; 302(5):1101-1117.
49. Choo Y, Isalan M. Advances in zinc finger engineering. Curr Opin Struct Biol 2000; 10(4):411-416.
50. Beerli RR, Barbas CF, 3rd. Engineering polydactyl zinc-finger transcription factors. Nat Biotechnol 2002; 20(2):135-141.
51. Choo Y, Klug A. Toward a code for the interactions of zinc fingers with DNA: selection of randomized fingers displayed on phage [published erratum appears in Proc Natl Acad Sci USA 1995; 92(2):646]. Proc Natl Acad Sci USA 1994; 91(23):11163-11167.
52. Rebar EJ, Pabo CO. Zinc finger phage: affinity selection of fingers with new DNA-binding specificities. Science 1994; 263(5147):671-673.
53. Isalan M, Choo Y. Engineering nucleic acid-binding proteins by phage display. Methods Mol Biol 2001; 148:417-429.
54. Wilson TE, Day ML, Pexton T et al. In vivo mutational analysis of the NGFI-A zinc fingers. J Biol Chem 1992; 267(6):3718-3724.
55. Choo Y, Klug A. A role in DNA binding for the linker sequences of the first three zinc fingers of TFIIIA. Nucleic Acids Res 1993; 21(15):3341-3346.
56. Elrod-Erickson M, Rould MA, Nekludova L et al. Zif268 protein-DNA complex refined at 1.6 A: a model system for understanding zinc finger-DNA interactions. Structure 1996; 4(10):1171-1180.
57. Laity JH, Dyson HJ, Wright PE. DNA-induced alpha-helix capping in conserved linker sequences is a determinant of binding affinity in Cys(2)-His(2) zinc fingers. J Mol Biol 2000; 295(4):719-727.
58. Foster MP, Wuttke DS, Radhakrishnan I et al. Domain packing and dynamics in the DNA complex of the N-terminal zinc fingers of TFIIIA. Nat Struct Biol 1997; 4(8):605-608.
59. Wu LC. ZAS: C2H2 zinc finger proteins involved in growth and development. Gene Expr 2002; 10(4):137-152.
60. Tseng H, Green H. Basonuclin: A keratinocyte protein with multiple paired zinc fingers. Proc Natl Acad Sci USA 1992; 89(21):10311-10315.
61. Ptashne M. A genetic switch: Gene control and phage λ. Cambridge: Cell Press & Blackwell Scientific Publications; 1986.
62. McCarty AS, Kleiger G, Eisenberg D et al. Selective dimerization of a C2H2 zinc finger subfamily. Mol Cell 2003; 11(2):459-470.
63. Wang BS, Grant RA, Pabo CO. Selected peptide extension contacts hydrophobic patch on neighboring zinc finger and mediates dimerization on DNA. Nat Struct Biol 2001; 8(7):589-593.
64. Tsai RY, Reed RR. Identification of DNA recognition sequences and protein interaction domains of the multiple-Zn-finger protein Roaz. Mol Cell Biol 1998; 18(11):6447-6456.
65. Shi Y, Lee JS, Galvin KM. Everything you have ever wanted to know about Yin Yang 1. Biochim Biophys Acta 1997; 1332(2):F49-66.
66. Thomas MJ, Seto E. Unlocking the mechanisms of transcription factor YY1: are chromatin modifying enzymes the key? Gene 1999; 236(2):197-208.
67. Song CZ, Keller K, Chen Y et al. Functional interplay between CBP and PCAF in acetylation and regulation of transcription factor KLF13 activity. J Mol Biol 2003; 329(2):207-215.
68. Song CZ, Keller K, Murata K et al. Functional interaction between coactivators CBP/p300, PCAF, and transcription factor FKLF2. J Biol Chem 2002; 277(9):7029-7036.
69. Yao YL, Yang WM, Seto E. Regulation of transcription factor YY1 by acetylation and deacetylation. Mol Cell Biol 2001; 21(17):5979-5991.
70. Dovat S, Ronni T, Russell D et al. A common mechanism for mitotic inactivation of C2H2 zinc finger DNA-binding domains. Genes Dev 2002; 16(23):2985-2990.
71. Cziferszky A, Mach RL, Kubicek CP. Phosphorylation positively regulates DNA binding of the carbon catabolite repressor Cre1 of Hypocrea jecorina (Trichoderma reesei). J Biol Chem 2002; 277(17):14688-14694.
72. Choo Y. Recognition of DNA methylation by zinc fingers. Nat Struct Biol 1998; 5(4):264-265.
73. Bell AC, West AG, Felsenfeld G. The protein CTCF is required for the enhancer blocking activity of vertebrate insulators. Cell 1999; 98(3):387-396.
74. Hark AT, Schoenherr CJ, Katz DJ et al. CTCF mediates methylation-sensitive enhancer-blocking activity at the H19/Igf2 locus. Nature 2000; 405(6785):486-489.
75. Prokhortchouk A, Hendrich B, Jorgensen H et al. The p120 catenin partner Kaiso is a DNA methylation-dependent transcriptional repressor. Genes Dev 2001; 15(13):1613-1618.
76. Daniel JM, Spring CM, Crawford HC et al. The p120(ctn)-binding partner Kaiso is a bi-modal DNA-binding protein that recognizes both a sequence-specific consensus and methylated CpG dinucleotides. Nucleic Acids Res 2002; 30(13):2911-2919.

CHAPTER 3

TFIIIA: A Sophisticated Zinc Finger Protein

Raymond S. Brown* and Jane Flint

Abstract

Transcription factor IIIA (TFIIIA) is widely regarded as the archetypal zinc finger protein. It is a member of a very large multigene family of eukaryotic DNA-binding proteins. More than two decades of research have been dedicated to understanding its interaction with the 5S ribosomal RNA gene (5S DNA). TFIIIA has nine tandem C_2H_2 zinc fingers along the peptide sequence. The three-dimensional structure of the N-terminal 6 zinc fingers bound to 31 base pairs of DNA shows that not all fingers are equal. Four of them make contacts located on both DNA strands while two fingers act as spacers. Individual fingers can recognize overlapping and interlocking base pair quartets. Side chains in the short alpha helices of fingers contact bases in the major groove. It is likely that the linker sequences connecting adjacent zinc fingers evolved to dictate which fingers bind to DNA. Signals essential for transcription initiation of 5S DNA, nuclear localization, and nucleocytoplasmic transport of 5S ribosomal RNA (5S rRNA) are located in the C-terminal part of TFIIIA.

Introduction

An RNA-binding protein accumulates in the oocytes of the ovaries of the African clawed frog (*Xenopus laevis*). This protein has a molecular mass of 40 kilodaltons (kD). About 10^{10} molecules of the protein are estimated to be present per oocyte and using fluorescent antibodies it can be seen that this protein is localized mainly outside of the nucleus.[1] Its partner in the cytoplasm was identified as 5S rRNA. About half of the 5S rRNA in the oocyte is associated with the 40 kD protein in a stable complex called the 7S storage particle (7S RNP).[2] Similar 7S RNPs are also present in the oocytes of other amphibians, teleost fish and maize embryos.[3]

In 1980 a transcription factor that is essential for RNA polymerase III (pol III) to transcribe *Xenopus* oocyte 5S rRNA genes was isolated. It was shown to be identical to the 40 kD RNA-binding protein. The protein isolated from the 7S RNP binds specifically to radioactively labeled 5S DNA.[4,5] This transcription factor was named TFIIIA. In fact this was the first eukaryotic transcription factor to be purified and used to transcribe an isolated gene. Surprisingly the promoter region that TFIIIA recognizes was found to be located inside the gene itself. Even more impressive was the discovery that a single TFIIIA protein binds without interruption to a long sequence of about 55 base pairs. This region of the 5S DNA was named the internal control region (ICR).

Transcription of 5S DNA

Expression of a fluorescent green protein/TFIIIA fusion protein shows that TFIIIA is localized as expected in the nucleolus[6] where rRNA genes are transcribed and ribosomes are assembled. Two other pol III transcription factors are also required for initiation of transcription of 5S DNA, namely TFIIIB and TFIIIC. In contrast to TFIIIA these factors contain a number of subunits. TFIIIB, which binds upstream of TFIIIA to 5S DNA, consists of three subunits, the TATA box-binding protein TBP, a single zinc finger protein Brf and a third subunit B. TFIIIC contains at least six subunits, some of which bind to sequences essential for transcription within the ICR (for a review see ref. 7).

5S rRNA is the product of transcription from 5S DNA. It is a component of the ribosome 60S subunit. The 120 nucleotide sequence of 5S rRNA is highly conserved among vertebrates. There are two major types of 5S rRNA genes in eukaryotes. *Xenopus* oocyte type (Xlo) and the somatic type (Xls) genes have sequences which differ at eight nucleotide positions between them.[8] Six of these sequence positions lie within the ICR region. The differences in sequence do not appear to significantly affect TFIIIA binding to 5S DNA in vitro.

TFIIIA Structure and Function

Internal Domains

In addition to the oocyte form of TFIIIA (oTFIIIA) there is a larger somatic protein, sTFIIIA, which has 22 additional amino acids at the N-terminus originating from translation at an upstream AUG codon.[9,10] The existence of sTFIIIA has been confirmed by precipitation with antibodies. Cellular levels of the two different forms of TFIIIA are probably related to switching from the production of oocyte 5S rRNA entirely to that of somatic 5S rRNA in the developing embryo.

The internal organization of oTFIIIA has been probed by limited digestion of 7S RNP with proteases such as elastase, trypsin or papain.[11] These enzymes can readily cut in exposed regions of a protein that would be expected to lie between the relatively resistant folded domains. In each case, proteolytic fragments are released from the C-terminus of the 40 kD protein in a step-wise fashion to give a 35 kD product first and then a 30 kD core fragment. The same oTFIIIA digestion pattern is obtained from either the 7S RNP or a TFIIIA-5S DNA complex. The 30 kD core

* Corresponding author. See list of "Contributors".

Zinc Finger Proteins: From Atomic Contact to Cellular Function, edited by Shiro Iuchi and Natalie Kuldell.

fragment remains stably bound to 5S rRNA or DNA. This property suggests that the structure of TFIIIA is organized in three domains that are revealed by protease digestion. The exact significance of these results with regard to the three-dimensional structure of intact TFIIIA remains unknown. After extensive digestion of the 7S RNP with chymotrypsin a ladder of small fragments which differ by about 3 kD are produced.[12] This curious result can be explained in terms of the sequence analysis described below.

Zinc Fingers

Addition of a metal chelating agent lowers the DNA-binding activity of TFIIIA. However, the protein function can be restored by addition of zinc.[13,14] The presence of integral bound zinc in the 7S RNP was shown by extended X-ray absorption fine structure (EXAFS).[15] Atomic absorption spectroscopy of the 7S RNP[12] and chemical analysis of purified oTFIIIA[16] established that there are about 9-11 moles of zinc in the protein. At the time, it was very surprising to find that a transcription factor with no enzymatic activity was a metalloprotein containing multiple zinc ions.

The amino acid sequence of the 344 residue *X. laevis* oTFIIIA was published in 1984.[17] Its organization, 9 alternating pairs of cysteines and pairs of histidines always separated by 12 residues, was unique. Such a repeating pattern can be seen particularly well in a DotPlot analysis,[18] in which a TFIIIA sequence is compared with itself. The nine internal repeats are revealed as regularly spaced, parallel, diagonal lines of scored points (see Fig. 1 in ref. 18). It is striking that repeat 6 has the lowest similarity to the others. The nine sequence repeat segments can be aligned using the invariant cysteines and histidines to reveal a consensus 30-34 amino acid motif.[12,18]

It was correctly proposed that consecutive pairs of cysteines and of histidines most likely combine together as four ligands binding a zinc ion.[12] Similar arrangements of cysteine and histidine ligands are common in metalloprotein enzymes. This Cys_2His_2 motif, often described as C_2H_2, has now come to be known as the TFIIIIA-like zinc finger. In contrast C_2H_2 zinc finger motifs connected with a highly conserved five residue H/C linker sequence (TGEKP) are often called Kruppel-like zinc fingers.[19] The nine zinc fingers in tandem that occupy nearly all of the N-terminal part of TFIIIA (residues 13-276) are required for transcriptional activity, together with the adjacent C-terminal region (residues 277-344).

C-Terminal Signals

The C-terminal part of oTFIIIA contains several amino acid sequences that are required for the different functions of the protein. A lysine-rich nuclear localization signal (NLS), EKRKLKEK (residues 274-286) is situated after finger 9. The presence of an NLS allows TFIIIA to migrate to the nucleus where it acts as a transcription factor. Domain swapping has demonstrated the presence of a functional leucine-rich nuclear export signal (NES), SLVLDKLTIQ (residues 335-344) close to the C-terminus of *Xenopus* oTFIIIA.[20] These signals suggest that TFIIIA might be a nuclear shuttle protein. However, although TFIIIA is certainly involved in the nuclear export of 5S rRNA, reentry of 5S rRNA into the nucleus takes place via a 5S RNP with ribosomal protein L5.[21] A further 15 amino acid sequence, KRSLASRLTGYIP (residues 291-304), has been identified as the transcription activating signal.[22] Deletion of the latter sequence inhibits transcription initiation.

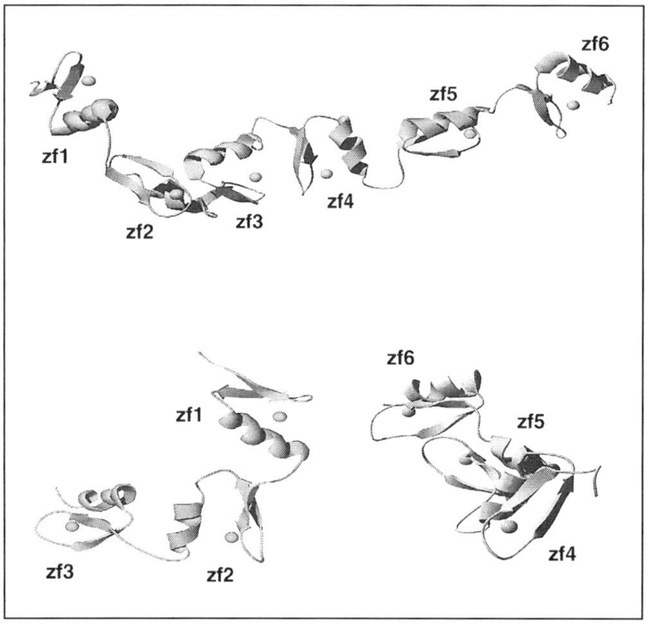

Figure 1. N-terminal zinc fingers 1-6 of TFIIIA. The crystal structure of a fragment of *X. laevis* oTFIIIA consisting of residues 10-188 is shown above.[48] Individual zinc fingers are numbered from zf1 to zf6. The secondary structure of the protein, showing the short alpha helices and beta strands, is represented as a ribbon and colored green. Zinc atoms in the fingers are represented as small spheres in violet. An endwise view which shows the right-handed helical conformation adopted by fingers 1-3 (residues 10-99) is seen on the lower left. In contrast fingers 4-6 (residues 100-188) have a different, irregular structure that is shown on the lower right. This figure was made with Swiss-PdbViewer and Pov-Ray™.[62,63]

Comparison of TFIIIA Sequences

The genomic DNA of *X. laevis* TFIIIA contains nine exons.[23,24] Some of these encode single zinc fingers suggesting that the protein was created through gene duplication. TFIIIA sequences containing nine zinc fingers are presently known for several amphibians; *Xenopus laevis*,[17] *Xenopus borealis*,[25] *Rana catesbeiana*,[25] *Bufo americanus* and *Rana pipiens*.[26] TFIIIA sequences have also been reported for channel catfish[27] and some mammmals; mouse,[28] rat[28] and human.[29,30] Bakers' yeast TFIIIA[31] has a long sequence inserted between fingers 8/9 and the fission yeast protein[32] has 10 zinc fingers with a long insert between fingers, 9/10. A plant sequence from *Aradopsis thaliana*[33] also has nine zinc fingers, but has long inserts between fingers 1/2 and between 4/5. Proteins with TFIIIA-like activity have also been identified in other plants, such as, tulip[34] and maize,[35] and in the amoeboid protozoon *Acanthamoeba castellanii*.[36]

Although TFIIIA is perhaps the best known example of a C_2H_2 zinc finger protein, it is actually quite atypical. Sequences from different species contain additional N-terminal and C-terminal extensions of varying lengths. There is little internal sequence similarity between any of the nine zinc fingers. The number of intervening residues found in a cysteine pair varies from 2 to 5, and from 3 to 5 for a histidine pair, in contrast to the more prevalent Cx_2C and Hx_3H that is typical of Kruppel-type zinc finger proteins. Sequence divergence within the protein has led to the proposal that TFIIIA may have evolved two differently adapted sets of zinc fingers, 1-3 that recognize DNA, and 4-7 that bind to 5S rRNA.[37,38]

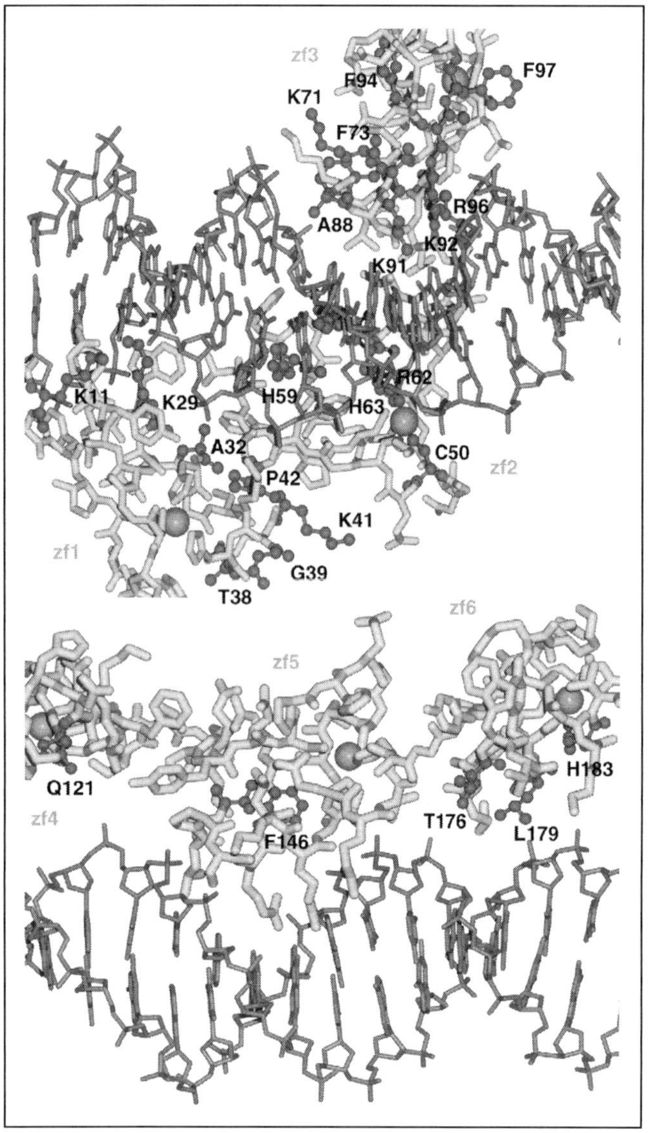

Figure 2. Sites of amino acid mutations in *Xenopus* oTFIIIA that lower DNA affinity. TFIIIA zinc fingers, zf1-zf3 (upper) and zf4-zf6 (lower), are colored green and shown binding to 5S DNA, +92 to +78 (upper) and +77 to +64 (lower), in blue.[48] Amino acids associated with loss of function by mutation are labeled and appear as ball-and-stick models in red.[41,55-59] In finger 1 these are; K11A (N-terminus), K29E and A32V (alpha helix), T38L, G39P, K41S and P42G (1/2 linker). Changes in finger 2 are; H59R, R62L, R62Q (alpha helix), H63L and C50Y (zinc ligands). In finger 3 DNA binding is decreased by point mutations; K71E (linker 2/3), F73L (beta strand), A88K, K91S, K92A, K92L, F94K, R96Y and F97L (alpha helix). Sites in finger 4 and 5 are Q121R (alpha helix) and F146S (beta strand) respectively. Mutations in finger 6 are T176I, L179S (alpha helix) and H183Y (zinc ligand). This figure was produced with RasMol and Pov-Ray™.[64, 63] Online color version at http://www.eurekah.com/chapter.php?chapid=1547&bookid=124&catid=30

TFIIIA Interaction with 5S DNA

DNA Footprint

The binding of TFIIIA to 5S DNA has been studied extensively with genetic, enzymatic and chemical probing techniques (for a review see ref. 39). Such approaches cannot provide the details of the atomic contacts between amino acid side chains of TFIIIA and individual base pairs in the target DNA. Nevertheless, they have established that the ICR contains three sequences essential for transcription of 5S DNA. These are called the box A sequence (+50 to +60), the intermediate element (+69 to +71) and the box C sequence (+80 to +89) (see Fig. 2).[40]

Protection of DNA by bound proteins can be probed with chemicals that methylate purine bases or ethylate phosphates and with hydroxyl radical reagents that cleave exposed regions of the phosphodiester chain. Accessibility can also be tested by digestion with DNase I, II and micrococcal nuclease. In the case of *Xenopus* TFIIIA, the DNase I footprint spans positions +42 to +96 (54 base pairs) of the plus (noncoding) strand of somatic 5S DNA. A similar span of the minus (coding strand) is also protected. This region corresponds closely to the ICR. A similar footprint from +42 to +96 can also be produced by binding of a truncated version of TFIIIA (fingers 1-9) synthesized in and purified from *Escherichia coli*.[41]

Distinct Regions within the TFIIIA Footprint

A prominent feature within the intact TFIIIA DNase I footprint is an accessible point at positions +62/+64 which divides the footprint into two distinct regions. This bipartite organization of the DNA binding site is mirrored in the protease digestion of the zinc finger domain into two parts. A smaller footprint extending from +62 to +92 can be produced by binding either the trypsin core fragment of oTFIIIA,[8,11] or a C-terminal truncated fragment containing zinc fingers 1-7 (residues 1-223).[41] From these data it is possible to conclude three things. First, the nine zinc fingers comprise the DNA-binding domain of TFIIIA. Second, the amino-to-carboxy terminal orientation of fingers 1-9 is in the 3' to 5' direction when bound to the noncoding strand. Third, the interaction is not completely uniform along the length of the protected 5S DNA region.

TFIIIA Fingers 1 to 7 and 5S DNA

An exonuclease approaches a DNA-protein complex from the direction of the flanking sequence until it reaches the binding site. At this point the exonuclease becomes obstructed by the protein and digestion will stop. Therefore the exonuclease III protection assay[42] complements the footprinting results obtained with DNase I, which may cut freely at any accessible point within the DNA sequence. We have used *X. laevis* oTFIIIA and Xlo 5S DNA to obtain detailed information about DNA recognition in a homologous system.

The trypsin core fragment was identified by standard N-terminal and C-terminal sequencing as comprising residues 5-218. This sequence contains zinc fingers 1-7. DNase I footprint results for the core fragment show that the protection extends from positions +61 to +94 on the coding strand and +65 to +90 on the noncoding strand of Xlo 5S DNA. The same increase in DNase I cutting at noncoding strand positions +62, +75, +90 and hypersensitivity at +93 is seen with both intact oTFIIIA and the 1-7 core fragment (unpublished data).

Digestion of both complexes was performed with exonuclease III/T4 DNA polymerase[43] under conditions where the unbound 5' end-labeled coding or noncoding DNA strands were completely degraded. The limits on the noncoding strand of Xlo 5S DNA were the same for both oTFIIIA and the 1-7 core protein with inhibition starting at +93/+94 to +89. On the coding strand pauses were found within the +44/+49 region for oTFIIIA and at +62/+63 for the core protein respectively (unpublished data). These observations indicate that the limits of protected 5S DNA are +62/+63 to +89 for fingers 1-7 whereas intact TFIIIA influences a region extending from +44 to +94.

DNA Interaction of C-Terminal Fingers

The binding site of the C-terminal part of TFIIIA can be deduced by subtracting the footprints of a series of C-terminal truncated mutants from each other.[41] By comparing the DNase I footprint of TFIIIA with those of fragments 1-9 and 1-8, zinc fingers 8-9 can be assigned to the box A sequence, base pairs +40 to +61 of the ICR.[44] The existence of direct contacts with base pairs is suggested by the TFIIIA-induced protection from methylation of guanine bases G53, G57 (coding strand) and G51, G59, G60 (noncoding strand).[45] The methylation of G51, G52 and G57 also interferes with TFIIIA binding to DNA.[46] Similarly, protection from hydroxyl radical cleavage occurs between noncoding positions +53 to +57.[47] This is further supported by resistance to exonuclease III digestion on coding strand +44 to +62. All these results are consistent with a hypothetical model in which zinc fingers 7-9 are in contact with certain guanine bases in the major groove, perhaps binding in a similar way to fingers 1-3.[46]

X-Ray Structure Determination

Crystallization

Making large amounts of double-stranded DNA for crystallization is not to be undertaken by the faint-hearted. Although it is technically possible to chemically synthesize a set of oligonucleotides (55-mers) required for crystallization screens, their low yields make it prohibitively costly. For this reason crystals were made using smaller fragments of *X. laevis* oTFIIIA synthesized in and refolded from *E. coli*. These fragments consisted of fingers 1-7 (residues 4-217), fingers 1-6 (residues 1-190) and fingers 7-9+C terminus (residues 191-344). Crystallization was successful with three reconstituted complexes containing 5S DNA duplexes with the sequences +62 to +90, +63 to +92[48] and +46 to +65 respectively (unpublished data).

Structure of Fingers 1-6 with DNA

An X-ray structure was published in 1998 of the TFIIIA N-terminal fingers 1-6 in a complex with a DNA duplex (31-mer, +63 to +92).[48] In complete agreement with the biochemical data, the amino-to-carboxy terminal orientation of fingers 1 to 6 is along the noncoding strand in the reverse direction (3' to 5'). As expected the six zinc fingers are all folded in a similar way around a zinc ion; they form a compact metal-linked beta-hairpin + alpha-helix structure (Fig. 1). Each zinc finger contains a short alpha helix about 12 residues in length. The variation in the number of residues between the pair of cysteines or the pair of histidines has little effect on the folding of individual zinc fingers.

Different Binding Modes

A novel feature of this three-dimensional structure is that the six zinc fingers are used in different ways to make an extended interaction with 31 base pairs of B-form double helical DNA. Zinc fingers 1-3 wrap around the DNA in a helical side-by-side fashion. This arrangement is similar to that seen in a nuclear magnetic resonance (NMR) structure of TFIIIA N-terminal fingers 1-3 (residues 11-101) binding to a DNA duplex (15-mer, +79 to +93).[49] Instead of continuing to wind around the double helix, the remaining fingers 4-6 have a different, extended configuration that runs along the DNA. Fingers 1, 2, 3 and 5 bind in the major groove in the classical way that is found in the crystal structures of other zinc finger proteins. In contrast, fingers 4 and 6 are positioned above the minor groove and act as spacers between their neighboring DNA-binding fingers. The two nonbinding fingers allow the protein to change direction and cross the minor groove twice. This results in a customized fit in which fingers 1-3 bind to the box C sequence and finger 5 binds to the intermediate element of the ICR.

In an X-ray structure at a modest resolution of 3.1 angstroms it is not possible to be absolutely certain of the protein-DNA interactions. However, consideration of atom positions in the model, the distances and angles between atoms, together with their chemical nature, identifies some if not all of the bonds. The DNA contacts span base pairs +68 to +93. Fingers 1-3 bind in the major groove to bases in the box C sequence (+81 to +91) xxAxGGGxxxx/xxTxCxCxxTG, in which only bases recognized by the fingers are specified, and finger 5 binds to bases in the intermediate element (+70 to +74) GGxxx/xxxxT. There are additional contacts to the 5'-phosphates of nucleotides 69, 79, 83, 84, 86, 87, 88 and to riboses 68, 81, 84 on the noncoding strand, and contacts to the 5'-phosphates of nucleotides 74, 86, 91, 92, 93 and riboses 75, 89, 92 on the coding strand.

Zinc Finger-DNA Recognition

As predicted correctly from sequence analysis, a zinc finger has a short DNA-binding alpha helix with amphipathic character.[50] Amino acid side chains in the alpha helices of TFIIIA fingers 1, 2, 3 and 5 make the majority of contacts to base pairs in the DNA major groove. In particular Arg62, Arg96, Arg151 and Arg154 contact guanines G85, G81, G71 and G70 respectively and Asn89 contacts adenine A83 (noncoding strand). Each amino acid makes two hydrogen bonds with a purine base, to the O6 and N7 atoms of guanine or the N6 and N7 atoms of adenine. There are also hydrophobic contacts between Leu148 and thymine T74, and Ala188 with cytidine C85 (coding strand).

Interactions made by TFIIIA fit well with the scheme suggested by Pabo based upon the Zif268-DNA complex where side chains of alpha helix positions -1, +3 and +6 make contacts with a triplet of bases on one strand of DNA in the 3'-5' direction.[51] These rules also apply to other zinc finger protein-DNA structures.[52] The pattern is more extended for TFIIIA DNA-binding fingers because the side chain of alpha helix position +2 often binds to an opposite base in the preceeding base pair on the coding strand. In addition, there are frequently contacts on the noncoding strand with the next two phosphates. Clearly, a single zinc finger recognizes a quartet of base pairs. The quartets can be overlapping and interlocking as in fingers 1-3 or separated as in the case of finger 5.

H/C Linker Structure

The H/C linkers that connect zinc fingers in *X. laevis* oTFIIIA have diverged from the typical Kruppel-type TGEKP sequence. The five linkers in fingers 1-6 (TGEKP, TGEKN, NIKICV, TQQLP and AG) vary in their lengths, sequences and three-dimensional structures. They form short loops that allow the individual fingers of TFIIIA to adopt the optimal DNA binding configuration. Consequently, no pair of adjoining zinc fingers is positioned in exactly the same way. However, linkers similar to the Kruppel-type sequence, TGEKP and TGEKN, still exist in TFIIIA and connect fingers 1/2 and 2/3. As fingers 1-3 wrap around the major groove of B-form DNA, their particular linkers may have evolved to support this kind of helical side-by-side packing.

The abrupt turn in the polypeptide chain before the linker is stabilized by hydrogen bonding interactions.[53] There is a H-bond between the main chain NH groups of Glu40 and Glu70 and Gln132 and the side chain gamma oxygens of threonines Thr38 and Thr86 and Thr130. Main chain H-bonds also exist between the carbonyl oxygen of the +5 residue of the preceding alpha helix and NH groups in the adjoining linker. As is the case in linker 1/2 where the carbonyl oxygen of Cys35 makes H-bonds to Gly39 and to Glu40 main chain NH groups. Similarly, in the 2/3 linker the Leu65 carbonyl oxygen makes H-bonds to Gly69 NH and to Glu70 NH groups. These interactions may stabilize the linker conformation so that a pair of fingers is correctly positioned for major groove DNA binding. The sequence variation in other TFIIIA linkers may have evolved to allow alternative spatial arrangements, or to dictate which of the fingers actually binds to RNA and DNA.

Mutagenesis and TFIIIA Activity

While there has been a greater effort concentrated in probing the 5S DNA, the activity of *Xenopus* TFIIIA has also been investigated by mutagenesis (see Fig. 2). The effects of amino acid substitutions and consequences of "finger swap" experiments upon DNA affinity have been investigated.[54] Replacement of linker sequences or mutations at the T, G, K or P positions within the 1/2 and 2/3 linkers cause a drastic decrease in DNA binding.[55,56] Substituting residue +6 in the alpha helix of finger 2 reduces the binding activity.[57] Mutations at positions +2, +5 or +6 in the alpha helix of finger 3,[58] at the +3 position of finger 4 or the -1 position of finger 6 are also detrimental to the binding activity.[41]

The exchange of finger 5 or finger 7 with their counterparts from another nine-zinc finger protein, p43, results in a decrease of TFIIIA affinity for DNA.[54] Many loss-of-function mutations in fingers 7-9, the C-terminus and zinc binding residues have been identified using a genetic assay in yeast cells.[59] The function of TFIIIA demands a highly extended and modular protein structure. Thus it is not surprising that mutations in the linkers, alpha helices and metal binding residues that may affect zinc finger structure also interfere with DNA-binding activity.

TFIIIA: A Sophisticated Protein

TFIIIA is a gene-specific transcription factor that has evolved with a number of very different cellular functions. It plays an important role in the formation of pol III transcription initiation complexes, the nuclear export and cytoplasmic storage of ribosomal 5S RNA in oocytes, and also interacts with other proteins.[60] Expression of the *Xenopus* TFIIIA gene is developmentally regulated, giving rise to short and longer forms of the protein specific to oocyte and somatic cells.

It is an unusual modular protein containing multiple zinc fingers that are highly adapted to recognize elements in a long promoter sequence, rather than bind to a short symmetrical DNA palindrome. Surprisingly it is not necessary for all of its fingers to exhibit DNA-binding activity although they are virtually indistinguishable in structure.[48] TFIIIA is a member of a group of fascinating proteins that "have it both ways" through a dual ability to interact with either double stranded DNA or folded RNA.[61]

References

1. Mattaj IW, Lienhard S, Zeller R et al. Nuclear exclusion of transcription factor IIIA and the 42S particle transfer RNA-binding protein in Xenopus oocytes: A possible mechanism for gene control? J Cell Biol 1983; 97(4):1261-1265.
2. Picard B, Wegnez M. Isolation of a 7S particle from Xenopus laevis oocytes: A 5S RNA-protein complex. Proc Natl Acad Sci USA 1979; 76(1):241-245.
3. Rincon-Guzman A, Beltran-Pena E, Ortiz-Lopez A et al. Ribonucleoprotein particles of quiescent maize embryonic axes. Plant Mol Biol 1998; 38(3):357-364.
4. Pelham HR, Brown DD. A specific transcription factor that can bind either the 5S RNA gene or 5S RNA. Proc Natl Acad Sci USA 1980; 77(7):4170-4174.
5. Honda BM, Roeder RG. Association of a 5S gene transcription factor with 5S RNA and altered levels of the factor during cell differentiation. Cell 1980; 22(1 Pt 1):119-126.
6. Mathieu O, Yukawa Y, Prieto J-L et al. Identification and characterization of transcription factor IIIA and ribosomal protein L5 from Arabidopsis thaliana. Nucleic Acids Res 2003; 31(9): 2424-2433.
7. Huang Y, Maraia RJ. Comparison of the RNA polymerase III transcription machinery in Schizosaccharomyces pombe, Saccharomyces cerevisiae and human. Nucleic Acids Res 2001; 29(13):2675-2690.
8. Xing YY, Worcel A. The C-terminal domain of transcription factor IIIA interacts differently with different 5S RNA genes. Mol Cell Biol 1989; 9(2):499-514.
9. Pelham HR, Wormington WM, Brown DD. Related 5S RNA transcription factors in Xenopus oocytes and somatic cells. Proc Natl Acad Sci USA 1981; 78(3):1760-1764.
10. Kim SH, Darby MK, Joho KE et al. The characterization of the TFIIIA synthesized in somatic cells of Xenopus laevis. Genes Dev 1990; 4(9):1602-1610.
11. Smith DR, Jackson IJ, Brown DD. Domains of the positive transcription factor specific for the Xenopus 5S RNA gene. Cell 1984; 37(2):645-652.
12. Miller J, McLachlan AD, Klug A. Repetitive zinc-binding domains in the protein transcription factor IIIA from Xenopus oocytes. EMBO J 1985; 4:1609-1614.
13. Hanas JS, Hazuda DJ, Bogenhagen DF et al. Xenopus transcription factor A requires zinc for binding to the 5S RNA gene. J Biol Chem 1983; 258(23):14120-14125.
14. Wingender E, Dilloo D, Seifart KH. Zinc ions are differentially required for the transcription of ribosomal 5S RNA and tRNA in a HeLa-cell extract. Nucleic Acids Res 1984; 12(23):8971-8985.
15. Diakun GP, Fairall L, Klug A. EXAFS study of the zinc-binding sites in the protein transcription factor IIIA. Nature 1986; 324(6098):698-699.
16. Brown RS. Structural investigation of the Xenopus 7S Ribonucleoprotein complex. PhD Thesis 1990. London University, Great Britain.
17. Ginsberg AM, King BO, Roeder RG. Xenopus 5S gene transcription factor, TFIIIA: characterization of a cDNA clone and measurement of RNA levels throughout development. Cell 1984; 39(3 Pt 2):479-489.
18. Brown RS, Sander C, Argos P. The primary structure of transcription factor TFIIIA has 12 consecutive repeats. FEBS Lett 1985; 186(2):271-274.
19. Rosenberg UB, Schroeder C, Preiss A et al. Structural homology of the product of the Drosophila Krueppel gene with Xenopus transcription factor IIIA. Nature 1986; 319:336-339.
20. Fridell RA, Fischer U, Luhrmann R et al. Amphibian transcription factor IIIA proteins contain a sequence element functionally equivalent to the nuclear export signal of human immunodeficiency virus type 1 Rev. Proc Natl Acad Sci USA 1996; 93(7):2936-2940.
21. Rudt F, Pieler T. Cytoplasmic retention and nuclear import of 5S ribosomal RNA containing RNPs. EMBO J 1996; 15(6):1383-91.
22. Mao X, Darby MK. A position-dependent transcription-activating domain in TFIIIA. Mol Cell Biol 1993; 13(12):7496-7506.

23. Tso JY, Van Den Berg DJ, Korn LJ. Structure of the gene for Xenopus transcription factor TFIIIA. Nucleic Acids Res 1986; 14(5):2187-2200.
24. Taylor W, Jackson IJ, Siegel N et al. The developmental expression of the gene for TFIIIA in Xenopus laevis. Nucleic Acids Res 1986; 14(15):6185-6195.
25. Gaskins CJ, Hanas JS. Sequence variation in transcription factor IIIA. Nucleic Acids Res 1990; 18(8):2117-2123.
26. Gaskins CJ, Smith JF, Ogilvie MK et al. Comparison of the sequence and structure of transcription factor IIIA from Bufo americanus and Rana pipiens. Gene 1992; 120(2):197-206.
27. Ogilvie MK, Hanas JS. Molecular biology of vertebrate transcription factor IIIA: Cloning and characterization of TFIIIA from channel catfish oocytes. Gene 1997; 203(2):103-112.
28. Hanas JS, Hocker JR, Cheng YG et al. cDNA cloning, DNA binding, and evolution of mammalian transcription factor IIIA. Gene 2002; 282(1-2):43-52.
29. Arakawa H, Nagase H, Hayashi N et al. Molecular cloning, characterization, and chromosomal mapping of a novel human gene (GTF3A) that is highly homologous to Xenopus transcription factor IIIA. Cytogenet Cell Genet 1995; 70(3-4):235-238.
30. Drew PD, Nagle JW, Canning RD et al. Cloning and expression analysis of a human cDNA homologous to Xenopus TFIIIA. Gene 1995; 159(2):215-218.
31. Archambault J, Milne CA, Schappert KT et al. The deduced sequence of the transcription factor TFIIIA from Saccharomyces cerevisiae reveals extensive divergence from Xenopus TFIIIA. J Biol Chem 1992; 267(5):3282-3288.
32. Schulman DB, Setzer DR. Identification and characterization of transcription factor IIIA from Schizosaccharomyces pombe. Nucleic Acids Res 2002; 30(13):2772-2781.
33. Mathieu O, Yukawa Y, Prieto JL et al. Identification and characterization of transcription factor IIIA and ribosomal protein L5 from Arabidopsis thaliana. Nucleic Acids Res 2003; 31(9):2424-2433.
34. Wyszko E, Barciszewska M. Purification and characterization of transcription factor IIIA from higher plants. Eur J Biochem 1997; 249(1):107-112.
35. Wyszko E, Radlowski M, Bartkowiak S et al. Maize TF IIIA—the first transcription factor IIIA from monocotyledons. Purification and properties. Acta Biochim Pol 1997; 44(3):579-589.
36. Polakowski N, Paule MR. Purification and characterization of transcription factor IIIA from Acanthamoeba castellanii. Gene 1997; 203(2):103-112.
37. Clemens KR, Wolf V, McBryant SJ et al. Molecular basis for specific recognition of both RNA and DNA by zinc finger protein. Science 1993; 260(5107):530-533.
38. Searles MA, Lu D, Klug A. The role of the central zinc fingers of transcription factor IIIA in binding to 5 S RNA. J Mol Biol 2000; 301(1):47-60.
39. Shastry BS. Transcription factor IIIA (TFIIIA) in the second decade. J Cell Sci 1996; 109(Pt 3):535-539.
40. Pieler T, Hamm J, Roeder RG. The 5S gene internal control region is composed of three distinct sequence elements, organized as two functional domains with variable spacing. Cell 1987; 48(1):91-100.
41. Clemens KR, Zhang P, Liao X et al. Relative contributions of the zinc fingers of transcription factor IIIA to the energetics of DNA binding. J Mol Biol 1994; 244(1):23-35.
42. Wu C. An exonuclease protection assay reveals heat-shock element and TATA box DNA-binding proteins in crude nuclear extracts. Nature 1985; 317(6032):84-87.
43. Nikolaev LG, Glotov BO, Belyavsky AV et al. Identification of sequence-specific DNA-binding factors by label transfer: Application to the adenovirus-2 major late promoter. Nucleic Acids Res 1988; 16(2):519-535.
44. Vrana KE, Churchill MEA, Tullius TD et al. Mapping functional regions of transcription factor TFIIIA. Mol Cell Biol 1988; 8(4):1684-1696.
45. Fairall L, Rhodes D, Klug A. Mapping of the sites of protection on a 5 S RNA gene by the Xenopus transcription factor IIIA. A model for the interaction. J Mol Biol 1986; 192(3):577-591.
46. Clemens KR, Liao X, Wolf V et al. Definition of the binding sites of individual zinc fingers in the transcription factor IIIA-5S RNA gene complex. Proc Natl Acad Sci USA 1992; 89(22):10822-10826.
47. Hayes JJ, Tullius TD. Structure of the TFIIIA-5 S DNA complex. J Mol Biol 1992; 227(2):407-417.
48. Nolte RT, Conlin RM, Harrison SC et al. Differing roles for zinc fingers in DNA recognition: Structure of a six-finger transcription factor IIIA complex. Proc Natl Acad Sci USA 1998; 95(6):2938-2943.
49. Wuttke DS, Foster MP, Case DA et al. Solution structure of the first three zinc fingers of TFIIIA bound to the cognate DNA sequence: determinants of affinity and sequence specificity. J Mol Biol 1997; 273(1):183-206.
50. Brown RS, Argos P. Fingers and helices. Nature 1986; 324(6094):215.
51. Pavletich NP, Pabo CO. Zinc finger-DNA recognition: Crystal structure of a Zif268-DNA complex at 2.1 A. Science 1991; 252(5007):809-817.
52. Wolfe SA, Nekludova L, Pabo CO. DNA recognition by Cys2His2 zinc finger proteins. Annu Rev Biophys Biomol Struct 2000; 29:183-212.
53. Laity JH, Dyson HJ, Wright PE. DNA-induced alpha-helix capping in conserved linker sequences is a determinant of binding affinity in Cys(2)-His(2) zinc fingers. J Mol Biol 2000; 295(4):719-727.
54. Hamilton TB, Turner J, Barilla K et al. Contribution of individual amino acids to the nucleic acid binding activities of the Xenopus zinc finger proteins TFIIIIA and p43. Biochemistry 2001; 40(20):6093-6101.
55. Choo Y, Klug A. A role in DNA binding for the linker sequences of the first three zinc fingers of TFIIIA. Nucleic Acids Res 1993; 21(15):3341-3346.
56. Ryan RF, Darby MK. The role of zinc finger linkers in p43 and TFIIIA binding to 5S rRNA and DNA. Nucleic Acids Res 1998; 26(3):703-709.
57. Hanas JS, Koelsch G, Moreland RM et al. Differential requirements of basic amino acids in transcription factor IIIA-5S gene interaction. Biochim Biophys Acta 1998; 1398(3):256-264.
58. Zang WQ, Veldhoen N, Romaniuk PJ. Effects of zinc finger mutations on the nucleic acid binding activities of Xenopus transcription factor IIIA. Biochemistry 1995; 34(47):15545-15552.
59. Bumbulis MJ, Wroblewski G, McKean D et al. Genetic analysis of Xenopus transcription factor IIIA. J Mol Biol 1998; 284(5):1307-1322.
60. Moreland RJ, Dresser ME, Rodgers JS et al. Identification of a transcription factor IIIA-interacting protein. Nucleic Acids Res 2000; 28(9):1986-1993.
61. Cassiday LA, Maher LJ 3rd. Having it both ways: Transcription factors that bind DNA and RNA. Nucleic Acids Res 2002; 30(19):4118-4126.
62. Guex N, Peitsch MC. SWISS-MODEL and the Swiss-PdbViewer: An environment for comparative protein modelling. Electrophoresis 1997; 18(15):2714-2723.
63. Pov-Ray™ (http://www.povray.org).
64. Bernstein HJ. Recent changes to RasMol, recombining the variants. Trends Biochem Sci 2000; 25(9):453-455.

CHAPTER 4

GAGA: Structural Basis for Single Cys_2His_2 Zinc Finger-DNA Interaction

G. Marius Clore* and James G. Omichinski

Abstract

The structural basis of sequence specific binding of the Cys_2His_2 single zinc finger DNA binding domain of the transcription factor GAGA is explored on the basis of the three-dimensional structure of a complex between the minimal DNA binding domain (DBD) of the GAGA factor (GAGA-DBD) and an oligonucleotide containing its GAGAG consensus binding site. The GAGA-DBD comprises a single classical Cys_2His_2 zinc finger core, and an N-terminal extension containing two highly basic regions, BR1 and BR2. The zinc finger core binds in the major groove and recognizes the first three GAG bases of the consensus in a manner very similar to that seen in other classical Cys_2His_2 zinc finger/DNA complexes. Unlike the latter which require tandem zinc finger repeats with a minimum of two units for high affinity binding, the GAGA-DBD makes use of only a single finger complemented by the two N-terminal basic regions BR1 and BR2. BR2 forms a helix that interacts in the major groove recognizing the last G of the consensus, while BR1 wraps around the DNA in the minor groove and recognizes the A in the fourth position of the consensus.

Introduction

The GAGA factor of *Drosophila melanogaster*, so-called because it binds to $(GA)_n$ rich sites, is a TFIIIA-like zinc finger protein which was originally identified on the basis of its ability to bind to $(GA)_n$ rich sites in the *Ultrabithorax* promoter.[1,2] The GAGA factor is encoded by the *Thrithorax-like* gene,[3] is required for normal expression of homeotic genes, and acts as a modifier of position-effect variegation.[4] Both in vivo[5,6] and in vitro[7-11] experiments suggest that the GAGA factor acts as an anti-repressor by helping to disrupt nucleosomes associated with gene regulatory sequences. Putative *Drosophila* target genes for the GAGA factor include genes such as the heat shock (hsp26 and hsp70), histone h3/h4, homeotic and housekeeping/constitutive genes.[4] The GAGA factor is 519 residues in length and comprises three domains: an N-terminal POZ/PTB protein interaction domain, a central DNA binding domain, and a polyglutamine-rich carboxy-terminus.[4] The minimal DNA binding domain (DBD) of the GAGA factor has been delineated and shown to bind specifically to DNA derived from the h3/h4 promoter and containing the sequence GAGAGAG with a dissociation constant of 5 nM.[12] The GAGA-DBD extends from residues 310 to 372 in the intact full length GAGA protein, and consists of a single classical (TFIIIA-like) Cys_2-His_2 zinc finger[13,14] preceded by two highly basic regions (BR1 and BR2) (Fig. 1A). (Note that there is also a basic region C-terminal to the zinc finger domain, but its presence or absence has no effect on DNA binding affinity[12]).

While the genes for several proteins containing only a single classical zinc finger have been identified,[12,15,16] the GAGA-DBD was the first single classical zinc finger containing protein for which sequence specific DNA binding by a Cys_2His_2 zinc finger region has been experimentally demonstrated. In other reported TFIIIA-like zinc finger proteins, a tandem array comprising a minimum of two zinc binding domains is required for sequence specific high-affinity DNA binding.[13,14] The ability of the single Cys_2His_2 zinc finger containing protein GAGA to specifically bind DNA suggests that all proteins containing a single classical zinc fingers are potential DNA-binding transcription factors. Over the last several years, several single zinc-finger domain proteins have been identified in a variety of plant species.[17] In each case, these plant proteins contain the highly conserved sequence QALGGH in their putative recognition helix within the single zinc-finger domain. In addition, the SUPERMAN protein from *Arabidopsis thalania* has been recently shown to be a sequence specific DNA-binding protein.[18] As with the DBD of the GAGA protein, the DBD of the SUPERMAN protein is composed of a single zinc-finger domain flanked by two basic rich regions. However, the DBD of SUPERMAN differs from the DBD of the GAGA protein in that one of the basic regions is C-terminal to the zinc finger while the second basic region is N-terminal to the zinc finger. This arrangement of a DBD containing a single zinc-binding domain flanked by basic regions on both termini has also been demonstrated with the N-terminal zinc-binding domains of chicken GATA-2 and GATA-3.[19] These DBDs consisting of a single zinc-binding domains flanked by basic rich regions represents a very versatile mode of specific DNA recognition that could be easily exploited in a number of different applications.

In this chapter, adapted from our previously published paper describing the three-dimensional solution NMR structure of the GAGA-DBD/DNA complex (PDB accession code 1YUI),[20] we summarize the structural basis for sequence specific recognition

* *Corresponding author. See list of "Contributors".*

Zinc Finger Proteins: From Atomic Contact to Cellular Function, edited by Shiro Iuchi and Natalie Kuldell.

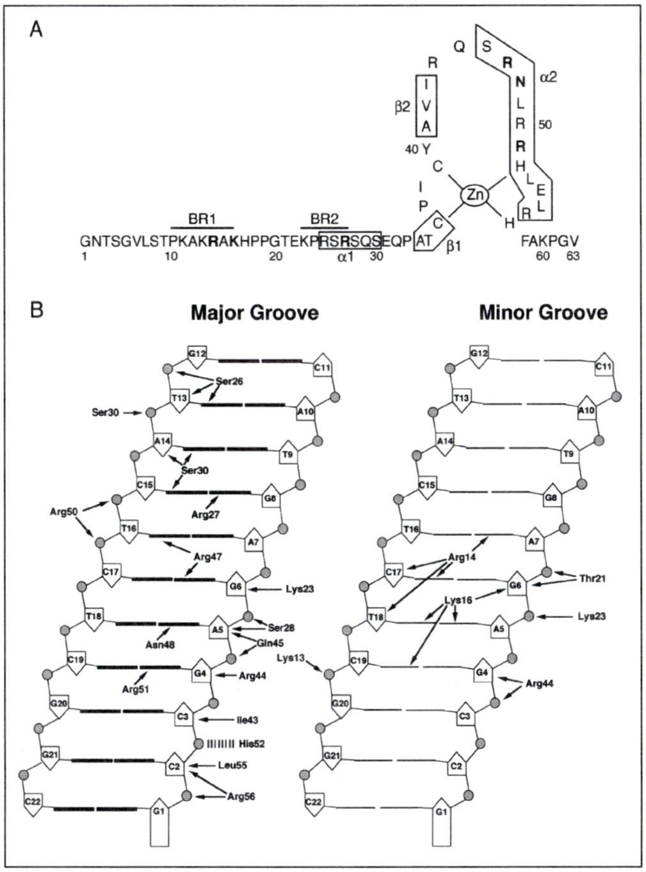

Figure 1. A) Sequence of the GAGA-DBD. The locations of the two basic regions, BR1 and BR2, and the secondary structure elements are indicated. Residues that make base specific contacts are shown in bold. Residues 1-63 of the GAGA-DBD correspond to residues 310-372 of the full length GAGA protein. B) Summary of the contacts between the GAGA-DBD and DNA. The DNA is represented as a cylindrical projection viewed from either the major or minor grooves as indicated. The bases are indicated as thick lines in the major groove and thin lines in the minor groove, with the deoxyribose sugar rings as pentagons and the phosphates as stippled circles. The hydrogen bond between the imidazole ring of His52 and the phosphate of C3 is shown as a broken line. (Adapted from ref. 20)

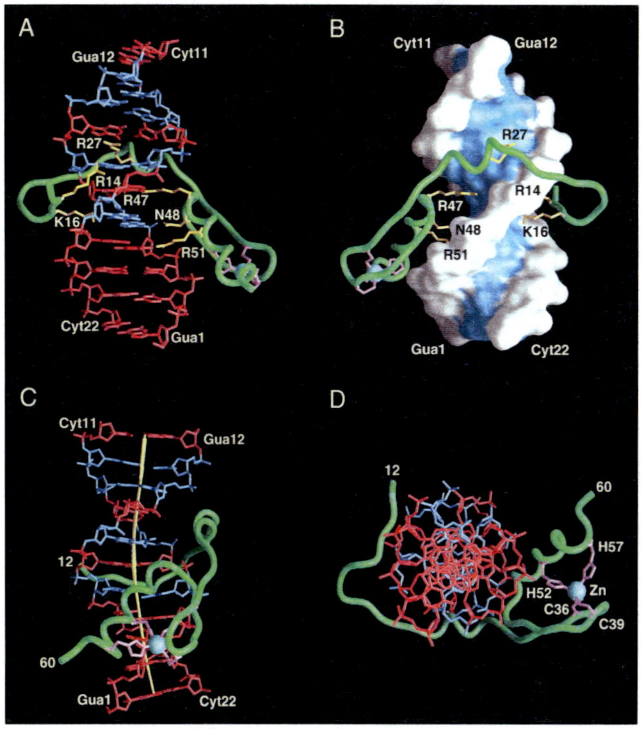

Figure 2. Views illustrating the interaction of the GAGA-DBD with DNA. The backbone of the protein is depicted as a green tube, side chains making base specific contacts are in yellow, the histidine and cysteine side chains coordinating the zinc (blue ball) in magenta, GC base pairs in red, and AT base pairs in blue. In (B) the DNA is depicted as a molecular surface with the bases colored in blue and the sugar-phosphate backbone in white. The path of the long axis of the DNA helix is shown in yellow in (C). (From ref. 20)

by the GAGA-DBD. The structure shows how a classical zinc finger complemented by an N-terminal extension comprising a basic helix and tail can recognize DNA in a sequence specific manner, making base specific contacts with every base in the pentanucleotide consensus sequence GAGAG. In addition, the presence of both major and minor groove contacts in the complex immediately suggests special constraints on interactions of the GAGA factor and DNA targets within nucleosomes.

Overall Descripton of the GAGA-DBD/DNA Structure

The key feature of the GAGA-DBD/DNA complex is the presence of base specific contacts to every base of the G_4AGAG_8 pentanucleotide consensus (Figs. 1B, 2 and 3; numbering is that of Figure 1B. Major groove recognition of G4, A5 and G6 is provided by the helix of the zinc finger, major groove recognition of G8 by the helix of BR2, and minor groove recognition of A7 by BR1. The DNA in the complex is essentially B-type and displays a very slight smooth bend of 10° (Fig. 2C).

The zinc finger core (residues 33-59) of the GAGA-DBD is centered around a tetrahedrally coordinated zinc atom, comprises a ββα motif (Figs. 2 and 3A), and is very similar to that of other classical zinc fingers with C^α atomic rms differences ranging from 0.8 Å relative to finger 2 of Tramtrack[21] to 1.6 Å relative to finger 3 of Zif268.[22] Arg51, Asn48 and Arg47 at positions 6, 3 and 2 of the zinc finger helix (α2, residues 46-56) contact the bases of G4, A5 and G6, respectively. The long axis of the helix α2 is oriented at approximately 60° to the long axis of the DNA. The guanidino group of Arg51 recognizes the O6 and N7 atoms of G4, the guanidino group of Arg47 recognizes the O6 atom of G6 and possibly the N7 atom of G6, and the side chain carboxyamide group of the Asn48 recognizes the 6-NH_2 group and N7 atom of A5 (Fig. 2A). The C^δ of Arg47 is also in close contact with the methyl group of T16 (Fig. 1B). The base specific interactions are supplemented by hydrophobic interactions with the sugars involving Arg56, Leu55, Arg44 and Ile43, and electrostatic interactions with the sugar phosphate backbone involving Arg56, His52, Arg50, Arg44 and Gln45. As in other classical zinc finger-DNA complexes[21-23] the $H^{\delta 1}$ proton of the imidazole ring of His52 is hydrogen-bonded to the phosphate of C3. BR2 forms a short helix (α1, residues 25-29) which is located in the major groove and oriented at 80° to the long axis of the DNA. The guanidinium group of Arg27 of the BR2 helix recognizes the base of G8, while Ser28, Ser30 and Ser26 anchor this helix to the phosphates of G6, A14 and T13, respectively (Fig. 3B). The orientation of the BR2 helix with respect to the zinc finger core is further stabilized by an electrostatic interaction between Ser46

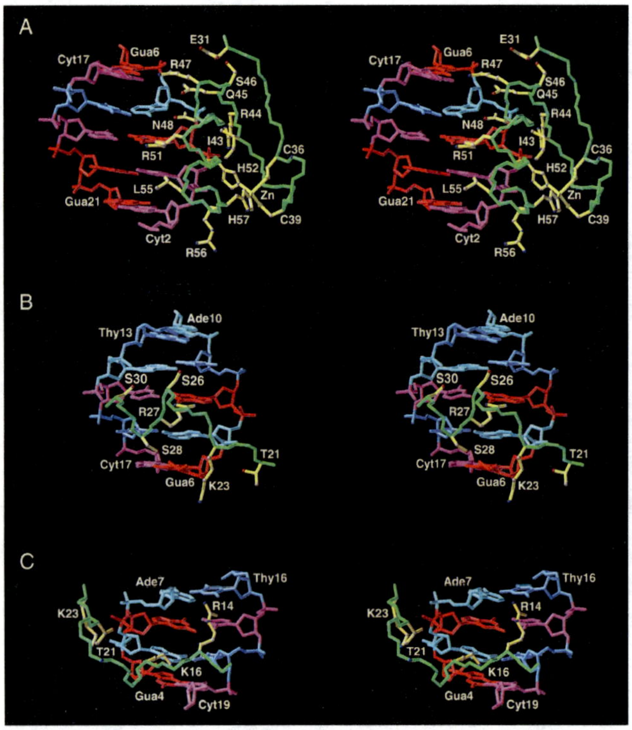

Figure 3. Stereoviews illustrating the interactions of the zinc finger core (A), the BR2 helix (B) and the BR1 N-terminal tail (C) of the GAGA-DBD with DNA. The protein backbone is in green, the side chains in yellow, T in dark blue, A in lighter blue, C in magenta and G in red. (From ref. 20.)

and Glu31 (Fig. 3A). In the minor groove, base specific contacts are made by Arg14 and Lys16 of BR1. The guanidino group of Arg14 interacts with the N3 and O2 atoms of A7 and C17, respectively, while Lys16 interacts with the O2 atoms of T18 and/or C19 (Fig. 3C).

Comparison with Other Classical Zinc Finger/DNA Complexes

At the time the NMR structure of the GAGA-DBD/DNA complex was published,[20] X-ray structures of three classical zinc finger/DNA complexes had been solved: Tramtrack,[21] Zif268[22] and GLI[23] which consist of a tandem array of two, three and five zinc finger repeats, respectively, separated by seven amino acids. The two zinc fingers of Tramtrack[21] and the three zinc fingers of Zif268[22] make specific contacts with the DNA, and the orientation of each finger with respect to the DNA is very similar in all cases. For GLI,[23] only four (fingers 2 to 5) of the five zinc fingers contact the DNA. The number of interactions between the zinc finger domains is small and each finger recognizes DNA in a largely independent manner.[14] Specific recognition is achieved by contacts between the helix of each zinc finger and the major groove of the DNA. The residues of the helix involved in specific DNA contacts are located principally at positions 6, 3, 2 or -1 (numbered relative to the N-terminus of the helix and corresponding to Arg51, Asn48, Arg47, and Gln45, respectively, of the GAGA-DBD; (Fig. 1A), with usually two of the four positions making one-to-one amino acid-to-base contacts with a base triplet.[13,14] The crystal structures of these three zinc finger/DNA complexes together with extensive data from site-directed mutagenesis, screening/selection and protein design experiments has led to a generalized consensus zinc finger recognition code in which the residues at position 6, 3, and either 2 or -1 recognizes the first, second and third base of the triplet (reading in the 5⟶to 3' direction), respectively.[13,14]

Recognition of the $G_4A_5G_6$ triplet by the zinc finger helix of the GAGA-DBD involves residues at positions 6 (Arg51), 3 (Asn48) and 2 (Arg47). This mode of interaction is identical to Finger 1 of Tramtrack with the difference that the residue at position 2 (Ser) recognizes a T.[24] Indeed, the side chain conformations and the nature of the contacts with the DNA bases for the Arg and Asn residues at positions 6 and 3 are essentially identical for the zinc finger of the GAGA-DBD and Finger 1 of Tramtrack[21] (Figs. 2A and 3A). It is also worth noting that the conformation of the Arg at position 6 in the GAGA-DBD, Finger 1 of Tramtrack,[21] and Fingers 1 and 3 of Zif268[22] are similar and in each case the guanidino group recognizes the O6 and N7 atoms of a G base with essentially the same hydrogen bonding geometry.

The number of intermolecular contacts, and hence the strength of the interaction of each zinc finger with DNA, is not sufficient in its own right to yield high affinity DNA binding.[13,14] The latter can be achieved in a number of ways. The simplest design approach, as exemplified by the three zinc finger/DNA crystal structures,[18-20] is to have a modular protein with a tandem array of two or more zinc fingers. Alternatively, a single zinc finger domain can be employed in conjunction with another DNA recognition motif. This is the path chosen by the GAGA-DBD where the additional contacts required for sequence specific binding are provided by an N-terminal extension comprising two basic regions, BR1 and BR2.

A direct comparison of the GAGA-DBD/DNA complex with the two zinc finger Tramtrack/DNA complex[21] is afforded in Figure 4 with the zinc finger core of the GAGA-DBD superimposed on Finger 2 of Tramtrack. The orientation of the zinc finger core of the GAGA-DBD and Finger 2 of Tramtrack with respect to the DNA are essentially identical. Moreover, the positioning of the BR2 helix of the GAGA-DBD approximately matches that of the helix of Finger 1 of Tramtrack. Unlike the Tramtrack-DBD/DNA complex, however, the GAGA-DBD/DNA complex has a significant minor groove component provided by the basic BR1 region which wraps around the minor groove and is involved in several base specific contacts (Figs. 1B, 2 and 3C).

N-terminal extensions to the classical zinc finger structure have been observed for the first of the zinc fingers of Tramtrack (a β-strand)[21] and SWI5 (a helix and β-strand).[24] In contrast to the N-terminal BR1 and BR2 regions of the GAGA-DBD, these extensions are in intimate contact with the zinc finger core and are thought to contribute to the stability of the domains. For SWI5, the N-terminal extension also enhances DNA binding, and although no structure of the SWI5 first finger/DNA complex has yet been published, model building suggests the involvement of the additional helix in contacts with the phosphate backbone.[24]

Correlation with Biochemical Data

The impact of the various base specific interactions with the GAGAG consensus in stabilizing the GAGA-DBD/DNA complex has also been investigated by gel retardation experiments.[20]

The most dramatic effects are seen for the G6⟶A,T,C substitutions which removes the hydrogen bond between the guanidino group of Arg47 and the O6 atom of G6. No complexes at all were observed for any of the G6⟶A,T,C substitutions, indicating a

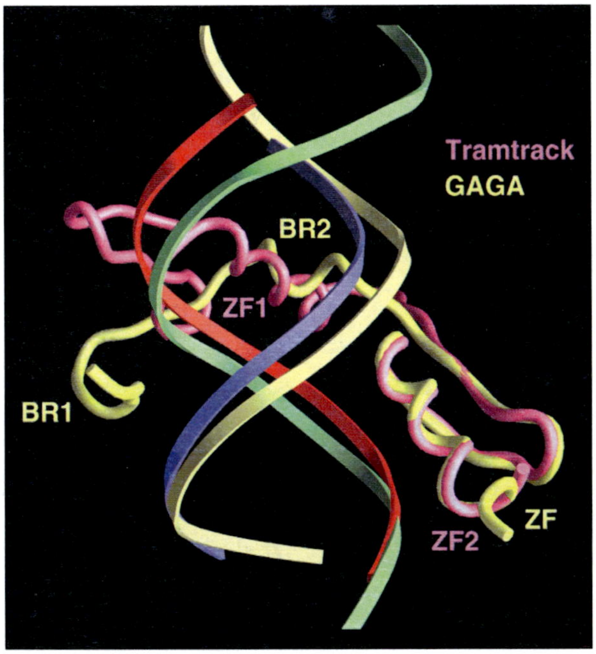

Figure 4. Comparison of the GAGA-DBD/DNA and Tramtrack-DBD/DNA complexes. The zinc finger core of the GAGA-DBD (residues 33-58) is superimposed on finger 2 (residues 140-165) of Tramtrack (C^{α} rms difference of 0.8 Å). The protein backbones are shown as tubes with the GAGA-DBD in yellow and the Tramtrack-DBD in magenta. The two DNA duplexes are shown as ribbons through the phosphate backbone; the shorter of the two duplexes with red and purple strands belongs to the GAGA-DBD/DNA complex, while the longer one with light green and pale yellow strands belongs to the Tramtrack-DBD/DNA complex. (From ref. 20).

reduction in affinity of at least an order of magnitude. This demonstrates the central importance of the Arg47-G6 interaction involving the major recognition helix of the zinc finger (Fig. 3A).

The A5→G,T,C substitutions, which decrease the affinity somewhat, remove one of the two hydrogen bonds between the carboxyamide of Asn48 and A5, specifically between the between the $O^{\delta1}$ atom of Asn48 and 6-NH_2 group of A5 (Fig. 3A). Consistent with the role of Asn48 is the observation that mutation of this residue to an Asp decreases the DNA binding affinity by more than 10-fold. This mutation removes the hydrogen bond between the side chain $N^{\delta2}H_2$ group of Asn48 and the N7 atom of A5, and introduces electrostatic repulsion between the negatively charged carboxylate of an Asp and the N7 atom of A5 which has partial negative charge (acting generally as a hydrogen bond acceptor).

Similarly, the G4→A,T,C substitutions remove one of the two hydrogen bonds between the guanidinium group of Arg51 and G4, specifically that involving the O6 atom of the G4 base. The effects of these substitutions are not as dramatic as those observed for the G6 substitutions, possibly because a hydrogen bond between Lys16 and the complementary base (C19→T,A,G) could still be formed in the minor groove (Fig. 3C).

The G8→T,C substitutions markedly reduce the binding, highlighting the importance of the G8-Arg27 interaction involving the BR2 helix (Fig. 3B). Interestingly, complex formation is still observed for the G8→A substitution, probably because one hydrogen bond could still be formed between the guanidinium group of Arg27 and the N7 atom of the A base.

The A7→G,T,C substitutions reflect the base interactions of Arg14 of BR1 in the minor groove (Fig. 3C). The A7→T,C substitutions still bind the GAGA-DBD, albeit with reduced affinity, since the O2 atoms of T and C occupy almost the same position as that of the N3 atom of A. The absence of complex formation for the A7→G substitution is probably due to steric hindrance and electrostatic repulsion arising from the presence of the bulky 2-NH_2 group in G which replaces the H2 proton of A in the minor groove. The importance of Arg14 for sequence specific recognition is further evidenced by the observation that a truncated GAGA-DBD starting at Ala15 exhibits no detectable complex formation as probed by gel shift experiments.

Relationship to Other DNA Binding Proteins

The general mode of DNA binding (as opposed to the specifics of side chain-base interactions and recognition) observed for the GAGA-DBD is reminiscent of that seen for the DBD of the GATA-1 transcription factor.[25] The GATA-1-DBD also clamps the DNA with a helix and loop interacting in the major groove, and a basic C-terminal tail wrapping around the minor groove. The size of the DNA binding site is similar, and, just as for the GAGA-DBD, the major groove interactions involve a zinc binding domain. The zinc binding module of GATA-1, however, is not a classical TFIIIA-like zinc finger but is structurally related to the amino-terminal zinc module of the glucocorticoid receptor.[26] Thus, both the GAGA-DBD and the GATA-1-DBD employ two structural motifs to interact with the DNA.

Relationship to Chromatin Remodeling

The structure of the GAGA-DBD/DNA complex shows how a single classical Cys_2His_2 zinc finger complemented by an N-terminal extension comprising a basic helix (BR2) and tail (BR1) can recognize DNA in a sequence specific manner. The requirement for binding of the zinc finger core and BR2 in the major groove, together with the simultaneous binding of BR1 in the minor groove opposite the recognition helix of the zinc finger core, suggests that there may be special constraints on interactions between the GAGA factor and DNA targets within nucleosomes, where a part of the binding site may be relatively inaccessible. Such interactions could favor (and be favored by) disruption of the nucleosome core by displacement or weakening of DNA-histone contacts, particularly in the presence of additional factors such as the nucleosome remodeling factor NURF.[9,27]

The minimal binding site for the GAGA-DBD comprises 9 basepairs with a central pentanucleotide G_4AGAG_8 motif (employing the numbering scheme shown in Figure 1B. Specific binding with only a relatively small decrease in affinity (i.e., less than a factor of 10), can be achieved with modifications, albeit one basepair at a time, of G4 and A5 to any other base, A7 to T or C, and G8 to A, with only G6 being invariant. The promoters that are known to be targets for the GAGA factor frequently contain a high density of GA repeats,[4] generating a large number of potential GAGA factor binding sites in close proximity. For example, the *hsp26-1* promoter contains at least 10 partially overlapping potential GAGA factor binding sites.[4,28] Likewise, a subset of highly repetitive DNA sequences found in heterochromatin of *Drosophila* are GA rich and found to be associated with the GAGA factor at all stages of the cell cycle, thereby possibly modifying heterochromatin in these regions.[8] This suggests that binding of multiple copies of the GAGA factor, possibly interacting with each other via protein-protein contacts, may be required to achieve efficient disruption of the nucleosome. It also suggests that the GAGA-factor may have the possibility of migrating along the DNA, diffusing along a linear lattice from one site to the next adjacent site. In the in vivo situation, the presence of such a large number of adjacent GAGA factor binding sites is likely to ensure that at least some sites would be located in a linker region on a nucleosome array, serving as an anchor point for GAGA factor binding. Consistent with that mechanism is the observation that the GAGA factor binds weakly to mononucleosomes, and even in the presence of the nucleosome remodeling factor NURF does not cause complete disruption of nucleosome structure under these conditions.[27] In contrast, the GAGA factor binds tightly to a nucleosome array containing multiple GAGA factor sites, even in the absence of NURF.[9,27]

The GAGA factor is able to alleviate the repression of transcription by linker histones (H1 or H5).[7] Conversely, the linker histone H1 decreases NURF/ATP dependent stimulation of GAGA factor binding.[9,29] The globular domain of the linker histones specifically recognizes and binds to the nucleosome core and is thought to be located asymetrically inside the superhelical gyre of DNA, just inside the nucleosome core region.[30,31] The globular domain of the linker histones belongs to the winged helix-turn-helix family of DNA binding proteins,[32] exemplified by HNF-3/*fork head*.[33] While the structure of a linker histone-DNA complex has not yet been determined, presumably because linker histones do not bind naked DNA in a sequence specific manner, it seems likely that the mode of binding of the linker histones would be similar to that seen in the complex of HNF-3/*fork head* with DNA. In the latter complex, the recognition helix of the helix-turn-helix motif binds in the major groove, and the two basic wings contact the adjacent minor grooves on either side of the interacting major groove.[33] The C-terminal wing also makes an arginine mediated base specific contact in the minor groove.[33] Thus, it is possible that the combination of major and minor groove binding by the GAGA-DBD, coupled with its significantly higher DNA sequence specificity relative to the linker histones, permits the GAGA-DBD to directly displace a linker histone under certain circumstances (e.g., when the linker histone is directly bound to a GAGA factor site).

As the GAGA-DBD effectively clamps the DNA by binding in the major and minor grooves, there are also limitations on the location of the N-terminal POZ domain and the polyglutamine-rich C-terminal tail in the DNA complex with the full-length GAGA factor. These must be free to interact with other components of the transcription machinery, including the nucleosome remodeling factor NURF which facilitates the complete displacement of the histone octamer from the complex in an ATP-dependent manner.[9]

Acknowledgements

We thank Gary Felsenfeld, Paolo Pedone and Angela Gronenborn for a highly fruitful collaboration on the structure of the GAGA-DBD/DNA complex.

References

1. Biggin MD, Tjian R. Transcription factors that activate the Ultrabithorax promoter in developmentally staged extracts. Cell 1988; 53:699-711.
2. Soeller WC, Oh CE, Kornberg TB. Isolation of cDNAs encoding the Drosophila GAGA transcription factor. Mol Cell Biol 1993; 13:7961-7970.
3. Farkas G, Gausz J, Galloni M et al. The trithorax-like gene encodes the Drosophila GAGA factor. Nature 1994; 371:806-808.
4. Granok H, Leibovitch BA, Shaffer CD et al. Ga-ga over GAGA factor. Curr Biol 1995; 5:238-241.
5. Lu Q, Wallrath LL, Elgin SCR. $(CT)_n \cdot (GA)_n$ repeats and heat shock elements have a distinct roles in chromatin structure and transcriptional activation of the Drosophila hsp26 gene. Mol Cell Biol 13:2802-2814.
6. O'Brien T, Wilkins RC, Giardina C et al. Distribution of GAGA protein on Drosophila genes in vivo. Genes & Dev 1995; 9:1098-1110.
7. Croston GE, Kerrigan LA, Lira M et al. Sequence specific antirepression of histone H1-mediated inhibition of basal RNA polymerase II transcription. Science 1991; 251:643-649.
8. Raff JW, Kellum R, Alberts B. The Drosophila GAGA transcription factor is associated with specific regions of heterochromatin throughout the cell cycle. EMBO J 1994; 13:5977-5983.
9. Tsukiyama T, Becker PB, Wu C. ATP-dependent nucleosome disruption at a heat-shock promoter mediated by binding of GAGA transcription factor. Nature 1994; 367:525-532.
10. Wall G, Varga-Weisz PD, Sandaltzopoulos R et al. Chromatin remodeling by GAGA factor and heat shock factor at the hypersensitive Drosophila hsp26 promoter in vivo. EMBO J 1995; 14:1727-1736.
11. Shopland LS, Hirayoshi K, Fernandes M et al. HSF access to heat shock elements in vivo depends critically on promoter architecture defined by GAGA factor, TFIID, and RNA polymerase II binding sites. Genes & Dev 1995; 9:2756-2769.
12. Pedone PV, Ghirlando R, Clore GM et al. The single Cys_2-His_2 zinc finger domain of the GAGA protein flanked by basic residues is sufficient for high-affinity specific DNA binding. Proc Natl Acad Sci USA 1996; 93:2822-2826.
13. Klug A, Schwabe JWR. Zinc fingers. FASEB J 1995; 9:597-604.
14. Berg JM, Shi Y. The galvanization of biology: A growing appreciation for the roles of zinc. Science 1996; 271:1081-1085.
15. Sakai H, Medrano LJ, Meyerowitz EM. Role of SUPERMAN in maintaining Arabdopsis floral whorl boundaries. Nature 1995; 378:199-203.
16. Tague BW, Goodman HM. Characterization of a family of Arabidopsis zinc finger protein cDNAs. Plant Mol Biol 1995; 28:267-279.
17. Isernia C, Bucci E, Leone M et al. NMR structure of the single QALGGH zinc finger domain from the Arabidopsis thaliana SUPERMAN protein. ChemBioChem 2003; 4:171-180.
18. Dathan N, Zaccaro L, Esposito S et al. The Arabidopsis SUPERMAN protein s able to specifically bind DNA through its single Cys2-His2 zinc finger motif. Nucl Acid Res 2002; 30:4845-4951.
19. Pedone P, Omichinksi JG, Nony P et al. The N-terminal fingers of chicken GATA-2 and GATA-3 are independent sequence-specific DNA binding domains. EMBO J 1997; 16:2874-2882.
20. Omichinski JG, Pedone PV, Felsenfeld G et al. The solution structure of a specific GAGA factor-DNA complex reveals a modular binding mode. Nature Struct Biol 1997; 4:122-132.
21. Fairall L, Schwabe JWR, Chapman L et al. The crystal structure of a two zinc-finger peptide reveals an extension to the rules of zinc-finger/DNA recognition. Nature 1993; 366:483-487.
22. Pavletich NP, Pabo CO. Zinc finger-DNA recognition: Crystal structure of a Zif268-DNA complex at 2.1 Å. Science 1991; 252:809-816.
23. Pavletich NP, Pabo CO. Crystal structure of a five-finger GLI-DNA complex: New perspectives on zinc fingers. Science 1993; 261:1701-1707.
24. Dutnall EN, Neuhaus D, Rhodes D. The solution structure of the first zinc finger domain of SWI5: A novel structural extension to a common fold. Structure 1996; 4:599-611.
25. Omichinski, JG, Clore GM, Schaad O et al. NMR structure of a specific DNA complex of a Zn-containing DNA binding domain of GATA-1. Science 1993; 261:438-446.
26. Luisi B, Xu WX, Otwinowski Z et al. Crystallographic analysis of the interaction of the glucocorticoid receptor with DNA. Nature 1991; 352:497-505.

27. Tsukiyama T, Wu C. Purification and properties of an ATP-dependent nucleosome remodelling factor. Cell 1995; 83:1011-1020.
28. Lis J, Wu C. Protein traffic on the heat shock promoter: Parking, stalling and trucking along. Cell 1993; 74:1-4.
29. Kingston RE, Bunker CA, Imbalzano AN. Repression and activation by multiprotein complexes that alter chromatin structure. Genes & Dev 1996; 10:905-920.
30. Hayes JJ. Site-directed cleavage of DNA by a linker histone-Fe(II) EDTA conjugate: Localization of a globular domain binding site within a nucleosome. Biochemistry 1996; 35:11931-11937.
31. Pruss D, Bartolomew B, Persinger J et al. An asymmetric model for the nucleosome: A binding site for linker histones inside DNA gyres. Science 1996; 274:614-617.
32. Ramakrishnan V, Finch JT, Graziano V et al. Crystal structure of globular domain of histone H5 and its implications for nucleosome binding. Nature 1993; 362:219-223.
33. Clark KL, Halay ED, Lai E et al. Cocrystal structure of the HNF-3/fork head DNA-recognition motif resmbles histone H5. Nature 1993, 364:412-420.

CHAPTER 5

The DNA-Binding Domain of GATA Transcription Factors—A Prototypical Type IV Cys_2-Cys_2 Zinc Finger

Angela M. Gronenborn

Abstract

The highly conserved DNA-binding domains (DBDs) of eukaryotic GATA factors comprise one or two zinc binding modules with four cysteines embedded in the sequence Cys-X_2-Cys-$X_{17/18}$-Cys-X_2-Cys and an adjacent basic region. The fold has been defined as a class IV zinc finger motif and belongs to the superfamily of glucocorticoid receptor-like DNA binding domains. Members of the GATA family are found in a wide range of organisms ranging from slime molds to fungi and plants to vertebrates and exhibit differing and complex roles in transcription regulation mediated by binding to regulatory DNA sequences of the form (A/T)GATA(A/G). Fungal GATA factors control nitrogen metabolism, light induction, siderophore biosynthesis and mating-type switching, while the prototypical animal factor GATA-1 is involved in the regulation of all erythroid cell-specific genes. A unique feature of the GATA DBDs is their ability to interact with DNA as well as with proteins, leading to a multitude of intricate functionalities. One important role is associated with dramatic chromatin rearrangement.

Introduction

The GATA transcription factors represent a major class of zinc-containing, regulatory proteins that are found in a wide range of organisms (for reviews see refs. 1 and 2). They are named after the consensus DNA sequence that they recognize, containing a six base-pair sequence with the invariant GATA core.[3,4] The amino acid sequence of the DNA binding domain (DBD) comprises a Cys_2-Cys_2 (C_2C_2) type IV zinc finger motif, followed by a basic terminal tail. The core of the structure consists of about 50 residues. At the N terminus the chain starts with a turn that is followed by two short, irregular antiparallel β sheets, an α helix and a long loop that contains a helical turn. At the C terminus of the core a Ω loop is found. β strands 1 and 2 form the first β sheet, and β strands 3 and 4 form the second β sheet. Following the Ω loop is a long, irregular tail which, in the free protein, is unstructured and highly flexible. The core is dominated by the Zn ion that is tetrahedrally coordinated to the Sγ atoms of the four cysteines. The first two coordinating cysteine residues are located on β strand 1, and the turn between β1 and β2, while the second pair resides on the α helix. The geometric constraints imposed by the Zn and its ligands therefore determine the orientation of the helix with respect to the remainder of the first antiparallel β sheet. The metazoan GATA DBDs always have a 17-residue stretch between the two cysteine pairs (Cys-X_2-Cys-X_{17}-Cys-X_2-Cys) and a leucine (an alanine in some *C. elegans* fingers) in the seventh position after the second cysteine. In vertebrates, two adjacent, homologous DNA-binding domains are present in tandem, separated by a few amino acids, with the carboxy-terminal domain being the physiological important DBD. The amino-terminal domain can modulate DNA binding specificity and/or participates in interactions with other proteins.[5,6] In invertebrates both two-domain and one-domain GATA factors occur. Plant GATA factors contain only one DBD and 18 residues between the two cysteine pairs. Putative plant GATA factors contain either leucine or glutamine in the seventh position. Fungal GATA factors come in two flavors: one is "animal-like" and the second is "plant-like." The residue at position seven after the second cysteine of plant-like fungal GATA factors is glutamic acid.[7] With one exception, fungal GATA factors contain only one DBD that resembles most closely the carboxy-terminal DBD of the vertebrate GATA factors.

GATA DBDs are most likely very ancient eukaryotic motifs and a sequence clearly related to GATA factors is present in the stalky gene of *Dictyostelium discoideum*.[8] A evolutionary relationship of GATA-binding motifs has been published[9] and a sequence comparison for GATA domains of different species is shown in Figure 1.

Structures of GATA DNA Binding Domains and DNA Recognition

High-resolution three-dimensional NMR structures have been determined for two GATA-binding domains complexed with their DNA recognition sequence: the chicken GATA-1 carboxy-terminal domain and the *Aspergillus nidulans* AreA domain.[10,11] The domain contains a compact core formed by two pairs of β sheets followed by an α helix and a loop (Fig. 2).

The structure is dominated by the tetrahedrally coordinated zinc via the cysteines and numerous additional hydrophobic contacts that determine the relative orientation of the two β sheets and the helix. The α helix and the loop connecting β strands β2 and β3 interact with the major groove of the DNA, mostly via hydrophobic amino acids. Methyl-groups of three leucines and

Zinc Finger Proteins: From Atomic Contact to Cellular Function, edited by Shiro Iuchi and Natalie Kuldell.

```
chicken GATA-1 (C)    AGTVCSNCQTSTTTLWRRSPMG-DPVCNACGLYYKLHQVNRPLTMRKDGIQTRNRKVSS-
chicken GATA-1 (N)    -ARECVNCGATATPLWRRDGTG-HYLCNACGLYHRLNGQNRPLIRPKKRLLVSKRAGTV-
human                 AGTQCTNCQTTTTTLWRRNASG-DPVCNACGLYYKLHQVNRPLTMRKDGIQTRNRKAS--
mouse                 AGTQCTNCQTTTTTLWRRNASG-DPVCNACGLYFKLHQVNRPLTMRKDGIQTRNRKAS--
rat                   AGTQCTNCQTTTTTLWRRNASG-DPVCNACGLYYKLHQVNRPLTMRKDGIQTRNRKAS--
fugu                  AGTCCANCQTTTTTLWRRNANG-DPVCNACGLYYKLHNVNRPLTMKKEGIQTPNRKMSN-
drosophila            AGLSCSNCHTTHTSLWRRNPAG-EPVCNACGLYYKLHSVPRPLTMKKDTIQKRKRK----
C.elegans             -GIECVNCRTNTTTLWRRNGEG-HPVCNACGLYFKLHKVERPITMKKDGIQTRNRKLSA-
yeast                 --TFCQNCATTNTPLWRRDESG-NPICNACGLYYKIHGVHRPVTMKKAIIKRRKRLV---
Neurospora            --TTCTNCFTQTTPLWRRNPDG-QPLCNACGLFLKLHGVVRPLSLKTDVIKKRNR-----
Arabidopsis           FQRRCSHCGTNNTPQWRTGPVGPKTLCNACGVRFKS-GRLCPEYRPADSPTFSNEIHSNL
Aspergillus(AreA)     GPTTCTNCFTQTTPLWRRNPEG-QPLCNACGLFLKLHGVVRPLSLKTDVIKKRNRSSANS
```

Figure 1. Amino acid sequences of several eukaryotic GATA DNA-binding domains. cGATA-1 (C) and (N), carboxy- and amino-terminal GATA-binding domains of the chicken GATA-1 factor (note that the last four residues of the amino-terminal domain are also the first four residues of the carboxy-terminal domain). *Drosophila melanogaster* Serpent, a one-finger GATA factor involved in regulating resistance to pathogens; C. elegans elt-1, a GATA like factor; Neurospora crassa WC1, a 'plant-like' GATA factor involved in photoinduction; Arabidopsis thaliana GATA-3, a factor of unknown function; AreA, a typical 'animal-like' fungal GATA-binding domain. The conserved cysteines involved in zinc binding are depicted in bold face.

one alanine provide the majority of contacts (Fig. 3). They are supplemented by several positively charged amino acids that form electrostatic interactions with the phosphate groups lining the outer edges of the major groove. In addition, residues in the carboxy-terminal tail contact the phosphate backbone. In the complex of the chicken GATA-1 the carboxy-terminal tail wraps around the DNA and crosses the minor groove, whereas the corresponding stretch of amino acids in the *A. nidulans* AreA runs along the edge of the minor groove, parallel to the phosphate backbone[11] (for example see Fig. 2A and Fig. 4A). The structure of the target DNA in the complex closely resembles classical B-type DNA in both complexes.

GATA Factor Activities

In vertebrates, GATA factors participate with other transcription factors in terminal differentiation. There are six GATA factors in mammals. The first discovered and best known, GATA-1, is essential for the differentiation of the erythroid and megaloblastoid cell lines.[1] Others participate in the differentiation of T cells, the endothelium of the gut, and in heart differentiation.[12] End-1 and Elt-1 are involved in the differentiation of the endoderm in *C. elegans*.[13] In *Drosophila* GATA factors regulate the immune response, bristle pattern and yolk protein synthesis.[14-16] Little is known of the function of GATA factors in plants. Although light-inducible promoters contain functionally significant GATA boxes, no involvement of a specific GATA factor has been demonstrated. An interesting case is the promoter of the nitrite reductase gene of *Arabidopsis thaliana*. This gene is induced both by light and nitrate. By analogy to fungi, GATA sites were found by in vivo footprinting in the nitrite reductase gene promoter.[17] To date, however, no GATA factor has been found to act as a specific regulator of nitrogen assimilation pathways in plants.

There are numerous GATA factor sequences in the available genomes, including metazoan genomes. Their number in the

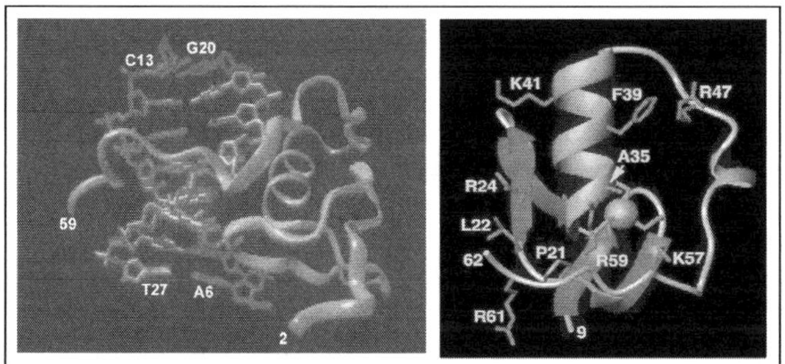

Figure 2. A) Structure of the C-terminal DBD of chicken GATA-1 complexed with the DNA target site. The protein backbone is shown in green and the color coding for the bases is as follows: A, red; T, lilac; G, dark blue; C, light blue. B) Core module of the DBD of AreA. The two anti-parallel β sheets are shown in red, the α helix in blue and the zinc ion as a purple sphere surrounded by the four coordinating cysteine residues in orange. Selected side-chains are shown in green. To view online color version, go to http://www.eurekah.com/chapter.php?chapid=1881&bookid=124&catid=30.

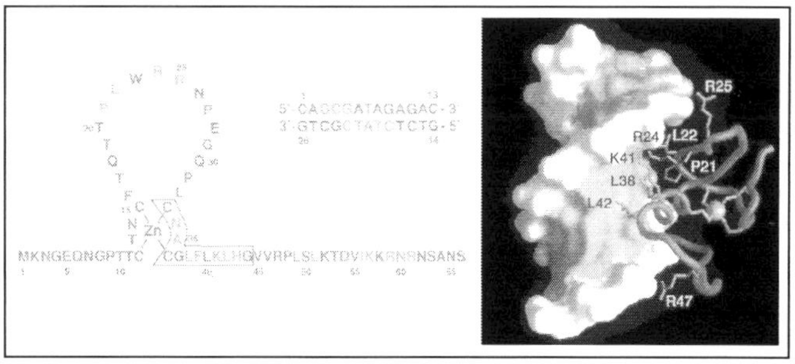

Figure 3. Amino acid and nucleotide sequences (left) and ribbon representation (right) of the DBD of the fungal GATA factor AreA complexed to the cognate CGATAG DNA target site. The protein backbone is depicted as a red tube and hydrophobic and hydrophilic amino acids that participate in DNA interactions in blue and green, respectively. The cysteine side-chains that coordinate the zinc ion (pink sphere) are shown in yellow. The major and minor groove surfaces of the DNA are shown in blue and orange, respectively. To view online color version, go to http://www.eurekah.com/chapter.php?chapid=1881&bookid=124&catid=30.

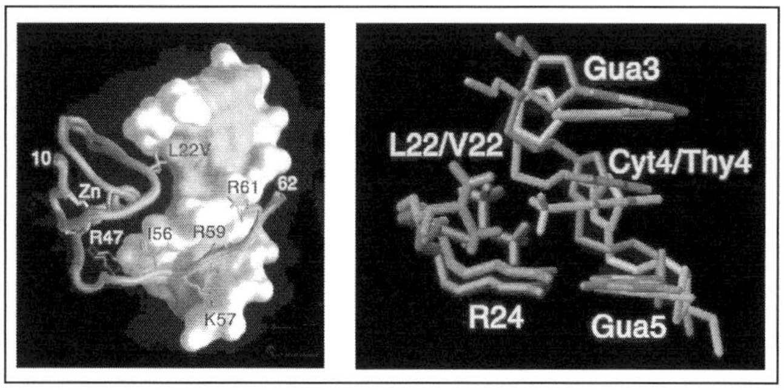

Figure 4. Comparison between the wild type and mutant AreA complexes with DNA. A) Tube representation of the protein backbones and space-filling representation of the DNA. Wild type AreA, red tube; mutant AreA, orange tube. The major and minor groove surfaces of the DNA are colored blue and orange, respectively. B) Detailed view of the interactions around the sites of mutation (residues 22-24). The wild type DBD backbone and side chains are shown in red and blue, respectively, while the mutant AreA backbone and side chains are shown in gold and green, respectively. The GCG element of the GCGATAG site is shown in red and the GTG element of the GTGATAG site is shown in tan. To view online color version, go to http://www.eurekah.com/chapter.php?chapid=1881&bookid=124&catid=30.

database is, however, considerably smaller than that for other transcription factors such as the classical zinc fingers. No functional conservation across the fungal and metazoan kingdoms or between vertebrates and invertebrates seems to have occurred. Although the DBD sequence motif and DNA binding specificity are conserved, the rest of the protein sequences are widely different. Thus, the same motif and binding specificity can serve different purposes in different contexts.

In ascomycetes the use of nitrogen sources involves transcriptional control at two levels: induction of specific genes and nitrogen metabolite repression.[18] For example, in order to induce nitrate and nitrite reductases, both the presence of nitrate and the absence of ammonium and glutamine are required. Significant chromatin rearrangement was observed that was independent of transcription and of the specific factor NirA, but strictly dependent on the GATA factor AreA.[19] The transcription of genes involved in the utilization of most nitrogen sources requires two transcription factors and each pathway is regulated specifically by one transcription factor of the zinc binuclear cluster class. The general regulator is always a GATA factor. The binding (or activation) of the specific factor is thought to require the presence of a specific inducer, the GATA factor, and a low intracellular concentration of glutamine. The overall process is very well understood and similar in *Aspergillus nidulans* and *Neurospora crassa*.

The master gene and GATA factor of nitrogen regulation is NIT2 in *N. crassa* and areA in *A. nidulans*.[18] The accepted model is that a negative-acting protein forms a complex with NIT2 (and AreA) in the presence of the repressing nitrogen sources, most likely mediated by glutamine. This negative regulator interacts both with residues in the DBD and in the carboxy terminus of AreA and NIT2, thereby preventing NIT2 and AreA binding to DNA. In the case of AreA, two other levels of regulation are present: first, transcription of *areA* is autoregulated and second, the stability of the *areA* mRNA is lower in mycelia grown on repressing nitrogen sources.[22]

In *S. cerevisiae* the basic pattern of regulation is the same but in this case four GATA factors are involved, all of which are interconnected in a complex regulatory network: Gln3p and Gat1p are transcriptional activators, while Dal80p and Gzf3p are repressors. The haploid genome of *S. cerevisiae* is most likely to be derived from an ancestral diploid and thus there have been two sets of genes able to diverge in the course of the evolution of the organism. The two filamentous fungi *A. nidulans* and *N. crassa* are rather 'wild' and represent probably the general situation in the ascomycetes, whereas baker's yeast is a very domesticated organism, therefore exhibiting the more complex regulatory mechanism.

"Vertebrate-Like" GATA Factors: Regulation of Siderophore Synthesis

The *urbs-1* gene of *Ustilago maydis* is involved in siderophore biosynthesis. This gene encodes a GATA factor with two typical GATA-DBDs. Unlike in vertebrates where the DNA-binding motifs of GATA factors are separated by only a few residues, the two motifs of Urbs-1 are separated by 92 amino acid residues.[23] As in the vertebrate factors, the DNA-binding finger is the carboxy-terminal domain. Regulators with similar sequences and functions have since been found in *N. crassa, A. nidulans* and *P. chrysogenum*[24] (but not in the genome of *S. cerevisiae*) implying a conservation of siderophore biosynthesis control.

"Plant-Like" GATA Factors: Light Regulation in *N. crassa*

Many functions in *N. crassa* are inducible by blue light, including conidiation, mycelial carotenoid biosynthesis and resetting of the circadian rhythm. The two loci involved, WC1 and WC2, encode "plant-like" GATA factors. These two factors contain typical PAS dimerization domains and act as dimers. A rapid light-induced turnover of WC1 in the WC1/WC2 complex is correlated with the induction of light-inducible promoters.

A Noncanonical Factor Involved in Mating-Type Switching

Although several unique features are present in its finger loop, Ash1p of budding yeast, clearly seems to be a GATA factor. Deletion of the gene results in massive daughter mating-type switching, whereas overexpression prevents mother switching.

The Leucine to Valine Mutant of AreA: A Paradigm for the Structural Basis of the Associated Phenotype

The Arst laboratory has mapped and characterized a myriad of mutations in the AreA DBD. As these mutations usually affect the growth of *A. nidulans* on all or some nitrogen sources, second site revertants can be isolated. Thus, the *A. nidulans* AreA system is a treasure trove, in which structural data can be uniquely correlated with functional studies both in vivo and in vitro.

The *areA*102 mutant grows exceptionally well on some amides but not at all on uric acid. The mutation responsible is a leucine to valine change at position 7 after the second cysteine (labeled L22V in the structures) (Fig. 4). This leucine in the seventh position of the zinc finger loop is conserved in almost all "animal-like" GATA factors (Fig. 1). Revertants with a valine to methionine change exhibit a "mirror image" mutant phenotype, thus they grow better on uric acid and worse on amides.[25] Specific suppressors map to the promoters of the uric acid permease genes. A number of these are mutations of CGATAG or AGATAG to TGATAG. All sites in the promoter region that are involved in utilization of amides are TGATAG sites. A detailed structural comparison of the interactions between the amino acids leucine and valine and their respective target bases for the wild type AreA/CGATG and the L22Vmutant/TGATAG complexes was carried out.[26] The global fold and the overall binding mode was found to be very similar for both DBDs (Fig. 4A). The positions for all sugars and bases are almost identical, with the exception of those directly surrounding the C/T mutation. The most significant perturbations to the protein structure occur around the leucine to valine mutation site. The protein backbone shifts slightly, exhibiting the largest displacement relative to the wild type coordinates. Also, the positioning of the valine side chain relative to the base is clearly altered and different $\chi 1$ angles are observed. As a result, the $\gamma 2$ methyl group of valine occupies a very similar position to that of the $\delta 1$ methyl of the wild type leucine. In the structure of the wild type AreA DBD with the wild type AGATAG DNA, the leucine side chain is involved in a hydrophobic contact with the first base, whether that be A, C, or T. In the mutant protein, the shorter valine side chain makes van der Waals contacts with the bulky methyl of the T. Thus, the missing methyl group of valine compared to leucine is compensated for by a methyl group now provided by the DNA (Fig. 4B). In another mutant, the longer methionine can interact well with an A or C, but clashes with a T. Therefore, the introduction of a T preceding the GATA sequence effectively compensates for the shorter valine side chain in the mutant, restoring a closed-packed hydrophobic interface. Based on these structural results with AreA, the DBDs of *C. elegans* and STALKY that contain an alanine and a threonine in the equivalent position should only recognize TGATAR sequences.

The GATA DBD Is Capable of More Than Just DNA Binding

Given the similarity of GATA DBDs, it is not easy to understand how promoter specificity is achieved. It has been established that GATA domains of mammalian and fungal proteins are able to interact with other proteins, including other zinc finger containing ones.[27] AreA/NIT2 can interact via its DBD with the NMR protein and the specific regulator NirA/NIT4. Whether further interactions with other specific regulators occur and whether the determinants for specificity reside in the promoters has not been established yet and requires further study.

GATA Factors and Chromatin Rearrangement

GATA-1 is stringently required in the maintenance of DNase I hypersensitive sites in the chicken- and human globin gene cluster enhancers[28,29] and it was demonstrated that GATA factors are capable of destabilizing nucleosomes in vitro.[30] It therefore seemed reasonable to suspect involvement of GATA factors in nucleosome maintenance. The nitrite and nitrate reductase bidirectional promoter of *A. nidulans* contains six positioned nucleosomes. In the presence of nitrate and the absence of ammonium, these nucleosomes are no longer observed, independent of transcription or the presence of the NirA-specific factor. AreA is absolutely required in this positioning.[20] Mutations in the AreA-binding domain result in constitutive chromatin rearrangement. Given that GATA-1 itself is acetylated,[31] it may well be that the vertebrate GATA-1 DBDs influence directly or indirectly histone acetylation.

Conclusions

GATA factors are widespread eukaryotic regulators most likely evolved from an ancestral GATA protein containing only a single type IV zinc finger. In vertebrates, evolution of GATA factors involved gene duplication whereas additional modular evolution in nonvertebrates has occurred. Vertebrate GATA factors stimulate gene transcription in hematopoietic development while in fungi they act either as transcription factors or repressors in a number of different processes, ranging from nitrogen source utilization to mating-type switching. The DBDs recognize DNA in a rather special fashion: most of the specific interactions with the bases in their respective target sites are hydrophobic. Mutational studies based on the powerful selection techniques possible in fungi, coupled with high-resolution structural work, provided a paradigm for the study of this type of side chain-base interactions. In addition, it was possible to elucidate interactions with other proteins and of the role of transcription factors in chromatin remodeling.

Acknowledgements

I am grateful to H. N. Arst Jr, C. Scazzocchio and members of the Felsenfeld laboratory for sharing results and/or helpful discussions.

References

1. Orkin SH. GATA-Binding transcription factors in hematopoietic cells. Blood 1992; 80:575-581.
2. Scazzocchio C. The fungal GATA factors. Curr Op Microbiol 2000; 3:126-131.
3. Merika M, Orkin SH. DNA-binding specificity of GATA family transcription factors. Mol Cell Biol 1993; 13:3999-4010.
4. Ko LJ, Engel JD. DNA-binding specificities of the GATA transcription factor family. Mol Cell Biol 1993; 13:4011-4022.
5. Yang HY, Todd E. Distinct roles for the two cGATA-1 finger domains. Mol Cell Biol 1992; 12:4562-4570.
6. Fox AH, Kowalski K, King GF et al. Key residues characteristic of GATA N-fingers are recognized by FOG. J Biol Chem 1998; 273:33595-33603.
7. Ballario P, Vittorioso P, Magrelli A et al. White collar-1, a central regulator of blue light responses in Neurospora, is a zinc finger protein. EMBO J 1996; 15:1650-1657.
8. Chang WT, Newell PC, Gross JD. Identification of the cell fate gene stalky in Dictyostelium. Cell 1996; 87:471-481.
9. Teackle GR, Gilmartin PM. Two forms of type IV zinc-finger motif and their kingdom-specific distribution between the flora, fauna and fungi. Trends Biochem Sci 1998; 23:100-102.
10. Omichinski JG, Clore GM, Schaad O et al. NMR structure of a specific DNA complex of Zn-containing DNA binding domain of GATA-1. Science 1993; 261:438-446.
11. Starich MR, Wikström M, Arst Jr HN et al. The solution structure of a fungal AREA protein-DNA complex: An alternative binding mode for the basic carboxyl tail of GATA factors. J Mol Biol 1998; 277:605-620.
12. Gao X, Sedgwick T, Shi YB et al. Distinct functions are implicated for the GATA-4,-5 and-6 transcription factors in the regulation of intestine epithelial cell differentiation. Mol Cell Biol 1998; 18:2901-2911.
13. Zhu J, Hill RJ, Heid PJ et al. end-1 encodes an apparent GATA factor that specifies the endoderm precursor in Caenorhabditis elegans embryos. Genes Dev 1997; 11:2883-2896.
14. Petersen UM, Kadalayil L, Rehorn KP et al. Serpent regulates Drosophila immunity genes in the larval fat body through an essential GATA motif. EMBO J 1999; 18:4013-4022.
15. Lossky M, Wensink PC. Regulation of Drosophila yolk protein genes by an ovary-specific GATA factor. Mol Cell Biol 1995; 15:6943-6952.
16. Haenlin M, Cubadda Y, Blondeau F et al. Transcriptional activity of Pannier is regulated negatively by heterodimerization of the GATA DNA-binding domain with a cofactor encoded by the u-shaped gene of Drosophila. Genes Dev 1997; 11:3096-3108.
17. Rastogi R, Bate NJ, Sivasankar S et al. Footprinting of the spinach nitrite reductase gene promoter reveals the preservation of nitrate regulatory elements between fungi and higher plants. Plant Mol Biol 1997; 34:465-476.
18. Arst Jr HN , Cove DJ. Nitrogen metabolite repression in Aspergilus nidulans. Mol Gen Genet 1973; 126:111-141.
19. Muro-Pastor MI, Strauss J, González R et al. The GATA factor AreA isessential for chromatin remodelling in an eucaryotic bidirectional promoter. EMBO J 1999; 18:1584-1597.
20. Marzluf GA. Genetic regulation of nitrogen metabolism in the fungi. Microbiol Mol Biol Rev 1997; 61:17-32.
21. Xiao X, Fu Y-H, Marzluf GA. The negative-acting NMR regulatory protein of Neurospora crassa binds to and inhibits the DNA-binding activity of the positive-acting nitrogen regulatory protein NIT2. Biochem 1995; 34:8861-8868.

22. Platt A, Langdon T, Arst Jr HN et al. Nitrogen metabolite signalling involves the C-terminus and the GATA domain of the Aspergillus transcription factor AREA and the 3' untranslated region of its mRNA. EMBO J 1996; 15:2791-2801.
23. Voisard C, Wang J, McEvoy JL et al. urbs1, a gene regulating siderophore biosynthesis in Ustilago maydis, encodes a protein similar to the erythroid transcription factor GATA-1. Mol Cell Biol 1993; 13:7091-7100.
24. Haas H, Zadra I, Stoffler G et al. The Aspergillus nidulans GATA factor SREA is involved in regulation of siderophore biosynthesis and control of iron uptake. J Biol Chem 1999; 274:4613-4619.
25. Kudla B, Caddick MX, Langdon T et al. The regulatory gene areA mediating nitrogen metabolite repression in Aspergillus, nidulans. Mutations, affecting specific gene activation alter a loop residue of a putative zinc finger. EMBO J 1990; 9:1355-1364.
26. Starich MR, Wikström M, Schumacher S et al. The solution structure of the Leu22Val mutant AREA DNA binding domain complexed with a TGATAG core element defines a role for hydrophobic packing in the determination of specificity. J Mol Biol 1998; 277:621-634.
27. Mackay JP, Crossley M. Zinc fingers are sticking together. Trends Biochem Sc 1998; 23:1-4.
28. Stamatoyannopoulos JA, Goodwin A, Joyce T et al. NF-E2 and GATA binding motifs are required for the formation of Dnase I hypersensitive site 4 of the human –globin locus control region. EMBO J 1995; 14:106-116.
29. Boyes J, Felsenfeld G. Tissue-specific factors additively increase the probability of the all-or-none formation of a hypersensitive site. EMBO J 1996; 15:2496-2507.
30. Boyes J, Omichinski J, Clark D et al. Perturbation of nucleosome structure by the erythroid transcription factor GATA-1. J Mol Biol 1998; 279:529-544.
31. Boyes J, Byfield P, Nakatani Y et al. Regulation of activity of the transcription factor GATA-1 by acetylation. Nature 1998; 396:594-598.

CHAPTER 6

MutM: Single C_2C_2 Zinc Finger-DNA Interaction

Ryoji Masui, Noriko Nakagawa and Seiki Kuramitsu*

Abstract

The prokaryotic MutM protein is a trifunctional DNA base excision repair enzyme that removes a wide range of oxidatively damaged bases, especially 8-oxoguanine, (N-glycosylase activity) and cleaves both the 3'- and 5'-phosphodiester bonds of the resulting apurinic/apyrimidinic site (AP lyase activity). This enzyme possesses a zinc finger motif (-Cys-X_2-Cys-X_{16}-Cys-X_2-Cys-) at the C terminus, which forms a β-hairpin loop. The positively charged Arg247 and Arg253 on the β-hairpin loop interact with the phosphate groups to pinch the backbone of the lesion-containing strand. This feature establishes that the zinc finger motif is essential for binding of MutM to DNA as well as its enzymatic activities.

Introduction

In aerobic organisms, cellular DNA is frequently damaged by activated oxygen species produced as a result of aerobic energy metabolism or oxidative stress. In particular, the 8-oxoguanine (GO) lesion is one of the most stable products of oxidative DNA damage.[1] GO can base pair with adenine as well as with cytosine, resulting in a G:C to T:A transversion.[2] The GO repair system composed of the MutM protein, also called formamidopyrimidine DNA glycosylase (Fpg), accompanied by MutY and MutT, prevents this mutation (Fig. 1).[3] The MutM protein is a trifunctional DNA base excision repair enzyme that removes a wide range of oxidatively damaged bases (N-glycosylase activity) and cleaves both the 3'- and 5'-phosphodiester bonds of the resulting apurinic/apyrimidinic site (AP lyase activity) (Fig. 2).[4]

The MutM protein is highly conserved across a wide range of organisms. Bacterial MutM proteins (Mr 30 kD) possess the invariant N-terminal sequence, PELPEV, two strictly conserved lysine residues, and a zinc finger motif (-Cys-X_2-Cys-X_{16}-Cys-X_2-Cys-) at the C terminus (Fig. 3). As MutM protein functions as a monomer, this enzyme belongs to the group possessing a single zinc finger motif of the C_2C_2-type. The length of the sixteen residues between two cysteines clusters is relatively short in comparison with other zinc finger motifs. This region is relatively rich in basic residues.

Recently, MutM homologues have been identified in the genome database of higher eukaryotes, including man.[5,6] These gene products are thought to constitute an alternative pathway for the repair of GO and other oxidative lesions, in addition to Ogg1, another eukaryotic 8-oxoguanine DNA glycosylase with activities identical to those of MutM. As MutY and MutT homologues have already been found in many eukaryotes, this finding demonstrates the generality of the GO system as a defense against oxidative damage.

Zinc in MutM

Atomic absorption spectrophotometric analysis has demonstrated that each MutM protein molecule possesses one zinc atom.[7] Extended X-ray absorption fine structure spectra show directly that the metal is coordinated to the sulfur atoms of four cysteine residues. The zinc atom can be displaced by other divalent metal cations. Cd^{2+}, $Cu^{2+}$and Hg^{2+} inhibit the activity of MutM,[8] although Co^{2+}-substituted MutM is still active.[9] The importance of the zinc finger motif for the activity has been indicated by site-directed mutagenesis and chemical modification of the cysteine residues within the zinc finger motif of the *E. coli* MutM protein.[8] Mutant MutM proteins with a single substitution of each cysteine residue by glycine show decreased capacity to bind DNA as well as decreased enzymatic activity; they still have a zinc ion, but bind the ion with much lower affinity than the wild type. The double mutant (C244S, C247S) lacks zinc without significant alteration of the secondary structure.[10] The cysteine residues within the zinc finger motif of the MutM protein are the primary sites of NO interaction, which irreversibly inhibits its activity.[11]

3D Structure of MutM

Precise coordination to a zinc ion in MutM proteins has been revealed by studies of the three-dimensional structure of *Thermus thermophilus* MutM.[12] The crystal structure was determined at 1.9 Å resolution with multiwavelength anomalous diffraction phasing using the intrinsic zinc ion of the zinc finger. MutM is composed of two domains connected by a flexible hinge (Fig. 4A). The N-terminal domain is a two-layered β-sandwich structure composed of nine antiparallel β strands. The C-terminal domain has four helices and two β strands that form a zinc finger motif. Most of the conserved residues including the catalytic residues are clustered in a large, electrostatically positive cleft between the domains. The two DNA-binding motifs, the helix-two turns-helix (H2TH) and zinc finger motifs, are also located in this region (Fig. 4C). The β-hairpin loop formed by the zinc finger motif is directed to the domain interface. The zinc

* Corresponding author. See list of "Contributors".

Zinc Finger Proteins: From Atomic Contact to Cellular Function, edited by Shiro Iuchi and Natalie Kuldell.

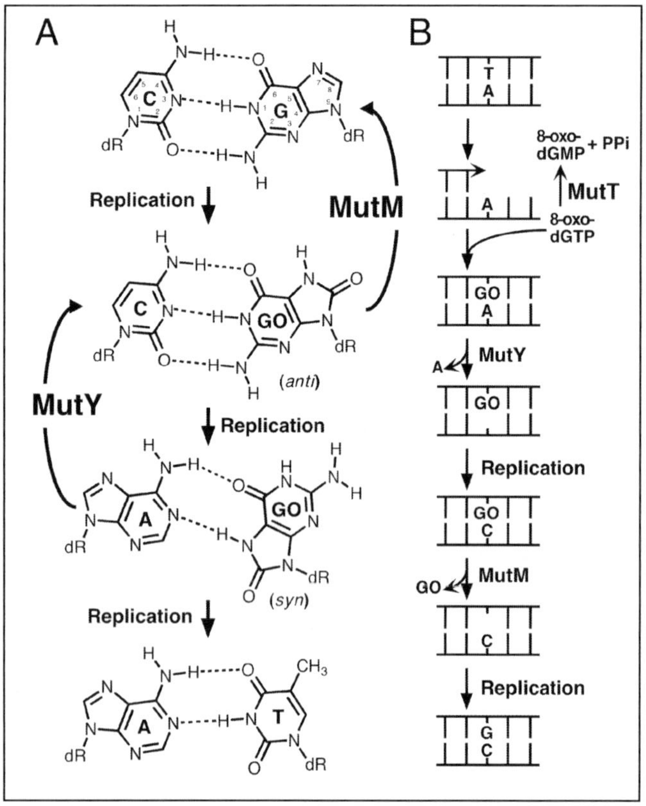

Figure 1. The GO system against mutations caused by GO. A) System acting against the occurrence of GO-containing base pairs in DNA. MutM removes an 8-oxoguanine base from the GO:C pair, whereas MutY removes an adenine base from the GO:A pair. These enzymes prevent G:C → T:A transversion caused by GO. B) System acting against incorporation of 8-oxo-dGTP into DNA upon replication. MutT catalyzes the degradation of 8-oxo-dGTP to 8-oxo-dGMP and pyrophosphate, which prevents T:A → G:C transversion.

finger motif of MutM is a single β-hairpin loop that extends into the inter-domain cleft.

To illustrate the features of DNA recognition by MutM, we first determined the 2.1 Å structure of *T. thermophilus* MutM bound to DNA containing GO opposite cytosine. A mutant MutM protein in which Glu2 was changed to alanine was employed for cocrystallization, as it has been shown to lack N-glycosylase activity but retain AP-lyase activity as well as its ability to bind DNA. In complex with a GO:C-containing 11 bp oligonucleotide, this E2A mutant protein yielded co-crystals

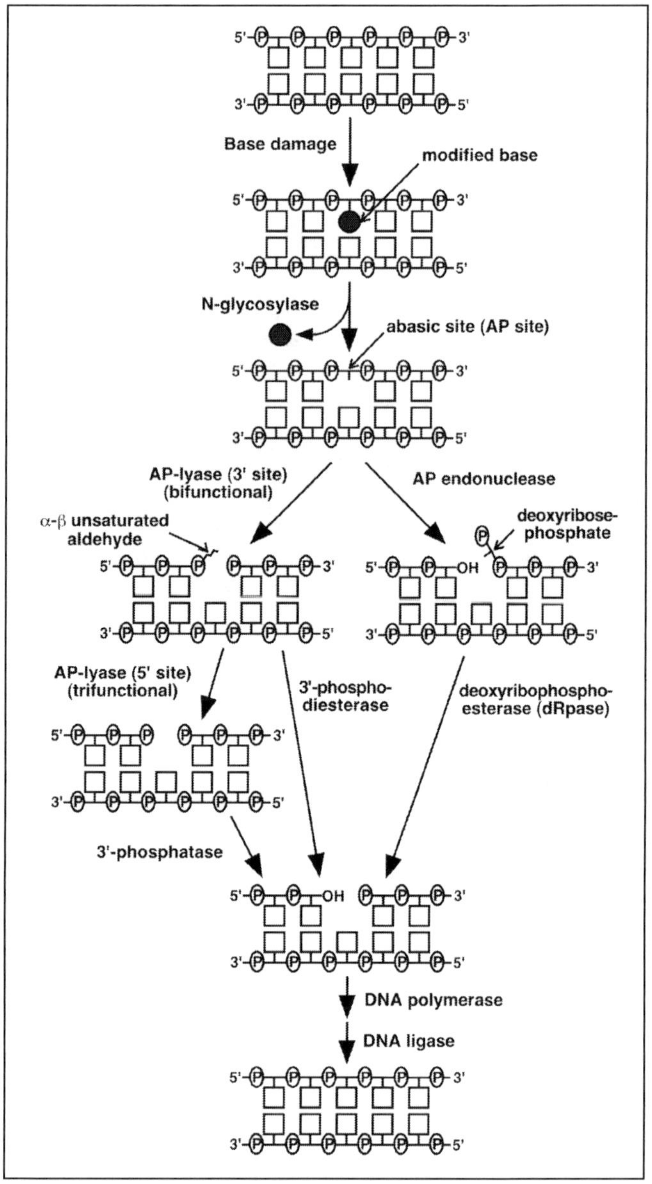

Figure 2. General pathways of base excision repair. Enzymic activities participating in each step are represented. For instance, MutM possesses N-glycosylase, AP-lyase (3' site), and AP-lyase (5' site) (and dRpase) activities. *E. coli* endonuclease IV and exonuclease III function as an AP endonuclease, but also have 3'-phosphodiesterase and 3'-phosphatase activities.

```
zinc finger motif                        β10      β11
                               240       250       260
Thermus thermophilus MutM       ..GLPCPACGRPVERRVVAGRGTHFCPTCQGEGP---- (266)
Escherichia coli MutM           ..GEPCRVCGTPIVATKHAQRATFYCRQCQK------- (268)
Synechococcus elongatus MutM    ..GEACRVCGTTIERLRLAGRSSHYCPQCQPLSSAIGK (283)
Bacillus stearothermophilus MutM ..GNPCKRCGTPIEKTVVAGRGTHYCPRCQR------- (273)
Listeria innocua MutM           ..GEPCVICGTPIEKIKLNGRGTHFCPHCQK------- (272)
                                    *  **          *    *  *
Arabidopsis thaliana MutM HI    ..PGKAFVDGKKIDFITAGGRTTAYVPELQKLYGKD.. (281)
Homo sapiens NEH1               ..PSRTRRAKRDLPKRTATQRPEGTSLQQDPEAPTV.. (354)
```

Figure 3. Aligned sequences of the zinc finger motifs from bacterial MutM proteins. The residue number of the *T. thermophilus* MutM sequence is shown in the uppermost line. Conserved residues are marked with asterisks.

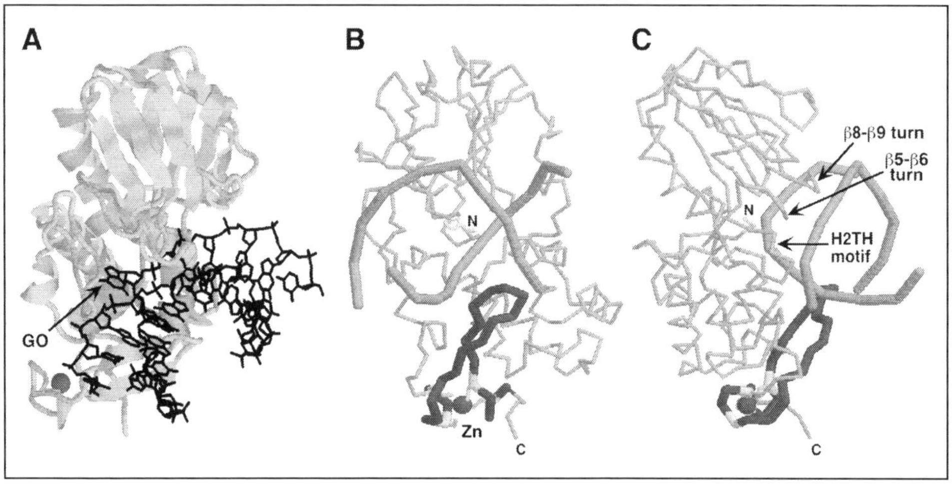

Figure 4. Overall structure of the MutM-DNA complex. A) Ribbon diagram of the complex. The DNA duplex is drawn by wireframes. A zinc atom is shown as a sphere. B) Wireframe diagram of the complex. DNA backbones are shown as thick wire-frames. A β-hairpin loop formed by the zinc finger motif is also shown as a thick wire-frame. Side chains of cysteine residues are diplayed as sticks. The letters N and C represent the N- and C-termini, respectively. C) Same as in A but viewed from a different perspective. The DNA interfaces besides the zinc finger motif are shown by arrows. Images were produced using RasMol.[15]

suitable for structure determination. In the resolved structure of the MutM-DNA complex, the N- and C-terminal domains mainly latch the backbone of the GO-containing DNA strand of the duplex (Fig. 4B). The DNA helix is sharply bent (~60°) at the locus of the GO site, but holds an almost canonical B-form conformation outside the lesion.

The substrate GO residue is fully extruded from the helix and is inserted deeply into an active site in the cleft on the enzyme (Fig. 4A). The catalytically important residues, Pro1, Glu2, and Lys52, are located near the GO residue in the active site. The space vacated by GO is filled with the side chains of Met70, Arg99, and Phe101 inserted from the minor groove side of the DNA. These residues interact with the opposite and neighboring bases, thus helping to stabilize the complex. The distorted conformation of the DNA is also enforced with extensive interactions between the protein and the backbone of the DNA. His67, Asn161, Tyr231, Arg247, and Arg253 form hydrogen bonds with nonbridging oxygens of phosphate moieties in the GO-containing strand, whereas His81 and Arg100 do so in the complementary strand.

Zinc Finger Motif of MutM

Four highly conserved regions are involved in DNA binding:

1. the loop between β5 and β6, containing His67 and Met70;
2. the loop between β8 and β9, containing Arg99, Arg100, and Phe101;
3. the H2TH motif, containing Asn161; and
4. the zinc finger motif, containing Arg247 and Arg253.

Among them, the zinc finger motif contacts the DNA helix from the side of the major groove, whereas the other three regions do so from the side of the minor groove (Fig. 4C). The positively charged Arg253, an invariant residue among MutM family proteins, on the edge of the b-hairpin loop, can connect the phosphate groups 5' and 3' to the GO residue (Fig. 5). This interaction causes backbone pinching of the GO-containing strand, which may help to extrude the GO moiety from the helix. Also, Arg253 interacts with the phosphate group on the 5' side of the residue neighboring GO (Fig. 5). Although Arg247 is not completely conserved among the MutM protein family, the corresponding lysine residue of *Bacillus stearothermophilus* MutM similarly interacts with the backbone phosphate (Fig. 3).[13] These interactions demonstrate that the zinc finger motif plays a direct role in the binding of MutM to substrate DNA.

Eukaryotic MutM homologues have no zinc finger motif at the C terminus, although they retain the same activities as MutM. Despite the lack of sequence homology in the C-terminal region, the eukaryotic MutM proteins possess a conserved arginine residue at the position corresponding to Arg253. This arginine may play a role similar to that of Arg253. Although no zinc

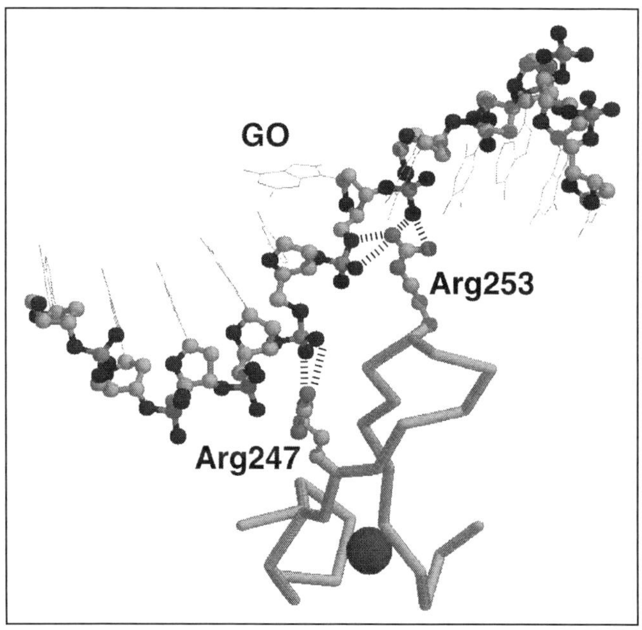

Figure 5. Interaction of two arginine residues with the GO-containing strand. The backbone of the DNA strand and side chains of Arg247 and Arg253 are represented as a ball-and-stick model. Short parallel lines denote hydrogen bonds. Images were produced using RasMol.[15]

finger motif has been demonstrated in other DNA glycosylases, a β-hairpin loop resembling the zinc finger motif of MutM has been reported.[14]

References

1. Dizdaroglu M. Formation of an 8-oxoguanine moiety in deoxyribonucleic acid on γ-irradiation in aqueous solution. Biochemistry 1985; 24:4476-4481.
2. Wood ML, Dizdaroglu M, Gajewski E et al. Mechanistic studies of ionizing radiation and oxidative mutagenesis: genetic effects of a single 8-hydroxyguanine (7-hydro-8-oxoguanine) residue inserted at a unique site in a viral genome. Biochemistry 1990; 290:7024-7032.
3. Michaels ML, Miller JH. The GO system protects organisms from the mutagenic effect of the spontaneous lesion 8-hydroxyguanine (7,8-dihydro-8-oxoguanine). J Bacteriol 1992; 174:6321-6325.
4. Bhagwat M, Gerlt JA. 3'- and 5'-strand cleavage reactions catalyzed by the Fpg protein from Escherichia coli occur via successive beta- and delta-elimination mechanisms, respectively. Biochemistry 1996; 35:659-665.
5. Hazra TK, Izumi T, Boldogh I et al. Identification and characterization of a human DNA glycosylase for repair of modified bases in oxidatively damaged DNA. Proc Natl Acad Sci USA 2002; 99:3523-3528.
6. Morland I, Rolseth V, Luna L et al. Human DNA glycosylases of the bacterial Fpg/MutM superfamily: an alternative pathway for the repair of 8-oxoguanine and other oxidation products in DNA. Nucleic Acids Res 2002; 30:4926-4936.
7. Boiteux S, O'Connor TR, Lederer F et al. Homogeneous Escherichia coli FPG protein. A DNA glycosylase which excises imidazole ring-opened purines and nicks DNA at apurinic/apyrimidinic sites. J Biol Chem 1990; 265:3916-3922.
8. O'Connor TR, Graves RJ, de Murcia G et al. Fpg protein of Escherichia coli is a zinc finger protein whose cysteine residues have a structural and/or functional role. J Biol Chem 1993; 268:9063-9070.
9. Buchko GW, Hess NJ, Bandaru V et al. Spectroscopic studies of zinc(II)- and cobalt(II)-associated Escherichia coli formamidopyrimidine-DNA glycosylase: extended X-ray absorption fine structure evidence for a metal-binding domain. Biochemistry 2000; 39:12441-12449.
10. Tchou J, Michaels ML, Miller JH et al. Function of the zinc finger in Escherichia coli Fpg protein. J Biol Chem 1993; 268:26738-26744.
11. Wink DA, Laval J. The Fpg protein, a DNA repair enzyme, is inhibited by the biomediator nitric oxide in vitro and in vivo. Carcinogenesis 1994; 15:2125-2129.
12. Sugahara M, Mikawa T, Kumasaka T et al. Crystal structure of a repair enzyme of oxidatively damaged DNA, MutM (Fpg), from an extreme thermophile, Thermus thermophilus HB8. EMBO J 2000; 19:3857-3869.
13. Fromme JC, Verdine GL. Structural insights into lesion recognition and repair by the bacterial 8-oxoguanine DNA glycosylase MutM. Nat Struct Biol 2002; 9:544-552.
14. Scharer OD, Jiricny J. Recent progress in the biology, chemistry and structural biology of DNA glycosylases. Bioessays 2001; 23:270-281.
15. Sayle R, Milner-White EJ. RASMOL: biomolecular graphics for all. Trends Biochem. Sci 1995; 20:374.

CHAPTER 7

Homing Endonuclease I-TevI: An Atypical Zinc Finger with a Novel Function

Patrick Van Roey,* Marlene Belfort and Victoria Derbyshire

Summary

I-TevI is a site-specific, sequence-tolerant homing endonuclease encoded by the *td* intron of bacteriophage T4. The enzyme consists of an N-terminal catalytic domain and a C-terminal DNA-binding domain that are connected by a long, flexible linker. The crystal structure of the DNA-binding domain of I-TevI, residues 130 to 245, complexed with the 20-bp primary binding region of its DNA target, reveals the presence of a zinc finger, comprising residues 151 to 167, that makes backbone contacts with the DNA from the minor groove. Biochemical data have shown that the zinc finger does not contribute to the DNA-binding affinity or to the specificity of the enzyme, but rather that it has a novel function and acts as a distance determinant that controls the relative positions of the catalytic and DNA-binding domains.

Introduction

Group I introns are intervening sequences within the coding sequence of genes that are excised by RNA splicing. They can also be mobile genetic elements at the DNA level.[1] If an intron-containing gene is in close proximity to an intron-less version of the same gene, there will be a gene conversion event whereby the intron-less copy of the gene becomes intron-plus. This process is called intron homing because the intron "goes home". Group I intron homing is brought about by the action of an endonuclease that is encoded by an ORF located in the RNA intron. The endonuclease recognizes the exon:exon junction of the intron-less allele—a contiguous sequence that is not present in the intron-plus version because of the presence of the intron. The endonuclease binds to the intron-less cognate allelle and introduces a double-stranded break. Then, the intron-plus gene aligns with the cleaved version, and recombination and repair events lead to copying of the intron, resulting in two intron-plus alleles.

Homing endonucleases are grouped into at least four different families that are defined on the basis of conserved sequence elements as the GIY-YIG, LAGLIDADG, H-N-H and His-Cys box families.[1,2] They are site-specific DNA endonucleases that recognize long DNA targets of 14-40 bp, with some degree of sequence tolerance. The biological function of homing endonucleases is to cleave intronless sites so that their host intron can be propagated. While a certain level of sequence specificity in both binding and cleavage is important to minimize random cleavage, precise cleavage is not essential, since the host's recombination and replication functions will repair the cleavage event as part of the homing process. In fact, the ability to cleave variant substrates is advantageous to both the endonuclease and the host intron.

GIY-YIG family members contain up to five conserved sequence motifs that make up the GIY-YIG module with essentially no similarity among the proteins beyond that.[3] In addition to the GIY-(9 to 10 residue)-YIG sequence, this module includes several other highly conserved residues, some of which have been shown to be critical for catalytic activity. These include Tyr1 Arg27, Glu75, and Asn90 (I-TevI sequence numbers).

I-TevI is encoded by a group I intron in the thymidylate synthase (*td*) gene of bacteriophage T4, and is the best characterized of the GIY-YIG enzymes and the first for which structural information is available (Fig. 1). I-TevI specifically recognizes its 37-bp DNA substrate, or homing site, as a monomer. Despite nanomolar affinity, the enzyme exhibits a high degree of sequence tolerance.[4]

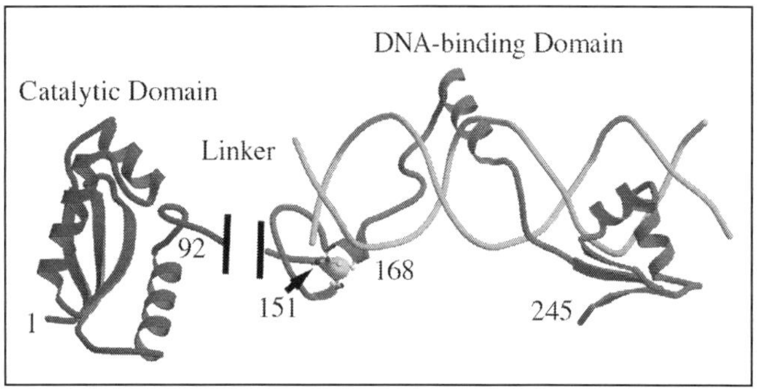

Figure 1. Overview of three-dimensional structural information currently available for I-TevI. The enzyme consists of an N-terminal catalytic domain joined to the C-terminal DNA-binding domain by a long, flexible linker that remains to be structurally characterized. The DNA-binding domain itself has an elongated structure and could only be crystallized in the presence of its DNA substrate. To view online color versions of images, go to http://www.eurekah.com/eurekahlogin.php?chapid=1734&bookid=124&catid=30.

* *Corresponding author. See list of "Contributors".*

Zinc Finger Proteins: From Atomic Contact to Cellular Function, edited by Shiro Iuchi and Natalie Kuldell.

No single nucleotide in the 37-bp target is essential for binding and cleavage, and many multiple substitutions are well tolerated.[5] Consistent with this sequence tolerance, ethylation and methylation interference studies indicated that most of the protein-DNA contacts are via the minor groove and the phosphate backbone.[4] The primary binding region of the enzyme is approximately 20 bp in length, spanning the intron insertion site (IS), with a second region of contact close to the cleavage site (CS), which is 23-25 bp upstream of the IS. In addition, I-TevI can tolerate insertions or deletions between the CS and IS, and still effect cleavage.[5]

I-TevI consists of two functionally distinct domains, an N-terminal catalytic domain and a C-terminal DNA-binding domain, that are separated by a long, flexible linker.[6] The catalytic domain contains the GIY-YIG module common to family members and forms a discrete structural domain contained within the first 92 amino acids of the protein.[3,7] This domain has an α/β-fold with a central three-stranded β-sheet surrounded by three α-helices. The GIY and YIG sequences are located on the first and second β-strands, respectively, and are important structurally as well as functionally. A surface composed primarily of the first and third α-helices and the second β-strand is thought to provide the main substrate interaction site. Three residues that are known to be important for catalytic activity, Tyr17, Arg27, and Glu75, are exposed on this surface. Evidence of a metal-binding site between Tyr17 and Glu75, as well as similarity in the arrangement of the three residues with that of the catalytically important residues of the His-Cys box homing endonuclease I-PpoI[8,9] (despite completely different protein folds) are indicative of similarities in the DNA cleavage activities of the GIY-YIG and His-Cys box families.[7]

The DNA-binding domain of I-TevI was initially defined to be contained between residues 130 and the C terminus of the 245-amino acid enzyme, based on footprinting and limited proteolysis experiments[6] The crystal structure[10] of this moiety, in complex with a 20-bp duplex DNA that corresponds to the primary binding region of the enzyme, shows that residues 151 to 245 consist of an extended modular structure that interacts with the full length of the DNA duplex (Fig. 2). The molecule consists of three distinct subdomains that are connected by elongated segments that lack secondary structure: a zinc finger (residues 151 to 167); a minor groove-binding α-helix (residues 183 to 194); and, a helix-turn-helix subdomain (residues 204 to 245). There are phosphate backbone contacts throughout the full length of the complex. The elongated segments, which are inserted in the minor groove, provide the only hydrogen bonding interactions with bases. However, due to the poor information content of the minor groove, these are of limited value with respect to substrate specificity. In contrast, the helix-turn-helix subdomain, inserted into the major groove, makes hydrophobic contacts with the C5-methyls of thymines and thus is selective for an AT-rich region.

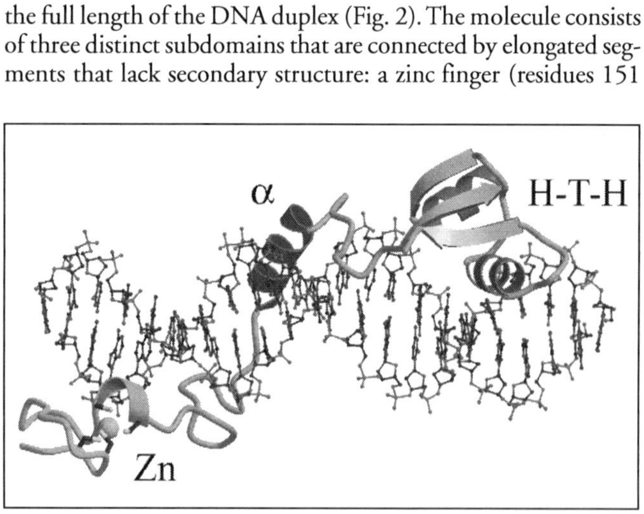

Figure 2. Structure of the DNA-binding domain of I-TevI. The DNA-binding domain has an extended fold that consists of three subdomains: a zinc finger (Zn), an α-helix inserted into the minor groove (α) and a helix-turn-helix (H-T-H). Segments that lack secondary structure (orange) connect the three subdomains but are important for DNA-binding as they are the only sites of direct hydrogen bonding contacts with the bases.

The biochemical and structural data present a picture of I-TevI as a catalytic GIY-YIG domain, common to all family members, connected by a long linker of more than 50 amino acids to a DNA-binding domain that itself is composed of several sub-domains. Sequence analysis indicates that the linker and the DNA-binding domain vary greatly among the different members of the family, with the exception of the C-terminal helix-turn-helix subdomain which appears to be more conserved, at least among a subset of family members. It is likely that the variations in the composition (and structure) of the linkers and of the DNA-binding domains are important for the recognition of the large and highly variable substrates. Furthermore, the GIY-YIG homing endonucleases appear to have different separations between their primary recognition sites and their cleavage sites, requiring differences in the relative positioning of the catalytic and DNA-binding domains, which can be achieved through alternative linker structures and domain interactions.

Structure of the Zinc Finger Subdomain

The determination of the crystal structure of the DNA-binding domain of I-TevI revealed the presence of a zinc finger.[10] This zinc finger was not anticipated from sequence analysis,[3] because it does not occur in any other GIY-YIG endonucleases and because of its unusual sequence, $CXCX_{10}CX_2C$. Sequence database searches indicate the presence of Cys_4-zinc fingers with similar separations of the cysteines in the nuclear factor NHR-63 of *Caenorhabditis elegans*[11] and in the β'-subunits of bacterial RNA polymerases. However, none of these zinc fingers have been structurally characterized. The structure of NHR-63 has not yet been reported while the zinc finger sequence lies just outside the resolved three-dimensional structure of the RNA polymerase of *Thermus aquaticus*.[12] The fact that I-TevI is the only GIY-YIG protein characterized to date that contains a zinc finger suggests that this element plays a specific role in the function of I-TevI, beyond mere DNA recognition.

The I-TevI zinc finger (Fig. 3) consists of two type-I β-turns, followed by a small loop and a single-turn α-helix. The first two cysteines, 151 and 153, are the first and third residues in the first β-turn, while the last two cysteines, 164 and 167, are the first and last residues of the helix. The zinc finger of I-TevI has been classified among the treble clef group,[11] although it is the smallest known member of the group and is quite degenerate. The treble clef zinc finger typically consists of a β-hairpin, a loop, and an α-helix. The first two cysteines are contributed by the zinc knuckle, a unique turn with a CPXCG consensus sequence that joins the two β-strands. In I-TevI, the corresponding sequence, CKCG, forms a tighter β-turn, bringing the two short (three residues each) β-strands closer together. Krishna and coworkers[11] further suggested that, due to the shortness of the α-helix, the I-TevI zinc finger could also be classified as a zinc ribbon, albeit one that is lacking a β-strand. Interestingly, several Cys_4-zinc fingers of the treble clef and zinc ribbon groups are known to be

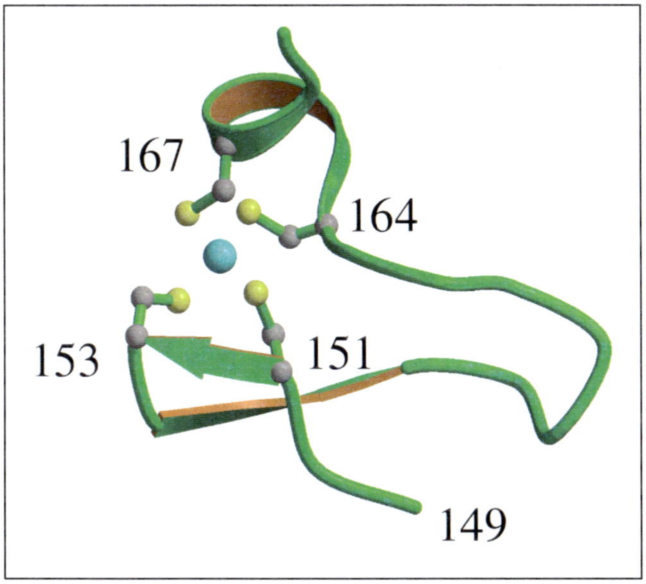

Figure 3. Fold of zinc finger of I-TevI. The zinc finger, CKCGVRIQTSAYTCSKC, is one of the smallest zinc fingers yet identified, and has a degenerate treble clef fold consisting of two β-strands and a single turn α-helix. Numbers indicate the N-terminus and the 4 cyteines that chelate Zn.

important for protein-protein interactions, rather than for DNA binding.

I-TevI Zinc Finger/DNA Interaction

In the crystal structure of the DNA-binding domain of I-TevI with a 20-bp substrate, the zinc finger makes few contacts with the DNA. Contacts are limited to hydrogen bonds between the main-chain nitrogen atoms of residues Tyr162 and Ser165 and the phosphates of the last two bases of the bottom strand (G50 and A51), respectively. No base-specific contacts are observed, but Oε1 of Gln158 is 3.6 Å from N1 of A31, the 5'-end overhanging base of the translationally related duplex that forms a base pair with T1 (Fig. 4). Although this distance is slightly beyond the normal range for hydrogen bonds, the residue is inserted into the major groove, where it could conceivably form a base-specific contact in a complex with a longer DNA substrate.

A Role As a Distance Determinant

The structural evidence that the I-TevI zinc finger does not contribute to the DNA-binding specificity of the enzyme, as well as the absence of zinc fingers in other GIY-YIG endonucleases, suggested that this element plays a role that is specific for the function of I-TevI and one that is likely to be more complex than basic protein-DNA interaction. This was confirmed on the basis of further biochemical experiments analyzing the role of the zinc finger in substrate interaction.[13] The DNA-binding affinities of several deletion derivatives of the DNA-binding domain were determined. A molecule consisting of all residues C-terminal to the zinc finger had a DNA-binding affinity similar to that of a molecule that includes the zinc finger. In contrast, removal of a segment that also includes the ten residues following the zinc finger, an unstructured segment that contacts four bases in the minor groove, did increase the K_d 10-fold. Consequently, it can be concluded that the zinc finger contributes little to the DNA-binding affinity, and should more appropriately be considered to be part of the linker.

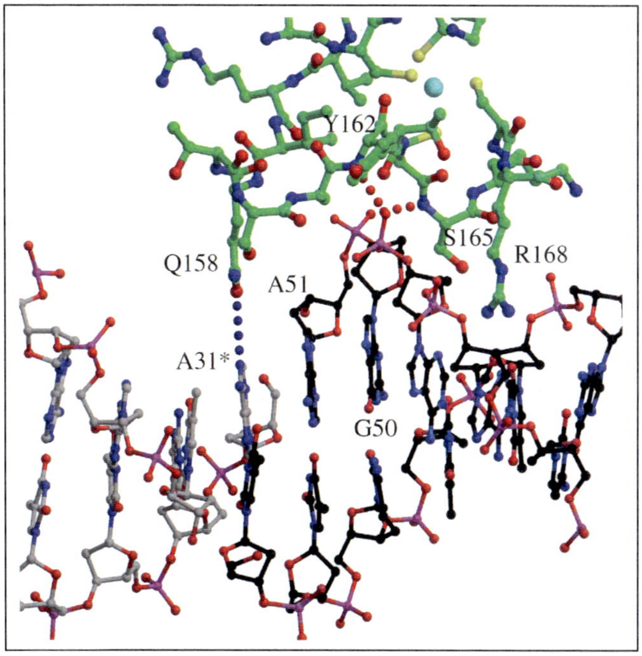

Figure 4. I-TevI zinc finger/DNA contacts. The zinc finger makes only two direct hydrogen bonding contacts (red dotted line) with the DNA, between the main-chain nitrogens of Tyr162 and Ser165 and the phosphate backbone. The DNA-binding domain of I-TevI was crystallized with a 20-bp DNA fragment with a one-base 5' overhang, T1 on the top strand and A31 on the bottom strand. In the crystal, these overhangs of translationally related molecules form a base pair, T1-A31*, resulting in pseudo-continuous DNA. The side chain of Gln158 is sufficiently close to the base of A31* to suggest that it could make an additional contact with the natural substrate.

In order to assess the role of the zinc finger in DNA cleavage, we constructed a series of full-length I-TevI derivatives with mutations in the zinc finger. These were either deletion derivatives with two or more cysteines deleted, including a 19-amino acid deletion that removed the whole zinc finger, or Cys-to-Ala substitutions of the cysteines that chelate the zinc moiety. The zinc finger mutant proteins were synthesized in vitro and were tested for their ability to cleave a plasmid containing the *td* homing site. The proteins were only modestly compromised in DNA-cleavage activity, by a factor of 4- to 7-fold as compared to the wild–type enzyme. This was a somewhat surprising result, given the severity of some of the mutations. Further analysis was carried out using homing site derivatives with insertions or deletions between the CS and the IS. Again, the zinc-finger mutants were competent to cleave these substrates. The most interesting findings came from experiments designed to determine the exact cleavage site for each I-TevI derivative on the mutant homing site substrates (Fig. 5). Wild-type I-TevI cleaves a wild-type substrate 23-25 bp upstream of the IS. With substrates containing modest insertions or deletions, the enzyme can move its catalytic domain back and forth to find the natural cleavage sequence. However, with substrates containing larger insertions or deletions, wild-type I-TevI prefers to revert to distance. Thus, for example, for a substrate containing a 10-bp insertion there is an approximately 10:1 bias for cleavage at the natural distance rather than at the natural sequence (Fig. 5, top).

For the zinc-finger mutants, the picture is quite different. In each case, these derivatives prefer to cleave at the natural sequence

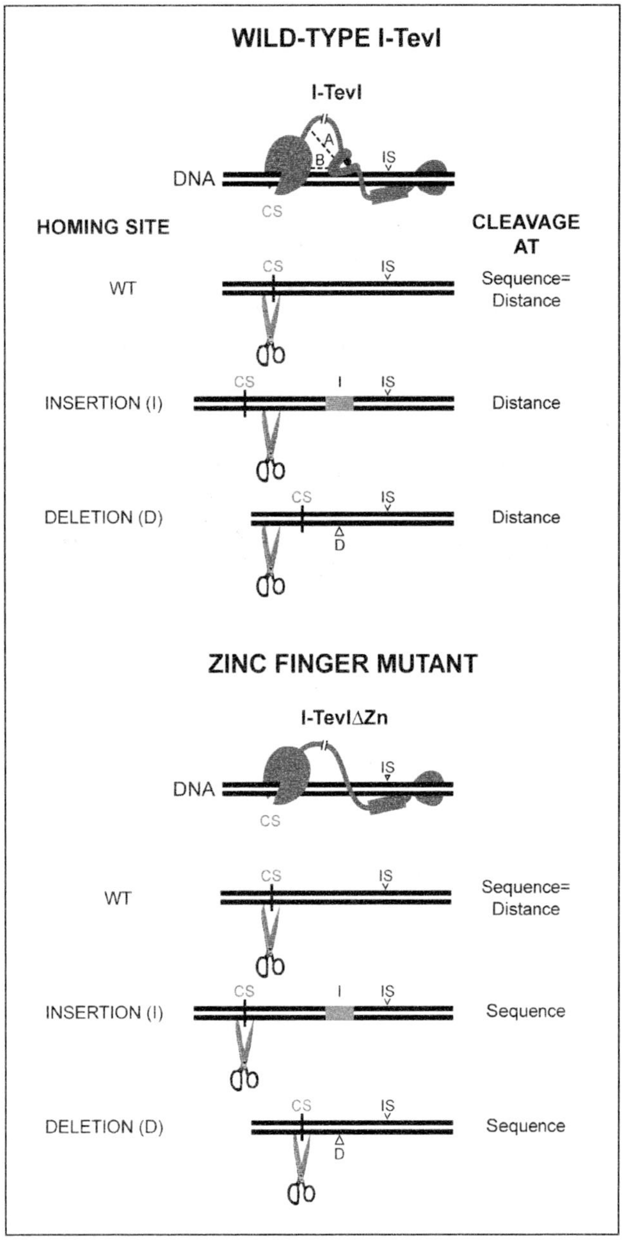

Figure 5. The I-TevI zinc finger acts as a distance determinant. The two different scenarios for cleavage preference of wild-type I-TevI at distance (top), and the zinc-finger mutant at sequence (bottom), are presented. Scissors represent site of preferred cleavage as described in the text. Linker organizer (A) and catalytic clamp (B) models for distance determination are designated on the wild-type protein, in the top panel.

site, even if that is displaced far from the IS. It seems that loss of the zinc finger correlates with loss of a distance determinant for cleavage. There are currently two models for how the zinc finger might act as a distance determinant that both involve putative protein-protein interactions between the zinc finger and other parts of the enzyme. In one model, the zinc finger acts as a "linker organizer", interacting with other parts of the lengthy linker region to reduce the potential for the catalytic domain to reach out to seek the natural cleavage sequence, and thus making it more likely to cleave at distance ("A" in top panel of Fig. 5). In the second model, the zinc finger is postulated to interact directly with the catalytic domain as a "catalytic clamp", again constraining the catalytic domain so that it is more likely to cleave the substrate at a set distance from the IS ("B" in top panel of Fig. 5). In fact, the data are most consistent with a function based on a combination of these two models.

How might the distance-constraining function of the zinc finger have evolved? From both structural and mechanistic standpoints, it would appear that I-TevI was formed by the joining of individual functional modules. Thus, a catalytic domain became attached to discrete DNA-binding units that provided substrate recognition for cleavage. It is a provocative fact that while the DNA binding units, i.e., the α-helix and helix-turn-helix subdomains with their connecting elongated segment, have a nanomolar K_d value, neither the zinc finger nor the catalytic domain contributes to the overall binding affinity of I-TevI. Indeed, the free catalytic domain has no demonstrable binding affinity. We therefore postulate that the zinc finger evolved to constrain the catalytic domain at the cleavage site, to increase its local concentration, and to facilitate cutting at the preferred sequences. Whether this function resulted from the adaptation of a zinc finger that originally contributed to DNA binding, or arose de novo by capture of a zinc finger with protein interaction potential is an interesting conundrum.

Acknowledgements

This research was supported by NIH grants GM56966 (PVR), GM39422 and GM44844 (MB).

References

1. Belfort M, Roberts RJ. Homing endonucleases: Keeping the house in order. Nucl Acids Res 1997; 25:3379-3388.
2. Chevalier BS, Stoddard BL. Homing endonucleases: Structural and functional insight into the catalysis of intron/intein mobility. Nucl Acids Res 2001; 29:3757-3774.
3. Kowalski JC, Belfort M, Stapleton MA et al. Configuration of the catalytic GIY-YIG domain of intron endonuclease I-TevI: Coincidence of computational and molecular findings. Nucl Acids Res 1999; 27:2115-2125.
4. Bryk M, Quirk SM, Mueller JE et al. The td intron endonuclease I-TevI makes extensive sequence-tolerant contacts across the minor groove of its DNA target. EMBO J 1993; 12:2141-2149.
5. Bryk M, Belisle M, Mueller JE et al. Selection of a remote cleavage site by I-TevI, the td intron-encoded endonuclease. J Mol Biol 1995; 247:197-201.
6. Derbyshire V, Kowalski JC, Dansereau JT et al. Two-domain structure of the td intron-encoded endonuclease I-TevI correlates with the two-domain configuration of the homing site. J Mol Biol 1997; 265:494-506.
7. Van Roey P, Meehan L, Kowalski JC et al. Catalytic domain structure and hypothesis for function of GIY-YIG intron endonuclease I-TevI. Nature Struct Biol 2002; 9:806-811.
8. Galburt EA, Chevalier B, Tang W et al. A novel endonuclease mechanism directly visualized for I-PpoI. Nature Structural Biology 1999; 6:1096-1099.
9. Galburt EA, Chadsey MS, Jurica MS et al. Conformational changes and cleavage by the homing endonuclease I-PpoI: A critical role for a leucine residue in the active site. J Mol Biol 2000; 300:877-887.
10. Van Roey P, Waddling CA, Fox KM et al. Intertwined structure of the DNA-binding domain of intron endonuclease I-TevI with its substrate. EMBO J 2001; 20:3631-3637.
11. Krishna SS, Majumdar I, Grishin NV. Structural classification of zinc fingers. Nucl Acids Res 2003; 31:532-550.
12. Campbell EA, Korzheva N, Mustaev A, et al. Structural mechanism for rifamycin inhibition of bacterial DNA polymerase. Cell 2001; 104:901-912.
13. Dean AB, Stanger M, Dansereau J et al. The zinc finger as a distance determinant in the flexible linker of intron endonuclease I-TevI. Proc Natl Acad Sci USA 2002; 99:8554-8561.

CHAPTER 8

Zinc Finger Interactions with Metals and Other Small Molecules

Jay S. Hanas,* Jason L. Larabee and James R. Hocker

Abstract

Zinc fingers encompass a wide variety of compact protein domains that are stabilized by a structural zinc ion which minimally interacts with a cysteine-rich coordination sphere. The selectivity for zinc ion binding is governed by coordinating amino acid side chains and by thermodynamic parameters. Since metal coordination spheres in zinc finger proteins are susceptible to chemical attack (principally at thiolates) and because zinc finger proteins have prominent roles in many cellular processes including the regulation of gene expression and signal transduction, an underlying mechanism for a number of cellular dysfunctions is likely to be the disruption of zinc coordination spheres by a variety of metals and other small molecules. For instance, a number of toxicity mechanisms are likely to be the consequence of zinc replacement by xenobiotic metals resulting in changes in polypeptide conformation and the concomitant loss of protein function. Zinc finger disruption could also occur by oxidation and modification of critical cysteine and histidine amino acids in the zinc coordination sphere resulting in zinc release and alteration of conformation. The chemical reactivity of metal coordination spheres of zinc finger proteins are utilized in normal physiological processes by providing regulatory sites for signal transduction via small molecules like nitric oxide and oxygen and their reactive intermediates. In addition, zinc finger proteins and their metal binding sites are promising targets for specific drug design to help ameliorate major diseases.

Introduction

The discovery that zinc has a structural role in eukaryotic gene regulatory proteins (as opposed to a functional role in enzyme catalysis) originated from studies which demonstrated the cysteine-rich transcription factor IIIA (TFIIIA) contained zinc and required the metal for specific binding to DNA.[1] That TFIIIA had zinc-dependent DNA binding domains was a novel observation at the time as most DNA binding proteins were believed to have helix-turn-helix motifs for DNA interaction similar to the bacterial repressor proteins. The elucidation of the amino acid sequence of TFIIIA[2] led Klug and colleagues to propose that repetitive cysteine and histidine residues present in the sequence could form modular domains of about 30 amino acids each, folded around a zinc atom.[3] A similar structure for TFIIIA was also proposed by Brown.[4] Klug termed this modular domain a "zinc finger" and proposed that each zinc-containing DNA binding module coordinated zinc via chelation with two cysteine and two histidine residues.[3] This cysteine-histidine coordination sphere for zinc in TFIIIA was later confirmed by EXAFS technology.[5] In 1989 it was demonstrated that the zinc in TFIIIA was responsible for holding the protein in the proper conformation for DNA binding as the removal of the metal resulted in an unfolding of the structure and loss of DNA binding specificity.[6]

In the original description of the zinc finger model for TFIIIA it was suggested this modular cysteine and histidine-rich structure folded around a zinc ion may be present in a wide variety of regulatory proteins as its compactness seemed an ideal structure for binding nucleic acids and for evolutionary duplication.[3] Berg identified this modular structure in a wide variety of bacterial, viral, and eukaryotic protein sequences.[7] The term "zinc finger" is now commonly used to identify a wide variety of compact protein domains stabilized by a structural zinc ion.[8-11] Zinc finger proteins encompass the largest class of proteins found in eukaryotic cells and have diverse roles in most cell processes including transcription and translation, DNA replication and repair, metabolism, signaling, cell proliferation, and apoptosis.[11] In most cases zinc ions are likely coordinated by both cysteine and histidine residues but some major families like nuclear receptors and GATA family members have only cysteine coordination. Zinc finger proteins are classified by the arrangement of the zinc coordinating amino acids (cysteines and histidines) and the secondary structures (fold groups) in which these amino acids are present. About 8 fold groups are currently specified.[11] One of the more common fingers and fold groups includes the C_2H_2 fold ($\beta\beta$-hairpin followed by an α-helix) characteristic of TFIIIA, the prototypical zinc finger protein. This fold contains about 30 amino acids. When the zinc coordinating cysteines are present in the sequence CPXCG, this structure is referred to as a "knuckle". This zinc knuckle structure is found in many zinc finger protein folds including the retroviral Gag (nucleocapsid) zinc finger fold.[12] The treble clef finger fold which consists of a zinc knuckle and a shortened α-helix is found in a wide variety of zinc finger proteins including RING fingers and the nuclear receptors.[11] Zinc ribbon folds which may represent the most diverse group of zinc

* Corresponding author. See list of "Contributors".

Zinc Finger Proteins: From Atomic Contact to Cellular Function, edited by Shiro Iuchi and Natalie Kuldell.

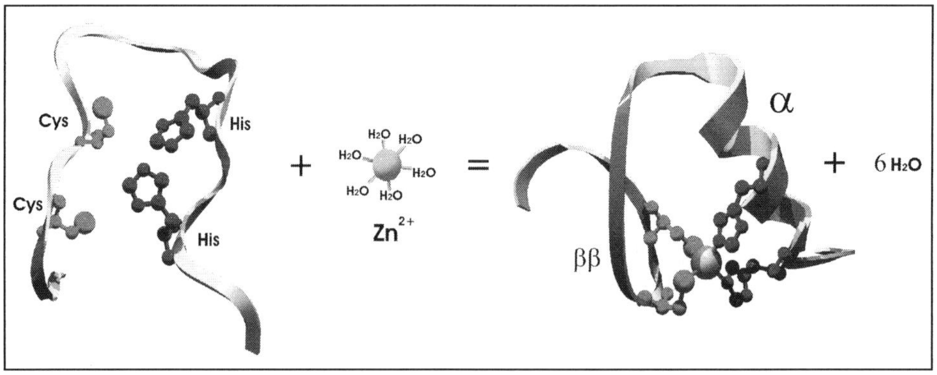

Figure 1. Zinc interaction induces the proper DNA binding conformation of the C_2H_2 finger polypeptide. By coordinating with the two cysteine side chains and the two histidine side chains, the zinc ion induces formation of the ββ turn structure in the N-terminal half of the peptide and the α helical structure of the C-terminal half of the peptide. Six water molecules are released in the peptide coordination reaction. Specific amino acids in the α helical structure form specific contacts with the DNA bases.

finger proteins are characterized by zinc coordination between two zinc knuckles separated by about 40 amino acids.[13] Cysteine-rich zinc clusters as well as zinc loop folds are characterized by the yeast Gal4 regulator and metallothioneins.[8] In general, zinc finger folds or domains function as interaction modules that bind to many large molecules including nucleic acids and other proteins and to a variety of small molecules including metals, redox agents, and therapeutic drugs. A number of X-ray and NMR structures are available for zinc fingers complexed with a variety of nucleic acid templates.[11]

Regulation of Zinc Finger Conformation by Zinc Ion Binding

Studies with TFIIIA correlated the loss of zinc with a significant change in protein conformation as well as with a concomitant loss of specific DNA binding ability.[1,6] These observations indicated that the role of zinc in TFIIIA was to hold the protein in the proper conformation allowing specific amino acid side chains to make specific contacts on the DNA binding site. Much has been learned about the role of zinc in maintaining zinc finger conformation from studying individual zinc fingers of about 30 amino acids. However, the specific DNA binding ability by these individual fingers in the presence of zinc has been difficult to demonstrate. In the absence of zinc, the finger polypeptide is in a random coil conformation as evidenced by circular dichroism (CD).[14,15] Added zinc induces a prominent α-helix in the finger structure as evidenced by the two negative CD Cotton peaks at 208 and 222 nm.[15] This α-helix is a prominent structural feature in the TFIIIA-type zinc finger (C_2H_2) as evidenced by NMR and X-ray crystallography.[16,17] The two zinc coordinating histidine residues located in this helix are three amino acids apart, except in the rare cases when they are four residues apart. This spacing is critical because it places these amino acids on the same face of the helix in an ideal position for zinc coordination. This helical structure allows specific noncoordinating amino acids located in invariant positions to be involved in formation of hydrogen bonds with specific nucleotide bases in the major groove of DNA.[17] Another prominent structure in the C_2H_2 zinc finger is the N-terminal anti-parallel ββ loop structure which is induced by zinc interaction with the two cysteine residues that are usually 2-4 amino acids apart (Fig. 1). The ββ structure is essential for overall zinc finger stability. There is some evidence that when zinc binds to the unfolded apo-finger the metal ion first interacts with the two coordinating cysteines followed by interaction with the histidines.[15,18,19]

The anti-parallel ββ sheet plays a role in making nonspecific contacts on the phosphodiester backbone; a conserved basic residue located two amino acids from the second coordinating cysteine in the second β strand makes a phosphodiester oxygen contact as does the protonated nitrogen on the first coordinating histidine.[17] NMR analysis using carboxymethylation of cysteines and protonation of imidazole nitrogens indicates zinc cation binding is blocked under these conditions.[20] In addition, carboxymethylation of cysteines in the absence of imidazole protonation did not result in proper α helical formation, again indicative of the importance of cysteine coordination in the overall finger folding process.[20] Loss of either cysteine still results in α helical formation but the helix is a looser and less extensive structure; loss of first histidine results in no helical structure whereas loss of the second histidine results in a wild-type CD spectrum.[15] It is suggested that the role of zinc is to bring the α helix and β sheet closer together to form a hydrophobic core which is essential for finger stability and that the first coordinating histidine is essential for hydrophobic core formation.[15] In vitro mutagenesis studies on TFIIIA demonstrated the importance of the first coordinating histidine residue for protein-DNA interactions.[21] Zinc binding to multiple finger domains appears to be more complex than binding to single finger domains. For example, zinc binding to the C_2C_2 DNA binding domain of the vitamin D receptor which contains two such fingers results in a minimal conformation change to the first finger and a much more significant conformational change to the second finger as assayed by CD and electrospray ionization mass spectrometry.[22]

Zinc Coordination Structures in Zinc Finger Proteins

For a number of reasons, zinc is an ideal cation for structural or catalytic roles in proteins. The zinc ion (Zn^{2+}) contains a filled d orbital (d^{10}) and therefore the metal cation does not participate in redox reactions. Rather it functions as a Lewis acid and will accept pairs of electrons.[23] This lack of redox behavior makes the zinc ion very stable and an ideal metal cofactor for enzymatic reactions that require a stable ion to function as a Lewis acid in catalytic mechanisms like proteolysis (carboxypeptidases) and carbon dioxide hydration (carbonic anhydrases).[24] As a structural cofactor in zinc finger proteins, which to a large degree interact with nucleic acids, this lack of redox potential is not deleterious to nucleic acid structure. Zinc cations interact with chemical ligands via chelation mechanisms governed by coordination chemistry whose bonding forces come from electron density transfers from the nucleophilic ligands (oxygen, nitrogen, sulfur) to the electrophilic metal cation. In metal-ligand chelation reactions, the polarizability or "hardness" of the metal and ligand electron shells help determine the stabilities of the metal-ligand interactions. Magnesium and calcium ions are considered "hard" and more readily interact with "hard" ligands in proteins, like carboxyl groups. Cadmium and lead ions are considered "soft" and more readily interact with "soft" ligands in proteins like sulfhydryl groups. Zinc ions are intermediate in this regard and have a preference for histidine residues and to a lesser extent cysteine residues.[25]

In a metal chelation reaction, two or more ligands bind the free metal ion and release water molecules from the hydrated metal in solution (Fig. 1). The increased entropy generated makes the metal-chelate complex very stable. The Irving-Williams effect predicts that increased chelation stability of metals occurs as the ionic radii decreases (calcium and magnesium chelation have the lowest stability and copper has the highest (over about a 10-fold log range). Zinc is anomalous in this series as it has a smaller radius than copper but binds less tightly.[26] Although in solution the zinc cation (Zn^{2+}) interacts with six water molecules, in proteins the zinc cation prefers the tetrahedral coordination sphere as this arrangement has the lowest strain and the shortest bond distances.[27] In fact, the electron structure of zinc likely directs its tetrahedral coordination since its vacant $4p4s^3$ orbitals can only accommodate four independent pairs of donated electrons.[28] The tetrahedral structure of protein-bound zinc regardless of the amino acid ligand attachments (cysteines, histidines, or carboxyls) is well suited for stabilizing a given zinc finger fold as higher coordination numbers might present an entropic barrier to protein folding as more polypeptide organization would be required with higher coordination number. The tetrahedral coordination preference for zinc also sets limits in searching protein sequence databases for potential zinc finger binding sites. In catalytic sites, the four coordination rule for zinc ions can be assumed by three amino acids and a fourth water molecule that is used in the hydrolysis reaction.[24]

At neutral pH in cells, cysteine and histidine coordination to zinc ions in proteins is likely to involve the deprotonated thiol (thiolate) as well as the uncharged imidazole nitrogen on histidine.[29,30] It is noted however that zinc coordination with a protonated thiol group is theoretically possible and this possibility should be noted in final structure determinations.[30] In addition, some thermodynamic evidence is presented that suggests partial thiol presence in the zinc coordination sphere.[29] For free amino acids, the pK_a of the cysteine thiol proton is 8.4 and the pK_a for protonation of the imidizole nitrogen is 6.0. Embedding these amino acids in a basic structure can lower their pK_as even further. So in the normal cellular milieu the unprotonated imidizole nitrogen with the free electron pair is readily available for zinc coordination. The cysteine thiol however may need to be deprotonated and the zinc ions themselves can assist in the deprotonization of the cysteine thiols.[32] Such thiol deprotonization would not be predicted for a hard metal such as magnesium and explains in part the preference of zinc for such coordination spheres.[32] The deprotonated thiolate anion of cysteine is a strong nucleophile and will accept hydrogen bonds from neighboring groups in the zinc finger structure to lower this nucleophilicity.[30] Core thiolates are predicted to be chemically reactive in zinc finger cores and screening of this structure by lysine/arginine groups may be necessary to preserve the overall structure.[32] This preservation is not complete however and, as will be discussed below, the zinc coordination sphere is chemically reactive to a number of electrophiles including xenobiotic metals and other small molecules. Such sensitivity likely plays a role in a number of disease processes.[33]

Table 1. Metal affinities of various zinc fingers

Zinc Fingers	Ions	K_d	Ref. No.
Consensus Prototypical C_2H_2 Sequence	Zn^{2+}	6.7 x 10^{-12} M	35
	Pb^{2+}	5.0 x 10^{-11} M	59
	Co^{2+}	6.3 x 10^{-8} M	35
	Ni^{2+}	1.6 x 10^{-6} M	43
	Fe^{2+}	2.5 x 10^{-6} M	43
	Mn^{2+}	> 10^{-5} M	43
Second C_2H_2 Zinc Finger of TFIIIA	Zn^{2+}	2.8 x 10^{-9} M	36
Third C_2H_2 Zinc Finger of Sp1	Zn^{2+}	6.0 x 10^{-10} M	37
	Co^{2+}	3.0 x 10^{-7} M	37
	Ni^{2+}	4.0 x 10^{-6} M	37
HIV-CCHC	Zn^{2+}	7.0 x 10^{-10} M	59
	Pb^{2+}	3.0 x 10^{-10} M	59
	Co^{2+}	9.0 x 10^{-8} M	59
Intact TFIIIA - 1 high affinity binding site	Zn^{2+}	1.0 x 10^{-8} M	38
	Ni^{2+}	2.3 x 10^{-5} M	38
	Cd^{2+}	2.8 x 10^{-6} M	38
Intact TFIIIA - 9 low affinity binding sites	Zn^{2+}	2.6 x 10^{-5} M	38
	Ni^{2+}	5.2 x 10^{-4} M	38
	Cd^{2+}	1.6 x 10^{-4} M	38
Intact metallothionein	Zn^{2+}	1.4 x 10^{-13} M	39
N-terminal β domain of metallothionein	Zn^{2+}	5.0 x 10^{-12} M	39

There is a great deal of interest in how zinc metalloproteins in general and zinc finger proteins in particular acquire and release zinc inside of the cell. The potential reversible nature of these interactions likely forms the basis for many regulatory processes for gene expression in the cell.[34] It is noted that the affinities for zinc by a number of individual zinc fingers, as well as zinc finger proteins like TFIIIA and metallothionein, can vary over many orders of magnitude. A consensus C_2H_2 zinc finger was found to bind zinc with a dissociation constant of 5.7 x 10^{-12} M whereas the second finger from the prototypical zinc finger protein TFIIIA had a dissociation constant for zinc binding of 2.8 x 10^{-9} M.[35,36] The third C_2H_2 zinc finger from the RNA polymerase II specificity protein Sp1 had a dissociation constant for zinc binding of 6.0 x 10^{-10} M.[37] Zinc binding to TFIIIA itself revealed one high affinity site (1x 10^{-8} M) and nine low affinity sites (2.6 x 10^{-5} M).[38] Human metallothionein II which contains 20 cysteines binds seven zinc cations with an overall affinity of 1.4 x 10^{-13} M (see Table 1) for a compilation of zinc finger affinities). This binding is dispersed among two domains, an N-terminal β domain with an affinity of 5.0 x 10^{-12} M for three zinc atoms and a C-terminal α domain which binds four zinc atoms with variable affinity.[39] The β domain appears to release one zinc atom much more readily than the α domain.[39] It is generally believed that the free zinc ion concentration in cells is very low and hence zinc metalloproteins do not acquire the metal from a pool of free metal ions.[40] Therefore, zinc uptake and transfer in the cell is possibly maintained by metal trafficking proteins like metallothioneins and ZIP proteins in higher organisms.[40,41] Redox reactions which have the potential to release zinc from a donor protein as well as direct transfer of zinc from a coordination sphere of the donor to a coordination sphere of the receptor protein are believed to be some of the mechanisms by which zinc metalloproteins including zinc finger proteins are populated with zinc in the cell. These observations suggest cells and metalloproteins have evolved complex mechanisms for zinc binding and zinc release, which are undoubtedly under regulatory control. The elucidation of these regulatory mechanisms will provide valuable information for understanding how cells function in normal and disease environments.

Zinc Finger Selectivity for Metal Ion Binding

The initial study demonstrating zinc dependency for the prototypical zinc finger protein, TFIIIA, also demonstrated the zinc specificity for this protein.[1] Brief EDTA treatment was used to abolish specific DNA binding by native TFIIIA and only zinc from the transition metal series at micromolar concentrations that included iron (Fe^{2+}), cobalt (Co^{2+}), nickel (Ni^{2+}), manganese (Mn^{2+}), and copper (Cu^{2+}) was able to restore DNA binding ability.[1] This observation is significant as it indicates that the C_2H_2 zinc coordination sphere exhibits exquisite specificity for zinc ions. This is especially important with respect to the redox active iron and cooper ions which if presented close to DNA in a finger structure could result in oxidative damage to DNA. Actual affinity measurements of zinc and nickel for TFIIIA indicated that two classes of metal binding sites exists on TFIIIA and that zinc binds with higher affinity; zinc ions bind with dissociation constants of 1.0×10^{-8} M and 2.3×10^{-5} M whereas nickel ions bind with dissociation constants of 2.6×10^{-5} M and 5.2×10^{-4} M.[38] Because of their ease of use and interpretation, many more metal binding studies are being performed with synthetic zinc finger polypeptides than with intact zinc finger-containing proteins.[42] Using synthetic polypeptides, Berg and colleagues have demonstrated that individual C_2H_2 zinc finger peptides readily bound zinc ions but do not readily bind iron, cobalt, nickel, or manganese ions.[43] The dissociation constants (K_d) for binding of this transition series of metals to a prototypical consensus C_2H_2 sequence was 2×10^{-12} M for Zn^{2+}, 5.0×10^{-8} M for Co^{2+}, 1.6×10^{-6} M for Ni^{2+}, 2.5×10^{-6} M for Fe^{2+}, and $>10^{-5}$ M for Mn^{2+} (also see Table 1).[43,44] At this time it is not clear why the affinities of the various transition metals are significantly lower for the intact TFIIIA protein than the individual synthetic C_2H_2 zinc fingers but it is possible the intact protein has a three-dimensional structure that limits access of metals to the coordination spheres.

There are relatively straight forward explanations as to why some of the more prevalent metal ions in cells (Mg, Ca, Na, and K) do not interact with the coordination spheres of cysteine and histidine-rich zinc finger proteins. These have to do with the "hard-soft acid-base" concepts as well as the Irving-Williams effect for chelation stabilities.[26] In general "hard" acids like sodium, potassium, magnesium, and calcium (metals are considered acids in that they are positive and will accept a negative species, that is a base) interact with "hard" bases or ligands like oxygen-rich ligands (for example, carboxylates). The "hardness" has to do with a lack of polarizability whereas "softness" refers to ease of polarizability. Soft acids have greater numbers of d electrons than hard acids and d-rich metals such as cadmium, lead, and mercury most readily interact with soft ligands like sulfhydryls. The hard-soft classification is not a sharp divide but rather borderline metals and ligands exist between these two extremes. Zinc ions are in this borderline classification and interact with nitrogen-based ligands like imidazoles but will also interact with sulfhydryls. The Irving-Williams series of metals refers to the stabilities of chelation complexes where chelation is defined as the coordination of two or more donor atoms from a single ligand to a central metal ion.[25] In general, for any oxygen, nitrogen, or sulfur chelate, the stabilities of the metal-chelate complex will increase with decreasing cation diameter. Therefore large diameter calcium and magnisium cation chelates are less stable than smaller diameter d-rich metals like cobalt, nickel, copper, and zinc.

A third stabilization factor helping to explain the zinc preference for the cysteine and histidine-rich metal coordination spheres characteristic of zinc finger proteins is ligand field stabilization energy (LFSE) which arises because of incompletely filled d orbitals in the transition metal series.[26] Because zinc has a completely filled d^{10} electronic structure, it does not lose any LFSE when it goes from an octahedral coordination sphere with water molecules in solution to a tetrahedral coordination sphere with sulfur and/or nitrogen ligands in proteins. The other transition metals (for example, nickel, cobalt, iron) experience a loss of LFSE when going from the octahedral hexaaqua form to the protein chelated form.[42] This loss of energy contributes to the lower affinity binding observed for these transition metals when binding zinc coordination spheres in proteins. The extra stability of the zinc-ligand interaction is also driven by the increased entropy resulting from water release from the zinc hydration shell when the water molecules are replaced by the stronger metal-ligand bonds (Fig. 1). In the case of zinc (Zn^{2+}), the metal is in the hexaaqua form and therefore six water molecules are available for release and entropic increase is necessary to substantially help offset the entropy loss upon zinc-dependent peptide folding. This tight binding by zinc to the finger structure is necessary for switching the apofinger into the correct conformation for specific DNA binding.

Alterations in Zinc Finger Structure and Function by Xenobiotic Metals

Most disease states ultimately result from alterations in gene expression. Zinc finger proteins have prominent roles in gene expression mechanisms in mammalian cells and up to 1% of mammalian genomes code for zinc finger proteins.[45] Since metal coordination spheres in zinc finger proteins are susceptible to chemical attack (principally at thiolates) and because zinc finger proteins have prominent roles in the regulation of gene expression, it was reasoned that zinc finger proteins may represent targets for a variety of toxic agents that could lead to different disease states including cancer.[46] Such toxic mechanisms could be the consequence of zinc replacement by other metals resulting in changes in conformation or result from chemical oxidation/modification of critical cysteine and histidine amino acids in the zinc coordination sphere. The elucidation of these toxic mechanisms is important since toxic and carcinogenic metal compounds are known to interfere with DNA metabolic processes like transcription and repair at low concentrations.[33,47-51]

Xenobiotic metals that disrupt zinc finger structure and function include cadmium, lead, mercury, gold, and arsenic. Cadmium ions (Cd^{2+}) have a known specificity for cysteine residues and are an obvious toxic metal to examine for inhibitory effects on cysteine-rich zinc finger proteins. Cadmium and arsenite ions, which also have specificity for reacting with thiol groups, were shown to block steroid hormone binding as well as DNA binding by the glucocorticoid receptor.[52] This protein is a member of the nuclear receptor superfamily and is characterized by two C_2C_2 zinc fingers. Micromolar concentrations of Cd^{2+} but not arsenite were shown to inhibit TFIIIA binding to the 5S RNA gene and the target of action was determined to be the zinc finger structure of the protein.[47] Cadmium ions did not exhibit any affinity for a synthetic peptide corresponding to zinc finger 3 of TFIIIA but exhibited high affinity for the third finger from activator protein Sp1 as well as a consensus C_2H_2 zinc finger.[53] The tumor suppressor protein p53, which is mutated in a large proportion of human carcinomas, requires zinc for specific DNA binding and has a CHC_2 metal binding domain.[54,55] Significantly, micromolar levels of cadmium ions (levels found in the blood of heavy smokers) were shown to inhibit specific DNA binding by tumor suppressor p53 in vitro and in vivo.[50] Through the use of conformation-specific monoclonal antibodies against p53, it was demonstrated that micromolar amounts of cadmium caused a conformational change in the protein in vivo which correlated with loss of specific DNA binding ability as well as loss of transactivation of target genes and loss of p53-dependent G_1 arrest in the cell cycle.[50] These results are significant since it is known that cadmium is carcinogenic and that normal p53 function in cells is a major impediment to tumor development. NMR results indicate that cadmium ions when bound to C_2H_2 zinc fingers in the human metalloregulatory transcription factor MTF-1, induce an altered conformation in the zinc fingers that can be restored to the $\beta\beta\alpha$ structure by added zinc ions.[56] DNA repair enzymes like the XPA protein (xeroderma pigmentosum A which contains a single C_4 zinc finger), poly (ADP-ribose) polymerase (which contains two C_2HC zinc fingers), are both inhibited in their DNA recognition mechanisms by cadmium and arsenite ions.[51] Since DNA repair is so vital for maintaining a normal cell phenotype, these results suggest additional mechanisms by which cadmium and arsenic may be exerting their carcinogenic activities.

Lead is the most prevalent toxic metal in our environment, and its propensity for interaction with protein sulfhydryl groups makes it likely to interact with zinc coordination spheres in zinc finger proteins. Using gel shift assays and cell extracts, lead ions (Pb^{2+}) were initially shown to alter the specific DNA binding ability of Sp1 in vitro and in vivo and were thus proposed to alter Sp1-dependent transcription in vivo.[57] Lead ions at micromolar concentrations were shown to inhibit the purified zinc finger domains of TFIIIA as well as Sp1 as assayed by DNase I protection.[58] The inhibition kinetics were rapid which suggested a direct involvement of lead with the DNA binding domains of these proteins. Thermodynamic studies reveal that lead ions bind zinc finger domains with dissociation constants ranging from 10^{-9} to 10^{-14} M depending upon the number of cysteine residues present in the domain.[59] In spite of this tight binding, circular dichroism and NMR studies indicate that the zinc finger domains are misfolded when bound to lead which likely explains the inhibitory nature of the lead-zinc finger interaction.[58,59] The lead ion has a much larger radius than the zinc ion and prefers six-coordinate ligation spheres rather than the four-coordinate spheres preferred by zinc. In addition, lead ions do not appear to coordinate histidine residues. These structural differences could account for the misfolding of zinc fingers when coordinated by lead ions. Therefore, disruption of zinc finger domains may be a common target mechanism that underlies many of the deleterious effects of lead.[60]

Other xenobiotic metals that were shown to alter the DNA binding functions of zinc finger proteins include mercuric and gold salts. Micromolar concentrations of mercury (Hg^{2+}) were found to inhibit specific DNA binding by TFIIIA and Sp1 but, in the case of TFIIIA, possibly at additional sites besides the protein zinc fingers.[61] This metal was also shown to inhibit transcription factor AP2 and restriction enzymes neither of which have zinc finger sites. Gold salts are known to have anti-inflammatory properties and exhibit high affinity for protein thiols. A thiomalate derivative of gold (Au^{1+}) was found to inhibit the C_2C_2 zinc finger-dependent DNA binding of the progesterone receptor in vitro and in vivo.[62] Thiomalate alone exhibited no such effects. Selenite, an oxyanion of the nonxenobiotic selenium, interacted with TFIIIA and Sp1 zinc fingers in rapid fashion and inhibited their DNA binding properties.[63] This inhibition was zinc finger specific as the AP2 transcription factor was not affected. Selenium compounds are proposed to be involved in redox coupling between thiol and zinc-bound states of zinc metalloproteins in cells.[64] This coupling may contribute to the anti-cancer and anti-inflammatory properties of selenium.

Interactions of Zinc Fingers with Signaling Molecules and Therapeutic Drugs

Zinc finger proteins play such prominent roles in gene expression and signal transduction that it is likely evolution has taken advantage of their chemical properties and evolved mechanisms for tuning these regulatory processes via interactions with proteins, small molecules, and other possible effectors. In addition, zinc finger proteins remain promising targets for drug design to help ameliorate major diseases. Protein interactions with zinc fingers are best exemplified with members of the nuclear receptor superfamily (steroid, glucocorticoid, retinoic acid, vitamin D, etc.). These receptors all have two C_2C_2 zinc fingers. The first finger region is responsible for specific DNA contacts and the second finger is responsible for protein-protein interactions forming homo- and heterodimer contacts with the second finger of the various classes of nuclear receptors. Such dimerization enables these transcription factors to "crosstalk" with a wide variety of receptor family members and greatly increases the regulatory capacity of this important family of regulatory proteins.

Further discussion will focus on zinc finger interactions with small molecules like nitric oxide and oxygen and their reactive intermediates, and with drugs of therapeutic interest. These interactions are based on the unique coordination chemistry of the metal binding spheres of zinc finger proteins. Nitric oxide (NO) is a major cellular mediator of signal transduction. NO release can S-nitrosate sulfhydryl groups under physiological conditions and cysteine-rich zinc finger domains were hypothesized to be molecular targets for this molecule and its derivatives.[65] Under aerobic conditions NO was demonstrated to release zinc from metallothionein by S-nitrosothiol modification of the zinc-chelating cysteine residues in the protein followed by disulfide bond formation.[66] In addition, NO was demonstrated to release zinc from metallothionein in endothelial cells.[67] Incubation of lymphocyte nuclear extracts with the NO donor molecule

S-nitrosocysteine (SNOC) resulted in the inhibition of the specific DNA binding ability of the RNA polymerase II activator zinc finger protein Sp1 which activates the IL-2 gene in these cells.[68] This inhibition of Sp1 DNA binding paralleled the loss of IL-2 gene transcription in activated lymphocytes. If lymphocytes were preactivated before NO treatment, no loss of Sp1 DNA binding was observed as well as no loss of IL-2 gene transcription.[68] These results collectively indicate that Sp1 once bound to the activator site on the IL-2 gene is resistant to NO-mediated disruption of the Sp1 zinc fingers. Sp1 also acts as a repressor of TNF-α gene expression and in this case NO was found to induce TNF-α by disrupting the repressor binding to the promoter of this gene.[69] These results provide sound evidence zinc finger transcription factors are physiological targets for NO-derived nitrosative stress in living cells.

Oxidation-reduction reactions at critical protein thiol groups play a major role in cellular responses to oxidative stress through alterations in gene expression.[70] Zinc finger proteins are prime targets for redox regulation since they contain cysteine-rich zinc coordination spheres. Evidence is presented that oxidative stress inhibits DNA binding by Sp1 in vitro and inhibits the in vivo transcription of the Sp1-dependent gene, dihydrofolate reductase (DHFR). This gene is a glycolytic housekeeping gene and is the type of gene believed to be initially shut down by a cell when encountering oxidative stress.[70] The tumor suppressor p53 contains a single CHC_2 zinc finger as well as four additional conserved cysteines which could serve as a secondary metal binding site.[71] Reactive oxygen intermediates (ROI) have a number of roles in the p53 anti-proliferative pathway including inducing DNA breaks, regulating DNA binding through oxidation of critical cysteines, and controlling several redox effector genes. With respect to the regulation of p53 DNA binding, oxidation of zinc-coordinating cysteines leads to loss of DNA binding.[72] In addition, thioredoxin which reduces protein thiols, activates p53 function in vivo which indicates that redox activation of p53 is physiologically relevant.[73] One of the highly conserved cysteines (Cys277) in p53 that does not have a direct role in zinc binding is believed to have a direct role in DNA binding.[74] Oxidation of this cysteine results in a protein that is unable to recognize a p53 binding site containing cytosines in the 3' end of the DNA recognition pentamer but will specifically bind to a pentamer with a thymine in this position.[74] Thus redox reactions are able to fine-tune the DNA binding specificity of p53.

Another example of a zinc finger protein whose activity is regulated by redox reactions is the heat shock protein (HSP) 33 which is a bacterial chaperone whose activity is stimulated by cellular oxidative stress.[75] This protein has a unique C_2C_2 zinc finger that binds zinc in the inactivated state. When a cell is subjected to oxidative stress, the zinc coordinating cysteines become oxidized resulting in zinc ejection. This cysteine oxidation and loss of zinc results in a protein conformational change that converts the protein to its activated chaperone conformation where it can now help maintain the proper conformation of other critical proteins necessary for cell survival. Zinc finger reactivities are a direct result of the unique coordination chemistry present at the conformational active sites of these proteins. These active centers represent a paradoxical state for the cell. Although the zinc ions themselves are not redox active, the cysteine residues and especially zinc-bound thiolates responsible for metal coordination and correct protein conformation are very redox active. The cell is in a constant balancing act with respect to enabling these centers to bind zinc as well as restoring that binding once oxidation and zinc release has occurred.

The reversibility and reactivity of the zinc binding and zinc coordination spheres in zinc finger proteins lends itself to the design of specific inhibitors that will oxidize these sites and/or eject zinc resulting in protein conformational changes and loss of protein function. The specificity of such agents can come from drug interactions with nonconserved amino acids in close proximity to the zinc coordinating center as well as from secondary and tertiary protein structure determinants. One human viral protein targeted for selective inactivation of a crucial zinc finger is the E6 oncoprotein from the human papillomavirus (HPV) which is the causative agent for cervical cancer. This protein has two C_2C_2 zinc fingers which are unique to this virus and have a role in forming a ternary complex with the viral E6-associated protein (E6AP) and the cellular p53 tumor suppressor. This complex results in the degradation of p53 via the ubiquitination-proteosome pathway. The uniqueness of these zinc fingers to the HPV virus makes them attractive candidates for therapeutic intervention. A large number of organic disulfide, azoic, and nitroso aromatic compounds with known sulfhydryl redox activity and zinc ejecting properties were screened for inhibitory activities against the human papillomavirus (HPV) oncoprotein E6, its zinc-dependent interaction with E6AP, and for growth inhibition of HPV transformed cells.[76] The organic disulfide 4,4' dithiodimorpholine was found to eject zinc from the E6 oncoprotein, inhibit its interaction with E6AP, and also activate p53 in an HPV transformed cell and induce apoptosis.[76] This agent is believed to form disulfides with the zinc–coordinating cysteine residues thus ejecting zinc and placing the protein in an inactive conformation. Another zinc finger protein being targeted for therapeutic intervention is the nucleocapsid 7 protein of the human immunodeficiency virus (HIV). Specific inactivation of the C_2HC zinc finger of this protein was accomplished using specific disulfide compounds as well as other organics like azodicarbonamide which are capable of undergoing redox reactions with zinc coordinating thiols resulting in zinc ejection and loss of activity.[77,78]

Although technically not a zinc finger structure because the zinc ion selected for drug design is involved in catalysis, the search for matrix metalloproteinases (MMPs) inhibitors is instructive in the discussion of zinc-targeted intervention therapies. Human macrophage elastase plays an important role in inflammatory processes and is believed to be involved in the destruction of elastase in lung alveolar tissues in emphysema. The crystal structure of a hydroxamic acid inhibitor with the zinc catalytic site of this enzyme reveals a number of structural requirements for such zinc-targeted therapies.[79] This structure reveals that positioning of the hydroxamic acid zinc chelator group of the inhibitor is dependent upon precise hydrogen bonding of the inhibitor polar groups with elastase amino acid side chains as well as interactions between a hydrophobic pocket in the enzyme active site with an aromatic group on the inhibitor.[79] This combination of unique protein-inhibitor interactions contributes to the strength and specificity of inhibitor binding and such information gleaned from this study is applicable to other zinc-targeted therapies as well.

References

1. Hanas JS, Hazuda DJ, Bogenhagen DF et al. Xenopus transcription factor A requires zinc for binding to the 5 S RNA gene. J Biol Chem 1983; 258:14120-14125.
2. Ginsberg AM, King BO, Roeder RG. Xenopus 5S gene transcription factor, TFIIIA: Characterization of a cDNA clone and measurement of RNA levels throughout development. Cell 1984; 39:479-489.
3. Miller J, McLachlan AD, Klug A. Repetitive zinc-binding domains in the protein transcription factor IIIA from Xenopus oocytes. EMBO J 1985; 4:1609-1614.
4. Brown RS, Sander C, Argos P. The primary structure of transcription factor TFIIIA has 12 consecutive repeats. FEBS Lett 1985; 186:271-274.
5. Diakun GP, Fairall L, Klug A. EXAFS study of the zinc-binding sites in the protein transcription factor IIIA. Nature 1986; 324:698-699.
6. Hanas JS, Duke AL, Gaskins C. Conformational states of Xenopus transcription factor IIIA. Biochemistry 1989; 28:4083-4088.
7. Berg JM. Potential metal-binding domains in nucleic acid binding proteins. Science 1986; 232:485-487.
8. Vallee BL, Coleman JE, Auld DS. Zinc fingers, zinc clusters, and zinc twists in DNA-binding protein domains. Proc Natl Acad Sci USA 1991; 88:999-1003.
9. Hanas JS, Gaskins CJ, Smith JF et al. Structure, function, evolution of transcription factor IIIA. Prog Nucleic Acid Res Mol Biol 1992; 43:205-239
10. Klug A, Schwabe JW. Protein motifs 5. Zinc fingers. FASEB J 1995; 9:597-604.
11. Krishna SS, Majumdar I, Grishin NV. Structural classification of zinc fingers. Nucleic Acids Res 2003; 31:532-550.
12. Klein DJ, Johnson PE, Zollars ES et al. The NMR structure of the nucleocapsid protein from the mouse mammary tumor virus reveals unusual folding of the C-terminal zinc knuckle. Biochemistry 2000; 39:1604-1612.
13. Zhu W, Zeng Q, Colangelo CM et al. The N-terminal domain of TFIIB from Pyrococcus furiosus forms a zinc ribbon. Nature Struct Biol 1996; 3:122-124.
14. Frankel AD, Berg JM, Pabo CO. Metal-dependent folding of a single zinc finger from transcription factor IIIA. Proc Natl Acad Sci USA 1987; 84:4841-4845.
15. Nomura A, Sugiura Y. Contribution of individual zinc ligands to metal binding and peptide folding of zinc finger peptides. Inorg Chem 2002; 41:3693-3698.
16. Lee MS, Gippert GP, Soman KV et al. Three-dimensional solution structure of a single zinc finger DNA-binding domain. Science 1989; 245:635-637.
17. Pavletich NP, Pabo CO. Zinc finger-DNA recognition: Crystal structure of a Zif268-DNA complex at 2.1 A. Science 1991; 252:809-817.
18. Shi Y, Berger RD, Berg JM. Metal binding properties of single amino acid deletion mutants of zinc finger peptides: Studies using cobalt (II) as a spectroscopic probe. Biophys J 1993; 64:749-753.
19. Miura T, Satoh T, Takeuchi H. Role of metal-ligand coordination in the folding pathway of zinc finger peptides. Biochim Biophys Acta 1998; 1384:171-179.
20. Parraga G, Horvath S, Hood L et al. Spectroscopic studies of wild-type and mutant "zinc finger" peptides: Determinants of domain folding and structure. Proc Natl Acad Sci USA 1990; 87:137-141.
21. Smith JF, Hawkins J, Leonard RE et al. Structural elements in the N-terminal half of transcription factor IIIA required for factor binding to the 5S RNA gene internal control region. Nucl Acids Res 1991; 19:6871-6876.
22. Veenstra TD, Johnson KL, Tomlinson AJ et al. Zinc-induced conformational changes in the DNA-binding domain of the vitamin D receptor determined by electrospray ionization mass spectrometry. J Am Soc Mass Spectrom 1998; 9:8-14.
23. Williams RJP. The biochemistry of zinc. Polyhedron 1987; 6:61-69.
24. McCall KA, Huang C-c, Fierke CA. Function and mechanism of zinc metalloenzymes. J Nutr 2000; 130:1437S-1446S.
25. Lippard SJ, Berg JM. Principals of Bioinorganic Chemistry. University Science Books. Mill Valley, CA: 1994:21-39.
26. Huheey JE, Keiter EA, Keiter JL. Inorganic chemistry: Principals of structure and reactivity. 4th ed. New York: HarperCollins College Publishers, 1993:578:346-355.
27. Dudev T, Lim C. Tetrahedral vs octahedral zinc complexes with ligands of biological interest: A DFT/CDM study. J Am Chem Soc 2000; 122:11146-11153.
28. Roe RR, Pang Y-P. Zinc's exclusive tetrahedral coordination governed by its electronic structure. J Mol Model 1999; 5:134-140.
29. Blasie CA, Berg JM. Structurebased thermodynamic analysis of a coupled metal binding-protein folding reaction involving a zinc finger peptide. Biochemistry 2002; 41:15068-15073.
30. Smith JN, Shirin Z, Carrano CJ. Control of thiolate nucleophilicity and specificity in zinc metalloproteins by hydrogen bonding: Lessons from model compound studies. J Am Chem Soc 2003; 125:868-869.
31. Simonson T, Calimet N. Cys_xHis_y-Zn^{2+} interactions: Thiol vs. thiolate coordination. Proteins 2002; 49:37-48.
32. Dudev T, Lim C. Factors governing the protonation state of cysteines in proteins: An Ab initio/CDM study. J Am Chem Soc 2002; 124:6759-6766.
33. Hartwig A. Zinc finger proteins as potential targets for toxic metal ions: Differential effects on structure and function. Antiox Redox Signal 2001; 3:625-634.
34. Berg JM, Shi Y. The galvanization of biology: A growing appreciation for the roles of zinc. Science 1996; 271:1081-1085.
35. Krizek BA, Merkle DL, Berg JM. Ligand variation and metal ion binding specificity in zinc finger peptides. Inorg Chem 1993; 32:937-940.
36. Berg JM, Merkle DL. On the metal ion specificity of "zinc finger" proteins. J Am Chem Soc 1989; 111:3759-3761.
37. Posewitz MC, Wilcox DE. Properties of the Sp1 zinc finger 3 peptide: Coordination chemistry, redox reactions, and metal binding competition with metallothionein. Chem Res Toxico 1995; 8:1020-1028.
38. Makowski GS, Sunderman Jr.FW. The interactions of zinc, nickel, and cadmium with Xenopus transcription factor IIIA, assessed by equilibrium dialysis. J Inorg Biochem 1992; 48:107-119.
39. Jiang L-J, Vasak M, Vallee BL et al. Zinc transfer potentials of the alpha- and beta- clusters of metallothionein are affected by domain interactions in the whole molecule. Proc Natl Acad Sci USA 2000; 97:2503-2508.
40. Finney LA, O'Halloran TV. Transition metal speciation in the cell: Insights from the chemistry of metal ion receptors. Science 2003; 300:931-936.
41. Jacob C, Maret W, Valee BL. Control of zinc transfer between thionein, metallothionein, and zinc proteins. Proc Natl Acad Sci USA 1998; 95:3489-3494.
42. Berg JM, Godwin HA. Lessons from zinc-binding peptides. Annu Rev Biophys Biomol Struct 1997; 26:357-371.
43. Krizek BA, Berg JM. Complexes of zinc finger peptides with Ni^{2+} and Fe^{2+}. Inorg Chem 1992; 31:2984-2986.
44. Krizek BA, Amann BT, Kilfoil VJ et al. A consensus zinc finger peptide: Design, high-affinity metal binding, a pH-dependent structure, and a His to Cys sequence variant. J Am Chem Soc 1991;113:4518-4523.
45. Mackay JP, Crossley M. Zinc fingers are sticking together. Trends Biochem 1998; 23:1-4.
46. Sunderman FW, Barber AM. Finger-loops, oncogenes, and metals. Claude Passmore Brown Memorial Lecture. Ann Clin Lab Sci 1988; 18:267-288.
47. Hanas JS, Gunn CG. Inhibition of transcription factor IIIA-DNA interactions by xenobiotic metal ions. Nucleic Acids Res 1996; 24:924-930.
48. Beyersmann D, Hechtenberg S. Cadmium, gene regulation, and cellular signalling in mammalian cells. Toxicol Appl Pharmacol 1997; 144:247-261.
49. Hartwig A. Carcinogenicity of metal compounds: Possible role of DNA repair inhibition. Toxicol Lett 1998; 102-103:235-239.
50. Meplan C, Mann K, Hainaut P. Cadmium induces conformational modifications of wild-type p53 and suppresses p53 response to DNA damage in cultured cells. J Biol Chem 1999; 274:31663-31670.

51. Hartwig A, Asmuss M, Blessing H et al. Interference by toxic metal ions with zinc-dependent proteins involved in maintaining genomic stability. Food Chem Toxicol 2002; 40:1179-1184.
52. Simons Jr SS, Chakraborti PK, Cavanaugh AH. Arsenite and cadmium(II) as probes of glucocorticoid receptor structure and function. J Biol Chem 1990; 265:1938-1945.
53. Petering DH, Huang M, Moteki S et al. Cadmium and lead interactions with transcription factor IIIA from Xenopus laevis: A model for zinc finger protein reactions with toxic metal ions and metallothionein. Mar Environ Res 2000; 50:89-92.
54. Pavletich NP, Chambers KA, Pabo CO. The DNA-binding domain of p53 contains the four conserved regions and the major mutation hot spots. Genes Dev 1993; 7:2556-2564.
55. Cho Y, Gorina S, Jeffrey PD et al. Crystal structure of a p53 tumor suppressor-DNA complex: Understanding tumorigenic mutations. Science 1994; 265:346-355.
56. Giedroc DP, Chen X, Pennella MA et al. Conformational heterogeneity in the C-terminal zinc fingers of human MTF-1. J Biol Chem 2001; 276:42322-42332.
57. Zawia NH, Sharan R, Brydie M et al. Sp1 as a target site for metal-induced pertubations of transcriptional regulation of developmental brain gene expression. Brain Res Dev Brain Res 1998; 107:291-298.
58. Hanas JS, Rodgers JS, Bantle JA et al. Lead inhibition of DNA-binding mechanism of Cys_2His_2 zinc finger proteins. Mol Pharmacol 1999; 56:982-988.
59. Payne JC, ter Horst MA, Godwin HA. Lead fingers: Pb^{2+} binding to structural zinc-binding domains determined directly by monitoring lead-thiolate charge-transfer bands. J Am Chem Soc 1999; 121:6850-6855.
60. Zawia NH, Crumpton T, Reddy GR et al. Disruption of the zinc finger domain: A common target that underlies many of the effects of lead. Neurotoxicol 2000; 2:1069-1080.
61. Rodgers JS, Hocker JR, Hanas RJ et al. Mercuric ion inhibition of eukaryotic transcription factor binding to DNA. Biochem Pharmacol 2001; 61:1543-1550.
62. Handel ML, deFazio A, Watts CKW et al. Inhibition of DNA binding and transcriptional activity of a nuclear receptor transcription factor by aurothiomalate and other metal ions. Mol Pharmacol 1991; 40:613-618.
63. Larabee JL, Hocker JR, Hanas RJ et al. Inhibition of zinc finger protein-DNA interactions by sodium selenite. Biochem Pharmacol 2002; 64:1757-1765.
64. Chen Y, Maret W. Catalytic selenols couple the redox cycles of metallothionein and glutathione. Eur J Biochem 2001; 268:3346-3353.
65. Kroncke K-D. Zinc finger proteins as molecular targets for nitric oxide-mediated gene regulation. Antiox Redox Signal 2001; 3:565-575.
66. Kroncke KD, Fehsel K, Schmidt T et al. Nitric oxide destroys zinc-sulfur clusters inducing zinc release from metallothionein and inhibition of the zinc finger-type yeast transcription activator LAC9. Biochem Biophys Res Commun 1994; 200:1105-1110.
67. Pearce LL, Gandley RE, Han W et al. Role of metallothionein in nitric oxide signalling as revealed by a green fluorescent protein fusion protein. Proc Natl Acad Sci USA 2000; 97:477-482.
68. Berendji D, Kolb-Bachofen V, Zipfel PF et al. Zinc finger transcription factors as molecular targets for nitric oxide-mediated immunosuppression: Inhibition of IL-2 gene expression in murine lymphocytes. Mol Med 1999; 5:721-730.
69. Wang S, Wang W, Wesley RA et al. A Sp1 binding site of the tumor necrosis factor promoter functions as a nitric oxide repressor element. J Biol Chem 1999; 274:33190-33193.
70. Wu X, Bishopric NH, Discher DJ et al. Physical and functional sensitivity of zinc finger transcription factors to redox change. Mol Cell Biol 1996; 16:1035-1046.
71. Meplan C, Richard M-J, Hainaut P. Redox signalling and transition metals in the control of the p53 pathway. Biochem Pharmacol 2000; 59:25-33.
72. Parks D, Bolinger R, Mann K. Redox state regulates binding of p53 to sequence-specific DNA, but not to nonspecific or mismatched DNA. Nucl Acids Res 1997; 25:1289-1295.
73. Casso D, Beach D. A mutation in a thioredoxin reductase homolog suppresses p53-induced growth inhibition in the fission yeast Schizosaccharomyces pombe. Mol Gen Genet 1996; 252:518-529.
74. Buzek J, Latonen L, Kurki S et al. Redox state of tumor suppressor p53 regulates its sequence-specific DNA binding in DNA-damaged cells by cysteine 277. Nucl Acids Res 2002; 30:2340-2348.
75. Jakob U, Eser M, Bardwell JCA. Redox switch of Hsp33 has a novel zinc-binding motif. J Biol Chem 2000; 275:38302-38310.
76. Beerheide W, Bernard H-U, Tan Y-J et al. Potential drugs against cervical cancer: Zinc-ejecting inhibitors of the human papillomavirus type 16 E6 oncoprotein. J Natl Cancer Inst 1999; 91:1211-1220.
77. Tummino PJ, Scholten JD, Harvey PJ et al. The in vitro ejection of zinc from human immunodeficiency virus (HIV) type I nucleocapsid protein by disulfide benzamides with cellular anti-HIV activity. Proc Natl Acad Sci USA 1996; 93:969-973.
78. Rice WG, Supko JG, Malspeis L et al. Inhibitors of HIV nucleocapsid protein zinc fingers as candidates for the treatment of AIDS. Science 1995; 270:1194-1197.
79. Nar H, Werle K, Bauer MMT et al. Crystal structure of human macrophage elastase (MMP-12) in complex with a hydroxamic acid inhibitor. J Mol Biol 2001; 312:743-751.

CHAPTER 9

Synthetic Zinc Finger Transcription Factors

Nicoletta Corbi, Valentina Libri and Claudio Passananti*

Abstract

The possibility of using designed transcription factors to control gene expression is highly appealing in view of a wide range of promising applications in research and biomedicine. In the last decade, the efforts of several research groups have clarified the molecular interactions between zinc finger proteins and DNA, generating a "recognition code" that relates amino acids in specific positions of the finger domain to its DNA target sequence. This chapter describes the methods to design artificial zinc finger transcription factors using the "code" and recent novel selection approaches. Efficacy and possible applications of artificial transcription factors are discussed.

Introduction

Despite the sequencing and the analysis of the genome of different organisms, the mechanisms by which a peculiar cell type coordinates gene regulation remain largely unknown. The complex process of eukaryotic gene expression requires integration of various events, accurately selected in million of years of evolution. These events include: the assembly of general transcription factors, the recruitment of theh transcription machinery to the promoter, utilization of cofactor-complexes involved in activation or repression and chromatin remodeling.[1-4] Transcription regulation is mainly achieved through the action of proteins known as transcription factors (TF), commonly organized in two separable modules: a DNA binding domain (DBD) capable of targeting specific DNA sequences and an effector domain (ED) capable of regulating at different levels the multiprotein machinery, that directs transcription. Transcription factors represent 6% of all the proteins in the human genome.[5] In the large class of DNA binding molecules the Cys_2His_2 (C_2H_2) zinc finger domain, with the 4,500 examples so far identified, represents the most frequently encoded motif in the human genome. In particular, C_2H_2 zinc finger proteins (ZFP) provide a simple and versatile framework favorable to design and select novel DNA binding proteins that recognize desired target sequences.[6-8] The C_2H_2 zinc finger domain is a compact motif of 28-30 amino acids, that comprises an α helix containing two invariant histidine residues coordinated through a zinc atom to two cysteines of a single β turn. The x-ray crystal structures of the three zinc finger domains of the transcription factor Zif268 bound to its DNA target site, reveal that an individual finger domain binds essentially three base pairs of double stranded DNA sequence with specific contacts through the amino-terminal part of the α helix.[9] In particular, a "code" that relates the amino acids of a single zinc finger to its associated DNA target has been proposed for a variety of zinc finger domains.[10-12] Using and expanding this "code" several research groups have recently engineered and/or selected peptides containing zinc finger domains able to recognize and to bind specific DNA sequences. These peptides represent DNA binding modules that can act as artificial transcription factors when fused to an appropriate effector domain.[13] Therefore, synthetic ZFP TFs represent the feasibility of "gene regulation on demand" in higher eukaryotes, opening great opportunities for genome analysis and control. Moreover, zinc finger peptides can act as RNA-binding modules, enlarging the number of the their possible applications in biotechnologies and molecular medicine.[14-17]

Evolution of the Recognition Code Concept

Despite the great variety of structural domains, the specificity of interactions observed between protein and DNA is most frequently due to the complementarity of the surfaces of a protein α helix and the major groove of DNA.[18] With regards to the recurring physical interaction of α helix and major groove, it has seemed feasible to describe contacts between particular amino acids and DNA bases by a simple set of rules that can be combined to form a stereochemical recognition code. Among the variety of DNA binding domains, the C_2H_2 zinc finger motif has represented an attractive motif as it is structurally well characterized and it presents distinct DNA binding properties.[19-21] ZFP domains display only minimal cooperativity, so each finger recognizes a DNA sequence with relative independence. An understanding of the interactions between ZFP domain and its DNA target sequence was initially guided by the analysis of sequence homology between individual fingers from the same protein, or from different, though related, proteins. Site directed mutagenesis experiments, that substituted the non homologous α helix residues, were able to change the specificity of the mutant finger.[6,22-25] Thus, the analysis of the available large database of zinc finger sequence and the rational mutagenesis revealed some base-contacting positions even before the solution of the Zif268 crystal structure. The determination of Zif268 protein/DNA complex at 2.1Å was the basis for many subsequent studies, and Zif268 has proved to be a very useful model system for understanding how C_2H_2 ZFPs recognize DNA.[21] The crystal structure shows

* *Corresponding author. See list of "Contributors".*

Zinc Finger Proteins: From Atomic Contact to Cellular Function, edited by Shiro Iuchi and Natalie Kuldell.

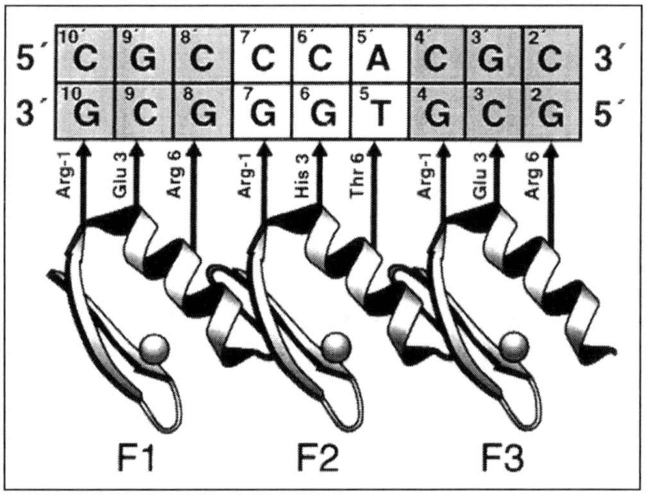

Figure 1. Interactions between Zif268 and its DNA target. Schematic diagram of modular recognition between the three zinc fingers of Zif268 and the triplet subsites of its DNA target, after the resolution at 2.1 Å.[21] Straight arrows indicate the stereochemical juxtapositioning of recognition residues with bases of the contacted sense DNA strand. The ZFP binds to DNA in an antiparallel mode since the N-terminal finger (F1) contacts the 3' end of the DNA. Reprinted from Isalan M, Choo Y, Klug A. Proc Natl Acad Sci USA 1997; 94(11):5617-21. Copyright (1997) National Academy of Sciences, U.S.A.

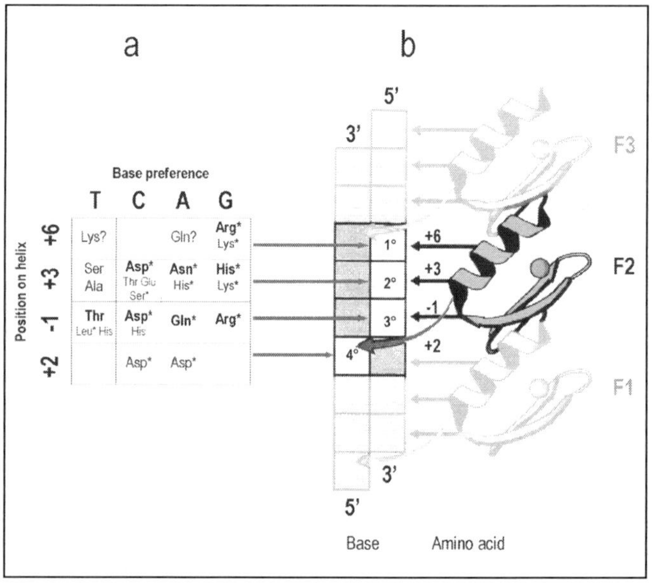

Figure 2. Table of recognition code. a) Amino acids located at the key positions in the zinc finger α helix (-1,+2,+3,+6) are listed in a matrix relating to the four bases at each position of a DNA subsite. Amino acid residues that arise recurrently from phage display selections are in bold, and asterisks indicate interactions observed in structural studies. Some correspondences are still uncertain (indicated by question mark) or poorly defined (left blank). This figure was adapted from reference 8. b) Schematic diagram of contacts between a ZFP and its DNA binding site including cross-strand interactions. Recognition contacts between each finger (position +2) and the antisense DNA strand (shown by curly arrows) mean that each finger binds overlapping 4-bp subsites.

the three Zif268 fingers wrap around the DNA in an antiparallel mode and each finger has a similar relation to the DNA with a simple pattern of interactions. In each finger three amino acids, in position -1, +3 and +6 of the α helix, contact three adjacent bases (three-base subsite) in one strand of DNA duplex. Contiguous zinc fingers recognize adjacent, but independent subsites (Fig. 1). This was the first simplest description of the mode of DNA recognition: a one-to-one interaction between amino acids and bases. Dejarlais and Berg were among the earliest to develop a zinc finger-DNA recognition code.[21,26] They performed binding studies on variants of the central zinc finger of Sp1 and combined in a matrix the correlations between the identity of amino acid in a contact position and the base pair recognized. They noted that the discrimination between different bases is rarely absolute thus introducing the concept of the degeneracy of the code. They distinguished between a prediction code that enables the prediction of the DNA sequence to which a given natural ZFP would optimally bind, and a simpler design code that would allow for the construction of a protein that would bind a desired DNA sequence. A certain grade of degeneracy in binding specificity may be tolerated or perhaps even desired for designing ZFPs.

A step forward in the construction of a code was obtained by the introduction of a selection system that allowed the isolation of desirable mutants with new specificity from a large pool of randomized ZFPs: the phage display system.[10-12,27-33] In this system, the zinc finger-binding domains are fused to a truncated version of gene III coat protein of phage M13, thus leading to their expression on the tip of the capsid. Phage, displaying the peptides, can be sorted by repeated binding and elution from affinity matrices containing different DNA target sequences. The Pabo group and Jamieson et al created a library of Zif268 variants randomizing the four positions of the first finger (-1, +2, +3, +6) and selected them against the Zif268 target with a single modified triplet. Led by the idea of the importance of the context, Choo and Klug performed more extensive randomization of the α helix (-1, +1, +2, +3, +5, +6, +8) of Zif268 central finger.[6]

Several rounds of selection, with modified Zif268 target sequence mutated in the middle triplet, returned a number of related zinc fingers that bind the given DNA target. Choo and Klug confirmed their data backwards performing a SELEX assay (Systematic Evolution of Ligands by Exponential Enrichment) in which zinc fingers, with fixed α helix positions, yield specific DNA sequence from a randomized pool. By comparing phage display and SELEX results, they found some statistically significant correlations summarized in a matrix. This code table was subsequently enriched to obtain the version shown in Figure 2a.[8;34] X-ray crystallographic studies of selected ZFPs have confirmed that some "coded contacts" occurs in the designed and selected ZFP variants. These data are consistent with physical basis of a stereochemical recognition code.[34-37]

Subsequent studies (other phage display selections, higher resolution of Zif268/DNA crystal structure and the resolution of other ZFP/DNA complexes) shaded light on the role of a fourth amino acid position in base specification, the position +2.[30-32,38-43] The amino acid in position +2 of one finger, usually an aspartic acid or a serine, makes a direct contact with the base at the 5' of adjoining triplet on the antisense strand. This observation involved a partial redefinition of binding subsites for each finger as an overlapping four base pair subsite (Fig. 2b), and therefore sequence specificity at the boundary between subsites arises from synergy between adjacent fingers.

The utility of the code is that new DNA binding proteins could be constructed by referring to a list of amino acid-base contacts simply substituting the appropriate residues in position -1, +2, +3, +6 in a standard ZFP scaffold. Designs based on the code have had a number of applications.[26,44-49] Corbi et al have

	T	C	A	G	
+6	Thr*	Glu*	Gln	Arg	1st base
+3	Ser	Asp	Asn	His	2nd base
-1	Thr	Glu*	Gln	Arg	3rd base
+2	Thr*	Asp	Asn	Ser*	4th base

Figure 3. Table of nondegenerate recognition code. The table allows identification of one specific amino acid at each of positions: -1, +2, +3, and +6 of the α-helical region of the zinc-finger domain from the overlapping 4-bp target sequence in a given DNA target. The depicted correspondences between amino acid and base are derived from observed X-ray crystal structures contacts and from potential amino acid-base interactions. The correspondences suggested exclusively in nondegenerate code (not reported in the degenerate code table) are indicated with asterisks.

published two papers in which ZFPs were rationally designed to bind arbitrary DNA sequences.[45,46] These ZFPs bind to the predicted sequence with a fairly strong affinity. As a proof of specificity, the proteins selected strictly related sequences among randomized ones in SELEX experiments. The results were encouraging as they validate some of the rules derived from the code. The design based on the code is a valid alternative to selection approach, even considering further level of complexity due to context-dependent effects such as synergy between adjacent fingers.[36,50-53] Sera and Uranga proposed a rapid creation scheme, based on a nondegenerate code (Fig. 3) to develop new ZFPs with satisfactory binding properties in a "high-through-put manner".[54] This method will enable manipulation of gene expressions by screening multiple ZFP TFs for multiple sites in a given promoter, providing an easy and rapid "shot-gun" approach even with no chromatin structural information on target genes.

Selection and Assembly of Single Zinc Finger Domains As Building Blocks

The modularity in both structure and function of zinc finger domain suggested that linking together single domains of known specificity could create novel DNA-binding proteins. Three successful construction methods have been described: parallel selection, sequential selection, and bipartite selection.[55-57]

Parallel Selection

This approach (Fig. 4a) preselects individual monomers by phage display and subsequently combines them into three-ZFPs. In particular, two Zif268 fixed fingers (anchor fingers), the first and the third one, ensure the correct positioning on target site, while the middle finger is randomized in the crucial positions.[29] To date, using parallel selection, Barbas group described zinc finger domains that recognize 30 different trinucleotides comprising the GNN and ANN subsets (where N is any one of the four nucleotides).[38,39,58-60] These selected "monomers" supply prefabricated building blocks that can be "stitched" together.[59] This method had several applications in the construction of artificial transcription factors that modulate the expression of diverse target genes even at the endogenous chromosomal site (Table 1).[7,13,39,58,61,62] The following two approaches address the problems dealing with target sites overlap and cooperativity in zinc finger-DNA recognition.

Sequential Selection

In this system (Fig. 4b) critical amino acid positions of individual zinc finger domains are randomized sequentially in the context of two "anchor" fingers.[32] Thus, finger three of Zif268 protein is substituted by the library represented by randomized finger one ($F1^R$). After the primary selection against the first three nucleotides, the N-terminal anchor finger is removed and the library of finger two ($F2^R$) is cloned at the C-terminus of the newly selected finger one ($F1^S$), which now is in the central position. The $F2^R$ library is selected to bind the next three nucleotides in the DNA sequence. After one additional round of

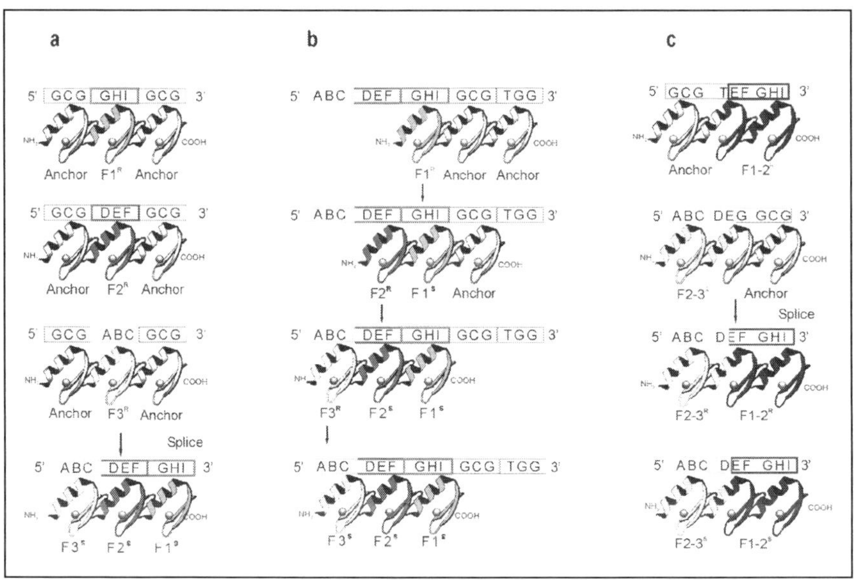

Figure 4. Strategies for selecting ZFPs with new DNA-binding specificities. Randomized fingers are indicated by superscript R (F^R). The fingers selected against the DNA subsites are indicated by superscript S (F^S). The hypothetical new binding site is indicated as 5'-ABC DEF GHI- 3'. a) Parallel selections of middle fingers from premade libraries are followed by a splicing step to make a new three-ZFPs. b) Sequential selections of zinc finger. Following each selection step, a new C-terminal finger library is cloned into the construct. In this way, a growing chain of new peptide "walks" across the gene of interest. c) In bipartite strategy, selections are carried out from two premade half libraries. Following which, the two selected zinc finger portions are spliced together and a further round of selection produces a new three-ZFP.[55]

Table 1. Genes regulated using designed ZFP TFs

Target Gene	Reporter	Endogenous	ZFP-DBD Used	Effector Domain	Activation	Repression	References
BCR-ABL	√		3-finger	VP16	√		Choo et al (1994)
BCR-ABL		√	3-finger	None		√	Choo et al (1994)
erB-2	√		3-finger and 6-finger	KRAB,ERD, SID		√	Beerli et al (1998)
Utrophin	√		3-finger	Gal4 and Vp16	√		Corbi et al (2000)
MDR1		√	5-finger	KRAB (two copies)		√	Bartsevich et al (2000)
Erytropoietin (EPO)	√	√	3-finger	VP16	√	√	Zhang et al (2000)
erB-2	√	√	6-finger	VP64 (VP16 x 4)	√		Beerli et al (2000) Dreier et al (2001)
erB-3	√	√	6-finger	VP64	√		Beerli et al (2000)
				(VP16 x 4)			Dreier et al (2001)
VEGF-√	√	√	3-finger	VP16 and NF-kBp65	√	√	Liu et al (2001)
PPARγ1,2		√	6-finger	KRAB		√	Ren et al (2002)
VEGF-A *		√	6-finger	VP16	√		Rebar et al (2002)
bax	√	√	3-finger and 5-finger	VP16	√		Falke et al (2003)
IE HSV-1	√	√	6-finger	KRAB		√	Papworth et al (2003)
HIV-1 Tat	√	√	3-fingers and 6-finger	KRAB		√	Reynolds et al (2003)
IGF2		√	3-finger	VP16 and NF-kBp65		√	Jouvenot et al (2003)
H19		√	3-finger	VP16 and NF-kBp65	√		Jouvenot et al (2003)

Features of artificial ZFP TFs and related target genes are listed. From the first column on the left: "Target genes", including both mammalian and viral genes; "Reporter", the transcription modulation has been tested, in cell lines, using transiently transfected reporter constructs; "Endogenous", the transcription modulation has been tested, in cell lines, on either stably integrated reporter constructs and/or on the chromosomal gene locus (asterisk indicates the unique example of whole mouse organism targeting); "ZFP-DBD used", number of zinc finger domains; "Effector Domain", type of functional domain; "Activation/Repression" of transcription

exchanges, a new three-finger protein is created in which all of the fingers ($F1^S$ $F2^S$ $F3^S$) have been selected in the context of the finger next to it.

Bipartite Selection

This method (Fig. 4c) combines the advantages of the two preceding methods and overcome incompatibilities at the interface between adjacent fingers.[63] According to this method selections are carried out in parallel from two premade "half" libraries, in which one-and-a-half fingers of the three-finger Zif268 are randomized. Selection of these two libraries is carried out at the same time against DNA sequences, in which half of the Zif268 target site is replaced by the sequence of interest. After these selections the two selected zinc finger portions are spliced together to make a novel three-ZFP.[64]

Alternative Strategies

Alternative strategies for selecting ZFPs have used a yeast one-hybrid and a bacterial two hybrid selection systems.[65,66] These systems show two useful features: the selection is performed in the context of living cells and the best proteins can be obtained through a one-step selection. In addition, a novel selection system using mammalian cultured cells has been recently described.[67]

Efforts in synthetic ZFPs design have also been focused on the assembly of ZFPs with more than three fingers.[14,57,68] In order to assembly such polydactyl ZFPs two strategies have been described:

connecting sets of fingers using linkers (covalent), or assembling sets of fingers using dimerization domains (noncovalent).[13,39,49,61,68-86] Computer graphics modeling, with Zif268/DNA complex, suggested the possibility of connecting two three-finger proteins using the conserved five-residue (TGEKP) linkers that join the fingers in many naturally occurring multi-finger domains.[69,71] Other designs have tested a longer, flexible or structured, linker to address the problem of the nonmatching helical periodicity of the zinc fingers versus B-DNA. Indeed, the periodicity gets progressively out of phase when there are more than three fingers.[70,77] These longer linkers appear to enhance binding specificity.[65,74,80] The noncovalent approach involves the use of dimerization domains to bring together two separate ZFPs. ZFPs fused either to Gal4 dimerization domain or to leucine zipper can dimerize and bind a longer DNA target.[83,85] Dimeric complexes can also form when ZFPs are fused to steroid hormone receptor ligand-binding domains (LBDs). The LBDs are inactive until ligand is added externally, thus allowing the chemical control activity of the ZFP TFs.

Affinity, Specificity and Efficacy of Artificial ZFP TFs

One of the crucial points in designing and selecting synthetic zinc finger peptides regards the calibration of correct binding affinity/specificity. In the cell environment a transcription factor is often able to bind multiple DNA targets with different degrees of affinity, exerting in this way an additional level of transcriptional regulation.[87] It is possible that binding too tightly is unfavourable for the transcription modulation.[13,88] It is likely that DNA binding specificity in relation to multiple DNA binding sites has been accurately calibrated during evolution so that any TF can correctly work in the proper pathways. Artificial TFs that are ideally realized to target a single promoter can by chance have multiple DNA targets, even if only a small portion of these potential sites in the genome will be accessible to TFs. In some favorable cases, the binding of a single ZFP to a DNA consensus site can lead to the simultaneous regulation of multiple desired genes. Assuming random base distribution, a 9 bp long DNA target sequence is present in the human genome ($\cong 3.5 \times 10^9$ bp) about 1.3×10^4 times.[13,47,61] Statistically, a target sequence must be at least 16 bp long to be unique in the human genome. For this reason Barbas group and other groups generated six finger polydactyl TFs that can specify 18 bp recognition.[13,61,69,89] The increased length of the target sequence in principle should determine a proportional increase of the DNA binding affinity. Nonetheless, on the basis of certain "mysterious" properties of ZFPs, in some cases in vitro affinities of three and six finger proteins, toward their respective binding sites, are practically equivalent. This could be due to an unfavourable cooperativity between certain fingers in binding long sequences.[90]

On the other hand, many other aspects of the artificial TFs must be simultaneously estimated to guarantee the optimum level of gene expression from a given target site in a given cell type. Aspects to consider include: (i) the nature of the effector domain, (ii) the position of the target site with respect to the target promoter, (iii) the accessibility of the factors to the target site in endogenous chromatin location.[88] The regulation of the correct concentration in the proper "latitude" of the artificial TFs is also desirable. For this purpose, various systems have been developed.[8] In conclusion, all these notions indicate that designer ZFP TFs need to be empirically tested for both the DNA binding properties and the trans-activation ability.[47]

Figure 5. Archetype structure of a designed ZFP TF. A specific antibody epitope (TAG) precedes a nuclear localization signal (NLS). A designed zinc finger DNA binding domain (DBD) is followed by a desired effector domain (ED).

Regulation of Gene Expression by Artificial ZFP TFs: Applications

There are several ways to alter gene expression in vivo. The transgenesis of cDNA encoding a gene of interest can lead to an increase of its mRNA level in the cell, homologous recombination methods can be used to ablate gene expression entirely, while the use of antisense reagents, rybozimes, or RNAi allow the lowering of mRNA levels to different degrees.[17,91-93] An alternative approach consists of the direct transcriptional regulation of a target endogenous locus. Several strategies can be followed for this purpose. The design of ZFPs with novel DNA-binding specifity is the most versatile to target any desired locus to either up-regulate or down-regulate gene expression.[8,57] To this end, the DNA binding domain can be fused to a repertoire of functional domains such as VP16, p65, for activation, KRAB, Sid for repression (Fig. 5).[17] The modular feature of ZFP-DNA recognition makes successful the "mix and match" approach, in which both natural fingers, with known site preferences, and synthetic fingers, with selected binding sites, are linked together to produce multifinger proteins with the proper affinity for the chosen DNA target.[13,17,94] Choo and Klug led the way in the realization of artificial ZFPs based on both coded contacts and selection.[48] In particular, they engineered a synthetic three-ZFP able to bind specifically the unique nine base pair region of the *BCR-ABL* fusion oncogene, generated by chromosomal breakpoints in acute lymphoblastic leukaemia. In transfected cell lines the artificial ZFP able to target *BCR-ABL* sequence, led to *BCR-ABL* mRNA reduction. The authors hypothesized that this is due to sequence specific binding of the ZFP and subsequent obstruction of the RNA polymerase II path. Since then, several studies have shown that designer ZFPs can regulate transcription of a target gene (Table 1).[13,16,17,47,48,61,62,65,95-101] Our group engineered several synthetic three-ZFPs on the basis of the recognition code selected by several research groups. One of these genes, named Jazz, was constructed to target the promoter of *utrophin*, a dystrophin related gene, with the aim to up-regulate the expression level of *utrophin* in Duchenne muscular dystrophy (DMD), complementing in this way the lack of dystrophin functions.[47] Chimeric proteins, generated by the fusion of the three-ZFP with both Gal4-activation domain and the natural Sp1 protein were able to drive the transcription of a test gene specifically from the human utrophin promoter.

Kang and Kim reported that ZFPs drive regulation of both transiently transfected and stably integrated reporter genes with comparable efficiency, despite important elements affecting transcription factor-DNA interaction, such as chromatin structure.[94]

Bartsevich and Juliano selected artificial ZFP TFs able to regulate the transcription of the *MDR1* multidrug resistance gene directly in the chromosomal context.[65] Beerli et al selected six-finger Sp1-based ZFPs able to target and to regulate the endogenous erB-2 and erB-3 protoncogenes, frequently overexpressed in human cancers.[13,38,61] Significantly, regulation of the two genes was highly specific, despite the fact that binding sites targeted by erB-2 and erbB-3 share 15 of 18 nucleotides. Moreover, they also obtained an inducible expression of the artificial ZFPs, which is particularly useful in both gene therapy and understanding of biological mechanisms.

Liu et al synthesized a panel of ZFPs to target the endogenous locus of vascular endothelial growth factor (*VEGF-A*), a specific inducer of new blood vessel growth, involved in a variety of medical conditions.[96] They demonstrated that combinations of three-ZFPs targeted to distinct-adjacent chromosomal sites linked to either VP16 or p65 were acting synergistically on *VEGF-A* activation. Moreover, they compared the behaviour of ZFPs targeted to accessible regions (identified by DNAase I hypersensitivity assay) with ZFPs targeted to inaccessible regions. While no evident differences were detected using naked reporter constructs, a clear difference in the activity of the endogenous *VEGF-A* gene was seen, thus emphasizing the importance of chromatin structure status. A similar approach was followed to activate the endogenous erythropoietin gene (EPO), suggesting the possibility of recruiting designed ZFPs to a site of interest in order to remodel the chromatin and regulate transcription.[95]

Peroxisome proliferator activated receptor-$\gamma 2$ (PPAR$\gamma 2$) has been used as an interesting example of knockdown expression. PPAR$\gamma 2$ is nuclear hormonal receptor, whose expression was selectively repressed by a designed ZFP. This ZFP distinguishes between the two isoforms $\gamma 1$ and $\gamma 2$, leading to determine an unique role of PPAR$\gamma 2$ isoform for a correct adipogenesis.[97] This kind of study could be functional for the design of a specific drug able to affect only the desired isoform.[14] Customized ZFP TFs were also able to inhibit transcription of viral DNA and therefore viral replication. Klug and colleagues designed ZFPs targeted to the viral immediate-early promoter (IE) of HSV-1 that were able to reduce the viral titer by about 90%.[99] Similarly, they obtained repression of the HIV-1 LTR promoter and inhibition of HIV-1 replication.[100] These results are promising for biotherapies to face viral infections.

Juliano et al reported data on an artificial transcription factor able to stimulate the expression of the endogenous pro-apoptotic gene bax, one of the p53 targets.[101] They propose the induction of bax gene as valuable tool in cancer chemotherapy, by diminishing survival of p53-deficient tumor cells.

All these data regarding the possibility to reprogram the expression of an endogenous gene target seem to predict that engineered transcription factors could be effective in a whole organism model. Indeed, expression of an artificial ZFP TF in a mouse ear model (delivered by adeno-vectors) led to induction of the *VEGF-A*, demonstrating for the first time the feasibility of regulating, in vivo, crucial biological processes as angiogenesis.[98] These data underline the potential therapeutic effects of designer TFs. In this context, Kim group proposed the elegant "GeneGrip" method based on screening of "natural" zinc fingers from human genome, in order to use them as modular building blocks in the generation of novel ZFPs.[102] A naturally occurring linker sequence was used to connect individual zinc fingers. Microarrays with cell lines stable transfected with engineered ZFPs permit to monitor the specificity of their action on a genome wide scale. More importantly, ZFPs constructed by shuffling naturally occurring human zinc fingers may be optimal in therapeutic applications in terms of reducing host immune response.

In addition, artificial ZFPs could be selected by ZFP libraries on the basis of an induced improved phenotype. Such ZFPs could also be used to regulate the expression of uncharacterized genes leading to identifying new genes and determining their role.[67]

ZFPs and RNA

The relevance of RNAs in many biological processes has generated great interest in designing sequence-specific RNA-binding molecules.[46,103,104] In particular, ZFPs selected to bind RNA, fused to appropriate modules, can regulate gene expression at different levels, interfering with alternative splicing, RNA trafficking and stability.[105]

Using a phage display approach Friesen et al reported on the amino acid requirements for the binding of TFIIIA to 5S RNA, demonstrating that some RNA/α helix contacts are identical to DNA contacts made by C_2H_2 zinc fingers.[106] Their results indicated that zinc finger recognition of RNA might be similar to DNA. However more than DNA, RNAs, on the basis of their rich repertoire of possible tertiary structures, display surfaces that resemble protein motifs. It has been observed that proteins and RNA can undergo peculiar conformational changes only in the context of the complex RNA/protein.[107] This characteristic gives the opportunity to design zinc fingers using structurebased strategies. McColl used this structure-based approach to design RNA-binding zinc fingers that recognize the HIV- REV response element (RRE).[108] An arginine-rich α helix from HIV REV was inserted into the middle zinc finger of Zif268 framework. The resulting hybrid zinc finger-REV was able to fold in a zinc dependent manner and binds specifically to the RRE. NMR experiments data provide evidences that monomeric zinc fingers can recognize specific nucleic acids demonstrating once again the modular nature of the zinc finger motif.[109] In particular, it seems to contact RNA into major groove through an extensive protein interface, involving probably not only the α helix, but also the β sheet of the zinc finger-binding surface. Moreover several studies have demonstrated that tethering of multiple binding modules can enhance RNA binding affinity and specificity, a strategy similarly used in DNA recognition.[104]

ZFPs and Plants

The manipulation of plants is an important issue with applications in basic plant biology research and agricultural biotechnologies.[110] Transgenesis has been the most common method to overexpress a gene in plants. Engineered ZFP TFs represent a novel promising tool to alter the expression of specific target genes, in order to get healthier or more nutritious crop plants.[111,112] Genes related to disease/stress resistance could be turned on, while genes related to pathologic status or to the production of anti-nutritive proteins could be turned off.[113] Importantly, to reduce or eliminate a gene expression in plants is not easy, since to date no natural modular repressors have been identified. However, foreign repressor domains may function in plants. The mammalian repressor domain KRAB appears to be a poor candidate for gene regulation in plants, since it is not found in organisms such as *Drosophila melanogaster*, *Caenorhabditis elegans*, *Saccharomyces cerevisiae*. By contrast, the Sid domain (Sin3A interaction domain) interacts with the highly conserved Sin3A and is therefore a promising effector domain to regulate endogenous plant genes.[111]

Several promising studies in the regulation of plant genes by ZFP TFs have recently reported.[114] A ZFP TF was able to activate a β-glucuronidase (GUS) reporter stably integrated in tobacco plants. Guan et al reported of ZFP TFs able to target the *Apetala 3 (Ape3)* promoter in *Arabidopsis* for both activation and repression. Importantly, in both cases activation and repression were inherited in subsequent generations.[79] In the future, libraries of ZFP TFs could be delivered using *Agrobacterium* transformation to obtain large libraries that could be screened for novel plant phenotypes.

Perspective

The design of efficient TFs, with a wide range of applications in both basic and applied sciences across many types of biotechnologies, remains a focal point of research in the post-genomic era.[14,115] At the moment, the delivery of artificial TFs into organisms and the ability to maintain the desired expression level of target genes are challenging objectives.[15]

Acknowledgements

This work was supported by Telethon A160 and Firb "Epigenetica e cromatina". N. Corbi is recipient of a FIRC fellowship.

References

1. Ptashne M, Gann A. Transcriptional activation by recruitment. Nature 1997; 386(6625):569-577.
2. Sauer F, Tjian R. Mechanisms of transcriptional activation: Differences and similarities between yeast, Drosophila, and man. Curr Opin Genet Dev 1997; 7(2):176-181.
3. Hampsey M. Molecular genetics of the RNA polymerase II general transcriptional machinery. Microbiol Mol Biol Rev 1998; 62(2):465-503.
4. Cremer T, Cremer C. Chromosome territories, nuclear architecture and gene regulation in mammalian cells. Nat Rev Genet 2001; 2(4):292-301.
5. Venter JC, Adams MD, Myers EW et al. The sequence of the human genome. Science 2001; 291(5507):1304-1351.
6. Desjarlais JR, Berg JM. Toward rules relating zinc finger protein sequences and DNA binding site preferences. Proc Natl Acad Sci USA 1992; 89(16):7345-7349.
7. Isalan M, Choo Y. Rapid, high-throughput engineering of sequence-specific zinc finger DNA-binding proteins. Methods Enzymol 2001; 340:593-609.
8. Pabo CO, Peisach E, Grant RA. Design and selection of novel Cys2His2 zinc finger proteins. Annu Rev Biochem 2001; 70:313-340.
9. Pavletich NP, Pabo CO. Zinc finger-DNA recognition: Crystal structure of a Zif268-DNA complex at 2.1 A. Science 1991; 252(5007):809-817.
10. Jamieson AC, Kim SH, Wells JA. In vitro selection of zinc fingers with altered DNA-binding specificity. Biochemistry 1994; 33(19):5689-5695.
11. Choo Y, Klug A. Toward a code for the interactions of zinc fingers with DNA: Selection of randomized fingers displayed on phage. Proc Natl Acad Sci USA 1994; 91(23):11163-7.
12. Choo Y, Klug A. Selection of DNA binding sites for zinc fingers using rationally randomized DNA reveals coded interactions. Proc Natl Acad Sci USA 1994; 91(23):11168-11172.
13. Beerli RR, Dreier B, Barbas III CF. Positive and negative regulation of endogenous genes by designed transcription factors. Proc Natl Acad Sci USA 2000; 97(4):1495-1500.
14. Jamieson AC, Miller JC, Pabo CO. Drug discovery with engineered zinc-finger proteins. Nat Rev Drug Discov 2003; 2(5):361-368.
15. Ansari AZ, Mapp AK. Modular design of artificial transcription factors. Curr Opin Chem Biol 2002; 6(6):765-772.
16. Urnov FD, Rebar EJ, Reik A et al. Designed transcription factors as structural, functional and therapeutic probes of chromatin in vivo. Fourth in review series on chromatin dynamics. EMBO Rep 2002; 3(7):610-615.
17. Urnov FD, Rebar EJ. Designed transcription factors as tools for therapeutics and functional genomics. Biochem Pharmacol 2002; 64(5-6):919-923.
18. Pabo CO, Sauer RT. Transcription factors: Structural families and principles of DNA recognition. Annu Rev Biochem 1992; 61:1053-1095.
19. Rhodes D, Klug A. An underlying repeat in some transcriptional control sequences corresponding to half a double helical turn of DNA. Cell 1986; 46(1):123-132.
20. Berg JM. Zinc finger domains: Hypotheses and current knowledge. Annu Rev Biophys Biophys Chem 1990; 19:405-421.
21. Pavletich NP, Pabo CO. Zinc finger-DNA recognition: Crystal structure of a Zif268-DNA complex at 2.1 A. Science 1991; 252(5007):809-817.
22. Desjarlais JR, Berg JM. Redesigning the DNA-binding specificity of a zinc finger protein: A data base-guided approach. Proteins 1992; 12(2):101-104.
23. Desjarlais, Berg JM. Redesigning the DNA-binding specificity of a zinc finger protein: A data base-guided approach. Proteins 1992; 13(3):272.
24. Nardelli J, Gibson TJ, Vesque C et al. Base sequence discrimination by zinc-finger DNA-binding domains. Nature 1991; 349(6305):175-178.
25. Nardelli J, Gibson T, Charnay P. Zinc finger-DNA recognition: Analysis of base specificity by site-directed mutagenesis. Nucleic Acids Res 1992; 20(16):4137-4144.
26. Desjarlais JR, Berg JM. Use of a zinc-finger consensus sequence framework and specificity rules to design specific DNA binding proteins. Proc Natl Acad Sci USA 1993; 90(6):2256-2260.
27. Rebar EJ, Pabo CO. Zinc finger phage: Affinity selection of fingers with new DNA-binding specificities. Science 1994; 263(5147):671-673.
28. Desjarlais JR, Berg JM. Length-encoded multiplex binding site determination: Application to zinc finger proteins. Proc Natl Acad Sci USA 1994; 91(23):11099-11103.
29. Choo Y, Klug A. Designing DNA-binding proteins on the surface of filamentous phage. Curr Opin Biotechnol 1995; 6(4):431-436.
30. Wu H, Yang WP, Barbas III CF. Building zinc fingers by selection: Toward a therapeutic application. Proc Natl Acad Sci USA 1995; 92(2):344-348.
31. Jamieson AC, Wang H, Kim SH. A zinc finger directory for high-affinity DNA recognition. Proc Natl Acad Sci USA 1996; 93(23):12834-12839.
32. Greisman HA, Pabo CO. A general strategy for selecting high-affinity zinc finger proteins for diverse DNA target sites. Science 1997; 275(5300):657-661.
33. Isalan M, Choo Y. Engineering nucleic acid-binding proteins by phage display. Methods Mol Biol 2001; 148:417-429.
34. Wolfe SA, Nekludova L, Pabo CO. DNA recognition by Cys2His2 zinc finger proteins. Annu Rev Biophys Biomol Struct 2000; 29:183-212.
35. Kim CA, Berg JM. A 2.2 A resolution crystal structure of a designed zinc finger protein bound to DNA. Nat Struct Biol 1996; 3(11):940-945.
36. Elrod-Erickson M, Benson TE, Pabo CO. High-resolution structures of variant Zif268-DNA complexes: Implications for understanding zinc finger-DNA recognition. Structure 1998; 6(4):451-464.
37. Choo Y, Klug A. Physical basis of a protein-DNA recognition code. Curr Opin Struct Biol 1997; 7(1):117-125.
38. Segal DJ, Dreier B, Beerli RR et al. Toward controlling gene expression at will: Selection and design of zinc finger domains recognizing each of the 5'-GNN-3' DNA target sequences. Proc Natl Acad Sci USA 1999; 96(6):2758-2763.
39. Dreier B, Beerli RR, Segal DJ et al. Development of zinc finger domains for recognition of the 5'-ANN-3' family of DNA sequences and their use in the construction of artificial transcription factors. J Biol Chem 2001; 276(31):29466-29478.

40. Elrod-Erickson M, Rould MA, Nekludova L et al. Zif268 protein-DNA complex refined at 1.6 A: A model system for understanding zinc finger-DNA interactions. Structure 1996; 4(10):1171-1180.
41. Pavletich NP, Pabo CO. Crystal structure of a five-finger GLI-DNA complex: New perspectives on zinc fingers. Science 1993; 261(5129):1701-1707.
42. Fairall L, Schwabe JW, Chapman L et al. The crystal structure of a two zinc-finger peptide reveals an extension to the rules for zinc-finger/DNA recognition. Nature 1993; 366(6454):483-487.
43. Houbaviy HB, Usheva A, Shenk T et al. Cocrystal structure of YY1 bound to the adeno-associated virus P5 initiator. Proc Natl Acad Sci USA 1996; 93(24):13577-13582.
44. Kim CA, Berg JM. Serine at position 2 in the DNA recognition helix of a Cys2-His2 zinc finger peptide is not, in general, responsible for base recognition. J Mol Biol 1995; 252(1):1-5.
45. Corbi N, Perez M, Maione R et al. Synthesis of a new zinc finger peptide; comparison of its 'code' deduced and 'CASTING' derived binding sites. FEBS Lett 1997; 417(1):71-74.
46. Corbi N, Libri V, Fanciulli M et al. Binding properties of the artificial zinc fingers coding gene Sint1. Biochem Biophys Res Commun 1998; 253(3):686-692.
47. Corbi N, Libri V, Fanciulli M et al. The artificial zinc finger coding gene 'Jazz' binds the utrophin promoter and activates transcription. Gene Ther 2000; 7(12):1076-1083.
48. Choo Y, Sanchez-Garcia I, Klug A. In vivo repression by a site-specific DNA-binding protein designed against an oncogenic sequence. Nature 1994; 372(6507):642-645.
49. McNamara AR, Ford KG. A novel four zinc-finger protein targeted against p190(BcrAbl) fusion oncogene cDNA: Utilisation of zinc-finger recognition codes. Nucleic Acids Res 2000; 28(24):4865-4872.
50. Isalan M, Choo Y, Klug A. Synergy between adjacent zinc fingers in sequence-specific DNA recognition. Proc Natl Acad Sci USA 1997; 94(11):5617-5621.
51. Wolfe SA, Greisman HA, Ramm EI et al. Analysis of zinc fingers optimized via phage display: Evaluating the utility of a recognition code. J Mol Biol 1999; 285(5):1917-1934.
52. Pabo CO, Nekludova L. Geometric analysis and comparison of protein-DNA interfaces: Why is there no simple code for recognition? J Mol Biol 2000; 301(3):597-624.
53. Wolfe SA, Grant RA, Elrod-Erickson M et al. Beyond the "recognition code": Structures of two Cys2His2 zinc finger/TATA box complexes. Structure (Camb) 2001; 9(8):717-23.
54. Sera T, Uranga C. Rational design of artificial zinc-finger proteins using a nondegenerate recognition code table. Biochemistry 2002; 41(22):7074-7081.
55. Choo Y, Isalan M. Advances in zinc finger engineering. Curr Opin Struct Biol 2000; 10(4):411-416.
56. Segal DJ, Barbas III CF. Design of novel sequence-specific DNA-binding proteins. Curr Opin Chem Biol 2000; 4(1):34-39.
57. Beerli RR, Barbas III CF. Engineering polydactyl zinc-finger transcription factors. Nat Biotechnol 2002; 20(2):135-141.
58. Dreier B, Segal DJ, Barbas III CF. Insights into the molecular recognition of the 5'-GNN-3' family of DNA sequences by zinc finger domains. J Mol Biol 2000; 303(4):489-502.
59. Segal DJ. The use of zinc finger peptides to study the role of specific factor binding sites in the chromatin environment. Methods 2002; 26(1):76-83.
60. Segal DJ, Beerli RR, Blancafort P et al. Evaluation of a modular strategy for the construction of novel polydactyl zinc finger DNA-binding proteins. Biochemistry 2003; 42(7):2137-2148.
61. Beerli RR, Segal DJ, Dreier B et al. Toward controlling gene expression at will: Specific regulation of the erbB-2/HER-2 promoter by using polydactyl zinc finger proteins constructed from modular building blocks. Proc Natl Acad Sci USA 1998; 95(25):14628-14633.
62. Jouvenot Y, Ginjala V, Zhang L et al. Targeted regulation of imprinted genes by synthetic zinc-finger transcription factors. Gene Ther 2003; 10(6):513-522.
63. Isalan M, Klug A, Choo Y. Comprehensive DNA recognition through concerted interactions from adjacent zinc fingers. Biochemistry 1998; 37(35):12026-12033.
64. Isalan M, Klug A, Choo Y. A rapid, generally applicable method to engineer zinc fingers illustrated by targeting the HIV-1 promoter. Nat Biotechnol 2001; 19(7):656-660.
65. Bartsevich VV, Juliano RL. Regulation of the MDR1 gene by transcriptional repressors selected using peptide combinatorial libraries. Mol Pharmacol 2000; 58(1):1-10.
66. Joung JK, Ramm EI, Pabo CO. A bacterial two-hybrid selection system for studying protein-DNA and protein-protein interactions. Proc Natl Acad Sci USA 2000; 97(13):7382-7387.
67. Blancafort P, Magnenat L, Barbas CF. Scanning the human genome with combinatorial transcription factor libraries. Nat Biotechnol 2003; 21(3):269-274.
68. Imanishi M, Hori Y, Nagaoka M et al. Design of novel zinc finger proteins: Towards artificial control of specific gene expression. Eur J Pharm Sci 2001; 13(1):91-97.
69. Liu Q, Segal DJ, Ghiara JB et al. Design of polydactyl zinc-finger proteins for unique addressing within complex genomes. Proc Natl Acad Sci USA 1997; 94(11):5525-5530.
70. Kim JS, Pabo CO. Getting a handhold on DNA: Design of poly-zinc finger proteins with femtomolar dissociation constants. Proc Natl Acad Sci USA 1998; 95(6):2812-2817.
71. Kamiuchi T, Abe E, Imanishi M et al. Artificial nine zinc-finger peptide with 30 base pair binding sites. Biochemistry 1998; 37(39):13827-13834.
72. Nagaoka M, Sugiura Y. Artificial zinc finger peptides: Creation, DNA recognition, and gene regulation. J Inorg Biochem 2000; 82(1-4):57-63.
73. Imanishi M, Hori Y, Nagaoka M et al. DNA-bending finger: Artificial design of 6-zinc finger peptides with polyglycine linker and induction of DNA bending. Biochemistry 2000; 39(15):4383-4390.
74. Moore M, Klug A, Choo Y. Improved DNA binding specificity from polyzinc finger peptides by using strings of two-finger units. Proc Natl Acad Sci USA 2001; 98(4):1437-1441.
75. Nagaoka M, Kaji T, Imanishi M et al. Multiconnection of identical zinc finger: Implication for DNA binding affinity and unit modulation of the three zinc finger domain. Biochemistry 2001; 40(9):2932-2941.
76. Nagaoka M, Nomura W, Shiraishi Y et al. Significant effect of linker sequence on DNA recognition by multi-zinc finger protein. Biochem Biophys Res Commun 2001; 282(4):1001-1007.
77. Moore M, Choo Y, Klug A. Design of polyzinc finger peptides with structured linkers. Proc Natl Acad Sci USA 2001; 98(4):1432-1436.
78. Liu Q, Xia Z, Zhong X et al. Validated zinc finger protein designs for all 16 GNN DNA triplet targets. J Biol Chem 2002; 277(6):3850-3856.
79. Guan X, Stege J, Kim M et al. Heritable endogenous gene regulation in plants with designed polydactyl zinc finger transcription factors. Proc Natl Acad Sci USA 2002; 99(20):13296-13301.
80. Arora PS, Ansari AZ, Best TP et al. Design of artificial transcriptional activators with rigid poly-L-proline linkers. J Am Chem Soc 2002; 124(44):13067-13071.
81. Imanishi M, Sugiura Y. Artificial DNA-bending six-zinc finger peptides with different charged linkers: Distinct kinetic properties of DNA bindings. Biochemistry 2002; 41(4):1328-1334.
82. Nomura W, Nagaoka M, Shiraishi Y et al. Influence of TFIIIA-type linker at the N- or C-terminal of nine-zinc finger protein on DNA-binding site. Biochem Biophys Res Commun 2003; 300(1):87-92.
83. Pomerantz JL, Wolfe SA, Pabo CO. Structure-based design of a dimeric zinc finger protein. Biochemistry 1998; 37(4):965-970.
84. Wang BS, Pabo CO. Dimerization of zinc fingers mediated by peptides evolved in vitro from random sequences. Proc Natl Acad Sci USA 1999; 96(17):9568-9573.
85. Wolfe SA, Ramm EI, Pabo CO. Combining structure-based design with phage display to create new Cys(2)His(2) zinc finger dimers. Structure Fold Des 2000; 8(7):739-750.

86. Wang BS, Grant RA, Pabo CO. Selected peptide extension contacts hydrophobic patch on neighboring zinc finger and mediates dimerization on DNA. Nat Struct Biol 2001; 8(7):589-593.
87. Berg JM. Sp1 and the subfamily of zinc finger proteins with guanine-rich binding sites. Proc Natl Acad Sci USA 1992; 89(23):11109-11110.
88. Yaghmai R, Cutting GR. Optimized regulation of gene expression using artificial transcription factors. Mol Ther 2002; 5(6):685-694.
89. Segal DJ, Barbas III CF. Custom DNA-binding proteins come of age: Polydactyl zinc-finger proteins. Curr Opin Biotechnol 2001; 12(6):632-627.
90. Ansari AZ. Fingers reach for the genome. Nat Biotechnol 2003; 21(3):242-243.
91. Thoma C, Hasselblatt P, Kock J et al. Generation of stable mRNA fragments and translation of N-truncated proteins induced by antisense oligodeoxynucleotides. Mol Cell 2001; 8(4):865-872.
92. Hutvagner G, Zamore PD. RNAi: Nature abhors a double-strand. Curr Opin Genet Dev 2002; 12(2):225-232.
93. Harris S, Foord SM. Transgenic gene knock-outs: Functional genomics and therapeutic target selection. Pharmacogenomics 2000; 1(4):433-443.
94. Kang JS, Kim JS. Zinc finger proteins as designer transcription factors. J Biol Chem 2000; 275(12):8742-8748.
95. Zhang L, Spratt SK, Liu Q et al. Synthetic zinc finger transcription factor action at an endogenous chromosomal site. Activation of the human erythropoietin gene. J Biol Chem 2000; 275(43):33850-33860.
96. Liu PQ, Rebar EJ, Zhang L et al. Regulation of an endogenous locus using a panel of designed zinc finger proteins targeted to accessible chromatin regions. Activation of vascular endothelial growth factor A. J Biol Chem 2001; 276(14):11323-11334.
97. Ren D, Collingwood TN, Rebar EJ et al. PPARgamma knockdown by engineered transcription factors: Exogenous PPARgamma2 but not PPARgamma1 reactivates adipogenesis. Genes Dev 2002; 16(1):27-32.
98. Rebar EJ, Huang Y, Hickey R et al. Induction of angiogenesis in a mouse model using engineered transcription factors. Nat Med 2002; 8(12):1427-1432.
99. Papworth M, Moore M, Isalan M et al. Inhibition of herpes simplex virus 1 gene expression by designer zinc-finger transcription factors. Proc Natl Acad Sci USA 2003; 100(4):1621-1626.
100. Reynolds L, Ullman C, Moore M et al. Repression of the HIV-1 5' LTR promoter and inhibition of HIV-1 replication by using engineered zinc-finger transcription factors. Proc Natl Acad Sci USA 2003; 100(4):1615-1620.
101. Falke D, Fisher M, Ye D et al. Design of artificial transcription factors to selectively regulate the pro-apoptotic bax gene. Nucleic Acids Res 2003; 31(3):e10.
102. Bae KH, Do KY, Shin HC et al. Human zinc fingers as building blocks in the construction of artificial transcription factors. Nat Biotechnol 2003; 21(3):275-280.
103. Friesen WJ, Darby MK. Specific RNA binding proteins constructed from zinc fingers. Nat Struct Biol 1998; 5(7):543-546.
104. Cheng AC, Calabro V, Frankel AD. Design of RNA-binding proteins and ligands. Curr Opin Struct Biol 2001; 11(4):478-484.
105. Cartegni L, Krainer AR. Correction of disease-associated exon skipping by synthetic exon-specific activators. Nat Struct Biol 2003; 10(2):120-125.
106. Friesen WJ, Darby MK. Phage display of RNA binding zinc fingers from transcription factor IIIA. J Biol Chem 1997; 272(17):10994-10997.
107. Frankel AD. Fitting peptides into the RNA world. Curr Opin Struct Biol 2000; 10(3):332-340.
108. McColl DJ, Honchell CD, Frankel AD. Structure-based design of an RNA-binding zinc finger. Proc Natl Acad Sci USA 1999; 96(17):9521-9526.
109. Friesen WJ, Darby MK. Specific RNA binding by a single C2H2 zinc finger. J Biol Chem 2001; 276(3):1968-1973.
110. Frommer WB, Beachy R. Plant biotechnology. A future for plant biotechnology? Naturally!. Curr Opin Plant Biol 2003; 6(2):147-149.
111. Stege JT, Guan X, Ho T et al. Controlling gene expression in plants using synthetic zinc finger transcription factors. Plant J 2002; 32(6):1077-1086.
112. Sanchez JP, Ullman C, Moore M et al. Regulation of gene expression in Arabidopsis thaliana by artificial zinc finger chimeras. Plant Cell Physiol 2002; 43(12):1465-1472.
113. Segal DJ, Stege JT, Barbas CF. Zinc fingers and a green thumb: Manipulating gene expression in plants. Curr Opin Plant Biol 2003; 6(2):163-168.
114. Ordiz MI, Barbas III CF, Beachy RN. Regulation of transgene expression in plants with polydactyl zinc finger transcription factors. Proc Natl Acad Sci USA 2002; 99(20):13290-13295.
115. Falke D, Juliano RL. Selective gene regulation with designed transcription factors: Implications for therapy. Curr Opin Mol Ther 2003; 5(2):161-166.

CHAPTER 10

TFIIIA and p43: Binding to 5S Ribosomal RNA

Paul J. Romaniuk

Abstract

TFIIIA and p43 are multifunctional zinc finger proteins that share a specific affinity for 5S ribosomal RNA. In this chapter I summarize over 25 years of research that highlights the similarities and differences in the RNA binding activity of these two proteins.

Introduction

In the immature oocytes of *Xenopus laevis*, other amphibia and teleost fish, 5S ribosomal RNA (rRNA) is transcribed in large amounts and stored in the cytoplasm as ribonucleoprotein particles (RNPs). In the case of *Xenopus*, approximately half of the cytoplasmic 5S rRNA is stored in a 7S RNP bound in a 1:1 complex with the protein TFIIIA, while the other half of the cytoplasmic 5S rRNA is stored in a 42S RNP bound in a 1:1 complex with the protein p43. Although TFIIIA and p43 both bind to 5S rRNA and each has 9 tandem zinc fingers, there is a remarkable lack of amino acid sequence homology between the two proteins outside of the conserved elements required to fold the zinc finger domains.

In addition to forming a 5S rRNA storage particle, TFIIIA also acts as an RNA polymerase III transcription factor by binding to the internal control region of the 5S rRNA gene. It therefore has a specific DNA binding activity as well as a specific RNA binding activity. The p43 protein has no known DNA binding activity. In the 42S RNP, p43 interacts with the p48 protein component of this particle as well as the 5S rRNA.

One of the questions that has occupied my research lab is "what are the similarities and differences in the mechanisms by which TFIIIA and p43 interact with nucleic acids?" Ours has been a very detailed approach to this question, using quantitative equilibrium binding assays and targeted mutagenesis of the nucleic acids and proteins in order to identify critical residues required for the formation of these specific protein-RNA complexes. In the case of TFIIIA, we have also been interested in understanding the similarities and differences in the binding of the protein to DNA and RNA. Before describing twenty years of research into these intriguing questions, I'll begin by describing the biology of the system we are investigating.

5S rRNA Synthesis and Storage during Oogenesis

Two strategies are used to ensure there are sufficient quantities of RNA components for the burst of ribosome assembly in the later stages of oogenesis (see review by Denis and le Maire).[1] The genes encoding the large rRNAs undergo 1000X amplification in order to provide the necessary transcriptional capacity. The expression of these amplified genes begins during the later, vitellogenic phase of oogenesis and provides the quantity of large rRNA required for ribosome assembly. The genes encoding the 5S rRNA component do not undergo amplification, but instead are completely derepressed and transcribed throughout oogenesis. During the previtellogenic stages of oogenesis, when expression of the large rRNA genes is low the excess 5S rRNA that accumulates is sequestered in the cytoplasm (Fig. 1).[2] The accumulated 5S rRNA is stored in two RNPs, one which sediments at 7S[3] and one which sediments at 42S.[4] As the vitellogenic phase of oogenesis proceeds, the 5S rRNA is imported to the nucleolus where it is assembled into the large ribosomal subunit.

The 7S RNP consists of a bimolecular complex of 5S rRNA bound to TFIIIA, the 5S rRNA gene-specific transcription factor.[5] Expression of the TFIIIA gene is under developmental regulation.[5-8] The levels of TFIIIA produced in immature oocytes (approximately 10^{12} molecules/oocyte) are sufficient to sequester the 5S rRNA in the cytoplasm. In the very late stages of oogenesis, the stored 5S rRNA is used for ribosome assembly and the released TFIIIA is degraded. After fertilization, the level of mRNA

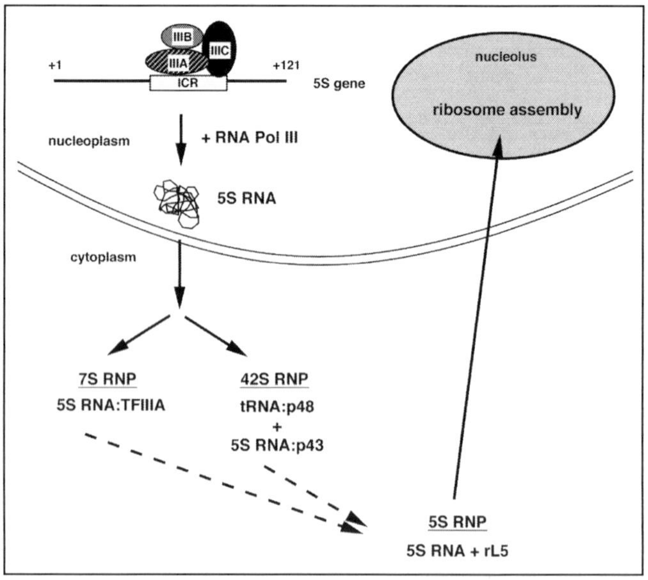

Figure 1. Metabolism of 5S rRNA during oogenesis in *Xenopus laevis*.

Zinc Finger Proteins: From Atomic Contact to Cellular Function, edited by Shiro Iuchi and Natalie Kuldell.

encoding TFIIIA drops dramatically during embryogenesis and the level of TFIIIA protein is reduced to approximately 10^4 molecules per somatic cell. This level of the transcription factor is in excess to the approximately 800 somatic 5S rRNA genes that are actively being transcribed, and the formation of a 7S RNP in somatic cells acts as part of a feedback regulatory mechanism that couples production of 5S rRNA and ribosomal protein L5.[8-13]

The 42S RNP consists of tRNA, 5S rRNA, and the proteins p43 and p48.[14] The 7S RNP contains about half of the oocyte 5S rRNA, while the 42S RNP contains the other half of the 5S rRNA and 90% of the oocyte tRNA. The components of the 42S RNP are found in the molar ratio 3:1:2:1 for tRNA:5S rRNA:p48:p43.[14] The p48 protein binds to tRNA and is a homologue of elongation factor EF-1α.[15,16] The p43 protein binds to 5S rRNA and was subsequently shown to be a zinc finger protein.[17] The 42S RNP is a tetramer of a fundamental 15S particle that contains the four protein and RNA components in the ratio given above.[1,14,18] In addition, the presence of minor protein components can account for approximately 10 to 30% of the total protein of the 42S RNP.[1,14]

While the sole biological role of the 7S RNP appears to be storage of 5S rRNA until it is required for ribosome assembly, the 42S RNP may fulfill several roles. The tRNA in the 42S RNP is aminoacylated, and there appears to be free exchange with tRNA in the cytoplasm. There is evidence that the 42S RNP carries aminoacyl tRNAs directly to ribosomes in the oocyte.[16] Therefore the 42S RNP is involved in regulating the pool of tRNA in the oocyte and may act as translation factor during protein biosynthesis.

In the latter stages of oogenesis, 5S rRNA stored in the cytoplasm must be imported into the nucleus for assembly into 60S ribosomal subunits. When bound to 5S rRNA, the nuclear import signal of TFIIIA is masked and 7S RNPs are excluded from the nucleus.[12,13] The form of 5S rRNA that is imported into the nucleolus and assembled into ribosomal subunits is a 5S RNP that consists of the RNA bound to ribosomal protein L5.[12,19-22] Importation of stored somatic 5S rRNA may occur more rapidly than oocyte 5S rRNA in the early phases of ribosome assembly.[23] Once there is sufficient L5 protein in the oocyte, newly synthesized 5S rRNA is likely incorporated directly into ribosomes, bypassing the cytoplasmic route entirely.

Interaction of TFIIIA with 5S rRNA and the 5S rRNA Gene

In 1980 TFIIIA was identified as the protein component of the 7S RNP, thus raising the question of how a single protein could form strong, specific interactions with the 5S rRNA gene and the 5S rRNA transcript.[5] The results of exhaustive experimental testing have demonstrated clearly that the modes of binding of TFIIIA to DNA and RNA are distinctly different. Different zinc fingers on the protein are involved in forming specific interactions with DNA versus RNA, and different bases and structural features in DNA and RNA are involved in forming the interactions with TFIIIA.

My lab started exploring the nature of TFIIIA binding to nucleic acids by considering the equilibrium binding properties of the protein. We developed a filter binding assay that could measure the specific interaction of TFIIIA with 5S rRNA and the 5S rRNA gene.[24,25] We used this assay to quantitatively measure the interaction of TFIIIA with both nucleic acid targets under a variety of binding conditions (Table 1). The protein has a slightly higher affinity for DNA compared to RNA. The 5S rRNA-TFIIIA

Table 1. Comparison of equilibrium binding parameters for the interaction of TFIIIA with the 5S rRNA gene and 5S RNA at 22° C

Parameter	5S rRNA Gene	5S rRNA
K_a	1.90×10^9 M^{-1}	1.4×10^9 M^{-1}
ΔG°	−12.4 kcal mol^{-1}	−12.1 kcal mol^{-1}
ΔH°	−12.1 kcal mol^{-1}	−8.3 kcal mol^{-1}
ΔS°	+1.23 cal deg^{-1} mol^{-1}	+13.1 cal deg^{-1} mol^{-1}
half life	15.3 min	25.7 min

complex dissociates more slowly than the complex of TFIIIA bound to the 5S rRNA gene, consistent with the biological roles of the two complexes. The RNA-protein complex is used to store excess 5S rRNA in the cytoplasm of oocytes and might be expected to be reasonably resistant to dissociation, while the DNA-protein complex should be more readily dissociated to allow for feedback regulation of transcription. The formation of the TFIIIA-5S rRNA complex at 22°C is favored by both enthalpy and entropy, while formation of the TFIIIA-DNA complex is favored by enthalpy. In addition to the parameters listed in Table 1, we also found that the pH optima for both interactions are very similar, but the DNA-protein complex is somewhat more sensitive to the monovalent salt concentration of the binding buffer than the RNA-protein complex. The establishment of these equilibrium binding assays and the understanding that we obtained about the similarities and differences in the equilibrium binding of TFIIIA to the two nucleic acids were important for the experiments that followed.

The binding site for TFIIIA on DNA lies within the coding part of the 5S rRNA gene.[26-28] Within this "internal control region" there are three distinct promoter elements (Fig. 2), with the intermediate and box C elements being essential for the binding of TFIIIA.[29,30] One of our initial goals for studying the TFIIIA-5S rRNA interaction was to determine the precise location of the binding site for the protein on the RNA. Footprinting proteins on RNA is not as straightforward as footprinting DNA-protein complexes for a number of reasons. The nuclease and chemical reagents used for footprinting RNA-protein complexes are influenced not only by their particular sequence specificities, but are also influenced by the secondary and tertiary structure of the RNA. As a result, it is difficult to get uniform cleavage patterns on "naked" RNA, which makes footprinting a protein bound to the RNA more difficult in turn. In addition, nicking an RNA molecule at a site outside the protein binding region may disrupt the structure of the RNA, leading to dissociation of the protein-RNA complex. Therefore it is necessary in many cases to select for RNA still bound to protein after treating with a chemical or enzymatic reagent in order to be able to detect the footprint. Another side effect of disrupting the structure of the RNA by the first nick is the fact that the RNA becomes much more susceptible to cleavage elsewhere, making it necessary to carefully titrate reagents and work out reaction conditions that ensure any RNA molecule is only cleaved once. To ensure each putative site represents a unique cleavage on a single RNA molecule, it is necessary to repeat the experiments with RNA that has been labeled at the 5' end in one case and the 3' end in the other.

Keeping these limitations in mind, we used two nucleases to footprint TFIIIA on 5S rRNA.[31] At the same time, other groups were using different reagents to footprint the TFIIIA-5S rRNA

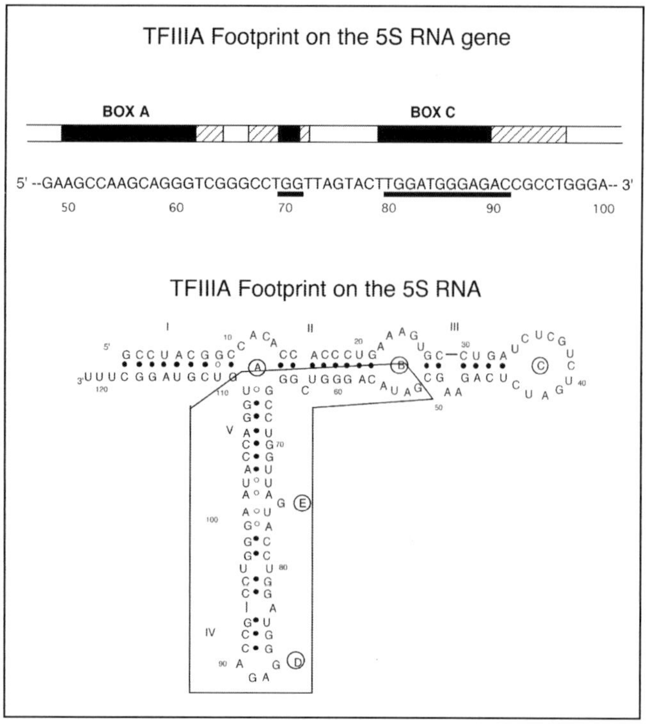

Figure 2. Footprint of TFIIIA on the 5S rRNA gene and the 5S rRNA. The box A, intermediate element and box C components of the polymerase III promoter are shown as solid boxes. The footprint on DNA extends from the start of box A to the end of the hatched region downstream of box C. Base sequence critical for TFIIIA binding is underlined.

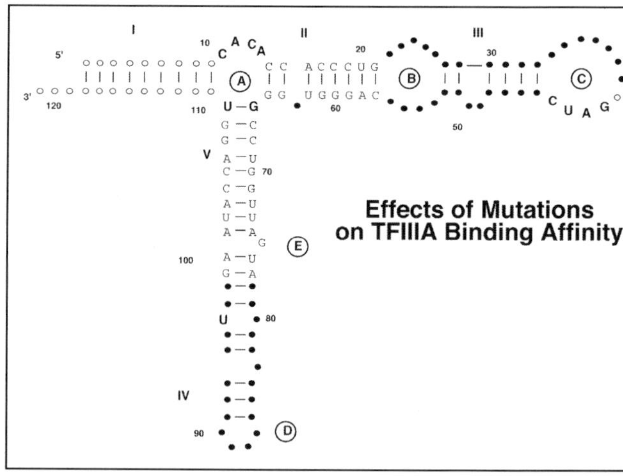

Figure 3. Effects of mutations in 5S rRNA on TFIIIA binding affinity. Open circles indicate nucleotides that were not tested for importance in TFIIIA binding. Closed circles represent nucleotides that could be substituted without significantly decreasing TFIIIA binding affinity. Nucleotides shown in normal script occur in areas where the structure (e.g., base paired stem) is important for TFIIIA binding, but the actual nucleotide sequence is irrelevant. Nucleotides shown in bold occur in regions where sequence and structure are important for TFIIIA binding.

complex.[32-36] The results obtained from these different studies were remarkably consistent. The binding site of TFIIIA on 5S rRNA as defined by these footprinting experiments is shown in Figure 2. It was apparent that the footprint of TFIIIA on the 5S rRNA involved roughly an identical region to the footprint of TFIIIA on the 5S rRNA gene. This similarity in the footprint areas provided additional motivation to identify the specific nucleotides in the RNA and base pairs in the DNA that are critical for the binding of TFIIIA.

In designing site directed mutations for an RNA ligand of a protein, the effects of any mutation on the overall structure of the RNA have to be taken into account. It is important that the mutation has a local, predictable effect on the RNA structure in order to draw accurate conclusions about the effect on protein binding. Additionally, mutations that disrupt a base paired stem may reduce the binding of a protein either because the specific nucleotide sequence that has been changed provides a direct interaction with the protein, or because the base paired stem is an important structural element for protein binding. These alternative effects can be addressed by making double, compensatory mutations that restore a base paired stem but alter the sequence of base pairs. In the case of 5S rRNA, our efforts to probe the requirement of specific sequence and structural elements for TFIIIA binding were greatly aided by the well known secondary structure of the RNA.

Over the course of several years, we constructed and expressed over 60 site-directed mutations of *Xenopus* 5S rRNA.[21,22,31,37-43] The effects of these mutations on the in vitro binding of TFIIIA, and the in vivo trafficking of 5S rRNA in the oocyte were assayed. Most of this work was carried out by Florence Baudin, Nik Veldhoen and Qimin You, three very talented graduate students. The results obtained with the filter binding assay are summarized in Figure 3. Although the effects observed were relatively small in magnitude, it is clear that the conserved structure of 5S rRNA is a main determinant in TFIIIA binding. The mutations with the greatest effect on TFIIIA binding are clustered in the junction between the three helical arms of the 5S rRNA.[38,43] The nucleotides in loop A are particularly critical, even though they do not fall within the footprint of TFIIIA on the RNA. Cleavage of the RNA backbone within loop A also significantly alters the TFIIIA-5S rRNA interaction.[44] A three way junction like this controls the coaxial stacking (colinearity in three dimensional space) of the three helical arms of an RNA (Fig. 4).[45-49] Particularly important are the specific base pairs that close the three helices at loop A, and the stacking of two adenine bases in the loop.[38] Based upon these results and other studies that compared the effects of some of the mutations on the solution structure of the 5S rRNA,[38,45,47 50] we proposed that TFIIIA likely forms a stronger interaction with a conformational state of the RNA in which helix II and helix V are colinear. The importance of loop A and region E sequences for the binding of TFIIIA was also observed by the groups of Tomas Pieler and Paul Huber using different site-directed mutants of 5S rRNA.[51,52]

The conformation of region E is also essential for TFIIIA binding (Fig. 3). This region of the 5S rRNA is noncomplementary and might be expected to be an internal loop of single-stranded nucleotides. However, studies we carried out in collaboration with Bernard and Chantal Ehresmann on the solution structure of the oocyte and somatic 5S rRNAs from *Xenopus* lead to the conclusion that region E is highly structured via the formation of noncanonical hydrogen bonding interactions.[45,46,50] NMR studies by Ignacio Tinoco Jr's group determined the specific network of noncanonical hydrogen bonding interactions that form to give region E its unique structure.[53]

Because TFIIIA binds to the coding sequence of the 5S rRNA gene, the set of 5S rRNA genes we constructed to use as templates to produce mutant 5S rRNAs could be used to probe the

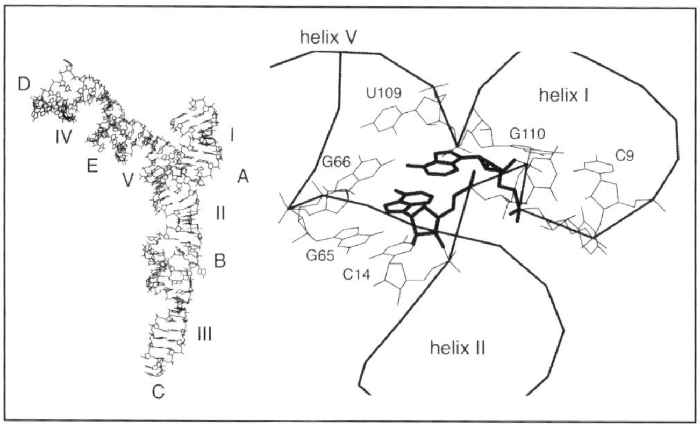

Figure 4. Loop A controls the conformation of the three helical arms of 5S rRNA. A wireframe model of the 5S rRNA is shown at left. The conformation of the loop A region of the RNA is shown at right. The three base pairs that close the loop are indicated. The two bold residues are the stacked A11 and A13 nucleotides.

sequence specificity of the TFIIIA-DNA interaction. In essence we were able to compare the effects of the same mutations on both nucleic acid binding activities of TFIIIA.[37,42] As the results summarized in Figure 5 show, there is no correlation between the mutations that reduce the binding of TFIIIA to the 5S rRNA, and those that reduce the binding of TFIIIA to the 5S rRNA gene. The results obtained in similar studies by other groups[52,54] are in general agreement with this conclusion. In the case of the 5S rRNA, some single mutations that disrupt a double helical structure reduced binding by TFIIIA.[41] In these cases, the binding could be restored by creating a double mutation that allowed for the formation of the double helical stem structure. Such compensatory effects were not observed when assaying the binding of TFIIIA to the same series of mutations in the 5S rRNA gene.[42] Thus TFIIIA relies upon different sequence and structural features of the DNA and RNA ligands in forming specific interactions with each. This conclusion was reinforced by studies of point mutations of the 5S rRNA gene,[40] and structural studies of TFIIIA-DNA complexes by NMR[55] and X-ray crystallography.[56] It seems quite clear that TFIIIA relies upon the overall three dimensional shape of the 5S rRNA to form a specific complex, while a specific sequence of base pairs is important in directing the binding of TFIIIA to the 5S rRNA gene.

With the cloning of the cDNA for *Xenopus* TFIIIA, and the ability to express and purify recombinant TFIIIA, it became possible to study the effects of site-directed mutations in the zinc fingers of the protein on the nucleic acid binding activities. Initially, truncated peptides of TFIIIA were assayed for the ability to bind specifically to the 5S rRNA gene and the 5S rRNA.[57-63] A peptide consisting of the first three fingers of TFIIIA retains much of the free energy of binding to the internal control region of the 5S rRNA gene.[64,65] A variety of footprinting techniques were used with truncation mutants and full length protein to map each of the zinc fingers of TFIIIA on the 5S rRNA gene.[57,58,60,66-69] Further insight into the interaction of TFIIIA with DNA came from careful equilibrium binding studies done by David Setzer's group with "broken finger" mutants of TFIIIA.[70] For each mutant in this series, a zinc-coordinating histidine in the targeted finger was replaced by an asparagine, leading to disruption of the folding of the finger. The energetic contribution of each finger was measured by a quantitative gel shift assay, and the approximate position of each finger on the DNA was measured by qualitative

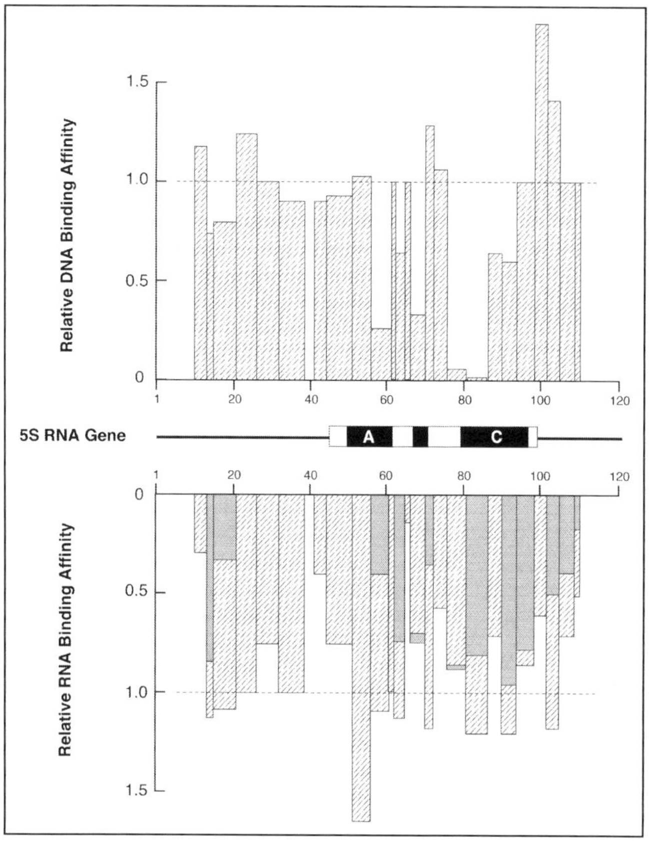

Figure 5. Comparison of the effects of mutations in the 5S rRNA gene and 5S RNA on TFIIIA binding. For the 5S RNA, the grey bars represent the effects of a single mutation in a base paired region, while the striped bar at the same location represents the effect of the compensating double mutation. Reprinted with permission from: You Q, Veldhoen N, Baudin F, Romaniuk PJ. Biochemistry 1991; 30: 2495-2500. ©1991 American Chemical Society.

DNase I footprinting. The results of these studies indicated that finger 3 made the most significant contribution to the binding of TFIIIA to DNA. Disruption of the other zinc fingers reduced the interaction of TFIIIA with DNA, although less severely than the finger 3 disruption. Further careful analysis of the effects of these mutations and truncation mutations on the energetics of TFIIIA binding to the 5S rRNA gene indicated that energetically unfavorable interactions occur between zinc fingers at opposite ends of the protein upon binding to DNA.[71] This observation has important implications for the binding of proteins with multiple zinc fingers to nucleic acids.

In my lab, Wei-Qing Zang, Nik Veldhoen and Tatyana Hamilton studied the importance of individual zinc fingers and amino acids in the binding of TFIIIA to DNA and RNA using a different approach. We took our cue from the classical studies of helix-turn-helix DNA binding proteins, in which "domain swap" mutants were created in order to determine protein features essential for binding to a specific sequence of base pairs in DNA. The concept behind this approach to creating site-directed mutants of a nucleic acid binding protein is that by using a protein from the same structural class, but with different specificities in the nucleic acid target sequence, it should be possible to use the binding domain of that protein as a donor for making specific amino acid sequence changes in the protein of interest without causing a significant disruption in the structure of the protein.

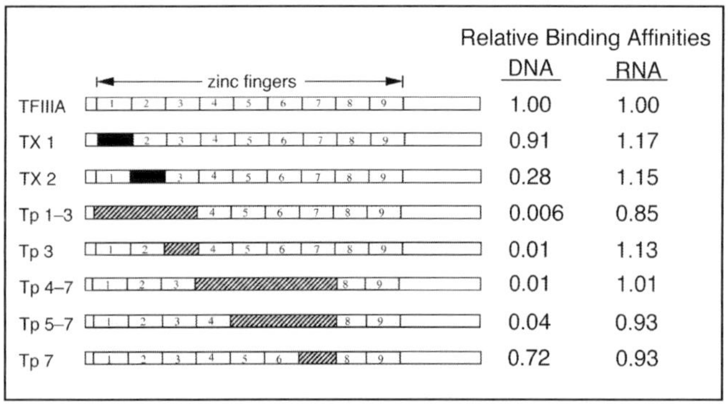

Figure 6. Effects of finger swapping mutations in fingers 1 through 7 on the binding of TFIIIA to the 5S rRNA gene and 5S RNA. Solid bars indicate that the donor fingers came from Xfin, while stripped bars indicate that the donor fingers came from p43.

With the structure intact, it becomes more straightforward to assign the underlying reasons for loss of function of specific mutant proteins.

We chose three proteins as possible donors of zinc finger sequences for creating mutants of TFIIIA. One obvious choice was p43, which has the same number of zinc fingers as TFIIIA, but binds only to 5S rRNA and not the 5S rRNA gene. We reasoned that p43 would provide us with valuable information on what fingers and amino acids were important for DNA binding by TFIIIA, but might not be informative on the RNA binding activity of TFIIIA. We also chose Xfin as donor, a large zinc finger protein of *Xenopus* which does not bind to either 5S rRNA or its gene, as well as the Wilms tumour suppressor protein WT1. Like TFIIIA, this protein binds to both DNA and RNA and its four finger domain likely has general features important for allowing interactions with two conformationally distinct nucleic acids.

By working our way through a deliberate series of mutations, starting with the swapping of multiple fingers of TFIIIA and then narrowing down the interesting target regions to ultimately study them by point mutagenesis (Fig. 6), we identified several amino acids within the α-helical region of finger 3 as being particularly important for the interaction of TFIIIA with DNA.[72] Contributions to DNA binding from fingers 2 and 5 were also identified from these studies. A great deal more insight into the specific amino acid-nucleotide interactions formed in the TFIIIA-DNA complex was obtained from NMR studies by Peter Wright's group[55] and X-ray crystallographic studies by Ray Brown.[56] Happily, the data from our mutagenesis experiments were consistent with the data from these structural studies.

The same mutants of TFIIIA that were used to study the interaction with DNA could also be assayed for 5S rRNA binding activity. Joel Gottesfeld's lab demonstrated that a peptide consisting of only fingers 4-7 of TFIIIA retained most of the free energy of binding to 5S rRNA.[61] N-terminal deletions beyond finger 4, or C-terminal deletions beyond finger 6 abolished binding to RNA. Proteolytic digests of TFIIIA-DNA and TFIIIA-5S rRNA complexes also demonstrated specific roles for fingers 1-3 in high affinity binding to DNA and fingers 4-7 in high affinity binding to 5S rRNA.[73] Comparison of the interaction of a finger 4-7 peptide and TFIIIA with mutants of 5S rRNA further supported the conclusion that the finger 4-7 peptide mimics the mechanism of RNA binding by full length TFIIIA.[74] Other groups studying truncation mutants of TFIIIA observed similar results for the most part,[44,51,59,75,76] although there was evidence that a peptide consisting of fingers 1-3 could bind to 5S rRNA with high affinity and high specificity.[44,75,77] By doing quantitative equilibrium binding studies, David Setzer's group showed that the mechanism of RNA binding by a finger 4-9 peptide appeared to be indistinguishable from the binding of full length TFIIIA on the basis of dissociation kinetics and half life of the RNA-protein complex.[44] In comparison, the complex formed between 5S rRNA and a peptide consisting of fingers 1-3 displayed rapid dissociation kinetics, suggesting that the mechanism of RNA binding for this peptide differed from that used by TFIIIA. It appears that inhibitory finger-finger interactions in the full length wild type protein may restrict the role of fingers 1-3 to DNA binding.

The essential role of fingers 4 to 7 of TFIIIA in binding to 5S rRNA was confirmed by David Setzer's group and their assays with broken finger mutants of TFIIIA.[44] The results of these studies showed that disrupting the structures of fingers 4, 5 and 6 in turn had the largest effects on the affinity of TFIIIA for 5S rRNA, although the magnitude of the effects were relatively small. In the case of the finger 5 and 6 mutants, this small decrease in affinity can be explained by the observation that the RNA-protein complexes formed had rapid dissociation kinetics similar to those observed for the finger 1-3 peptide. This result suggests that loss of the "normal" RNA binding mechanism by the broken finger 5 and 6 mutants was compensated for by a gain of an "alternative" RNA binding mechanism attributable to fingers 1-3. In work with peptides encompassing fingers 4 to 6 or 4 to 7, where a compensating mechanism for RNA binding cannot come into play, Joel Gottesfeld, Martyn Darby and Tomas Pieler independently showed key roles for specific amino acids in finger 4 (lysine 118, glutamine 121) and finger 6 (threonine 176, tryptophan 177, threonine 178).[51,61,75] The TWT sequence motif in finger 6 is found in TFIIIA from a variety of species and is also found in finger 6 of p43, making it a sequence of particular interest.

My lab built upon these observations by working to identify the amino acids within full length TFIIIA that are critical for binding to 5S rRNA. Once again, finger swapping mutants were constructed covering the finger 4 to 7 region of TFIIIA, using p43 and WT1 as donor proteins (Fig. 7).[78] The p43 finger swap mutants established that finger 5 plays a significant role in DNA binding. The WT1 finger swap mutants had a number of interesting properties. Completely replacing fingers 4 to 7 of TFIIIA with the four zinc fingers of WT1 resulted in a mutant TFIIIA protein that was no longer able to bind to 5S rRNA or the 5S rRNA gene. This hybrid protein did have the RNA binding specificity displayed by WT1, that is, it had an affinity for a specific RNA aptamer that binds to the WT1 zinc finger domain. However, although the isolated WT1 zinc finger domain binds to a 12 base pair sequence of DNA with high affinity, the mutant TFIIIA protein did not display this DNA binding activity. This result suggests the intriguing possibility that finger–finger interactions within the mutant TFIIIA, and possibly wild type TFIIIA, restrict the contribution of certain fingers within the 4-7 central region to a role only in RNA binding.

By creating a refined set of mutations in the α–helical region of finger 4, we determined that Lys118 and Gln121 play a minor role at best in the binding of full length TFIIIA to 5S rRNA.[78] It is possible that these residues are more important in the binding of a finger 4-7 peptide to 5S rRNA because they form interactions that help anchor one of the terminal fingers of the peptide to the RNA. Consistent with this concept is our observation that mutations in the 5S rRNA that fall within the proposed binding sites for finger 4 and finger 7 have much larger effects on the

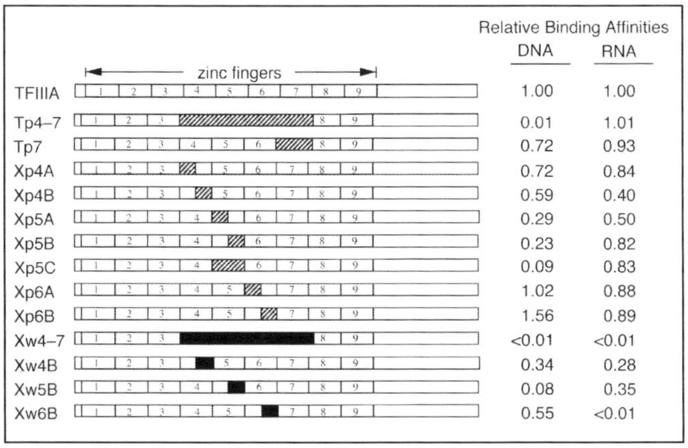

Figure 7. Effects of finger swapping mutations in fingers 4 through 7 on the binding of TFIIIA to the 5S rRNA gene and 5S rRNA. Striped bars indicate that the donor fingers came from p43, while solid bars indicate that the donor fingers came from WT1.

binding of a finger 4–7 peptide than they do on the binding of full length TFIIIA.[74]

When we assayed a set of mutant TFIIIA proteins that had substitutions in the α–helical region of finger 6, we observed that replacement of the TWT motif with the amino acids RSD resulted in a loss of affinity for 5S rRNA similar to that observed when the entire finger 4 through 7 region had been substituted (Fig. 8). This result is consistent with those of other labs,[51,61,75] and indicates that one or more of these amino acids plays a key role in 5S rRNA binding. It is interesting that this mutation has a much larger effect on RNA binding affinity than was observed for the broken finger mutation of finger 6.[44] It suggests that other sequence and structural elements of finger 6 in the TWT>RSD mutant may be involved in putative finger–finger interactions within TFIIIA that prevent the use of an alternative RNA binding mechanism involving fingers 1 to 3.

Mutants of 5S rRNA and TFIIIA have also been used as tools to map the approximate binding sites for individual zinc fingers on the RNA, and to define a "minimal" RNA binding site for TFIIIA. We worked in collaboration with Joel Gottesfeld's group to study the effects of mutations in 5S rRNA on the binding of a TFIIIA peptide containing fingers 4–7.[74] Mutations in the core region of the 5S rRNA structure (loop A, helix II, loop B, helix V and loop E) reduce the binding of both the full length TFIIIA and the finger 4–7 peptide (Fig. 9). Mutations extending beyond this core region that reduced the binding of full length TFIIIA had no effect on the binding of the peptide to 5S rRNA. Combining these observations with the results of nuclease footprinting

		Relative 5S RNA Binding Affinities
TFIIIA	PCKKDDSCSFVGKTWTLYLKHVAECHQD	1.00
Xw6B	-------------RSDHLKT-ΔTRT-TG	<0.01
Xw6B-I	-------------RSD-----------	0.09
Xw6B-II	----------------HLKT--------	0.43
Xw6B-III	---------------------ΔTRT-TG	0.85
	← β-sheet → ← α-helix →	

Figure 8. The effect of targeted mutations in finger 6 on TFIIIA binding to 5S rRNA

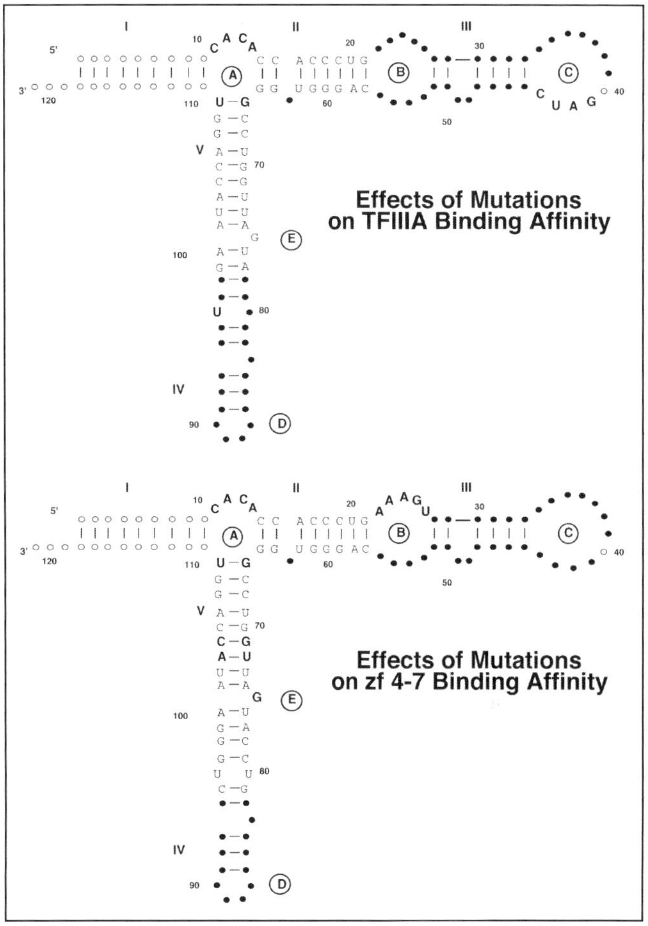

Figure 9. Comparison of the effects of site directed mutations in 5S rRNA on the binding of TFIIIA and a finger 4–7 peptide. See legend to (Fig. 3) for details on the symbols used.

experiments, we proposed binding sites on 5S rRNA for the zinc fingers of TFIIIA. Other groups have mapped different sets of TFIIIA mutants on 5S rRNA using similar techniques.[59,76,79] The current view of zinc finger distribution along the 5S rRNA resulting from all of these studies is shown in Figure 10.

Having established that the central zinc fingers of TFIIIA bind to the core region of the 5S rRNA, Gottesfeld's group turned to the question of what would represent a "minimal" RNA ligand for TFIIIA. The key to these experiments was to ensure that fragments derived from the 5S rRNA retained a three-way junction and were structurally stable.[74,80] Ultimately, they found that a 55 nucleotide fragment of the 5S rRNA retains high affinity binding to peptides that consist of fingers 4–7 or 4–6 of TFIIIA. This minimal RNA ligand is derived from 5S rRNA but consists of a complete deletion of helix II, loop B, helix III and loop C, and a shortened helix IV (Fig. 11). A 61 nucleotide fragment of 5S rRNA with a similar structure was found by Aaron Klug's group to act similarly as a high affinity RNA ligand for TFIIIA.[76]

Interaction of p43 with 5S rRNA

Once the cDNA had been cloned and sequenced for the p43 protein component of the 42S RNP, it became apparent that another zinc finger protein in *Xenopus* bound specifically to 5S rRNA.[17] Comparison of the amino acid sequence of p43 and TFIIIA showed that both proteins have 9 zinc fingers, with seven of identical size. However, amino acid sequence homology

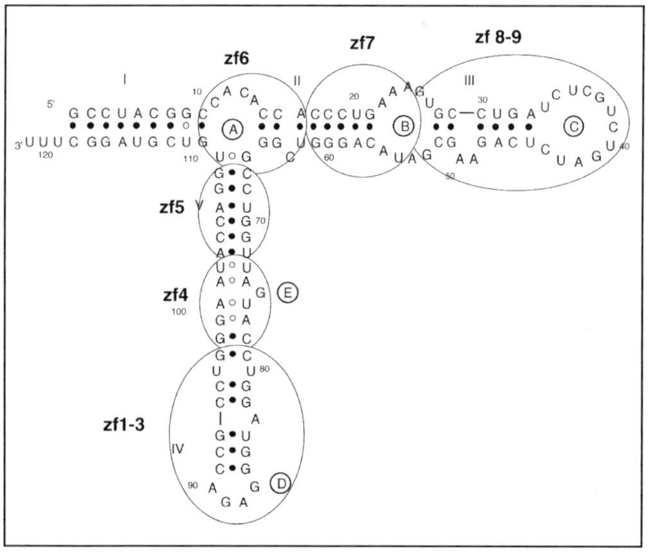

Figure 10. Proposed binding sites for individual fingers of TFIIIA on 5S rRNA.

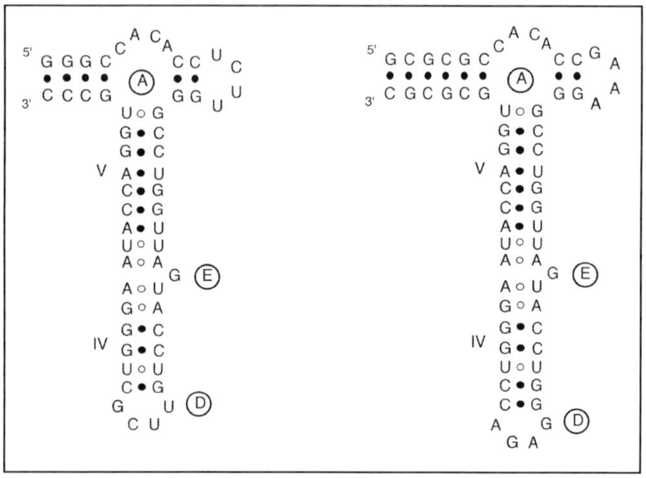

Figure 11. Minimal RNA ligands for TFIIIA. The ligand on the left was developed by Joel Gottesfeld's lab and the ligand on the right was developed by Aaron Klug's lab.

between the two proteins is primarily restricted to the conserved residues required for folding of the zinc fingers. Unlike TFIIIA, p43 is unable to bind to the 5S rRNA gene and appears to be exclusively an RNA binding protein.

A natural starting point in characterizing the RNA binding activity of p43 was to draw comparisons with what was known about the binding of TFIIIA to 5S rRNA. Nuclease footprinting experiments with both proteins demonstrated that each bind to the same region of the 5S rRNA.[81] However, there were intriguing differences in the specific nuclease protection patterns observed with the two proteins, suggesting there might be significant mechanistic details unique to each. A zinc finger peptide that includes fingers 5 through 9 of TFIIIA has strong affinity for 5S rRNA, while the same peptide from p43 demonstrates no binding to the RNA.[63] A peptide with fingers 1–4 of p43 binds strongly to 5S rRNA, while the comparable peptide from TFIIIA does not. This observation further strengthened the view that p43 and

Table 2. Comparison of equilibrium binding parameters for the interaction of p43 and TFIIIA with 5S rRNA at 22° C

Parameter	p43	TFIIIA
K_a	1.61×10^9 M^{-1}	1.4×10^9 M^{-1}
ΔG°	−12.2 kcal mol^{-1}	−12.1 kcal mol^{-1}
ΔH°	+2.1 kcal mol^{-1}	−8.3 kcal mol^{-1}
ΔS°	+48.5 cal deg^{-1} mol^{-1}	+13.1 cal deg^{-1} mol^{-1}

TFIIIA are mechanistically distinct in their RNA binding properties.

Using the methods and reagents we had generated in studying the interaction of TFIIIA with 5S rRNA, a graduate student in my lab, Wei-Qing Zang set out to study the interaction of p43 with 5S rRNA.[82] By carefully characterizing the equilibrium binding of p43 to 5S rRNA, and comparing the results to those for TFIIIA, Wei-Qing showed that the mechanism of RNA binding by p43 is very distinct from that of TFIIIA (Table 2). Although the two proteins have similar affinity for the RNA, the binding of p43 to 5S rRNA is an entropy-driven process that is quite resistant to monovalent salt concentration. These observations are consistent with a mode of binding that involves hydrophobic and van der Waals interactions rather than the formation of hydrogen bonds. Using our library of site-directed mutants of 5S rRNA, Wei-Qing was able to demonstrate that p43 relies upon different sequences and structures in the 5S rRNA for high affinity binding than those required for TFIIIA binding (Fig. 12).[82]

As mentioned above, there is relatively little sequence homology between p43 and TFIIIA outside of the conserved residues required to fold the zinc fingers. One of the few regions of homology occurs in finger 6, where the TWT amino acid motif critical for the binding of TFIIIA to 5S rRNA is conserved in finger 6 of p43. We compared the effects of substituting RSD for TWT in finger 6 of TFIIIA and p43. While this substitution in TFIIIA greatly decreased affinity for 5S rRNA, the same substitution in p43 had little effect on 5S rRNA binding.[78] However, finger 6 of p43 can functionally substitute for finger 6 of TFIIIA in a hybrid TFIIIA protein.[63,78] These observations lead to an interesting paradox: why is a conserved amino acid motif in two proteins that bind to the same RNA only required by one protein for that interaction?

Future Directions

The summary I've provided demonstrates the depth to which the TFIIIA–5S rRNA interaction has been characterized. Because the 5S rRNA is a common ligand to TFIIIA, p43 and ribosomal protein L5 the efforts of a number of groups to probe the nature of these three RNA–protein interactions has generated a wealth of data, all of which begs the question: are there future directions to be pursued in studying the interaction of 5S rRNA with TFIIIA and p43?

Indeed, there are still a number of compelling questions that beckon. Of particular interest to my lab is the question of what specific fingers and residues in p43 are required for binding to 5S rRNA. With that information in hand, a more rational comparative study of TFIIIA and p43 can be carried out to understand at a more detailed level the mechanisms of RNA binding by these

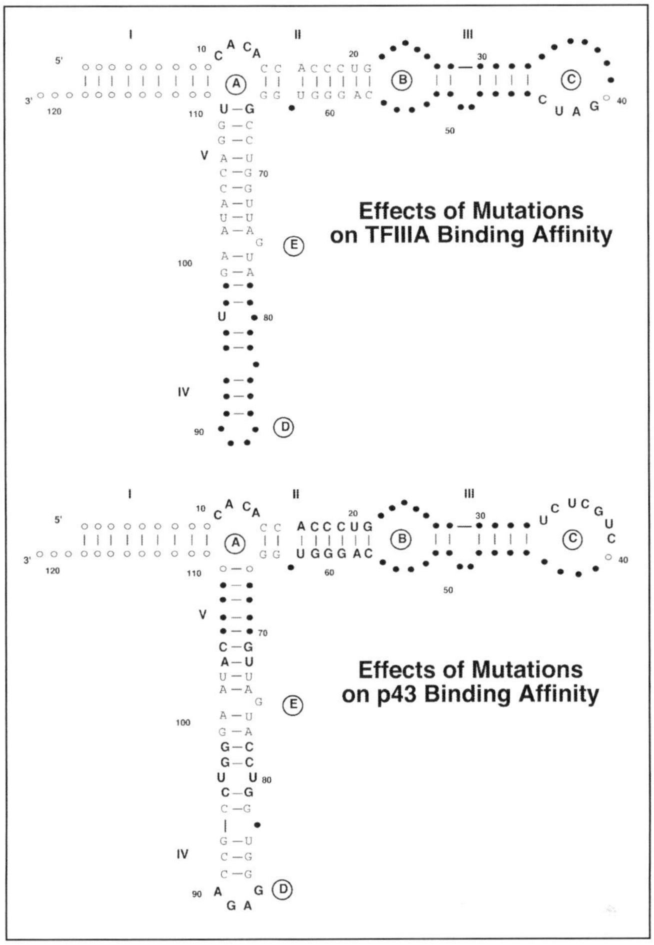

Figure 12. Comparison of the effects of site directed mutations in 5S rRNA on the binding of TFIIIA and p43. See legend to Figure 3 for details on the symbols used.

two proteins. Another open question concerns what aspects of protein structure determine which fingers in these proteins participate in binding to 5S rRNA. It is particularly intriguing that fingers 1 to 3 in TFIIIA can bind specifically to 5S rRNA when expressed as a peptide, but apparently do not contribute significantly to the RNA binding activity of full length TFIIIA. Another pressing question is why the TWT motif found in finger 6 of both TFIIIA and p43 is required for 5S rRNA binding only in TFIIIA. The model proposed by David Setzer of unfavorable finger-finger interactions determining the role of individual zinc fingers in nucleic acid binding deserves considerably more study. Investigating these questions may lead to important insights into how proteins with more than 3 or 4 zinc fingers form specific complexes with one or more ligands.

Acknowledgements

During the time I've been involved in studying the RNA–protein interactions of TFIIIA and p43, I have had the great fortune to work with a talented group of graduate students, and I thank them for their enthusiasm and dedication. Together, we also had the support of a number of excellent research assistants, whose contributions have been recognized by co-authorship on our primary publications. I would like to thank my collaborators for their generous contributions to our joint work: Lizabeth Allison, Bernard and Chantal Ehresmann, Joel Gottesfeld, David Setzer and Eric Westhof. I have also benefited tremendously from in-depth discussions of our work with other colleagues, particularly Ray Brown, Martyn Darby, Debbie Johnson, Paul Huber and Tomas Pieler. Finally, I thank the Natural Sciences and Engineering Research Council of Canada for their continued funding of this work.

Note Added in Proof

Since the submission of this manuscript, a three dimensional structure of a subset of TFIIIA zinc fingers bound to a modified 5S RNA ligand has been reported (Lu D et al. Nature 2003; 426:96-100).

References

1. Denis H, le Maire M. Thesaurisomes, a novel kind of nucleoprotein particle. Subcell Biochem 1983; 9:263-297.
2. Ford PJ. Noncoordinated accumulation and synthesis of 5S ribonucleic acid by ovaries of Xenopus laevis. Nature 1971; 233:561-564.
3. Picard B, Wegnez M. Isolation of a 7S particle from Xenopus laevis oocytes: A 5S RNA-protein complex. Proc Natl Acad Sci USA 1979; 76:241-245.
4. Denis H, Mairy M. Recherches biochimiques sur l'oogenese. Eur J Biochem 1972; 25:524-534.
5. Honda BM, Roeder RG. Association of a 5S gene transcription factor with 5S RNA and altered levels of the factor during cell differentiation. Cell 1980; 22:119-126.
6. Taylor W, Jackson IJ, Siegel N et al. The developmental expression of the gene for TFIIIA in Xenopus laevis. Nucl Acids Res 1986; 14:6185-6195.
7. Shastry BS, Honda BM, Roeder RG. Altered levels of a 5S gene-specific transcription factor (TFIIIA) during oogenesis and embryonic development of Xenopus laevis. J Biol Chem 1984; 259:11373-11382.
8. Pelham HRB, Wormington WM, Brown DD. Related 5S RNA transcription factors in Xenopus oocytes and somatic cells. Proc Natl Acad Sci USA 1981; 78:1760-1764.
9. Rollins MB, Del Rio S, Galey AL et al. Role of TFIIIA zinc fingers in vivo - analysis of single-finger function in developing Xenopus embryos. Mol Cell Biol 1993; 13:4776-4783.
10. Pittman RH, Andrews MT, Setzer DR. A feedback loop coupling 5S rRNA synthesis to accumulation of a ribosomal protein. J Biol Chem 1999; 274:33198-33201.
11. Andrews MT, Brown DD. Transient activation of oocyte 5S RNA genes in Xenopus embryos by raising the level of the trans-acting factor TFIIIA. Cell 1987; 51:445-453.
12. Mattaj IW, Lienhard S, Zeller R et al. Nuclear exclusion of transcription factor IIIA and the 42S particle transfer RNA-binding protein in Xenopus oocytes: A possible mechanism for gene control? J Cell Biol 1983; 97:1261-1265.
13. Guddat U, Bakken AH, Pieler T. Protein-mediated nuclear export of RNA: 5S rRNA containing small RNPs in Xenopus oocytes. Cell 1990; 60:619-628.
14. Picard B, le Maire M, Wegnez M et al. Biochemical research on oogenesis. Composition of the 42-S storage particles of Xenopus laevis oocytes. Eur J Biochem 1980; 109:359-368.
15. Viel A, Djé MK, Mazabraud A et al. Thesaurin a, the major protein of Xenopus laevis previtellogenic oocytes, present in the 42 S particles, is homologous to elongation factor EF-1 alpha. FEBS Lett 1987; 223:232-236.
16. Viel A, Le Maire M, Philippe H et al. Structural and functional properties of thesaurin a (42Sp50), the major protein of the 42-S particles present in Xenopus laevis previtellogenic oocytes. J Biol Chem 1991; 266:10392-10399.

17. Joho KE, Darby MK, Crawford ET et al. A finger protein structurally similar to TFIIIA that binds exclusively to 5S RNA in Xenopus. Cell 1990; 61:293-300.
18. Kloetzel P-M, Whitfield W, Sommerville J. Analysis and reconstitution of an RNP particle which stores 5S RNA and tRNA in amphibian oocytes. Nucl Acids Res 1981; 9:605-621.
19. Steitz JA, Berg C, Hendrick JP et al. A 5S rRNA/L5 complex is a precursor to ribosome assembly in mammalian cells. J Cell Biol 1988; 106:545-556.
20. Wormington WM. Developmental expression and 5S rRNA binding activity of Xenopus laevis ribosomal protein L5. Mol Cell Biol 1989; 9:5281-5288.
21. Allison LA, North MT, Murdoch KJ et al. Structural requirements of 5S rRNA for nuclear transport, 7S ribonucleoprotein particle assembly, and 60s ribosomal subunit assembly in Xenopus oocytes. Mol Cell Biol 1993; 13:6819-6831.
22. Allison LA, Romaniuk PJ, Bakken AH. RNA-protein interactions of stored 5S RNA with TFIIIA and ribosomal protein L5 during Xenopus oogenesis. Dev Biol 1991; 144:129-144.
23. Allison LA, North MT, Neville LA. Differential binding of oocyte-type and somatic-type 5S rRNA to TFIIIA and ribosomal protein L5 in Xenopus oocytes: Specialization for storage versus mobilization. Dev Biol 1995; 168:284-295.
24. Romaniuk PJ. Characterization of the RNA binding properties of transcription factor IIIA of Xenopus laevis oocytes. Nucl Acids Res 1985; 13:5369-5387.
25. Romaniuk PJ. Characterization of the equilibrium binding of Xenopus transcription factor IIIA to the 5S RNA gene. J Biol Chem 1990; 265:17593-17600.
26. Sakonju S, Bogenhagen DF, Brown DD. A control region in the center of the 5S RNA gene directs specific initiation of transcription: I. The 5' border of the region. Cell 1980; 19:13-25.
27. Sakonju S, Brown DD, Engelke D et al. The binding of a transcription factor to deletion mutants of a 5S rRNA gene. Cell 1981; 23:665-669.
28. Sakonju S, Brown DD. Contact points between a positive transcription factor and the Xenopus 5S RNA gene. Cell 1982; 31:395-405.
29. Pieler T, Hamm J, Roeder RG. The 5S gene internal control region is composed of three distinct sequence elements, organized as two functional domains with variable spacing. Cell 1987; 48:91-100.
30. Pieler T, Oei S-L, Hamm J et al. Functional domains of the Xenopus laevis 5S gene promoter. EMBO J 1985; 4:3751-3756.
31. Romaniuk PJ, de Stevenson IL, Wong H-HA. Defining the binding site of Xenopus transcription factor IIIA on 5S RNA using truncated and chimeric 5S RNA molecules. Nucl Acids Res 1987; 15:2737-2755.
32. Darsillo P, Huber PW. The use of chemical nucleases to analyze RNA-protein interactions - the TFIIIA-5S rRNA complex. J Biol Chem 1991; 266:21075-21082.
33. Andersen J, Delihas N, Hanas JS et al. 5S RNA structure and interaction with transcription factor A. 2. Ribonuclease probe of the 7S RNP from Xenopus laevis immature oocytes and RNA exchange properties of the 7S particle. Biochemistry 1984; 23:5759-5766.
34. Andersen J, Delihas N, Hanas JS et al. 5S RNA structure and interaction with transcription factor A. 1. Ribonuclease probe of the 7S particle from Xenopus laevis immature oocytes and RNA exchange properties of the 7S particle. Biochemistry 1984; 23:5752-5759.
35. Huber PW, Wool IG. Identification of the binding site on 5S rRNA for the transcription factor IIIA: Proposed structure of a common binding site on 5S rRNA and on the gene. Proc Natl Acad Sci USA 1986; 83:1593-1597.
36. Christiansen J, Brown RS, Sproat BS et al. Xenopus transcription factor IIIA binds primarily at junctions between double helical stems and internal loops in oocyte 5S RNA. EMBO J 1987; 6:453-460.
37. Baudin F, Romaniuk PJ. A difference in the importance of bulged nucleotides and their parent base pairs in the binding of transcription factor IIIA to Xenopus 5S RNA and 5S RNA genes. Nucl Acids Res 1989; 17:2043-2056.
38. Baudin F, Romaniuk PJ, Romby P et al. Involvement of hinge nucleotides of Xenopus laevis 5S rRNA in the RNA structural organization and in the binding of transcription factor TFIIIA. J Mol Biol 1991; 218:69-81.
39. In: Romaniuk PJ, de Stevenson IL, You Q, eds. The specificity of the RNA binding activity of Xenopus transcription factor IIIA. New York: Alan R. Liss, Inc.; 1989. Cech TR, ed. Molecular Biology of RNA.
40. Veldhoen N, You QM, Setzer DR et al. Contribution of individual base pairs to the interaction of TFIIIA with the Xenopus 5S RNA gene. Biochemistry 1994; 33:7568-7575.
41. You Q, Romaniuk PJ. The effects of disrupting 5S RNA helical structures on the binding of Xenopus transcription factor IIIA. Nucl Acids Res 1990; 18:5055-5062.
42. You Q, Veldhoen N, Baudin F et al. Mutations in 5S DNA and 5S RNA have different effects on the binding of Xenopus transcription factor IIIA. Biochemistry 1991; 30:2495-2500.
43. Romaniuk PJ. The role of highly conserved single-stranded nucleotides of Xenopus 5S RNA in the binding of transcription factor IIIA. Biochemistry 1989; 28:1388-1395.
44. Setzer DR, Menezes SR, Del Rio S et al. Functional interactions between the zinc fingers of Xenopus transcription factor IIIA during 5S rRNA binding. RNA 1996; 2:1254-1269.
45. Romaniuk PJ, de Stevenson IL, Ehresmann C et al. A comparison of the solution structures and conformational properties of the somatic and oocyte 5S rRNAs of Xenopus laevis. Nucl Acids Res 1988; 16(5):2295-2312.
46. Westhof E, Romby P, Romaniuk PJ et al. Computer modeling from solution data of spinach chloroplast and of Xenopus laevis somatic and oocyte 5S rRNAs. J Mol Biol 1989; 207:417-431.
47. Romby P, Baudin F, Brunel C et al. Ribosomal 5S RNA from Xenopus laevis oocytes: Conformation and interaction with transcription factor IIIA. Biochimie 1990; 72:437-452.
48. Shen Z, Hagerman PJ. Conformation of the central, three-helix junction of the 5S ribosomal RNA of Sulfolobus acidocaldarius. J Mol Biol 1994; 241:415-430.
49. Ha T, Zhuang X, Kim HD et al. Ligand-induced conformational changes observed in single RNA molecules. Proc Natl Acad Sci USA 1999; 96:9077-9082.
50. de Stevenson IL, Romby P, Baudin F et al. Structural studies on site-directed mutants of domain 3 of Xenopus laevis oocyte 5S ribosomal RNA. J Mol Biol 1991; 219:243-255.
51. Theunissen O, Rudt F, Pieler T. Structural determinants in 5S RNA and TFIIIA for 7S RNP formation. Eur J Biochem 1998; 258:758-767.
52. Rawlings SL, Matt GD, Huber PW. Analysis of the binding of Xenopus transcription factor IIIA to oocyte 5 S rRNA and to the 5 S rRNA gene. J Biol Chem 1996; 271:869-877.
53. Wimberly B, Varani G, Tinoco I. The conformation of loop E of eukaryotic 5S Ribosomal RNA. Biochemistry 1993; 32:1078-1087.
54. Sands MS, Bogenhagen DF. TFIIIA binds to different domains of 5S RNA and the Xenopus borealis 5S RNA gene. Mol Cell Biol 1987; 7:3985-3993.
55. Wuttke DS, Foster MP, Case DA et al. Solution structure of the first three zinc fingers of TFIIIA bound to the cognate DNA sequence: Determinants of affinity and sequence specificity. J Mol Biol 1997; 273:183-206.
56. Nolte RT, Conlin RM, Harrison SC et al. Differing roles for zinc fingers in DNA recognition: Structure of a six finger transcription factor IIIA complex. Proc Natl Acad Sci USA 1998; 95:2938-2943.
57. Vrana KE, Churchill MEA, Tullius TD et al. Mapping functional regions of transcription factor TFIIIA. Mol Cell Biol 1988; 8:1684-1696.

58. Fiser-Littell RM, Duke AL, Yanchick JS et al. Deletion of the N-terminal region of Xenopus transcription factor IIIA inhibits specific binding to the 5S RNA gene. J Biol Chem 1988; 263:1607-1610.
59. Theunissen O, Rudt F, Guddat U et al. RNA and DNA binding zinc fingers in xenopus TFIIIA. Cell 1992; 71:679-690.
60. Clemens KR, Liao XB, Wolf V et al. Definition of the binding sites of individual zinc fingers in the transcription factor IIIA-5S RNA gene complex. Proc Natl Acad Sci USA 1992; 89:10822-10826.
61. Clemens KR, Wolf V, McBryant SJ et al. Molecular basis for specific recognition of both RNA and DNA by a zinc finger protein. Science 1993; 260:530-533.
62. Clemens KR, Zhang PH, Liao XB et al. Relative contributions of the zinc fingers of transcription factor IIIA to the energetics of DNA binding. J Mol Biol 1994; 244:23-35.
63. Darby MK, Joho KE. Differential binding of zinc fingers from Xenopus TFIIIA and p43 to 5S RNA and the 5S RNA Gene. Mol Cell Biol 1992; 12:3155-3164.
64. Liao XB, Clemens KR, Tennant L et al. Specific interaction of the first three zinc fingers of TFIIIA with the internal control region of the Xenopus 5S RNA gene. J Mol Biol 1992; 223:857-871.
65. Liggins JR, Privalov PL. Energetics of the specific binding interaction of the first three zinc fingers of the transcription factor TFIIIA with its cognate DNA sequence. Proteins 2000; Suppl 4:50-62.
66. Hansen PK, Christensen JH, Nyborg J et al. Dissection of the DNA binding domain of Xenopus laevis TFIIIA: Quantitative DNase I footprinting analysis of specific complexes between a 5S RNA gene fragment and N-terminal fragments of TFIIIA containing 3 zinc finger, 4 zinc finger or 5 zinc finger domains. J Mol Biol 1993; 233:191-202.
67. Churchill MEA, Tullius TD, Klug A. Mode of interaction of the zinc finger protein TFIIIA with a 5S RNA gene of Xenopus. Proc Natl Acad Sci USA 1990; 87:5528-5532.
68. Hayes JJ, Clemens KR. Locations of contacts between individual zinc fingers of Xenopus laevis transcription factor IIIA and the internal control region of a 5S RNA gene. Biochemistry 1992; 31:11600-11605.
69. Hayes JJ, Tullius TD. Structure of the TFIIIA-5S DNA complex. J Mol Biol 1992; 227:407-417.
70. Del Rio S, Menezes SR, Setzer DR. The function of individual zinc fingers in sequence specific DNA recognition by transcription factor IIIA. J Mol Biol 1993; 233:567-579.
71. Kehres DG, Subramanyan GS, Hung VS et al. Energetically unfavorable interactions among the zinc fingers of transcription factor IIIA when bound to the 5S rRNA gene. J Biol Chem 1997; 272:20152-20161.
72. Zang WQ, Veldhoen N, Romaniuk PJ. Effects of zinc finger mutations on the nucleic acid binding activities of Xenopus transcription factor IIIA. Biochemistry 1995; 34:15545-15552.
73. Bogenhagen DF. Proteolytic footprinting of transcription factor TFIIIA reveals different tightly binding sites for 5S RNA and 5S DNA. Mol Cell Biol 1993; 13:5149-5158.
74. McBryant SJ, Veldhoen N, Gedulin B et al. Interaction of the RNA binding fingers of Xenopus transcription factor IIIA with specific regions of 5S ribosomal RNA. J Mol Biol 1995; 248:44-57.
75. Friesen WJ, Darby MK. Phage display of RNA binding zinc fingers from transcription factor IIIA. J Biol Chem 1997; 272:10994-10997.
76. Searles MA, Lu D, Klug A. The role of the central zinc fingers of transcription factor IIIA in binding to 5S RNA. J Mol Biol 2000; 301:47-60.
77. Ryan RF, Darby MK. The role of zinc finger linkers in p43 and TFIIIA binding to 5S rRNA and DNA. Nucl Acids Res 1998; 26:703-709.
78. Hamilton TB, Turner J, Barilla K et al. Contribution of individual amino acids to the nucleic acid binding activities of the Xenopus zinc finger proteins TFIIIA and p43. Biochemistry 2001; 40:6093-6101.
79. Giel-Pietraszuk M, Barciszewska MZ. Additivity of interactions of zinc finger motifs in specific recognition of RNA. J Biochem 2002; 131:571-578.
80. Neely LS, Lee BM, Xu J et al. Identification of a minimal domain of 5S ribosomal RNA sufficient for high affinity interactions with the RNA-specific zinc fingers of transcription factor IIIA. J Mol Biol 1999; 291:549-560.
81. Sands MS, Bogenhagen DF. Two zinc finger proteins from Xenopus laevis bind the same region of 5S RNA But with different nuclease protection patterns. Nucl Acids Res 1991; 19:1797-1803.
82. Zang WQ, Romaniuk PJ. Characterization of the 5S RNA binding activity of Xenopus zinc finger protein p43. J Mol Biol 1995; 245:549-558.

CHAPTER 11

RNA Binding by Single Zinc Fingers

Martyn K. Darby

Abstract

RNA-protein complexes have important functions in gene expression and regulation. Zinc fingers of the Cys-Cys-His-His (C_2H_2) class that bind RNA do so via contacts with amino acid side chains in the α-helical portion of the zinc finger, similar to their interaction with DNA. In general, two or more tandem zinc fingers are present in naturally occurring zinc finger proteins. In vitro selection and recombination techniques have isolated single zinc fingers that bind complex RNA structures with high affinity and specificity. These zinc fingers may ultimately find use as pharmaceutical or agricultural agents designed specifically to modify the function of cellular or viral RNA.

Introduction

RNA is key to gene expression. It serves important functions as a template, catalyst and amino-acid acceptor for protein synthesis, a cofactor for mRNA splicing, a sequence guide for RNA editing and modification, a structural component for the ribosome and spliceosome, and also serves as the genome of many plant and animal viruses.[1] In all of these varied functions, complexes of RNA with proteins are essential. The structures of proteins that bind RNA are as diverse as the functions of RNA, but can be classified through a few characteristic protein motifs. The most common of these are the RNA recognition motif (RRM),[2] the KH domain,[3] the dsRNA binding domain,[4] the arginine-rich motif,[5] and three of the major classes of zinc finger.[6] These zinc finger classes are distinguished structurally by the arrangement of cysteine and histidine zinc ligands, and with the examples known at present, also fall into functional classes based on their predominant roles in living cells (see Table 1 for descriptions). The first class, C_2H_2 zinc fingers, are by far the most abundant nucleic acid binding motif in the human genome, and have the unique ability among zinc finger proteins to mediate binding to both RNA and DNA. A second class, the CCHC zinc fingers of retroviral nucleocapsids bind and package genomic RNA through tandem zinc fingers or "knuckles", and the third class, CCCH zinc fingers, are present in proteins that regulate mRNA stability through AU-rich elements.

The function of C_2H_2 zinc finger proteins in DNA binding has been illuminated by detailed structural studies of DNA-zinc finger complexes at the atomic level, and subsequent application of the principles of protein-DNA interaction learned from these structures has led to the design of specific DNA binding proteins and transcription factors.[7-10] Hopes for future gene therapies, based on the controlled manipulation of gene expression, are founded on these synthetic proteins.[11,12] RNA-zinc finger complexes are yet to be solved at an atomic level, yet some of the features characteristic of DNA-zinc finger recognition appear to apply to the interaction of zinc fingers with RNA and have been successfully exploited to isolate RNA binding zinc fingers with novel sequence specificity.[13] Like their DNA binding counterparts, therapeutic or agricultural applications for such RNA-binding proteins, particularly as anti-viral agents, are envisioned. However, there is unlikely to be a straightforward parallel between the nature of zinc finger interaction with DNA and RNA, since there are fundamental differences in structure that appear to influence the principles governing protein binding to RNA rather than DNA. DNA is frequently bound by protein structures, including the α helix of C_2H_2 zinc fingers, that recognize the unique pattern of hydrogen bond donors and acceptors found within the major groove of the DNA chain.[14] In contrast, the major groove of duplex RNA is too narrow to accommodate an α helix and in addition, the rich tertiary structure of RNA contributes significantly to the formation of protein-RNA complex. Furthermore, most RNA complexes upon formation exhibit substantial conformational changes in both RNA and protein (Fig. 1).[15,16]

This chapter explores efforts to develop synthetic RNA binding C_2H_2 zinc fingers that bind specific RNAs, from early observations that determined the role of zinc fingers in RNA binding, to contemporary phage display approaches for constructing zinc fingers for developing future applications.

RNA Binding C_2H_2 Zinc Finger Proteins

The RNA binding properties of C_2H_2 zinc fingers have been known since the identification of the repeated zinc finger sequence, yet the number of C_2H_2 zinc finger proteins shown to bind RNA since that time remains small in comparison to the number shown to bind DNA (Table 1). Why are there so many confirmed DNA binding C_2H_2 zinc fingers and so few RNA specific? The answer may lie in the lack of an identifying sequence characteristic other than the zinc finger consensus. A search of the PIR annotated protein sequence database identified more than 4000 protein matches to a consensus C_2H_2 zinc finger sequence (Cys-X_{2-5}-Cys-X_{12}-His-X_{3-6}-His) and approximately one third of these were connected with the amino acids TGEKP (TGERP) or Krüppel-like linkers,[17] characteristic of DNA binding zinc fingers. Given that about half of the currently confirmed DNA binding C_2H_2 proteins have the Krüppel linker, up to two-thirds of the consensus matches may be DNA binding proteins. The remaining third may well be RNA binding proteins that await

Zinc Finger Proteins: From Atomic Contact to Cellular Function, edited by Shiro Iuchi and Natalie Kuldell.

Table 1. RNA binding and RNA associated zinc finger proteins*

Finger Type	Protein	Comments
C_2H_2	TFIIIA	9 zinc fingers, binds 5S RNA (7S RNP) and 5S RNA gene. Transcription factor.
Cys-X_{3-5}-Cys-X_{12}-His-X_{4-6}-His	P43	9 zinc fingers, binds 5S RNA isolated from 42S RNP in Xenopus oocytes
	WT1	4 zinc fingers, +KTS isoform associated with splicing factors. Transcription factor.
	Wig-1	3 zinc fingers, p53 induced, localized to nucleoli
	dsRBP-Zfa	7 zinc fingers, binds dsRNA through 3 N-terminal fingers or 3 C-terminal zinc fingers
	JAZ	4 spaced zinc fingers, binds dsRNA or DNA/RNA hybrids.
	MOK2	2 zinc fingers, binds RNA and DNA (brain and testis)
	NUFIP	1 zinc finger associates with FMR RNA binding protein
	XFO-5, XFG-6	5 – 15 zinc fingers, members of FAR family proteins
	ZNF74	12 zinc fingers, deleted in DiGeorge syndrome, binds poly(U)
	PEP	4 zinc fingers, co-precipitates in complex including hnRNPK
	SL100	18 zinc fingers, binds to SLBP/SL RNA complex using fingers 2-8
	Xfin	31 zinc fingers, binds poly(G) preferentially
CCHC (C_2HC) Cys-X_2-Cys-X_4-His-X_4-Cys	HIV p7	2 zinc knuckle, retroviral nucleocapsid protein
CCHC (C_2HC) Cys-X_2-Cys-X_{12}-His-X_{10}-Cys-X_7-Cys-X_2-Cys-X_7-His-X_4-Cys	Nanos	2 zinc fingers, translational regulator, Cys and His spacing unlike nucleocapsid knuckles, non-specific RNA binding.
CCCH (C_2CH)	Nup475/ TTP	2 zinc fingers, binds to AU-rich mRNA 3′ UTR
Cys-X_8-Cys-X_5-Cys-X_3-His	CPSF	5 zinc fingers and 1 CCHC zinc fingers, cleavage and polyadenylation factor
	Pos1	2 zinc fingers, translational repressor, binds mRNA 3′ UTR complex with SPN4
C_2C_2 Cys-X_{2-4}-Cys-X_{10}-Cys-X_2-Cys	ZNF265	2 zinc fingers, splicing factor, binds cyclin B1 mRNA

* Consensus sequences are given for each zinc finger class. 'X' indicates any amino acid.

discovery. Perhaps more likely, the conformational flexibility of RNA may account for the lack of consensus amino acid sequences within zinc fingers that define an RNA-binding zinc finger and there may not be a sequence motif, but rather a structural motif. Of the C_2H_2 proteins listed in Table 1 only TFIIIA, p43 and WT1 exhibit sequence specific interaction with RNA. RNA selection experiments in vitro such as SELEX have been used to isolate specific RNA aptamers that bind to zinc finger proteins.[18,19] However, given the extraordinary power of RNA aptamers to distinguish even small chemical modifications, it is not surprising that RNAs specific for zinc fingers can be isolated in this manner. Support for RNA binding by C_2H_2 zinc fingers in some cases is indirect and inferred from subcellular location in the cytoplasm with splicing complexes or association with other RNA binding proteins.

TFIIIA and p43

TFIIIA and p43 are the clearest examples of sequence-specific C_2H_2 RNA-binding zinc finger proteins. The archetypal C_2H_2 zinc finger protein, TFIIIA, was demonstrated to be an RNA binding protein,[20] soon after its identification as a transcription factor.[21] In the *Xenopus oocyte*, where many proteins and RNAs are stockpiled to fuel the rapid cycles of cell division characteristic of early embryonic development, TFIIIA was found in abundance complexed with 5S rRNA in a 7S RNP particle.[20,22] Thus the notion of the zinc finger protein as an RNA binding protein was established early in the infancy of zinc finger biochemistry. The role of the zinc fingers in DNA binding was established later through deletion mutagenesis[23] and their involvement in RNA binding was suggested (retrospectively) by a residual N-terminal 20 kD RNA bound proteolytic fragment, including the zinc finger region, of the 7S RNP.[24]

Identification of a TFIIIA-related protein from another oocyte storage particle further implicated zinc fingers in RNA binding.[25] The 42S ribonucleoprotein particle accumulates in immature *Xenopus oocytes* prior to ribosome biogenesis. This particle comprises two proteins, p43 and p50, together with 5S rRNA and amino acyl-tRNAs.[26] Similar to TFIIIA, p43 is complexed with 5S rRNA in a 7S sub-particle, but unlike TFIIIA, p43 fails to stimulate 5S RNA gene transcription or to bind DNA, suggesting that it interacts only with 5S RNA.[25] Thus, the role of the C_2H_2 zinc finger as a motif that could interact exclusively with RNA was firmly established.

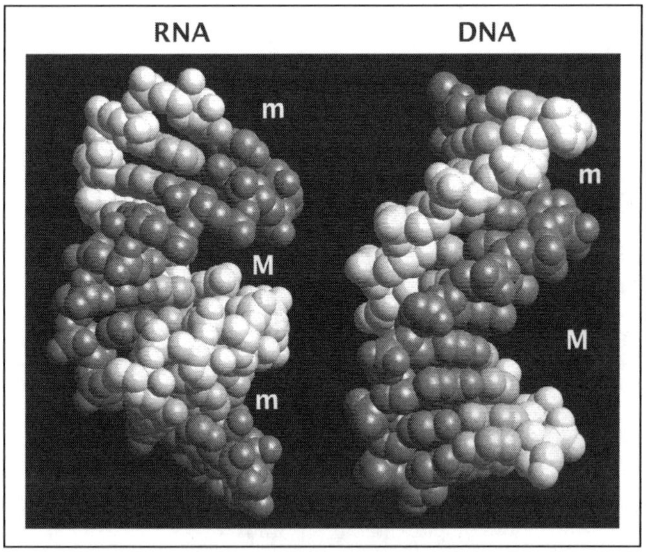

Figure 1. Comparison of RNA and DNA structure. Duplex forms of RNA (pdb: 1SRA) and DNA (pdb: 1BNA) are shown with backbone phosphodiester chains in gray and white. The major groove (M) of the RNA duplex is deeper and narrower than that of DNA and prohibits access by a protein α-helix unless distorted. In contrast, the minor groove (m) is accessible, but possible H-bond contacts limit sequence distinction to AT or GC base-pairs.

RNA Binding by Sub-Groups of Zinc Fingers in Multi-Finger Proteins

Much of our understanding of the interaction of C_2H_2 zinc fingers with RNA derives from deletion and site-directed mutagenesis experiments and reveals that subgroups of zinc fingers within multi-finger proteins can perform independent functions such as RNA or DNA binding, although not always with affinity comparable to the intact protein (Fig. 2).[27] The close structural relationship between TFIIIA and p43, in contrast to their differing nucleic acid binding properties, has made this pair of proteins a favorite choice for comparative structure-function studies.[25,27-31] By analyzing groups of three or four zinc fingers from TFIIIA, DNA binding activity was localized principally to the amino terminal zinc fingers, specifically fingers 1-3 (Fig. 2 light gray boxes).[29,32,33] These zinc fingers in TFIIIA are connected by the Krüppel linkers and this linker sequence is notably absent from p43, which binds only RNA. However, conversion of p43 linkers to those in TFIIIA does not confer DNA binding activity, suggesting that the Krüppel linker sequence, while indicative of a DNA binding protein, is insufficient.[29,31,34] Central zinc fingers from TFIIIA (fingers 4 to 7) and amino terminal zinc fingers from p43 have the highest RNA affinity of isolated zinc finger fragments (Fig. 2, blue boxes).[29,32,33,35,36] Single finger alanine-scanning revealed an important role for zinc fingers 4 and 6 of TFIIIA when expressed as a four finger fragment (fingers 4 to 7).[37] Finger 6 of both TFIIIA and p43 contains a conserved amino acid triplet, Thr-Trp-Thr, which plays a significant role for TFIIIA interaction with 5S rRNA, but not for p43.[27] It is worth noting that single zinc fingers derived from either protein have no measurable affinity for either nucleic acid and that proteins containing a single finger are a rarity within any class of zinc finger protein found in nature.

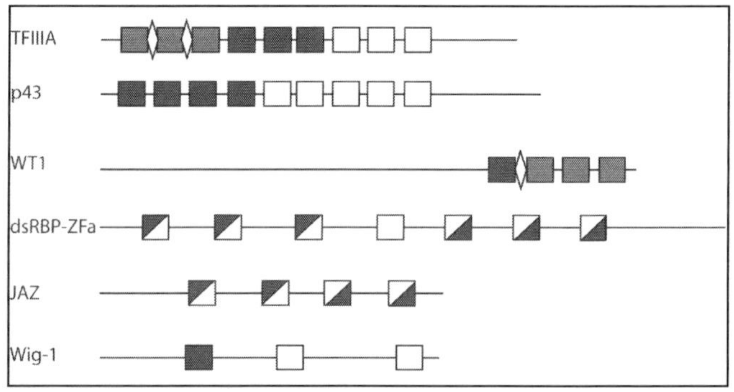

Figure 2. Scale comparison of zinc finger arrangements in RNA binding zinc finger proteins. The distribution of minimal RNA binding (dark gray) and DNA binding (light gray) zinc fingers within multi-finger C_2H_2 zinc finger proteins. Half-filled boxes are used where two separate groups of zinc fingers have been identified to have RNA binding properties. Krüppel-like linkers are indicated by the white diamonds.

Phage Display of RNA Binding Zinc Fingers

The role of individual amino acids in RNA binding by zinc fingers has been sought through a variety of mutagenesis experiments in which individual amino acids or groups thereof have been changed from wild type. Displaying a library of mutated zinc fingers on the surface of bacteriophage screens the effects of many amino acid changes simultaneously. Using this approach important amino acid contacts between RNA and TFIIIA zinc finger 4 were identified and the utility of a phage display approach was established.[37]

Phage display presents multiple copies of a single mutant on the surface of a filamentous phage such as fd or M13 (Fig. 3).[38] Binding of a phage library to an RNA substrate (selector RNA) is carried out in solution and complexes formed between phage and biotinylated RNA are recovered by affinity purification on streptavidin-coated microtiter plates. Phage that do not form complexes with RNA are removed by washing under conditions of varied stringency. In principle, a single round of affinity selection could yield specific RNA-binding phage. However, in practice, initial selections are impure and the phage are amplified in bacteria after each round of selection and then in an iterative process reselected against the same RNA. Successful isolation of RNA-binding phage is judged by the fraction of input phage recovered from each round of selection. DNA sequencing of recovered phage is used to reveal the amino acid sequence of the zinc fingers that bind the selector RNA.

Alanine scanning mutagenesis of each of the central zinc fingers of TFIIIA (fingers 4 to 7), identified zinc fingers 4 and 6 as important RNA binding fingers. In this experiment, four amino acid positions previously shown to make contact with DNA in crystal structures -1, +2, +3 and +6 with respect to the amino-terminal end of the α helix, were substituted by alanine. Since alanine at these positions does not support RNA binding, screening all amino acid substitutions at these positions through phage display reveals any amino acid combinations that can support 5S RNA binding (Fig. 4). A cDNA library representing all possible permutations of amino acids at these four positions in TFIIIA zinc finger 4 was constructed using degenerate oligonucleotides and cloned into a phage display vector that expressed the TFIIIA zinc fingers 4 through 7 as an N-terminal fusion of fd coat protein III. Phage recovery of 3% of input was observed

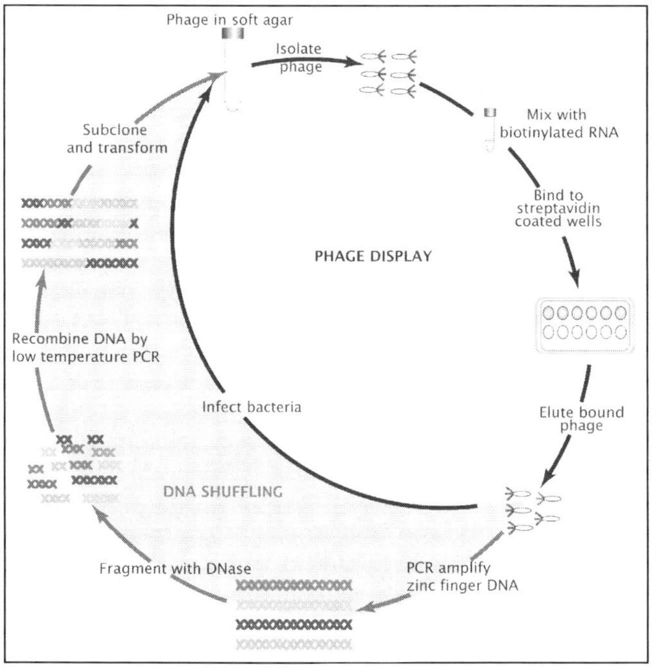

Figure 3. Phage display and shuffling protocol: In vitro evolution of zinc finger function. The flow of methodology used in a standard phage display experiment is represented by the black oval. Phage recovered from affinity selection against RNA are directly used to re-infect bacteria for subsequent rounds of selection. The gray portion of the flow diagram shows an additional strategy for isolating zinc fingers. Zinc finger cDNA is amplified from selected phage and is fragmented and recombined before reintroduction to bacteria, in this case by transformation rather than electroporation.

after 3 rounds of selection, which represented a 1500-fold enrichment over a control selection without RNA (0.002%). In all of the phage sequenced, position -1, immediately N-terminal to the zinc finger 4 α helix, was lysine, suggesting a critical role for this amino acid. To confirm this observation through standard site-directed mutagenesis, lysine at -1 was changed to alanine. This mutation reduced the RNA affinity by 43 fold. In contrast, changing the other three mutated positions (+2, +3 and +6) to alanine caused only a 7 fold reduction in RNA binding affinity. Interestingly, arginine was not found at position -1 in any of the zinc finger phage sequenced, despite its greater representation in the library, suggesting a specific requirement for a positive charge provided by the less bulky lysine side chain. These experiments established the validity of the phage display approach and encouraged a wider sequence search for zinc fingers that bind novel RNA substrates, that is, for any chosen RNA.

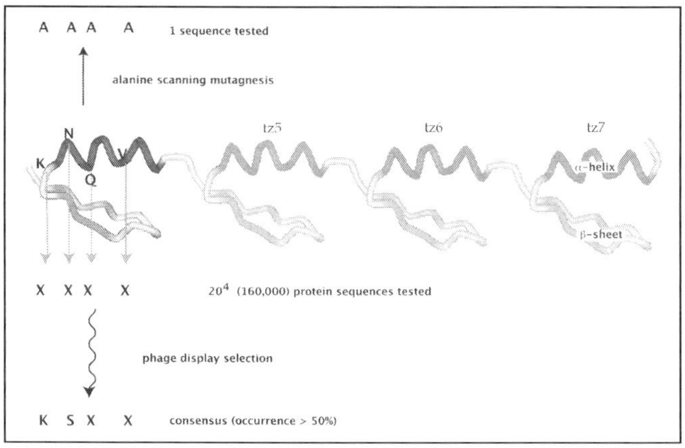

Figure 4. Assessment of the contribution of amino acids in TFIIIA derived zinc fingers by alanine scanning and phage display of TFIIIA zinc finger 4. Traditional site-directed mutagenesis determines the effect of individual changes in amino acid sequence to test hypotheses regarding the role of specific amino acids, whereas phage display screens up to 10 million sequences simultaneously to determine a consensus amino acid sequence that supports binding.

Selecting RNA Binding Proteins with Novel Specificity

The modular nature of C_2H_2 zinc finger binding to DNA was clearly established through finger swapping experiments. A single finger binds a base-pair triplet and rearranging the order of zinc fingers produces a protein that binds a corresponding rearrangement of cognate triplets.[39] Therefore, combining zinc fingers selected by phage display to specific nucleotide triplets creates a protein with novel and predictable specificity.[40-42] A similar correspondence between RNA nucleotides and zinc fingers is unlikely, based upon RNAse protection experiments with TFIIIA and p43 that show important interactions at disparate junctions of RNA helices and bulged nucleotides in 5S rRNA.[43-47] However, as few as two zinc fingers from TFIIIA or p43 bind RNA.[29] Therefore, it seemed feasible that a two zinc finger phage display library could be used to isolate RNA binding proteins. Our C_2H_2 design approach used a randomized α helix as the principle recognition element within the RNA-binding zinc finger framework of tandem copies of TFIIIA zinc finger 4.

While the α helix is the zinc finger structural element that makes significant contacts with 5S RNA, its size and shape restricts the RNA structures that may be bound. The major groove of A-form helical RNA is too narrow to accommodate an α helix and the minor groove, which is accessible, has a pattern of potential hydrogen bonds that allows discrimination simply for the presence or absence of A or G.[14] In contrast to duplex DNA, RNA forms complex tertiary structures, which are used to direct specific association with proteins. Proteins that bind RNA specifically recognize RNA structures that widen the major groove by various means and provide access to a richer pattern of potential base contacts as well as an elaborate binding pocket.

In looking for an existing paradigm to test the hypothesis that C_2H_2 zinc fingers with novel specificity could be selected, we noted the role of a positively charged α helix in the binding of HIV Rev protein (regulator of expression of viral proteins) to the Rev response element (RRE) of the RNA genome.[48,49] Thus it seemed feasible that a zinc finger phage display library containing randomized alpha helices could be used to isolate a specific binding protein for RRE. Zinc finger proteins with sufficient affinity to block Rev binding to RRE could block a key regulatory point in the HIV life cycle and hence inhibit virion production.

HIV-1 Rev protein is responsible for a switch between early and late phases of HIV mRNA expression.[50] In the early phase of HIV gene expression, all mRNA is fully spliced to yield RNA encoding regulatory proteins. Later, as Rev protein accumulates and binds to the RRE present in HIV introns, unspliced and partially spliced mRNA bound to Rev, are exported to the cytoplasm.[51] These mRNAs encode proteins that are essential for the formation of the viral coat and thus production of new virions. Rev binds to its high affinity site in the RRE as a single monomer, causes conformational changes in the RNA structure and promotes binding of additional Rev (to a total of 8 or more) molecules via protein-protein and protein-RNA contacts.[52] Rev functions on any heterologous RNA containing an RRE and is independent of the splicing process.[53]

The interaction between Rev and RRE has been studied intensively and functional amino acids and nucleotides have been identified. Deletion studies have shown that an unusually arginine rich 17 amino acid fragment of Rev binds to stem loop IIB of the RRE (RRE-IIB) with an affinity comparable to that of Rev.[48,54] The affinity of this peptide for RRE-IIB-RNA correlates with its helical content. The peptide structure is inherently unstable, but the helix can be stabilized through addition of a negatively charged group at the N terminus and amidation of the C terminus together with introducing an AAAAR segment. Alanine scanning identified several amino acids (Arg35, Arg38, Arg39, Arg44, Thr34 and Asn40) that upon mutation strongly reduced RNA binding.[54]

Selection of Rev Response Element (RRE) Binding Zinc Fingers

The sequence pool from which zinc fingers with affinity for RRE were isolated was constructed as a zinc finger phage display library based on tandem zinc fingers of TFIIIA RNA binding zinc finger 4 (Fig. 5). Amino acids at nine positions, between -2 and +10 of the zinc finger α helix, were randomized to match as closely as possible a subset of amino acids determined to be present in α helices of zinc fingers from a search of the PIR database. In addition, a single amino acid position in the linker between fingers was randomized to permit variations in relative finger orientations, as suggested by the observations that relative finger orientation is critical for DNA zinc finger interaction.[55] To maintain the structural integrity of the zinc finger within the α-helical region, two histidines (+7 and +11 with respect to the start of the α helix) required for zinc coordination were included, as were two additional residues, phenylalanine (-3) and leucine (+4) essential for the hydrophobic core of the zinc finger. The antiparallel β sheet region of zinc finger 4 was unchanged. cDNA encoding a randomized finger 4 was constructed from degenerate oligonucleotides and two copies of this finger were ligated to make a library of tandem zinc finger proteins. This library was displayed on the surface of bacteriophage fd as a coat protein III fusion and zinc finger expressing phage that bound to RRE-IIB RNA, full-length RRE RNA or 5S rRNA were selected (Fig. 3).[13]

Ten unique tandem zinc fingers were identified among 34 sequenced phage recovered from six rounds of selection against HIV RRE-IIB RNA. Strikingly, two zinc finger sequences, named RR1 and RR2 accounted for 47% of all sequences. The affinity of these zinc fingers for the selector RNA was highly specific. For example, RR1 bound the RRE-IIB RNA with a K_d of 7.9 nM, but bound an unrelated RNA, 5S rRNA, nonspecifically. Similarly, a 5S rRNA selected zinc finger bound 5S rRNA with a K_d of 0.35 nM and failed to bind RRE-IIB RNA at concentrations up to 2.4 μM. Furthermore, all zinc finger peptides bound only the appropriate target RNA in the presence of a 1000-fold molar excess of competitor tRNA or poly[A] RNA. In these experiments a very large number of zinc finger sequences (10^{18}) were potentially encoded by the library, which therefore cannot be comprehensively screened with 10^7 transformants. However, this limitation can be overcome in part by recombining (shuffling) zinc finger cDNA sequences from the output of the phage display experiment, i.e., from sequences for zinc fingers that already make one or more productive interactions with RNA.

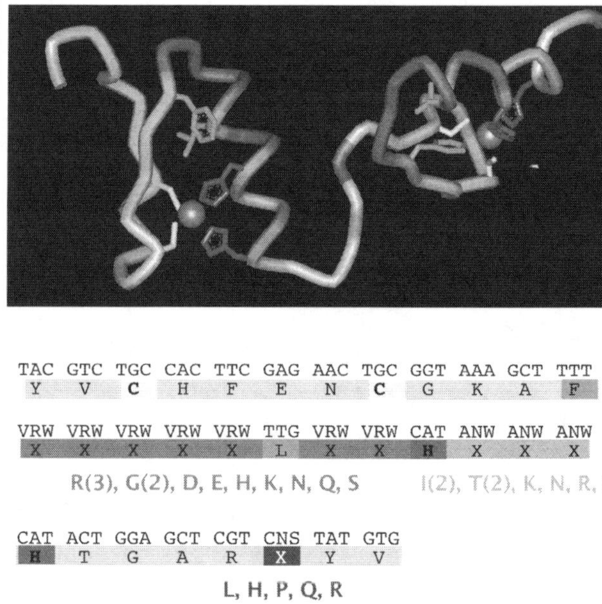

Figure 5. Design of a two C_2H_2 zinc finger phage display library for isolation of novel RNA binding zinc finger proteins. Tandem zinc fingers were expressed using the degenerate DNA sequence shown at bottom. Three different degenerate codons (indicated by shading (see web version for color at http://www.eurekah.com/chapter.php?chapid=2046&bookid=124&catid=30)) were used to encode different sets of amino acid at different positions along the zinc finger α-helix or linker as indicated in the model (based on TFIIIA zinc fingers 4 and 5, pdb: 1TF6). The choice of amino acids was as close as possible to a consensus of more than 4000 zinc finger sequences present in the PIR protein database. Amino acids essential for maintenance of the zinc finger structure are shown in stick representation: Zinc ligands cysteine, histidine and the hydrophobic core formed by leucine and phenylalanine.

DNA shuffling is essentially an in vitro recombination technique that uses low temperature PCR to recombine fragments of related DNA.[56,57] Shuffling zinc finger cDNA in this way creates a new library from a pool of sequences that already have binding affinity for the selector RNA, which can be displayed on phage for selection. Using this approach with zinc fingers selected against RRE-IIB, the most common recombination event joined whole zinc fingers, most likely because the linker and β sheet between α helices of neighboring zinc fingers is the longest region of identical sequence shared between selected proteins. Significantly, RNA binding affinity was doubled by using this method. When shuffling was incorporated during several rounds of the selection procedure for full-length RRE, zinc fingers with sub-nanomolar dissociation constants were recovered. Naturally occurring two zinc finger proteins with this degree of RNA affinity or specificity are yet to be described. Full length RRE selected zinc fingers bound the RRE-IIB RNA fragment with a significantly lower affinity than RRE-IIB selected zinc fingers, suggesting that their mode of binding or binding sites differ.[58]

In Vivo RRE Binding Zinc Fingers

The binding activity of selected zinc fingers in living cells was confirmed using zinc finger-Tat fusion proteins.[54,59] Tat protein is an RNA binding transcriptional activator, which functions through a binding site in nascent RNA by stimulating RNA

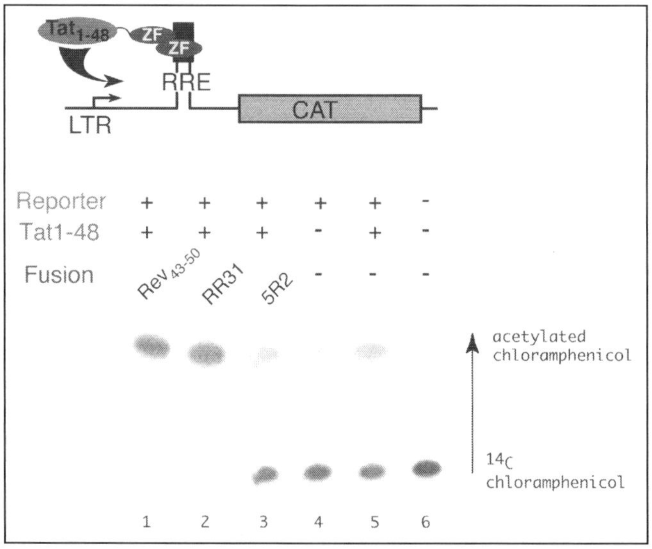

Figure 6. An in vivo assay for RRE-IIB RNA binding. To test RNA binding, zinc fingers are fused with HIV Tat transcriptional activation domain and cotransfected with a reporter gene containing the RNA binding site and a chloramphenicol acetyl transferase (CAT) coding sequence under control of the HIV LTR promoter. Zinc finger binding to the target RNA sequence in the nascent transcript stimulates transcription and the production of CAT. The lower panel shows a typical result. Conversion of chloramphenicol to its acetylated form by CAT activity is determined by thin layer chromatography and is indicated by a faster migrating radioactive spot. Rev_{43-50}, is a positive control fusion of Tat activation domain with the Rev RNA binding peptide (lane 1). A negative control of the Tat activation domain alone is shown in lane 5. Binding by a zinc finger selected to RRE-IIB (RR31) is confirmed by CAT activation (lane 2), whereas a zinc finger selected to an unrelated RNA (5S RNA) does not activate CAT activity (lane 3).

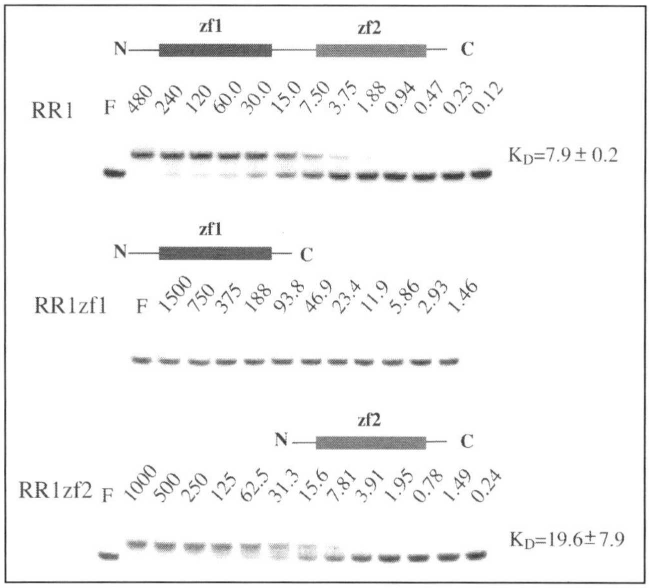

Figure 7. RNA binding by a single selected zinc finger peptide. An RNA mobility shift assay with a selected tandem zinc finger protein (RR1) is shown in the top panel. The zinc fingers of RR1 were separately expressed and assayed in the two lower panels. Only the C-terminal peptide zf2 retains significant RNA binding activity.

polymerase elongation.[60] Fusing the activation domain of Tat with an RNA binding domain creates a transcriptional activator and if the cognate binding site is present in the nascent transcript, transcription rates are elevated. Therefore, protein binding to RNA can be readily measured by standard transfection and reporter gene activity assays. Using this approach, RRE-IIB selected zinc fingers fused to Tat activator domain stimulated transcription of a reporter gene construct in which the Tat binding site was replaced by RRE-IIB (Fig. 6).[13] Control 5S RNA selected zinc fingers fused to Tat activation domain do not stimulate transcription from an RRE-IIB containing template. Attempts to use RRE-selected zinc fingers to inhibit competitively Rev activity in cultured cells have as yet been unsuccessful and could be a consequence of the very high apparent affinity of Rev binding to RRE due to the cooperative nature of the Rev interaction.

RNA Binding by Single C_2H_2 Zinc Fingers of RRE

RNA binding can be specified in a sequence specific manner by single C_2H_2 zinc fingers. Sequence inspection of the zinc fingers selected against RRE-IIB revealed that the C-terminal zinc finger (zf2) is more conserved and biased towards basic residues, suggesting that each zinc finger in the selected protein contributed differently to RNA binding (Fig. 7).[58] Individual zinc fingers were expressed and purified and their affinity for RRE-IIB RNA was determined by protein titration gel shift analysis. RR1 N-terminal zinc finger (zf1) did not bind ^{32}P labeled RRE-IIB RNA. However RR1-zf2 bound RRE-IIB RNA with a K_D of 26.1 nM, which is just 3 fold lower than RR1, demonstrating that the observed binding of the two finger construct RR1 is mostly attributable to a single zinc finger (RR1-F2). The specificity, as determined by titration with nonspecific competitor or binding to mutated RRE-IIB RNA was unchanged.

A synthetic zinc finger peptide, ZNF28,[61] corresponding to RR1-zf2 also binds to RRE-IIB with a dissociation constant of approximately 6 nM (Fig. 8). A model for the RNA binding finger complexed with RRE-IIB-tr RNA compared with binding of the Rev peptide is shown in Figure 9. The α helix of ZNF28 may be oriented parallel to Rev, that is with N-terminal ends together, or anti-parallel, with the C terminus of one neighboring the N terminus of the other. In the anti-parallel orientation of the zinc finger α helix an asparagine in ZNF28 can be aligned with a critical asparagine within the Rev peptide.

Rev α Helix Zinc Finger Framework

Another approach to designing a RRE-IIB specific zinc finger protein has embedded the Rev α helix in a zinc finger.[62] As detailed above, the affinity of the isolated recognition peptide of Rev is dependent on α-helical content. Thus, the stability of the C_2H_2 zinc finger secondary structure was natural choice of structural framework with which to stabilize the Rev α helix. To accommodate Rev peptide in the second zinc finger from Zif268, two nonessential arginine residues were replaced with histidine (Fig. 10). The resulting peptide was shown to fold in a zinc-dependent manner by circular dichroism of a cobalt-peptide complex and its competition with zinc. Gel shift assays in vitro show similar affinities for zf2-rev or a Rev peptide stabilized through flanking amino acids binding to RRE-IIB. Zinc finger structural requirements were confirmed through mutation of the zinc-coordinating cysteines to serine or omission of zinc during folding and resulted in a 3 and 7 fold reduction in affinity, respectively. Substrate specificity was confirmed through mutations in RRE-IIB that narrow the RNA major groove at the Rev binding

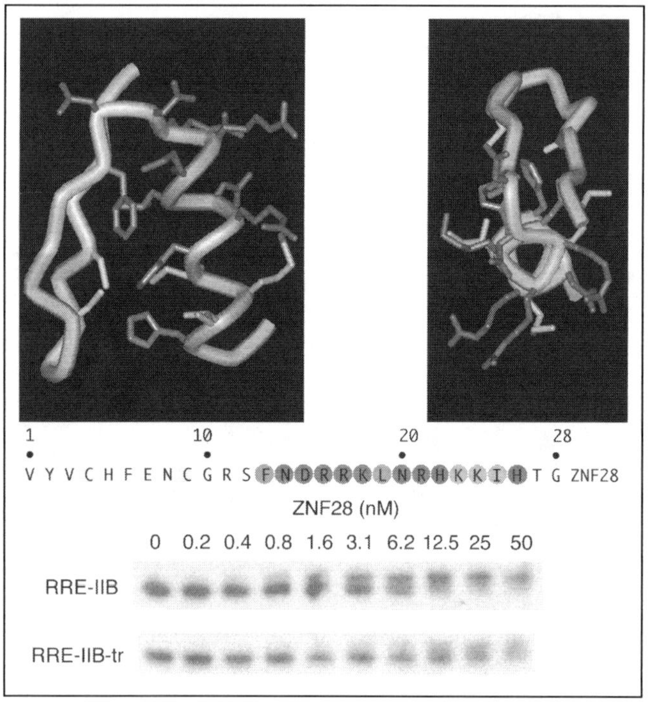

Figure 8. A synthetic peptide, ZNF28, corresponding to RR1-zf2 binds RRE-IIB RNA. A synthetic zinc finger peptide designed for structural studies binds to the RRE-IIB RNA used to select the parent molecule RR1 and to a truncated version of RRE-IIB, RRE-IIB-tr, used in earlier NMR studies of the Rev-IIB RNA interaction and lacks a portion of the loop. A model for the peptide, ZNF28, is shown in two views at top. The left hand panel views the side of the a-helix, on the right the view is from the finger tip along the axis of the helix. The amino acid positions subjected to selection are circled.

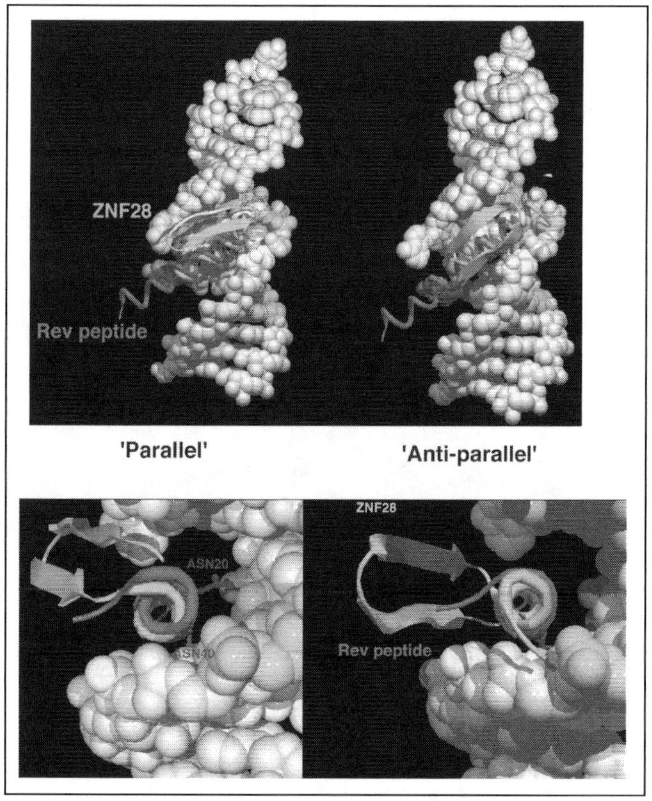

Figure 9. Models for ZNF28 binding to RRE-IIB. This model was derived by aligning the Rev α-helix (pdb: 1ETF) with the α-helix of ZNF28 (blue). Asn40 of Rev, which makes essential base-specific contacts with the G-A base-pair in RRE-IIB, and zinc finger Asn20 are indicated. The zinc finger is oriented either parallel with the Rev helix (left) or antiparallel (right) which best aligns Asn20 and Asn40 residues. A color coded version is available on line. See web version for color at http://www.eurekah.com/chapter.php?chapid=2046&bookid=124&catid=30.

site (G71A), or else change the sequence of the Rev binding pocket (C46-G74). A five fold reduction in affinity for C46-G74 compared to wild type is consistent with major groove binding by the zinc finger. Significantly, more marked differences in binding affinities were reported with the RRE-IIB mutations in cultured cells using fusion of zf2-rev with Tat activation domain assay.

Zinc Fingers That Bind RNA Mismatched Base Pair

An interesting, systematic approach for elucidating a potential recognition code for RNA binding zinc fingers has made use of a combination of phage display and DNA-RNA hybrids.[63] The central zinc finger of a 3 finger DNA binding domain from Zif268 was randomized at nucleotides within the α helix coding region to create a library. Phage display was used to select proteins that would bind an eleven base pair DNA-RNA hybrid molecule in which the central three base pairs were RNA and the flanking base-pairs DNA. Each base pair combination was tested at a single position within the RNA triplet. Interestingly, attempts to isolate Watson-Crick RNA base pairs failed, consistent with the notion that the α helix recognition motif of zinc fingers requires an RNA structural feature to open the major groove. The selected peptide (E4.1) fused to maltose binding protein bound the G•A mismatch specifically and has little affinity for either dG•A or Watson-Crick base pairs.

Future Directions For Designing RNA Binding Zinc Fingers

The use of zinc fingers as a tool for constructing RNA binding proteins of predetermined specificity has great potential for growth. The next steps will require an understanding of how nature has taken advantage of zinc fingers in the evolution of RNA binding proteins, namely in the use of multi-finger proteins and combinations of RNA binding motifs.

Selection of Multi Zinc Finger Proteins

Although the core of TFIIIA and p43 binding to 5S RNA resides in a few zinc fingers, all fingers clearly make some contacts with nucleotides, which permits close binding of a large RNA structure (see chapter 10). In part this mode of binding reflects the role of TFIIIA and p43 as protective storage particles. Higher affinity or recognition of more extensive RNA could be achieved by combining RNA binding zinc fingers as has already been accomplished with polydactyl zinc finger DNA binding proteins. The knowledge that a single zinc finger library can be used to select peptides that bind to a specific RNA will make construction of multi-finger RNA binding proteins feasible. Phage display can be used to build sequentially upon a specific zinc finger

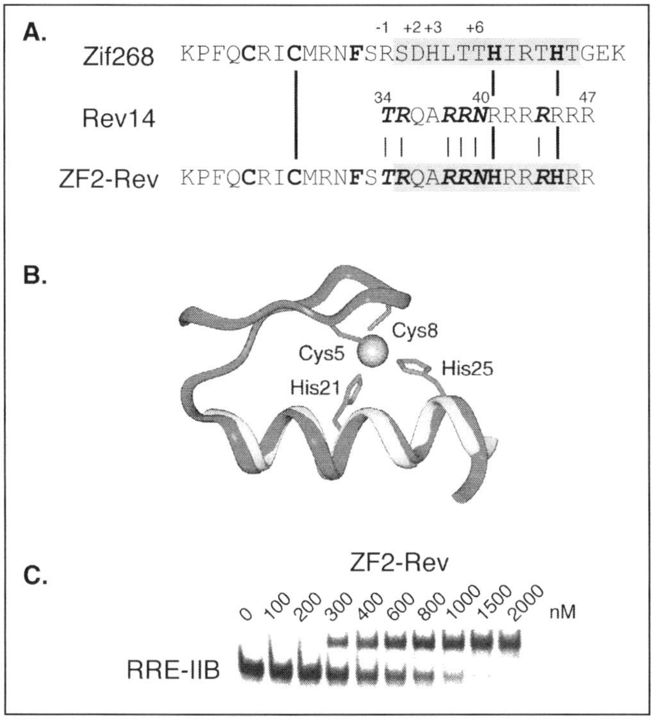

Figure 10. Rev binding helix embedded in a zinc finger framework. A) An alignment of Rev RNA recognition helix with zinc finger 2 of the DNA binding protein zif268. Amino acids essential for zinc finger formation are indicated in bold type, while those important for Rev binding to RNA are given in bold-italic. The zinc finger recognition helix is boxed. B) Overlap between structures for zif268 (white, pdb 1ZAA) and Rev peptide (gray, pdb 1ETF). C) RRE-IIB gel shift with indicated concentrations of ZF2-Rev peptide. Adapted with permission from reference 62. National Academy of Sciences, U.S.A. © 1999.

from either a known protein or one that has been selected against the RNA target. This 'founder' zinc finger is ligated to a randomized zinc finger and selected again against the target RNA. The two finger proteins recovered from this selection are ligated to a randomized finger and so on. In this manner a multi-finger protein that binds large RNA targets could be constructed. A similar approach for DNA binding protein design has been described.[64]

CCCH Zinc Fingers: Recognition of Short Motifs

Other finger structures may be more versatile for construction of RNA binding proteins, most notably the CCHC nucleocapsid zinc fingers and CCCH zinc finger of ARE-binding proteins. The latter is of particular interest, because ARE-binding proteins recognize very short nucleotide sequences. Tristetraprolin (TTP/NUP475/GOS24) binds a short RNA sequence UUAUUUAUU, that is the core of the AU-rich element (ARE) found in the 3' untranslated region of a number of mRNAs, and leads to the degradation of the mRNA (see Chapter 13).[65] High affinity RNA binding (dissociation constant 19 nM for a 24 residue RNA) is mediated through tandem CCCH zinc fingers and a single finger is unable to bind RNA in a gel-shift assay.[66] However, using a more sensitive assay, weak binding by a single CCCH zinc finger to the subsequence, UUUAUUU, can be detected (dissociation constant of 5 μM).[67] An NMR spectrum has recently been acquired for a peptide representing the tandem zinc fingers and may explain the two finger requirement.[68] The N-terminal zinc finger is highly structured, but a spectrum for the C-terminal zinc finger is induced by nine base ribonucleotide, indicative of induced fit mechanism, in which both RNA and protein components of the complex undergo conformational changes. Several key amino acids in the zinc finger have been identified through site-directed mutagenesis for RNA binding[69] and could be modified and presented to RNA as a phage display library.

In summary, single zinc fingers can provide a specific RNA binding motif. The probability for isolating specific RNA binding peptides using a single zinc finger library phage display library is much improved over the tandem finger libraries used to date, since the smaller number of potential sequences can be screened more comprehensively. Novel zinc fingers proteins could be used to target RNA either as competitive inhibitors of RNA dependent processes, as inhibitors of animal or plant RNA viruses, or as a tool be used to visualize RNA in vivo, or localize RNA to a particular subcellular location.

Acknowledgements

I would like to acknowledge the important contributions of Drs. Westley Friesen, Robert Ryan and Michele Shields for the work done in my laboratory. Dr. Markus Germann determined the model for a single zinc finger. My laboratory was supported by grants from the National Science Foundation (MCB-9206873) and the National Institutes of Health (GM50846 and GM/AI47459).

References

1. Simons RW, Grunberg-Manago M. RNA structure and function. Plainview. NY: Cold Spring Harbor Laboratory Press, 1998.
2. Query CC, Bentley RC, Keene JD. A common RNA recognition motif identified within a defined U1 RNA binding domain of the 70K U1 snRNP protein. Cell Apr 7 1989; 57(1):89-101.
3. Siomi H, Matunis MJ, Michael WM et al. The premRNA binding K protein contains a novel evolutionarily conserved motif. Nucleic Acids Res 1993; 21(5):1193-1198.
4. St Johnston D, Brown NH, Gall JG et al. A conserved double-stranded RNA-binding domain. Proc Natl Acad Sci USA 1992; 89(22):10979-10983.
5. Calnan BJ, Tidor B, Biancalana S et al. Arginine-mediated RNA recognition: The arginine fork. Science 1991; 252(5010):1167-1171.
6. Laity JH, Lee BM, Wright PE. Zinc finger proteins: New insights into structural and functional diversity. Curr Opin Struct Biol 2001; 11(1):39-46.
7. Pavletich NP, Pabo CO. Zinc finger-DNA recognition: Crystal structure of a Zif268-DNA complex at 2.1 A. Science 1991; 252(5007):809-817.
8. Rebar EJ, Pabo CO. Zinc finger phage: Affinity selection of fingers with new DNA-binding specificities. Science 1994; 263(5147):671-673.
9. Choo Y, Klug A. Physical basis of a protein-DNA recognition code. Curr Opin Struct Biol 1997; 7(1):117-125.
10. Choo Y, Sanchez-Garcia I, Klug A. In vivo repression by a site-specific DNA-binding protein designed against an oncogenic sequence. Nature 1994; 372(6507):642-645.
11. Jamieson AC, Miller JC, Pabo CO. Drug discovery with engineered zinc-finger proteins. Nat Rev Drug Discov 2003; 2(5):361-368.
12. Ordiz MI, Barbas 3rd CF, Beachy RN. Regulation of transgene expression in plants with polydactyl zinc finger transcription factors. Proc Natl Acad Sci USA 2002; 99(20):13290-13295.
13. Friesen WJ, Darby MK. Specific RNA binding proteins constructed from zinc fingers. Nat Struct Biol 1998; 5(7):543-546.
14. Steitz T. Similarities and Differences between RNA and DNA Recognition by Proteins. In: Gesteland R, Atkins J, eds. The RNA World. Vol Monograph 24. Cold Spring Harbor: CSHL Press, 1993:219-237.

15. Leulliot N, Varani G. Current topics in RNA-protein recognition: Control of specificity and biological function through induced fit and conformational capture. Biochemistry 2001; 40(27):7947-7956.
16. Williamson JR. Induced fit in RNA-protein recognition. Nat Struct Biol Oct 2000;7(10):834-837.
17. Schuh R, Aicher W, Gaul U et al. A conserved family of nuclear proteins containing structural elements of the finger protein encoded by Kruppel, a Drosophila segmentation gene. Cell 1986; 47(6):1025-1032.
18. Bardeesy N, Pelletier J. Overlapping RNA and DNA binding domains of the wt1 tumor suppressor gene product. Nucleic Acids Res 1998; 26(7):1784-1792.
19. Zhai G, Iskandar M, Barilla K et al. Characterization of RNA aptamer binding by the Wilms' tumor suppressor protein WT1. Biochemistry 2001; 40(7):2032-2040.
20. Pelham HR, Brown DD. A specific transcription factor that can bind either the 5S RNA gene or 5S RNA. Proc Natl Acad Sci USA 1980; 77(7):4170-4174.
21. Engelke DR, Ng SY, Shastry BS et al. Specific interaction of a purified transcription factor with an internal control region of 5S RNA genes. Cell 1980; 19(3):717-728.
22. Picard B, Wegnez M. Isolation of a 7S particle from Xenopus laevis oocytes: A 5S RNA-protein complex. Proc Natl Acad Sci USA 1979; 76(1):241-245.
23. Vrana KE, Churchill ME, Tullius TD et al. Mapping functional regions of transcription factor TFIIIA. Mol Cell Biol 1988; 8(4):1684-1696.
24. Smith DR, Jackson IJ, Brown DD. Domains of the positive transcription factor specific for the Xenopus 5S RNA gene. Cell 1984;37(2):645-652.
25. Joho KE, Darby MK, Crawford ET et al. A finger protein structurally similar to TFIIIA that binds exclusively to 5S RNA in Xenopus. Cell 1990;61(2):293-300.
26. Picard B, le Maire M, Wegnez M et al. Biochemical Research on oogenesis. Composition of the 42-S storage particles of Xenopus laevix oocytes. Eur J Biochem 1980; 109(2):359-368.
27. Hamilton TB, Turner J, Barilla K et al. Contribution of individual amino acids to the nucleic acid binding activities of the Xenopus zinc finger proteins TFIIIIA and p43. Biochemistry 2001; 40(20):6093-6101.
28. Sands MS, Bogenhagen DF. Two zinc finger proteins from Xenopus laevis bind the same region of 5S RNA but with different nuclease protection patterns. Nucleic Acids Res 1991; 19(8):1797-1803.
29. Darby MK, Joho KE. Differential binding of zinc fingers from Xenopus TFIIIA and p43 to 5S RNA and the 5S RNA gene. Mol Cell Biol 1992; 12(7):3155-3164.
30. Zang WQ, Romaniuk PJ. Characterization of the 5 S RNA binding activity of Xenopus zinc finger protein p43. J Mol Biol 1995; 245(5):549-558.
31. Ryan RF, Darby MK. The role of zinc finger linkers in p43 and TFIIIA binding to 5S rRNA and DNA. Nucleic Acids Res 1998; 26(3):703-709.
32. Theunissen O, Rudt F, Guddat U et al. RNA and DNA binding zinc fingers in Xenopus TFIIIA. Cell 1992;71(4):679-690.
33. Clemens KR, Wolf V, McBryant SJ et al. Molecular basis for specific recognition of both RNA and DNA by a zinc finger protein. Science 1993; 260(5107):530-533.
34. Choo Y, Klug A. A role in DNA binding for the linker sequences of the first three zinc fingers of TFIIIA. Nucleic Acids Res 1993; 21(15):3341-3346.
35. Searles MA, Lu D, Klug A. The role of the central zinc fingers of transcription factor IIIA in binding to 5 S RNA. J Mol Biol 2000;301(1):47-60.
36. Neely LS, Lee BM, Xu J et al. Identification of a minimal domain of 5 S ribosomal RNA sufficient for high affinity interactions with the RNA-specific zinc fingers of transcription factor IIIA. J Mol Biol 1999; 291(3):549-560.
37. Friesen WJ, Darby MK. Phage display of RNA binding zinc fingers from transcription factor IIIA. J Biol Chem 1997; 272(17):10994-10997.
38. Smith GP, Scott JK. Libraries of peptides and proteins displayed on filamentous phage. Methods Enzymol 1993; 217:228-257.
39. Nardelli J, Gibson TJ, Vesque C et al. Base sequence discrimination by zinc-finger DNA-binding domains. Nature 1991; 349(6305):175-178.
40. Wolfe SA, Greisman HA, Ramm EI et al. Analysis of zinc fingers optimized via phage display: Evaluating the utility of a recognition code. J Mol Biol 1999; 285(5):1917-1934.
41. Pabo CO, Peisach E, Grant RA. Design and selection of novel Cys2His2 zinc finger proteins. Annu Rev Biochem 2001; 70:313-340.
42. Liu Q, Segal DJ, Ghiara JB et al. Design of polydactyl zinc-finger proteins for unique addressing within complex genomes. Proc Natl Acad Sci USA 1997; 94(11):5525-5530.
43. Christiansen J, Brown RS, Sproat BS et al. Xenopus transcription factor IIIA binds primarily at junctions between double helical stems and internal loops in oocyte 5S RNA. Embo J 1987; 6(2):453-460.
44. Pieler T, Erdmann VA, Appel B. Structural requirements for the interaction of 5S rRNA with the eukaryotic transcription factor IIIA. Nucleic Acids Res 1984; 12(22):8393-8406.
45. Theunissen O, Rudt F, Pieler T. Structural determinants in 5S RNA and TFIIIA for 7S RNP formation. Eur J Biochem 1998; 258(2):758-767.
46. Setzer DR, Menezes SR, Del Rio S et al. Functional interactions between the zinc fingers of Xenopus transcription factor IIIA during 5S rRNA binding. RNA 1996; 2(12):1254-1269.
47. Darsillo P, Huber PW. The use of chemical nucleases to analyze RNA-protein interactions. The TFIIIA-5 S rRNA complex. J Biol Chem 1991; 266(31):21075-21082.
48. Kjems J, Calnan BJ, Frankel AD et al. Specific binding of a basic peptide from HIV-1 Rev. EMBO Journal 1992; 11(3):1119-1129.
49. Pollard VW, Malim MH. The HIV-1 Rev protein. Annu Rev Microbiol 1998;52:491-532.
50. Sodroski J, Goh WC, Rosen C et al. A second post-transcriptional trans-activator gene required for HTLV- III replication. Nature 1986; 321(6068):412-417.
51. Hope T, Pomerantz RJ. The human immunodeficiency virus type 1 Rev protein: A pivotal protein in the viral life cycle. Curr Top Microbiol Immunol 1995; 193:91-105.
52. Malim MH, Cullen BR. HIV-1 structural gene expression requires the binding of multiple Rev monomers to the viral RRE: Implications for HIV-1 latency. Cell 1991; 65(2):241-248.
53. Pasquinelli AE, Ernst RK, Lund E et al. The constitutive transport element (CTE) of Mason-Pfizer monkey virus (MPMV) accesses a cellular mRNA export pathway. Embo J 1997; 16(24):7500-7510.
54. Tan R, Chen L, Buettner JA et al. RNA recognition by an isolated alpha helix. Cell 1993; 73(5):1031-1040.
55. Wuttke DS, Foster MP, Case DA et al. Solution structure of the first three zinc fingers of TFIIIA bound to the cognate DNA sequence: Determinants of affinity and sequence specificity. J Mol Biol 1997; 273(1):183-206.
56. Stemmer WP. DNA shuffling by random fragmentation and reassembly: In vitro recombination for molecular evolution. Proc Natl Acad Sci USA 1994; 91(22):10747-10751.
57. Stemmer WP. Rapid evolution of a protein in vitro by DNA shuffling. Nature 1994; 370(6488):389-391.
58. Friesen WJ, Darby MK. Specific RNA binding by a single C2H2 zinc finger. J Biol Chem 2001;276(3):1968-1973.
59. Southgate C, Zapp ML, Green MR. Activation of transcription by HIV-1 Tat protein tethered to nascent RNA through another protein. Nature 1990; 345(6276):640-642.
60. Brigati C, Giacca M, Noonan DM et al. HIV Tat, its TARgets and the control of viral gene expression. FEMS Microbiol Lett 2003; 220(1):57-65.
61. Darby MK, Germann MW. Unpublished.
62. McColl DJ, Honchell CD, Frankel AD. Structure-based design of an RNA-binding zinc finger. Proc Natl Acad Sci USA 1999; 96(17):9521-9526.
63. Blancafort P, Steinberg SV, Paquin B et al. The recognition of a noncanonical RNA base pair by a zinc finger protein. Chem Biol 1999; 6(8):585-597.
64. Greisman HA, Pabo CO. A general strategy for selecting high-affinity zinc finger proteins for diverse DNA target sites. Science 1997; 275(5300):657-661.

65. Blackshear PJ. Tristetraprolin and other CCCH tandem zinc-finger proteins in the regulation of mRNA turnover. Biochem Soc Trans 2002; 30(Pt 6):945-952.
66. Lai WS, Carballo E, Strum JR et al. Evidence that tristetraprolin binds to AU-rich elements and promotes the deadenylation and destabilization of tumor necrosis factor alpha mRNA. Mol Cell Biol 1999; 19(6):4311-4323.
67. Michel SL, Guerrerio AL, Berg JM. Selective RNA binding by a single CCCH zinc-binding domain from Nup475 (Tristetraprolin). Biochemistry 2003; 42(16):4626-4630.
68. Blackshear PJ, Lai WS, Kennington EA et al. Characteristics of the interaction of a synthetic human tristetraprolin tandem zinc finger peptide with AU-rich element-containing RNA substrates. J Biol Chem 2003 (Please add the missing information)
69. Lai WS, Carballo E, Thorn J et al. Interactions of CCCH zinc finger proteins with mRNA. Binding of tristetraprolin-related zinc finger proteins to Au-rich elements and destabilization of mRNA. J Biol Chem 2000; 275(23):17827-17837.
70. Shastry BS. Transcription factor IIIA (TFIIIA) in the second decade. J Cell Sci 1996; 109 (Pt 3):535-539.
71. Caricasole A, Duarte A, Larsson SH et al. RNA binding by the Wilms tumor suppressor zinc finger proteins. Proc Natl Acad Sci USA 1996; 93(15):7562-7566.
72. Mendez-Vidal C, Wilhelm MT, Hellborg F et al. The p53-induced mouse zinc finger protein wig-1 binds double-stranded RNA with high affinity. Nucleic Acids Res 2002; 30(9):1991-1996.
73. Finerty Jr PJ, Bass BL. A Xenopus zinc finger protein that specifically binds dsRNA and RNA-DNA hybrids. J Mol Biol 1997; 271(2):195-208.
74. Yang M, May WS, Ito T. JAZ requires the double-stranded RNA-binding zinc finger motifs for nuclear localization. J Biol Chem 1999; 274(39):27399-27406.
75. Arranz V, Harper F, Florentin Y et al. Human and mouse MOK2 proteins are associated with nuclear ribonucleoprotein components and bind specifically to RNA and DNA through their zinc finger domains. Mol Cell Biol 1997; 17(4):2116-2126.
76. Bardoni B, Schenck A, Mandel JL. A novel RNA-binding nuclear protein that interacts with the fragile X mental retardation (FMR1) protein. Hum Mol Genet 1999; 8(13):2557-2566.
77. Klocke B, Koster M, Hille S et al. The FAR domain defines a new Xenopus laevis zinc finger protein subfamily with specific RNA homopolymer binding activity. Biochim Biophys Acta 1994; 1217(1):81-89.
78. Grondin B, Bazinet M, Aubry M. The KRAB zinc finger gene ZNF74 encodes an RNA-binding protein tightly associated with the nuclear matrix. J Biol Chem 1996; 271(26):15458-15467.
79. Amero SA, Elgin SC, Beyer AL. A unique zinc finger protein is associated preferentially with active ecdysone-responsive loci in Drosophila. Genes Dev 1991; 5(2):188-200.
80. Hovemann BT, Reim I, Werner S et al. The protein Hrb57A of Drosophila melanogaster closely related to hnRNP K from vertebrates is present at sites active in transcription and coprecipitates with four RNA-binding proteins. Gene 2000; 245(1):127-137.
81. Dominski Z, Erkmann JA, Yang X et al. A novel zinc finger protein is associated with U7 snRNP and interacts with the stem-loop binding protein in the histone premRNP to stimulate 3'-end processing. Genes Dev 2002; 16(1):58-71.
82. Koster M, Kuhn U, Bouwmeester T et al. Structure, expression and in vitro functional characterization of a novel RNA binding zinc finger protein from Xenopus. Embo J 1991; 10(10):3087-3093.
83. Gorelick RJ, Henderson LE, Hanser JP et al. Point mutants of Moloney murine leukemia virus that fail to package viral RNA: Evidence for specific RNA recognition by a "zinc finger-like" protein sequence. Proc Natl Acad Sci USA 1988; 85(22):8420-8424.
84. Curtis D, Treiber DK, Tao F et al. A CCHC metal-binding domain in Nanos is essential for translational regulation. Embo J 1997; 16(4):834-843.
85. Worthington MT, Amann BT, Nathans D et al. Metal binding properties and secondary structure of the zinc-binding domain of Nup475. Proc Natl Acad Sci USA 1996; 93(24):13754-13759.
86. Barabino SM, Hubner W, Jenny A et al. The 30-kD subunit of mammalian cleavage and polyadenylation specificity factor and its yeast homolog are RNA-binding zinc finger proteins. Genes Dev 1997; 11(13):1703-1716.
87. Ogura K, Kishimoto N, Mitani S et al. Translational control of maternal glp-1 mRNA by POS-1 and its interacting protein SPN-4 in Caenorhabditis elegans. Development 2003; 130(11):2495-2503.
88. Tabara H, Hill RJ, Mello CC et al. pos-1 encodes a cytoplasmic zinc-finger protein essential for germline specification in C. elegans. Development 1999; 126(1):1-11.
89. Plambeck CA, Kwan AH, Adams DJ et al. The structure of the zinc finger domain from human splicing factor ZNF265 fold. J Biol Chem 2003; 278(25):22805-22811.

CHAPTER 12

Wig-1, a p53-Induced Zinc Finger Protein that Binds Double Stranded RNA

Cristina Mendez-Vidal, Fredrik Hellborg, Margareta T. Wilhelm, Magdalena Tarkowska and Klas G. Wiman*

Abstract

The *wig-1* gene (for Wild type p53-Induced Gene 1) is a direct transcriptional target of wild type p53. Wig-1 is a highly conserved unusual nuclear zinc finger protein that binds double stranded RNA. Overexpression of Wig-1 inhibits cell growth. Wig-1 has been implicated in neuronal apoptosis and stem cell proliferation/differentiation. Elucidation of the biological function of Wig-1 should provide new insights into the p53 tumor suppressor pathway.

Introduction

The p53 tumor suppressor protein is activated by different stress signals such as irradiation, genotoxic drugs, hypoxia, and oncogenic activation. This leads to p53 accumulation and induction of cell cycle arrest, differentiation, apoptosis and/or senescence. p53 orchestrates these biological responses through transcriptional regulation of a set of target genes. This involves binding to a conserved sequence site that consists of two copies of the motif 5'-PuPuPuC(A/T) (T/A)GPyPyPy-3' separated by 0-13 nucleotides. Induction of the p21 and GADD45 genes triggers cell cycle arrest whereas transactivation of proapoptotic genes, for instance Bax, Fas, KILLER/DR5, Noxa, and PUMA, induces apoptosis. p53-mediated repression of the antiapoptotic Bcl-2 and IGF1 receptor genes may also contribute to the apoptotic response. The p53-induced MDM2 protein regulates p53 stability by binding and ubiquitinating p53, resulting in proteasome-mediated degradation of p53 in the cytoplasm. This ensures low levels of p53 in the absence of p53-activating stimuli. p53 induces several proteins that regulate MDM2, including cyclin G, PTEN and WISP1 (reviewed by Vousden and Lu).[1]

p53 is inactivated by point mutation or deletion in around 50% of all human tumors. Most point mutations cluster in the p53 core domain that interacts specifically with DNA, leading to inactivation of p53 DNA binding and failure to transactivate p53-regulated genes. This makes mutant p53 an interesting target for cancer therapy. Several novel strategies for p53-based cancer therapy have been designed and in some cases tested clinically, including p53 gene therapy and reactivation of mutant p53 by small molecules.[2]

To learn more about the p53 tumor suppressor pathway, it is important to identify and study the function of p53-regulated genes. *wig-1* is a novel p53 target gene that encodes a zinc finger protein with unusual characteristics and unknown function.

Discovery of Mouse *wig-1* and Its Rat Ortholog *pag608*

The mouse *wig-1* gene was identified in J3D mouse T lymphoma cells carrying a temperature sensitive Val135 mutant p53 construct (tsp53). This construct is expressed as mutant p53 at 37°C but temperature shift to 32°C induces wild type p53 expression, triggering cell cycle arrest and massive apoptosis. *wig-1* was identified by differential display analysis as one of several mRNAs expressed at 32°C but not 37°C in J3D-tsp53 cells. Wig-1 is a 290 amino acid protein with an apparent molecular weight of 32 kD. Northern blotting using the *wig-1* coding region as a probe revealed two main transcripts of 7.6 kb and 2.2 kb. Both transcripts are induced by gamma irradiation in mouse fibroblasts in a wild type p53-dependent manner. Moreover, *wig-1* is strongly induced in brain, testis, kidney, lungs and spleen of irradiated mice, and also expressed at detectable levels in unirradiated brain and testis.[3]

The rat homolog of *wig-1*, PAG608 (p53-Activated Gene 608) was independently identified by Israeli et al.[4] At the protein level, PAG608 is 97% identical to Wig-1 and is induced in tsp53-carrying cells in the same manner as *wig-1*. The protein localizes to the nucleus and possibly to nucleoli. A weak apoptotic effect of PAG608 was observed in transient transfection experiments.

Wig-1/PAG608 is a zinc finger protein that contains three Cys_2-His_2 (C_2H_2) zinc finger motifs and a nuclear localization signal between the second and the third zinc finger (Fig. 1). The two N-terminal zinc fingers show strong homology to each other, whereas the third zinc finger differs significantly.

Human *wig-1*

Human Wig-1 shows 87% amino acid sequence identity to mouse Wig-1.[5] The zinc fingers are almost perfectly conserved,

* *Corresponding author. See list of "Contributors".*

Zinc Finger Proteins: From Atomic Contact to Cellular Function, edited by Shiro Iuchi and Natalie Kuldell.

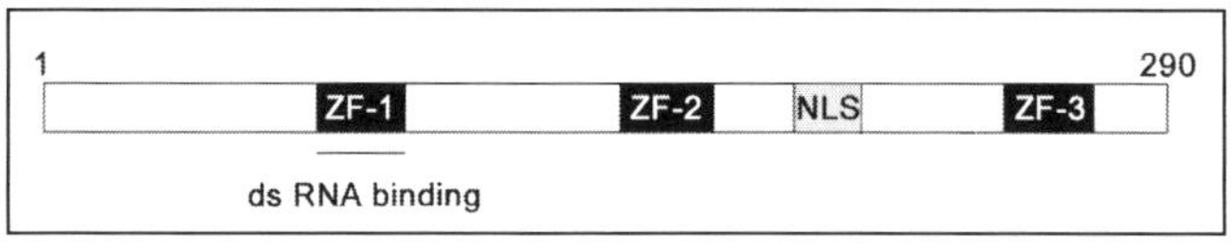

Figure 1. Schematic representation of the mouse Wig-1 protein. Zinc finger regions (ZF1-3) are indicated by black boxes. The nuclear localization signal (NLS) is in gray and the region implicated in double-stranded RNA binding is underlined.

with only one amino acid mismatch in the third zinc finger between the human and the mouse protein. The distances between the zinc fingers are also conserved (Fig. 2). Human *wig-1* is transcribed into four mRNA species detectable by Northern blotting. The significance of the multiple transcripts is unknown. Human Wig-1 is expressed as two distinct bands on Western blots. This could be due to posttranslational modification such as phosphorylation and/or initiation of translation from different ATG codons. Human *wig-1* is located to the short arm of chromosome 3 (3q26.3-27), a region frequently amplified in a variety of tumors. Ectopically expressed human Wig-1 inhibits cell growth in colony formation assays, although less efficiently than p53 itself.

Regulation of *wig-1*

wig-1 mRNA is upregulated within 2-4 hours following p53 activation[3] suggesting that *wig-1* is a direct transcriptional target of p53. To prove this, the promoter region of both mouse and human *wig-1* has been characterized. In mouse *wig-1* three motifs corresponding to the consensus p53 binding motif have been identified upstream of the putative transcription start site. Two of these formed DNA-protein complexes with recombinant p53, and were also found to confer p53-dependent activation to a luciferase reporter plasmid.[6] In human *wig-1,* only one motif corresponding to the consensus p53 binding sequence has been found, located in the intron downstream of exon 1. This DNA motif forms complexes with recombinant p53 in vitro (M. Wilhelm and F. Hellborg, unpublished results), but its functionality in living cells remains to be experimentally proven. Although Wig-1 is highly conserved between mammalian species the location of the p53 binding sites differs. Nevertheless, the direct regulation by p53 is conserved.

The Wig-1 Protein Binds Double-Stranded RNA

Several proteins bind to double-stranded (ds) RNA structures through a conserved dsRNA-binding motif (dsRBM).[7] This motif is composed of approximately 70 amino acid residues and confers sequence-independent dsRNA-binding and sometimes binding to DNA-RNA hybrids. Recently, a novel class of dsRNA-binding proteins lacking the consensus dsRBM has been described.[8] This novel class of proteins contains unusual C_2H_2 zinc finger motifs separated by long linkers. It is believed that this peculiar distribution of zinc finger motifs is responsible for the binding to dsRNA of this group of proteins. The first example of a zinc finger protein lacking a dsRBM and showing a preference for binding to dsRNA came with the identification of dsRBP-ZFa.[9] This 55.6 kD protein containing seven zinc finger motifs was originally found during the screening of a *Xenopus* cDNA expression library with radiolabeled dsRNA. It primarily localizes to the nucleus of *Xenopus* oocytes and its function is currently unknown. The identification of dsRBP-ZFa as a dsRNA-binding protein first indicated that there are proteins that can bind dsRNA in a way distinct from those utilizing the consensus dsRBM. Since then, two other dsRNA-binding zinc finger proteins have been reported. JAZ (for Just Another Zinc finger protein) is a protein with four zinc finger domains that are required for its nuclear localization. JAZ induces apoptosis when overexpressed in murine fibroblasts.[10] Wig-1 is the third nuclear zinc finger protein described so far showing a preferential binding to dsRNA and a similar zinc finger distribution as that found in dsRBP-ZFa and JAZ.[11] Further, the first zinc finger domain in Wig-1 has been demonstrated to be necessary for efficient dsRNA-binding.

Wig-1, dsRBP-ZFa and JAZ share several key features (Table 1). The linkers connecting adjacent zinc fingers are unusually long (34-75 residues as opposed to 6-8 in other zinc finger proteins)

```
human    1 MILLQHAVLPPPKQPSPSPPMSVATRSTGTLQLPPQKPFGQEASLPLAGEEELSKGGEQDC 61
rat      1 MILLQQVWLPLPNRPSTSPPMSVAARSTGTLQLPPQKAFGQEASLPLAGEEDLAKRGEPDS 61
mouse    1 MILLQQVWLPLPNRPSTSPPMSVAARSTRTLQLPPQKAFGQEASLPLAGEEDLAKRGEPDS 61
           *****   ** *  ** ******* *** ******** ************* * * ** *

human   62 ALEELCKPLYCKLCNVTLNSAQQAQAHYQGKNHGKKLRNYYAANSCPPPARMSN-VVEPAA 121
rat     62 ALEELCKPLFCKLCNVTLNSAQQAQAHYQGKNHGKKLRNYYAANSCPPPARMSS-VAEPVA 121
mouse   62 ALEELCKPLFCKLCNVTLNSAQQAQAHYQGKNHGKKLRNYYAANSCPPPARVSSVVAEPVA 122
           ********* ***************************************** *  * ** *

human  122 TPVVPVPPQMGSFKPGGRVILATENDYCKLCDASFSSPAVAQAHYQGKNHAKRLRLAEAQS 182
rat    122 TPLVPVPPQVGSCKPGGRVILATENDYCKLCDASFSSPAVAQAHYQGKNHAKRLRLAEAQS 182
mouse  123 TPLVPVPPQVGSCKPGGRVILATENDYCKLCDASFSSPAVAQAHYQGKNHAKRLRLAEAQS 183
           ** ****** ** ************************************************

human  183 NSFSESSELGQRRARKEGNEFKMMPNRRNMYTVQNNSGPYFNPRSRQRIPRDLAMCVTPSG 243
rat    183 HSFSDSAEAGQRRTRKEGSEFKMVTTRRNMYTVQSNSGPYFNARSRQRIPRDLAMCVTPSG 243
mouse  184 HSFSDSAEAGQRRTRKEGSEFKMVATRRNMNPVQSNSGPYFNARSRQRIPRDLAMCVTPSG 244
            *** * * **** **** ****   ****  ** ******* ******************

human  244 QFYCSMCNVGAGEEMEFRQHLESKQHKSKVSEQRYRNEMENLGYV  288
rat    244 QFYCSMCNVGAGEEVEFRQHLESKQHKSKVSEQRYRSEMENLGYVQ 289
mouse  245 QFYCSMCNVGAGEEVEFRQHLESKQHKSKVSEQRYRSEMENLGYVQ 290
           ************** ********************* ********
```

Figure 2. Alignment of human, mouse and rat Wig-1 protein sequences. Zinc fingers are indicated in bold. The nuclear localization signal between zinc finger 2 and 3 is underlined.

Table 1. Comparison of dsRBP-ZFa, JAZ and Wig-1

Protein	Identification	ZF Moties	DNA Binding	DNA-RNA Binding	Subcellular Localization	Biological Function
dsRBP-ZFa	Screening of Xenopus cDNA library with dsRNA	7	No	Yes	Nuclear (Xenopus oocytes)	Unknown
JAZ	Isolation of proteins that interact with PKR in mouse cDNA library	4	No	Yes	Nuclear/ nucleolar	Unknown/ apoptosis
Wig-1	Isolation of p53-induced mRNAs in mouse J3D-tsp53 cells	3	No	Yes	Nuclear/ nucleolar	Unknown/ cell growth inhibition

and lack the consensus linker sequence TG(Q/E)KP found in DNA-binding zinc finger proteins. It has been suggested that proteins harboring widely spaced fingers may show greater versatility in its binding specificity, strength and span.[12,13] Furthermore, zinc finger proteins normally have four amino acids separating the histidines involved in zinc coordination. In contrast, these three proteins have inter-histidine spaces of five amino acids. The linker regions show no sequence similarities. Interestingly, Wig-1, dsRBP-ZFa and JAZ all show a characteristic binding to DNA-RNA hybrids and lack DNA binding activity. Like the binding to dsRNA through the consensus dsRBM, the binding of this novel group of proteins to dsRNA and DNA-RNA hybrids is apparently sequence-independent. However, they show a clear preference for binding to an A-form helix versus the B-form helix recognized by most DNA-binding proteins. No function has so far been attributed to dsRBP-ZFa, JAZ and Wig-1, although JAZ and Wig-1 have been implicated in inhibition of cell growth and/or apoptosis.

In addition, it is worth mentioning the identification of two other proteins sharing structural similarities to Wig-1, JAZ and dsRBP-ZFa but these proteins are less well characterized regarding their nucleic acids binding properties. ZFR (zinc finger RNA binding) is a nuclear protein highly expressed in testis, ovary and brain that was isolated in a screen for RNA binding proteins expressed during mouse spermatogenesis.[14] ZFR also exists in worms, flies and humans and contains three widely spaced C_2H_2 zinc finger motifs, two putative nuclear localization signals and a 314 amino acids region conserved in several dsRNA-binding proteins harboring dsRBMs. Human *zfr* is mostly expressed in brain.[15] In nucleic acids binding assays, recombinant mouse ZFR protein displayed affinity for RNA known to contain many duplexed regions but also for DNA.[14] Homozygous deletion of the *zfr* gene in the mouse leads to both increased programmed cell death and decreased mitotic index and ultimately to early (E-9.5-10.5) embryonic lethality.[16] Hzf is another protein containing three zinc finger domains similar to the zinc finger domains found in Wig-1, JAZ and dsRBP-ZFa. It was isolated as a novel gene regulated during hematopoietic development from embryonic stem (ES) cells containing a random gene trap insertion and induced to differentiate into hematopoietic cells.[17] Hzf is predominantly expressed in megakaryocytes suggesting a role in the differentiation and/or maturation of this cell lineage. Indeed, studies on Hzf-deficient mice showed an essential function of Hzf in megakaryopoiesis.[18] Importantly, the Hzf putative peptide has been shown to exhibit sequence similarities (22-32%) with dsRBP-ZFa and Wig-1.

In conclusion, dsRNA binding may be a property of any protein containing zinc fingers of the type and with the distribution found in Wig-1, JAZ and dsRBP-ZFa. The possible functional relationship between proteins in this group needs further investigation.

Wig-1 and Neuronal Apoptosis

Rat *wig-1* (PAG608) has been reported to be constitutively expressed throughout the hippocampus, and has also been implicated in neuronal cell death following cerebral ischemia. In a rat model for transient global cerebral ischemia, increased *wig-1/ pag608* mRNA and protein levels were observed in the vulnerable pyramidal neurons of the hippocampal CA1 region.[19] This region is known to undergo apoptosis in response to ischemia, suggesting a role for wig-1 in neuronal cell death. Moreover, p53 and Wig-1 expression were increased by 6-hydroxydopamine (6-OHDA) treatment in catecholaminergic neuronal cell lines.[20] 6-OHDA is a known neurotoxin that causes degeneration of tyrosine hydroxylase-positive neurons, and has been used to generate animal models of degenerative diseases such as Parkinson's disease. Interestingly, the p53 induction following 6-OHDA treatment and the subsequent apoptosis-related morphological changes could be inhibited with antisense wig-1 cDNA. In addition, a Wig-1 zinc finger 1 deletion-mutant also inhibited cell death. This implies not only that Wig-1 is an effector of neuronal apoptosis but that the dsRNA-binding domain of Wig-1 is crucial for the apoptotic effect. Furthermore, Wig-1 was observed to increase p53 protein expression in neuronal cells, suggesting a positive feedback loop in which 6-OHDA induces p53 that in turn induces Wig-1 that increases p53 levels, triggering neuronal apoptosis.

Wig-1 and Stem Cell Proliferation

Another important clue as to the function of *wig-1* has come from the study of mechanisms that regulate self-renewal of haematopoitetic stem cells (HSC). Bmi-1 is a member of the PcG family of transcriptional repressors that control development by the regulation of cell growth and differentiation genes. Bmi-1-deficient mice developed hypocellular bones and died less than 2 months after birth due to no self-renewal of adult HSC.[21] When comparing gene expression profiles of bone marrow mononuclear cells obtained from wild type and Bmi-1 deficient mice, upregulation of *wig-1* mRNA was observed in HSC lacking Bmi-1. Moreover, analysis of transcriptional profiles for three different types of stem cells, i.e., embryonic stem cells, neural stem cells, and haematopoietic stem cells, showed that *wig-1* mRNA was enriched in all three types of stem cells.[22] These results suggest that *wig-1* has an important role in the control of stem cell proliferation and/or differentiation.

Acknowledgments

The authors are supported by the Swedish Cancer Society (Cancerfonden), the King Gustaf V Jubilee Fund, the Ingabritt & Arne Lundberg Research Foundation, the Robert Lundberg Foundation, and the Karolinska Institute Funds.

References

1. Vousden KH, Lu X. Live or let die: The cell's response to p53. Nat Rev Cancer 2002; 2(8):594-604.
2. Lane DP, Lain S. Therapeutic exploitation of the p53 pathway. Trends Mol Med 2002; 8(4 Suppl):S38-42.
3. Varmeh-Ziaie S, Okan I, Wang Y et al. Wig-1, a new p53-induced gene encoding a zinc finger protein. Oncogene 1997; 15(22):2699-2704.
4. Israeli D, Tessler E, Haupt Y et al. A novel p53-inducible gene, PAG608, encodes a nuclear zinc finger protein whose overexpression promotes apoptosis. EMBO J 1997; 16(14):4384-4392.
5. Hellborg F, Wang Q, Méndez-Vidal C et al. Human *wig-1*, a *p53*-target gene that encodes a growth inhibitory zinc finger protein. Oncogene 2001; 20:5466-5474.
6. Wilhelm MT, Mendez-Vidal C, Wiman KG. Identification of functional *p53*-binding motifs in the mouse *wig-1* promoter. FEBS Lett 2002; 524(1-3):69-72.
7. Fierro-Monti I, Mathews MB. Proteins binding to duplexed RNA: One motif, multiple functions. Trends Biochem Sci 2000; 25(5):241-246.
8. Iuchi S. Three classes of C2H2 zinc finger proteins. Cell Mol Life Sci 2001; 58(4):625-635.
9. Finerty Jr PJ, Bass BL. A Xenopus zinc finger protein that specifically binds dsRNA and RNA-DNA hybrids. J Mol Biol 1997; 271(2):195-208.
10. Yang M, May WS, Ito T. JAZ requires the double-stranded RNA-binding zinc finger motifs for nuclear localization. J Biol Chem 1999; 274(39):27399-27406.
11. Mendez-Vidal C, Wilhelm MT, Hellborg F et al. The p53-induced mouse zinc finger protein wig-1 binds double-stranded RNA with high affinity. Nucleic Acids Res 2002; 30(9):1991-1996.
12. Reuter G, Giarre M, Farah J et al. Dependence of position-effect variegation in Drosophila on dose of a gene encoding an unusual zinc-finger protein. Nature 1990; 344(6263):219-223.
13. Fasano L, Roder L, Core N et al. The gene teashirt is required for the development of Drosophila embryonic trunk segments and encodes a protein with widely spaced zinc finger motifs. Cell 1991; 64(1):63-79.
14. Meagher MJ, Schumacher JM, Lee K et al. Identification of ZFR, an ancient and highly conserved murine chromosome-associated zinc finger protein. Gene 1999; 228(1-2):197-211.
15. Kleines M, Gartner A, Ritter K et al. Cloning and expression of the human single copy homologue of the mouse zinc finger protein zfr. Gene 2001; 275(1):157-162.
16. Meagher MJ, Braun RE. Requirement for the murine zinc finger protein ZFR in perigastrulation growth and survival. Mol Cell Biol 2001; 21(8):2880-2890.
17. Hidaka M, Caruana G, Stanford WL et al. Gene trapping of two novel genes, Hzf and Hhl, expressed in hematopoietic cells. Mech Dev 2000; 90(1):3-15.
18. Kimura Y, Hart A, Hirashima M et al. Zinc finger protein, Hzf, is required for megakaryocyte development and hemostasis. J Exp Med 2002; 195(7):941-952.
19. Tomasevic G, Shamloo M, Israeli D et al. Activation of p53 and its target genes p21(WAF1/Cip1) and PAG608/Wig-1 in ischemic preconditioning. Brain Res Mol Brain Res 1999; 70(2):304-313.
20. Higashi Y, Asanuma M, Miyazaki I et al. The p53-activated gene, PAG608, requires a zinc finger domain for nuclear localization and oxidative stress-induced apoptosis. J Biol Chem 2002; 277(44):42224-42232.
21. Park IK, Qian D, Kiel M et al. Bmi-1 is required for maintenance of adult self-renewing haematopoietic stem cells. Nature 2003; 423(6937):302-305.
22. Ramalho-Santos M, Yoon S, Matsuzaki Y et al. "Stemness": Transcriptional profiling of embryonic and adult stem cells. Science 2002; 298(5593):597-600.

CHAPTER 13

Tandem CCCH Zinc Finger Proteins in mRNA Binding

Perry J. Blackshear,* Ruth S. Phillips and Wi S. Lai

Summary

A small family of mammalian zinc finger proteins containing an unusual putative tandem zinc finger motif was identified approximately 13 years ago. The tandem zinc finger domain was characterized by two hypothetical fingers with identical Cx8Cx5Cx3H spacing, with exactly 18 amino acids between the carboxyl terminal H of the first zinc finger and the amino terminal C of the second zinc finger. The two fingers also shared a characteristic amino-terminal lead-in sequence of RYKTEL or a close variant. Although first thought to be transcription factors, these proteins are becoming better understood as the result of experiments with knockout mice for tristetraprolin (TTP), currently the best-studied member of the family. These mice developed a systemic inflammatory syndrome found to be secondary to elevations of tumor necrosis factor alpha (TNF) and possibly granulocyte-macrophage colony-stimulating factor (GM-CSF). These elevations were found to be due to stabilized mRNAs for these cytokines, and subsequent work found that TTP could bind to the AU-rich elements within the 3'-untranslated region of these mRNAs and destabilize them, apparently by initiating a process of 3'-5' deadenylation. This chapter will summarize some of our current thinking about this small but interesting protein family, including binding site and binding domain characterization, and recent developments in mutagenesis and structure determination.

Introduction and Background

Zinc fingers are small, independently folding protein motifs that require one or more zinc ions to maintain structural stability. Zinc finger-containing proteins are among the most abundant in the eukaryotic genome and are most commonly classified as transcription factors containing the well-described Cys_2His_2 zinc finger structure (reviewed in ref. 1). Recently, a more unusual class of zinc finger proteins has become recognized. In these Cys_3H or CCCH zinc finger proteins, the zinc ion is coordinated by three cysteines and one histidine.

To our knowledge, the first description of a novel family of tandem CCCH zinc finger proteins was made by Brown and colleagues in their initial description of the cMG1 protein, published in Oncogene in July, 1990.[2] The cDNA encoding this protein was cloned as a gene induced by epidermal growth factor in rat intestinal epithelial cells. Gomperts et al[2] were apparently the first to describe the internal cysteine-rich repeat sequence that we now refer to as the tandem zinc finger (TZF) domain; they noted that this was found both in their clone of cMG1 and also in a partial clone of TIS11 (tristetraprolin or TTP) previously published by Herschman and colleagues.[3] As stated by Gomperts et al[2] "... it is noteworthy that each of the repeated elements in TIS11 and cMG1 contains three cysteines and one histidine residue, suggesting a possible coordination of these four amino acids with a metal ion in a manner analogous to zinc-finger structures." It is this TZF domain that will be the focus of this review.

We became involved in the study of the CCCH TZF proteins more than 10 years ago, when one of us (WSL) cloned an insulin-induced gene from a library from an insulin-sensitive fibroblast cell line by differential hybridization. This gene was rapidly induced by insulin, serum and other growth factors, and encoded a hypothetical protein that we called tristetraprolin or TTP because of three PPPPG motifs in its primary sequence. We described this sequence and the "immediate early response gene" characteristics of its induction in the September 25, 1990 issue of the *Journal of Biological Chemistry*.[4] A partial clone of this cDNA (TIS11) had been identified earlier in a screen for phorbol ester-responsive genes, and was described in the January, 1989 issue of *Oncogene*;[3] these authors described the corrected sequence in the July, 1991 issue of *Oncogene*.[5] The same gene was also cloned nearly simultaneously as a serum-induced gene (*Nup475*) by DuBois et al;[6] this clone was described in the November 5, 1990, issue of the Journal of Biological Chemistry.

A third mammalian cDNA was cloned by the Herschman group in a study seeking to identify cDNAs with similar TZF domains; this protein was labeled TIS11d, and was described in the March, 1991, issue of Molecular and Cellular Biology.[7] The protein translated from their original TIS11d sequence (GenBank accession number M58564.1) was truncated in the amino terminus and had a reading frame shift in the carboxyl terminus when compared to the current human protein RefSeq (NP_008818), as we have discussed previously.[8,9] We have recently confirmed that the extended amino terminus is expressed as part of the protein in the mouse (S.B. Ramos and P.J. Blackshear, unpublished data).

* *Corresponding author. See list of "Contributors".*

Zinc Finger Proteins: From Atomic Contact to Cellular Function, edited by Shiro Iuchi and Natalie Kuldell.

The Human Genome Nomenclature Committee has recently formalized the nomenclature for these three genes in man as *ZFP36* (encoding TTP, TIS11, Nup475 or G0S24), *ZFP36L1* (for cMG1, TIS11b, Berg 36, ERF1 and BRF1), and *ZFP36L2* (for TIS11D, ERF2 and BRF2). Other aspects of the genomics of the human genes, their chromosomal localization and common variants, can be found in a recent review.[10] We will refer to the proteins in this chapter as TTP, ZFP36L1 and ZFP36L2.

As far as we know, these are the three mammalian family members of the CCCH TZF proteins. A fourth protein has been identified in frogs[8] and fish.[11] This transcript appears to be expressed only in oocytes, ovary and eggs in these species, and contains not only the characteristic tandem CCCH zinc fingers but also two more degenerate fingers of this type. Although there has been recent speculation that this protein, known as C3H-4 in *Xenopus*, is actually a family of related proteins,[12] the recent completion of the small *Fugu* genome suggests that there is only one representative of this gene in this species, as well as the *Fugu* representative of the three mammalian CCCH TZF proteins (see below).

Because of its purported nuclear localization, immediate early response gene induction characteristics, and presumed zinc finger structure, TTP, the best studied of the mammalian CCCH TZF proteins, was presumed to be a transcription factor. In fact, in addition to its well-known "squelching" effect on overall gene transcription,[13] TTP has been implicated as a transcription factor in more recent studies. That this remains a possibility is suggested by the fact that it can shuttle between nucleus and cytoplasm.[9] However, the best-documented function of the protein, in our view, is its ability to bind to AU-rich elements in certain mRNA molecules and destabilize those mRNAs, apparently by initiating a process of deadenylation. These studies were based on the fact that TTP knockout mice chronically over-expressed tumor necrosis factor alpha (TNF), due in part to their chronic over-expression of TNF mRNA.[14,15] This was found to be stabilized in macrophages derived from the TTP-deficient mice.[13] In transfection systems, TTP (and its other family members) can destabilize this mRNA after binding directly to AU-rich elements in its 3'-untranslated region (3'UTR).[16-18] That the tandem zinc finger (TZF) domain is responsible for the binding was demonstrated by the ability of single amino acid mutations in any of the canonical cysteines or histidines to greatly decrease binding[16] as well as in the ability of both biosynthetic and synthetic TZF domain peptides to bind to the ARE probe with high affinity.[19] Similar conclusions about a physiological role for TTP in the regulation of granulocyte macrophage colony stimulating factor (GM-CSF) have been drawn from studies in bone-marrow derived stromal cells from TTP knockout mice and their wild-type littermates.[20,21] Although TTP can bind to AREs in the mRNAs encoding interleukin 3 and other mRNAs, the only mRNAs shown to be affected so far in the knockout mice or cells derived from them are the mRNAs encoding TNF and GM-CSF. Thus, we will focus this review on the TZF domains of these proteins as ARE binding domains.

Binding Sites for TTP

To our knowledge, the only physiologically relevant binding sites for TTP that have been proven from experiments in cells derived from knockout mice are those in the mRNAs encoding TNF and GM-CSF. TTP can bind to other AU-rich elements in mRNAs, and can be shown to exert its usual deadenylating ability on these mRNAs, in cell transfection experiments; examples include the mRNA for IL-3, and for TTP itself. However, we will focus this discussion on the two known physiological substrates.

In early experiments,[13] we showed by UV light-induced crosslinking that TTP could bind to the ARE-containing portion of the TNF mRNA. We then showed that this could be narrowed down to a core ARE of 24 b, that was still compatible with TTP binding and TTP-dependent destruction in cotransfection experiments.[16,17] However, when one or more of the key A residues within this core ARE was changed to C or G residues, binding could be almost completely abrogated (although this depended somewhat on the placement of these residues within the core – see below).

The naturally occurring TNF ARE contains a number of sequences resembling so-called class II AREs that may be capable of TTP binding, raising the possibility of multiple TTP binding sites within the same ARE. To explore the natural variation within the TNF ARE, we performed an alignment of the ARE portion of the known mammalian TNF mRNAs using ClustalW (Fig. 1). This was done by searching known mRNA sequences with the human ARE portion of the TNF ARE using the nr database from GenBank and DNA-DNA blast with the repetitive elements filter disabled; we also searched the mammalian EST collection using the same program. As will be apparent from this figure, the general gestalt of

Alignment of mammalian TNFα AREs

```
Human     CACUUGUGAUUAUUUAUUAUUUAUUUAUUAUUUAUUUAUUUACAGAUGAAUGUAUUUAUUUGGGAG 66
Baboon    CACUUGUGAUUAUUUAUUAUUUAUUUAUUAUUUAUUUAUUUACCGAUGAAUGUAUUUAUUUGGGAG 66
Beluga    CACUUGUGAUUAUUUAUUAUUUAUUUAUUAUUUAUUUAUUUACUAAUGAAUGUAUUUAUUCAGGAG 66
Pig       CACUUGGCAUUAUUAUUUAUUUAUUUAUUUAUUAUUUAUUUACUAGUGAAUGUAUUUAUUCAGGAG 66
Dolphin   CACUUGUGAUUAUUUAUUAUUUAUUUAUUAUUUAUUUAUUUACUAAUGAAUGUAUUUAUUCAGGAG 66
Cow       CACUUGUGAUUAUUUAUUAUUUAUUUAUUAUUUAUUUAUUUACUAAUGAAUGUAUUUAUUCAGGAG 66
Zebu      CACUUGUGAUUAUUUAUUAUUUAUUUAUUAUUUAUUUAUUUACUAAUGAAUGUAUUUAUUCAGGAG 66
Sheep     CACUUAUGAUUAUUUAUUAUUUAUUUAUUAUUUAUUUAUUUACUAAUGAAUGUAUUUAUUCAGGAG 66
Goat      CACUUAUGAUUAUUUAUUAUUUAUUUAUUAUUUAUUUAUUUACUAAUGAAUGUAUUUAUUCAGGAG 66
Horse     CACUUGAGAUUAUUUAUUAUUUAUUUAUUAUUUAUUUAUUU------------------------- 41
Woodchuck CACUUAUAAUUAUUUAUUACUUAUUUAUUAUUUAUUUAUUUACUGAUGAAUGUAUUUAUUUAGAAG 66
Mouse     CACUUAUGAUUAUUUAUUAUUUAUUUAUUAUUUAUUUAUUUGCUUAUGAAUGUAUUUAUUUGGAAG 66
Rat       CACCUGUGACUAUUUAUUUAUUAUUUAUUAUUUAUUUAUUUGCAGAUGAAUGUAUUUAUUUGGGAG 66
Rabbit    GAUUAUUGAUUAUUUAUUAUUUAUUUAAUAUUUAUUUAUUUGCCGAUGAAUGUAUUUAUUUGGAAG 66
           *      * ****  **  ******* *  ********** *   **************  * **
```

Figure 1. Alignment of known mammalian TNF AREs. These represent all the known mammalian TNF AREs at the time of this writing, based on known cDNA sequences as well as uncharacterized ESTs that were thought to represent TNF from these species. They were aligned with ClustalW. The asterisks represent identical residues at that position in all species. The horse sequence had not been extended at the time of this writing. See the online color version of this and subsequent figures at http://www.eurekah.com/chapter.php?chapid=2045&bookid=124&catid=30.

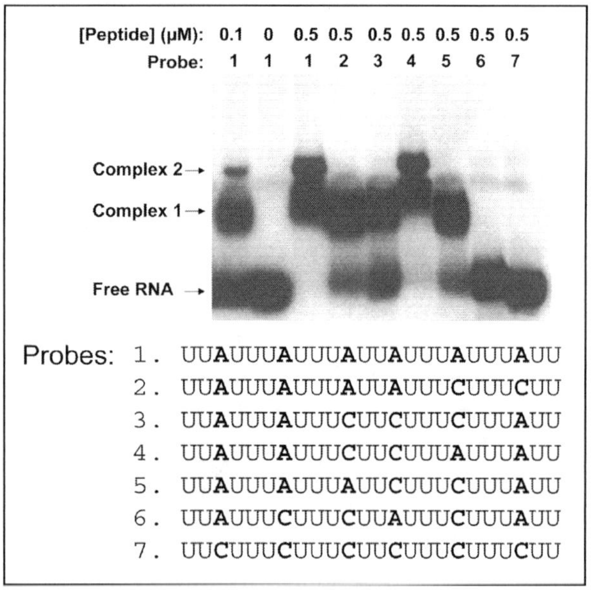

Figure 2. Binding of synthetic human TTP73 peptide to normal and mutant ARE sequences. Gel shift assays were performed with radiolabeled wild-type and mutant ARE probes; the sequences of the ARE portion of the probes are shown, with the original A residues and C substitutions shown in bold. The numbers of the probes correspond to the lane numbers at the top of the gel. The probes were bound to the TTP73 peptide at the concentrations shown at the top of the gel. Complexes 1 and 2, and the positions of the free RNA probes, are indicated. Taken from ref. 19 with permission.

these ARE elements is similar in all of these mammals, although the alignment is not perfect throughout the whole length of the ARE. It should be noted that the alignment of the TNF mRNAs from these species is much less good on either side of the ARE. We will return to this alignment after a brief discussion of the tolerable binding sites within the core human ARE.

Using the human TNF ARE sequences as a starting point, we recently evaluated the minimal binding sequence in the ARE for human TTP and a synthetic 73 amino acid TZF peptide derived from it (TTP73;[19]). The critical figure from that paper is shown as Figure 2. The parental binding site consisted of 24 b from the human TNF ARE, representing bases 17-40 in the first line of Figure 1, i.e., $U^1U^2A^3U^4U^5U^6A^7U^8U^9U^{10}A^{11}U^{12}U^{13}A^{14}U^{15}U^{16}U^{17}A^{18}U^{19}U^{20}U^{21}A^{22}U^{23}U^{24}$. When this sequence was radiolabeled and used to bind to the synthetic peptide TTP73 at low peptide concentrations (0.1 μM, lane 1 in Fig. 2), the probe was shifted into the gel as two main bands, a predominant lower complex (Complex 1) and lesser amounts of a secondary complex (Complex 2); under these conditions, a considerable amount of probe remained "unshifted" at the bottom of the gel in lane 1. The effect of the buffer alone on the probe is shown in lane 2, with no gel shift occurring. When a higher concentration of peptide (0.5 μM) was used, then essentially all of the probe was shifted into the gel and formed complexes 1 and 2 with approximately equal intensity, with no evidence for the formation of complexes of lower gel mobility, even at longer exposures of the autoradiograph. This result was compatible with either two mol per mol TZF peptide occupancy of the 24-mer probe, or perhaps monomer and dimer formation, both conformations permitting probe binding.

We explored these possibilities by selectively mutating the A residues in this target binding site to C residues (Fig. 2). When all six of these As were mutated to Cs, as in probe 7, there was no detectable probe binding to the peptide at 0.5 μM (last lane in Fig. 2). This is in keeping with previous results using C and G replacements of the As in longer probes.[22] A very informative probe was probe 3, in which A11, A14 and A18 were changed to C. In this situation, only a single shifted complex 1 was formed when the peptide was used at 0.5 μM (lane 5 in Fig. 2). In contrast, when only the two middle As (A11 and A14) were mutated to Cs (probe 4), then both binding complexes 1 and 2 were formed with almost the same intensity as with the wild-type probe. These data strongly suggested that the 5' portion of probe 3 (UUAUUUAUUU) was able to bind to peptide, but that the 3' portion (UUUCUUUAUU) was not; in contrast, when the 3' end of the probe was changed to UUUAUUUAUU, as in probe 4, then apparently normal binding was permitted.

These data suggested that, at a minimum, two A residues would be required for a full binding site, surrounded by at least 2 Us and with Us between them. This was further studied in additional mutations, as shown in Figure 2. For example, probe 2 also contained only a single binding site, despite having a more extensive run of wild-type sequence: UUAUUUAUUUAUUAUUUC. This was reduced in size somewhat in probe 5: UUAUUUAUUUAUUC, which again permitted single complex formation as with probe 2. Finally, probe 6, in which the 5', middle, and 3' terminal ends contained single As, did not bind probe at all under these conditions. Thus, the pentamers UUAUUU and UUUAUU were not sufficient to bind peptide under these conditions.

These data strongly suggested that tandem occupancy of the 24 b ARE occurred, with one mol of peptide binding to each binding site, with a minimal effective size of 10 b in the form of UUAUUUAUUU or UUUAUUUAUU. We found in subsequent work that oligonucleotides 9 bases and larger were sufficient to change the NMR conformation of a biosynthetic peptide based on TTP73, whereas probes of 7 b or fewer were not sufficient to cause full conformational changes.[19] Both types of experiments indicated that the minimal completely effective binding site was the 9mer, UUAUUUAUU. A similar conclusion was drawn in an independent set of studies.[23]

When the sequences aligned in Figure 1 were evaluated for the presence of this optimal nonamer, several new patterns emerge. For example, the human sequence is the prototype for a large group that includes whales, monkeys, ruminants, and mice, in which there are five overlapping nonamers (Fig. 3). The horse sequence was not complete at the 3'-end at the time of this writing, but is identical in known residues to the other members of this group. According to the data from,[19] this would probably permit either two or three mol of TTP or its TZF peptide to bind to the overall ARE in tandem. This is in keeping with data from that paper, in which three complexes appeared to be present when a long ARE probe was allowed to bind to the TZF peptide, probably representing single, double and triple peptide occupancy of the probe.

This pattern is somewhat different for the rat, pig, rabbit and woodchuck (Fig. 3). In the case of the rat sequence, there are four overlapping nonamer binding sites; we assume because of steric interference that only two mol per mol of TTP or peptide could bind to this ARE, but this remains to be validated experimentally. In the case of the rabbit, there are a total of three nonamer binding sites, but since two overlap we predict that only two mol per mol of protein or peptide should bind to this element. The

```
Human      GAUUAUUUAUUAUUUAUUUAUUAUUUAUUUAUUUAC
Baboon     GAUUAUUUAUUAUUUAUUUAUUAUUUAUUUAUUUAC
Beluga     GAUUAUUUAUUAUUUAUUUAUUAUUUAUUUAUUUAC
Dolphin    GAUUAUUUAUUAUUUAUUUAUUAUUUAUUUAUUUAC
Cow        GAUUAUUUAUUAUUUAUUUAUUAUUUAUUUAUUUAC
Zebu       GAUUAUUUAUUAUUUAUUUAUUAUUUAUUUAUUUAC
Sheep      GAUUAUUUAUUAUUUAUUUAUUAUUUAUUUAUUUAC
Goat       GAUUAUUUAUUAUUUAUUUAUUAUUUAUUUAUUUAC
Horse      GAUUAUUUAUUAUUUAUUUAUUAUUUAUUUAUUU--
Mouse      GAUUAUUUAUUAUUUAUUUAUUAUUUAUUUAUUUGC

Rat        GACUAUUUAUUUAUUAUUUAUUAUUUAUUUAUUUGC

Rabbit     GAUUAUUUAUUAUUUAUUUAAUAUUUAUUUAUUUGC

Pig        CAUUAUUAUUUAUUUAUUUAUUUAUUAUUUAUUUAC

Woodchuck  AAUUAUUUAUUACUUAUUUAUUAUUUAUUUAUUUAC
```

Figure 3. Presence of TTP binding nonamer in TNF AREs. The portions of the AREs shown in Figure 1 were analyzed for the presence of the putative canonical TTP binding nonamer, UUAUUUAUU. The lines above the sequences indicate the presence of the nonamer. The horse sequence was not complete at the time of this writing.

pig sequence contains five overlapping sites that are somewhat different from the majority of mammals; again, assuming steric interference, we predict two mol per mol of protein or peptide binding, although since the overlap of the nonamers is less than in the other mammals there may be some tandem "triple" binding. Finally, the woodchuck should permit at least three mol/mol peptide/RNA binding based on these assumptions (Fig. 3). It will be interesting to determine whether the rat, rabbit and pig mRNAs behave differently in assays of mRNA deadenylation and turnover than those of the other mammals, since they appear to support only two mol/mol of protein/RNA binding, based on this analysis. It is interesting that this analysis predicts at least two mol/mol of protein/RNA binding in all known mammalian sequences, raising the interesting possibility that at least two binding sites are optimal for TTP to accomplish its physiological function.

Obviously, if the consensus binding site is only slightly looser than the strict nonamer sequence that we have been discussing, then other binding sites will be present in the TNF mRNAs, and in other mRNAs as well. It will be important to determine which "AREs" occurring in nature are truly responsive to TTP and its related proteins in knockout experiments, since binding of TTP to degenerate ARE sequences can be demonstrated readily in cell-free systems but may not be physiologically relevant because of, for example, lower binding affinity. Although we know from direct experimental testing that substitution of all the A residues in a binding site with either C or G abrogates binding, we do not yet know whether modest tampering of the UUU internal sequence affects binding in a physiologically relevant manner.

Relatively little is known about TNF in nonmammalian species, although TTP and members of its family have been cloned from fish and frogs. We searched the GenBank database for fish cDNAs encoding TNF, and sequences that were long enough to include a 3' ARE like sequence are shown in Figure 4. In this case, the sequences were so unrelated that a consensus alignment was not possible. Interestingly, of these four fish species only the catfish sequence contained a single nonamer TTP binding site, and none of the fish contained tandem sites in their "AREs". The seabream sequence has recently been shown to bind both human and zebrafish TTP.[24] Despite evidence for TTP binding, however, TTP from either human and zebrafish was not capable of causing deadenylation or degradation of the seabream TNF mRNA in cotransfection experiments. This supports the speculation that one or more "real" TTP binding sites are necessary for full TTP "activity" in terms of deadenylating and degrading mRNA. Clearly, considerable future work will be necessary to fully describe the characteristics of TTP targets, preferably in the context of TNF action in fish and other nonmammalian species.

The other known physiological target of TTP is GM-CSF. This mRNA is greatly stabilized in bone-marrow derived stromal cells from TTP knockout mice; furthermore, the stable mRNA exists largely in the form of a completely polyadenylated transcript, whereas in the normal cells the mRNA exists as two approximately equally represented fully adenylated and deadenylated

```
FLOUNDER UAUGUAUUUAUUUAUAUUUAUUUAUAUUUAAAAUGGUGGGGUGAGGUGGUAACUAUUAUA

CATFISH  UAUGAUUUAAUGCAUAUUUAUUAUAAACGUAUUUAUUUAAUUAUUUAUUCUAUAUUUAUU

SEABREAM UAUUUAUUUGUAUUUAUUUAUAUUUAAAUUGUUGGGAUAAGGUGUUGAAAGAUUUAUAUUUAUAU

TROUT    UAUUUAUAUGGAAAUACAUAUCACUUUAAAAAAUGAUAUUUAUUAUUUAUAUGAAAAUACA
```

Figure 4. "AREs" from fish TNF sequences. Fish TNF sequences from GenBank were searched for the most ARE like sequence within the 3'-UTR, and these are shown here. The

```
HUMAN      UAUUUAUAUAUUUAUAUUUUUAAAAUAUUUAUUUAUUUAUUUAUUUAAGUUCAUAUUCCA
HORSE      UAUUUAUAUAUUUAUGUAUUUUAA-UAUUUAUUUAUUUAUUUAUUUAAGCUCAUACUCCA
MOUSE      UAUUUAUAUAUUUAUAUUUUUUAAAUAUUUAUUUAUUUAUUUAUUUAA
WOODCHUCK  UAUUUAUAUAUUUAUACUUUUAAAAUAUUUAUUUAUUUAUUUAUUG
COW        UAUUUAUAUAUUUAUGUAUUUUAA-UAUUUAUUUAUUUAUUUAUUUAAACUCAUACCCCA
DOG        UAUUUAUAUAUUUAUGUAUUUUAA-UAUUUAUUUAUUUAUUUAUUUAAGAUCAUACUCUG

PIG        UAUUAACCUAUUUAUGUAUUUUAA-UAUUUAUUUAUUUAUUUAUCUAUUUAUUUAUUUA
           ****  *   *******    ***  **  ******************
```

Figure 5. Alignments of mammalian GM-CSF AREs. These are taken from known mammalian GM-CSF mRNA sequences, as well as ESTs, from GenBank. The asterisks at the bottom indicate base identity at those sites, as determined by the ClustalW alignment program. The overlines indicate the presence of the TTP binding nonamer.

```
humanTNF       ----------------------GAUUAUUUAUUAUUUAUUUAU-UAUUUAUUUAUUUAC- 36
BaboonTNF      ----------------------GAUUAUUUAUUAUUUAUUUAU-UAUUUAUUUAUUUAC- 36
CowTNF         ----------------------GAUUAUUUAUUAUUUAUUUAU-UAUUUAUUUAUUUAC- 36
DolphinTNF     ----------------------GAUUAUUUAUUAUUUAUUUAU-UAUUUAUUUAUUUAC- 36
BelugaTNF      ----------------------GAUUAUUUAUUAUUUAUUUAU-UAUUUAUUUAUUUAC- 36
ZebuTNF        ----------------------GAUUAUUUAUUAUUUAUUUAU-UAUUUAUUUAUUUAC- 36
GoatTNF        ----------------------GAUUAUUUAUUAUUUAUUUAU-UAUUUAUUUAUUUAC- 36
MouseTNF       ----------------------GAUUAUUUAUUAUUUAUUUAU-UAUUUAUUUAUUUAC- 36
SheepTNF       ----------------------GAUUAUUUAUUAUUUAUUUAU-UAUUUAUUUAUUUAC- 36
WoodchickTNF   ----------------------AAUUAUUUAUUACUUAUUUAU-UAUUUAUUUAUUUAC- 36
HorseTNF       ----------------------GAUUAUUUAUUAUUUAUUUAU-UAUUUAUUUAUUU--- 34
RabbitTNF      ----------------------GAUUAUUUAUUAUUUAUUUAA-UAUUUAUUUAUUUGC- 36
RatTNF         ----------------------GACUAUUUAUUUAUUAUUUAU-UAUUUAUUUAUUUGC- 36
PigTNF         ----------------------CAUUAUUAUUUAUUUAUUUAU-UUAUUAUUUAUUUAC- 36
PigGM          UAUUAACCUAUUUAUGUAUUUUAAUAUUUAUUUAUUUAUUUAUCUAUUUAUUUAUUUA-- 58
HumanGM        -----------UAUUUAUAUAUUUAUAUUUUAAAAUAUUUAUUUAUUUAUUUAUUUAAG 49
WoodchuckGM    -----------UAUUUAUAUAUUUAUACUUUUAAAAUAUUUAUUUAUUUAUUUAUUG--- 46
HorseGM        -----------UAUUUAUAUAUUUAUGUAUUUUAA-UAUUUAUUUAUUUAUUUAUUUAAG 48
CowGM          -----------UAUUUAUAUAUUUAUGUAUUUUAA-UAUUUAUUUAUUUAUUUAUUUAAA 48
DogGM          -----------UAUUUAUAUAUUUAUGUAUUUUAA-UAUUUAUUUAUUUAUUUAUUUAAG 48
MouseGM        -----------UAUUUAUAUAUUUAUAUUUUUUAAAUAUUUAUUUAUUUAUUUAUUUAA- 48
                                        *    ******  *  *********
```

Figure 6. ClustalW alignment of AREs from mammalian TNF and GM-CSF sequences. The AREs shown in Figures 1 and 5 were subjected to an unbiased alignment with the ClustalW program, and the results are shown here, with asterisks under the bases indicating identity at each site. The hyphens were inserted by the program to maximize the alignment.

transcripts.[20,21] We therefore analyzed the potential ARE TTP binding sites from the known mammalian GM-CSF mRNAs long enough to contain a full ARE; these are illustrated in Figure 5. With the exception of the pig sequence, all of the mammalian GM-CSF mRNAs illustrated in Figure 5 contain overlapping nonamer TTP binding sites. In the case of most of the mammalian sequences shown in Figure 5, there would only be space for a single mol/mol of TTP binding to the ARE if no overlap is permitted; however, since the Us at the extreme 5' and 3' ends of the binding site nonamer overlap by only one base, then tandem protein or peptide occupancy might be permitted. The other possibility for multiple protein binding is of course if a loosened consensus sequence is permitted.

Given that the TNF and GM-CSF mRNAs are the only ones known to be regulated by TTP in normal mammalian physiology, we aligned the AREs of the known mammalian orthologues using a completely unbiased ClustalW alignment (Fig. 6). Remarkably, there were only two regions of perfect conservation: A more 3' perfect UUAUUUAUU nonamer, and a more 5' hexamer UAUUUA. Closer inspection of the alignment showed that the 5' hexamer was in the middle of a canonical nonamer in 14 of the 21 AREs aligned. In the other cases, the nonamer was altered by an A residue in place of the first U of the canonical nonamer (6 sequences) and/or an AU instead of the 3'terminal UU of the classical nonamer (rabbit TNF only) and/or a 3' terminal UC in the pig GM-CSF sequence. Attempts to confirm the rabbit TNF and the pig GM-CSF mRNA sequences by EST searching revealed that there were no corresponding ESTs in the database, so that these differences in consensus should be viewed only as potential differences. In addition, inspection of the sequences in Figure 6 shows that both the rabbit TNF and pig GM-CSF AREs have canonical nonamers further 5' of the conserved pentamer in Figure 6. These data support a model in which two binding sites for TTP and related proteins are present in the AREs of known mammalian mRNA targets, and thus may be necessary for full activity of these proteins on mRNA deadenylation and turnover. These two binding sites may require enough separation from each other to permit tandem occupancy of two protein molecules per molecule of RNA.

Restricting the definition of potential TTP targets in this way would greatly decrease the number of potential targets in normal physiology, and would also markedly decrease the number of potential targets in the AU-rich element database (ARED; http://rc.kfshrc.edu.sa/ared/New_ARED.htm), for example. From the forgoing discussion, a combination of experimentally derived consensus sequences, coupled with phylogenetic comparisons, should allow us at some point in the relatively near future to describe the sequence requirements for physiologically relevant TTP binding, and thereby delineate the set of possible physiological targets in the "transcriptome".

The Mammalian CCCH TZF RNA Binding Domain

A similar phylogenetic argument can be made concerning the CCCH TZF binding domain. We will consider this TZF domain as the entity necessary for high affinity binding to RNA, based on our initial studies in which mutation of a single key histidine or cysteine almost completely abrogated RNA binding (see ref. 17). However, it is important to note that mutation of a single cysteine or histidine in one of the two zinc fingers permitted a small amount of binding to occur, at least in some experiments using expression of the normal and mutant proteins in 293 cells, followed by gel shift assays using radiolabeled RNA probes with the expressed proteins.[13] We assume that this will be reflected in a major decrease in binding affinity, and that the physiological action of TTP and related proteins relies on the presence of both intact zinc fingers; nonetheless, the interaction of a single zinc finger with RNA may occur with measurable affinity, as suggested by the gel shift data cited above as well as by recent work by Berg and colleagues.[25]

One approach to the elucidation of key residues within the TZF domain is to determine conserved residues within the three known mammalian proteins, then mutate them selectively and look for changes in RNA binding, as well as changes in intact protein-dependent mRNA deadenylation and degradation. Figure 7 shows an alignment of the known mammalian representatives of the TTP family of CCCH TZF domain proteins; some were translated from EST sequences in GenBank, and all were aligned and assigned names based on their closest homology to the human proteins. For the purposes of this initial discussion, we have used the mammalian proteins only, since the functions of TNF, for example, may not be analogous in nonmammalian vertebrates. As shown in Figure 7, there are a number of areas of perfect conservation among the three proteins from the various mammalian species, taking as given the complete conservation of the CCCH canonical zinc finger motifs and the perfect intra- and inter-finger spacing. Other conserved regions and residues include: The complete RYKTELC leading sequence leading into the first C of the first zinc finger (this C is represented as C109 in the human TTP protein, RefSeq NP_003398); the R110 residue immediately following C109; E114; G116; KCQFAHG, containing canonical C124 and H128; ELR at E132; RHPKYKTELC at the lead-in sequence before the start of the second zinc finger at C147; F150; G154; CPYG following C156; and RCHFIH including the last C162 and H166. There are also a number of, for example, TTP-specific residues within the mammalian alignments, e.g., L152 instead of T; however, the significance of these protein-specific residues is not yet clear.

```
                     109      118   124 128                147      156   162 166
                      .        .     .   .                  .        .     .   .
HumanTTP        RYKTELCRTFSESGRCRYGAKCQFAHGLGELRQANRHPKYKTELCHKFYLQGRCPYGSRCHFIH
MouseTTP        RYKTELCRTYSESGRCRYGAKCQFAHGLGELRQANRHPKYKTELCHKFYLQGRCPYGSRCHFIH
RatTTP          RYKTELCRTYSESGRCRYGAKCQFAHGPGELRQANRHPKYKTELCHKFYLQGRCPYGSRCHFIH
CowTTP          RYKTELCRTFSESGRCRYGAKCQFAHGLGELRQASRHPKYKTELCHKFYLQGRCPYGSRCHFIH
HorseTPP        RYKTELCRTFSESGRCRYGAKCQFAHGLGELRQASRHPKYKTELCHKFYLQGRCPYGSRCHFIH
PigTTP          RYKTELCRTFSESGRCRYGAKCQFAHGLGELRQASRHPKYKTELCHKFYLQGRCPYGSRCHFIH

HumanZFP36L1    RYKTELCRPFEENGACKYGDKCQFAHGIHELRSLTRHPKYKTELCRTFHTIGFCPYGPRCHFIH
HorseZFP36L1    RYKTELCRPFEENGACKYGDKCQFAHGIHELRSLTRHPKYKTELCRTFHTIGFCPYGPRCHFIH
SheepZFP36L1    RYKTELCRPFEENGACKYGDKCQFAHGIHELRSLTRHPKYKTELCRTFHTIGFCPYGPRCHFIH
CowZFP36L1      RYKTELCRPFEENGACKYGDKCQFAHGIHELRSLTRHPKYKTELCRTFHTIGFCPYGPRCHFIH
RatZFP36L1      RYKTELCRPFEENGACKYGDKCQFAHGIHELRSLTRHPKYKTELCRTFHTIGFCPYGPRCHFIH
MouseZFP36L1    RYKTELCRPFEENGACKYGDKCQFAHGIHELRSLTRHPKYKTELCRTFHTIGFCPYGPRCHFIH

HumanZFP36L2    RYKTELCRPFEESGTCKYGEKCQFAHGFHELRSLTRHPKYKTELCRTFHTIGFCPYGPRCHFIH
MouseZFP36L2    RYKTELCRPFEESGTCKYGEKCQFAHGFHELRSLTRHPKYKTELCRTFHTIGFCPYGPRCHFIH
RatZFP36L2      RYKTELCRPFEESGTCKYGEKCQFAHGFHELRSLTRHPKYKTELCRTFHTIGFCPYGPRCHFIH
CowZFP36L2      RYKTELCRPFEESGTCKYGEKCQFAHGFHELRSLTRHPKYKTELCRTFHTIGFCPYGPRCHFIH
PigZFP36L2      RYKTELCRPFEESGTCKYGEKCQFAHGFHELRSLTRHPKYKTELCRTFHTIGFCPYGPRCHFIH

Consensus       RYKTELCR   E G C YG KCQFAHG  ELR   RHPKYKTELC  F   G CPYG RCHFIH
```

Figure 7. Alignment of the TZF domains from mammalian CCCH proteins. These were made using the program ClustalW, and include the amino-terminal lead-in sequences from all known cDNA and EST sequences corresponding to mammalian CCCH TZF proteins. The consensus sequence of amino acid identities is shown at the bottom of the figure. The amino acids corresponding to the numbered dots at the top of the figure represent residues of the CCCH zinc fingers, with the numbering system derived from the human TTP RefSeq, NP_ 003398.

We have begun to systematically mutate these conserved residues and evaluate their effect on protein function.[26] To date, all mutations that have affected binding of the intact protein to a TNF ARE probe have had similar effects on transfection-based assays of TTP-dependent mRNA deadenylation and degradation, so for the purposes of this discussion we will focus mainly on the binding assay. An example of this type of experiment is shown in Figure 8. These studies used wild-type and mutant human TTP expressed in human 293 cells, which conveniently lack TTP, as the source of ARE binding proteins; they also used probes derived from TNF (A) and GM-CSF (B) as gel-shift targets. Panel C demonstrates the roughly equivalent expression of the epitope-tagged wild-type and mutant TTP proteins present in the 293 cell extracts used for the binding assay. As readily seen when comparing (Fig. 8A), lanes 1 and 2, there were at least three endogenous probe binding proteins in 293 cell extracts when used in this assay at this protein concentration (labeled I, II and III); the identities of these proteins are not known. However, when the 293 cell extracts containing wild-type TTP were used as the protein source in this gel-shift assay, there was the characteristic formation of two typical TTP-probe complexes (lane 3) that were not present in the extracts from cells transfected with vector alone (lane 2). As can be seen at the bottom of this gel, most but not all of the probe was shifted into the gel to form the TTP-probe complexes in lane 3. Essentially identical findings were seen when a different ARE probe was used (panel B, same lanes). However, when mutant TTP proteins were used in the same assay, it is apparent that there was essentially no detectable RNA binding activity under these conditions. This is demonstrated by the lack of the presence of the TTP probe complexes in lanes 4-8; the presence of the endogenous 293 cell ARE binding proteins in the same extracts; and the normal amount of probe remaining at the bottom of the gels. Similar findings were present when the second probe was used (panel B), and panel C demonstrates roughly equivalent concentrations of expressed proteins in these extracts.

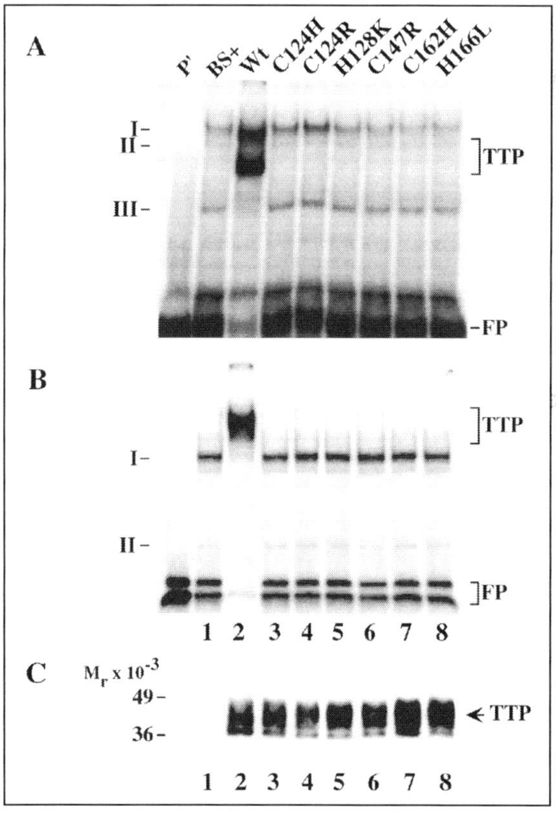

Figure 8. The effect of mutations on residues within the TZF domain upon binding of ARE probes to the intact TTP protein. Gel shift assays were performed on cytosolic extracts from 293 cells expressing wild-type TTP or the indicated mutant forms. P' is a lane containing probe alone; BS+ is a lane containing an equivalent amount of a 293 cell extract transfected with the BS+ vector alone; Wt is wild-type TTP; the other lanes are labeled by the appropriate amino acid substitution in the numbered residues from the TTP TZF domain. A represents gel shifts using a TNF ARE probe; the numbered complexes formed are indicated to the left. B represents gel shifts using a GM-CSF ARE probe, with the same protein extracts as in A. C indicates the TTP protein expression in the extracts used, determined by western blotting with an antibody against the epitope tag of TTP. FP= free probe. Taken from ref. 26 with permission.

These experiments demonstrated no detectable binding of single zinc finger residue mutants under these conditions, in which the mutations tested were fairly radical modifications of the canonical C and H residues within the two zinc fingers of human TTP (using the same numbering system as in Fig. 7). We then applied the same type of assay to mutations of some of the other conserved residues described in Figure 7 (Table 1; ref. 26). This

Table 1. Summary of point mutant effects on TTP binding to an ARE probe and destabilizing an ARE-containing mRNA

Interval	Amino Acid Change	ARE Binding	Decrease TNF mRNA?	Increase TNF mRNA?
CX_8C				
	R110E(L)	++	++++	
	H148E(L)	++++	++++	
	F112Y	++++	++++	
	F112Q	++	++	
	F150Y	++++	++++	
	G116E	+++	++++	
	G154K	++++	++++	
	R117E(L)	++++	++++	
	R155L	++++	++++	
	C147R	0	0	Yes
CX_5C				
	Y120F	++++	++++	
	Y158E	0	0	Yes
	Y158Q	0	0	Yes
	G121T(R)	++	++	
	G159A(E)(K)	++	++	
	K123D	++++	++++	
	R161L	+	+	
	C124H(R)	0	0	Yes
	C162H	0	0	Yes
CX_3H				
	F126Y	++++	++++	
	F126N	0	0	Yes
	F164Y	++++	++++	
	F164L	0	0	Yes
	H128K	0	0	Yes
	H166L	0	0	Yes

These data summarize the results from three to four individual experiments. "Interval" represents the distance between each of two key amino acids in the two CCCH motifs, and *X* represents any amino acid. "Amino acid change" refers to the results of mutations encoding conserved amino acids of the CCCH motifs, as illustrated in Figure 1. "ARE binding" means the ability of expressed hTTP mutants to bind to a GM-CSF mRNA ARE probe as assessed by gel mobility shift assays. The binding activity of wild-type TTP is indicated by ++++, whereas no binding activity is indicated by 0. "Decrease TNF mRNA?" represents the ability of a mutant TTP construct to cause breakdown of the TNF mRNA in cotransfection experiments in 293 cells, indicated as 0 (no effect) to ++++ (equivalent to wild-type TTP). "Increased TNF mRNA?" refers to the ability of a given mutant to cause an increase in the steady-state concentration of TNF or other ARE containing mRNAs in the 293 cell cotransfection studies. Modified from ref. 26 with permission.

table does not include mutations within the highly conserved lead-in sequences for both fingers or the interfinger spacer which also has a number of conserved residues. Some of these results were somewhat surprising. For example, mutation of the perfectly conserved R110 to an acidic E residue had only a modest effect on RNA binding, and did not seem to affect the ability of TTP to promote TNF mRNA degradation in the co-transfection studies. Similarly, mutation of G116 or G154 also had only a modest effect on TTP binding activity. However, mutation of other conserved residues, such as Y158 to either E or Q, completely abolished RNA binding ability. Another example of this type of profound change was the F residue in the final intra-finger sequence in either zinc finger: mutation of F126 or F164 to Y had no effect on binding, whereas mutation of F126 to N or F164 to L abolished binding.

Many of these highly conserved residues that are critical for binding have been found to be in significant positions within the recently described three-dimensional structure of this binding domain (see below). Similar studies are underway for a variety of other of the conserved residues within the zinc fingers themselves; within the lead-in sequences; within the intra-finger spacer; and finally, "spacers" are being introduced into the highly conserved inter-finger and intra-finger distances, to determine how well changes in these distances can be tolerated. This is of particular importance when evaluating the numerous members of this extended protein family in *C. elegans* (see below), where there is considerable variability in the distances between key residues in the two zinc fingers.

In addition to the RNA binding properties of the TZF domain, it has emerged within the past few years that the TZF domain contains sequences that seem to be capable of nuclear localization. TTP has long known to be capable of moving between the nuclear and cytosolic compartments,[27] and recent studies from our laboratory have identified nuclear export sequences within the amino terminus of TTP and the carboxyl-termini of ZFP36L1 and L2.[9] However, the nature of the nuclear localization signals was not clear. Recently, Murata and colleagues demonstrated that two arginine residues within the inter-finger spacer were important for nuclear localization of the protein.[28] In our own work, the TZF domain alone seems to be entirely nuclear, i.e., when lacking the nuclear export sequences. Mutation of the same two arginine residues in the inter-finger linker of the whole TTP and ZFP36L1 proteins caused the expressed proteins to be completely cytosolic, compatible with an important role for these two residues as a nuclear localization motif. However, when these mutant proteins were expressed in 293 cells and used in a gel shift assay, they were unable to bind to an ARE probe, suggesting that these conserved basic residues within the inter-finger linker may be critical for the three dimensional configuration of the protein and are necessary for binding to occur. This in turn suggests that these residues may not constitute true nuclear localization signals, since the fact that the mutant protein can no longer bind RNA suggests that a fairly fundamental change in structure has occurred that may prevent the nuclear import of the protein, independent of true primary sequence nuclear localization motifs. Distinguishing among these possibilities will require further experimentation.

Conserved Residues within Vertebrate CCCH TZF Domain Proteins

As mentioned above, a fourth member of this family of proteins was identified in frogs and fish.[8,11] Although an analysis of the *Xenopus* and *Silurana tropicalis* ESTs encoding this fourth protein, known as *Xenopus* C3H-4, suggested that it might represent a family of closely related sequences,[12] we recently determined by blasting the *Fugu* genome that there probably is only a single representative of this protein in the *Fugu* genome, as well as three additional proteins representing the *Fugu* versions of TTP, ZFP36L1 and ZFL36L2. Therefore, for the purposes of this discussion, vertebrate genomes seem to express only four members of the TTP family of TZF CCCH finger proteins, using the fairly strict spacing definition used here. We recently searched GenBank for all known members of this family in vertebrates, in general using tblastn to search nr, EST and genome databases for the various vertebrates. The results of this overall alignment using ClustalW are shown in Figure 9. In this alignment, the prototype TZF domain belonged to mouse TTP, and the distance from mouse was directly proportional to the difference in sequence within the TZF domain. The representation of these differences in a phylogenetic tree are shown in Figure 10. Interestingly, Figure 9 demonstrates that all three of the *Xenopus* orthologues of TTP, ZFP36L1 and ZFP36L2, known as C3H-1, C3H-2, and C3H-3, respectively, are closer to mouse TTP than any of the fish members of the family. In fact, the fish members are often difficult to assign to an orthologous family member, and so were arbitrarily numbered 1-4. In particular, the boundaries between the fish orthologues of ZFP36L1 and ZFP36L2 were often blurred, so it seems possible that these proteins will be found to be interchangeable in the fish genome, with biological activity possibly more dependent upon expression pattern than primary sequence.

This alignment allows for the identification of amino acids that are conserved among these various disparate groups of vertebrate animals, as shown at the bottom of Figure 9. Remarkably, many of the conserved residues found in the mammalian alignment are still conserved among all vertebrates, and occasional differences, such as the equivalent of E145 to Q change in zebrafish "TTP" as the only vertebrate with this change, are possibly due to EST sequencing errors or polymorphisms. Nonetheless, these conserved residues will continue to represent mutation targets for understanding the RNA binding domain and nuclear localization functions of the proteins.

```
MouseTTP          RYKTELCRTYSESGRCRYGAKCQFAHGLGELRQANRHPKYKTELCHKFYLQGRCPYGSRCHFIH
RatTTP            RYKTELCRTYSESGRCRYGAKCQFAHGPGELRQANRHPKYKTELCHKFYLQGRCPYGSRCHFIH
HumanTTP          RYKTELCRTFSESGRCRYGAKCQFAHGLGELRQANRHPKYKTELCHKFYLQGRCPYGSRCHFIH
CowTTP            RYKTELCRTFSESGRCRYGAKCQFAHGLGELRQASRHPKYKTELCHKFYLQGRCPYGSRCHFIH
HorseTPP          RYKTELCRTFSESGRCRYGAKCQFAHGLGELRQASRHPKYKTELCHKFYLQGRCPYGSRCHFIH
PigTTP            RYKTELCRTFSESGRCRYGAKCQFAHGLGELRQASRHPKYKTELCHKFYLQGRCPYGSRCHFIH
XenopusC3H1       RYKTELCRTFSETGTCKYGAKCQFAHGKIELREPNRHPKYKTELCHKFYLYGECPYGSRCNFIH
HorseZFP36L1      RYKTELCRPFEENGACKYGDKCQFAHGIHELRSLTRHPKYKTELCRTFHTIGFCPYGPRCHFIH
ChickenZFP36L1    RYKTELCRPFEENGACKYGDKCQFAHGIHELRSLTRHPKYKTELCRTFHTIGFCPYGPRCHFIH
SheepZFP36L1      RYKTELCRPFEENGACKYGDKCQFAHGIHELRSLTRHPKYKTELCRTFHTIGFCPYGPRCHFIH
CowZFP36L1        RYKTELCRPFEENGACKYGDKCQFAHGIHELRSLTRHPKYKTELCRTFHTIGFCPYGPRCHFIH
RatZFP36L1        RYKTELCRPFEENGACKYGDKCQFAHGIHELRSLTRHPKYKTELCRTFHTIGFCPYGPRCHFIH
MouseZFP36L1      RYKTELCRPFEENGACKYGDKCQFAHGIHELRSLTRHPKYKTELCRTFHTIGFCPYGPRCHFIH
HumanZFP36L1      RYKTELCRPFEENGACKYGDKCQFAHGIHELRSLTRHPKYKTELCRTFHTIGFCPYGPRCHFIH
HumanZFP36L2      RYKTELCRPFEESGTCKYGEKCQFAHGFHELRSLTRHPKYKTELCRTFHTIGFCPYGPRCHFIH
MouseZFP36L2      RYKTELCRPFEESGTCKYGEKCQFAHGFHELRSLTRHPKYKTELCRTFHTIGFCPYGPRCHFIH
RatZFP36L2        RYKTELCRPFEESGTCKYGEKCQFAHGFHELRSLTRHPKYKTELCRTFHTIGFCPYGPRCHFIH
CowZFP36L2        RYKTELCRPFEESGTCKYGEKCQFAHGFHELRSLTRHPKYKTELCRTFHTIGFCPYGPRCHFIH
PigZFP36L2        RYKTELCRPFEESGTCKYGEKCQFAHGFHELRSLTRHPKYKTELCRTFHTIGFCPYGPRCHFIH
ChickenZFP36L2    RYKTELCRPFEESGACKYGEKCQFAHGFHELRSLTRHPKYKTELCRTFHTIGFCPYGPRCHFIH
XenopusC3H3       RYKTELCRPFEENGACKYGEKCQFAHGFHELRSLTRHPKYKTELCRTFHTIGFCPYGPRCHFIH
XenopusC3H2       RYKTELCRPFEENGSCKYGDKCQFAHGIHELRSLTRHPKYKTELCRTFHTIGFCPYGPRCHFIH
ZebrafishZFP36L1  RYKTELCRPFEENGACKYGDKCQFAHGIHELRSLSRHPKYKTELCRTFHTIGFCPYGPRCHFIH
Fugu4             RYKTELCRPFEENGSCKYGDKCQFAHGIHELRSLSRHPKYKTELCRTFHTIGFCPYGPRCHFIH
TroutZFP36L1      RYKTELCRPFEENGTCKYGDKCQFAHGMHELRSLSRHPKYKTELCRTFHTIGFCPYGPRCHFIH
SalmonZFP36L1     RYKTELCRPFEENGTCKYGDKCQFAHGMQELRSLSRHPKYKTELCRTFHTMGFCPYGPRCHFIH
Fugu2             RYKTELCRTYEESGTCKYGTKCQFAHGLDELRGISRHPKYKTELCRTFHTIGFCPYGARCHFVH
Fugu3             RYKTELCRSFTESGFCKYGGKCQFAHGAEELRDLNRHPKYKTEPCRTFHTIGFCPYGVRCHFVH
SiluranaTTP       RYKTELCRTFHEIGSCKYGSKCQFAHGSDELRGLNRHPKYKTELCRTYHTIGFCPYGSRCHFIH
CatfishTTP        RYKTELCRSFQENGSCKYGNKCQFAHGESELRGLYRHPKYKTEACRTFYNFGYCPYGSRCHFIH
TroutTTP          RYKTELCRSFQENGSCKYGSKCQFAHGEPELRGLYRHPKYKTEACRTFYNFGYCPYGARCHFIH
ZebrafishTTP      RYKTELCRSFQEHGSCKYGAKCQFAHGENELRGLYRHPKYKTQACRTFYQFGYCPYGSRCHFIH
SticklebackTTP    RYKTELCRGFQETGSCKYGSKCQFAHGEAELRGLFRHPKYKTEPCRTFYNFGYCPYGSRCHFIH
MedakaTTP         RYKTELCRGFQETGSCKYGSKCQFAHGEAELRGLYRHPKYKTEPCRTFYNFGYCPYGPRCHFIH
XenopusC3H4       RYKTELCSRYAESGFCAYRNRCQFAHGLSELRPPVQHPKYKTELCRSFHVLGTCNYGLRCLFIH
Silurana1         RYKTELCTRYAESGFCAYRNRCQFAHGLSELRPPVQHPKYKTELCRSFHVLGTCNYGLRCLFIH
Carp1             RYKTELCSRYAETGTCKYAERCQFAHGLHDLHVPSRHPKYKTELCRTYHTAGYCVYGTRCLFVH
Zebrafish1        RYKTELCSRYAETGTCKYAERCQFAHGLHDLHVPSRHPKYKTELCRTYHTAGYCVYGTRCLFVH
Trout2            RYKTELCSRYADTGFCKYAERCQFAHGLHDLHVQSRHPKYKTELCRTFHTAGYCVYGTRCLFAH
Fugu1             RYKTELCTSYSATGFCKYGERCQFAHGLHELHIPFHHPKYKTELCRSYHTTGYCYYGSRCLFVH
                  *******    :    * * *  :******  :*:    :******: *:.::  * * ** ** * *
```

Figure 9. Alignment of all known vertebrate CCCH protein TZF domains. These were taken from blast searches of vertebrate cDNAs and ESTs, and then aligned by ClustalW. The asterisks indicate amino acid residues that were identical at that site in all proteins; the colons indicate amino acid similarity at that position; and the single dots indicate lesser degrees of amino acid similarity. Although the frog sequences could be assigned to orthologous mammalian proteins by their overall sequence identities, the fish sequences were not; these were either numbered randomly or assigned possible orthologous identities. The top to bottom sequence order is related to the extent of amino acid identity to mouse TTP, shown at the top, with the least related sequences at the bottom.

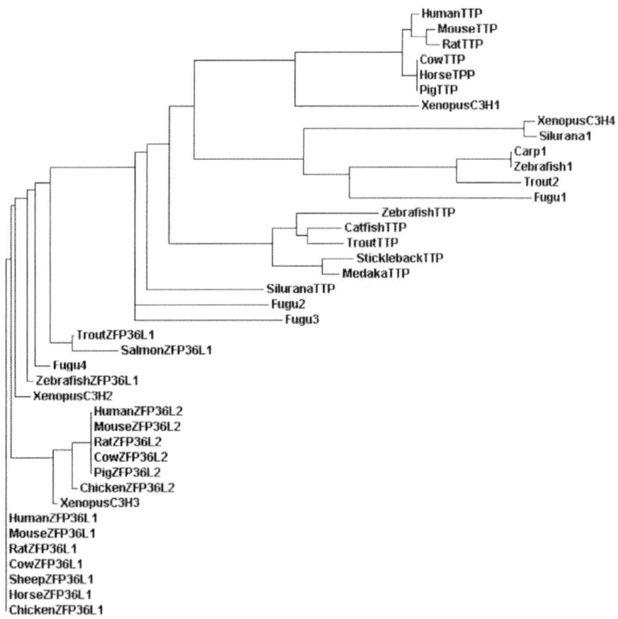

Figure 10. Phylogram of the sequences shown in Figure 9. This figure shows the hypothetical relationships among the sequences shown in Figure 9. Note that this tree was calculated based on the alignments of the TZF domains alone, as shown in Figure 9, rather than on the alignment of the whole proteins.

Two other alignments may be of interest. In the first, we blasted the plant and invertebrate animal sequences in GenBank (ESTs, nr and genomes) with human TTP, and looked for the canonical CCCH motifs and identical spacing. The results of this search are shown in Figure 11. There are representative proteins containing versions of this domain from many invertebrates, including two in *S. cerevesiae*, one in *S. pombe*, *Drosophila*, *Bombyx mori* (silkworm), the *Anopheles* mosquito, and *Plasmodium falciparum*, and still more distantly related family members in plants. To date, exploration of the function of these proteins in invertebrates has been disappointing; for example, disruption of both genes in yeast yielded a completely viable organism.[29] Nonetheless, as the structure of the prototype member of this family, presumably the human TTP TZF domain, is determined, it will be of interest to determine the corresponding structures of some of these invertebrate forms and work out their function in their respective organisms.

```
A.
11   SYKTVICQAWLE-SKTCAFAENCRFAHGEEELRPSLIEA-RQNNKYRTKLCDKYTT--TGLCPYGKRCLFIH
12   SYKTVICQAWLE-SKTCTFAENCRFAHGEEELRPAKLES-RQNNKYKTKLCDKYTT--TGLCPYGKRCLFIH
10   SYKTVICQAWLE-SKTCSFADNCRFAHGEEELRPTFVEP-LQNNKYKTKLCDKYTT--TGLCPYGKRCLFIH
TTP  RYKTELCRTFSE-SGRCRYGAKCQFAHGLGELR--QANR---HPKYKTELCHKFYL--QGRCPYGSRCHFIH
1    LYKTELCRSWMD-HGRCNYGERCQYAHGELEKR--PVPR---HPKYKTEACQSFHQ--SGYCPYGPRCHFIH
6    AFKTSLCLSHKR-GKTCIYGEACKFAHGVHELRCQQTTR--NHRNYKTVLCDKFTT--TGYCKYGARCQFIH
7    AFKTALCLSHKR-GKTCIYGEQCKFAHGVHELRCQQAKK--NHRNYKTVLCDKFTT--TGYCKYGIRCQFIH
8    AFKTSLCGFHRR-GQKCAYGEKCKFAHSVHELRFPQTKR--NHRNYKTVLCNNFST--TGHCKYGIRCQFIH
9    AFKTALCGFQRR-GQKCIYGEQCKFAHSVHELRFTQAKK--THRNYKTVLCDKFST--TGYCKYGARCQFIH
4    AFKTALCKTFQL-TRACSYGEQCKFAHSVEELQLKQKNRGVNHPKYKTVLCDNFSR--TGHCKYGTKCQFIH
5    AFKTALCKTFQL-TKACSYGEQCKFAHSVEELQLKHQNLGINNPKYKTVLCDNFST--TGHCKYGTKCQFIH
3    AFKTALCDAYKR-SQACSYGDQCRFAHGVHELRLPMNPRGRNHPKYKTVLCDKFSM--TGNCKYGTRCQFIH
2    AFKTALCDAFKR-SGSCPYGEACRFAHGENELRMPSQPRGKAHPKYKTQLCDKFSN--FGQCPYGPRCQFIH
14   NFKTRLCMTHAAGINPCALGARCKFAHGLKELRASDIPTRYPNNKYKTKLCKNFARGGSGVCPYGLRCEFVH
15   NFKTRLCMTHAAGINPCALGARCKFAHGLKELRASDIPTRYPNNKYKTKLCKNFARGGSGVCPYGLRCEFVH
13   NYKTRLCMMHASGIKPCDMGARCKFAHGLKELRATDAPARYPNNKYKTKLCKNFARGGTGFCPYGLRCEFVH
      :** :*           *  .  *::**.  * :             : :*:*  *..:      * * ** :* *:*

B.
TTP     RYKTELCRTFSESGRCRYGAKCQFAHGLGELRQANRHPKYKTELCHKFYLQGRCPYGSRCHFIH 64
1       LYKTELCRSWMDHGRCNYGERCQYAHGELEKRPVPRHPKYKTEACQSFHQSGYCPYGPRCHFIH 64
        *******:: : ***.** :**:***  * * . ******** *:.*: .* ****.******
```

Figure 12. *C. elegans* CCCH TZF proteins. Shown in (A) is an alignment of CCCH TZF domains from all known *C. elegans* "TZF" proteins with that of human TTP. The hyphens were placed by the ClustalW alignment program to permit loosening of the spacing conventions of the mammalian proteins. The amino acid identities and similarities are indicated by the symbols described in the legend to Figure 9. In (B) is shown the alignment of TTP with the single *C. elegans* protein that maintains the exact vertebrate spacing conventions. Outside of the TZF domain, this protein bears very little resemblance to any of the vertebrate CCCH TZF proteins.

Figure 11 shows only a single representative of this group in *C. elegans*, but if the *C. elegans* genome, nr and EST databases are blasted with human TTP and the spacer conventions are not insisted upon, then at least 13 related *C. elegans* proteins are produced (Fig. 12A). Some of these already have known functions or attributes. For example, the closest sequence to human TTP is sequence 1 in Figure 12A, shown aligned individually with human TTP using ClustalW in Figure 12B. This protein, referred to as a pair of hypothetical proteins F38B7.1a and F38B7.1b (see GenBank accession number NP_505926), or its mRNA, is expressed at a high level during all stages of development. According to the RefSeq citation, RNA interference

```
Oryza sativa                    MFKTELCNKWEETGDCPYGDQCQFAHGVTELRPVIRHPRYKTAVCRMVLAGDVCPYGHRCHFRH
Hordeum vulgare                 MFKTELCNKWEETGACPYGDQCQFAHGVSELRPVIRHPRYKTEVCRMVLNGEVCPYGHRCHFRH
Vitis vinifera                  MFKTELCNKWQESGTCPYGDHCQSAHGIEELRPVIRHPRYKTEVCRMVLAGDACPYGHRCHFRH
2 Arabidopsis thaliana          MTKTELCNKWQETGTCPYGDHCQFAHGIKELRPVIRHPRYKTEVCRMVLAGDNCPYGHRCHFRH
1 Arabidopsis thaliana          MMKTELCNKWQETGACCYGDNCQFAHGIDELRPVIRHPRYKTEVCRMMVTGAMCPYGHRCHFRH
Bombyx mori                     RYKTELCRPFEEAGVCKYGDKCQFAHGVRELRNLQRHPKYKTELCRTFHSVGFCPYGPRCHFVH
Anopheles gambiae               RYKTELCRPYEEAGECKYGDKCQFAHGMQELRNLQRHPKYKTELCRTFHSVGFCPYGPRCHFVH
Drosophila melanlogaster        RYKTELCRPFEEAGECKYGEKCQFAHGSHELRNVHRHPKYKTEYCRTFHSVGFCPYGPRCHFVH
Crassostrea virginica           RYKTELCRPFEESGHCKYGDKCQFAHGAHELRNLNRHPKYKTELCRTFHTIGFCPYGPRCHFIH
C Elegans                       LYKTELCRSWMDHGRCNYGERCQYAHGELEKRPVPRHPKYKTEACQSFHQSGYCPYGPRCHFIH
Encephalitozoon cuniculi        LYKTEMCRSHTEIGYCRYGDKCQFAHSKAELRYVQRHPKYKTETCKTFWEEGSCPYGKRCCFIH
Schizosaccharomyces pombe       LYKTEPCKNWQISGTCRYGSKCQFAHGNQELKEPPRHPKYKSERCRSFMMYGYCPYGLRCCFLH
CTH1 Saccharomyces cerevesiae   LYKTELCESFTIKGYCKYGNKCQFAHGLNELKFKKKSNNYRTKPCINWSKLGYCPYGKRCCFKH
CTH2 Saccharomyces cerevesiae   LYKTELCESFTLKGSCPYGSKCQFAHGLGELKVKKSCKNFRTKPCVNWEKLGYCPYGRRCCFKH
                                  *** *.      * * **..** **.  * :      .:::  *        **** ** * *
```

Figure 11. Sequence alignments of invertebrate CCCH TZF domain proteins. These peptide sequences are derived from cDNAs and ESTs from GenBank and were aligned without any vertebrate sequences to anchor the alignment. The extent of amino acid identity or similarity is indicated at the bottom of the alignment, as described in the legend to Figure 9. The *C. elegans* sequence used was from the single *C. elegans* protein of this class that maintains the vertebrate spacing conventions (see Fig. 12). The dendrogram at the bottom is a cladogram, that is, it represents a the extent of the sequence similarity among the sequences rather than the evolutionary relationships. This was based on the sequences of the TZF domain alone.

studies did not yield an obvious phenotype. When the entire sequence was blasted against the human protein database in GenBank, the only region of homology determined was with the TZF domain, so this protein probably should not be considered an orthologue of any of the known vertebrate proteins, at least at this stage in our knowledge.

More is known about some of the other *C. elegans* proteins with differences in spacing compared to the vertebrate TZF domains. For example, a number of the sequences are from Muscle Excess (MEX) proteins, also known as Maternal Effect Lethal (MEL) proteins. Mex-1, shown in Figure 12A as sequence 2, is embryonic lethal 80-100% of the time when decreased by RNA interference; the surviving progeny are sterile and often have ruptured vulvas (see RefSeq NP_496551). There are other MEX family members, and other groups of proteins include the Posterior Segregation (POS) proteins, Maturing Oocyte Expressed (MOE) proteins, and others. All are presumed to be nucleic acid binding proteins. Probably the best known member of this family is Pharynx and Intestine in Excess (PIE-1), which is not shown in Figure 12 because of its much greater inter-finger distance than the putative TZF domain proteins shown in Figure 12. Its current status and recent references are referred to in GenBank RefSeq record NP_499619; these data suggest that it is thought to act as a transcriptional repressor as well as a possible translational activator. This once again raises the interesting possibility that the mammalian proteins will also be found to be DNA binding proteins, in addition to their apparent RNA binding protein abilities. The PIE-1 protein and other *C. elegans* family members are not apparently orthologues of any of the vertebrate proteins, outside of the TZF domain similarities, and it will be interesting to determine how the *C. elegans* proteins divide up into RNA and/or DNA binding protein families.

Structure of TZF Domains

Obviously, a structure of the TZF domain in complex with its RNA target sequence would represent an important advance in our knowledge in this area. It may also provide clues to the mechanism of action of TTP and its related proteins; clues to the process by which RNA binding apparently induces a conformational change in the binding protein; clues to the possible change in RNA conformation induced by the protein binding, perhaps making it a better substrate for nucleases; and finally, a structure might inform the development of inhibitors and activators, both of which might have therapeutic implications in certain settings. Unfortunately, TTP and its relatives behave badly when expressed in bacteria, generally falling out of solution during purification. However, two recent approaches have begun to yield informative data. We recently found that a synthetic 73 amino acid peptide containing the amino-terminal lead-in sequence, the complete TZF domain and a small carboxyl-terminal extension of TTP was soluble and capable of binding various ARE substrates containing one or more copies of the proposed canonical nonamer.[19] A biosynthetic, labeled version of this was also soluble and capable of binding ARE substrates, down to and including the same nonamer. An interesting aspect of this binding is that the amino terminal zinc finger appeared to change configuration upon RNA binding, as assessed by NMR. In addition, the second zinc finger, which was previously undetectable in the peptide in solution, became "visible" to NMR upon RNA binding. These studies suggested that RNA binding confers a significant conformational change upon the TZF domain. The data also indicated that the conformational change that occurred with RNA binding began to deteriorate as the RNA binding site oligonucleotide was shortened from the 9 b core ARE binding motif.

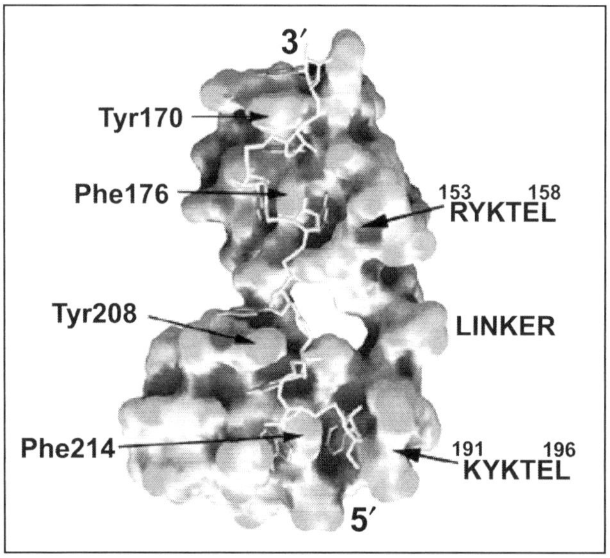

Figure 13. CCCH TZF domain structure. The figure represents a surface diagram of the TZF domain peptide from ZFP36L2 (TS11D) complexed with the RNA nonamer UUAUUUAUU. The RNA chain is shown in white, with rings; the 3' and 5' ends of the RNA oligo are indicated, as are hydrophobic pockets formed by the RYKTEL and KYKTEL lead-in sequences, and the positions of various aromatic side chains important for stacking interactions with the bases. See the text for further details. From ref. 30 by permission of the Nature Publishing Group.

Very recently, Hudson et al[30] determined the NMR structure of the TZF domain from human ZFP36L2 (TIS11D), in complex with the same nonamer binding site, UUAUUUAUU. This novel structure has many interesting features, some of which can be appreciated from a surface diagram taken from their paper (Fig. 13), including:

(1) There was a conformational change in the peptide structure upon RNA binding; (2) The RNA oligonucleotide did not appear to exhibit any significant secondary structure in the complex; (3) The two zinc fingers were very similar in structure, with the amino terminal zinc finger bound to the 3' UAUU half-site, and the carboxyl terminal zinc finger bound to the 5' half-site; (4) Base U1, the 5'-most base in the nonamer, was not structured in the complex. Other information suggested that the 3'-most U might also be replaced by another base with retention of the structure; (5) The peptide-RNA interactions appeared to be in the form of hydrogen bonding and electrostatic interactions; (6) The "lead-in" sequences (R/K)YKTEL were critical in that key aromatic side chains appeared to form stacking interactions with the side chains of the bases, in structural features that the authors referred to as "hydrophobic pockets" (Fig. 13).

Given the very high percentage of identical amino acids in the ZFP36L2 and TTP TZF domains, it seems likely that this structure will be very similar in all members of this family, at least in the mammalian proteins. We recently modeled the human TTP TZF domain based on the ZFP36L2 structure described by Hudson et al[30] and found that the TTP residues directly in contact with the RNA were identical to those in the ZFP36L2 peptide.[31] Thus, we can make the assumption that a nearly identical situation will exist for all the mammalian family members in complex with the RNA binding site. Further challenges will include additional structure function studies on many conserved and variable residues within the different TZF domains; the elucidation of the mechanisms by which the intact proteins promote deadenylation after this binding complex has formed; the identity of the bona fide physiological binding partners for the other family members; the numerous regulatory possibilities afforded by phosphorylation, nuclear-cytoplasmic shuttling, transcriptional regulation and protein binding events; and finally, the possibility of using this novel structural information to develop new types of therapies for diseases caused in part by elevated levels of circulating TNF.

References

1. Laity JH, Lee BM, Wright PE. Zinc finger proteins: New insights into structural and functional diversity. Curr Opin Struct Biol 2001; 11(1):39-46.
2. Gomperts M, Pascall JC, Brown KD. The nucleotide sequence of a cDNA encoding an EGF-inducible gene indicates the existence of a new family of mitogen-induced genes. Oncogene 1990; 5(7):1081-1083.
3. Varnum BC, Lim RW, Sukhatme VP et al. Nucleotide sequence of a cDNA encoding TIS11, a message induced in Swiss 3T3 cells by the tumor promoter tetradecanoyl phorbol acetate. Oncogene 1989; 4(1):119-120.
4. Lai WS, Stumpo DJ, Blackshear PJ. Rapid insulin-stimulated accumulation of an mRNA encoding a proline-rich protein. J Biol Chem 1990; 265(27):16556-16563.
5. Ma Q, Herschman HR. A corrected sequence for the predicted protein from the mitogen-inducible TIS11 primary response gene. Oncogene 1991; 6(7):1277-1278.
6. DuBois RN, McLane MW, Ryder K et al. A growth factor-inducible nuclear protein with a novel cysteine/histidine repetitive sequence. J Biol Chem 1990; 265(31):19185-19191.
7. Varnum BC, Ma QF, Chi TH et al. The TIS11 primary response gene is a member of a gene family that encodes proteins with a highly conserved sequence containing an unusual Cys-His repeat. Mol Cell Biol 1991; 11(3):1754-1758.
8. De J, Lai WS, Thorn JM et al. Identification of four CCCH zinc finger proteins in Xenopus, including a novel vertebrate protein with four zinc fingers and severely restricted expression. Gene 1999; 228(1-2):133-145.
9. Phillips RS, Ramos SB, Blackshear PJ. Members of the tristetraprolin family of tandem CCCH zinc finger proteins exhibit CRM1-dependent nucleocytoplasmic shuttling. J Biol Chem 2002; 277(13):11606-11613.
10. Blackshear PJ. The CCCH tandem zinc finger domain as a novel nucleic acid binding element. In: Iuchi S, KN, ed. Zinc Finger Proteins. 2003:in press.
11. te Kronnie G, Stroband H, Schipper H et al. Zebrafish CTH1, a C3H zinc finger protein, is expressed in ovarian oocytes and embryos. Dev Genes Evol 1999; 209(7):443-446.
12. Blackshear PJ. Xenopus laevis genomic biomarkers for environmental toxicology studies. In: WsaS WA, ed. Biomarkers of environmentally associated diseae. LLC: CRC Press, 2002:339-353.
13. Carballo E, Lai WS, Blackshear PJ. Feedback inhibition of macrophage tumor necrosis factor-alpha production by tristetraprolin. Science 1998; 281(5379):1001-1005.
14. Taylor GA, Carballo E, Lee DM et al. A pathogenetic role for TNF alpha in the syndrome of cachexia, arthritis, and autoimmunity resulting from tristetraprolin (TTP) deficiency. Immunity 1996; 4(5):445-454.
15. Carballo E, Gilkeson GS, Blackshear PJ. Bone marrow transplantation reproduces the tristetraprolin-deficiency syndrome in recombination activating gene-2 (-/-) mice. Evidence that monocyte/macrophage progenitors may be responsible for TNFalpha overproduction. J Clin Invest 1997; 100(5):986-995.
16. Lai WS, Carballo E, Strum JR et al. Evidence that tristetraprolin binds to AU-rich elements and promotes the deadenylation and destabilization of tumor necrosis factor alpha mRNA. Mol Cell Biol 1999; 19(6):4311-4323.
17. Lai WS, Carballo E, Thorn JM et al. Interactions of CCCH zinc finger proteins with mRNA. Binding of tristetraprolin-related zinc finger proteins to Au-rich elements and destabilization of mRNA. J Biol Chem 2000; 275(23):17827-17837.
18. Lai WS, Blackshear PJ. Interactions of CCCH zinc finger proteins with mRNA: Tristetraprolin-mediated AU-rich element-dependent mRNA degradation can occur in the absence of a poly(A) tail. J Biol Chem 2001; 276(25):23144-23154.
19. Blackshear PJ, Lai WS, Kennington EA et al. Characteristics of the interaction of a synthetic human tristetraprolin tandem zinc finger peptide with AU-rich element-containing RNA substrates. J Biol Chem 2003; 278(22):19947-19955.
20. Carballo E, Lai WS, Blackshear PJ. Evidence that tristetraprolin is a physiological regulator of granulocyte-macrophage colony-stimulating factor messenger RNA deadenylation and stability. Blood 2000; 95(6):1891-1899.
21. Carballo E, Blackshear PJ. Roles of tumor necrosis factor-alpha receptor subtypes in the pathogenesis of the tristetraprolin-deficiency syndrome. Blood 2001; 98(8):2389-2395.
22. Lai WS, Kennington EA, Blackshear PJ. Tristetraprolin and its family members can promote the cell-free deadenylation of AU-rich element-containing mRNAs by poly(A) ribonuclease. Mol Cell Biol 2003; 23(11):3798-3812.
23. Worthington MT, Pelo JW, Sachedina MA et al. RNA binding properties of the AU-rich element-binding recombinant Nup475/TIS11/tristetraprolin protein. J Biol Chem 2002; 277(50):48558-48564.
24. Garcia-Castillo J. Interactions between tristetraprolin and the TNFa mRNA of the Marine Fish Gilthead Seabream. Paper presented at: International Society for Developmental and Comparative Immunology (ISDCI) - 9th International Congress. Scotland: University of St Andrews, 2003.
25. Michel SL, Guerrerio AL, Berg JM. Selective RNA binding by a single CCCH zinc-binding domain from Nup475 (Tristetraprolin). Biochemistry 2003; 42(16):4626-4630.
26. Lai WS, Kennington EA, Blackshear PJ. Interactions of CCCH zinc finger proteins with mRNA: Nonbinding tristetraprolin mutants exert an inhibitory effect on degradation of AU-rich element-containing mRNAs. J Biol Chem 2002; 277(11):9606-9613.
27. Taylor GA, Thompson MJ, Lai WS et al. Mitogens stimulate the rapid nuclear to cytosolic translocation of tristetraprolin, a potential zinc-finger transcription factor. Mol Endocrinol 1996; 10(2):140-146.
28. Murata T, Yoshino Y, Morita N et al. Identification of nuclear import and export signals within the structure of the zinc finger protein TIS11. Biochem Biophys Res Commun 2002; 293(4):1242-1247.
29. Thompson MJ, Lai WS, Taylor GA et al. Cloning and characterization of two yeast genes encoding members of the CCCH class of zinc finger proteins: Zinc finger-mediated impairment of cell growth. Gene 1996; 174(2):225-233.
30. Hudson BP, Martinez-Yamout MA, Dyson HJ et al. Recognition of the mRNA AU-rich element by the zinc finger domain of TIS11d. Nat Struct Mol Biol 2004; 11(3):257-264.

CHAPTER 14

Ribosomal Zinc Finger Proteins: The Structure and the Function of Yeast YL37a

John Dresios, Yuen-Ling Chan and Ira G. Wool*

Abstract

Zinc finger motifs are common in ribosomal proteins: they are widely distributed in nature, having been found amongst the proteins of both subunits of the ribosomes of all species examined in the three kingdoms; the motif is always of the C_2C_2 variety and occurs only once in a protein. Despite wide distribution there is neither strict conservation of the ribosomal proteins with the motif nor of the entire motif in homologous proteins. A comprehensive genetic, biochemical, and structural analysis has been made of the contribution of the zinc finger to the function of yeast ribosomal protein YL37a, to date the only study of its kind. Replacement, one at a time, of the cysteines with serines in the motif in YL37a revealed that all four cysteines are required for the binding of zinc; nonetheless, cells with mutations in three of the four cysteines do not suffer a significant impairment of growth, nor is the binding to rRNA of the mutant proteins materially affected. It is possible that the zinc finger motif in ribosomal proteins are the vestiges, biological fossils if you will, of a former function, and that the motif has been preserved despite the ribosomal proteins having come to use alternate amino acid sequences and/or structures to bind to rRNA as has been shown to be the case for YL37a.

Introduction

Ribosomes are large ribonucleoprotein complexes that mediate protein synthesis in all organisms in the biosphere; they function to link the genotype to the phenotype. The amino acid sequences of the 79 mammalian[1,2] and of the 78 yeast[3] ribosomal proteins (rps) have been determined. The sequences are needed for a solution of the structure of eukaryotic ribosomes, and they should help to understand the evolution of rps, to unravel their function, and to define the rules that govern their interaction with ribosomal RNAs (rRNAs). An unexpected finding was that some rps had amino acid sequences that were first associated with binding to DNA.[1,2,4-6] These motifs include helix-turn-helix,[7] basic region leucine-zippers,[5,6] and zinc fingers.[1,2,5]

Zinc finger motifs are present in the rps of species from the three kingdoms.[4] Seven rat (R) rps have zinc finger motifs of the C_2C_2 variety; three are in the small (S) and four are in the large (L) ribosomal subunit; they are RS27, RS27a, RS29, RL36a, RL37, RL37a, and RL40 (ref. 2) (Fig. 1); moreover,

YS27	37	C	PG	C	LNITTVFSHAQTAVT	C	ES	C
YS27a	45	C	SNPT	C	GAGVFLANHKDRLY	C	GK	C
YS29	21	C	RV	C	SSHTGL(I/V)RKY(G/D)LNI	C	RQ	C
YL36a	22	C	KGKT	C	RKHTQHK...43 aa..VVLRLE	C	VK	C
YL37	19	C	NR	C	GRRSFHVQKKT	C	SS	C
YL37a	39	C	SF	C	GKKTVKRGAAGIWT	C	SC	C
YL40	20	C	RK	C	YARLPPRATN	C	RKRK	C

Figure 1. Zinc finger motifs in *S. cerevisiae* ribosomal proteins. The numbers in the second column are the position in the amino acid sequence of the first cysteine residue in the motif.

analyses by atomic absorption spectroscopy indicated that mammalian ribosomes have approximately six moles of zinc.[8] In rps the zinc finger motif occurs only once; this is in contrast to DNA binding proteins where there are often tandem repeats of the motif.[9]

There is a crystal structure of the large ribosomal subunit (50S) from archaebacterium *Haloarcula marismortui* (*Hm*) resolved to 2.4 Å;[10] the structure confirms the presence of zinc finger motifs in rps:[1,2] *Hm*L37e (a homolog of RL37); *Hm*L37ae (RL37a); *Hm* L24e (RL24); and *Hm*L44e (RL36a). The rat homolog of *Hm*L24e (RL24) lacks the internal two cysteines and, therefore, is unlikely to coordinate zinc. The rat homolog of *Hm*L44e (RL36a) has an unusually large insert (56 amino acids) between the two internal cysteines of the motif and was overlooked in the original analysis.[2] In sum, of the four zinc finger proteins in the large subunit, one is unique to rat, RL40, and one to *H. marismortui*, *Hm*L24e.

In the crystal structure of the 50S subunit of the ribosomes of the eubacterium *Deinococcus radiodurans* (*Dr*) there are two rps with zinc finger motifs, *Dr*L32 and *Dr*L36;[11] both lack eukaryotic homologs. In the structure of the 30S subunit from the eubacterium *Thermus thermophilus* (*Tt*),[12] a zinc atom is coordinated by four cysteine residues in rps *Tt*S4 (a homolog of RS9; RS9 lacks the zinc coordinating residues and thus the motif) and in *Tt*S14 (homolog of RS29).

Thus, in archaebacterial and eukaryotic ribosomes, the large subunit contains four rps with zinc finger motifs; in contrast, the eubacterial 50S subunit has only two. The similarity of the zinc finger rps of eukaryotic and of archaebacterial ribosomes accords

* Corresponding author. See list of "Contributors".

Zinc Finger Proteins: From Atomic Contact to Cellular Function, edited by Shiro Iuchi and Natalie Kuldell.

```
             1          11         21
YL37a        MAKRTKKVGI TGKYGVRYGS SLRRQVKKLE
HmL37ae               R TGRFGPRYGL KIRVRVADVE

             31         41         51
YL37a        IQQHARYDCS FCGKKTVKRG AAGIWTCSCC
HmL37ae      IKHKKKHKCP VCGFKKLKRA GTGIWMCGHC

             61         71         81
YL37a        KKTVAGGAYT VSTAAAATVR STIRRLREMV EA
HmL37ae      GYKIAGGCYQ PETVAGKAV- ---------M KA
```

Figure 2. The sequence of amino acids in *S. cerevisiae* ribosomal protein YL37a and in *H. marismortui HmL37ae*. The residues at the amino-terminus (positions 1-18) and in the zinc finger motif (positions 39-60) of YL37a that were mutated are designated in bold.

with the finding that, in general, the amino acid sequences of their rps are close, closer than either are to bacteria. Nonetheless, it is important to note that there is neither strict conservation of the rps with zinc finger motifs nor of the motif in homologous rps (see later as well).

Little is known of the role, if any, of the motif or of the metal in the structure and the function of the zinc finger rps. But, the importance of metals in general and of zinc in particular for many cellular processes inclines one to take seriously the possibility that the rps with zinc fingers might be critical for the assembly or for the function of the ribosome, perhaps because the motif has a role in interactions with rRNA.

The YL37a Zinc Finger Motif

The sequence of nucleotides (nts) in the two genes encoding yeast (Y) rp YL37a (Fig. 2) had been determined,[13,14] and the amino acid sequence derived from the nt sequence had established that YL37a has a zinc finger motif of the C_2C_2 type. What had not been determined at the time was whether YL37a binds zinc and, if it does, whether the structure formed by the coordination of the metal is necessary for the contribution of the protein to the function of the ribosome.

YL37a is essential for cell viability.[14] The lethal phenotype of a strain in which both copies of the *YL37a* gene were disrupted could be complemented with a copy of either the *YL37a-A* or the *YL37a-B* gene.[15] Complementation of the null phenotype of the yeast strain *yl37a* with the open reading frame encoding the homologous rat RL37a indicated that the portions of the amino acid sequences of the yeast and of the rat proteins most critical for function are conserved.[14]

To address the question of the importance of the zinc finger motif for YL37a function, both chromosomal copies of the *YL37a* gene were disrupted. The efficiency of the complementation of the lethal *yl37a* null phenotype with plasmid-encoded copies of a YL37a gene in which the cysteine residues of the zinc finger motif were replaced, one at a time, with the codons for serine was assessed. The effect of the mutations on the binding of zinc to YL37a was also determined.[14] Later the role in the function of YL37a of the amino acids in the intervening region of the zinc finger motif and of the residues at the amino-terminus of the protein was evaluated.[15] In addition, the YL37a binding sites in 26S rRNA were identified and the role of the zinc finger motif and of residues in the amino-terminal region on the interaction of the protein with the nucleic acid was investigated.[15]

Table 1. *The effect of mutations in the zinc finger motif of YL37a on yeast growth and on zinc binding*

Mutation	Doubling Time in YPD[a] (min)	Zn Bound to Protein (μmol/μmol)
MBP	–	0.15
MBP-YL37a-wild type	102	1.00
MBP-YL37a-C39S	lethal	0.27
MBP-YL37a-C42S	109	0.16
MBP-YL37a-C57S	106	0.16
MBP-YL37a-C59S	102	0.93
MBP-YL37a-C60S	141	0.20

[a] YPD liquid media: 1% yeast extract, 2% bactopeptone, 2% glucose.

The Phenotype of Cells with Mutations in the Cysteine Residues of the YL37a Zinc Finger Motif

There is a great deal of data that suggests that the tetrahedral coordination of zinc initiates the folding of the finger motif.[16-18] In general, mutations of the amino acids that coordinate zinc abolish the binding of the metal and *parri passu* lead to the loss of function of the protein. Hence, the initial prediction was that each cysteine residue in the YL37a zinc finger motif was essential for tetrahedral coordination of the metal, for the folding of the module, and for the function of the protein. These assumptions were tested.[14] Single replacements of individual cysteine residues with serine were constructed in a *YL37a-B* gene in a high copy plasmid. The mutations were: C39S, C42S, C57S, C59S, and C60S (Fig. 2).

A plasmid bearing a mutation of the cysteine at position 39 (C39S) in a *YL37a-B* gene was unable to rescue the *yl37a* null strain (Table 1), i.e., cysteine 39 is essential for the function of YL37a. However, in contradistinction to expectation, plasmids with mutations C42S, C57S, C59S, or C60S in *YL37a-B* were able to complement the null strain. There are at least two explanations of these results. First, that the coordination of zinc is not be required for the function of YL37a nor of ribosomes and, hence, for the growth of yeast cells. This explanation assumes that all four cysteine residues are required for the coordination of zinc; i.e., that mutation of any one of the cysteines in the motif abolishes metal binding. A second explanation is that YL37a with three cysteines and a serine at the fourth position is able to coordinate zinc. Serine differs from cysteine only in that the γ-oxygen of the former is replaced by a sulfur atom in the latter. Serine does participate in the coordination of zinc in several enzymes.[19] However, canonical zinc fingers are not known to use amino acids other than cysteine or histidine to coordinate the metal. To investigate the two possibilities an in vitro assay for the binding of zinc to YL37a was developed.

Zinc Binding to Purified YL37a-MBP

YL37a binds zinc; the binding is stoichiometric, specific, and most importantly requires all four of the cysteine residues in the motif.[14] YL37a was expressed in *E. coli* fused to maltose binding protein (MBP). The purified recombinant YL37a-MBP was resolved by electrophoresis and transferred to nitrocellulose paper. The fusion protein bound radioactively labeled zinc; whereas MBP alone did not. Each mole of YL37a-MBP bound 0.8 mole of zinc. The radioactive zinc bound to YL37a could be replaced by an excess of nonradioactive zinc, and by excesses of cobalt, cadmium, or nickel, cations that are known to bind to zinc finger motifs. In contrast, zinc could not be replaced by even greater excesses of calcium or of magnesium, cations that do not bind to the motif. In addition, the cysteine-specific reagent p-hydroxymercuriphenylsulfonate released radioactive zinc from YL37a, a further indication that the binding of zinc to the motif depends on the cysteine residues in the protein.[14]

The effect of mutations of the cysteine residues of the zinc finger motif in YL37a on the binding of the metal was determined quantitatively (Table 1).[14] The replacement of the nonconserved cysteine residue at position 59 with serine did not affect binding. Replacement, one at a time, with serine of the conserved cysteine residues in the motif (at positions 39, 42, 57, and 60) reduced the binding of zinc to mutant MBP-YL37a to levels that did not appreciably exceed background. The conclusion was clear: all four conserved cysteine residues are essential for the binding of zinc in vitro, but only one, C39, is essential for YL37a function in vivo. Thus, YL37a can bind zinc, but the function of the rp, whatever that might be, does not depend on the coordination of the metal.

Sequences in YL37a Zinc Finger Motif between the Chelating Cysteines 2 and 3

Several observations had suggested that the amino acids in the intervening region between the internal cysteines of the zinc finger motifs might be important for the function of rps.[4-6] One observation, especially noteworthy here, concerns members of the eubacterial rp S14 family which are related to the rat zinc finger rp RS29. Homologous S14 proteins from several bacterial species have the entire element; whereas *Escherichia coli* (*Ec*) S14 has a degenerate form of the motif. In an alignment with the others there is a cysteine in *Ec*S14 at what would be the initial position; the other three are absent. The hydrophilic and hydrophobic character of the linker sequence, however, is preserved in *Ec*S14 and 8 of the 12 residues share identity with amino acids at the same positions in *Bacilus subtilis (Bs)* S14. The spin given to this finding is that *Ec*S14 once had a full C_2C_2 motif and that parts were lost during divergent evolution. There is, of course, no direct evidence for this bias.

It is likely that the same amino acid sequences and/or structures in the *B. subtilis* and *Ec*S14 proteins are used to bind to 16S rRNA since the binding site is conserved.[20] If this assumption is correct then the conclusion is that the proteins do not use the canonical secondary and tertiary structures that are formed by association with zinc in binding to rRNA, since *Ec*S14 presumably lacks the capacity to bind the metal and, hence, to form the structure. One reconciliation is that the binding to rRNA employs the side chains of the conserved basic, hydrophobic, and aromatic amino acids in the linker region rather than in the finger structure per se. Basic and aromatic amino acids are often involved in RNA recognition.[21,22]

Table 2. The doubling times of* S. cerevisiae *strains with mutations in the zinc finger motif and in the amino-terminal region in a plasmid-borne copy of the* YL37a-B *gene

Mutation	Doubling Time in YPD (min)
Zinc Finger Motif	
YL37a-Wild-type	102
YL37a-K44A	124
YL37a-K45A	111
YL37a-K48A	122
YL37a-R49A	131
YL37a-K44A-K45A	136
YL37a-K48A-R49A	610
YL37a-W55A	Lethal
N-terminal region	
YL37a-K6A-K7A	141
YL37a-K13A	173
YL37a-Y14A	Lethal
YL37a-R17A	201
YL37a-Y18A	155

It seemed possible then that one or more of the four conserved basic residues (K44, K45, K48, and R49) and/or the aromatic residue (W55) in the linker sequence is critical for the binding of YL37a to rRNA and for its function. To test the possibility, nts encoding these residues in a plasmid encoded *YL37a-B* gene were replaced with those for alanine and the rescue of the null strain was assessed (Table 2). Substitution of alanines for lysines 44, 45, or 48 or for arginine 49 in the intervening region consistently increased the doubling times of cells, although the increase was relatively small. The possibility that a region of charge density in YL37a is needed for interaction with the RNA was tested by constructing mutations in pairs of basic amino acids. The doubling time of cells with mutations of lysines 44 and 45 (YL37a-K44A-K45A) was increased approximately 30%. A greater effect was seen in a second double mutant, YL37a-K48A-R49A, where the viability of the cells was retained but their doubling time was increased six-fold (Table 2). It should be noted that of the several mutations in basic residues, the largest negative effect on growth appeared when alanine was substituted for arginine 49: the R49A mutation decreased growth more than the K44A, K45A, or K48A single mutations; and the K48A-R49A double mutation decreased growth far more than a K44A-K45A variant (Table 2). Thus, although all three lysines in the intervening region of the zinc finger motif contribute to YL37a function, the most critical basic residue is arginine 49. This may reflect in part the great propensity of arginine for interaction with nucleic acids. In protein-RNA interactions, the predominant proton donor is arginine; moreover, arginine is more often than lysine involved in the recognition of the 2'-hydroxyl of ribose and of phosphate oxygens in the backbone.[22]

Finally, it is noteworthy that the mutation of the aromatic amino acid tryptophan at position 55 to alanine was lethal (Table 2).

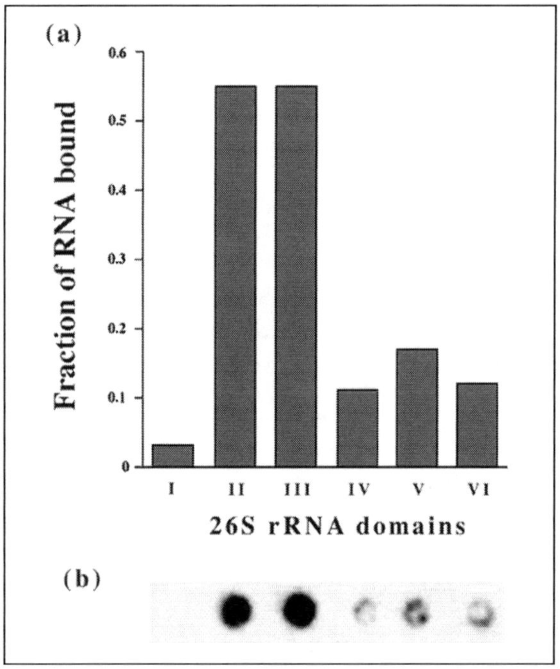

Figure 3. The binding of the fusion protein MBP-YL37a to the individual domains of 26S rRNA. The binding of 1 μM of MBP-YL37a to 1 nM of each of the six radioactive 26S rRNA domains was determined by spotting the reaction mixture (10 μl) in a single drop on a nitrocellulose filter. In (a), a histogram of the fraction of the input of the separate RNA domains in complexes with MBP-YL37a. In (b), autoradiographs of the complexes of MBP-YL37a and the radioactive 26S rRNA domain fragments captured on nitrocellulose filters.

YL37a Binding Sites in 26S rRNA

It is all but certain that it is the rRNAs that are most critical for the chemistry of protein synthesis: mRNA, aminoacyl- and peptidyl-tRNA, and the accessory initiation, elongation, and termination factors bind to sites in rRNA; it is likely that nts in rRNA catalyze peptide bond formation; and changes in the conformation of the rRNAs may underlie translocation. However, the importance of rps is not to be gainsaid since functional ribosomes cannot be formed in their absence. Rps serve as chaperones facilitating the folding of the rRNAs during biogenesis and stabilize an optimal conformation in the mature particle.[10,20] Mutations in rps that produce a phenotype often are in rRNA binding sites.[23] Thus, the chemistry of RNA-protein recognition is critical to a determination of the structure of the particle and to the unraveling of the molecular details of its function.

It was assumed that the function of YL37a, and hence the phenotype of mutations in the protein, must derive from its interaction with RNA. For that reason it seemed important to first identify the rRNA binding sites for YL37a and then to evaluate the effects of mutations in YL37a on binding to the sites. Towards this end, a set of six rDNA fragments that encode the domains in 26S rRNA was generated by PCR amplification from a plasmid that had a *Saccharomyces cerevisiae* 35S rDNA transcription unit.[24] The rDNA domain fragments were subcloned in appropriate plasmid vectors for the preparation of radioactive transcripts using T7 RNA polymerase.[15] An in vitro assay to quantitate binding was developed and the affinity of MBP-YL37a for each of the six 26S rRNA domains was determined (Fig. 3). Domains II and III have the primary sites to which YL37a associates.[15]

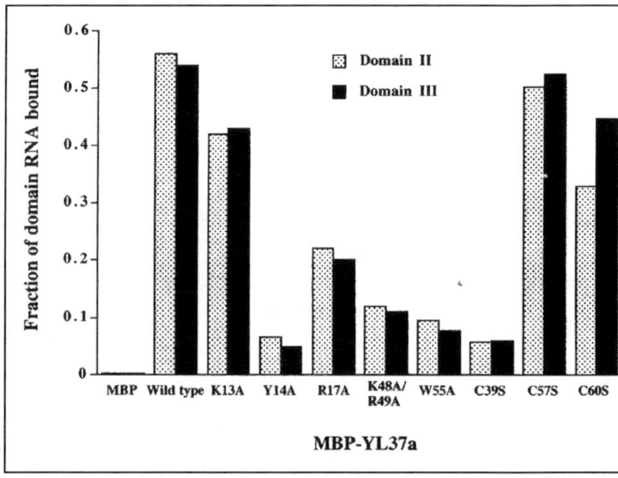

Figure 4. The effects of mutations in YL37a on the binding of the ribosomal protein to domain II or III of 26S rRNA. Histograms of the fraction of radioactive domain II or domain III of 26S rRNA in complexes with MBP-YL37a having mutations in the zinc finger motif or in the amino-terminal region of the ribosomal protein.

YL37a Zinc Finger Binding to 26S rRNA Domains II and III

To delineate the contribution of the zinc finger to the interaction of YL37a with domains II and III of 26S rRNA, the effect of mutations in the cysteines and in the intervening sequence of the motif was assessed. In short, there is a close correspondence between the effect of the mutations on the growth of cells and on the binding to 26S rRNA. A substitution of serine for the first cysteine in the motif (MBP-YL37a-C39S), a mutation that is lethal for yeast cells,[14] all but abolished binding to domains II and III (Fig. 4). More importantly, for the YL37a-C57S and YL37a-C60S mutations, there is a clear dissociation between their inability to bind zinc on the one hand, and the lack of an affect on cell growth or on binding to rRNA on the other. The C57S and C60S mutations had abolished the binding of zinc to YL37a, but the former did not affect the growth of yeast cells and the latter caused only a small increase in the doubling time (Table 1). The same mutations either did not affect binding of the protein to domain II or III or caused only a small decrease (Fig. 4). There was negligable binding of the double mutant MBP-YL37a-K48A-R49A to either domain II or III (Fig. 4). Finally, there was no binding to 26S rRNA of MBP-YL37a-W55A, a mutation that was lethal for yeast cells (Table 2 and Fig. 4).

Interaction of 26S rRNA with YL37a Amino-Terminal Region

Once again, it was possible that the side chains of conserved amino acids in YL37a apart from the zinc finger motif, contributed to the binding to rRNA. This interpretation is supported by the observation that K16 near (but not in) the zinc finger motif in *Bacillus stearothermophilus* S14 can be cross-linked to 16S rRNA.[25] Regions rich in basic amino acids are a common feature of RNA-binding proteins. In the amino-terminal region of YL37a six of the fifteen residues between positions 3 and 17 are basic and there are, as well, two tyrosines (Fig. 2). In all eukaryotic organisms for which an amino acid sequence of YL37a is available, six of these eight residues (K6, K7, K13, Y14, R17, and Y18) are either invariant or the changes are conservative.

Mutations to alanine, one at a time, of these six residues, in each instance, decreased cell growth; a Y14 mutation was lethal (Table 2). Moreover, as in the case of mutations in the zinc finger motif, the effect of mutations in the amino-terminal region of YL37a on binding to 26S rRNA mirrors their effects on cell growth and viability (Fig. 4).

Structural Data

The genetic and the biochemical experiments do not unambiguously define the nature of the participation of individual amino acids in YL37a in the function of the protein, since the results do not distinguish between a direct contribution of a residue to the free energy of binding to rRNA and an indirect contribution because the residue is critical for the maintenance of the three-dimensional structure. A decision between these alternatives requires correlation of the results of the genetic and biochemical experiments with the three-dimensional structure of the protein-RNA complex. On the other hand, hydrogen bonds, nonpolar contacts, and electrostatic interactions between amino acids and nts can only be inferred from the X-ray structure;[21] biochemistry and genetics are needed to define the subset of proximities that have functional significance.

An atomic structure of the 50S subunit of *H. marismortui* ribosomes that has the archaebacterial homolog of YL37a, *Hm*L37ae, has been determined by X-ray diffraction of crystals at 2.4 Å resolution.[10] The C_2C_2 motif in *Hm*L37ae (Fig. 5) coordinates a zinc atom and the finger is folded into a form similar to a zinc ribbon motif. There is conservation of the secondary structures of *H. marismortui* 23S and of *S. cerevisiae* 26S rRNAs and of the amino acid sequences of *Hm*L37ae and YL37a (Fig. 2). The proteins have 41% amino acid identity (51% similarity) albeit *Hm*L37ae lacks 10 of the amino acids that are near the carboxyl-terminus of YL37a. Thus, the structure of *Hm*L37ae can be used, with some confidence, to interpret the biochemical and genetic results obtained with YL37a.

*Hm*L37ae (Fig. 5) has a globular body (residues 35 to 73) with two α-helical extensions. The globular body is formed by the folded zinc finger; the motif is comprised of four anti-parallel β-strands and five loops. One α-helical extension is near the amino-terminus (residues 21 to 34) and the other is at the carboxyl-terminus (residues 74 to 81). There is also a 3_{10}-helix (residues 11 to 13) closer to the amino-terminus. One surface of the carboxyl-terminal helix is packed against *Hm*L2 (YL8 in yeast ribosomes). Basic and aromatic amino acids form an extended surface on the side of *Hm*L37ae that is in contact with rRNA. These residues include K46, K48, and R49 in the intervening region of the zinc finger; R13, R17, and Y18 at the amino-terminus; and K34 and K36 as well.

That YL37a binds to domains II and III of 26S rRNA[15] accords with the crystal structure of the *H. marismortui* 50S subunit where *Hm*L37ae is in proximity to domains II, III, IV of 23S rRNA.[10] None of the cysteines in the zinc finger motif of *Hm*L37ae are within hydrogen-bonding distance (2.4 to 3.4 Å) of a nt in 23S rRNA. For the mutations of three of the cysteines in YL37a (C42, C57, and C60) this is compatible with a lack of a significant growth phenotype or an effect on binding to rRNA. The lethal phenotype of the C39S mutation can only be ascribed to a perturbation of the structure of YL37a, since the mutation abolishes binding to rRNA even though it is unlikely that the amino acid makes a contact to any nt. Indeed, the homologous cysteine in *Hm*L37ae is not within hydrogen-bonding distance of a nt (the closest nt is more than 10 Å away). In support of this interpretation, it should be noted that there is not a single instance, in a database of the structures of 45 protein-RNA complexes, of a cysteine being implicated in the recognition of a nt.[22]

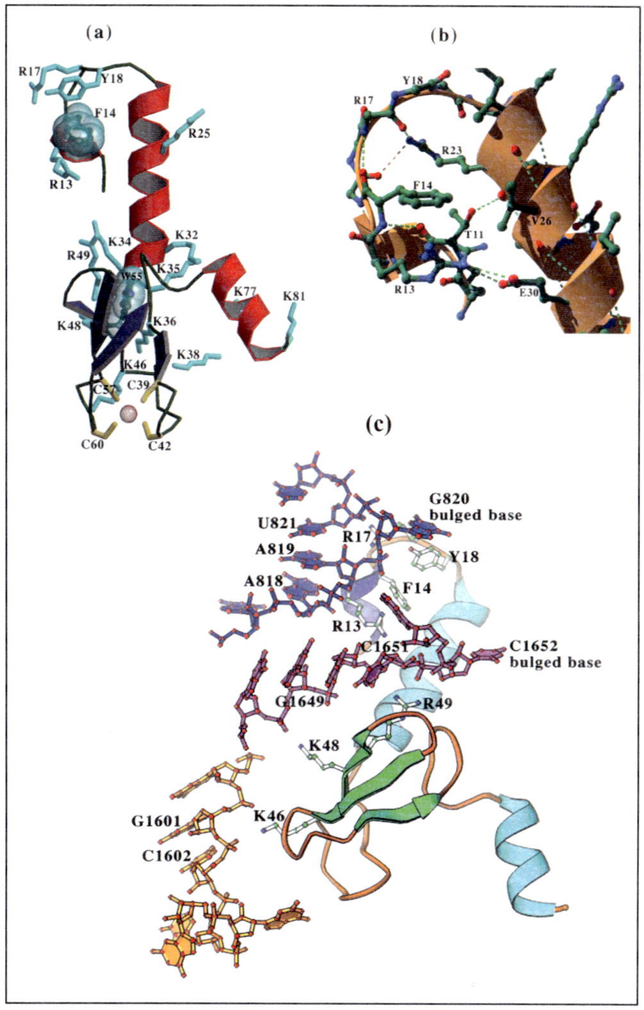

Figure 5. The structure of the ribosomal protein *Hm*L37ae taken from the crystal structure of the 50S ribosomal subunit.[10] In (a), a ribbon diagram of *Hm*L37ae. α-Helices are red; β-strands are dark blue; loops are green; and side chains of amino acids are light blue. Two aromatic residues, F14 and W55, are represented by van der Waals spheres. The four cysteines (yellow) that coordinate a zinc atom (grey-red ball) in the finger motif are at the bottom. In (b), the interactions of amino acids at the amino-terminus of *Hm*L37ae. In (c), interactions between amino acids in *Hm*L37ae and nucleotides in 23S rRNA. Domain II nucleotides (817-822) are lavender; domain III nucleotides are either orange (1600-1605) or magenta (1648-1653).

The results suggest that the two basic residues, K44 and K45 in YL37a, combine to create a region of positive charge, since the double mutation, YL37a-K44A-K45A, slowed the growth of yeast cells and the binding to rRNA domains more than either mutation alone. Of the residues in the intervening region of the zinc finger that were mutated, K45 (K44 in YL37a) is not within hydrogen-bonding distance of a nt; whereas the side chain of K46 (K45 in YL37a) makes contacts to the O1 phosphate of a guanosine and the O2 phosphate of an adjacent cytosine in domain III (Table 3).

Table 3. Atomic contacts between amino acids in **HmL37ae** *and nucleotides in 23S rRNA of* **H. marismortui**

Amino Acids HmL37ae	YL37a[a]	23S rRNA Nucleotide	Atomic Contact Amino Acid[b]	Nucleotide[b]	Distance (Å)
Zinc finger motif					
K46	K45	G1601	Nζ	O1P	3.2
		C1602	Nζ	O2P	3.2
K48	K48	G1649	Nζ	O1P	3.1
R49	R49	C1651	$NH\eta^1$	O2P	3.0
		C1651	$NH\eta^1$	O5'	3.4
		C1652	$NH\eta^1$	O2P	3.0
		C1652	$NH\eta^2$	O2P	2.7
		C1652	**Cζ**	**O2P**	**3.3**
N-terminal region					
R13	K13	A818	**Cδ**	**O2'**	**2.9**
R17	R17	A819	$NH\eta^2$	O3'	3.2
		G820	$NH\eta^2$	O1P	2.6
		G820	Nε	O1P	3.3
		G820	**Cζ**	**O1P**	**3.3**
		U821	**$NH\eta^1$**	**C4'**	**3.2**
Y18[c]	Y18	G820	OHη	O1P	2.7

The contacts are from the 2.4 Å resolution crystal structure of the *H. marismortui* 50S ribosomal subunit.[10] [a] The homologous amino acids in YL37a (Fig. 2). The amino acids that were mutated (Y14, C39, C42, C57, C60, K44, and W55), but whose homologs in *Hm*L37ae are not within 3.4 Å of a 23 S rRNA nucleotide are not listed; K6 and K7 were mutated but were not resolved in the diffraction data. [b] The side chain atoms are designated by Greek letters; van der Waals contacts are in bold. [c] Y18 also makes backbone contacts to the base of A1829 and to the sugar of A1885 in domain IV.

The YL37a-K48A-R49A double mutation increased the doubling time of yeast cells six-fold and markedly affected its affinity for rRNA. Once again, the phenotype of the double mutation was far more severe than the sum of the separate K48A or R49A mutations. The ionic nature of the interaction is supported by the observation that arginine or lysine to alanine substitutions at positions K48 and R49 respectively reduced binding to rRNA. Indeed, the side chain of K48 in *Hm*L37ae makes a contact to an O1 phosphate of a guanosine in domain III of 23S rRNA; whereas, the side chain amino groups of R49 make multiple contacts to O2 phosphates and to a sugar oxygen of adjacent cytosines also in domain III, as well as a van der Waals contact to an O2 phosphate of the second cytosine (Table 3). Despite the multiple contacts, the functional importance of R49 is in some way amplified by K48. Of course, the relationship between the two residues can be put the other way around. In any case, the two basic amino acids either form an essential region for electrostatic interaction or the fine structure of the region is dependent on the two. It may be that we have here an illustration of a theme for the chemistry of RNA-protein recognition: the participation of clusters of two or more basic amino acids in the recognition of RNA binding sites. Clusters of basic residues are common in proteins that bind to RNA, especially in rps.[1] The results suggest that it is the overall charge density of the intervening sequence that is important for binding to 26S rRNA rather than specific residues.

A W55A mutation in YL37a was lethal. In *Hm*L37ae, W55, the only solvent inaccessible residue in the protein, occupies a cavity in the hydrophobic core of the globular body formed by the β-strands of the zinc finger motif (Fig. 5); W55, however, is not close to an rRNA nt (the closest is more than 10 Å away). Because alanine is smaller than tryptophan it is unlikely that the former would fill the cavity ordinarily occupied by the latter and the mutation might well cause the collapse of the fold. Thus, the contribution of W55 to the function of YL37a is likely to derive from a role in the maintenance of the structure of the protein rather than from participation in binding to rRNA.

Among the several basic residues at the amino-terminus of YL37a that were surveyed, a K13A mutation slowed growth. The side chain of R13 (the homolog in *Hm*L37ae of K13 in YL37a) makes a van der Waals contact to the 2' oxygen of the sugar of an adenine in domain II (Table 3). The side chain of a second basic residue, R17, makes contacts to three consecutive nts in domain II: van der Waals contacts to the O1 phosphate of a guanosine and to a sugar carbon of a uridine; two electrostatic contacts to the O1 phosphate of the guanosine; and finally a contact to the 3' oxygen of a sugar in an adenosine (Table 3). The R17A mutation in YL37a caused a marked decrease in the growth of yeast cells, presumably because these contacts were compromised; indeed, the binding of YL37a-R17A to 26S rRNA was all but abolished. The mutation YL37a-Y18 retards the growth of yeast cells, and this could be explained by the observation that the hydroxyl of the aromatic ring of Y18 in *Hm*L37ae makes a contact to an O1 phosphate of a guanosine.

The closest nt to F14 in *Hm*L37a (Y14 in YL37a) is more than 10 Å away; hence, the presumption has to be that the Y14A mutation was lethal because the residue is essential to the maintenance of the stability of the amino-terminal region, thereby, preserving the orientation of the contacts of nearby amino acids, particularly R17 and Y18, to rRNA. The amino-terminal region of *Hm*YL37ae forms a 3_{10}-helix-turn-helix (residues 10-34)

(Fig. 5). R13 forms two hydrogen bonds with E30 converting a hook into a closed loop. The side chains of R17 and Y18 are on the outer surface of the loop where they make contacts to nts (Table 3). The interior of the loop has a network of interactions between protein backbone carbonyls and amides and side chains of conserved amino acids in that way forming a stable fold having a cavity filled by the bulky side chain of F14; 84 % of F14 is buried. Presumably, alanine in the YL37a-Y14A mutant would not fill the cavity and, hence, would not maintain the integrity of the structure; this in turn might prevent the proper presentation of the side chains of R17 and Y18 for binding to rRNA, a failure that could rationalize the lethal phenotype of the Y14A mutation.

The contacts presumed to have been altered by mutations in YL37a, and to have affected binding to rRNA, are predominantly between amino acid side chains and nt oxygen phosphates or sugars. That the main amino acid contacts are not to nt bases suggests that recognition by YL37a is of the shape of its rRNA binding site rather than a readout of sequence. Inspection of the three-dimensional structure of the relevant regions of 23S rRNA supports this interpretation. The conformation of the rRNA binding pocket for *Hm*L37ae is complex with surfaces that have many deviations from A-form. Guanosine 820, which has three separate contacts to R17 and one to Y18, is bulged out of a helix distorting the A-form geometry (Fig. 5). In a like manner, cytosine 1652, which has three contacts to R49, is also bulged out of a helix, once again distorting the A-form geometry (Fig. 5). Nts that distort A-form geometry, usually because they form noncanonical pairs, can widen the otherwise inaccessible major groove, or they can reveal distinctly shaped features in the minor groove that proteins can recognize.[21,26] It seems most likely that R17 and Y18 in the amino-terminal region of YL37a, and R49 in the zinc finger motif, is recognizing the unique shape of the rRNA in its binding site, and that G820 and C1652 are the identity elements in rRNA for the binding of the rp.

The contacts of the zinc finger motif residues K45, K48, and R49 are to nts in domain III; whereas, the contacts of residues at the amino-terminus, K13, R17, and Y18, are predominantly to domain II. Yet, mutations in these six residues affect binding to both domains more or less equally (Fig. 4). Binding to the two domains is specific since a forty-fold excess of nonradioactive domain II rRNA did not affect binding of MBP-YL37a to radioactive domain III rRNA; and nonradioactive domain III rRNA did not inhibit binding to domain II.[15] Still the original observation is a paradox that implies cooperativity in the binding of YL37a to the two rRNA domains. The possibility has been tested and there is cooperativity but it is not reciprocal. An excess of nonradioactive domain III RNA increased the binding of MBP-YL37a to radioactive domain II RNA 1.7 fold. In the reciprocal experiment domain II RNA had no effect on binding of MBP-YL37a to radioactive domain III.[15] It appears that binding of residues in the zinc finger motif to domain III alters the conformation of the amino-terminus of YL37a so as to favor association with domain II.

If we assume that the rps only condition the folding and the end state conformation of rRNA, then the corollary that derives from this study is that the sites in domains II and III to which YL37a binds must contribute directly or indirectly to one or another of the partial reactions of protein synthesis.

Biological Significance of YL37a Zinc Finger Motif

It is a tenet of the faith of ribosomologists that the "proto-ribosome" had only RNA. This is a belief that is supported by the mounting evidence that the rRNAs are responsible for the basic biochemistry of protein synthesis. The rps are now viewed as a later evolutionary embellishment and are deemed to facilitate the folding of the rRNA and the maintenance of an optimal configuration, in this way endowing protein synthesis with speed and accuracy. This may be too severe a restriction on the role of the proteins in ribosome function, nonetheless, it is likely that RNA preceded the proteins and hence it is pertinent to ask from whence the latter came.

The occasion for the transition during evolution from a ribosome that had only RNA to a RNP machine may have coincided with, or been a response to, the appearance of nucleases, which would have put an RNA ribosome at risk. There are at least two possible story lines for the origin of ribosomal proteins. That they were designed specifically for the ribosome, or that they were coopted from amongst a set of preexistent proteins that already had defined functions. The two possibilities are by no means exclusive nor is it likely that all the proteins were added at one time. If the latter conjecture has substance, the preexistent proteins most likely to have been recruited would have been those that already had the capacity to bind to nucleic acids. It is possible then that zinc finger motifs in rps are the vestiges of a former function, perhaps, of binding to DNA or to other RNAs, and that the motif has been preserved despite the rps having come to use alternate structures or alternate amino acids for associating with rRNA.[4-6] This interpretation fits the phenotype of the cysteine mutations in the zinc finger motif in YL37a: their replacement with serines yields variants that lack the capacity to bind zinc but, nonetheless, support growth and bind to rRNA; instead basic and aromatic residues in the intervening sequence and at the amino-terminus of the protein are important in binding to rRNA. It has not yet been determined, however, whether the dispensability of zinc binding is a common feature of rps with finger motifs, or whether it is unique to YL37a.

Mutations to serine of three of the four cysteines in the zinc finger motif of YL37a (C42S, C57S, and C60S) abolish tetrahedral coordination of zinc; nonetheless, cells with the mutations do not suffer a significant impairment of growth; and binding to rRNA of the mutant proteins is not affected. The results force one to conclude, either that the mutations do not materially affect the structure of the motif despite not binding zinc, or that the motif fold is not essential for the function of YL37a. Whether the motif can fold into its characteristic three-dimensional conformation in the absence of the coordination of the metal is not known. It is possible that zinc only stabilizes the folded state and that binding to rRNA is an alternate way to maintain the conformation of the protein, a conformation determined in the first instance by the amino acid sequence. Based on the results of the experiments with YL37a and of the natural experiments during the evolution of, for example, the S14 family of rps (see earlier), we favor the secondary possibility: that the fold of the zinc finger motif in YL37a and perhaps in other rps is not essential for function.

Acknowledgements

The work was supported by a grant from the National Institutes of Health (GM 33702).

References

1. Wool IG, Chan YL, Glück A. Structure and evolution of mammalian ribosomal proteins. Biochem Cell Biol 1995; 73:933-947.
2. Wool IG, Chan YL, Glück A. Mammalian ribosomes: The structure and the evolution of the proteins. In: Hershey JWB, Mathews MF, Sonnenberg N, eds. Translational Control. Cold Spring Harbor, New York: Cold Spring Harbor Press, 1996:685-732.
3. Mager WH, Planta RJ, Ballesta JPG et al. A new nomenclature for the cytoplasmic ribosomal proteins of Saccharomyces cerevisiae. Nucleic Acids Res 1997; 25:4872-4875.
4. Wool IG. The bifunctional nature of ribosomal proteins and speculations on their origins. In: Nierhaus KH, Franceschi F, Subramanian AR, Erdmann VA, Wittmann-Liebold B, eds. The Translational Apparatus. New York: Plenum Press, 1993:727-737.
5. Wool IG. Extraribosomal functions of ribosomal proteins. In: Green R, Schroeder R, eds. Ribosomal RNA and Group I Introns. Austin, TX: RG Landes, 1996:153-178.
6. Wool IG. Extraribosomal functions of ribosomal proteins. Trends Biochem Sci 1996; 21:164-165.
7. Rice PA, Steitz TA. Ribosomal protein L7/L12 has a helix-turn-helix motif similar to that found in DNA-binding regulatory proteins. Nucleic Acids Res 1989; 17:3757-3762.
8. Chan Y-L, Suzuki K, Olvera J, Wool IG. Zinc finger-like motifs in rat ribosomal proteins S27 and S29. Nucleic Acids Res 1993; 21:649-655.
9. Schwabe JWR, Rhodes D. Beyond zinc fingers: Steroid hormone receptors have a novel structural motif for DNA recognition. Trends Biochem Sci 1991; 16:291-296.
10. Ban N, Nissen P, Hansen J et al. The complete atomic structure of the large ribosomal subunit at 2.4 Å resolution. Science 2000; 289:905-920.
11. Harms J, Schluenzen F, Zarivach R et al. High resolution structure of the large ribosomal subunit from a mesophilic eubacterium. Cell 2001; 107:679-88.
12. Wimberly BT, Brodersen DE, Clemons WM et al. Structure of the 30S ribosomal subunit. Nature. 2000; 407:327-348.
13. Cherry JM, Adler C, Ball C et al. SGD: Saccharomyces Genome Database. Nucleic Acids Res 1998; 26:73-79.
14. Rivlin AA, Chan YL, Wool IG. The contribution of a zinc finger motif to the function of yeast ribosomal protein YL37a. J Mol Biol 1999; 294:909-919.
15. Dresios J, Chan YL, Wool IG. The role of the zinc finger motif and of the residues at the amino-terminus in the function of yeast ribosomal protein YL37a. J Mol Biol 2002; 316:475-488.
16. Boysen RI, Hearn MT. The metal binding properties of the CCCH motif of the 50S ribosomal protein L36 from Thermus thermophilus. J Pept Res 2001; 57:19-28.
17. Fowle DA, Stillman MJ. Comparison of the structures of the metal-thiolate binding site in Zn(II)-, Cd(II)-, and Hg(II)-metallothioneins using molecular modeling techniques. J Biomol Struct Dyn 1997; 14:393-406.
18. Parraga G, Horvath SJ, Eisen A et al. Zinc-dependent structure of a single-finger domain of yeast ADR1. Science 1988; 241:1489-1492.
19. Vallee BL, Auld DS. Cocatalytic zinc motifs in enzyme catalysis. Proc Natl Acad Sci USA 1993; 90:2715-2718.
20. Stern S, Powers T, Changchien LM et al. RNA-protein interactions in 30S ribosomal subunits: Folding and function of 16S rRNA. Science 1989; 244:783-790.
21. Draper DE. Themes in RNA-protein recognition. J Mol Biol 1999; 293:255-270.
22. Allers J, Shamoo Y. Structurebased analysis of protein-RNA interactions using the program ENTANGLE. J Mol Biol 2001; 311:75-86.
23. Ramakrishnan V, White SW. Ribosomal protein structures: Insights into the architecture, machinery and evolution of the ribosome. Trends Biochem Sci 1998; 23:208-212.
24. Nogi Y, Yano R, Nomura M. Synthesis of large rRNAs by RNA polymerase II in mutants of Saccharomyces cerevisiae defective in RNA polymerase I. Proc Natl Acad Sci USA 1991; 88:3962-3966.
25. Urlaub H, Kruft V, Bischof O et al. Protein-rRNA binding features and their structural and functional implications in ribosomes as determined by cross-linking studies. The EMBO J 1995; 14:4578-4588.
26. Misra VK, Hecht JL, Sharp KA et al. Salt effects on protein-DNA interactions. The cI repressor and Eco RI endonuclease. J Mol Biol 1994; 238:264-280.

CHAPTER 15

LIM Domain and Its Binding to Target Proteins

Algirdas Velyvis and Jun Qin*

Abstract

LIM domain is a unique double-zinc finger motif found in a variety of proteins such as homeodomain transcription factors, kinases, and adaptors. The LIM-containing proteins are involved in diverse biological processes including cytoskeleton organization, cell lineage specification and organ development. Dysfunctions of LIM domains induce pathological effects including muscle detachment, embryonic lethality, and oncogenesis. Acting as a protein-protein interaction motif, the LIM domain has a conserved scaffold but highly variable mode in recognizing diverse target proteins. This chapter describes the structure and function of LIM domain proteins and discusses the molecular basis by which the domain mediates protein-protein interactions.

Introduction

LIM domains were discovered as a novel double zinc finger sequence motif, $C\text{-}X_2\text{-}C\text{-}X_{17\text{-}19}\text{-}H\text{-}X_2\text{-}C\text{-}X_2\text{-}C\text{-}X_2\text{-}C\text{-}X_{15\text{-}19}\text{-}C$, with conserved distribution of cysteine and histidine residues in *lin-11*, *Isl-1* and *mec-3* (hence - LIM) gene products.[1,2] LIM domains are found in proteins from varied branches of eukaryotes: plants, animals, fungi and mycetozoa (*D. discoideum*) and have been classified as A, B, C or D types based on their sequence similarity.[3,4] Several domains fall outside of these classes due to pronounced sequence divergence.

LIM domains have been also classified into three groups, 1, 2 and 3. LIM domains from classes A and B are most frequently found fused to other functional domains such as kinase domain. But nuclear LIM-only (LMO) proteins, which contain two LIM domains only, are also important members of this group, called group 1.[5] Class C domains are called group 2. This group's proteins often contain two copies of LIM domain per protein molecule where the domains are more similar to each other than those from classes A and B. There are no other domains in association with the C-class LIM domains. It was suggested that class C-containing proteins arose via internal duplication of LIM domains.[4] Class D serves as a sorting bin for LIM domains lacking homology to other classes and with little similarity among themselves. Given that sequence similarity of LIM domains in classes A - C predicts rather well the global architecture of the entire protein, perhaps it is not surprising that the motley collection of LIM domains in class D appears in dissimilar proteins. These are called group 3 and harbor from one to five LIM domains, either with or without additional functional domains or motifs.

Diverse Functions of LIM-Containing Proteins

Proteins containing LIM domains are involved in a variety of biological processes including regulation of gene transcription, cytoskeleton organization, cell lineage specification and organ development. LIM-containing proteins can have significantly different functions even in the same groups. Three commonly known subclasses of group 1 LIM-containing proteins are: (i) LIM-homeodomain (LIM-HD) transcription factors, which participate in activation of transcription and development of the nervous system.[6] (ii) LMO proteins that are thought to act as molecular adapters involved in development and oncogenesis.[7] The LMO-1 and LMO-2 were first discovered at translocation breakpoints of T-cell leukemia patients.[8,9] The LMO family has been now extended to LMO-3 and LMO-4.[10] (iii) LIM kinases, which are involved in cytoskeleton establishment and regulation. LIM kinases (Limk1 and Limk2) are known to phosphorylate and thereby inhibit cofilin.[11,12] Cofilin functions by depolymerizing F-actin, hence activation of LIM-kinases by the upstream small GTPase dependent PAK1 or ROCK kinases leads to accumulation of actin filaments.[13]

A functionally well-characterized subclass of group 2 LIM proteins is cysteine-rich proteins (CRP1 - CRP3), which are prominent in myogenesis and muscle structure. Expression of all three is upregulated at distinct stages of embryo development and myogenesis.[13] Antisense disruption of CRP3 blocks terminal differentiation in myoblasts.[14] Interestingly, in CRP3 knock-out the development appears to be normal, but the arrangement of striated muscle myofibrils is disrupted, arguing for a structural or structureregulating role of CRP3. A structural function for all three CRPs in fibroblast cells is implied by robust intracellular staining at actin filaments[15] and in focal adhesions (at least for CRP1).[16] Given that these proteins bind to α-actinin (a well known actin bundle cross-linking protein) and to zyxin (a focal adhesion marker protein), such intracellular localization is expected.[13] CRP3 is also reported to bind to βI-spectrin in muscle cells and colocalizes with spectrin at the plasma membrane where it overlies Z- and M- lines of striated muscle.[17] The composition and function of these regions of muscle cells is rather similar to focal adhesions, therefore CRP proteins have similar structural roles in different cell types.

* *Corresponding author. See list of "Contributors".*

Zinc Finger Proteins: From Atomic Contact to Cellular Function, edited by Shiro Iuchi and Natalie Kuldell.

The class D LIM domains that belong to group 3 are exemplified by zyxin (3 LIM domains), paxillin (4 LIM domains), and PINCH (5 LIM domains) - three adaptor proteins that are classical markers of focal adhesion (FA) plaques.[18,19] All are thought to act as adaptor proteins that orchestrate assembly of FAs. These proteins regulate cell shape change and spreading via distinct LIM-mediated protein-protein interactions.

Zinc Coordination in LIM Domains

The presence of a conserved pattern of cysteine and histidine residues in LIM domains suggested that LIM domains likely contain structural Zn^{2+} ions and this is indeed so. From the current data, the consensus pattern appears to be C - X_2 - C - $X_{16\text{-}23}$ - H/ C - X_2 - C/H - X_2 - C - X_2 - C - $X_{15\text{-}30}$ - C - $X_{1\text{-}3}$ - C/H/D (/ indicates alternative amino acid residue). Metal coordination by LIM domains was extensively studied on two members of the family, CRP1 and CRIP proteins. These proteins contain two and one copies of LIM domains respectively, with metal coordination consensus residues CCHCCCCC, which all belong to class C. It was established that these LIM domains, expressed in heterologous bacterial hosts or isolated from natural sources, contain approximately 2 equivalents of zinc per equivalent of protein.[20] Bound zinc was also demonstrated for native preparations of zyxin.[16] Metal ions can be stripped from LIM domains by denaturation with low pH or guanidinium hydrochloride. A number of other metals- Cd^{2+}, Co^{2+}, Cu(I),- can be reconstituted into LIM domains upon refolding, again at two ions per domain as established by atomic absorption spectroscopy[16] and by titration.[21] UV spectroscopy shows bands consistent with formation of S-metal bonds in the complex.[21] A strong CoS_4 band dominates the Co^{2+} spectrum. Interestingly, the two metal sites appear to have different affinities for metal ions. This observation was craftily exploited by preparing a LIM domain with the high affinity site occupied by Zn^{2+} and the second site by Co^{2+}.[21] This preparation yielded a spectrum consistent with CoS_3N type complex. 1D ^{113}Cd NMR spectra of Cd^{2+}-substituted LIM domains provided further information on the coordination. Chemical shifts of two signals per LIM domain observed in Cd^{2+} NMR are consistent with Cd-S_4 coordination.[21] The chemical shift of one of the signals, however, is in the region consistent with both S_4 and S_3N coordination. The possibility for S_3N coordination was strongly supported by observation of J-coupling between Cd^{2+} and hydrogen atoms in the approximately 7 parts per million region, which can be explained only by the presence of Cd^{2+} coordinated to a histidine imidazole ring.[21] However, none of the above studies determined which of the 7 conserved cysteines form the S_4 and the S_3N site respectively. Based on a generally valid assumption that individual C-X_2-C/H units coordinate the same metal ion, Beckerle and coworkers used mutagenesis to address this question.[22] Results of Co^{2+} UV-visible and ^{113}Cd NMR spectroscopy on the mutant proteins indicated that the N-terminal 4 residues in the protein sequence (CCHC) form the S_3N site, while the S_4 site is composed of the 4 C-terminal cysteines. On a side note, among the mutants tested was a cysteine to aspartic acid substituted domain, in which the C-terminal CCCC zinc finger was transformed into CCCD. Aspartic acid was predicted to be a zinc binding residue in many LIM domains among LIM-HD proteins, but generally it is not common as a Zn^{2+}-coordinating group. The cysteine to aspartic acid mutant folded and ^{113}Cd NMR and Co UV-visible spectra are consistent with conversion of the S_4 to the (apparently perfectly functional) S_3O Zn^{2+}-coordination sphere. Finally, the arrangements of metal coordinating residues in the LIM domains were derived from NMR structure determination.[23,24]

LIM Domain As Protein Interaction Motif

Currently, the only function ascribed to LIM domains is protein-protein recognition. Two recent reviews[10,13] present an impressive list of about 30 known protein-protein interaction pairs for LIM domains. Given that the LIM domain count is in the hundreds and the consensus opinion in the field that LIM domains function by recognizing target proteins, it is clear that this is but the tip of the iceberg. The following sections present some well-characterized examples of LIM domain-mediated protein-protein interactions.

LIM-HD and LMO Proteins

Well-known partners for LIM domains, discovered simultaneously for LIM-HD[25] and LMO[26] proteins, are proteins from the Ldb (LIM domain binding) family. For LIM-HD proteins, in nearly all cases protein-protein interactions depend on LIM domains.[6] It is the LIM-Ldb interaction mediated formation of different heterodimers of LIM-HD proteins that is thought to determine distinct identities for motor neurons during development.[27] Ldb1 specifically binds to nuclear group 1 LIM proteins: LMO1, LMO2, Lmx1, Lhx1, Isl1, Mec3, but not LIM kinase, paxillin, zyxin, enigma or CRP1.[26] Deletion mutagenesis was used to determine the LIM-interacting part of Ldb1. A 39-residue stretch within Ldb1 (residues 300-338 of mouse protein) was identified that is sufficient to bind LIM domains.[28,29] This region is termed "LID" – LIM interaction domain. The LID region contains both highly hydrophobic and hydrophilic stretches of residues and encompasses a 34 residue sequence where human and *C. elegans* proteins share 23 identical residues (additionally, 7 residues are conservatively substituted), which implies strong evolutionary pressure to preserve the LIM - LID interaction. It is not clear so far what features of LIM domains are responsible for recognition of LID: in some instances it is LIM1 (first LIM), while in others the LIM2 (second LIM) domain of the LIM-HD protein is implicated. Also, some conflicting data exist for *X. laevis* Xlim-1 where both LIM1 and LIM2 appear to bind to Ldb1.[29] LID of Ldb1 appears to be functionally independent from the Ldb1 dimerization domain,[28] and monomeric behaviour of LID – LIM (of LMO2 or LMO4 proteins) fusion proteins suggests 1:1 binding stoichiometry.[30] The importance of LIM domain-mediated interactions for LIM-HD proteins is underscored by the finding that Y116C mutation in a human LIM-HD protein (Lhx3) causes a genetic disease, combined pituitary hormone deficiency.[31] Residues equivalent to Tyr116 play a central role in hydrophobic cores of known LIM domain structures, hence this mutation is predicted to cause unfolding and loss of function of the LIM domain. Indeed, mutant Lhx3 behaves consistently with such prediction.[32] While binding to Ldb by LIM-HD and LMO proteins is important for the activation of transcription, a recently discovered RLIM protein seems to have a negative regulatory role[33] by binding to Lhx3 via its basic region of about 100 residues.[34]

Zyxin

A number of binding partners are known for zyxin: CRP proteins,[35,16] α-actinin,[36] proto-oncogene Vav,[37] p130^{Cas},[38] and members of the Ena/VASP family of proteins.[39] LIM1 of zyxin is necessary and sufficient for binding to CRP, as demonstrated by blot overlay and GST pull-down assays.[40] Interestingly, although two zinc fingers in LIM domains form a single domain, individual fingers appear to have some functional independence.[22,41] This is also implied by the study that the N-terminal and not the C-terminal zinc finger of zyxin LIM1 is responsible for CRP1 recognition.[42] The C-terminal zinc finger in LIM1 of zyxin could be swapped with the C-terminal zinc finger from another LIM domain (but not deleted!) with no apparent effect on binding to CRP1. In CRP1, LIM1 and LIM2 appear to cooperate in binding to zyxin, so that deletion of either reduces the extent of binding more than twenty-fold in blot overlay assay.[43] The entire LIM_1 is required, but the C-terminal zinc finger in LIM2 is disposable. The linker between the two LIM domains is inert. Taken together these studies present an interesting picture—LIM domains in CRP1 are linked by a long disordered segment and structurally seem to be entirely independent.[44] Zyxin was also suggested to bind to actin filaments via α-actinin and play a regulatory rather than a structural role in FAs. Ample biochemical data support this notion. Interaction with Ena/VASP proteins is important in targeting zyxin to lamellipodia,[45] although under normal cellular situations the vast majority of zyxin resides in focal adhesions.[46] Ena/VASP proteins are known to bind profilin—an actin polymerizing protein. In an important development, it was shown that zyxin-VASP complex is able to initiate actin polymerization.[47]

Paxillin

Paxillin is one of the central proteins in FAs acting as a scaffold for a myriad of binding partners.[48] The LIM domain-containing C-terminal half of paxillin is known to bind to protein tyrosine phosphatase – PEST[49,50] and this binding is postulated to bring PTP-PEST into the vicinity of its target, phosphorylated p130^{Cas}.[51] PTP-PEST is required for disassembly of FAs possibly by dephosphorylating p130^{Cas} and paxillin[52] and therefore modulating their FA targeting regions. The LIM domain region is responsible for targeting paxillin to Fas,[53] via a yet unknown binding event. It was recently shown that paxillin LIM domains bind to α- and γ-tubulin as well,[54] which led to a proposal that paxillin is responsible for the interplay between actin filaments and microtubules. A fragment containing both the C-terminal LIM3 and LIM4 of paxillin is competent to bind to PTP-PEST, while LIM3 or LIM4 in isolation are not.[51] Deletions of LIM3 or LIM4 from an otherwise intact paxillin or its fragments abolish binding, as does disruption of the N- or C-terminal zinc fingers within LIM3 and the N-terminal finger in LIM4 with point mutagenesis. Comprehensive deletion of PTP-PEST showed that a part of the protein that centers on the 2nd (of five total) proline-rich region of PTP-PEST is the binding epitope. The structure of this PTP-PEST peptide is not known, but the 50% content of proline likely precludes formation of a globular fold. It is interesting that a close homologue of paxillin, Hic-5 protein, also binds to PTP-PEST.[55] Broadly the same region of the phosphatase, rich in prolines, is involved. Here, in contrast to paxillin, LIM2, LIM3 and LIM4 are involved in binding, with LIM3 being absolutely necessary for detectable binding, while the presence of LIM2 and LIM4 clearly augments the extent of binding.[55] Given that the LIM domains of paxillin and His-5 share 68% sequence similarity,[51] it is not obvious whether the protein interface is the same for both members of the family, or only the net result, formation of the complex with PTP-PEST, is conserved.

Enigma Protein

A relatively well-characterized system is Enigma protein that contains three LIM domains at its C terminus. Yeast two-hybrid screening revealed that enigma binds to the insulin receptor (InsR) internalization motif.[56] Point mutants in two internalization-driving tetrapeptides within this segment somewhat diminished (weaker peptide) and thoroughly blocked (stronger peptide) interaction with full-length enigma and LIM3 construct. The interaction is specific since the internalization segments of IGF1, EGF, sequence-optimized EGF, Transferrin and LDL receptors failed to bind. However, the paired sequence-optimized EGF receptor motif (the tandem repeat of the internalization motif of EGF that has been optimized to display high activity in the internalization assay) did bind, albeit to a lesser extent. The relevance of the last observation is not clear, since LIM2 in Mec-3, a nuclear LIM-HD protein which has no known function in receptor endocytosis also binds to the paired sequence-optimized EGFR motif. Furthermore, this paired sequence-optimized motif binds equally well to LIM2, LIM3 and the C-terminally truncated LIM3 of enigma.[56] Yeast two-hybrid screening was also instrumental in discovering that the LIM domain fragment of enigma also binds to Ret Receptor tyrosine kinase.[57] Mutating the most C-terminal tyrosine at position 1062 to phenylalanine or deleting twenty-three C-terminal residues (including Tyr1062) from Ret abolished the interaction. In fact, a 61 C-terminal residue segment of Ret is perfectly capable of binding to enigma[58] in a GST pull-down assay. Their association is not phosphorylation-dependent, and is specific for enigma, as zyxin and CRP1 did not bind. Enigma-Ret binding was confirmed by the observation that coinjection of DNA for the LIM domain-containing part of enigma with the Ret construct into fibroblasts abolished the mitogenic effect of Ret in a dominant-negative style.[57] Unlike InsR, Ret peptide is specifically recognized by enigma LIM2,[58] while the LIM domains from LIM-HD proteins, zyxin, CRP1 or paxillin do not bind Ret or InsR peptides. Which residues are important for Ret - LIM2 recognition was further assessed with a competition assay.[58] Wild type (NKLY1062) and mutant (AKLA and NKLF) peptides within an otherwise intact 20-residue segment of Ret Tyr1062 region were used as competitors for Ret in GST-pull down. While AKLA peptide and InsR LIM3-recognizing peptide failed to inhibit the interaction, NKLY and NKLF peptides were inhibitors. The capacity of NKLF peptide to act as an inhibitor, however, contradicts previous data[57] that the Y1062F mutation disrupts binding. Also, the GST pull-down titration experiment showed that lower amounts of Y1062F mutant Ret are retained by GST-LIM_2. The apparent discrepancy may arise from the fact that the amount of observed binding in GST-pull down assay depends on times and volumes of the binding and washing steps, in addition to the affinity of interaction. Also, 20-residue peptides may be suboptimal for binding, since high concentrations were required to document competition.[58] A random peptide library was screened to identify sequence requirements for peptides recognized by enigma LIM3.[57] A screen with fixed tyrosine which is assigned at position 0, and is prominent in both internalization-targeting peptides in InsR and recognized by LIM3, identified strong preference for proline at positions –1 and +2. A second screen with a fixed PXXP feature yielded a consensus GP-Hyd1-GP-Hyd2-Y/F-A, where Hyd1 is Met, Phe, Tyr or Ile and Hyd2 is Ile, Met or Val. This compares nicely with the sequence naturally present in InsR: GPLGPLYA. Moreover, GPLY is present, which is the strong InsR internalization-driving motif whose mutation blocks binding to LIM3.[56]

PINCH

PINCH (Particularly Interesting New Cys-His) protein was discovered in an immunological screen for senescent erythrocyte antigens.[59] Nearly the entire protein sequence of PINCH is contained in the LIM domains, with very short interdomain linker peptides, and a C-terminal extension with numerous positive charges. A yeast two-hybrid screen of the human lung library revealed that PINCH interacts with the bait that consisted of the N-terminal ankyrin repeat domain of integrin-linked kinase (ILK).[60] Paxillin or zyxin LIM domain fragments failed to bind to ILK. MBP-PINCH is able to pull down ILK from cell lysates and ILK is retained on an anti-PINCH immunoglobulin column. Binding was confirmed in vitro with an ELISA assay between recombinant MBP-PINCH and GST-ILK. This establishes that even though binding in the cells might be mediated by bridging proteins, it is likely a direct association. Deletion mutagenesis with a subsequent yeast two-hybrid assay established that LIM_1 of PINCH is necessary and sufficient for binding. LIM1 was also competent to bind to ILK in pull-down and ELISA[61,60] experiments. The ankyrin repeat domain of ILK is solely responsible for binding to PINCH.[60,61] The LIM4 domain of PINCH was found to interact with Nck2 protein that is also an adaptor.[62] Deletion mutagenesis revealed that the third SH3 domain of Nck2 is responsible for binding to LIM4,[60] however, the affinity of the interaction was found to be very weak.[24]

In summary, unlike short peptide-recognizing SH3 or SH2 domains, LIM domains bind to a wide variety of partners, which make it difficult to pinpoint their binding specificities. Hence, despite the success in the case of Enigma LIM3, combinatorial peptide synthesis and modification techniques are of limited value for research in the field of LIM domains. Structure determination, which would define conformations of specificity-defining groups at domain surfaces and would directly reveal the nature of the interfaces between the molecules in the complex, seems to be the most robust way to uncover the biochemical basis for recognition of targets by LIM domains.

Structures of LIM Domains

LIM2 of cCRP1 is the first LIM domain for which the 3D structure has been determined.[63] The structure revealed two contiguous zinc fingers. In each zinc finger, two of the Zn^{2+}-coordinating cysteine residues are located at the turns connecting individual β strands in β hairpins. The NOE patterns in these turns closely resemble patterns observed in the cysteine-containing turns (also metal-coordinating) in *P. furiosus* rubredoxin (Rd) and hence these turns have been called "rubredoxin knuckles" (Fig. 1). Moreover, these turns contain a unique feature—H^N-S hydrogen bonds from backbone amide protons to metal chelating sulfur atoms, as demonstrated for Rd itself. Heteronuclear Spin-Echo difference (HSED) spectrum on the Cd^{2+}-substituted LIM2 sample showed clearly the existence of partially covalent H^N-S H-bonds. ^1H-^{113}Cd HMQC experiment on Cd^{2+}-substituted LIM_2 unequivocally established metal coordination in LIM_2. The two zinc fingers in cCRP1 LIM2 stack together (Fig. 2) to form a single domain with several bulky hydrophobic residues forming a small core. The core is dominated by stacking of the aromatic rings of Trp138 and Phe143 (Fig. 3) that reside in the very C terminus of the N-terminal zinc finger and the inter-finger dipeptide, respec-

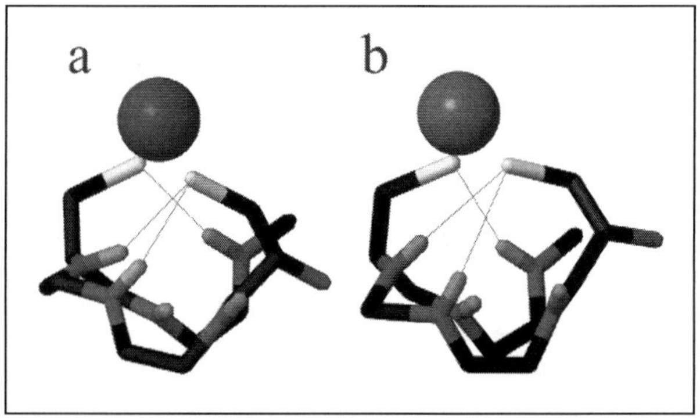

Figure 1. Rubredoxin (Rd) turns. a) N-terminal turn in *P. furiosus* Rd (PDB ID: 1brf). b) N-terminal turn in qCRP2 LIM2 (PDB ID 1CTL). Spheres – Fe^{3+} (a) and Zn^{2+} (b). H^N-S H-bonds are denoted by lines.

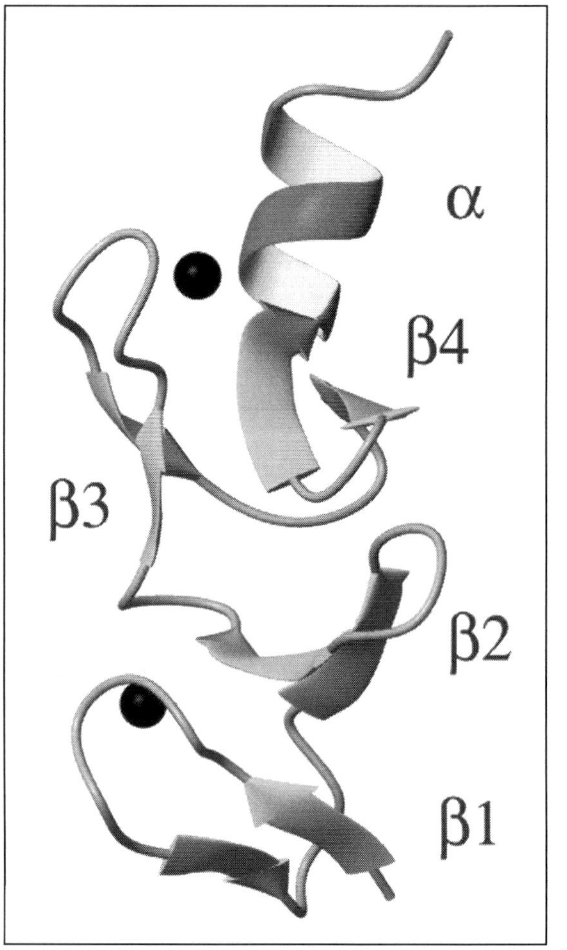

Figure 2. Ribbon representation of cCRP1 LIM2, with secondary structure elements labeled and Zn^{2+} ions in spheres.

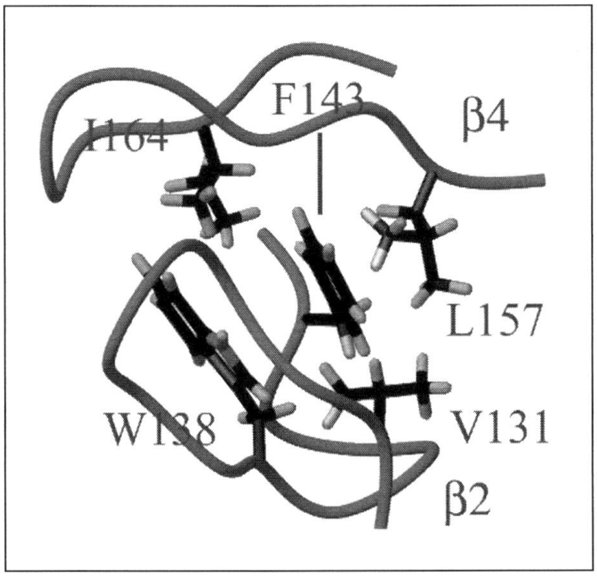

Figure 3. Hydrophobic core in cCRP2 LIM2. Backbone is shown as a tube. The hydrophobic residues in the core are labeled.

tively. The importance of these residues is reflected in their conservation: among currently known LIM domain sequences, the position equivalent to equivalent to Phe143 is occupied by phenylalanine (80.4%), leucine (10.8%), or tyrosine, valine, tryptophan or isoleucine (6.9%) (total 98.1%). It is quite possible that the preference for aromatic residues at these two positions is due to strengthening of the hydrophobic cluster by additional aromatic π - π interaction. Other hydrophobic residues (also highly conserved) pile on the sides of these two: Val131 from the N-terminal zinc finger and Leu152, Leu157, Ieu164 from the C-terminal finger. Interestingly, Perez-Alvarado et al[63] also discovered a rather good superposition between the C-terminal zinc finger of LIM2 and the DNA-binding zinc fingers in the glucocorticoid receptor and in the GATA-1 transcription factor (backbone rmsds 2.0 and 1.7 Å, respectively), which suggested a possible a role for LIM domains as DNA-binding molecules. However, so far this hypothesis is not supported by experimental evidence.

Since the report of cCRP LIM2 structure, numerous structures of other LIM domains have been determined by NMR spectroscopy. These include all three groups of LIM domains: (i) LIM1 domains of LMO2 and LMO4,[64] which belong to group 1. (ii) LIM domains in group 2 family including LIM2 of cCRP1 as mentioned above, CRIP protein (one LIM domain),[65] full-length cCRP1 (LIM1 and LIM2),[43] LIM1 and LIM2 of qCRP2.[66,67] In addition, the structure of an isolated N-terminal zinc finger of LIM domain in Lasp-1 has been determined.[41] (iii) LIM1 and LIM4 domains of PINCH, which belong to group 3.[23,24] The main structural features of these LIM domains were found to be the same as cCRP LIM2: four short β hairpins with a C-terminal α helix; two "rubredoxin knuckles"; small hydrophobic core contributed by both zinc fingers. Although the global fold is similar among all known LIM domain structures, there are some differences, notably the relative orientation of the two zinc fingers varies about the long axis of the domain (Fig. 4). Different orientations between the fingers in LIM domains may be important for protein-protein recognition. Each zinc finger appears to be an independent folding unit despite the interactions between the two fingers. This is evidenced by the determination of the structure of isolated N-terminal CCHC half of the LIM domain of Lasp-1 protein.[41] Clearly, four zinc

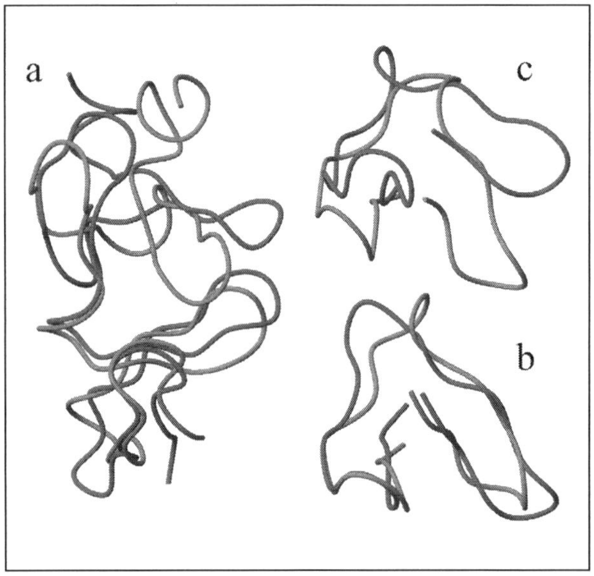

Figure 4. Overlay of CRIP LIM domain (gray, PDB ID 1IML) with cCRP1 LIM2 (black), illustrating the relative twist of the C-terminal Zn-finger in CRIP. Backbone residues 1-23 of CRIP were best fit superimposed onto corresponding residues of LIM2. a) Overall view; b and c) – top view of individual fingers. b) N-terminal Zn-fingers, used for superimposition. c) C-terminal fingers, showing rotation in CRIP relative to LIM_2.

binding residues are the key structural determinants, which orchestrate the fold of this fragment. Other residues in Lasp-1 also adopt conformations close to those observed in the N-terminal halves of full-length LIM domains.

A key issue about the LIM domain is its structural basis in recognizing target proteins. Recent chemical shift mapping studies on PINCH LIM domains have provided important insights into this issue:[23,24] For PINCH LIM1, it was shown that C-terminal hydrophobic patch and several adjoining residues are most perturbed upon binding to ankyrin repepat (ANK) domain of ILK (Fig. 5a) whereas residues from N-terminal basic region

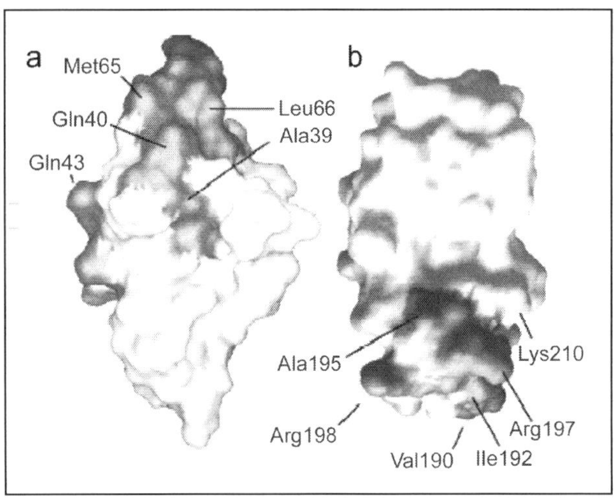

Figure 5. Comparison of PINCH LIM1 and LIM4 binding modes to target proteins: a) the major binding surface (shaded area with labeled residues) of LIM1 (PDB ID 1G47) to ILK ANK domain; b) the major binding surface (shaded area with labeled residues)

of PINCH LIM4 appear to be responsible for binding to SH3 domain (Fig. 5b). The structure of LMO-Ldb1 complex revealed a rather extensive binding interface involving both zinc finger 1 and zinc finger 2.[64] Hence, LIM domains do not appear to have a single, well defined binding epitope, but rather present on their surfaces clusters of different functional groups, individually tailored for diverse binding partners. Such diverse recognition mode has also been observed in PTB and SH2 domains.[68,69] Clearly, more structures of LIM domains in complex with target proteins are necessary for a thorough understanding of this fascinating protein interaction motif.

Acknowledgments

The work was supported by NIH grants to J.Q. (HL58758, GM62823).

References

1. Karlsson O, Thor S, Norberg T et al. Insulin gene enhancer binding protein Isl-1 is a member of a novel class of proteins containing both a homeo- and a Cys-His domain. Nature 1990; 344:879-882.
2. Freyd G, Kim SK, Horvitz HR. Novel cysteine-rich motif and homeodomain in the product of the Caenorhabditis elegans cell lineage gene lin-11. Nature 1990; 344:876-879.
3. Taira M, Evrard JL, Steinmetz A et al. Classification of LIM proteins. Trends Genet 1995; 11:431-432.
4. Dawid IB, Toyama R, Taira M. LIM domain proteins. CR Acad Sci III 1995; 318:295-306.
5. Dawid IB, Breen JJ, Toyama R. LIM domains: Multiple roles as adapters and functional modifiers in protein interactions. Trends Genet 1998; 14:156-162.
6. Hobert O and Westphal H. Functions of LIM-homeobox genes. Trends Genet 2000; 16:75-83.
7. Rabbitts, T.H. LMO T-cell translocation oncogenes typify genes activated by chromosomal translocations that alter transcription and developmental processes. Genes Dev 1998; 12(17):2651-7
8. Larson RC, Osada H, Larson TA et al. The oncogenic LIM protein Rbtn2 causes thymic developmental aberrations that precede malignancy in transgenic mice. Oncogene 1995; 11(5):853-62
9. Larson RC, Lavenir I, Larson TA et al. Protein dimerization between Lmo2 (Rbtn2) and Tal1 alters thymocyte development and potentiates T cell tumorigenesis in transgenic mice. EMBO J 1996; 15(5):1021-7
10. Bach I. The LIM domain: Regulation by association. Mech. Dev 2000; 91:5-17.
11. Arber S, Barbayannis FA, Hanser H et al. Regulation of actin dynamics through phosphorylation of cofilin by LIM-kinase. Nature 1998; 39:805-809.
12. Yang,N, Higuchi,O, Ohashi,K et al. Cofilin phosphorylation by LIM-kinase 1 and its role in Rac-mediated actin reorganization. Nature 1998; 393:809-812.
13. Khurana,T, Khurana, B, Noegel, AA. LIM proteins: Association with the actin cytoskeleton. Protoplasma 2002; 219:1-12.
14. Arber S, Halder G, Caroni P. Muscle LIM protein, a novel essential regulator of myogenesis, promotes myogenic differentiation. Cell 1994; 79:221-231.
15. Louis HA, Pino JD, Schmeichel KL et al. Comparison of three members of the cysteine-rich protein family reveals functional conservation and divergent patterns of gene expression. J Biol Chem 1997; 272:27484-27491.
16. Sadler I, Crawford AW, Michelsen JW et al. Zyxin and cCRP: Two interactive LIM domain proteins associated with the cytoskeleton. J Cell Biol 1992; 119:1573-1587.
17. Flick MJ, Konieczny SF. The muscle regulatory and structural protein MLP is a cytoskeletal binding partner of betaI-spectrin. J Cell Sci 2000; 113(Pt 9):1553-1564.
18. Wu C, Dedhar S. Integrin-linked kinase (ILK) and its interactors: A new paradigm for the coupling of extracellular matrix to actin cytoskeleton and signaling complexes. J Cell Biol 2001; 155:505-510.
19. Wang Y, Gilmore TD. Zyxin and paxillin proteins: Focal adhesion plaque LIM domain proteins go nuclear. Biochim Biophys Acta 2003; 1593(2-3):115-20.
20. Michelsen JW, Schmeichel KL, Beckerle MC et al. The LIM motif defines a specific zinc-binding protein domain. Proc Natl Acad Sci USA 1993; 90:4404-4408.
21. Kosa JL, Michelsen JW, Louis HA et al. Common metal ion coordination in LIM domain proteins. Biochemistry 1994; 33:468-477.
22. Michelsen JW, Sewell AK, Louis HA et al. Mutational analysis of the metal sites in an LIM domain. J Biol Chem 1994; 269:11108-11113.
23. Velyvis A, Yang Y, Wu C et al. Solution structure of the focal adhesion adaptor PINCH LIM1 domain and characterization of its interaction with the integrin-linked kinase ankyrin repeat domain. J Biol Chem 2001; 276:4932-4939.
24. Velyvis A, Vaynberg J, Yang Y et al. Structural and functional insights into PINCH LIM4 domain-mediated integrin signaling Nat Struct Biol 2003; 10:558-564.
25. Agulnick AD, Taira M, Breen JJ et al. Interactions of the LIM-domain-binding factor Ldb1 with LIM homeodomain proteins. Nature 1996; 384:270-272.
26. Jurata LW, Kenny DA, Gill GN. Nuclear LIM interactor, a rhombotin and LIM homeodomain interacting protein, is expressed early in neuronal development. Proc Natl Acad Sci USA 1996; 93:11693-11698.
27. Jurata LW, Pfaff SL, Gill GN. The nuclear LIM domain interactor NLI mediates homo- and heterodimerization of LIM domain transcription factors. J Biol Chem 1998; 273:3152-3157.
28. Jurata LW, Gill GN. Functional analysis of the nuclear LIM domain interactor NLI. Mol Cell Biol 1997; 17:5688-5698.
29. Breen JJ, Agulnick AD, Westphal H et al. Interactions between LIM domains and the LIM domain-binding protein Ldb1. J Biol Chem 1998; 273:4712-4717.
30. Deane JE, Sum E, Mackay JP et al. Design, production and characterization of FLIN2 and FLIN4: The engineering of intramolecular ldb1: LMO complexes. Protein Eng 2001; 14:493-499.
31. Netchine I, Sobrier ML, Krude H et al. Mutations in LHX3 result in a new syndrome revealed by combined pituitary hormone deficiency. Nat Genet 2000;25:182-186.
32. Howard PW, Maurer RA. A Point Mutation in the LIM Domain of Lhx3 Reduces Activation of the Glycoprotein Hormone alpha -Subunit Promoter. J Biol Chem 2001; 276:19020-19026.
33. Bach I, Rodriguez-Esteban C, Carriere C et al. RLIM inhibits functional activity of LIM homeodomain transcription factors via recruitment of the histone deacetylase complex. Nat Genet 1999; 22:394-399.
34. Ostendorff HP, Peirano RI, Peters MA et al. Ubiquitination-dependent cofactor exchange on LIM homeodomain transcription factors. Nature 2002; 416:99-103.
35. Crawford AW, Pino JD, Beckerle MC. Biochemical and molecular characterization of the chicken cysteine-rich protein, a developmentally regulated LIM-domain protein that is associated with the actin cytoskeleton. J Cell Biol 1994; 124:117-127.
36. Crawford AW, Michelsen JW, Beckerle MC. An interaction between zyxin and alpha-actinin. J Cell Biol 1992; 116:1381-1393.
37. Hobert O, Schilling JW, Beckerle MC et al. SH3 domain-dependent interaction of the proto-oncogene product Vav with the focal contact protein zyxin. Oncogene 1996; 12:1577-1581.
38. Yi J, Kloeker S, Jensen CC et al. Members of the Zyxin family of LIM proteins interact with members of the p130Cas family of signal transducers. J Biol Chem 2002; 277:9580-9589.
39. Drees B, Friederich E, Fradelizi J et al. Characterization of the interaction between zyxin and members of the Ena/ vasodilator-stimulated phosphoprotein family of proteins. J Biol Chem 2000; 275:22503-22511.
40. Schmeichel KL, Beckerle MC. The LIM domain is a modular protein-binding interface. Cell 1994; 79:211-219.

41. Hammarstrom A, Berndt KD, Sillard R et al. Solution structure of a naturally-occurring zinc-peptide complex demonstrates that the N-terminal zinc-binding module of the Lasp-1 LIM domain is an independent folding unit. Biochem 1996; 35:12723-12732.
42. Schmeichel KL, Beckerle MC. Molecular dissection of a LIM domain. Mol Biol Cell 1997; 8:219-230.
43. Schmeichel KL, Beckerle MC. LIM domains of cysteine-rich protein 1 (CRP1) are essential for its zyxin-binding function. Biochem J 1998; 331(Pt 3):885-892.
44. Yao X, Perez-Alvarado GC, Louis HA et al. Solution structure of the chicken cysteine-rich protein, CRP1, a double-LIM protein implicated in muscle differentiation. Biochem 1999; 38:5701-5713.
45. Nix DA, Fradelizi J, Bockholt S et al. Targeting of zyxin to sites of actin membrane interaction and to the nucleus. J Biol Chem 2001; 276:34759-34767.
46. Beckerle MC. Zyxin: Zinc fingers at sites of cell adhesion. Bioessays 1997; 19:949-957.
47. Fradelizi J, Noireaux V, Plastino J et al. ActA and human zyxin harbour Arp2/3-independent actin-polymerization activity. Nat Cell Biol 2001; 3:699-707.
48. Turner CE. Paxillin and focal adhesion signaling. Nat Cell Biol 2000; 2:E231-E236.
49. Cote JF, Turner CE, Tremblay ML. Intact LIM 3 and LIM 4 domains of paxillin are required for the association to a novel polyproline region (Pro 2) of protein-tyrosine phosphatase-PEST. J Biol Chem 1999; 274:20550-20560.
50. Shen Y, Schneider G, Cloutier JF et al. Direct association of protein-tyrosine phosphatase PTP-PEST with paxillin. J Biol Chem 1998; 273:6474-6481.
51. Cote JF, Turner CE, Tremblay ML. Intact LIM 3 and LIM 4 domains of paxillin are required for the association to a novel polyproline region (Pro 2) of protein-tyrosine phosphatase-PEST. J Biol Chem 1999; 274:20550-20560.
52. Shen Y, Lyons P, Cooley M et al. The noncatalytic domain of protein-tyrosine phosphatase-PEST targets paxillin for dephosphorylation in vivo. J Biol Chem 2000; 275:1405-1413.
53. Brown MC, Perrotta JA, Turner CE. Identification of LIM3 as the principal determinant of paxillin focal adhesion localization and characterization of a novel motif on paxillin directing vinculin and focal adhesion kinase binding. J Cell Biol 1996; 135:1109-1123.
54. Herreros L, Rodriguez-Fernandez JL, Brown MC et al. Paxillin localizes to the lymphocyte microtubule organizing center and associates with the microtubule cytoskeleton. J Biol Chem 2000; 275:26436-26440.
55. Nishiya N, Iwabuchi Y, Shibanuma M et al. Hic-5, a Paxillin Homologue, Binds to the Protein-tyrosine Phosphatase PEST (PTP-PEST) through Its LIM 3áDomain. J Biol Chem 1999; 274:9847-9853.
56. Wu RY, Gill GN. LIM domain recognition of a tyrosine-containing tight turn. J Biol Chem 1994; 269:25085-25090.
57. Durick K, Wu RY, Gill GN et al. Mitogenic Signaling by Ret/ptc2 Requires Association with Enigma via a LIM Domain. J Biol Chem 1996; 271:12691-12694.
58. Wu RY, Durick K, Songyang Z et al. Specificity of LIM Domain Interactions with Receptor Tyrosine Kinases. J Biol Chem 1996; 271:15934-15941.
59. Rearden A. A new LIM protein containing an autoepitope homologous to "senescent cell antigen." Biochem Biophys Res Commun 1994; 201:1124-1131.
60. Tu Y, Li F, Goicoechea S et al. The LIM-only protein PINCH directly interacts with integrin-linked kinase and is recruited to integrin-rich sites in spreading cells. Mol Cell Biol 1999; 19:2425-2434.
61. Li F, Zhang Y, Wu C. Integrin-linked kinase is localized to cell-matrix focal adhesions but not cell-cell adhesion sites and the focal adhesion localization of integrin-linked kinase is regulated by the PINCH-binding ANK repeats. J Cell Sci 1999; 112(Pt 24):4589-4599.
62. Tu Y, Li F, Wu C. Nck-2, a novel Src homology2/3-containing adaptor protein that interacts with the LIM-only protein PINCH and components of growth factor receptor kinase-signaling pathways. Mol Biol Cell 1998; 9:3367-3382.
63. Perez-Alvarado GC, Miles C, Michelsen JW et al. Structure of the carboxy-terminal LIM domain from the cysteine rich protein CRP. Nat Struct Biol 1994; 1:388-398.
64. Deane JE, Mackay JP, Kwan AH et al. Structural basis for the recognition of ldb1 by the N-terminal LIM domains of LMO2 and LMO4. EMBO J 2003, 22(9):2224-33
65. Perez-Alvarado GC, Kosa JL, Louis HA et al. Structure of the cysteine-rich intestinal protein, CRIP. J Mol Biol 1996; 257:153-174.
66. Konrat R, Weiskirchen R, Krautler B et al. Solution structure of the carboxyl-terminal LIM domain from quail cysteine-rich protein CRP2. J Biol Chem 1997; 272:12001-12007.
67. Kontaxis G, Konrat R, Krautler B et al. Structure and intramodular dynamics of the amino-terminal LIM domain from quail cysteine- and glycine-rich protein CRP2. Biochem 1998; 37:7127-7134.
68. Li SC, Zwahlen C, Vincent SJ et al. Structure of a Numb PTB domain-peptide complex suggests a basis for diverse binding specificity. Nat Struct Biol 1998; 5:1075-83.
69. Mallis RJ, Brazin KN, Fulton DB et al. Structural characterization of a proline-driven conformational switch within the Itk SH2 domain. Nat Struct Biol 2002; 9:900-5.

CHAPTER 16

RING Finger-B Box-Coiled Coil (RBCC) Proteins As Ubiquitin Ligase in the Control of Protein Degradation and Gene Regulation

Kazuhiro Ikeda, Satoshi Inoue and Masami Muramatsu*

Abstract

The protein family harboring the RING finger motif, defined as a linear array of conserved cysteines and histidines, has grown enormously in the last decade. The members of the family are involved in various biological processes including growth, differentiation, apoptosis, transcription and also in diseases and oncogenesis. It has been postulated that the RING finger domains have crucial roles in these phenomena themselves, in some cases, working with other domains in other proteins, although the precise mechanisms and common features of RING finger function have not been fully elucidated. However, most recently, an accumulating body of evidence has revealed that some of the RING finger proteins work as E3 ubiquitin ligases in ubiquitin-mediated specific protein degradation pathway. In this review, we focus on the RING finger protein with special reference to E3 ligase.

Structure of RING Finger

The RING finger protein sequence motif was first identified in the human gene RING1—Really Interesting New Gene 1—which is located proximal to the major histocompatibility region on chromosome 6.[1,2] The RING finger motif can be defined as a unique linear series of conserved cysteine and histidine residues: Cys-X_2-Cys-$X_{11\text{-}16}$-Cys-X-His-X_2-Cys -X_2-Cys-$X_{7\text{-}74}$-Cys-X_2-Cys (RING-CH or C_3HC_4 type), where X can be any amino acid (Fig. 1). So far, three-dimensional structures of RING domains from human PML (for promyelocytic leukemia protein),[3] immediate early equine herpes virus (IEEHV) protein,[4] human recombination-activating gene 1 protein (RAG1),[5] human MAT1 (for menage a trois-1 protein)[6] and human Cbl (for Casitas B-lineage lymphoma protein)[7] with a cognate ubiquitin-conjugating enzyme (E2) have been solved at atomic resolution. These studies have confirmed that the RING finger binds zinc ions in a similar manner as the classical zinc finger motif. Particularly, the RING finger is composed of a unique 'cross-brace' arrangement with two zinc ions and folds into a compact domain comprising a small central β sheet and an α helix. There are subfamilies of RING fingers which have Cys5 substituted with histidine (RING-H2) and a cysteine or histidine substituted with other metal binding residues such as aspartic acid and threonine.[8,9] Although the RING domain was initially found in only a few genes, more than 3000 proteins harboring the RING finger domain have been detected from diverse eukaryotes in the SMART database as of July 2003. Because of this evolutional conservation and variation in loop lengths, the RING domain appears to have a considerable flexibility within the rigid structure.

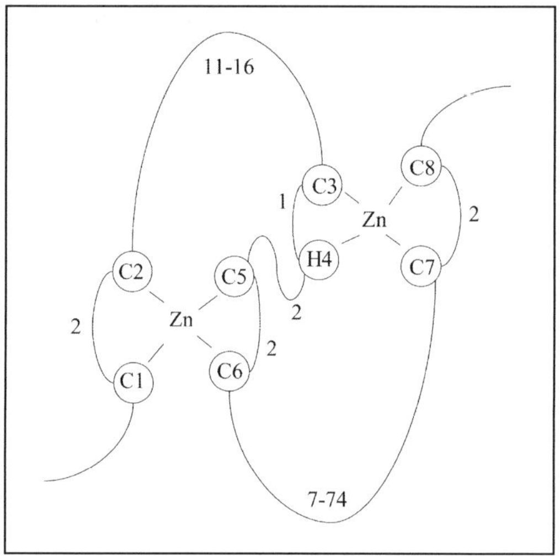

Figure 1. Schematic representation of the structure of RING finger domain. The metal-ligand residues, either cysteine (C) or histidine (H), are shown as numbered spheres. The numbers next to the loops connecting the metal-ligand residues indicate the minimum and maximum number of loop residues.

Family of RING Finger Protein

The RING fingers and their variants are generally located close to an amino or carboxyl terminus though there are no fixed rules. Most of the RING finger is associated with certain protein domains to form larger conserved motifs which may define the function of the protein, thus the family being divided into subfamilies along with the associated domains (Fig. 2). A similar domain

* Corresponding author. See list of "Contributors".

Zinc Finger Proteins: From Atomic Contact to Cellular Function, edited by Shiro Iuchi and Natalie Kuldell.

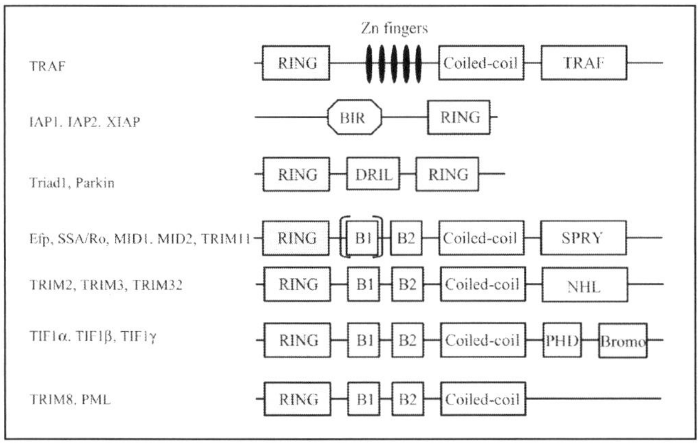

Figure 2. Structures of the RING finger protein family. Representative RING finger proteins with frequently associated domains are presented.

architecture often corresponds with a similar function. For instance, TRAFs (for tumor necrosis factor (TNF) receptor-associated factors) 2-5 have an N-terminal RING domain followed by five zinc fingers, a coiled coil, and a C-terminal TRAF domain.[10] TRAF1 has all of these domains except for the RING. Members of TRAF family have been shown to be involved in TNF-related cytokine signal transduction through interactions between their TRAF domains and the intracytoplasmic parts of receptors of the TNF receptor family which are suicide receptors to transfer apoptotic signals into the cells.[11-13]

The inhibitors of apoptosis gene family, IAP1, IAP2 and XIAP, have a RING domain at their C termini and BIR (baculovirus IAP repeat) domain at their N termini. The BIR domains of the proteins bind and inhibit caspase.[14,15] Interestingly, the RING fingers of XIAP and IAP2 possess E3 ubiquitin ligase activity and are thought to be responsible for self-degradation when an apoptotic signal is transduced.[16] In addition, the anti-apoptotic activity of the protein is lost when the RING domain is mutated.[17]

There are interesting subfamilies uniquely possessing two RING fingers. Triad1 (for two RING fingers and DRIL1) and parkin have two RING finger domains separated by the double RING finger linked (DRIL) domain. Triad1 was identified as a nuclear RING finger protein, which is up-regulated during retinoic acid induced granulocytic differentiation of acute leukemia cells.[18] Parkin is a responsible gene for familial autosomal recessive Parkinson's disease.[19,20] Parkin binds to the E2 ubiquitin-conjugating enzymes through its C-terminal RING finger and has ubiquitin–protein ligase activity.[21] Parkin ubiquitinates and promotes the degradation of a putative G protein-coupled transmembrane polypeptide, Pael (parkin-associated endothelial-like) receptor, the insoluble form of which is accumulated in the brains of Parkinson's disease.[22] The insoluble parkin overexpressed in cells causes unfolded protein-induced cell death, whereas coexpression of Parkin suppresses the accumulation of Pael receptor and subsequent cell death.[21] Parkin also ubiquitinates and promotes the degradation of CDCrel-1 (for cell division cycle related-1) and itself.[23] Familial-linked mutations disrupt the ubiquitin–protein ligase function of Parkin and impair Parkin and CDCrel-1 degradation.

Function of RING Finger

It has been shown in early studies that the RING finger proteins have crucial roles in the growth, differentiation, transcription, signal transduction and oncogenesis.[24] For example, PML is fused to the retinoic acid receptor α (RARα) in acute promyelocytic leukemia (APL) translocation,[25-27] BRCA1 is mutated in early-onset breast cancer and ovarian cancer,[28] TIF1α is a positive cofactor of nuclear hormone receptors[29] and TRAF transduces signals from members of the TNF receptor superfamily to the transcription factor NF-κB.[14] Although those studies appear to show some essential roles played by the RING finger domains in the function of these proteins, the general function of the RING finger domain has not been resolved. However, recently, it was uncovered that the RING finger proteins are involved in the ubiquitin-mediated protein degradation pathway.

The ubiquitin-dependent protein degradation is a specific and sophisticated mechanism in which a target protein to be destroyed is tagged with the ubiquitin. Ubiquitination is accomplished by a complex process involving ubiquitin-activating enzyme (E1), ubiquitin-conjugating enzyme (E2) and ubiquitin ligase (E3).[30] Ubiquitin ligase mediates the transfer of ubiquitin from E2 to a substrate, marking it for degradation by the 26S proteosome. Therefore, E3 enzyme is thought to be important for the specific recognition of the substrate in the ubiquitination pathway. There is accumulating evidence that RING finger domains are identified in E3 complexes and proteins, suggesting the broad use of these domains for ubiquitination. As mentioned above, RING finger domain has the conserved cysteine and histidine residues. The C_3HC_4 type RING finger is found in several E3 proteins including Cbl,[31] BRCA1,[32] Efp (for estrogen-responsive finger protein)[33] and Mdm2 (for murine double minute 2).[34] The RING-H2 subtype is found in Rbx1 (for RING box protein 1) and Apc11 (for anaphase promoting complex (APC) subunit 11) in SCF (Skp1-Cullin-F-box) and APC E3 complexes,[35] respectively, and other ubiquitin ligases. Thus, evidence is accumulating that the RING finger proteins has crucial roles as an E3 ubiquitin ligase in diverse biological functions and diseases. Cbl is one of the initially identified E3 ligase which is involved in the regulation of various tyrosine kinase-linked receptors such as growth factor receptors (for example EGF and PDGF receptors), cytokine receptors and immuno-receptors (for example T-cell, B-cell and Fc-receptors).[36] Cbl recognizes activated protein tyrosine kinases and recruits E2 ubiquitin conjugating enzymes through its SH2 and RING finger domain, respectively. For EGF and PDGF receptors, increased recruitment of Cbl to the activated receptor complex leads to enhanced ubiquitination and degradation of the activated receptor. In contrast, oncogenic mutation in the Cbl RING finger which fails to bind E2 ubiquitin conjugating enzymes abrogates Cbl-mediated EGF receptor ubiquitination and degradation.[37] Thus, it appears that Cbl functions as an adapter to recruit the ubiquitination machinery to activated tyrosine kinase-linked receptors and stimulates receptor ubiquitination and degradation. This causes enhanced down-regulation of the receptor from the cell surface and attenuation of growth factor receptor signaling.

The RING finger protein Mdm2 is identified as an E3 ubiquitin ligase of the tumor-suppressor protein p53 which is a transcription factor and a potent inhibitor of the cell cycle. Mdm2 can bind to p53 and promote its ubiquitination and subsequent degradation by the proteasome.[38,39] It is also known that Mdm2 can ubiquitinate itself, suggesting that some of E3s self-regulate their own stability. The RING finger of Mdm2 is necessary for both p53 ubiquitin and Mdm2 auto-ubiquitination. Substitution of the Mdm2 RING finger domain with the RING finger from another RING protein maintains the autoubiquination and degradation of Mdm2 but is not able to stimulate p53 ubiquitination. Moreover, mutations in the RING finger domain do not impair binding capacity between Mdm2 and p53. These

Gene		Structure				
TRIM1	MID2, FXY2	R	B1	B2	CC	SPRY
TRIM2	NARF	R		B2	CC	NHL
TRIM3	BERP, RNF22	R		B2	CC	NHL
TRIM4		R		B2	CC	SPRY
TRIM5		R		B2	CC	SPRY
TRIM6	IFP1 (long)	R		B2	CC	SPRY
TRIM7	GNIP1	R		B2	CC	SPRY
TRIM8	GERP, RNF27	R	B1	B2	CC	
TRIM9		R	B1	B2	CC	SPRY
TRIM10	HERF1, RNF9	R		B2	CC	SPRY
TRIM11	BIA1	R		B2	CC	SPRY
mTRIM12		R		B2	CC	
TRIM13	RFP2	R		B2	CC	
TRIM14				B2	CC	SPRY
TRIM15		R		B2	CC	SPRY
TRIM16	BEEP		B1	B2	CC	SPRY
TRIM17	TERF	R		B2	CC	SPRY
TRIM18	MID1, FXY	R	B1	B2	CC	SPRY
TRIM19	PML	R	B1	B2	CC	
TRIM21	SSA/Ro	R		B2	CC	SPRY
TRIM22	STAF50	R		B2	CC	SPRY
TRIM23	ARD1	R	B1	B2	CC	ARF
TRIM24	TIF1α	R	B1	B2	CC	PHD, BROMO
TRIM25	EFP	R	B1	B2	CC	SPRY
TRIM26	AFP	R		B2	CC	SPRY
TRIM27	RFP	R		B2	CC	SPRY
TRIM28	TIF1β, KAP1	R	B1	B2	CC	PHD, BROMO
TRIM29	ATDC		B1	B2	CC	
mTRIM30	mRPT1	R		B2	CC	SPRY
TRIM31	RING	R		B2	CC	
TRIM32	HT2A	R	B1		CC	NHL
TRIM33	TIF1γ	R	B1	B2	CC	PHD, BROMO
TRIM34	IFP1 (middle)	R		B2	CC	SPRY
TRIM35	mNC8	R		B2	CC	SPRY
TRIM36		R	B1	B2	CC	SPRY
TRIM37	MUL, TEF3	R	B1	B2	CC	TRAF
TRIM38	RoRet, RNF15	R		B2	CC	SPRY
TRIM39	TFP	R		B2	CC	SPRY
TRIM40	RNF35	R			CC	mSPRY
TRIM41				B2	CC	SPRY
TRIM42		R	B1	B2	CC	FN3
TRIM43		R		B2	CC	SPRY
TRIM44	DIPB			B2		
TRIM45		R	B1	B2	CC	Filamin
TRIM46		R	B1	B2	CC	FN3
TRIM47	GOA	R	B1	B2	CC	SPRY
TRIM48		R		B2	CC	SPRY

Figure 3. RBCC/TRIM family genes. The gene names are listed in numerical order of *TRIM* genes followed by commonly used names in the second column. Their domain structures are schematically shown to the right.

observations suggest that the RING finger domain appears to be required for specific recognition of substrates in some degree, but is not generally involved in substrate binding.[40]

RBCC/TRIM Subfamily

Frequently, the RING is associated with cysteine-rich B-box domains followed by a predicted coiled coil domain. The B-box domain can be defined as a series of conserved cystein and histidine residues: B1 [Cys-X_2-Cys-$X_{7\text{-}10}$-Cys-X_2-Cys-$X_{4\text{-}5}$-Cys-X_2-Cys/His-$X_{3\text{-}6}$-His-$X_{2\text{-}8}$-His] and B2 [Cys-$X_{2\text{-}4}$-His/Cys-$X_{4\text{-}9}$-Cys-X_2-Cys/His-X_4-Cys/His-X_2-His/Cys] where X can be any amino acid. Structural analysis revealed that it consists of 15 or fewer β-strands.[41] The coiled coil is a common protein motif involving a number of α-helices wound around each other in a highly organized manner and is often used to control oligomerization.[42] These RING, one or two B-boxes and a coiled coil domain motifs are

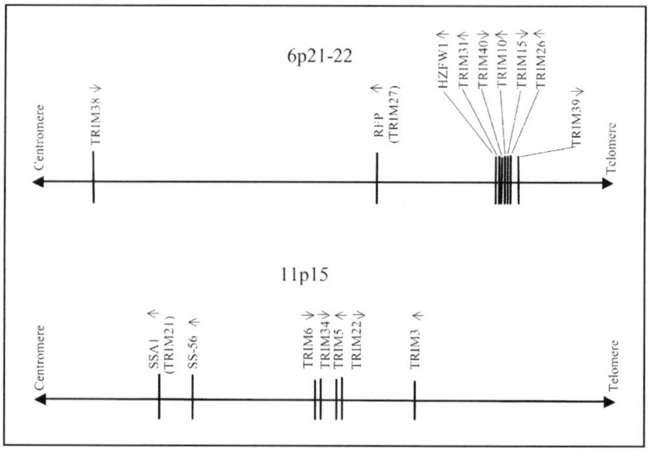

Figure 4. Clustered localization of the *RBCC/TRIM* genes in 6p21-22 and 11p15. *TRIM22, SSA1, TRIM34, TRIM6, TRIM5, TRIM3* and *SS-56* are clustered in 11p15, whereas *RFP, TRIM31, TRIM10, TRIM15, TRIM26, TRIM38, TRIM39, TRIM40* and *HZFW1* are clustered in 6p21-22. The sense strand orientations of each gene are indicated by the arrows.

called RBCC or tripartite motif (TRIM) (Fig. 3).[43] This largest subfamily was first identified in a putative transcriptional regulator, Xenopus XNF7[24] and about 50 members have been identified since then. Though either one or two B-boxes are present, the spacing between the RING, B-boxes and the coiled coil is highly conserved with 38-40 residues between the RING and first B-box, and less than 10 amino acids between the second B-box and the coiled coil. There is no apparent homology among these separating sequences. According to the recent progress of genomic analysis, it was revealed that the chromosomal localization of the RBCC/TRIM subfamily genes has an intriguing feature. Although the genes encoding the RBCC/TRIM family members are dispersed throughout the genome, there are two distinct clusters on chromosomes 11p15 and 6p21-22 (Fig. 4). *TRIM22, SSA1/TRIM21, TRIM34, TRIM6, TRIM5, TRIM3* and *SS-56* are clustered in 11p15. *RFP/TRIM27, TRIM31, TRIM10, TRIM15, TRIM26, TRIM38, TRIM39, TRIM40* and *HZFW1* are clustered in 6p21-22. In particular, mRNAs for *TRIM 21, 22* and *34* are shown to be up-regulated by interferons.[44-46] These findings suggest that duplication of an ancestral *RBCC/TRIM* gene at these genomic loci may have occurred and, regulation and function may be conserved to some extents in these genes.[43]

The RBCC/TRIM proteins are associated with certain domains such as B30.2-like or SPRY, NHL, PHD and BROMO domains at their C-terminus. These additional domains may contribute to the function of the subfamily. The B30.2-like domain is a series of 160-170 amino acids containing three highly conserved motifs (LDP, WEVE and LDYE) which is named after the B30-2 exon within the human class I histocompatibility complex locus. The SPRY domain, which was originally named from SPla and the RYanodine Receptor, is composed of around 140 amino acids containing the latter two conserved motifs in B30.2-like domain. At present, the B30.2-like domain is considered as a subclass of the SPRY domain family. The SPRY domain is contained in many of the RBCC/TRIM subfamily including XNF-7, RPT-1, SSA-Ro and STAF50. The significantly conserved SPRY domains imply the biological importance of this gene, however the function of the SPRY domain is not known. The NHL domain name was derived from the three founding members:

NCL-1, HT2A, and LIN-41.[47] NCL is involved in rRNA metabolism. HT2A was identified as an interacting partner of the HIV Tat protein and Lin41 is involved in posttranscriptional regulation of mRNA. The NHL motif has a slight homology with WD40 domain, suggesting a protein-protein interaction. The C-terminal PHD fingers and bromodomains are found in TIF1α and KAP1/TIF1β. The PHD domain is a motif characteristically defined by seven cysteines and a histidine that are highly homologous to the RING motif and is contained in some transcription factors. The PHD domain of MEKK1 (MEK kinase 1) exhibited E3 ubiquitin ligase activity toward ERK (extracellular signal-regulated protein kinase) 2, suggesting a negative regulatory mechanism for decreasing ERK1/2 activity.[48] The bromodomain is also found in transcription factors, can bind histones with acetylated lysines and appears to be involved in chromatin remodeling.[49] KAP1/TIF1β and TIF1α are involved in transcriptional regulation. Genes belonging to this RBCC/TRIM family are implicated in a variety of processes such as development and cell growth and are involved in several human diseases. PYRIN,[50] MID1 (Midline 1)[51] and MUL (for mulibrey nanism proteins)[52] are mutated in familial Mediterranean fever, X-linked Opitz/GBBB syndrome and mulibrey nanism, respectively, whereas PML, RFP (ret finger protein) and TIF1α acquire oncogenic activity when fused to RARα, RET or B-raf, respectively.

It has been shown that the RBCC/TRIM proteins can oligomerize through their coiled coil domains. In homodimerization of RFP proteins, the coiled coil region with the B-box but not the RING finger is required.[53] In this case, while the B-box is not an interacting interface itself, the mutation of conserved cysteine residues within the B-box affects the ability of RFP to multimerize, suggesting that its structural integrity is necessary for this interaction to occur.[53] The coiled coil domain of RFP is also necessary for interaction with Enhancer of polycomb protein (EPC) to repress gene transcription.[54] The homodimerization and binding with EPC occurs with the proximal coil in RFP protein. RFP also directly interacts and colocalizes with PML in a subset of the PML NBs (nuclear bodies).[55] This interaction is mediated by the RFP B-box and the distal two coils. The association of RFP with the PML NBs is altered by mutations that affect RFP/PML interaction and in APL patients-derived cells. These results indicate that RFP have an important role in regulating cellular growth and differentiation. MID1 protein, which is mutated in patients with Opitz GBBB syndrome, and the highly related gene MID2 also make both homo- and hetero-dimers mediated by the coiled coil motifs. The dimerization is a prerequisite for the association of MID and Alpha 4 (a regulatory subunit of PP2-type phosphatases) and the complex formation with microtubules which seems important for normal midline development.[56] In contrast, it has been shown that the entire RBCC/TRIM domain is required for hetero-oligomerization or binding natural ligands. The RBCC region of KAP1/TIF1β associates with the KRAB (Krüppel-associated box) transcriptional repressor domain of KOX-1.[57] From extensive studies of the interaction, it has been revealed that the interaction is specific for the KAP1 RBCC/TRIM domain. Namely, when each RBCC/TRIM motifs of KAP1 was swapped with other corresponding ones of MID1, KAP1 did not bind the KRAB domain any more. Therefore, each domain of the RBCC may function as an independent functional unit and have important roles in the specific recognition of interacting partners or oligomers formation. In other RBCC/TRIM proteins, only one copy of the B-box motif is present, but inspection of the whole family reveals a conserved residue spacing between the RING, B-box and coiled coil domains.[43] This strongly suggests that the overall architecture of the RBCC/TRIM motif is highly conserved, perhaps relating to the motif acting as a scaffold for higher-order protein-protein interactions.[57] Molecular modeling suggests that the position and orientation of the B-box (adjacent to the coiled coil) would be critical for the correct alignment of the α-helices that form the coiled coil. Interestingly, unlike the RING and coiled coil motifs, the B-box is only found in RBCC/TRIM family members suggesting that it is a critical determinant of the overall motif and its function.[43]

As mentioned above, the latest findings of RING fingers in E3 ubiquitin ligases imply that the members of this RBCC/TRIM subfamily are potential candidates for specific regulators/adopters in ubiquitin-dependent protein degradation. In fact, some genes belonging to the subfamily has been proven to act as E3 ligase. We next discuss such genes focusing on the recent findings.

Ring Fingers That Act As E3 Ligases

Efp

Estrogen-responsive finger protein (Efp) is a member of the RING-finger, B1 and B2-boxes, coiled coil and SPRY (RBCC-SPRY) subfamily in the RING finger family. Efp was isolated as an estrogen-responsive gene by genomic binding-site cloning using a recombinant estrogen receptor (ER) protein.[58] The estrogen-responsive element (ERE) to which ER can bind is found at the 3'-untranslated region (UTR) in the *Efp* gene and the gene's expression is predominantly detected in female reproductive organs including uterus, ovary and mammary gland[59] and in breast and ovarian cancers.[60] Estrogen-induced expression is found in the uterus, brain and mammary gland cells. *Efp* knockout mice have an underdeveloped uterus and estrogen responses of uterin cells from knockout mice are markedly attenuated, suggesting that Efp is necessary for estrogen-induced cell growth.[61] Moreover, tumor growth of breast cancer MCF7 cells implanted in female athymic mice has been demonstrated to be reduced by treatment with antisense Efp oligonucleotide. In contrast, Efp-overexpressing MCF7 cells in ovariectomized athymic mice generate tumors in the absence of estrogen.[33] These results indicate that Efp mediates estrogen-dependent growth in breast cancer cells. We identified 14-3-3σ which is responsible for reduced cell growth, as a binding factor to Efp and found an accumulation of 14-3-3σ in *Efp* knockout mouse embryonic fibroblasts. Furthermore, it has been revealed that Efp is an ubiquitin ligase (E3) that targets proteolysis of 14-3-3σ. Specifically, the RING which preferentially bound to ubiquitin-conjugating enzyme UbcH8 has been shown to be essential for the ubiquitination of 14-3-3σ. Our findings provide an insight into the cell-cycle machinery and tumorigenesis of breast cancer by identifying 14-3-3σ as a target for proteolysis by Efp, leading to cell proliferation. The degradation of 14-3-3σ is subsequently followed by dissociation of the protein from cyclin–Cdk complexes, leading to cell cycle progression and tumor growth (Fig. 5).

MID1, MID2

Opitz GBBB syndrome (OS; Opitz syndrome) is a genetically and phenotypically complex disorder defined by characteristic facial anomalies, structural heart defects, as well as anal and genital anomalies.[62,63] A positional cloning approach has revealed a candidate gene designated *MID1*[51] which is a member of the RBCC-SPRY family. Most of the mutations identified so far in

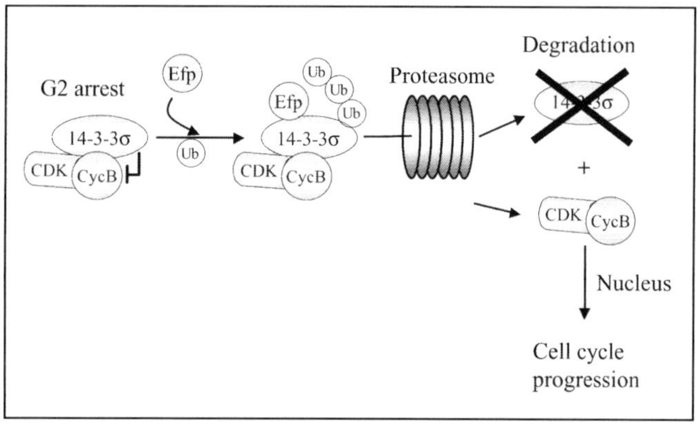

Figure 5. Models of Efp action as E3 ligase. Estrogen-induced RING finger protein Efp recognizes a cell cycle inhibitor 14-3-3σ which keeps Cyclin B in cytoplasm. Efp modifies 14-3-3σ with ubiquitin and the resulting ubiquitinated 14-3-3σ is recruited to 26S proteosome to be destroyed. The dissociated cyclin B is now capable of entering the nucleus where it drives cell cycle.

patients with Opitz syndrome cluster in the SPRY domain of MID1. It has been shown that MID1 associates with microtubules, whereas mutant forms of MID1 do not.[64] These results suggest that MIDI has a physiological role in microtubule dynamics.

Recently, the α4 protein, a regulatory subunit of protein phosphatase 2A (PP2A)[65] was isolated by yeast two-hybrid screening with MID1 as bait. It was demonstrated that the B-box 1 is sufficient for a strong interaction with α4. MID2,[66] which is highly similar to MID1, also binds α4. Cellular localizations of MID1 and α4 are coincident with cytoskeletal structures and MID1 with a mutation at the C terminus that mimics the mutant protein of some individuals with OS results in the formation of cytoplasmic clumps containing both proteins. The identified substrate for E3 ligase activity of MID1 is a cytosolic PP2A. In contrast, addition of a proteasome inhibitor to OS-derived fibroblasts expressing dysfunctional MID1 does not cause either enrichment of PP2A or accumulation of the enzyme's polyubiquitinated forms,[67] suggesting that MID1 mutations result in decreased proteolysis of the C subunit of PP2A in individuals with OS.

PML

PML also belongs to a subfamily of proteins containing a RBCC/TRIM motif.[43,68] PML has been implicated in the pathogenesis of acute promyelocytic leukemia that arises following a reciprocal chromosomal translocation that fuses the *PML* gene located on chromosome 15 with the retinoic acid receptor alpha (*RARα*) gene located on chromosome 17. The resulting PML-RARα fusion protein preserves most of the functional domains of both PML and RARα, but it lacks C-terminus of PML and N-terminus of RARα. The fusion protein shows cell type- and promoter-specific differences from the wild type RARα,[25,26,69] while it maintains a responsiveness to retinoic acid. Overexpression of PML-RARα inhibits vitamin D3 and transforming growth factor β-induced differentiation and also reduces serum starvation-induced apoptosis in U937 cells.[70] In addition, dimerization of PML with PML-RARα is required to block differentiation.[71] Thus, PML-RARα is considered to function as a dominant negative protein by interfering with the function of PML and RARα.

In normal cells, cellular distribution of PML is found to form a discrete subnuclear compartment (nuclear body, NB)[72,73] or PML oncogenic domain.[74] Other proteins containing Sp100[75] and PLZF (for promyelocytic leukemia zinc finger)[76] have been reported to localize to the NBs. Interestingly, PLZF-RARα fusion protein is also found in a rare form of APL.[77] It is shown that the nuclear bodies were dispersed into a microspeckle pattern in APL cells but reformed with retinoic acid treatment by which APL cells differentiated into granulocytes. In addition, the NB is the preferred site where the early steps of transcription and replication of DNA virus occurs.[78] Therefore, the regulation of NB formation is thought to be involved in the pathogenesis of APL. Recently, PML is shown to be covalently modified by SUMO-1 (Small Ubiquitin-like Modifier-1) of ubiquitin-like proteins.[79] Mutations in the PML RING finger disrupt the nuclear body formation in vivo[3,69] and cause a failure of growth suppression,[80,81] apoptosis and anti-viral activities[82] of PML. The dependence on an intact RING finger for PML NBs formation implies specific protein interactions regulated by the RING structure. Recent studies have shown that PML RING interacting with the SUMO-1 E2 enzyme UBC9 is SUMO modified and the sumolylation of PML has an important role in regulating the formation of NBs.[83]

PML has two B-boxes (B1 and B2) adjacent to the RING domain. Mutations of conserved zinc-chelating residues in B1 and/or B2 boxes collapsed PML NB formation, whereas they did not affect PML oligomerization.[84] PML B-boxes are also involved in growth suppression.[80] It has been revealed that PML is sumoylated in B1 box which is responsible for binding of the 11S proteasomal subunit to PML NBs.[85]

The coiled coil region in PML is indispensable for multimerization or heterodimerization with PML-RARα,[69,71,86] formation of PML NB and growth suppression activity.[80] Notably, the important role of the coiled coil domain for the complex formation is also suggested from the studies of other RBCC/TRIM subfamily.[43,57]

TRIM8

TRIM8, a member of RBCC subfamily, is shown to interact with SOCS-1 (suppressor of cytokine signaling-1) which is induced by cytokines and inhibits cytokine signaling by binding to downstream signaling molecules such as JAK (Janus kinase) kinases.[87-89] The B-box coiled coil region of TRIM8 is sufficient for efficient interaction with SOCS-1, but the RING portion of the protein is not required for the binding. By contrast, both the SOCS box and the SH2 domain in SOCS-1 appear to be necessary for the interaction between SOCS-1 and TRIM8. It was found that exogenous coexpression of TRIM8/GERP with SOCS-1 decreased the stability of SOCS-1 protein and TRIM8 restored the IFN-γ-mediated transcription which was inhibited by the expression of SOCS-1.[90] These results suggest that TRIM8 is the putative E3 ligase for SOCS-1 and inhibits SOCS-1 function by targeting it for proteasomal degradation.

TRIM11

TRIM11 is a member of the protein family composed of a RING finger domain, which is a putative E3 ubiquitin ligase, a B-box domain, a coiled coil domain and a SPRY domain. A recent experiment with yeast two-hybrid screening has revealed that TRIM11 can interact with Humanin[91] which is a newly identified anti-apoptotic peptide that specifically suppresses Alzheimer's disease (AD)-related neurotoxicity. It is known that Bax (Bcl2-associated X protein) has a crucial role in apoptosis. In response to death stimuli, Bax protein changes the conformation exposing membrane-targeting domains, translocates to mitochondrial membrane and releases the cytochrome c and other apoptogenic proteins. Indeed, Humanin is shown to bind with Bax and prevents the translocation of Bax from cytosol to mitochondria.[92] Moreover, Humanin blocks Bax association with isolated mitochondria and suppresses cytochrome c release. Therefore, Humanin seems to exert its anti-apoptotic effect by interfering the Bax function.

The coiled coil domain of TRIM11 is indispensable for the interaction with Humanin. The SPRY domain also contributes to the recognition of Humanin, whereas SPRY domain alone cannot. It was found that the intracellular level of Humanin was drastically reduced by the coexpression of TRIM11, and mutation of the RING finger domain or treatment with proteasome inhibitor attenuates the effect of TRIM11 on the intracellular level of Humanin.[91] These results suggest that the TRIM11 participates in the ubiquitin-mediated degradation of Humanin as an E3 ligase.

SSA/Ro (SSA1, TRIM21)

Sjögren syndrome is an autoimmune disease in which exocrine glands including salivary and lacrimal glands develop a chronic inflammation, and whose symptoms are dry eyes, dry mouth and fatigue. Autoantibodies to Ro recognize a ribonucleoprotein complex composed of small single-stranded RNAs and of one or more peptides. Although the Ro autoantigen is heterogeneous and found in most tissues and cells with differences in structure and quantity across tissues, it is detected in 35 to 50% of patients with systemic lupus erythematosus and in up to 97% of patients with Sjögren syndrome.[93] The 60-kD protein (Ro60) and the 52-kD protein (Ro52) were identified[94] and, another novel 56-kD protein (Ro56/SS-56) has been identified, recently.[95] Ro52 and Ro56 proteins belong to RBCC-SPRY subfamily. It is thought that the Ro autoantigen is involved in the regulation of transcription because it possesses functional domains associated with gene-regulation and binds to nucleic acids.[93] Its precise function is not understood, however. In a study, Ro52 was reported to be ubiquitinated in the cell.[96] The observation suggests that Ro52 may be downregulated by the ubiquitin-proteasome pathway in vivo. Interestingly, sera from patients with Sjögren syndrome showed heterogeneity in their reactivity to poly-ubiquitinated Ro52, probably because of their differing antigenic determinants. This heterogeneity of the reactivity may be associated with the varying clinical features found in Sjögren patients.

Conclusion

Here, we summarized the structural characteristics and functions of RING finger proteins specifically in terms of the E3 ligase activity. However, relatively few proteins have been really proven to function as E3 ligase. Thus, most RING finger proteins remain to be further investigated. Investigation of the RING finger proteins as a novel E3 ligase family will elucidate important mechanisms of cellular protein degradation and provide new insight into the physiological and pathophysiological roles of the pathway. Particularly, the molecular mechanisms of specific substrate recognition by E3 with the RING and other associated domains must be determined. Moreover, the RING finger proteins such as PML may possess unknown functions other than E3 ligase. Functional analysis of the RING finger proteins will help to understand biological roles of the family including the ubiquitin-mediated protein degradation pathway.

References

1. Freemont PS, Hanson IM, Trowsdale J. A novel cysteine-rich sequence motif. Cell 1991; 64(3):483-484.
2. Lovering R, Hanson IM, Borden KL et al. Identification and preliminary characterization of a protein motif related to the zinc finger. Proc Natl Acad Sci USA 1993; 90(6):2112-2116.
3. Borden KL, Boddy MN, Lally J et al. The solution structure of the RING finger domain from the acute promyelocytic leukaemia proto-oncoprotein PML. EMBO J 1995; 14(7):1532-1541.
4. Barlow PN, Luisi B, Milner A et al. Structure of the C3HC4 domain by 1H-nuclear magnetic resonance spectroscopy. A new structural class of zinc-finger. J Mol Biol 1994; 237(2):201-211.
5. Bellon SF, Rodgers KK, Schatz DG et al. Crystal structure of the RAG1 dimerization domain reveals multiple zinc-binding motifs including a novel zinc binuclear cluster. Nat Struct Biol 1997; 4(7):586-591.
6. Gervais V, Busso D, Wasielewski E et al. Solution structure of the N-terminal domain of the human TFIIH MAT1 subunit: New insights into the RING finger family. J Biol Chem 2001; 276(10):7457-7464.
7. Zheng N, Wang P, Jeffrey PD et al. Structure of a c-Cbl-UbcH7 complex: RING domain function in ubiquitin-protein ligases. Cell 2000; 102(4):533-539.
8. Freemont PS. The RING finger. A novel protein sequence motif related to the zinc finger. Ann N Y Acad Sci 1993; 684:174-192.
9. Saurin AJ, Borden KL, Boddy MN et al. Does this have a familiar RING? Trends Biochem Sci 1996; 21(6):208-214.
10. Takeuchi M, Rothe M, Goeddel DV. Anatomy of TRAF2. Distinct domains for nuclear factor-kappaB activation and association with tumor necrosis factor signaling proteins. J Biol Chem 1996; 271(33):19935-19942.
11. Sato T, Irie S, Reed JC. A novel member of the TRAF family of putative signal transducing proteins binds to the cytosolic domain of CD40. FEBS Lett 1995; 358(2):113-118.
12. Cheng G, Cleary AM, Ye ZS et al. Involvement of CRAF1, a relative of TRAF, in CD40 signaling. Science 1995; 267(5203):1494-1498.
13. Rothe M, Sarma V, Dixit VM et al. TRAF2-mediated activation of NF-kappa B by TNF receptor 2 and CD40. Science 1995; 269(5229):1424-1427.
14. Takahashi R, Deveraux Q, Tamm I et al. A single BIR domain of XIAP sufficient for inhibiting caspases. J Biol Chem 1998; 273(14):7787-7790.
15. Liston P, Fong WG, Kelly NL et al. Identification of XAF1 as an antagonist of XIAP anti-Caspase activity. Nat Cell Biol 2001; 3(2):128-133.
16. Yang Y, Fang S, Jensen JP et al. Ubiquitin protein ligase activity of IAPs and their degradation in proteasomes in response to apoptotic stimuli. Science 2000; 288(5467):874-877.
17. Clem RJ, Miller LK. Control of programmed cell death by the baculovirus genes p35 and iap. Mol Cell Biol 1994; 14(8):5212-5222.
18. van der Reijden BA, Erpelinck-Verschueren CA, Lowenberg B et al. TRIADs: A new class of proteins with a novel cysteine-rich signature. Protein Sci 1999; 8(7):1557-1561.
19. Kitada T, Asakawa S, Hattori N et al. Mutations in the parkin gene cause autosomal recessive juvenile parkinsonism. Nature 1998; 392(6676):605-608.
20. Morett E, Bork P. A novel transactivation domain in parkin. Trends Biochem Sci 1999; 24(6):229-231.
21. Imai Y, Soda M, Inoue H et al. An unfolded putative transmembrane polypeptide, which can lead to endoplasmic reticulum stress, is a substrate of Parkin. Cell 2001; 105(7):891-902.
22. Imai Y, Soda M, Hatakeyama S et al. CHIP is associated with Parkin, a gene responsible for familial Parkinson's disease, and enhances its ubiquitin ligase activity. Mol Cell 2002; 10(1):55-67.
23. Zhang Y, Gao J, Chung KK et al. Parkin functions as an E2-dependent ubiquitin-protein ligase and promotes the degradation of the synaptic vesicle-associated protein, CDCrel-1. Proc Natl Acad Sci USA 2000; 97(24):13354-13359.
24. Reddy BA, Etkin LD, Freemont PS. A novel zinc finger coiled-coil domain in a family of nuclear proteins. Trends Biochem Sci 1992; 17(9):344-345.

25. Kakizuka A, Miller Jr WH, Umesono K et al. Chromosomal translocation t(15;17) in human acute promyelocytic leukemia fuses RAR alpha with a novel putative transcription factor, PML. Cell 1991; 66(4):663-674.
26. de The H, Lavau C, Marchio A et al. The PML-RAR alpha fusion mRNA generated by the t(15;17) translocation in acute promyelocytic leukemia encodes a functionally altered RAR. Cell 1991; 66(4):675-684.
27. Goddard AD, Borrow J, Freemont PS et al. Characterization of a zinc finger gene disrupted by the t(15;17) in acute promyelocytic leukemia. Science 1991; 254(5036):1371-1374.
28. Miki Y, Swensen J, Shattuck-Eidens D et al. A strong candidate for the breast and ovarian cancer susceptibility gene BRCA1. Science 1994; 266(5182):66-71.
29. Le Douarin B, Zechel C, Garnier JM et al. The N-terminal part of TIF1, a putative mediator of the ligand-dependent activation function (AF-2) of nuclear receptors, is fused to B-raf in the oncogenic protein T18. EMBO J 1995; 14(9):2020-2033.
30. Hershko A, Ciechanover A. The ubiquitin system. Annu Rev Biochem 1998; 67:425-479.
31. Joazeiro CA, Wing SS, Huang H et al. The tyrosine kinase negative regulator c-Cbl as a RING-type, E2-dependent ubiquitin-protein ligase. Science 1999; 286(5438):309-312.
32. Lorick KL, Jensen JP, Fang S et al. RING fingers mediate ubiquitin-conjugating enzyme (E2)-dependent ubiquitination. Proc Natl Acad Sci USA 1999; 96(20):11364-11369.
33. Urano T, Saito T, Tsukui T et al. Efp targets 14-3-3 sigma for proteolysis and promotes breast tumour growth. Nature 2002; 417(6891):871-875.
34. Zhang Y, Xiong Y. Control of p53 ubiquitination and nuclear export by MDM2 and ARF. Cell Growth Differ 2001; 12(4):175-186.
35. Seol JH, Feldman RM, Zachariae W et al. Cdc53/cullin and the essential Hrt1 RING-H2 subunit of SCF define a ubiquitin ligase module that activates the E2 enzyme Cdc34. Genes Dev 1999; 13(12):1614-1626.
36. Galisteo ML, Dikic I, Batzer AG et al. Tyrosine phosphorylation of the c-cbl proto-oncogene protein product and association with epidermal growth factor (EGF) receptor upon EGF stimulation. J Biol Chem 1995; 270(35):20242-20245.
37. Thien CB, Walker F, Langdon WY. RING finger mutations that abolish c-Cbl-directed polyubiquitination and downregulation of the EGF receptor are insufficient for cell transformation. Mol Cell 2001; 7(2):355-365.
38. Haupt Y, Maya R, Kazaz A et al. Mdm2 promotes the rapid degradation of p53. Nature 1997; 387(6630):296-299.
39. Kubbutat MH, Jones SN, Vousden KH. Regulation of p53 stability by Mdm2. Nature 1997; 387(6630):299-303.
40. Geyer RK, Yu ZK, Maki CG. The MDM2 RING-finger domain is required to promote p53 nuclear export. Nat Cell Biol 2000; 2(9):569-573.
41. Chen Y, Chen CF, Riley DJ et al. Aberrant subcellular localization of BRCA1 in breast cancer. Science 1995; 270(5237):789-791.
42. Lupas A. Coiled coils: New structures and new functions. Trends Biochem Sci 1996; 21(10):375-382.
43. Reymond A, Meroni G, Fantozzi A et al. The tripartite motif family identifies cell compartments. EMBO J 2001; 20(9):2140-2151.
44. Tissot C, Mechti N. Molecular cloning of a new interferon-induced factor that represses human immunodeficiency virus type 1 long terminal repeat expression. J Biol Chem 1995; 270(25):14891-14898.
45. Der SD, Zhou A, Williams BR et al. Identification of genes differentially regulated by interferon alpha, beta, or gamma using oligonucleotide arrays. Proc Natl Acad Sci USA 1998; 95(26):15623-15628.
46. Orimo A, Tominaga N, Yoshimura K et al. Molecular cloning of ring finger protein 21 (RNF21)/interferon-responsive finger protein (ifp1), which possesses two RING-B box-coiled coil domains in tandem. Genomics 2000; 69(1):143-149.
47. Slack FJ, Ruvkun G. A novel repeat domain that is often associated with RING finger and B-box motifs. Trends Biochem Sci 1998; 23(12):474-475.
48. Lu Z, Xu S, Joazeiro C et al. The PHD domain of MEKK1 acts as an E3 ubiquitin ligase and mediates ubiquitination and degradation of ERK1/2. Mol Cell 2002; 9(5):945-956.
49. Dhalluin C, Carlson JE, Zeng L et al. Structure and ligand of a histone acetyltransferase bromodomain. Nature 1999; 399(6735):491-496.
50. Ancient missense mutations in a new member of the RoRet gene family are likely to cause familial Mediterranean fever. The International FMF Consortium. Cell 1997; 90(4):797-807.
51. Quaderi NA, Schweiger S, Gaudenz K et al. Opitz G/BBB syndrome, a defect of midline development, is due to mutations in a new RING finger gene on Xp22. Nat Genet 1997; 17(3):285-291.
52. Avela K, Lipsanen-Nyman M, Idanheimo N et al. Gene encoding a new RING-B-box-Coiled-coil protein is mutated in mulibrey nanism. Nat Genet 2000; 25(3):298-301.
53. Cao T, Borden KL, Freemont PS et al. Involvement of the rfp tripartite motif in protein-protein interactions and subcellular distribution. J Cell Sci 1997; 110(Pt 14):1563-1571.
54. Shimono Y, Murakami H, Hasegawa Y et al. RET finger protein is a transcriptional repressor and interacts with enhancer of polycomb that has dual transcriptional functions. J Biol Chem 2000; 275(50):39411-39419.
55. Cao T, Duprez E, Borden KL et al. Ret finger protein is a normal component of PML nuclear bodies and interacts directly with PML. J Cell Sci 1998; 111(Pt 10):1319-1329.
56. Trockenbacher A, Suckow V, Foerster J et al. MID1, mutated in Opitz syndrome, encodes an ubiquitin ligase that targets phosphatase 2A for degradation. Nat Genet 2001; 29(3):287-294.
57. Peng H, Begg GE, Schultz DC et al. Reconstitution of the KRAB-KAP-1 repressor complex: A model system for defining the molecular anatomy of RING-B box-coiled-coil domain-mediated protein-protein interactions. J Mol Biol 2000; 295(5):1139-1162.
58. Inoue S, Orimo A, Hosoi T et al. Genomic binding-site cloning reveals an estrogen-responsive gene that encodes a RING finger protein. Proc Natl Acad Sci USA 1993; 90(23):11117-11121.
59. Orimo A, Inoue S, Ikeda K et al. Molecular cloning, structure, and expression of mouse estrogen-responsive finger protein Efp. Colocalization with estrogen receptor mRNA in target organs. J Biol Chem 1995; 270(41):24406-24413.
60. Ikeda K, Orimo A, Higashi Y et al. Efp as a primary estrogen-responsive gene in human breast cancer. FEBS Lett 2000; 472(1):9-13.
61. Orimo A, Inoue S, Minowa O et al. Underdeveloped uterus and reduced estrogen responsiveness in mice with disruption of the estrogen-responsive finger protein gene, which is a direct target of estrogen receptor alpha. Proc Natl Acad Sci USA 1999; 96(21):12027-12032.
62. Opitz JM. G syndrome (hypertelorism with esophageal abnormality and hypospadias, or hypospadias-dysphagia, or "Opitz-Frias" or "Opitz-G" syndrome)—perspective in 1987 and bibliography. Am J Med Genet 1987; 28(2):275-285.
63. Robin NH, Opitz JM, Muenke M. Opitz G/BBB syndrome: Clinical comparisons of families linked to Xp22 and 22q, and a review of the literature. Am J Med Genet 1996; 62(3):305-317.
64. Gaudenz K, Roessler E, Quaderi N et al. Opitz G/BBB syndrome in Xp22: Mutations in the MID1 gene cluster in the carboxy-terminal domain. Am J Hum Genet 1998; 63(3):703-710.
65. Jerome LA, Papaioannou VE. DiGeorge syndrome phenotype in mice mutant for the T-box gene, Tbx1. Nat Genet 2001; 27(3):286-291.
66. Cainarca S, Messali S, Ballabio A et al. Functional characterization of the Opitz syndrome gene product (midin): Evidence for homodimerization and association with microtubules throughout the cell cycle. Hum Mol Genet 1999; 8(8):1387-1396.
67. Short KM, Hopwood B, Yi Z et al. MID1 and MID2 homo- and heterodimerise to tether the rapamycin-sensitive PP2A regulatory subunit, alpha 4, to microtubules: Implications for the clinical variability of X-linked Opitz GBBB syndrome and other developmental disorders. BMC Cell Biol 2002; 3(1):1.
68. Borden KL. RING fingers and B-boxes: Zinc-binding protein-protein interaction domains. Biochem Cell Biol 1998; 76(2-3):351-358.

69. Kastner P, Perez A, Lutz Y et al. Structure, localization and transcriptional properties of two classes of retinoic acid receptor alpha fusion proteins in acute promyelocytic leukemia (APL): Structural similarities with a new family of oncoproteins. EMBO J 1992; 11(2):629-642.
70. Grignani F, Ferrucci PF, Testa U et al. The acute promyelocytic leukemia-specific PML-RAR alpha fusion protein inhibits differentiation and promotes survival of myeloid precursor cells. Cell 1993; 74(3):423-431.
71. Grignani F, Testa U, Rogaia D et al. Effects on differentiation by the promyelocytic leukemia PML/RARalpha protein depend on the fusion of the PML protein dimerization and RARalpha DNA binding domains. EMBO J 1996; 15(18):4949-4958.
72. Weis K, Rambaud S, Lavau C et al. Retinoic acid regulates aberrant nuclear localization of PML-RAR alpha in acute promyelocytic leukemia cells. Cell 1994; 76(2):345-356.
73. Koken MH, Puvion-Dutilleul F, Guillemin MC et al. The t(15;17) translocation alters a nuclear body in a retinoic acid-reversible fashion. EMBO J 1994; 13(5):1073-1083.
74. Dyck JA, Maul GG, Miller Jr WH et al. A novel macromolecular structure is a target of the promyelocyte-retinoic acid receptor oncoprotein. Cell 1994; 76(2):333-343.
75. Szostecki C, Guldner HH, Netter HJ et al. Isolation and characterization of cDNA encoding a human nuclear antigen predominantly recognized by autoantibodies from patients with primary biliary cirrhosis. J Immunol 1990; 145(12):4338-4347.
76. Koken MH, Reid A, Quignon F et al. Leukemia-associated retinoic acid receptor alpha fusion partners, PML and PLZF, heterodimerize and colocalize to nuclear bodies. Proc Natl Acad Sci USA 1997; 94(19):10255-10260.
77. Chen Z, Brand NJ, Chen A et al. Fusion between a novel Kruppel-like zinc finger gene and the retinoic acid receptor-alpha locus due to a variant t(11;17) translocation associated with acute promyelocytic leukaemia. EMBO J 1993; 12(3):1161-1167.
78. Ishov AM, Maul GG. The periphery of nuclear domain 10 (ND10) as site of DNA virus deposition. J Cell Biol 1996; 134(4):815-826.
79. Kamitani T, Nguyen HP, Kito K et al. Covalent modification of PML by the sentrin family of ubiquitin-like proteins. J Biol Chem 1998; 273(6):3117-3120.
80. Fagioli M, Alcalay M, Tomassoni L et al. Cooperation between the RING + B1-B2 and coiled-coil domains of PML is necessary for its effects on cell survival. Oncogene 1998; 16(22):2905-2913.
81. Mu ZM, Chin KV, Liu JH et al. PML, a growth suppressor disrupted in acute promyelocytic leukemia. Mol Cell Biol 1994; 14(10):6858-6867.
82. Regad T, Saib A, Lallemand-Breitenbach V et al. PML mediates the interferon-induced antiviral state against a complex retrovirus via its association with the viral transactivator. EMBO J 2001; 20(13):3495-3505.
83. Kamitani T, Kito K, Nguyen HP et al. Identification of three major sentrinization sites in PML. J Biol Chem 1998; 273(41):26675-26682.
84. Borden KL, Lally JM, Martin SR et al. In vivo and in vitro characterization of the B1 and B2 zinc-binding domains from the acute promyelocytic leukemia protooncoprotein PML. Proc Natl Acad Sci USA 1996; 93(4):1601-1606.
85. Lallemand-Breitenbach V, Zhu J, Puvion F et al. Role of promyelocytic leukemia (PML) sumolation in nuclear body formation, 11S proteasome recruitment, and As2O3-induced PML or PML/retinoic acid receptor alpha degradation. J Exp Med 2001; 193(12):1361-1371.
86. Le XF, Yang P, Chang KS. Analysis of the growth and transformation suppressor domains of promyelocytic leukemia gene, PML. J Biol Chem 1996; 271(1):130-135.
87. Chen XP, Losman JA, Rothman P. SOCS proteins, regulators of intracellular signaling. Immunity 2000; 13(3):287-290.
88. Haque SJ, Harbor PC, Williams BR. Identification of critical residues required for suppressor of cytokine signaling-specific regulation of interleukin-4 signaling. J Biol Chem 2000; 275(34):26500-26506.
89. Terstegen L, Maassen BG, Radtke S et al. Differential inhibition of IL-6-type cytokine-induced STAT activation by PMA. FEBS Lett 2000; 478(1-2):100-104.
90. Toniato E, Chen XP, Losman J et al. TRIM8/GERP RING finger protein interacts with SOCS-1. J Biol Chem 2002; 277(40):37315-37322.
91. Niikura T, Hashimoto Y, Tajima H et al. A tripartite motif protein TRIM11 binds and destabilizes Humanin, a neuroprotective peptide against Alzheimer's disease-relevant insults. Eur J Neurosci 2003; 17(6):1150-1158.
92. Guo B, Zhai D, Cabezas E et al. Humanin peptide suppresses apoptosis by interfering with Bax activation. Nature 2003; 423(6938):456-461.
93. Sibilia J. Ro(SS-A) and anti-Ro(SS-A): An update. Rev Rhum Engl Ed 1998; 65(1):45-57.
94. Itoh Y, Reichlin M. Autoantibodies to the Ro/SSA antigen are conformation dependent. I: Anti-60 kD antibodies are mainly directed to the native protein; anti-52 kD antibodies are mainly directed to the denatured protein. Autoimmunity 1992; 14(1):57-65.
95. Billaut-Mulot O, Cocude C, Kolesnitchenko V et al. SS-56, a novel cellular target of autoantibody responses in Sjogren syndrome and systemic lupus erythematosus. J Clin Invest 2001; 108(6):861-869.
96. Fukuda-Kamitani T, Kamitani T. Ubiquitination of Ro52 autoantigen. Biochem Biophys Res Commun 2002; 295(4):774-778.

CHAPTER 17

Structure and Function of the CBP/p300 TAZ Domains

Roberto N. De Guzman, Maria A. Martinez-Yamout, H. Jane Dyson and Peter E. Wright*

Abstract

The TAZ (transcriptional adapter zinc binding) domains in the transcriptional coactivators CBP/p300 are zinc-binding motifs that are primarily involved in protein-protein recognition. The activation domains of more than 30 transcription factors have been reported to bind to the TAZ domains, and each TAZ domain generally binds a different subset of transcription factors. Structurally, the TAZ domains contain three zinc binding sites and four alpha helices that are packed against each other to form a hydrophobic core. Each zinc atom is bound to one histidine and three cysteine residues that are located at the end of one helix, in the interconnecting loop, and at the beginning of a second helix, forming a characteristic HCCC-type zinc binding motif. In the absence of zinc, the proteins are unfolded. The role of the zinc atoms is to organize and stabilize the four amphipathic helices into a global fold, which then serves as the site of protein-protein interaction. The TAZ1 domains of CBP and p300 binds other proteins mostly by hydrophobic interactions in the grooves and canyons created by the helical interfaces. The first three helices of TAZ1 and TAZ2 are structurally homologous; the fourth helix, however, is in a different relative orientation. This major structural difference between the two domains appears to play an important role in determining the protein-binding specificity of the TAZ domain.

Introduction

The protein CBP (cyclic-AMP response element binding protein (CREB) binding protein)[1] and its paralog p300[2] are large multifunctional transcriptional coactivator proteins that bind to the activation domains of numerous transcription factors.[3-8] Structural and biological studies of CBP/p300 paint a picture of a large protein (2441 residues for human CBP) containing several globular domains that are independently capable of interacting with different sets of transcription factors. Three dimensional structures have been reported for the KIX domain[9-11] and the nuclear coactivator binding domain (CABD).[12] In addition, CBP and p300 contain four zinc-binding domains that map to three cysteine/histidine-rich regions, initially designated in the literature as CH1, CH2, and CH3.[13] An alternative nomenclature was proposed by Ponting et al.[14] The CH1 region corresponds to the TAZ1 (for transcriptional adaptor zinc-binding) domain, the CH2 region forms a zinc-binding domain similar to the PHD zinc-finger motif, and the CH3 region forms two independently-folded zinc-binding domains, the ZZ domain, and the TAZ2 domain.

The two TAZ domains are important sites of protein-protein interaction in CBP/p300, and to date, about 30 different transcription factors have been reported to bind to TAZ1 and TAZ2 (Fig. 1). The zinc-binding motifs of TAZ1 and TAZ2 share a high degree of homology, each consisting of one histidine and three cysteine residues (Fig. 2). Sequence homology searches indicate that the TAZ domains are currently found only in CBP/p300-related proteins, and each protein contains two TAZ domains. The TAZ2 domain is always associated with the much larger acetyltransferase module of CBP/p300, an arrangement evolutionarily conserved even in plant proteins.[15,16] Each TAZ domain contains three zinc-binding sites with the consensus sequence His1-X_3-Cys2-$X_{4\text{-}12}$-Cys3-$X_{2\text{-}4}$-Cys4, where X is any

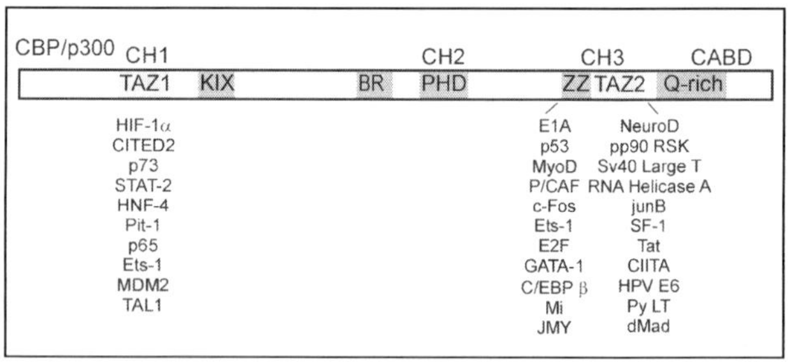

Figure 1. CBP/p300 contains four zinc-binding domains that map to the three cyteine/histidine-rich regions, CH1, CH2, and CH3. The first cysteine/histidine-rich region, CH1, is the TAZ1 domain, and the CH3 region is made of two zinc-binding domains, ZZ and TAZ2. CH2 forms another zinc-binding domain that is homologous to the PHD zinc finger motif. The relative location of the CREB-binding domain (KIX), Bromodomain (BR), glutamine-rich region, and the C-terminal activator binding domain (CABD) of CBP/p300 are indicated. Shown below are the protein factors that have been reported to interact with TAZ1 and the CH3 region.

* *Corresponding author. See list of "Contributors".*

Zinc Finger Proteins: From Atomic Contact to Cellular Function, edited by Shiro Iuchi and Natalie Kuldell.

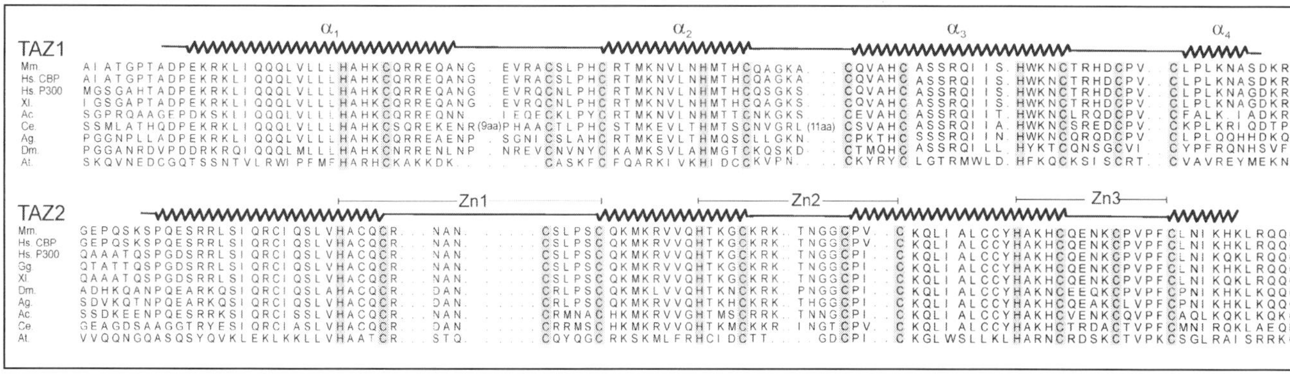

Figure 2. Sequences of TAZ1 and TAZ2 aligned at the three zinc binding clusters (Zn1, Zn2, and Zn3). Secondary structures are indicated as wavy lines for helices and solid lines for loops. The abbreviations and accession numbers are: Mm. (*Mus musculus*/mouse AAL87531); Hs. (*Homo sapiens*/human Q92793 (CBP) / Q09472 (P300)); Xl. (*Xenopus laevis*/frog AAH44677); Ac. (*Aplysia californica*/sea hare AAL54859); Ce. (*Caenorhabditis elegans*/worm P34545); Ag (*Anopheles gambiae*/mosquito EAA06516); Dm. (*Drosophila melanogaster*/fly AAB53050); Gg. (*Gallus gallus*/chicken TC65031); At. (*Arabidopsis thaliana*/plant AAF79331).

amino acid and the subscript denotes the number of residues between the zinc ligands.

Although the zinc-binding motifs of TAZ domains are homologous, TAZ1 and TAZ2 domains have been reported to bind a different subset of transcription factors, as indicated in Figure 1 and Tables 1 and 2. Sequence analyses of the activation domains of their binding partners also do not reveal an obvious sequence homology. CBP and p300 do not bind nucleic acids, and no catalytic activity has been reported for regions containing the TAZ domains. Thus, their primary role appears to be in protein-protein recognition, a common function for a growing list of zinc-binding domains. Structural results from NMR spectroscopy reveal that TAZ1[17-20] and TAZ2[21] form a novel zinc-binding motif and show how they participate in protein-protein interaction. Comparison of the structures of TAZ1 and TAZ2 (see later discussion) illustrates how a structural change in the orientation of one helix allows them to discriminate between two different subsets of binding partners.

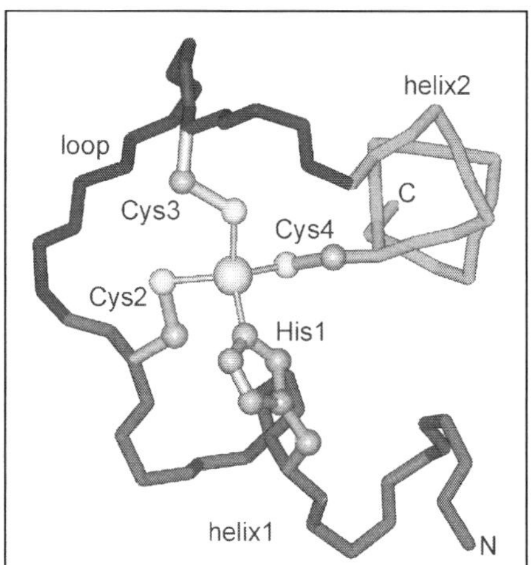

Figure 3. A zinc-binding motif of the TAZ domains consists of four zinc-binding ligands His1, Cys2, Cys3, and Cys4, located at the C-terminal end of a helix, the interconnecting loop, and the N-terminal end of the next helix.

The Zinc-Binding Site

The basic building block of a TAZ domain is a single zinc-binding site, as illustrated in Figure 3. Each zinc binding motif consists of two helices joined by a connecting loop, with one histidine and three cysteine ligands that are arranged sequentially to form an HCCC-type zinc-binding motif.[21] The first two zinc ligands (His1 and Cys2) are located at the end of a helix, the third ligand (Cys3) is in the loop following the helix, and the fourth ligand (Cys4) is at the beginning of the next helix. The zinc atom is coordinated to the histidine Nε2 atom; the other possible metal-binding site, Nδ1, was excluded based on chemical shift analysis. Packing interactions between the two helices are stabilized by hydrophobic contacts mediated by nonpolar residues.

TAZ1 and TAZ2 Structures

The basic zinc-binding cluster in Figure 3 is repeated three times to form a complete TAZ domain, as shown in Figure 4. The NMR structure of the TAZ2 domain from CBP[21] shown in Figure 4 is currently the only available structure of a free TAZ domain. The structure of the free CBP TAZ1 has not yet been reported; the structure shown in Figure 4 was taken from the TAZ1/HIF-1α complex.[17] The free TAZ1 domain from p300 was recently reported to be only partially folded in the presence of $Zn^{2+}$and to behave like a molten globule. Binding of the free p300 TAZ1 to HIF-1α was required for it to form a stable three dimensional structure.[22] This conclusion was based on the poor dispersion of the NMR proton-nitrogen correlation spectra (2D HSQC) of p300 TAZ1 in the presence of Zn^{2+}. We have found, however, that the free TAZ1 domain from CBP has a well-dispersed 2D HSQC spectra in the presence of Zn^{2+} ions, indicating that the CBP domain is folded even in the absence of binding partners. Preliminary NMR structure calculations for the free CBP TAZ1 domain indicate a folded structure that is very similar to that found in the TAZ1/HIF-1α complex (R. N. De Guzman, S.A. Dames, M. A. Martinez-Yamout, J. Wojciak, H.J. Dyson, P.E. Wright, unpublished data).

In TAZ1 and TAZ2, the three zinc-binding motifs, Zn1-Zn3, are distributed approximately at the apices of an irregular triangle. This arrangement maximizes the hydrophobic interactions between the four helices (α_1-α_4), forming a hydrophobic core in the middle of the domain, and directing the connecting loops outwards. The four helices are generally amphipathic in character,

Table 1. List of proteins that have been reported to bind the TAZ1 domain

	Protein	Function	Ref.
HIF-1α	Hypoxia inducible factor	regulates cellular oxygen level	26
CITED2	CBP/p300-interacting transactivator with ED-rich tail	inhibits HIF-1α activity	27
SCL/TAL-1	T-cell acute lymphoblastic leukemia transcription factor	bHLH transcription factor involved chromosomal translocation in T-cell acute lymphoblastic leukemia	45
p73	homolog of p53	apoptosis, growth suppression	46
TBP	TATA-binding protein	basal transcription machinery	47-49
Ets-1	transcription factor	T- and B-cell function and development, Ras signalling	50
Ets-2	transcription factor	cellular transformation, Ras signalling	51
RXR	retinoic acid receptor		
p65		component of NF-κB; inflammation, immune response, control of cell division and apoptosis	52,53
Pit-1	pituitary-specific transcription factor	pituitary cell development, growth hormone expression	54
STAT2	signal transduction and activator of transcription	antiviral defense; IFN-α response	55
HNF-4	hepatocyte nuclear factor	transcriptional regulation of glucose, cholesterol, and fatty acid metabolism	56
PML	promyelocytic leukemia transcription factor	stimulates nuclear receptor transcriptional activity	57

with the hydrophobic residues facing the core of the domain, and the nonpolar residues solvent exposed. The connecting loops are mostly polar. The first three helices, α_1-α_3, are structurally homologous in the TAZ domains. Although the TAZ1 α_1 helix is longer than the corresponding helix in TAZ2 α_1, the relative position of the zinc atom in both domains is very similar.

The major structural difference between TAZ1 and TAZ2 is in the orientation of the fourth helix, α_4 (red in Fig. 4). In TAZ1, both α_1 and α_4 are on the same face of α_3, while in TAZ2, α_3 is sandwiched between α_1 and α_4. The difference in the orientation of the fourth helix also results in a major shift in the position of the zinc atom in Zn3, relative to Zn1 and Zn2.[17] This structural difference is also reflected in the number of inter-helical contacts: there are more inter-helical NOEs between α_1 and α_4 in TAZ1 than in TAZ2. In TAZ2, the arrangement of the four helices results in a more globular structure, approximating a pyramidal-like overall shape. In TAZ1, the overall shape is more of a flatter triangle, or a pie-shaped structure, due to the positioning of α_4. This structural difference will be more evident below in the discussion of the protein-protein complexes formed by TAZ1.

Preparation of TAZ Domains

Some bacterially expressed TAZ1 aggregates in the insoluble fraction, but there is a significant amount (up to ~ 60%) that remains in the soluble fraction. This soluble fraction increases to more than 90% when TAZ1 is coexpressed with binding partners such as HIF-1α or CITED2.[17,20] TAZ1 can be expressed and purified under denaturing conditions, reduced in 10-20 mM dithiothreitol at pH 7.5-8.5, and folded with addition of three equivalents of Zn^{2+} ions.[17,20] The NMR ^{15}N-HSQC spectra for refolded TAZ1 and for TAZ1 purified under native conditions are identical, indicating that the refolding procedure does not alter the biologically relevant structure.[17] In the absence of Zn^{2+} or with an excess of EDTA, TAZ1 and TAZ2 are unstructured in solution, and circular dichroism (CD) spectra indicate that there is little or no residual helical character.

Bacterially expressed TAZ2 is less soluble than TAZ1, forming inclusion bodies[21] that require solubilization in 6 M guanidinium hydrochloride or 8 M urea. Coexpression with its binding partners E1A or p53, also resulted in the formation of inclusion bodies, but some soluble protein is formed. As for TAZ1, three equivalents of Zn^{2+} were needed to fold TAZ2 under reducing conditions.[21] A large excess of Zn^{2+} resulted in a decrease of the helical content in the CD spectra of TAZ2. This probably indicates nonspecific metal binding since TAZ2 contains additional metal binding residues (four cysteines and two histidines).

Protein-Protein Interactions of the TAZ1 Domains

Recent NMR structures of the complexes formed by CBP and p300 TAZ1 reveal the structural determinants of protein recognition by TAZ1.[17-20] Figure 5 shows the structures of complexes formed by the CBP TAZ1 with the C-terminal activation domains (CAD) of HIF-1α[17] and CITED2.[20] The heterodimeric transcription factor HIF-1 (Hypoxia Inducible Factor) plays an important role in the cellular response to low oxygen conditions,

Table 2. List of proteins that have been reported to bind the TAZ2 domain

	Protein	Function	Ref.
E1A	adenovirus early 1A oncoprotein	cellular transformation	2,36-38
HPV E6	human papillomavirus E6 oncoprotein	repression of p53 transcriptional activity	58
SV40 LT	simian virus type 40 large T antigen	simian virus transforming activity	59
p53	tumor suppressor p53 protein	apoptosis, cell cycle control	30-34
Mad	effector of drosophila *Dpp* gene	TFGβ signaling, drosophila Dpp signalling	60
ARIX1/Phox2	homeodomain protein	neuron differentiation	61
Py LT	polyomavirus large T antigen	viral replication and cellular transformation	62
CIITA	class II transactivator	antigen presentation regulates cytokine expression	63
Tax	human T-cell leukemia virus type 1 transactivator	human T-cell leukemia virus transcription	64
Tat	human immunodeficiency virus transactivator	HIV replication	65
SF-1	steroidogenic factor 1	development of the adrenal gland	66
E2F1	transcription factor	cell cycle control	43,67
Ets-1	transcription factor	cell growth and development	50
JunB	transcription factor	cell differentiation	68
p68	RNA helicase	gene transcription	69,70
C/EBP β	CCAAT-box/enhancer binding protein	blood cell development	71
GATA-1	erythroid transcription factor	erythroid cell differentiation	39
BETA-2/Neuro D	beta-cell enriched factor	transcriptional regulation of insulin gene expression	72
Mi	Microphthalmia transcription factor	melanocyte development	73,74
TFIIB	transcription factor II B	basal transcription machinery	75
PCAF	p300/CBP-associated factor	histone acetyltransferase	44
MyoD	myogenic factor	muscle cell differentiation	40
PP90RSK	ribosomal S6 kinase	Ras signaling	76
c-Fos	protooncogene	binds with c-jun to mediate signal transduction, cell differentiation	41
YY1	Ying Yang transcription factor	cell growth and differentiation	42

and thus, is a major protein involved in the pathology of cancer, heart disease and stroke.[23-25] Under conditions of oxygen deprivation, the HIF-1α subunit recruits CBP/p300 to direct expression of genes involved in vascularization and glycolysis to ensure cellular survival.[26] HIF-1 also controls the expression of another nuclear protein, CITED2 (CBP/p300 Interacting Transactivator with glutamic acid (E) and aspartic acid (D) rich Tail) at the transcriptional level. CITED2 competes with HIF-1α for binding to the TAZ1 domain, effectively blocking HIF-1 transcriptional activity.[27] This suggests regulation during hypoxia by a feedback loop mediated by protein-protein interactions of HIF-1α and CITED2with TAZ1. The structures of TAZ1/HIF-1α[17,18] and TAZ1/CITED2[19,20] complexes provide a detailed picture of the protein-protein interactions mediated by a zinc-binding domain during hypoxia.

Structure of the TAZ1/HIF-1α Complex

The structure of the CBP TAZ1 domain in complex with the HIF-1α CAD is shown in Figure 5A. The HIF-1α CAD is unstructured in isolation, but, upon complex formation, folds and wraps around TAZ1, forming three helices, α_A, α_B, and α_C, that bind in the grooves formed by the interfaces of TAZ1 helices.[17,18] The α_A helix is poorly defined and flexible, and interacts mainly with TAZ1 α_1. Although the HIF-1α CAD lacking this N-terminal region is capable of tight binding to TAZ1 (K_d = 46 nM)[17,20] (and indeed this region was not included in the structure determination of the p300/HIF-1α complex),[18] these additional residues increase the affinity by about five-fold (K_d = 10 nM).[17,20] The poorly structured N-terminal α_A helical region is followed by an extended region containing the sequence LPQL, which binds tightly at the junction of the TAZ1 $\alpha_1/\alpha_2/\alpha_3$ interface. Following the LPQL loop, is a short helix, α_B, which binds at the TAZ1 α_1/α_3 interface. This helix contains the so-called hypoxic switch, an asparagine residue in position 803 that when hydroxylated, impairs the interaction of Hif-1α with CBP or p300.[28] Amino acid analysis and NMR has defined the site of hydroxylation at the pro-S position of the C^β atom of Asn803.[29] Finally, the HIF-1α CAD forms the C-terminal α_C, which binds tightly at the TAZ1 α_1/α_4 interface. The TAZ1/CITED2 Complex.

The NMR structures of the p300[19] and CBP[20] TAZ1/CITED2 complexes provide another detailed picture of how TAZ1 serves as a scaffold for protein-protein interaction. The structural features, four helices (α_1-α_4), three zinc binding clusters, and the overall global fold of TAZ1 are preserved in the two complexes. There are minor differences in the TAZ1 structure between the Hif-1α and CITED2 complexes: a slight bending of α_1 in the CITED2 complex,[19,20] and a slight increase in the length of the α_4 in the CBP TAZ1/CITED2 complex.[20] The main difference between the

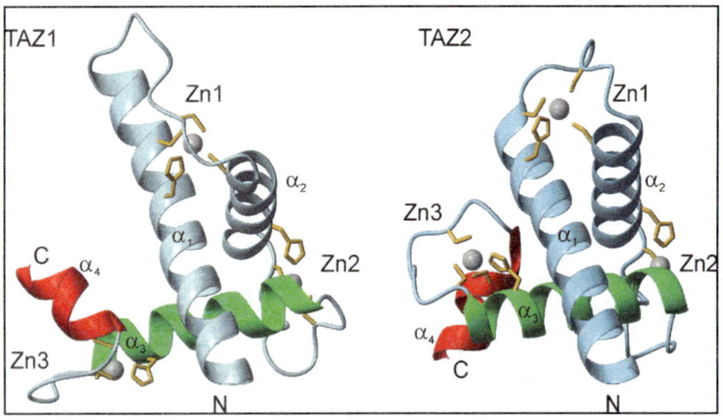

Figure 4. Ribbon diagram of the structures of TAZ1 and TAZ2. A TAZ domain is made of four helices (α_1-α_4) and three zinc-binding clusters (Zn1-Zn3). α_1, α_2 and α_3 (blue and green) are structurally homologous, while α_4 (red) is in opposite relative orientation. α_1 is also longer in TAZ1 than in TAZ2.

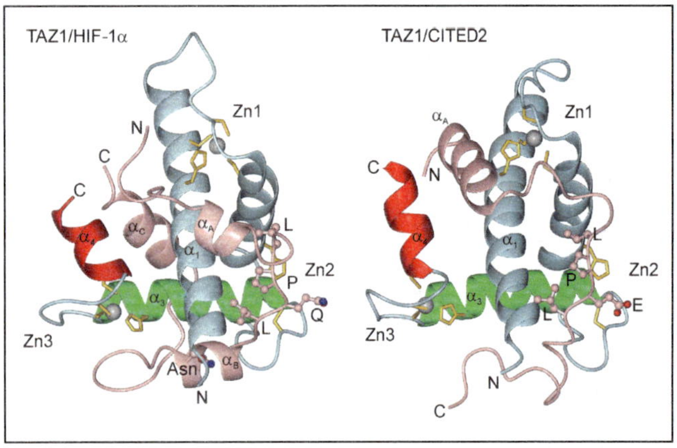

Figure 5. Structures of the protein-protein complexes of the CBP TAZ1 bound to activation domains of HIF-1α (left) and CITED2 (right). The four TAZ1 helices α_1 (blue), α_2 (blue), α_3 (green) , α_4 (red), and the three zinc clusters (Zn1-Zn3) are indicated. The bound proteins, HIF-1α and CITED2, are colored pink and the secondary structures are labeled (α_A-α_B for HIF-1α, and α_A for CITED2). The four residues of the LPQL (HIF-1α) and LPEL (CITED) loops are shown. The hypoxic switch in HIF-1α (Asn803) is located at the C-terminus of α_B.

HIF-1α and CITED2 complexes is in the structures of the two bound ligands. Like the HIF-1α CAD, the free CITED2 activation domain is unstructured and folds on binding to TAZ1,[20] forming a tight-binding helix, α_A, and an extended region that loops around TAZ1 (Fig. 5). The CITED2 α_A binds on the same face of TAZ1 α_1 as the disordered N-terminal loop and helix α_A of HIF-1α.[20] This helix also contacts the TAZ1 α_1/α_4 interface, but on the opposite side and at a more oblique angle compared to the tight-binding helix α_C of HIF-1α.[19,20] The CITED α_A helix is followed by a loop that contains the LPEL motif (see following paragraph). There is no equivalent structure in the CBP TAZ1/CITED2 complex to the HIF-1α helix α_B,[20] although the structure of the p300 TAZ1/CITED2 complex[19] showed a 3-residue 3_{10}-helix in this region. This difference in the structures of the CBP and p300 CITED2 complexes is probably due to the different CITED2 constructs used. There is no equivalent of HIF-1α helix α_C in either the CBP or p300 TAZ1 complexes with CITED2. The residues at the C-terminus of CITED2 are flexible and do not make contact with the surface of TAZ1.[20] The two protein complexes shown in Figure 5 have very similar dissociation constants, suggesting that HIF-1α and CITED2 exert their biological activities by competing with a limited amount of CBP/p300 and are able to displace each other during the hypoxic response.

The HIF-1α and CITED2 activation domains do not show an obvious sequence homology to each other, but after the structures of the complexes became available, it became evident that the two proteins do share a limited structural and sequence homology. Both domains contain a four-residue sequence, LPQL in HIF-1α and LPEL in CITED2 (shown as a ball-and-stick diagram in Fig. 5) that bind in the same manner at the TAZ1 $\alpha_1/\alpha_2/\alpha_3$ interface.[19,20] This apparent homology of sequence and structure does not appear to be general: sequence analysis of the various transcription factors that bind to TAZ1 reveals that the LP(Q/E)L sequence is not a common motif for binding.

TAZ2 Protein-Protein Interactions

The CBP/p300 TAZ2 domain also participates in protein-protein interaction, with a different subset of transcription factors. A list of proteins that have been reported to bind the TAZ2 domain of CBP/p300 is given in Table 2. Our laboratory has explored the binding of the activation domains of the p53 tumor suppressor protein and the adenoviral E1A oncoprotein to TAZ2. The p53 tumor suppressor plays an important role in cell cycle control and apoptosis; defects in the *p53* gene have been implicated in many types of human cancers. Binding of the N-terminal transactivation domain of p53 with CBP/p300[6] is necessary for the full transcriptional activity of p53 and for p53-mediated control of the cell growth cycle.[30-34] Although no structural information has been reported for the 100-residue transactivation domain of p53, a peptide within this region containing residues involved in transactivation was reported to form an amphipathic helix on binding to MDM2.[35] NMR results indicate that the TAZ2 domain within the CH3 region of CBP/p300 interacts with the p53 transactivation peptide,[21] and chemical shift mapping indicates that the TAZ2 residues mostly affected by p53 binding map to the TAZ2 $\alpha_1/\alpha_2/\alpha_3$ interface.[21]

Another important protein that binds the CH3 region of CBP/p300 is the adenoviral E1A oncoprotein.[2,36-38] Indeed, the CH3 region is often referred to as the E1A-binding domain. The interaction between CBP/p300 and E1A is necessary for the transforming ability of adenovirus, and when E1A is coexpressed with other CBP/p300-binding proteins, it abrogates the interaction between CBP/p300 and the other binding partners.[39-44] This suggests that the CH3 domain plays an important role in the protein-protein interactions mediated by CBP. Although sequence analysis predicts helical regions in the activation domains of p53 and E1A, it is not yet known whether they form helices upon binding to TAZ2.

Conclusions

It is clear that the zinc-binding domains of transcriptional coactivators such as p300 and CBP are of considerable importance in the function of these proteins. Unique zinc-binding folds have been identified for the TAZ domains, and structural studies have been able to provide rationales for the differences in the binding behavior of the TAZ1 and TAZ2 domains. Work continues in a number of laboratories on this fascinating new field.

Acknowledgments

We thank members of the Wright/Dyson lab for continuing helpful discussions and comments. RND is supported by a fellowship from the Leukemia and Lymphoma Society. This work was supported by grant CA96865 from the National Institutes of Health.

References

1. Chrivia JC, Kwok RP, Lamb N et al. Phosphorylated CREB binds specifically to nuclear protein CBP. Nature 1993; 365:855-859.
2. Eckner R, Ewen ME, Newsome D et al. Molecular cloning and functional analysis of the adenovirus E1A- associated 300-kD protein (p300) reveals a protein with properties of a transcriptional adaptor. Genes Devel 1994; 8:869-884.
3. Blobel GA. CBP and p300: Versatile coregulators with important roles in hematopoietic gene expression. J Leukoc Biol 2002; 71:545-556.
4. Janknecht R. The versatile functions of the transcriptional coactivators p300 and CBP and their roles in disease. Histol Histopathol 2002; 17:657-668.
5. Chan HM, La Thangue NB. p300/CBP proteins: HATs for transcriptional bridges and scaffolds. J Cell Sci 2001; 114:2363-2373.
6. Grossman SR. p300/CBP/p53 interaction and regulation of the p53 response. Eur J Biochem 2001; 268:2773-2778.
7. McManus KJ, Hendzel MJ. CBP, a transcriptional coactivator and acetyltransferase. Biochem Cell Biol 2001; 79:253-266.
8. Goodman RH, Smolik S. CBP/p300 in cell growth, transformation, and development. Genes Devel 2000; 14:1553-1577.
9. Goto NK, Zor T, Martinez-Yamout M et al. Cooperativity in transcription factor binding to the coactivator CREB-binding protein (CBP). The mixed lineage leukemia protein (MLL) activation domain binds to an allosteric site on the Kix domain. J Biol Chem 2002; 277:43168-43174.
10. Radhakrishnan I, Pérez-Alvarado GC, Parker D et al. Solution structure of the KIX domain of CBP bound to the transactivation domain of CREB: A model for activator:Coactivator interactions. Cell 1997; 91:741-752.
11. Zor T, Mayr BM, Dyson HJ et al. Roles of phosphorylation and helix propensity in the binding of the KIX domain of CREB-binding protein by constitutive (c-Myb) and inducible (CREB) activators. J Biol Chem 2002; 277:42241-42248.
12. Demarest SJ, Martinez-Yamout M, Chung J et al. Mutual synergistic folding in recruitment of CBP/p300 by p160 nuclear receptor coactivators. Nature 2002; 415:549-553.
13. Borrow J, Stanton Jr VP, Andresen JM et al. The translocation t(8;16)(p11;p13) of acute myeloid leukaemia fuses a putative acetyltransferase to the CREB-binding protein. Nat Genet 1996; 14:33-41.
14. Ponting CP, Blake DJ, Davies KE et al. ZZ and TAZ: New putative zinc fingers in dystrophin and other proteins. Trends Biochem Sci 1996; 21:11-13.
15. Pandey R, Muller A, Napoli CA et al. Analysis of histone acetyltransferase and histone deacetylase families of Arabidopsis thaliana suggests functional diversification of chromatin modification among multicellular eukaryotes. Nucleic Acids Res 2002; 30:5036-5055.
16. Chakravarti D, Ogryzko V, Kao HY et al. A viral mechanism for inhibition of p300 and PCAF acetyltransferase activity. Cell 1999; 96:393-403.
17. Dames SA, Martinez-Yamout M, De Guzman RN et al. Structural basis for Hif-1 alpha /CBP recognition in the cellular hypoxic response. Proc Natl Acad Sci USA 2002; 99:5271-5276.
18. Freedman SJ, Sun ZY, Poy F et al. Structural basis for recruitment of CBP/p300 by hypoxia-inducible factor-1alpha. Proc Natl Acad Sci USA 2002; 99:5367-5372.
19. Freedman SJ, Sun ZY, Kung AL et al. Structural basis for negative regulation of hypoxia-inducible factor-1alpha by CITED2. Nat Struct Biol 2003; 10:504-512.
20. De Guzman RN, Martinez-Yamout M, Dyson HJ et al. Interaction of the TAZ1 domain of CREB-binding protein with the activation domain of CITED2: Regulation by competition between intrinsically unstructured ligands for nonIdentical binding sites. J Biol Chem 2004; 279:3042-3049.
21. De Guzman RN, Liu HY, Martinez-Yamout M et al. Solution structure of the TAZ2 (CH3) domain of the transcriptional adaptor protein CBP. J Mol Biol 2000; 303:243-253.
22. Dial R, Sun ZY, Freedman SJ. Three conformational states of the p300 CH1 domain define its functional properties. Biochemistry 2003; 42:9937-9945.
23. Semenza GL. HIF-1 and human disease: One highly involved factor. Genes Devel 2000; 14:1983-1991.
24. Semenza GL. Hypoxia-inducible factor 1: Oxygen homeostasis and disease pathophysiology. Trends Mol Med 2001; 7:345-350.
25. Pugh CW, Ratcliffe PJ. Regulation of angiogenesis by hypoxia: Role of the HIF system. Nat Med 2003; 9:677-684.
26. Arany Z, Huang LE, Eckner R et al. An essential role for p300/CBP in the cellular response to hypoxia. Proc Natl Acad Sci USA 1996; 93:12969-12973.
27. Bhattacharya S, Michels CL, Leung MK et al. Functional role of p35srj, a novel p300/CBP binding protein, during transactivation by HIF-1. Genes Devel 1999; 13:64-75.
28. Lando D, Peet DJ, Whelan DA et al. Asparagine hydroxylation of the HIF transactivation domain: A hypoxic switch. Science 2002; 295:858-861.
29. McNeill LA, Hewitson KS, Claridge TD et al. Hypoxia-inducible factor asparaginyl hydroxylase (FIH-1) catalyses hydroxylation at the beta-carbon of asparagine-803. Biochem J 2002; 367:571-575.
30. Avantaggiati M, Ogryzko V, Gardner K et al. Recruitment of p300/CBP in p53-dependent signal pathways. Cell 1997; 89:1175-1184.
31. Gu W, Shi XL, Roeder RG. Synergistic activation of transcription by CBP and p53. Nature 1997; 387:819-823.
32. Lee CW, Sorensen TS, Shikama N et al. Functional interplay between p53 and E2F through coactivator p300. Oncogene 1998; 16:2695-2710.
33. Lill NL, Grossman SR, Ginsberg D et al. Binding and modulation of p53 by p300/CBP coactivators. Nature 1997; 387:823-827.
34. Scolnick DM, Chehab NH, Stavridi ES et al. CREB-binding protein and p300/CBP-associated factor are transcriptional coactivators of the p53 tumor suppressor protein. Cancer Res 1997; 57:3693-3696.
35. Kussie PH, Gorina S, Marechal V et al. Structure of the MDM2 oncoprotein bound to the p53 tumor suppressor transactivation domain. Science 1996; 274:948-953.
36. Stein RW, Corrigan M, Yaciuk P et al. Analysis of E1A-mediated growth regulation functions: Binding of the 300-kilodalton cellular product correlates with E1A enhancer repression function and DNA synthesis-inducing activity. J Virol 1990; 64:4421-4427.
37. Arany Z, Newsome D, Oldread E et al. A family of transcriptional adaptor proteins targeted by the E1A oncoprotein. Nature 1995; 374:81-84.
38. Lundblad JR, Kwok RPS, Laurance ME et al. Adenoviral E1A-associated protein p300 as a functional homologue of the transcriptional coactivator CBP. Nature 1995; 374:85-88.
39. Blobel GA, Nakajima T, Eckner R et al. CREB-binding protein cooperates with transcription factor GATA-1 and is required for erythroid differentiation. Proc Natl Acad Sci USA 1998; 95:2061-2066.
40. Yuan W, Condorelli G, Caruso M et al. Human p300 protein is a coactivator for the transcription factor MyoD. J Biol Chem 1996; 271:9009-9013.
41. Bannister AJ, Kouzarides T. CBP-induced stimulation of c-Fos activity is abrogated by E1A. EMBO J 1995; 14:4758-4762.
42. Lee JS, Galvin KM, See RH et al. Relief of YY1 transcriptional repression by adenovirus E1A is mediated by E1A-associated protein p300. Genes Devel 1995; 9:1188-1198.
43. Trouche D, Cook A, Kouzarides T. The CBP coactivator stimulates E2F1/DP1 activity. Nucleic Acids Res 1996; 24:4139-4145.
44. Yang X-J, Ogryzko VV, Nishikawa J et al. A p300/CBP-associated factor that competes with the adenoviral oncoprotein E1A. Nature 1996; 382:319-324.

45. Huang SM, Qiu Y, Stein RW et al. P300 functions as a transcriptional coactivator for the TAL1/SCL oncoprotein. Oncogene 1999; 18:4958-4967.
46. Zeng X, Li X, Miller A et al. The N-terminal domain of p73 interacts with the CH1 domain of p300/CREB binding protein and mediates transcriptional activation and apoptosis. Mol Cell Biol 2000; 20:1299-1310.
47. Abraham SE, Lobo S, Yaciuk P et al. p300, and p300-associated proteins, are components of TATA-binding protein (TBP) complexes. Oncogene 1993; 8:1639-1647.
48. Dallas PB, Yaciuk P, Moran E. Characterization of monoclonal antibodies raised against p300: Both p300 and CBP are present in intracellular TBP complexes. J Virol 1997; 71:1726-1731.
49. Kwok RPS, Lundblad JR, Chrivia JC et al. Nuclear protein CBP is a coactivator for the transcription factor CREB. Nature 1994; 370:223-226.
50. Yang C, Shapiro LH, Rivera M et al. A role for CREB binding protein and p300 transcriptional coactivators in Ets-1 transactivation functions. Mol Cell Biol 1998; 18:2218-2229.
51. Jayaraman G, Srinivas R, Duggan C et al. p300/cAMP-responsive element-binding protein interactions with Ets-1 and Ets-2 in the transcriptional activation of the human stromelysin promoter. J Biol Chem 1999; 274:17342-17352.
52. Gerritsen ME, Williams AJ, Neish AS et al. CREB-binding protein p300 are transcriptional coactivators of p65. Proc Natl Acad Sci USA 1997; 94:2927-2932.
53. Zhong H, Voll RE, Ghosh S. Phosphorylation of NF-κB p65 by PKA stimulates transcriptional activity by promoting a novel bivalent interaction with the coactivator CBP/p300. Mol Cell Biol 1998; 1:661-671.
54. Kishimoto M, Okimura Y, Yagita K et al. Novel function of the transactivation domain of a pituitary-specific transcription factor, Pit-1. J Biol Chem 2002; 277:45141-45148.
55. Bhattacharya S, Eckner R, Grossman S et al. Cooperation of Stat2 and p300/CBP in signalling induced by interferon-α. Nature 1996; 383:344-347.
56. Dell H, Hadzopoulou-Cladaras M. CREB-binding protein is a transcriptional coactivator for hepatocyte nuclear factor-4 and enhances apolipoprotein gene expression. J Biol Chem 1999; 274:9013-9021.
57. Doucas V, Tini M, Egan DA et al. Modulation of CREB binding protein function by the promyelocytic (PML) oncoprotein suggests a role for nuclear bodies in hormone signaling. Proc Natl Acad Sci USA 1999; 96:2627-2632.
58. Zimmermann H, Degenkolbe R, Bernard HU et al. The human papillomavirus type 16 E6 oncoprotein can down-regulate p53 activity by targeting the transcriptional coactivator CBP/p300. J Virol 1999; 73:6209-6219.
59. Eckner R, Ludlow JW, Lill NL et al. Association of p300 and CBP with simian virus 40 large T antigen. Mol Cell Biol 1996; 16:3454-3464.
60. Waltzer L, Bienz M. A function of CBP as a transcriptional coactivator during Dpp signalling. EMBO J 1999; 18:1630-1641.
61. Swanson DJ, Adachi M, Lewis EJ. The homeodomain protein Arix promotes protein kinase A-dependent activation of the dopamine beta -hydroxylase promoter through multiple elements and interaction with the coactivator cAMP-response element-binding protein-binding protein. J Biol Chem 2000; 275:2911-2923.
62. Cho S, Tian Y, Benjamin TL. Binding of p300/CBP coactivators by polyoma large T antigen. J Biol Chem 2001; 276:33533-33539.
63. Sisk TJ, Gourley T, Roys S et al. MHC class II transactivator inhibits IL-4 gene transcription by competing with NF-AT to bind the coactivator CREB binding protein (CBP)/p300. J Immunol 2000; 165:2511-2517.
64. Suzuki T, Yoshida M. Tax protein of HTLV-1 inhibits CBP/p300-mediated transcription by interfering with recruitment of CBP/p300 onto DNA element of E-box or p53 binding site. Oncogene 1999; 18:4137-4143.
65. Hottiger MO, Nabel GJ. Interaction of human immunodeficiency virus type 1 Tat with the transcriptional coactivators p300 and CREB binding protein. J Virol 1998; 72:8252-8256.
66. Ito M, Yu RN, Jameson JL. Steroidogenic factor-1 contains a carboxy-terminal transcriptional activation domain that interacts with steroid receptor coactivator-1. Mol Endocrinol 1998; 12:290-301.
67. Trouche D, Kouzarides T. E2F1 and E1A(12S) have a homologous activation domain regulated by RB and CBP. Proc Natl Acad Sci USA 1996; 93:1439-1442.
68. Lee JS, See RH, Deng T et al. Adenovirus E1A downregulates cJun- and JunB-mediated transcription by targeting their coactivator p300. Mol Cell Biol 1996; 16:4312-4326.
69. Rossow KL, Janknecht R. Synergism between p68 RNA helicase and the transcriptional coactivators CBP and p300. Oncogene 2003; 22:151-156.
70. Nakajima T, Uchida C, Anderson S et al. RNA helicase A mediates association of CBP with RNA polymerase II. Cell 1997; 90:1107-1112.
71. Mink S, Haenig B, Klempnauer KH. Interaction and functional collaboration of p300 and C/EBPbeta. Mol Cell Biol 1997; 17:6609-6617.
72. Qiu Y, Sharma A, Stein R. p300 mediates transcriptional stimulation by the basic helix-loop-helix activators of the insulin gene. Mol Cell Biol 1998; 18:2957-2964.
73. Sato S, Roberts K, Gambino G et al. CBP/p300 as a cofactor for the Microphthalmia transcription factor. Oncogene 1997; 14:3083-3092.
74. Price ER, Ding HF, Badalian T et al. Lineage-specific signaling in melanocytes - c-Kit stimulation recruits p300/CBP to microphthalmia. J Biol Chem 1998; 273:17983-17986.
75. Felzien LK, Farrell S, Betts JC et al. Specificity of cyclin E-Cdk2, TFIIB, and E1A interactions with a common domain of the p300 coactivator. Mol Cell Biol 1999; 19:4241-4246.
76. Nakajima T, Fukamizu A, Takahashi J et al. The signal-dependent coactivator CBP is a nuclear target for $pp90_{RSK}$. Cell 1996; 86:465-474.

CHAPTER 18

A Zinc Ribbon Motif Is Essential for the Formation of Functional Tetrameric Protein Kinase CK2

Odile Filhol, Maria José Benitez and Claude Cochet*

Abstract

Protein kinase CK2 plays an essential role in the regulation of many cellular functions. The enzyme is an heterotetrameric complex formed by the association of two catalytic α/α' subunits with two regulatory β subunits. High-resolution structure of the CK2β subunit revealed the presence of a zinc binding motif made by three-stranded antiparallel β sheets and two "knuckles" (Cys-X4-Cys and Cys-X2-Cys) contained in the invariant motif CPX3C-X22-CPXC. This zinc binding motif belongs to the sub-family of zinc ribbon domains. CK2β exist as a dimer in which the zinc ribbon motif makes many hydrophobic interactions with the zinc ribbon motif of the other monomer forming the protein-protein interface. Importantly, functional and biochemical studies have indicated that the integrity of the zinc binding motif which is pivotal in the formation of the CK2β homodimer, is also instrumental for the regulatory functions of this important protein.

Protein phosphorylation is a major mechanism for the regulation of fundamental cellular processes.[1] Among the hundred of protein kinases encoded by the genome of eukaryotic cells, the Ser/Thr protein kinase CK2 (formely called casein kinase II) represents an essential component of this family of regulatory enzymes.[2-5] The broad spectrum of protein substrates which are phosphorylated by CK2 underscores the functional importance of this protein kinase in the regulation of many cellular functions.[5]

Protein kinase CK2 exibits a high degree of conservation of amino acid sequence between evolutionarily distant organisms such as yeast, Drosophila and human.[2] When purified from many sources, CK2 appears as a tetrameric enzyme (with a molecular mass of 130 kDa), composed of two types of structurally analogous catalytic α (42 to 44 kDa) and α'(38 kDa) subunits and two regulatory β (27 kDa) subunits which associate to form stable $\alpha_2\beta_2$, $\alpha\alpha'\beta_2$, or $\alpha'_2\beta_2$ structures. The lethal phenotype resulting from disruption of the α subunit in yeast can be rescued by expression of Drosophila α showing that the catalytic subunit is functionnaly interchangeable between distantly related species.[6] In male mice, CK2α' is preferencially expressed in late stages of spermatogenesis and disruption of its gene results in infertility owing to defective spermatogenesis.[7] However, the apparently normal embryonic development of CK2α' knockout mice argues again for a strong functional overlap of both CK2 isoforms. RNA-mediated interference in Caenorhabditis elegans has shown that inhibition of the CK2β gene leads to embryonic lethality.[8] Similarly, disruption of the CK2β subunit in mice leads to a cell-autonomous defect and early embryonic lethality.[9] Thus, the production of both the α and β subunits of CK2 appears to be required for cell survival.

Mounting evidence suggest that CK2 is a component of regulatory protein kinase networks that are involved in various aspects of transformation and cancer. Targetted overexpression of CK2α in T cells and mammary glands of transgenic mice leads to lymphomagenesis and mammary tumorigenesis, respectively.[10,11] A number of studies have also shown that in all human cancers that have been examined, CK2 activity is consistently enhanced suggesting that the kinase plays a role in cell proliferation but also participates in the transduction of cell survival signals.[12] Finally, the involvement of CK2 in viral infection and oncogenesis has been demonstrated.[3]

Structural and Regulatory Features

Several recent reports have described specific structural aspects of the CK2 subunits and their functional properties providing long-awaited insights into a potential modus operandi of this pivotal protein kinase.

In many organisms as well as in humans several isoforms of the catalytic subunit of CK2 have been identified.[13-16] Although closely related, they are in fact the products of different genes. The crystal structure of the catalytic α subunit from *Zea mays* has shown the common bilobal architecture described for protein kinases. More importantly, this study provided clues for the apparent constitutive activity of the enzyme and for the fact that this kinase can use GTP as well as ATP as phosphate donors.[17]

Distinct isoforms of CK2 regulatory subunits exist in *S. cerevisiae*, *A. thaliana* and in *D. melanogaster*. Surprisingly, in mammals, this subunit is encoded by a single gene. On the basis of sequence comparisons, CK2β does not share similarity with any known protein. However, CK2β exhibits remarkable conservation between species suggesting that this subunit plays important roles in cellular functions. This contention is illustrated by the demonstration that the expression of CK2β is essential for cell proliferation and survival in *D. melanogaster*[18] and in mice.[9]

* *Corresponding author. See list of "Contributors".*

Zinc Finger Proteins: From Atomic Contact to Cellular Function, edited by Shiro Iuchi and Natalie Kuldell.

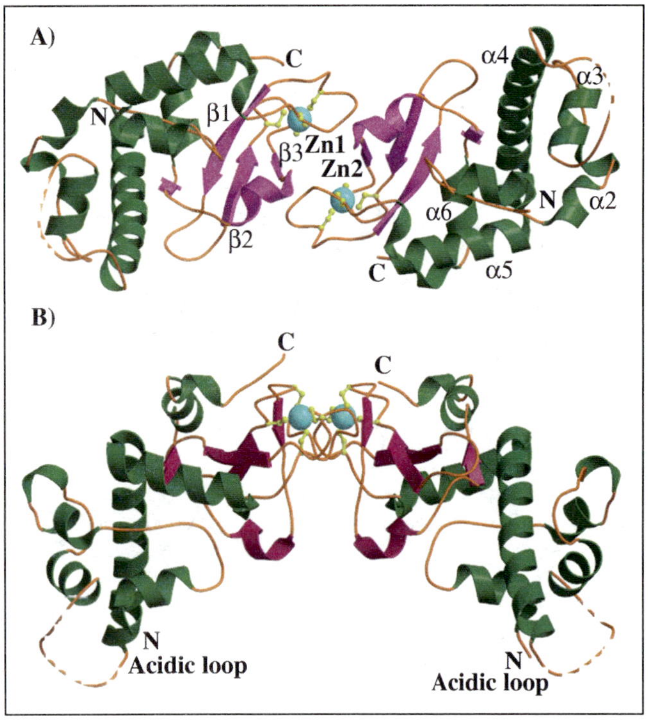

Figure 1. Ribbon representations of the CK2β dimer. The β strands are shown as pink arrows. The α helices are represented as green coils, loops as brown lines and the acidic loop as dashed brown lines. The zinc atoms and cysteine ligands are shown in ball-and-stick, blue and yellow respectively. A) view along the 2-fold axis. B) The view in (A) was rotated by 90°, the two-fold axis is vertical. (adapted from ref. 19).

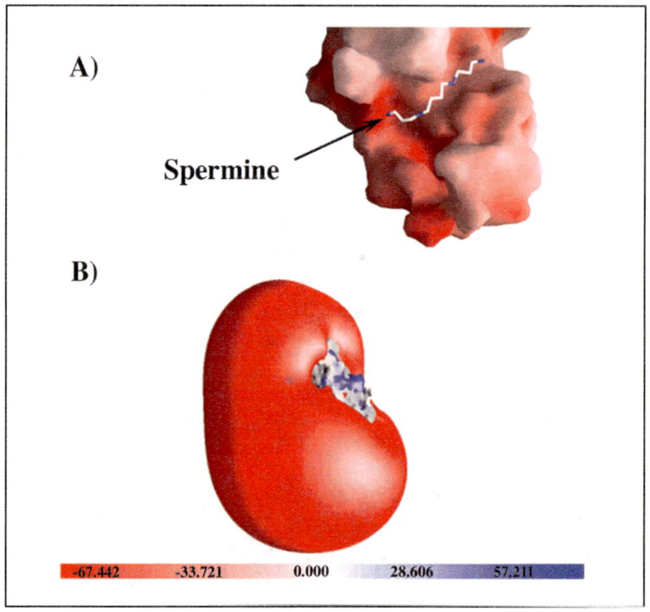

Figure 2. The electrostatic potential of the molecular surface of the N-terminal region of CK2β defines an acidic groove. A) Modelization of polyamine-CK2β contacts. The interaction of a spermine molecule with the CK2β molecule was modelized by manual docking. B) The anisotropic distribution of the electric field around the CK2β dimer is coloured from deep blue (+) to red (-). The molecule shows a striking asymmetric charge distribution which generates a monolobol potential field enveloping nearly all the molecule.

Overall Folding Properties of CK2β

The 1.7 Å resolution structure of a C-terminally truncated form of CK2β that was determined by X-ray crystallography demonstrated that two different crystal forms of the protein contain two monomers in the asymetric unit (Fig. 1A). The number of contacts and extent of interactions between monomers reflect a physiological self-association and the crystal structure is in accord with the native dimeric state of CK2β in solution.[19] Each monomer is an ovoid-shaped molecule containing six α helices and three β strands. The overall topology showed no structural homology with other known proteins. However the global fold comprises two separately identifiable domains that are packed closely together. Domain I (residues 1-104) is entirely α-helical (α1, α2, α3, α4 and α5). Helices α4 and α5 are forming a 95° angle of an unusual arrangement which ressembles the "L" letter. In contrast, helices α1, α2, α3, α4 wrap around helix forming a protuding acidic loop. Domain II (residues 105-161) exhibits a three-stranded antiparallel β-sheet (β1, β2 and β3) in which a zinc ion located between the α5 and α6 helices is tetrahedrally coordinated by four cysteines, (Cys 109, Cys 114, Cys 137 and Cys 140). Helix α6 (residues 163-170) makes contacts with both β1 and helix α5. The structure lack the C-terminal domain that contains the major site of association with the catalytic subunit. However, residues 170-177 which are part of the α-β subunit interface extend away from the bulk of the protein (see below). Two CK2β monomers pack against each other generating a crab-shaped molecule in which the zinc finger motif present in domain II mediates the highly stable dimerization of the protein (Fig. 1B).

The Acidic Binding Groove

A representation of the molecular surface of CK2β shows that the two domain I of the dimer generate an extended acidic groove formed by helices α1 and α3, an acidic loop (residues 55-64) and the N-terminus of helix α4. The groove is 35Å long, 7 Å wide and 4.5 Å deep (Fig. 2A). Remarkably, the size of the groove and the acidic side chains that line it creates a continuous negatively charged surface around the CK2β dimer, suggesting that this protuding region represents a binding pocket for basic ligands (Fig. 2B). Interestingly, it has been observed that this isolated region of CK2β exhibited an autonomous binding activity for polyamines reflecting a functional folding of this region of the protein.[20] Based on its similarity to the clusters of acidic amino acids that are typically observed in CK2 substrates, it has been proposed that this acidic segment $_{55}$DLEPDEELED$_{64}$ is reminiscent of the autoinhibitory sequences that have been identified in a number of other protein kinases.[4] Indeed, this acidic stretch was previously proposed to represent a regulatory region involved in the down regulation of CK2 activity,[21] and in its stimulation by polybasic ligands.[20]

The Zinc Ribbon Motif

Among the six cysteines present in the primary structure of the human CK2β subunit, only Cys 109, Cys 114, Cys 137 and Cys 140 located in domain II are invariant in different species. The crystal structure shows that these cysteine residues are indeed involved in a zinc binding motif. There are several types of zinc binding motifs categorized by the nature and spacing of their Zn^{2+}-chelating residues.[22-24] In the CK2β structure, the motif is made by three-stranded antiparallel β sheets (β1, β2, and β3) but

1	2	3		4
CK2β	Hs	103	QGDFGYCPRV-YCENQP--MLPIGLSDIPG----EAMVKLYCP--------KCMDVYT	145
CK2β	Dm	103	TGDFGHCPRV-YCESQP--MLPLGLSDIPG----EAMVKTYCP--------KCIDVYT	145
CK2β	Ce	102	DHDFGVCPRV-YCENQP--MLPIGLSDVPG----EAMVKLYCP--------RCNMVFV	144
CK2β	At	191	NYDFGRCPRV-YCCGQP--CLPVGQSDIPR----ASTVKIYCP--------KCEDVYY	233
CK2β	Sp	106	KCDFGHCPRV-LCNGQP--MLPVGLSDIAH----AKSVKLYCP--------RCEDVYT	148
CK2β	Sc	151	HKEFGTCPRY-YCNGMQ--LLPCGLSDTVG----KHTVRLYCP--------SCQDLYL	193
RPB9	Dm	84	RTEDHACP---KCSHR--EAVFFKAQTRRA--EEEMRLYYVCTN------QNCTHRWTE	129
TFIIS	Hs	6	QTDLFTCG---KCKKK--NCTYTQVQTRSA--DEPMTTFVVCN--------ECGNRWKF	49
TFIIβ	Sc	18	LNIVLTCP---ECKVYPPKIVERFS-----------EGDVVCA--------LCGLVLSD	54
TOP3	Hs	652	PEPIRK P---Q NK---DMVLKTKKN--------GGFYLS MGF-----PE RSAVWL	691

Figure 3. Structure-based sequence alignement of zinc ribbon domains in representative CK2β proteins and in members of the topoisomerase family. 1) Protein name: CK2β= CK2β regulatory subunit; RPB9= DNA-directed RNA polymerase II subunit 9; TFIIS= transcriptional elongation factor SII; TFIIB= transcription initiation factor IIB; TOP3= topoisomerase III. 2) Organism name: At= *Arabidopsis thaliana*; Ce= *Caenorhabditis elegans*; Dm= *Drosophila melanogaster*; Hs= *Homo sapiens*; Sc=, *Saccharomyces cerevisiae*; Sp= *Schizosaccharomyces pombe*. 3) Sequence number of the first residue shown in the alignment. 4) Sequence number of the last residue shown in the alignement. Cys residues in Zn-binding sites are highlighted in red and are shown in white. Uncharged residues in hydrophobic sites of β-strands are highlighted in yellow. Secondary structure consensus for β-strand residues is shown below the alignment by black arrows.

unlike the classical zinc finger, no α helix is observed. The Zn^{2+} binding site of CK2β is well ordered and contains two non-canonical zinc finger loops termed "knuckles" (Cys-X4-Cys and Cys-X2-Cys) contained in the invariant motif $CP_{X3}C$-$_{X22}$-CP_XC. The two knuckles are placed at almost 90° to each other, resulting in tetrahedral geometry of zinc ion. Based on pronounced structural similarity and residual sequence similarity, the zinc binding motif present in CK2β belongs to the sub-family of zinc ribbon domains. This motif resembles the Zn-ribbon structure of transcription factors TFIIS,[25] TFIIB,[26] the RNA polymerase II subunit 9 RPB9,[27] and Topoisomerase I and III[28] (Fig. 3). Interestingly, the three β strands and the P loop that form the core of the Zn^{2+} binding motif of CK2β show a striking similar topology to the structure depicted for TFIIS (Fig. 4). The knuckles of the Zn^{2+} binding site of CK2β are similar to those described for TFIIS and overall, the zinc ribbon motif of CK2β aligns structurally with an RMSD of 1.13 Å (96 atoms) with TFIIS. The β sheet surface of the Zn ribbon motif in CK2β is remarkable for its hydrophobicity, with 13 hydrophobic side chains defining the central surface. Unlike TFIIS which is a highly soluble nucleic-acid binding domain, the zinc ribbon motif of CK2β makes many hydrophobic interactions with the zinc ribbon motif of the other monomer forming the protein-protein interface. Most of the interactions between two CK2β monomers are contacts between residues that are localized at the edge of the Cys109-X4-Cys114 element and on the β3 strand. In this core region, several conserved hydrophobic residues make van der Waals interactions that form several important contacts between the two monomers (Fig. 5). The most significant contacts are Pro 110, Tyr

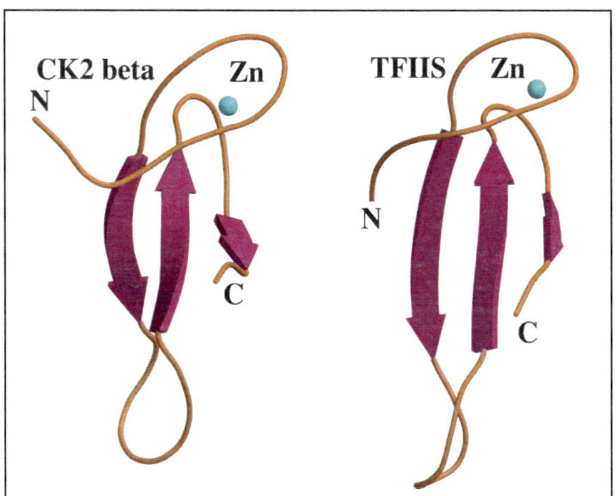

Figure 4. Ribbon representations of the zinc ribbon domains of CK2β and of transcription elongation factor TFIIS. The β strands are shown as pink arrows, loops as brown lines and zinc as blue sphere. (adapted from ref. 19).

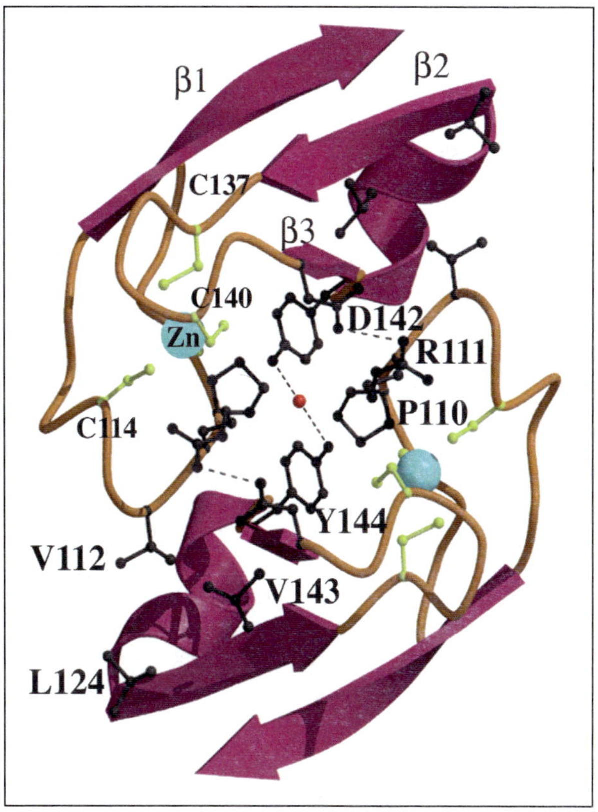

Figure 5. Ribbon representation of the dimer interface. The β strands are shown as pink arrows,loops as brown lines and zinc as a blue sphere. Amino acid chains and the water molecule involved in the protein-protein interaction, and the zinc ligands are represented in ball-and-stick in black, red and yellow respectively. (Adapted from ref. 19).

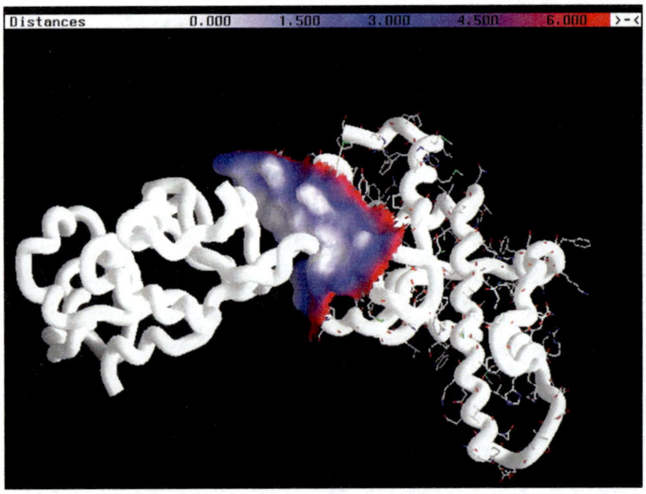

Figure 6. Worm representation of the CK2β dimer with the contact interface between monomers. The surface is color coded according to the difference between the distance and sum of van der Waals radii of the two atoms forming the interface. The distance is coloured from white (0 Å), blue (3 Å) to red (6 Å).

144, and Lys 147 of one monomer which make hydrophobic contacts with their counterpart residues on the other monomer. Away from the two-fold axis, important hydrophobic interactions can be found such as Val A 112 with Val B 143 and Leu B 124. The Cys109-X4-Cys114 element and the β3 strand form a continuous interface that buries a total area of 540 $Å^2$ for each monomer and defines the interface between two CK2β molecules (Fig. 6). In contrast, the unvariant region located between the β1 and β2 sheets of the zinc finger motif (Gly123-Ileu 127) is not involved in the formation of the interface between two CK2β monomers. The conformationally prominent residues in this region extend onto the molecular surface of the CK2β homodimer generating potential sites of interaction with other molecules.

Implications for CK2 Functions

The regulatory CK2β subunit alone has no known catalytic activity, but it does associate with the catalytic CK2α subunit to generate a stable holoenzyme complex. Several studies suggest that the regulatory subunit modulates the ability of CK2α to interact with and to phosphorylate substrate proteins.[29-31] Thus, CK2β appears as a crucial mediator of cellular functions of CK2. Furthermore, CK2β has been reported to be an interacting partner and activator of the mammalian A-Raf kinase[21] and an inhibitor of the *Xenopus laevis* c-Mos serine/threonine kinase.[32] In addition, the CK2β3 isoform present in *Arabidopsis thaliana* has been identified as an interacting partner of the circadian clock-associated 1 protein and interferes with regulation of circadian rhythms upon overexpression.[33] Taking together, these observations suggest a regulatory function for CK2β in signaling networks.

It has been demonstrated that in living cells, the CK2β subunit is synthetized in excess of CK2α and interacts slowly with it.[34] Therefore a vast majority of CK2α can potentially interact with CK2β. The β-β dimerization which can occurs in the absence of CK2α is a prerequisite for the incorporation of catalytic CK2 subunits into tetrameric complexes.[35] The high-resolution structure of the human CK2 holoenzyme (Fig. 7) revealed that the CK2β homodimer is the building block of this molecular hetero-complex bridging the two catalytic subunits.[36] Modification of sulfhydryl groups of all cysteinyl residues of recombinant CK2β using p-chloromercuribenzoid acid abrogated the ability of the protein to form either homodimer or canonical $\alpha_2\beta_2$ heterotetramers.[37] A more direct approach to study the functional significance of the zinc ribbon motif of CK2β used CK2β mutants in which Cys 109 and Cys 114 were exchanged to serines.[38] As expected, these mutations disrupted the zinc binding motif and resulted in loss of interactions between CK2β subunits. Interestingly, these mutants also failed to interact with catalytic CK2 subunits. The high-resolution structure of tetrameric CK2 (Fig. 7) also shows how stable binding of the catalytic subunits requires interactions with both C-terminal tails of the CK2β dimer

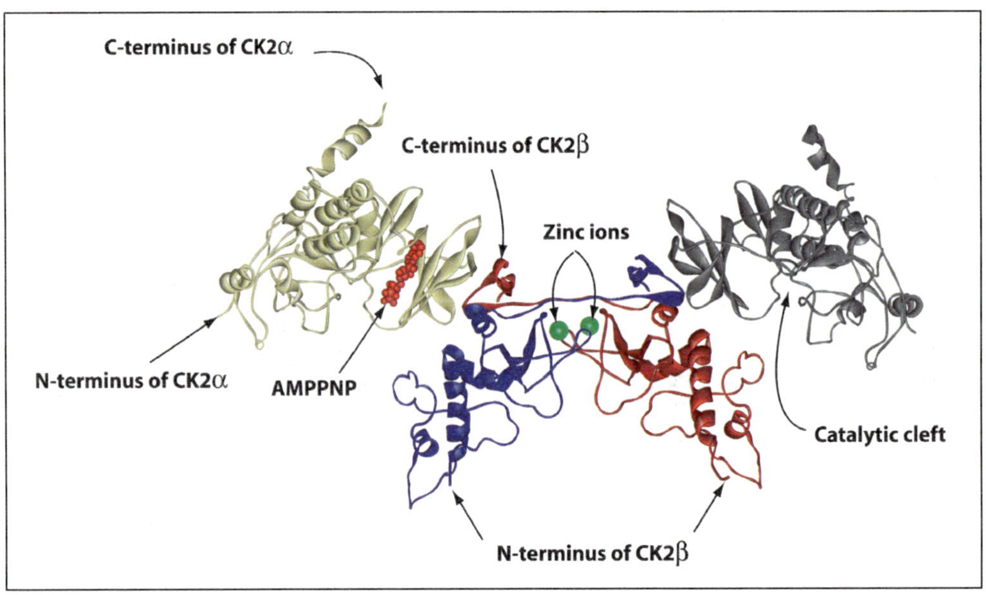

Figure 7. Overall shape of the human CK2 holoenzyme showing that the CK2β homodimer is the building block of this hetero-complex, bridging the two catalytic subunits. The view is perpendicular to the C2 axis. The two CK2β chains are drawn in blue and red, the two CK2α subunits in yellow and grey. The non-hydrolysable analog AMPPNP bound to the ATP binding site of one catalytic subunit is illustrated in red. (Adapted from ref. 36).

in a manner such that dimerization-deficient CK2β mutants are not able to interact with CK2α.[36] The importance of CK2β dimerization for the in vivo CK2β functions was demonstrated in *D. melanogaster* by expression of mutagenised *DmCK2*β transgenes in a *DmCK2*β null mutant background.[18] Mutations of either cysteinyl residue pair (109/114 or 137/140) involved in Zn^{2+} binding resulted in a CK2β protein which was unable to rescue the lethality of the CK2β null mutant. The failure of these mutants to substitute for the loss of endogenous DmCK2β function could be due to misregulation of the catalytic CK2 subunit, a loss of interaction with a critical CK2β binding partner or an accelerated protein turnover of the mutated proteins.[38] Importantly, these studies emphasize the absolute requirement of the β-β dimerisation motif for CK2β function.

The regulatory CK2β subunit has been shown to exert control over the catalytic activity of CK2 at a number of possible levels, i.e., in enhancing the catalytic activity and stability of CK2, and in the modulation of its substrate selectivity. The contribution of the CK2β subunit to the catalytic activity of CK2α was evaluated comparing the phosphorylation in vitro of different CK2 substrates by the isolated CK2α subunit or the tetrameric holoenzyme.[39] It was observed that CK2β modulates the ability of the catalytic CK2α subunit to interact with and phosphorylate substrate proteins demonstrating that CK2β plays a key role in the targeting of CK2 substrates. We have extended this study, analyzing the contribution of the zinc ribbon motif of CK2β in the regulation of the substrate specificity of catalytic CK2α subunit. CK2β mutants were constructed in which two of the conserved zinc ribbon residues, Pro110 and Val143, located at the edge of the Cys109-X4-Cys114 element and on the β3 strand respectively, were mutated to aspartic acid (Fig. 5). Reconstitution experiments showed that the recombinant mutant proteins behave as wild type CK2β to form homodimers (Fig. 8A). They were also fully competent to interact with the catalytic subunits of CK2 to generate the canonical multi-subunit holoenzyme (Fig. 8B). Polyamine binding activity of the mutant CK2β proteins were also unchanged (not shown). Unexpectedly, we observed that the substrate specificity of the reconstituted mutant holoenzymes was very similar to the one observed for the isolated CK2α catalytic subunit indicating that both mutants were defective in regulating the associated catalytic CK2α subunit (Fig. 9). Overall, these results show that the zinc ribbon mutants are fully competent to reconstitute an apparently normal tetrameric CK2. However, the enzymatic activity of the reconstituted CK2 holoenzyme complex was severely affected by the mutations in the zinc ribbon motif.

Based on our results, we postulate that mutations in this domain of residues that form important contacts between the CK2β monomers would induce conformational changes that alleviate the regulatory function of this subunit on CK2 activity. Collectively, these studies show the functional importance of the zinc ribbon motif of CK2β. A correct relative positioning of the sites of contact between the two CK2β monomers is likely to be important for forming a functional CK2β homodimer. The loss of function for the corresponding CK2β mutants suggests that even though Pro110 and Val143 residues play no direct role in catalysis, the structural integrity of the β-β protein interface is important for the regulation of the CK2 catalytic activity.

Beside the complex nature of the interaction between the catalytic and regulatory subunits of CK2, there is a growing body of evidence to suggest that CK2β also performs functions that are distinct from its role as a regulatory subunit of CK2 and the notion that the catalytic subunits of CK2 exist outside the holoenzyme

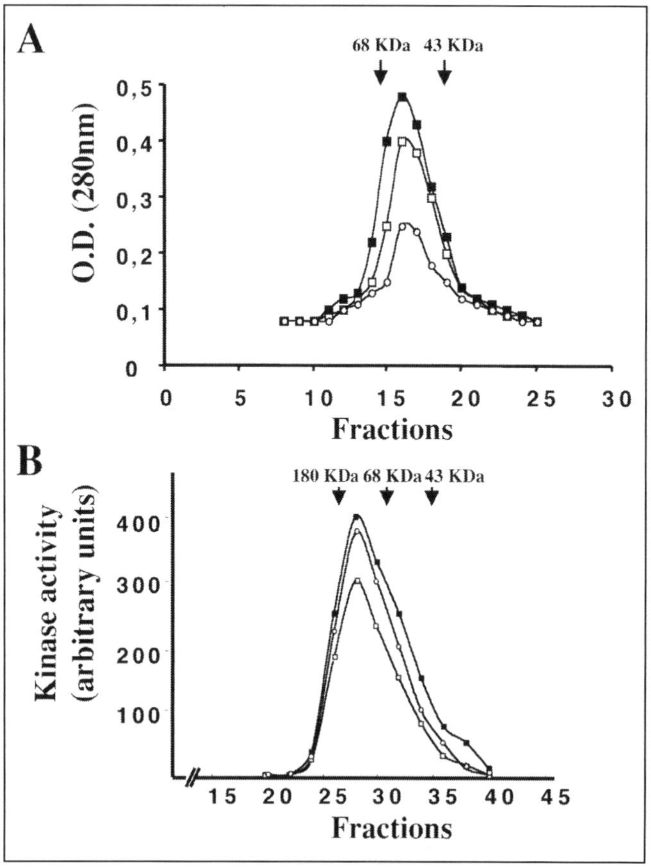

Figure 8. Biochemical characterization of mutant CK2β subunits. A) Gel exclusion chromatography of wt CK2β and mutants $CK2\beta^{P110D}$ and $CK2\beta^{V143D}$. Proteins were chromatographed on a Ultrogel ACA44 column equilibrated in 20mM Tris, HCl pH 7.4, 0.2M NaCl, 2% glycerol. Fractions were collected for adsorbance determination at 280nm. (■) wt CK2β; (□) $CK2\beta^{P110D}$ (○) $CK2\beta^{V143D}$. B) Gel exclusion chromatography of CK2 holoenzymes reconstituted by combining CK2α subunit with wt and mutated CK2β subunits. Proteins were loaded on a Sephacryl S-200 HR column equilibrated in 20mM Tris. HCl pH 7.4, 0.2M NaCl, 2% glycerol. Fractions were collected for protein kinase activity determinations. (■) wt CK2β; (□) $CK2\beta^{P110D}$; (○) $CK2\beta^{V143D}$ Elution volumes for marker proteins of known molecular mass were determined separately. (Unpublished results from our group.)

complex was brought by several studies. In this respect, the crystal structure of tetrameric CK2 highlighted that the limited surface of the α-β contacts suggests an inter-subunit flexibility compatible with an association-dissociation in vivo.[36] In addition, recent evidence from live-cell fluorescence imaging have revealed the independent movements of the catalyic and regulatory CK2 subunits within cells.[40] It was observed in this study, that both CK2 subunits were separately translocated into the nucleus of growing cells and that CK2β was not targeted to the nucleus by virtue of its stable association with CK2α. In contrast, a deletion mutant unable to interact with CK2α but containing the zinc ribbon motif was efficiently targeted to the nucleus (Fig.10). Interestingly, GFP-CK2β mutant in which Cys109 and Cys114 were replaced by serine residues, did not accumulate within the nucleus but remained in the cytoplasm (Fig. 10). These observations suggest that the integrity of the zinc ribbon motif, which is pivotal in the CK2β homodimer structure, is also instrumental for the correct nuclear targeting of the CK2β subunit in living cells.

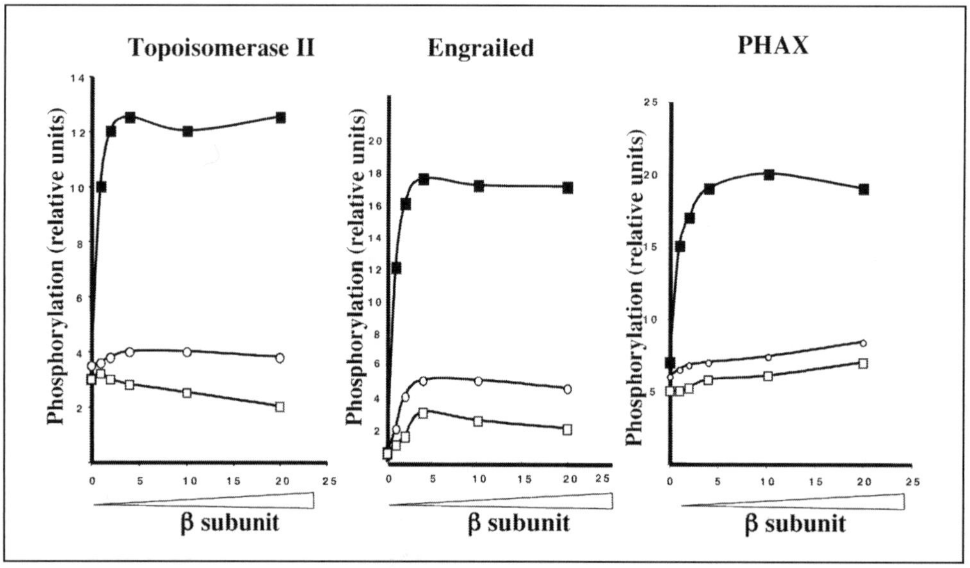

Figure 9. Stimulation of CK2α subunit catalytic activity by wt and mutated CK2β subunits. Known CK2 protein substrates (Topoisomerase II, Engrailed, PHAX) were phosphorylated by CK2α in the presence of increasing amount of wt CK2β (■), CK2β^{P110D} (□), or CK2β^{V143D} (○). Phosphorylated proteins were separated by SDS-PAGE electrophoresis and ^{32}P incorporation was determined by autoradiography. (unpublished results from our group).

Conclusion

As a result of the structural work highlighted here and other biochemical studies, we have now dramatically improved our evaluation of the structural conformation of the isolated CK2 subunits and their arrangement within tetrameric complexes.[19,36,41] The structure of the CK2β homodimer revealed the presence of a zinc binding motif mediating the dimerization of the protein. Thus, the regulatory subunit of CK2 joins the expanding subset of self-associating zinc ribbon proteins in which the Zn binding motif mediates protein-protein interactions. The zinc ribbons which are the largest fold group of zinc fingers frequently display limited sequence and structural similarity, mainly restricted to the zinc ligands and the zinc knuckle motifs. However, the zinc binding motif of CK2β shows a clear structural similarity with the zinc binding sites found in representative members of the classical zinc ribbon fold group such as the transcription factors TFIIS, TFIIB, RPB9 and Topoisomerase I and III. In the case of CK2β, the zinc ribbon motif has been now recognized as an important structural

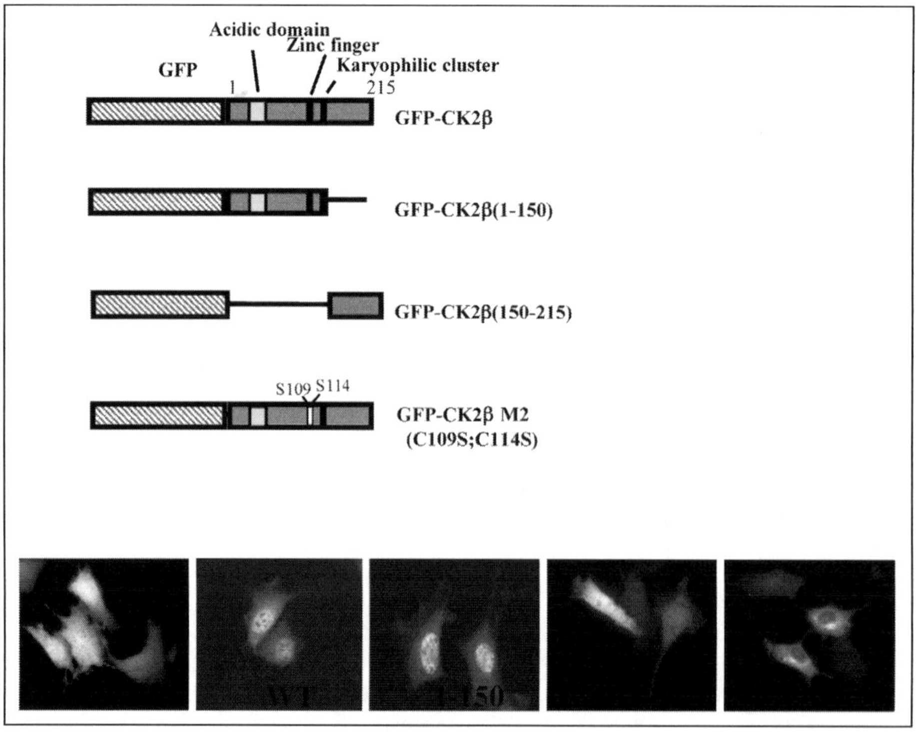

Figure 10. Nuclear import of mutant CK2β subunits. NIH3T3 cells were transiently transfected with different GFP-CK2β constructs as illustrated in the diagram. Transfected cells were observed for protein localization. (Adapted from ref. 40).

feature of this regulatory protein. The integrity of this motif appears crucial not only for the correct molecular architecture of this important protein, but also for its regulatory functions.

Acknowledgements

We acknowledge L. Chantalat and O. Dideberg for help in generating structural diagrams. We thank A. Larsen for providing us with Topoisomerase II, A. Joliot for Engrailed, and C. Mazza for the PHAX protein. Work on CK2 in our group is supported by grants from the INSERM, the CNRS (contrat 8BC06G), the Commissariat à l'Energie Atomique, the Association pour la Recherche contre le Cancer (réseau ARECA), and the Ligue Nationale contre le Cancer.

References

1. Hunter T. Signaling-2000 and beyond. Cell 2000; 100(1):113-27.
2. Allende JE, Allende CC. Protein kinases. 4. Protein kinase CK2: an enzyme with multiple substrates and a puzzling regulation. FASEB J 1995; 9(5):313-23.
3. Guerra B, Issinger OG. Protein kinase CK2 and its role in cellular proliferation, development and pathology. Electrophoresis 1999; 20(2):391-408.
4. Litchfield DW. Protein kinase CK2: structure, regulation and role in cellular decisions of life and death. Biochem J 2003; 369(Pt 1):1-15.
5. Meggio F, Pinna LA. One-thousand-and-one substrates of protein kinase CK2? FASEB J 2003;17(3):349-68.
6. Padmanabha R, Chen-Wu JL, Hanna DE et al. Isolation, sequencing, and disruption of the yeast CKA2 gene: casein kinase II is essential for viability in Saccharomyces cerevisiae. Mol Cell Biol 1990; 10(8):4089-99.
7. Xu X, Toselli PA, Russell LD et al. Globozoospermia in mice lacking the casein kinase II alpha' catalytic subunit. Nat Genet 1999; 23(1):118-21.
8. Fraser AG, Kamath RS, Zipperlen P et al. Functional genomic analysis of C. elegans chromosome I by systematic RNA interference. Nature 2000; 408(6810):325-30.
9. Buchou T, Vernet M, Blond O et al. Disruption of the regulatory Beta subunit of protein kinase CK2 in mice leads to a cell-autonomous defect and early embryonic lethality. Mol Cell Biol 2003; 23(3):908-15.
10. Landesman-Bollag E, Song DH, Romieu-Mourez R et al. Protein kinase CK2: signaling and tumorigenesis in the mammary gland. Mol Cell Biochem 2001; 227(1-2):153-65.
11. Seldin DC, Leder P. Casein kinase II alpha transgene-induced murine lymphoma: relation to theileriosis in cattle. Science 1995; 267(5199):894-7.
12. Ahmed K, Gerber DA, Cochet C. Joining the cell survival squad: an emerging role for protein kinase CK2. Trends Cell Biol 2002; 12(5):226-30.
13. Lozeman FJ, Litchfield DW, Piening C et al. Isolation and characterization of human cDNA clones encoding the alpha and the alpha' subunits of casein kinase II. Biochemistry 1990; 29(36):8436-47.
14. Maridor G, Park W, Krek W et al. Casein kinase II. cDNA sequences, developmental expression, and tissue distribution of mRNAs for alpha, alpha', and beta subunits of the chicken enzyme. J Biol Chem 1991; 266(4):2362-8.
15. Shi X, Potvin B, Huang T et al. A novel casein kinase 2 alpha-subunit regulates membrane protein traffic in the human hepatoma cell line HuH-7. J Biol Chem 2001; 276(3):2075-82.
16. Xu X, Rich ES, Jr, Seldin DC. Murine protein kinase CK2 alpha': cDNA and genomic cloning and chromosomal mapping. Genomics 1998; 48(1):79-86.
17. Niefind K, Guerra B, Pinna LA et al. Crystal structure of the catalytic subunit of protein kinase CK2 from Zea mays at 2.1 A resolution. EMBO J 1998; 17(9):2451-62.
18. Jauch E, Melzig J, Brkulj M et al. In vivo functional analysis of Drosophila protein kinase casein kinase 2 (CK2) beta-subunit. Gene 2002; 298(1):29-39.
19. Chantalat L, Leroy D, Filhol O et al. Crystal structure of the human protein kinase CK2 regulatory subunit reveals its zinc finger-mediated dimerization. EMBO J 1999; 18(11):2930-40.
20. Leroy D, Heriche JK, Filhol O et al. Binding of polyamines to an autonomous domain of the regulatory subunit of protein kinase CK2 induces a conformational change in the holoenzyme. A proposed role for the kinase stimulation. J Biol Chem 1997; 272(33):20820-7.
21. Boldyreff B, Meggio F, Pinna LA et al. Reconstitution of normal and hyperactivated forms of casein kinase-2 by variably mutated beta-subunits. Biochemistry 1993; 32(47):12672-7.
22. Krishna SS, Majumdar I, Grishin NV. Structural classification of zinc fingers: survey and summary. Nucleic Acids Res 2003; 31(2):532-50.
23. Laity JH, Lee BM, Wright PE. Zinc finger proteins: new insights into structural and functional diversity. Curr Opin Struct Biol 2001; 11(1):39-46.
24. Mackay JP, Crossley M. Zinc fingers are sticking together. Trends Biochem Sci 1998; 23(1):1-4.
25. Qian X, Gozani SN, Yoon H et al. Novel zinc finger motif in the basal transcriptional machinery: three-dimensional NMR studies of the nucleic acid binding domain of transcriptional elongation factor TFIIS. Biochemistry 1993; 32(38):9944-59.
26. Zhu W, Zeng Q, Colangelo CM et al. The N-terminal domain of TFIIB from Pyrococcus furiosus forms a zinc ribbon. Nat Struct Biol 1996; 3(2):122-4.
27. Wang B, Jones DN, Kaine BP et al. High-resolution structure of an archaeal zinc ribbon defines a general architectural motif in eukaryotic RNA polymerases. Structure 1998; 6(5):555-69.
28. Grishin NV. C-terminal domains of Escherichia coli topoisomerase I belong to the zinc-ribbon superfamily. J Mol Biol 2000; 299(5):1165-77.
29. Filhol O, Cochet C, Wedegaertner P et al. Coexpression of both alpha and beta subunits is required for assembly of regulated casein kinase II. Biochemistry 1991; 30(46):11133-40.
30. Marin O, Meggio F, Sarno S et al. Physical dissection of the structural elements responsible for regulatory properties and intersubunit interactions of protein kinase CK2 beta-subunit. Biochemistry 1997; 36(23):7192-8.
31. Meggio F, Boldyreff B, Marin O et al. Role of the beta subunit of casein kinase-2 on the stability and specificity of the recombinant reconstituted holoenzyme. Eur J Biochem 1992; 204(1):293-7.
32. Chen M, Cooper JA. The beta subunit of CKII negatively regulates Xenopus oocyte maturation. Proc Natl Acad Sci USA 1997; 94(17):9136-40.
33. Sugano S, Andronis C, Ong MS et al. The protein kinase CK2 is involved in regulation of circadian rhythms in Arabidopsis. Proc Natl Acad Sci U S A 1999; 96(22):12362-6.
34. Luscher B, Litchfield DW. Biosynthesis of casein kinase II in lymphoid cell lines. Eur J Biochem 1994; 220(2):521-6.
35. Graham KC, Litchfield DW. The regulatory beta subunit of protein kinase CK2 mediates formation of tetrameric CK2 complexes. J Biol Chem 2000; 275(7):5003-10.
36. Niefind K, Guerra B, Ermakowa I et al. Crystal structure of human protein kinase CK2: insights into basic properties of the CK2 holoenzyme. EMBO J 2001; 20(19):5320-31.
37. Meggio F, Ruzzene M, Sarno S et al. pCMB treatment reveals the essential role of cysteinyl residues in conferring functional competence to the regulatory subunit of protein kinase CK2. Biochem Biophys Res Commun 2000; 267(1):427-32.
38. Canton DA, Zhang C, Litchfield DW. Assembly of protein kinase CK2: investigation of complex formation between catalytic and regulatory subunits using a zinc-finger-deficient mutant of CK2beta. Biochem J 2001; 358(Pt 1):87-94.
39. Martel V, Filhol O, Nueda A et al. Dynamic localization/association of protein kinase CK2 subunits in living cells: a role in its cellular regulation? Ann N Y Acad Sci 2002; 973:272-7.
40. Filhol O, Nueda A, Martel V et al. Live-cell fluorescence imaging reveals the dynamics of protein kinase CK2 individual subunits. Mol Cell Biol 2003; 23(3):975-87.
41. Filhol O, Martiel JL, Cochet C. Protein kinase CK2: A new view of an old molecular complex. EMBO Reports 2004; 4:351-355.

CHAPTER 19

The FYVE Finger: A Phosphoinositide Binding Domain

Harald Stenmark

Abstract

The FYVE finger is an evolutionarily conserved double-zinc binding domain with structural similarity to RING and PHD fingers. It consists of two β-hairpins that are stabilized by a C-terminal α-helix and the coordination of two Zn^{2+} ions. The most characteristic feature of the FYVE finger is a R(R/K)HHCR motif associated with the first β-strand. This motif mediates coordination of the ligand, phosphatidylinositol 3-phosphate (PI3P), via contacts with the phosphate and inositol hydroxyl groups of PI3P. PI3P, a rare lipid formed on endosomes and phagosomes by phosphorylation of phosphatidylinositol, is crucial for phagosome maturation and endocytic trafficking; and FYVE finger proteins are important effectors of PI3P as it has been shown that several FYVE finger proteins among the 27 human proteins play central roles in endocytic and phagocytic membrane trafficking.

Introduction

The FYVE finger was originally discovered as a double-zinc finger required for the association of the early-endosome autoantigen EEA1 with endosome membranes. Its name derives from four of the first proteins shown to contain this motif (Fab1, YOTB, Vac1, EEA1).[1] Genomic analyses show that the FYVE finger is highly conserved in yeast, nematodes, flies, plants and mammals, and that typically about 0.1% of the genes encode FYVE finger proteins.[2] Thus, *Saccharomyces cerevisiae* contains 5 FYVE finger genes, whereas the human genome contains 27 such genes. With few exceptions, FYVE fingers occur in only one copy per protein. While most zinc fingers bind to large ligands (nucleic acids and proteins), the FYVE finger highly specifically binds to a membrane lipid, phosphatidylinositol 3-phosphate (PI3P).[3-5] We will here review the structural basis and functional implications of this interaction.

Formation and Functions of Phosphatidylinositol 3-Phosphate

Phosphatidylinositol 3-phosphate (PI3P) is formed by 3'-phosphorylation of the abundant membrane phospholipid, phosphatidylinositol (Fig. 1). The reaction is catalyzed by phosphatidylinositol 3-kinases (PI 3-kinases), notably the class III enzymes.[6] The latter are the most conserved among PI 3-kinases and are found both in yeast (Vps34) and in humans (hVps34). The catalytic subunit is associated with a regulatory subunit (named Vps15 and p150 in yeast and mammals, respectively) which has serine/threonine protein kinase activity.[7,8] Together with several other proteins, the catalytic and regulatory subunits are found in high-molecular weight complexes.[9] Unlike class I and II PI 3-kinases, class III PI 3-kinases are active even in unstimulated cells, and the levels of PI3P in most mammalian cell types remain fairly constant, typically amounting to about 0.2% of the amount of phosphatidylinositol.[10] In certain cell types, such as phagocytes, PI3P levels may increase more than two-fold upon cell stimulation, and this is likely to have important functional implications.[11,12] PI3P is mainly associated with limiting membranes of early (sorting) endosomes and intraluminal vesicles of multivesicular endosomes.[13] This highly specific localization probably results from the fact that the early-endosomal GTPase Rab5 interacts with p150 and thereby directs formation of PI3P on early endosomes only.[14,15]

Metabolism of PI3P can occur via three pathways.[16] Quantitatively, the most important metabolic pathway appears to be degradation by lipases within the lumen of multivesicular endosomes.[13,17] However, it is also clear that PI3P can be dephosphorylated by PI3P 3-phosphatases of the myotubularin family[18-20] or phosphorylated by the PI3P 5-kinase Fab1/PIKfyve (Fig. 1).[21,22]

Evidence from genetic and pharmacological studies indicates that PI3P is crucial for endocytic membrane trafficking and phagosome maturation.[23,24] Indeed, the catalytic and regulatory subunits of the class III PI 3-kinase in yeast, Vps34 and Vps15, were both isolated in a screen for vacuolar protein sorting mutants (the vacuole is the yeast counterpart of mammalian lysosomes).[7,25] Yeast cells devoid of Vps34 or Vps15 do not display any overt defects in processes other than vacuolar protein sorting and autophagy, indicating that in this organism PI3P is exclusively required for these processes. The picture is a little more complicated in mammalian cells, in which the role of PI3P has been studied with PI 3-kinase inhibitors and with microinjected antibodies against hVps34. Since pharmacological inhibitors of PI 3-kinase also inhibit class I PI 3-kinases (which produce phosphoinositides other than PI3P),[6] pharmacological studies alone do not unequivocally establish a role for PI3P in cellular processes. However, when studies using microinjected anti-hVps34 or synthetic PI3P are also taken into account, we can conclude that the following membrane trafficking processes in mammalian cells are controlled by PI3P: Endocytic membrane

Zinc Finger Proteins: From Atomic Contact to Cellular Function, edited by Shiro Iuchi and Natalie Kuldell.

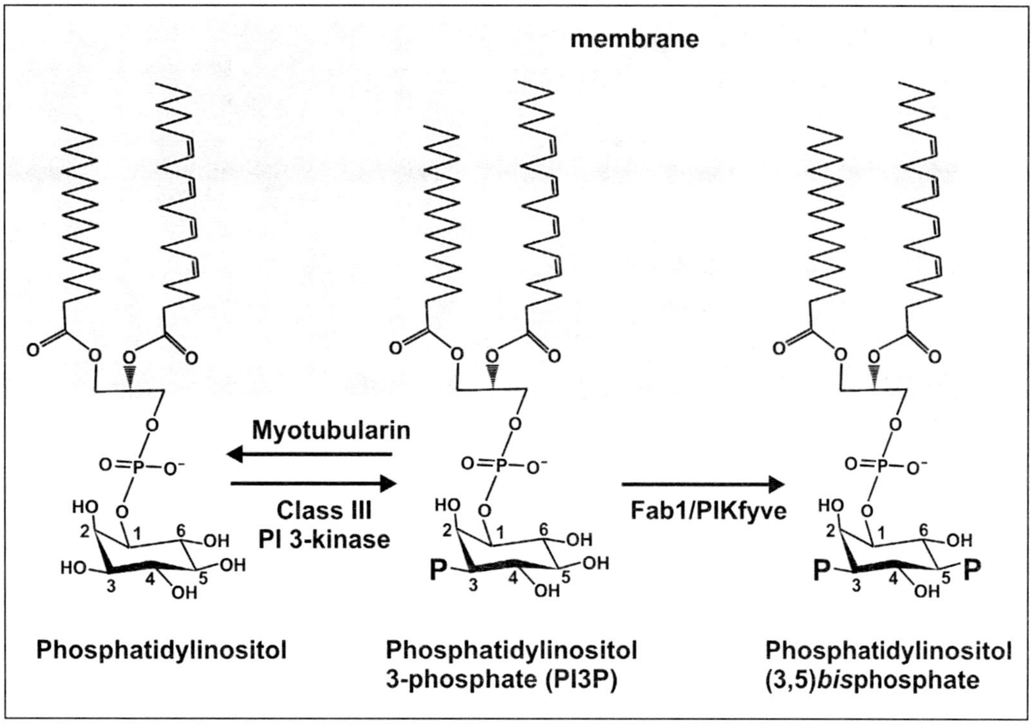

Figure 1. Formation and metabolism of PI3P. PI3P is formed by phosphorylation of phosphatidylinositol by a class III PI 3-kinase. Turnover of PI3P may occur by lysosomal lipases (not shown), dephosphorylation by members of the myotubularin family, or 5-phosphorylated by Fab1/PIKfyve.

fusion,[26-28] endosomal invagination,[29,30] phagosome maturation[12,31] and autophagy.[32,33] For all these processes, FYVE finger containing effectors of PI3P have already been identified (see below and Table 1).

Structure and Ligand Binding of the FYVE Finger

The FYVE finger typically consists of 60-70 amino acid residues and shares both sequence and structural similarities with RING and PHD fingers.[2] The structure is built around a pair of β-hairpins which are stabilized by a C-terminal α-helix and two Zn^{2+}-binding clusters.[34] Each Zn^{2+}-binding cluster consists of two pairs of CXXC motifs (in single-letter amino acid code; X is any amino acid), in which the cysteines coordinate Zn^{2+}. In a few cases, the fifth cysteine is replaced by a histidine residue. Similar to RING and PHD fingers, the two cysteine rich clusters bind two Zn^{2+} ions in a cross-braced fashion. Beside the 8 conserved cysteines, 3 short sequence motifs distinguish the FYVE finger: An N-terminal WXXD motif, a R(R/K)HHCR motif associated with the β1-strand, and a C-terminal RVC motif.[2] Three high-resolution crystal structures and one solution structure of FYVE fingers are available,[34-37] and the crystal structure of the EEA1 FYVE finger bound to inositol(1,3)bisphosphate,[36] reveals that all the three conserved sequence motifs contribute to ligand binding (Fig. 2A). In particular, the R(R/K)HHCR motif creates a shallow basic pocket that mediates PI3P binding (Fig. 2B). The EEA1 FYVE structure also explains why the FYVE finger only binds PI3P and not related phosphoinositides phosphorylated in the 4'- and/or 5'- positions of the inositol ring. Binding of the latter ligands would cause sterical problems and disrupt hydrogen bonds formed with inositol ring hydroxyl groups.

Alanine mutagenesis of conserved residues of the EEA1 FYVE finger has shown that most of the conserved residues contribute to the structural integrity of the FYVE finger.[38] Only two mutations were found to inhibit ligand binding without causing structural distortion, those of the underlined arginine residues in the RRHHCR motif. The equivalent residues of other FYVE fingers should therefore be regarded as primary candidates for mutagenesis whenever a functional analysis that investigates the role of PI3P binding is required.

Monomeric FYVE fingers bind the PI3P headgroup with too low affinity (K_D values in the high micromolar range) to explain the targeting of these proteins to endosomal membranes.[39] Binding to such membranes is strengthened by several mechanisms. First, a conserved hydrophobic "turret" loop preceding the β1-strand is predicted to insert into the lipid bilayer,[34-36] (see Fig. 2A) and biophysical studies using purified FYVE fingers and liposome membranes support this idea.[37,40] Thus, the K_D value for binding of a monomeric Hrs FYVE domain to PI3P-containing membranes has been estimated to be about 2.5 μM, which is a more than 50-fold increase in affinity over the isolated PI3P headgroup.[39] Second, avidity for PI3P-containing membranes can be enhanced through multimerization. This has been well documented in the case of EEA1, which forms a parallel dimer due to coiled-coil interactions N-terminal to the FYVE finger (Fig. 2B) and which significantly increases the avidity for PI3P-containing membranes.[36] Likewise, while the isolated FYVE finger of Hrs is cytosolic, a tandem Hrs FYVE finger is efficiently targeted to endosomes.[13] Third, the association of FYVE finger containing proteins with membranes may be strengthened through interactions with additional molecules. For example, a domain adjacent to the FYVE finger of EEA1 binds to the endosomal GTPase Rab5,[41] and whereas the monomeric EEA1 FYVE finger is not targeted to endosome membranes, constructs that contain additional dimerization/Rab5-binding domains are.[1]

Even though sequences in addition to the FYVE finger are required for endosomal targeting of EEA1 and Hrs, there are a

Table 1. Examples of mammalian FYVE finger containing proteins

Protein	Proposed Function	Localization	References
EEA1	Rab5/Rab22 effector. Tethering of endosomal membranes. Phagosome maturation.	Early endosomes Early phagosomes	41,51,74
Rabenosyn-5 (Vac1)	Rab4/Rab5 effector. Tethering of endosomal membranes. Organization of endosomal microdomains.	Early endosomes	53
Rabip4/RUFY	Rab4 effector. Endocytic recycling.	Early endosomes	55,56
Hrs (Vps27)	Sorting of ubiquitinated membrane proteins from endosomes to lysosomes.	Early endosomes MVEs	59-61
SARA	Scaffolding of Smad2 in TGFb signalling. Regulation of endocytic trafficking?	Early endosomes	42,69,70
MTMR3	PI3P and PI(3,5)P_2 3-phosphatase. Regulator of autophagy?	Cytosol Autophagosomes?	20,75
PIKfyve (Fab1)	PI3P 5-kinase. Membrane recycling from vacuoles/lysosomes. Protein sorting into MVEs.	Early endosomes	21,63
Fgd1*	GDP/GTP exchange factor for Cdc42. Regulation of the subcortical actin cytoskeleton. Activation of G1 progression.	Plasma membrane	46,76

Yeast homologs are indicated in parentheses. For a more exhaustive list see ref. 2. *The FYVE finger of Fgd1 has low affinity for PI3P and also binds PI5P.[39]

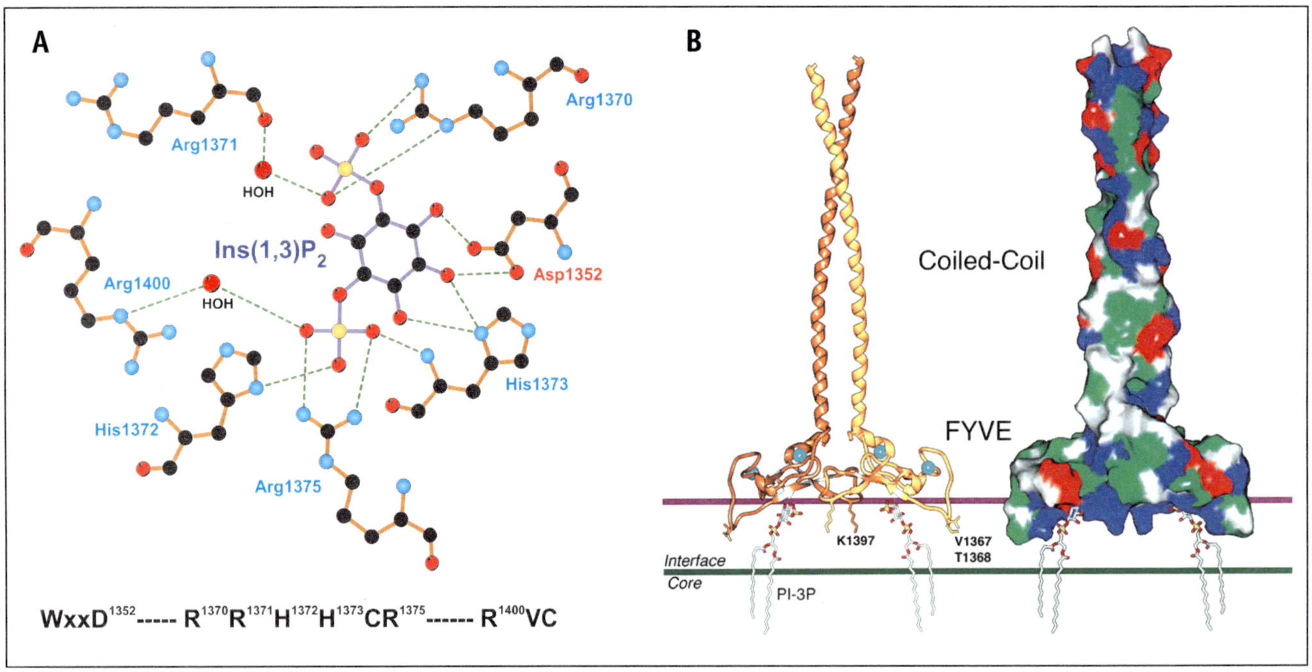

Figure 2. Structural basis of the interaction of the EEA1 FYVE domain with PI3P. A) The indicated side groups of the EEA1 FYVE domain interact with inositol 1,3-bisphosphate, the headgroup of PI3P. Note that all the three conserved motifs of the FYVE domain (bottom line; residue nubmering according to the published sequence of EEA1[77]) are involved in ligand binding. Black= carbon; red= oxygen; blue= nitrogen; yellow= sulfur. Based on the published crystal structure of the ligand bound FYVE domain of EEA1[36] and adapted from ref. 2. B) model for the binding of dimeric EEA1 to a PI3P-containing membrane. The molecular surfaces are shown in the same orientation as the schematic drawing and are colored blue (basic), red (acidic), green (hydrophobic), and white (uncharged polar). The two Zn^{2+} ions are shown as light blue spheres. The structure is docked to a membrane, with the polar interface and hydrophobic core regions of the membrane shown to scale. Residues predicted to be involved in nonspecific hydrophobic and electrostatic interactions with membrane lipids other than PI3P are shown. Myristoyl tails on PI3P have been modeled. Based on the published crystal structure of the ligand bound FYVE domain of EEA1[36] and reproduced with kind permission from J. Hurley.[78]

few examples of FYVE fingers that by themselves are sufficient for endosomal targeting. The FYVE fingers of SARA, Endofin and FENS-1 are efficiently targeted to endosomal membranes without the need for additional domains.[42-45] It is not known if this reflects an inherent propensity of these FYVE fingers to dimerize or whether they interact with additional endosomal molecules.

Cellular Functions of FYVE Finger Proteins

Most of the FYVE finger proteins expressed in yeast and humans have been functionally characterized to some extent (see examples in Table 1). Since PI3P is mainly localized to endosomes,[13] it is hardly surprising that many FYVE finger proteins are found on these organelles. However, certain FYVE finger proteins, such as the Cdc42 GDP/GTP exhange factor Fgd1 and the myotubularin-related PI3P phosphatase MTMR3 (which has been implicated in autophagy) are localized to other organelles. The reason why these proteins are not localized to endosomes could be that their FYVE domains bind PI3P with low affintiy or that they contain additional targeting domains that override the endosomal targeting conferred by the FYVE domain.[20,46]

Several FYVE finger proteins bind to the GTP-bound form (and thus serve as effectors) of endosomal Rab GTPases (Table 1). Both EEA1 and Rabenosyn-5 are FYVE finger containing Rab5 effectors that are required for Rab5-dependent membrane fusion in the endocytic pathway.[41,47,48] This includes homotypic fusion between early endosomes as well as heterotypic fusion between endocytic or Golgi-derived vesicles and early endosomes. Yeast cells defective of Vac1, the yeast homolog of Rabenosyn-5, have impaired Golgi to endosome trafficking, presumably because Golgi-derived vesicles do not fuse with early endosomes.[49,50] In addition to binding to Rab5, EEA1 and Rabenosyn-5 also bind to other endosomal Rab GTPases. EEA1 has a C-terminal Rab5 binding domain (adjacent to the FYVE finger) and an N-terminal C_2H_2 zinc finger which can bind both Rab5 and Rab22, an endosomal GTPase with unknown function.[51] Since the two Rab binding domains are spaced apart by a long coiled-coil region, it is possible that simultaneous binding to Rab GTPases on two apposing membranes may serve to tether vesicles/endosomes prior to fusion. Both Rabenosyn-5 and Vac1 contain an N-terminal C_2H_2 zinc finger similar to that found in EEA1. Surprisingly, while the C_2H_2 finger of Rabenosyn-5 binds to Rab5, the similar zinc finger of Vac1 does not bind to the yeast Rab5 homolog Vps21.[52] In addition to the N-terminal Rab5 binding domain, Rabenosyn-5 also has a C-terminal Rab5 binding domain.[53] Thus, like EEA1, this protein might potentially serve as a tether between two Rab5 positive membranes, even though it does not contain the long coiled-coil regions found in EEA1. However, Rabenosyn-5 in addition contains a Rab4 binding domain in the central part of the molecule, just C-terminal to the FYVE finger. Since overexpression of Rabenosyn-5 causes the apparent merging of otherwise separate Rab4 and Rab5 containing endosomal microdomains, it has been proposed that Rabenosyn-5 may serve as an organizer of such microdomains.[53] It is not yet understood how the same protein may serve to regulate both membrane tethering/fusion and domain organization.

Studies using microinjected antibodies indicate that hVPS34 and hence PI3P is required for maturation of phagosomes, a process that is crucial for efficient inactivation of phagocytosed microorganisms.[12,54] Interestingly, EEA1 associates with early phagosomes in a PI3P dependent manner, and antibodies against EEA1 inhibit phagosome maturation.[54] The requirement for EEA1 in phagosomal maturation could possibly reflect the need for Golgi- or endosome-derived vesicles to fuse with the phagosome as part of the maturation process.[24]

In addition to EEA1 and Rabenosyn-5, several other FYVE finger containing proteins serve as regulators of endocytic trafficking, albeit at distinct steps. Rabip4/RUFY is an effector of the early-endosomal Rab GTPase Rab4, which, in contrast to Rab5, regulates recycling of endocytosed molecules to the plasma membrane. Overexpression studies suggest that Rabip4 indeed functions as a regulator of endocytic recycling, possibly by regulating vesicle formation on endosomal membranes.[55,56] Other endosomal FYVE finger proteins are involved in cargo sorting and endosomal involution. Hrs (hepatocyte growth factor regulated tyrosine kinase substrate)[57] is a clathrin-associated protein[58] which binds to ubiquitinated membrane proteins and targets these to lysosomes.[59-62] Besides its sorting function, this protein is also required for endosomal invagination.[21,62] Interestingly, another FYVE finger protein, PIKfyve also seems to play a role in endosome invagination, although its possible relationship with Hrs is not known.[63] Unlike most other FYVE finger proteins, PIKfyve has enzymatic activity -it is a PI3P 5-kinase which phosphorylates PI3P into $PI(3,5)P_2$. Evidence primarily derived from studies of Fab1, the yeast homolog of PIKfyve, have indicated that $PI(3,5)P_2$ is crucial for the formation of intraluminal endosomal vesicles, and for membrane trafficking from the vacuole (lysosome) to other compartments.[21,64,65] Surprisingly, while EEA1, Rabenosyn-5, Hrs and PIKfyve are targeted to endosomes by virtue of their FYVE domains.[1,47,66,67] endosomal targeting of Rabip4 appears to depend on an N-terminal RUN domain. The FYVE domain of this protein might nevertheless play a role in its recruitment into specific membrane microdomains.[68]

Although several FYVE finger containing proteins serve as regulators of endocytic membrane trafficking, at least one FYVE finger protein functions in signal transduction. SARA (Smad anchor for receptor activation) localizes to early endosomes via PI3P binding by its FYVE domain, and here it serves as a scaffold for Smad2, a crucial signal transducer downstream of the activated transforming growth factor-β (TGFβ) receptor.[42,69] Phosphorylation of Smad2 by the ligand-bound receptor causes Smad2 to dissociate from endosomes, and to associate with Smad4. The Smad2/Smad4 complex then enters the nucleus in order to activate transcription of TGFβ–responsive genes. Interestingly, Smad2 phosphorylation and nuclear translocation depend on endocytosis of the receptor/ligand complex and its localization to EEA1/SARA-positive early endosomes.[43] This provides a clear example of compartmentalized receptor signalling, and SARA plays a crucial role in this compartmentalization by restricting Smad2 phosphorylation to early endosomes. An alternative internalization pathway of TGFβ receptors into caveolin-positive (but EEA1-negative) vesicles ("caveosomes") causes the association of the receptor with a ubiquitin ligase and results in its rapid turnover.[70] Thus, segregation of TGFβ receptors into distinct endocytic compartments regulates Smad activation and receptor turnover. It is worth noting that SARA may not exclusively function in signal transduction: Similar to several other FYVE finger proteins, this protein also appears to play a role in the regulation of endocytic membrane traffic.[71] One might conclude that the most common function of FYVE finger proteins is to regulate membrane trafficking into or out of the endosome.

Conclusion

The FYVE finger is a rare but important zinc finger that serves to recruit a small subset of proteins to endosomal membranes by binding to PI3P. Such recruitment ensures specificity in endocytic membrane fusion and receptor sorting and creates an environment for compartmentalized receptor signaling. Some variant FYVE fingers lack several of the residues that mediate PI3P binding.[2,72] These FYVE fingers bind PI3P with too low affinity to be recruited to endosomes, and their functions will have to be revealed by future studies. Even though the FYVE finger is principally a lipid binding domain, some FYVE fingers may well be involved in protein-protein interactions as well.[73]

References

1. Stenmark H, Aasland R, Toh BH et al. Endosomal localization of the autoantigen EEA1 is mediated by a zinc-binding FYVE finger. J Biol Chem 1996; 271:24048-24054.
2. Stenmark H, Aasland R, Driscoll PC. The phosphatidylinositol 3-phosphate-binding FYVE finger. FEBS Lett 2002; 513:77-84.
3. Gaullier J-M, Simonsen A, D'Arrigo A et al. FYVE fingers bind PtdIns(3)P. Nature 1998; 394:432-433.
4. Patki V, Lawe DC, Corvera S et al. A functional PtdIns(3)P-binding motif. Nature 1998; 394:433-434.
5. Burd CG, Emr SD. Phosphatidylinositol(3)-phosphate signaling mediated by specific binding to RING FYVE domains. Mol Cell 1998; 2:157-162.
6. Vanhaesebroeck B, Leevers SJ, Ahmadi K et al. Synthesis and function of 3-phosphorylated inositol lipids. Annu Rev Biochem 2001; 70:535-602.
7. Stack JH, DeWald DB, Takegawa K et al. Vesicle-mediated protein transport: Regulatory interactions between the Vps15 protein kinase and the Vps34 PtdIns 3-kinase essential for protein sorting to the vacuole in yeast. J Cell Biol 1995; 129:321-334.
8. Volinia S, Dhand R, Vanhaesebroeck B et al. A human phosphatidylinositol 3-kinase complex related to the yeast Vps34p-Vps15p protein sorting system. EMBO J 1995; 14:3339-3348.
9. Kihara A, Noda T, Ishihara N et al. Two distinct Vps34 phosphatidylinositol 3-kinase complexes function in autophagy and carboxypeptidase Y sorting in Saccharomyces cerevisiae. J Cell Biol. 2001 ;152:519-530.
10. Stephens L, McGregor A, Hawkins P. Phosphoinositide 3-kinases: Regulation by cell-surface receptors and function of 3-phosphorylated lipids. In: Cockcroft S, ed. Biology of phosphoinositides. Oxford: Oxford University Press, 2000:32-108.
11. Ellson CD, Anderson KE, Morgan G et al. Phosphatidylinositol 3-phosphate is generated in phagosomal membranes. Curr Biol 2001; 11:1631-1635.
12. Vieira OV, Botelho RJ, Rameh L et al. Distinct roles of class I and III phosphatidylinositol 3-kinases in phagosome formation and maturation. J Cell Biol 2001; 155:19-25.
13. Gillooly DJ, Morrow IC, Lindsay M et al. Localization of phosphatidylinositol 3-phosphate in yeast and mammalian cells. EMBO J 2000; 19:4577-4588.
14. Christoforidis S, Miaczynska M, Ashman K et al. Phosphatidylinositol-3-OH kinases are Rab5 effectors. Nature Cell Biol 1999; 1:249-252.
15. Murray JT, Panaretou C, Stenmark H et al. Role of Rab5 in the recruitment of hVps34/p150 to the early endosome. Traffic 2002; 3:416-427.
16. Stenmark H, Gillooly DJ. Intracellular trafficking and turnover of phosphatidylinositol 3-phosphate. Semin Cell Dev Biol 2001; 12:193-199.
17. Wurmser AE, Emr SD. Phosphoinositide signaling and turnover: PtdIns(3)P, a regulator of membrane traffic, is transported to the vacuole and degraded by a process that requires lumenal vacuolar hydrolase activities. EMBO J 1998; 17:4930-4942.
18. Taylor GS, Maehama T, Dixon JE. Myotubularin, a protein tyrosine phosphatase mutated in myotubular myopathy, dephosphorylates the lipid second messenger, phosphatidylinositol 3-phosphate. Proc Natl Acad Sci USA 2000; 97:8910-8915.
19. Laporte J, Liaubet L, Blondeau F et al. Functional redundancy in the myotubularin family. Biochem Biophys Res Commun 2002; 291:305-312.
20. Walker DM, Urbé S, Dove SK et al. Characterization of MTMR3: an inositol lipid 3-phosphatase with novel substrate activity. Curr Biol 2001; 11:1600-1605.
21. Odorizzi G, Babst M, Emr SD. Fab1p PtdIns(3)P 5-kinase function essential for protein sorting in the multivesicular body. Cell 1998; 95:847-858.
22. Sbrissa D, Ikonomov OC, Shisheva A. PIKfyve, a mammalian ortholog of yeast Fab1p lipid kinase, synthesizes 5-phosphoinositides. J Biol Chem 1999; 274:21589-21597.
23. Gillooly DJ, Simonsen A, Stenmark H. Cellular functions of phosphatidylinositol 3-phosphate and FYVE domain proteins. Biochem J 2001; 355:249-258.
24. Gillooly DJ, Simonsen A, Stenmark H. Phosphoinositides and phagocytosis. J Cell Biol 2001; 155:15-17.
25. Schu PV, Takegawa K, Fry MJ et al. Phosphatidylinositol 3-kinase encoded by yeast VPS34 gene essential for protein sorting. Science 1993; 260:88-91.
26. Jones AT, Clague MJ. Phosphatidylinositol 3-kinase activity is required for early endosome fusion. Biochem J 1995; 311:31-34.
27. Li G, D'Souza-Schorey C, Barbieri MA et al. Evidence for phosphatidylinositol 3-kinase as a regulator of endocytosis via activation of Rab5. Proc Natl Acad Sci USA 1995; 92:10207-10211.
28. Siddhanta U, McIlroy J, Shah A et al. Distinct roles for the p110α and hVPS34 phosphatidylinositol 3'- kinases in vesicular trafficking, regulation of the actin cytoskeleton, and mitogenesis. J Cell Biol 1998; 143:1647-1659.
29. Fernandez-Borja M, Wubbolts R, Calafat J et al. Multivesicular body morphogenesis requires phosphatidylinositol 3-kinase activity. Curr Biol 1999; 14:55-58.
30. Futter CE, Collinson LM, Backer JM et al. Human VPS34 is required for internal vesicle formation within multivesicular bodies. J Cell Biol 2001; 155:1251-1263.
31. Fratti RA, Backer JM, Gruenberg J et al. Role of phosphatidylinositol 3-kinase and Rab5 effectors in phagosomal biogenesis and mycobacterial phagosome maturation arrest. J Cell Biol 2001; 154:631-644.
32. Blommaart EFC, Krause U, Schellens JPM et al. The phosphatidylinositol 3-kinase inhibitors wortmannin and LY294002 inhibit autophagy in isolated rat hepatocytes. Eur J Biochem 1997; 243:240-246.
33. Petiot A, Ogier-Denis E, Blommaart EFC et al. Distinct classes of phosphatidylinositol 3'-kinases are involved in signaling pathways that control macroautophagy in HT-29 cells. J Biol Chem 2000; 275:992-998.
34. Misra S, Hurley JH. Crystal structure of a phosphatidylinositol 3-phosphate-specific membrane-targeting motif, the FYVE domain of Vps27p. Cell 1999; 97:657-666.
35. Mao Y, Nickitenko A, Duan X et al. Crystal structure of the VHS and FYVE tandem domains of Hrs, a protein involved in membrane trafficking and signal transduction. Cell 2000; 100:447-456.
36. Dumas JJ, Merithew E, Sudharshan E et al. Multivalent endosome targeting by homodimeric EEA1. Mol Cell 2001; 8:947-958.
37. Kutateladze T, Overduin M. Structural mechanism of endosome docking by the FYVE domain. Science 2001; 291:1793-1796.
38. Gaullier J-M, Rønning E, Gillooly DJ et al. Interaction of the EEA1 FYVE finger with phosphatidylinositol 3-phosphate and early endosomes. Role of conserved residues. J Biol Chem 2000; 275:24595-24600.
39. Sankaran VG, Klein DE, Sachdeva MM et al. High-affinity binding of a FYVE domain to phosphatidylinositol 3-phosphate requires intact phospholipid but not FYVE domain oligomerization. Biochemistry 2001; 40:8581-8587.
40. Stahelin RV, Long F, Diraviyam K et al. Phosphatidylinositol 3-phosphate induces the membrane penetration of the FYVE domains of Vps27p and Hrs. J Biol Chem 2002; 277:26379-26388.

41. Simonsen A, Lippé R, Christoforidis S et al. EEA1 links PI(3)K function to Rab5 regulation of endosome fusion. Nature 1998; 394:494-498.
42. Panopoulou E, Gillooly DJ, Wrana JL et al. Early endosomal regulation of Smad-dependent signaling in endothelial cells. J Biol Chem 2002; 277:18046-18052.
43. Hayes S, Chawla A, Corvera S. TGF beta receptor internalization into EEA1-enriched early endosomes: Role in signaling to Smad2. J Cell Biol 2002; 158:1239-1249.
44. Seet LF, Hong W. Endofin, an endosomal FYVE domain protein. J Biol Chem 2001; 276:42445-42454.
45. Ridley SH, Ktistakis N, Davidson K et al. FENS-1 and DFCP-1 are FYVE-domain containing proteins with distinct functions in the endosomal and Golgi compartments. J Cell Sci 2001; 114:3991-4000.
46. Nagata K, Driessens M, Lamarche N et al. Activation of G_1 progression, JNK mitogen-activated protein kinase, and actin filament assembly by the exchange factor FGD1. J Biol Chem 1998; 273:15453-15457.
47. Nielsen E, Christoforidis S, Uttenweiler-Joseph S et al. Rabenosyn-5, a novel Rab5 effector, is complexed with hVPS45 and recruited to endosomes through a FYVE finger domain. J Cell Biol 2000; 151:601-612.
48. Rubino M, Miaczynska M, Lippé R et al. Selective membrane recruitment of EEA1 suggests a role in directional transport of clathrin-coated vesicles to early endosomes. J Biol Chem 2000; 275:3745-3748.
49. Weisman LS, Wickner W. Molecular characterization of VAC1, a gene required for vacuole inheritance and vacuole protein sorting. J Biol Chem 1992; 267:618-623.
50. Tall GG, Hama H, DeWald DB et al. The phosphatidylinositol 3-phosphate binding protein Vac1p interacts with a Rab GTPase and a Sec1p homologue to facilitate vesicle-mediated vacuolar protein sorting. Mol Biol Cell 1999; 10:1873-1889.
51. Kauppi M, Simonsen A, Bremnes B et al. The small GTPase Rab22 interacts with EEA1 and controls endosomal membrane trafficking. J Cell Sci 2002; 115:899-911.
52. Merithew E, Stone C, Eathiraj S et al. Determinants of Rab5 interaction with the N terminus of early endosome antigen 1. J Biol Chem 2003; 278:8494-8500.
53. de Renzis S, Sonnichsen B, Zerial M. Divalent Rab effectors regulate the sub-compartmental organization and sorting of early endosomes. Nat Cell Biol 2002; 4:124-133.
54. Fratti RA, Backer JM, Gruenberg J et al. Role of phosphatidylinositol 3-kinase and Rab5 effectors in phagosomal biogenesis and mycobacterial phagosome maturation arrest. J Cell Biol 2001; 154:631-644.
55. Cormont M, Mari M, Galmiche A et al. A FYVE-finger-containing protein, Rabip4, is a Rab4 effector involved in early endosomal traffic. Proc Natl Acad Sci USA 2001; 98:1637-1642.
56. Yang J, Kim O, Wu J et al. Interaction between tyrosine kinase Etk and a RUN domain- and FYVE domain-containing protein RUFY1. A possible role of ETK in regulation of vesicle trafficking. J Biol Chem 2002; 277:30219-30226.
57. Komada M, Kitamura N. Growth factor-induced tyrosine phosphorylation of Hrs, a novel 115-kilodalton protein with a structurally conserved putative zinc finger domain. Mol Cell Biol 1995; 15:6213-6221.
58. Raiborg C, Bache KG, Mehlum A et al. Hrs recruits clathrin to early endosomes. EMBO J 2001; 20:5008-5021.
59. Raiborg C, Bache KG, Gillooly DJ et al. Hrs sorts ubiquitinated proteins into clathrin-coated microdomains of early endosomes. Nat Cell Biol 2002; 4:394-398.
60. Shih SC, Katzmann DJ, Schnell JD et al. Epsins and Vps27p/Hrs contain ubiquitin-binding domains that function in receptor endocytosis. Nat Cell Biol 2002; 4:389-393.
61. Bilodeau PS, Urbanowski JL, Winistorfer SC et al. The Vps27p Hse1p complex binds ubiquitin and mediates endosomal protein sorting. Nat Cell Biol 2002; 4:534-539.
62. Lloyd TE, Atkinson R, Wu MN et al. Hrs regulates endosome invagination and receptor tyrosine kinase signaling in Drosophila. Cell 2002; 108:261-269.
63. Ikonomov OC, Sbrissa D, Shisheva A. Mammalian cell morphology and endocytic membrane homeostasis require enzymatically active phosphoinositide 5-kinase PIKfyve. J Biol Chem 2001; 276:26141-26147.
64. Yamamoto A, DeWald DB, Boronenkov IV et al. Novel PI(4)P 5-kinase homologue, Fab1p, essential for normal vacuole function and morphology in yeast. Mol Biol Cell 1995; 6:525-539.
65. Gary GD, Wurmser AE, Bonangelino CJ et al. Fab1p is essential for PtdIns(3)P 5-kinase activity and the maintenance of vacuolar size and membrane homeostasis. J Cell Biol 1998; 143:65-79.
66. Raiborg C, Bremnes B, Mehlum A et al. FYVE and coiled-coil domains determine the specific localisation of Hrs to early endosomes. J Cell Sci 2001; 114:2255-2263.
67. Sbrissa D, Ikonomov OC, Shisheva A. Phosphatidylinositol 3-phosphate-interacting domains in PIKfyve. Binding specificity and role in PIKfyve. Endomenbrane localization. J Biol Chem 2002; 277:6073-6079.
68. Mari M, Macia E, Marchand-Brustel Y et al. Role of FYVE-finger and the run domain for the subcellular localization of RABIP4. J Biol Chem 2001; 276:42501-42508.
69. Tsukazaki T, Chiang TA, Davison AF et al. SARA, a FYVE domain protein that recruits Smad2 to the TGFβ receptor. Cell 1998; 95:779-791.
70. Di Guglielmo GM, Le Roy C, Goodfellow AF et al. Distinct endocytic pathways regulate TGF-β receptor signalling and turnover. Nat Cell Biol 2003; 5:410-421.
71. Hu Y, Chuang JZ, Xu K et al. SARA, a FYVE domain protein, affects Rab5-mediated endocytosis. J Cell Sci 2002; 115:4755-4763.
72. Stenmark H, Aasland R. FYVE-finger proteins - effectors of an inositol lipid. J Cell Sci 1999; 112:4175-4183.
73. Lawe DC, Patki V, Heller-Harrison R et al. The FYVE domain of early endosome antigen 1 is required for both phosphatidylinositol 3-phosphate and Rab5 binding. J Biol Chem 2000; 275:3699-3705.
74. Christoforidis S, McBride HM, Burgoyne RD et al. The Rab5 effector EEA1 is a core component of endosome docking. Nature 1999; 397:621-626.
75. Schaletzky J, Dove SK, Short B et al. Phosphatidylinositol-5-phosphate activation and conserved substrate specificity of the myotubularin phosphatidylinositol 3-phosphatases. Curr Biol 2003; 13:504-509.
76. Pasteris NG, Cadle A, Logie LJ et al. Isolation and characterization of the faciogenital dysplasia (Aarskog-Scott syndrome) gene: a putative Rho/Rac guanine nucleotide exchange factor. Cell 1994; 79:669-678.
77. Mu FT, Callaghan JM, Steele-Mortimer O et al. EEA1, an early endosome-associated protein. EEA1 is a conserved alpha-helical peripheral membrane protein flanked by cysteine "fingers" and contains a calmodulin-binding IQ motif. J Biol Chem 1995; 270:13503-13511.
78. Misra S, Miller GJ, Hurley JH. Recognizing phosphatidylinositol 3-phosphate. Cell 2001; 107:559-562.

CHAPTER 20

The BTB Domain Zinc Finger Proteins

Gilbert G. Privé, Ari Melnick, K. Farid Ahmad and Jonathan D. Licht*

Abstract

The BTB/zinc finger proteins have a wide range of functions in development and homeostasis and a wide range of interactions. The BTB domain appears essential for these proteins to dimerize, facilitate DNA looping, form specific multi-protein structures in the nucleus and interact with co-repressor molecules. The BTB domains of the proteins differ in their affinity for co-repressors and contribution to the transcriptional activity of the proteins. The BTB domain also allows interaction with other BTB proteins perhaps by forming higher order multimers and a network of BTB interactions likely exists. All of the BTB/zinc finger proteins have other important functional domains. PLZF and Bcl6 have a second repression domain, while Miz-1 has an important activation domain internal to the protein. In addition, the zinc fingers of these proteins can interact with co-factors implicated in transcriptional activity as well as nuclear cytoplasmic shuttling. Lastly given the recent information indicating the importance of the BTB domain in ubiquitylation pathways the BTB/zinc finger proteins may also play a role in degradation of specific proteins in the cell. Whether this is related to or distinct from their transcriptional functions remains to be discovered.

Introduction

The BTB (Broad Complex, Tramtrack and Bric à Brac) domain, also known as POZ (Poxvirus and Zinc finger) domain, is a highly conserved, widely distributed motif first identified in *Drosophila* and poxvirus proteins.[1-5] Examples of genes encoding BTB domain proteins have since been found in all eukaryotes studied to date. The domain consists of 90-120 residues, and is typically present as a single copy at the extreme N-terminus of proteins that often contain other motifs. BTB domains are protein-protein interaction motifs, and were shown to mediate homo-oligomerization,[6-8] hetero-oligomerization[5,9-12] and interactions with non-BTB containing proteins.[13-27]

While there are only four examples of BTB genes in *S. cerevisiae*, the number is increased dramatically in plants and in each of the major animal lines (nematodes, arthropods and vertebrates)[28] (Table 1). The expansion of BTB domain proteins in multi-cellular lineages suggests that these proteins many play crucial roles in cellular differentiation and development, and these expansions appear to be species-specific. For example, in the *C. elegans* genome, the majority of BTB domain proteins also contain a MATH domain,[29] while in *D. melanogaster* and the vertebrates, most BTB domains are most often found in proteins that contain either kelch repeats,[30] voltage-gated ion channels[31] (where the BTB domain is commonly known as the T1 domain), or Krüppel-type C_2H_2 zinc finger domains (Fig. 1).

The human genome encodes 205 BTB proteins, and these are roughly equally distributed among the three subclasses of proteins. There are 47 BTB/zinc finger proteins, 53 BTB/kelch proteins and 57 BTB(T1)/ion channels. The remaining ~20% of human BTB domain proteins are associated with a variety of other domains, or with no identified domain such as the Skp1/ElonginC proteins. One recently identified subfamily are RhoBTB proteins consisting of a GTPase domain, a proline rich region and a tandem arrangement of two BTB domains at the C-terminus (Fig. 1). This family of BTB proteins is largely uncharacterized, but other Rho GTPase containing proteins have been shown to be involved in the regulation of a number of cellular processes, including cytoskeletal remodelling, cell cycle progression, morphogenesis and apoptosis.[32] There is also evidence linking Rho-regulated signal transduction pathways with tumorigenesis and metastasis. For instance, DBC2 (deleted in breast cancer) is a recently identified RhoBTB protein homozygously deleted in 3.5% of breast tumours.[33]

The subclass of BTB/zinc finger proteins are generally thought to be transcription factors, and several of these have received considerable attention because of their involvement in human cancers. Two of the most studied human BTB domain proteins are PLZF and BCL-6, which are transcriptional repressors implicated in t(11:17)-associated acute promyelocytic leukemia (APL) and in diffuse large cell lymphoma (DLCL), respectively. The crystal structure of residues 6-126 of PLZF serves to identify the structural features of the BTB domain in the zinc finger and kelch repeat proteins.[6] Briefly, the structure is a butterfly-shaped homodimer with a tightly interwound peptide chain (Fig. 2). The first ~30 residues form strand $\beta 1$ and helix $\alpha 1$, and these elements make up an integral part of the dimer interface. The remaining core of the domain is mostly α-helical, with a short 3 stranded β-sheet exposed to the surface. The dimer interface is hydrophobic and makes up one quarter of the monomer surface, suggesting that the PLZF BTB domain is an obligate homodimer.

Structural alignments of BTB-fold proteins reveal a core region of 90 amino acids common to all examples of the domain, including the T1 and Skp1 subfamiles of the fold. The pairwise

* *Corresponding author. See list of "Contributors".*

Zinc Finger Proteins: From Atomic Contact to Cellular Function, edited by Shiro Iuchi and Natalie Kuldell.

Table 1. BTB domain-containing proteins

Genome	Number	Rank
H. sapiens	205	12
C. elegans	183	8
D. melanogaster	147	6
S. cerivisiae	5	> 100

Data compiled at http://xtal.uhnres.utoronto.ca/prive/btb_database.html. Rank indicates the prevalence in the given genome in the Interpro database.[200]

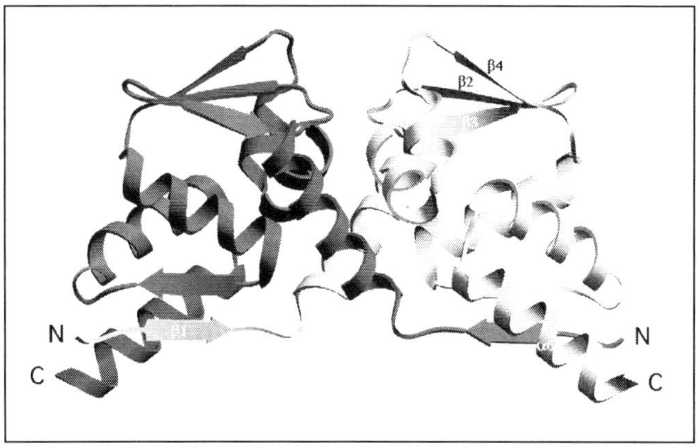

Figure 2. Ribbon diagram of the PLZF BTB domain dimer. One monomer is colored dark gray, the light gray. The N- and C-termini are labeled, as well as the secondary structure elements.

sequence identity between the core of these different classes of BTB domains is less than 20%, and the homologies are at the limit of detection by sequence-only methods.[28] While all of these proteins share a common core fold, these associate into different types of quaternary arrangements. Thus, the BTB domains from PLZF and BCL-6 form homodimers with major interchain contacts from the N-terminal extension, while the ion channel T1 domain forms homotetramers[31] and the Skp1/ElonginC proteins forms heterotrimers[34,35] utilizing distinctly different oligomerization interfaces. The 30 amino acid N-terminal extension is seen only in the zinc finger and kelch-repeat proteins (Fig. 3), while the Skp1/ElonginC proteins have C-terminal extentions.

The biological properties of BTB-containing proteins are largely a result of the protein-protein interactions made by the domain. The primary function of the BTB domain, as found in zinc finger and kelch-repeat proteins, is to mediate homodimerization, as this role is critical to proteins that contain the domain. For example, in PLZF the BTB domain is required for dimerization,[6,8] transcriptional repression,[36] formation of high molecular weight DNA-protein complexes.[37] growth suppression[38] and localization of the protein to nuclear speckles.[39] Mutations that disrupt the interface and abrogate dimerization of the PLZF BTB domain or outright removal of the PLZF BTB domain results in completely nonfunctional proteins.[40]

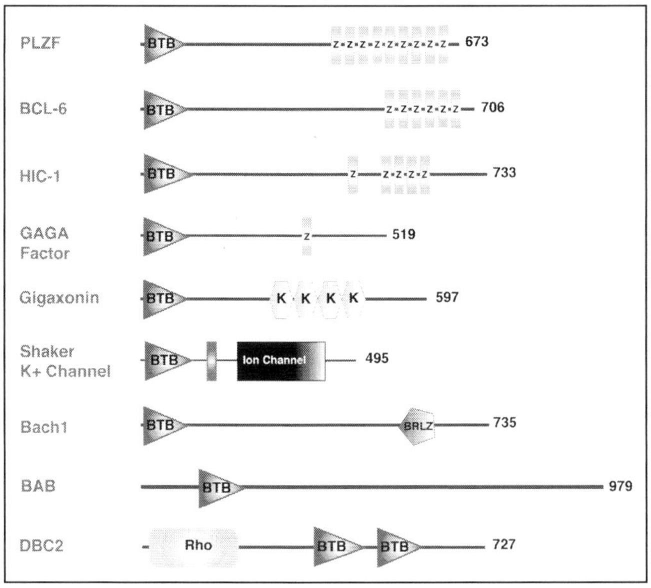

Figure 1. Schematic representation of selected BTB proteins. All, except GAGA factor and BAB, are human proteins. Z represents Kruppel-type C_2H_2 zinc finger domains, K indicates the Kelch Motif, BRLZ represents the basic-leucine zipper domain, Rho represents the Rho GTPase domain. The long rectangle (in the Shaker K+ channel) represents a transmembrane domain. The lengths of the proteins are indicated on the right.

BTB Domain Oligomerization

BTB domain dimers can potentially form higher order complexes, such as those observed with the head-to-tail polymerization of the SAM domains of the transcriptional repressors TEL[41] and polyhomeotic.[42] The structural main evidence for this is the observation of a tight contact between PLZF BTB homodimers involving the β1-β5' strands at the "bottom" of the PLZF BTB protein.[6,8] Such an interaction has been observed in two different crystal lattices, further supporting the biological relevance of this dimer-dimer interaction. If the homodimers associate as a repeating chain, it suggests that the full-length PLZF protein may be able to assemble into a large complex with multiple-zinc finger

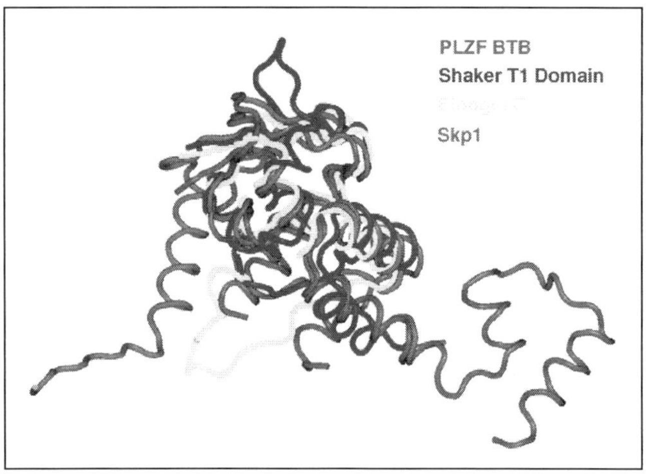

Figure 3. Superposition of the PLZF BTB domain, the Shaker T1 tetramerization domain, elongin C and skp1. The PLZF BTB domain has an approximately 30-amino acid N-terminal extension, while skp1 has a two-helix C-terminal extension.

modules. In many BTB transcription factors, the domain is separated from the C-terminal DNA-binding region by several hundred residues, which are predicted to have little fixed structure. This leads to a model in which an oligomeric chain of BTB domains is linked to a series of DNA reading heads by long flexible linkers, which can in turn interact with multiple DNA recognition sites. It has been demonstrated that the ability PLZF to form a high molecular weight DNA-protein complex is dependent on its BTB domain. Furthermore, the BTB domain is necessary for directing PLZF,[39] BCL-6[43] and ZID[5] to a punctate expression pattern in the nucleus, suggesting that these proteins are part of large assemblies.

In the case of *Drosophila* GAGA factor, polyvalent binding to natural operator sequences was also dependent on the formation of higher order protein complexes mediated by the N-terminal BTB domain.[44,45] GAGA factor is a constitutively expressed *Drosophila* transcription factor encoded by the essential *Trithorax-like* gene[46] and it binds to GA rich DNA sequences through its single zinc finger.[47] Unlike many BTB domain proteins, however, GAGA factor has been shown to activate transcription.[48] Electron microscopy (EM) experiments have directly shown that GAGA factor binds multiple DNA elements as a large oligomer and that the DNA in this complex is bent and might be wrapped around the surface of a GAGA factor multimer.[45] It is thought that, after transient chromatin remodeling by NURF,[49] the association of a GAGA factor oligomer with multiple binding sites within a promoter constrains the DNA trajectory, consequently reorganizing the promoter DNA and maintaining it in an open conformation.[45] This is correlated with in vivo evidence that have shown that GAGA-binding elements coincide with DNase-hypersensitive sites at promoters.[50] Similar higher order complexes have also been observed for the murine BTB-containing transcription factor Bach1, for which polyvalent binding to natural operator sequences was dependent on the presence of the BTB domain.[51]

Chain formation may also be a mechanism for heteromeric associations between different BTB proteins. Some examples of heteromeric pairs formed between BTB domains include PLZF and BCL-6,[9,10] PLZF and FAZF,[11] BCL-6 and BAZF[12] and GAGA factor and TTK.[5,52] However, there appears to be a strict specificity to heterodimer formation, as it was shown that the following pairs of domains do not interact: PLZF and LRF,[9] BCL-6 and LRF,[9] BCL-6 and kaiso,[10] TTK and ZID,[5] TTK and ZF5.[5] It is important to note that such interactions have been observed by yeast two-hybrid, co-immunoprecipitation and in vitro co-expression experiments only. In fact, when non-tagged and tagged ZID derivatives of differing sizes were mixed together rather than co-translated, co-immunoprecipitation was inefficient, indicating that limited exchange took place.[5] Therefore the notion that BTB heterodimers may form is an attractive but as yet unproven possibility. For the DNA-binding members of the BTB family, heterodimerization might increase the number of gene promoters that proteins family binds can bind to, increasing the potential for functional variation. However, the biological relevance of the observed heterodimers remains to be clarified. Based on their three-dimensional crystal structures, we have proposed that the PLZF and BCL-6 BTB domains form obligate homodimers.[6] This is due to the extensive hydrophobic interface between monomers, and on thermal denaturation studies on the PLZF BTB domain showing a single step transition between folded dimers and unfolded monomers.[6-8] It is possible, nonetheless, that BTB heteromers can form through homodimer-homodimer contacts, generating much larger BTB oligomers.

Interactions of the BTB Domain with Co-repressors

The BTB domain is strongly implicated in the repression of gene expression through the local control of chromatin conformation. Several BTB domains were shown to interact with components of a histone deacetylase complex, such as the co-repressors silencing mediator of retinoid and thyroid hormone receptor (SMRT), the nuclear receptor co-repressor (N-CoR), Sin3A and histone deacetylase 1 (HDAC-1).[13-22] Repression of gene transcription by BTB containing transcription factors such as Bach2,[53] PLZF,[14-18,21,22] FAZF,[54] ZID[13] and BCL-6[13,17,19,20] is mediated by the recruitment of this complex to the chromatin, where the removal of acetyl groups from the histone lysine tails leads to changes in chromatin structure. Furthermore, transcriptional repression mediated by these particular BTB domains is reversible with trichostatin A, a specific inhibitor of HDAC activity.[55] Disruption of the BTB fold of PLZF and BCL-6 by point mutation abrogates the interaction of the BTB domain with these co-repressors, further verifying the critical role of BTB dimerization.[40,54] Recruitment of a histone deacetylase complex, however, is not a universal property of BTB domain containing transcriptional repressors. The HIC-1 (hypermethylated in cancer) gene encodes a BTB domain with 13 amino acid alanine rich insert in the region analogous to an alpha helical region of the solved BTB domain structures.[56] This domain does not bind to N-Cor and SMRT and repression by this domain is not inhibited by HDAC inhibitors.[55] Furthermore the BTB domains of Bric à Brac 1 (BAB1) and Bric à Brac 2 (BAB2) interact with BIP2, a *Drosophila* TATA-box protein associated factor (TAF_{II}),[57] raising the possibility that some BTB domains may repress transcription by interfering with the basal transcriptional machinery.

The Emerging Role of BTB Domains in Ubiquitylation

Very recently the BTB domains was identified as a substrate adaptor for cullin 3 (Cul3) E3 ligase complexes.[23,24,26,27] Several groups found that BTB domain of the *C. elegans* protein Mel-26 interacts directly with cullin 3 (Cul3) and targets the microtubule-severing protein MEI-1 for degradation by the Cul3 E3-ubiquitin ligase complex.[23,24,27] In yeast, *S. pombe* Cul3 (Pcu3p) forms a complex with all three BTB proteins in this species.[26] Complexes of human Cul3 with BTB/kelch and BTB/MATH proteins, as well as with the BTB/zinc finger protein PLZF were found, strongly suggesting that a large number of BTB proteins may be regulated by the Cul3/ubiqutin ligase complex or regulate the stability of other proteins through this complex.[27]

Skp1 and Elongin C are small proteins that consist almost exclusively of the core BTB fold, and these are part of the related SCF (Skp1-Cul1-F-box)[34,58] and ECS (Elongin C-Cul2-SOCS box)[59] E3-ligases, respectively. The latter complex is well known because of the interactions with the VHL tumor suppressor, a SOCS-box protein associated with the von Hippel-Lindau cancer syndrome and kidney cancer. Skp1 and Elongin C interact with F-box or SOCS-box adaptor proteins such as Cdc4 or VHL to recognize target substrates. In the newly described Cul3-E3 ligase complexes, however, BTB domain proteins may have a "built-in" substrate-recognizing functionality in their C-terminal downstream domains.[25] For example, the N-terminal

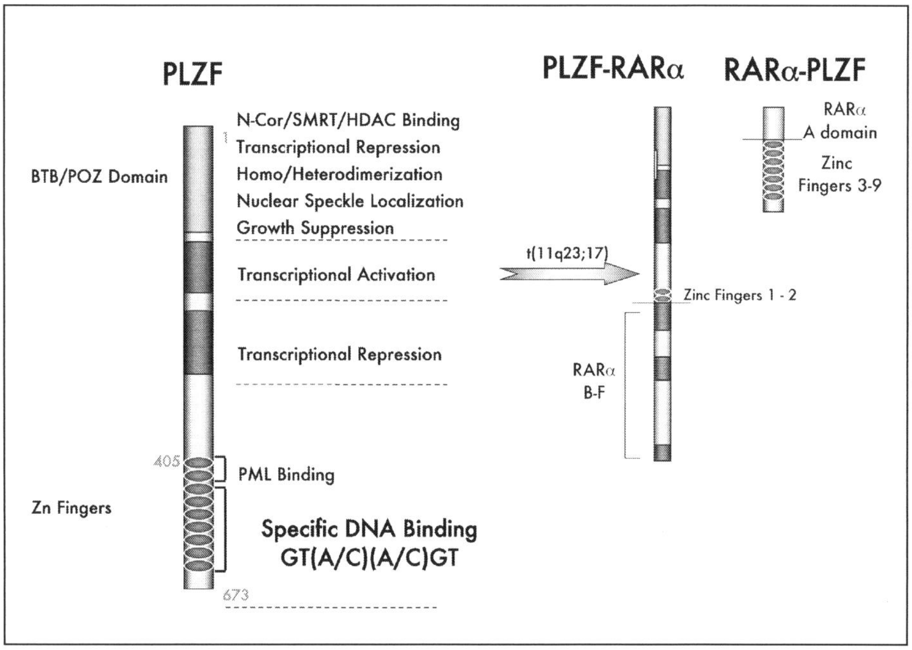

Figure 4. Domain structure of PLZF and the PLZF-RARa fusion proteins of acute promyelocytic leukemia.

BTB domain of Mel-26 associates with Cul3, while the C-terminal kelch repeats bind directly to the target MEI-1.[24,27] Thus, BTB domain proteins may carry out both the cullin-interacting and substrate-recognizing functions that are achieved by two proteins in the SCF and ECS systems. Given that zinc fingers are also protein-protein interaction domains it also offers potential additional non-transcriptional functions for the BTB/zinc finger proteins.

The Role of PLZF in Leukemia

The PLZF Gene and Its Fusion with RARα

The *promyelocyticleukemia zinc finger (PLZF)* gene was identified by its rearrangement in a case of APL with t(11;17)(q23;q21) translocation.[60,61] Since then 16 cases of APL with a fusion between the *PLZF* and *RARα* genes were described,[62-65] representing less than 1% of cases of APL. The *PLZF* gene is located on chromosome 11q23 about 1 MB centromeric to the *MLL (mixed lineage leukemia)* gene.[66] The *PLZF* gene was completely sequenced and found to be contained in six exons spread over 201 kb of DNA. The resulting 8kb mRNA has a long 6kB 5' untranslated region of unknown function. In (t:11;17) associated APL, the *PLZF* gene is rearrange with the RARα gene. The breakpoint in *PLZF* is between exons two and three except in one case where the breakpoint was more 3' in the *PLZF* gene between exons three and four. The chromosomal rearrangement is reciprocal, leading to two fusion gene products PLZF-RARα and RARα-PLZF. PLZF RARα contains the N-terminal portion of PLZF, which contains the transcriptional effector domains of PLZF lined to domains B-F of RARα including the DNA binding and ligand binding domain of RARα. The reciprocal fusion RARα-PLZF links the A1 or A2 domain of RARα resulting from the alternative promoter's of the *RARα* gene to the last 7 of nine zinc fingers of PLZF. The PLZF-RARα fusion is an aberrant form of the retinoic acid receptor while RARα-PLZF is an abnormal form of PLZF (Fig. 4). In contrast to the more frequent t(15;17) APL, this translocation is generally associated with a poor prognosis and lack of response to all-*trans*retinoic acid (ATRA), in vivo and in vitro.[62] However, more recently several cases responsive to combination therapies such as ATRA and G-CSF [65] or ATRA and chemotherapy,[63,67] were reported indicating that the resistance of this syndrome is not absolute. However in one of these cases only PLZF-RARα and not the reciprocal fusion gene was expressed.[67] This and another case[64] suggested that PLZF-RARα may be sufficient to generate disease and suggesting that the reciprocal RARα-PLZF protein might play a role in the refractoriness of t(11;17)-APL.

The PLZF Protein

The PLZF zinc finger transcription factor[60] contains nine C-terminal Krüppel-like C_2H_2 zinc finger motifs and a N-terminal BTB domain (Fig. 4). PLZF is a 673 amino acid nuclear serine/threonine phosphorylated protein[68] that binds to DNA specifically to a consensus sequence of GTAC$^{T}/_{A}$GTAC. The last three zinc fingers of PLZF which are retained in the RARα-PLZF fusion are sufficient to recognize the PLZF cognate binding site.[36,68,69] In fact RARα-PLZF can bind to a site within the *Hoxb2* promoter more strongly than PLZF itself[70] suggesting that the N-terminal BTB domain, as demonstrated originally by Bardwell[71] may negatively regulate DNA binding. The highly related BTB zinc finger FAZF[72] also known as TZFP[73] or ROG[74] also has a N-terminal BTB domain and only three C-terminal zinc fingers which are very similar to those of PLZF. The three FAZF zinc fingers recognize the PLZF binding site.[72,75] The FAZF gene is localized on chromosome 19q13.1 adjacent to the *MLL2* gene. This information suggests that the genes encoding PLZF and MLL may have arisen from an ancient duplication of genetic material and the more simple FAZF protein may have been the precursor of the larger PLZF. What the function of the more N-terminal zinc fingers of PLZF might be is not yet certain. One publication suggested that these fingers might be important for transcriptional repression function of the PLZF- RARα fusion.[76] The first three zinc fingers of PLZF can mediate interaction with the RAR as well as estrogen receptor in vitro and overexpression of these zinc fingers or full-length PLZF can inhibit the transcriptional activity of those receptors and interfere with RARα/RXR interaction.[77] Whether this is a significant interaction in vivo remains unclear however. A recent paper showed that zinc fingers 5-8 interacted with the intracellular fragment of proHB-EGF and mediated nuclear-cytoplasmic shuttling of PLZF.[78]

The PLZF protein is localized to the nucleus[79,80] in a pattern of ~50 small nuclear speckles,[80] which are dependent on the presence of the BTB domain.[76] PLZF is co-localized or adjacent to the PML protein which is disrupted in the more common t(15;17) associated form of APL in a large multi-protein complex called the PML oncogenic domain (POD) or nuclear body (NB)[81,82] suggesting a functional interplay. In t(15;17) and t(5;17) APL, PLZF is de-localized into a microspeckled pattern similar to that of PML-RARα[21,81,83] suggesting that delocalization of PLZF might be important in the pathogenesis of APL.[83,84]

PLZF in Development

PLZF is expressed in undifferentiated myeloid cell lines and at lower levels in more differentiated erythroleukemia, promyelocytic and monocytic cell lines as well as in peripheral blood mononuclear cells.[60,79,80] PLZF is down-regulated during differentiation of NB4 and HL60 cells[60] but is up-regulated in the MDS cell line after treatment with calcium ionophore, perhaps recapitulating some aspect of monocyte development.[80] CD34+ human progenitor cells could be immunostained with PLZF antisera in a distinct nuclear speckled pattern.[79] When such cells were placed into culture and allowed to differentiate, PLZF levels declined with terminal differentiation.[85] PLZF expression may be important for the maintenance or survival of hematopoietic stem cells and or early progenitors, down-regulated with differentiation and re-expressed in monocytes where it plays a role in regulating monocytic differentiation through direct interaction with the vitamin D_3 receptor.[86] PLZF expression increases through megakaryocytic development and overexpression of PLZF in an erythro-megakaryocytic cell line induced megakaryocytic markers and the thrombopoietin receptor suggesting a role of PLZF in megakaryocytic development.[87] This appeared to be mediated through interaction between PLZF and the GATA-1 transcription factor. PLZF may also modulate hematopoietic development through interaction with the GATA-2 factor.[88] The BTB domain of PLZF mediates interaction with the zinc finger domains of GATA-2 and PLZF as well as PLZF-RARα can block trans-activation by GATA-2, potentially acting to modulate hematopoetic cell division.

During mouse embryogenesis, PLZF is expressed in a segmental pattern[89] in the nervous system in a pattern reciprocal to the Krox 20 zinc finger gene and the Hoxb2 homeobox gene. In co-transfection experiments PLZF can repress the Hoxb2 promoter through a cognate PLZF binding site as well as an AT rich region.[70] PLZF expression is also notable in the limb buds where *Hox* genes help to guide the development[89] and mice lacking *PLZF*[90] display homeotic limb defects. This phenotype was associated with a more anterior pattern of expression of some *HoxD* genes in the developing hind limb bud, suggesting that PLZF could inhibit the expression of this class of pattern formation genes. Moreover, Barna et al showed that PLZF directly binds to and represses *HoxD* genes. *PLZF* null mice have not exhibited an obvious hematopoetic phenotype, nor have they developed leukemia or other tumors. These data do indicate, however, that as suspected and found for many otherleukemia-associated proteins that *HOX* genes may be major targets of PLZF and its fusion protein with RARα.

PLZF may play roles in other tissues as well. Recently PLZF was identified as a gene whose expression was stimulated in response to glucocorticoid and progesterone treatment of endometrial stromal and myometrial cells.[91] In addition PLZF was identified as a gene up-regulated in breast cancer cells in response to glucocorticoids and less so in response to progestins.[92] PLZF may in part mediate growth suppression by these agents. PLZF was detected in the uterus in the mid and late secretory phase of the menstrual cycle but not in the proliferative phase, consistent with the general negative effect of PLZF on cell growth. These results warrant examination of the reproductive tract of PLZF null animals for possible developmental changes.

Biological Effects of PLZF and PLZF Target Genes

Enforced expression of PLZF was achieved in myeloid 32D cells and U937 cells as well as non-hematopoetic cells NIH 3T3 cells. In all cases expression of PLZF led to G1 arrest and a block in the ability of cells to traverse the S-phase.[38,78,86,93,94] Similarly expression of the highly related FAZF protein induces growth arrest virtually identical to PLZF in U937 cells.[85] Prolonged expression of PLZF or FAZF eventually leads to apoptosis,[85,94] however somewhat paradoxically PLZF expressing cells could survive two weeks of withdrawal from IL3.[93] The ability of 32D cells to differentiate in the presence of G-CSF or GM-CSF was blocked by PLZF. This was not simply due to the fact that PLZF blocks cell division, since even before CSF treatment, PLZF expressing cells showed elevated expression of the primitive Sca1 marker and reduced expression of the mature granulocyte marker Gr1. PLZF expression in monocytic U937 cells inhibited the ability of vitamin D3 (VD_3) to stimulate monocytic differentiation of these cells while leaving differentiation in response to retinoic acid intact.[86] Similarly enforced PLZF expression inhibited VD_3-mediated differentiation in HL60 cells.[95] The ability of PLZF into inhibit the action of the VDR was associated with the ability of the BTB domain of PLZF to bind and inhibit transcriptional activation by the Vitamin D receptor.

PLZF can repress cell growth by binding to the promoters and inhibiting the expression of cell cycle regulators such as cyclin A and c-myc.[38,93,94] Enforced expression of cyclinA or c-myc could reverse growth suppression by PLZF, and both of these promoters bound PLZF in vivo as demonstrated by chromatin immunoprecipitation, validating them as PLZF targets.[94] Furthermore about 25% of genes identified as regulated by PLZF in a microarray analysis are known or suspected c-myc targets. PLZF regulation of Hox genes[96] may also play a role in regulation of cell growth and differentiation. Through a combination of microarray analyses, bioinformatics and chromatin immunoprecipitation studies we plan to more comprehensively identify PLZF targets important in development and differentiation. Currently our working model suggests that PLZF, by direct repression of these genes as well as potentially through its action on GATA-2 is associated with relative cell quiescence. In addition through interaction with other transcription factors such as the VDR, PLZF inhibits differentiation.[86] Furthermore the BTB domain of PLZF mediates interaction with that of FAZF[72] and additional BTB dimer/dimer interactions by PLZF and the PLZF-RARα fusion can be envisioned that may lead to regulation of additional sets of target genes.

Scheduled down-regulation of PLZF during hematopoiesis may allow for controlled cell division and differentiation. Hence PLZF might be considered a tumor suppressor disrupted by the t(11;17) translocation in APL. The PLZF-RARα fusion proteins resulting from this translocation may act as dominant negative inhibitors of PLZF. In fact, heterodimerization between PLZF-RARα and PLZF seems preferential to the formation of PLZF homodimeric complexes, in vitro.[76] Therefore, high-level expression of PLZF-RARα in t(11;17)(q23;q21) blasts might sequester PLZF from binding to its natural target genes and/or bind to limiting quantities of PLZF transcriptional co-factors.[97]

Growth factor-mediated signalling may also modulate PLZF-mediated gene repression. We found that transcriptional repression by PLZF was inhibited by co-expression of a mutant, constitutively active form of the tyrosine kinase receptor Flt3. Flt3 is the most common mutation identified in humanleukemia and mediates signalling through the ras-MAP kinase pathway as well as through the JAK-STAT pathway.[98] Co-expression of constitutively activated ras or stress kinases a well as mutant Flt3 blocks transcriptional repression by PLZF.[99,100] Signaling by active Flt3 leads to the export of the critical PLZF co-repressor SMRT from the nucleus to the cytoplasm. While signalling decreases PLZF-SMRT interaction it increases interaction between SMRT and the p65 subunit of NFκB, which can then shuttle SMRT to the cytoplasm (Takahashi and Licht Submitted). This may represent a general way in which signalling may influence multiple transcriptional repressors that utilize the SMRT co-factor. A mechanism more specific to PLZF was recently identified, potentially important in skin development. Upon proteolytic cleavage of the heparin-binding EGF-like growth factor (ectodomain shedding) the EGF-like ligand to bind to its cognate receptor. Meanwhile a small ~7kD fragment found within the cytoplasm of the cell translocates to the nucleus where it can bind to PLZF through the N-terminal zinc fingers of PLZF which are not critical for sequence specific DNA binding, leading to the export of PLZF from the cell nucleus. This interaction was verified to occur naturally in keratinocytes. Nuclear export of PLZF was associated with reversal of PLZF-mediated suppression of cell growth and cyclin A expression.[78] The phorbol ester TPA stimulates HB-EGF cleavage and treatment of murine skin with this agent induces keratinocytes hyperplasia, possibly by inhibiting the ability of PLZF to control cell growth. Hence signalling through receptor tyrosine kinases may, through two different mechanisms defeat PLZF-mediated transcriptional and growth regulation. PLZF can associate with another cytoplasmic factor Epsin1, that is associated with clathrin-coated pits.[101] Epsin1 has a structure similar to β-catenin a well-known transcriptional co-factor. As noted below Kaiso, another BTB-zinc finger protein interacts with a catenin-like factor, suggesting a more general interplay between nuclear BTB-zinc finger proteins and cytoplasmic co-factors.[102] PLZF interacts with Epsin1 through the zinc finger motifs as well as sequences N-terminal to the zinc fingers and can recruit Epsin1 to the nucleus. The effect of Epsin1 on PLZF transcriptional function and nuclear localization remains to be defined.

Transcriptional Repression by PLZF and PLZF Partner Proteins

Two different portions of PLZF can act as repression domains when tethered to the GAL4 DNA binding domain. The first corresponding to the BTB/POZ domain interacts with nuclear receptor co-repressors N-COR, SMRT and sin3A as well as histone deacetylases.[15,16,18,22,103,104] PLZF co-localizes to the same nuclear sub-location as the SMRT co-repressor.[104] Dimerization by the BTB domain appears to be required for its ability to mediate repression however our group performed a detailed mutational analysis of the BTB domain of PLZF and found that a charged pocked at the "top" of the domain was necessary for repression. These mutations in the isolated BTB domain tethered to GAL4 changes the domain from a repressor to an activator and when the mutations are made in full-length PLZF the protein is inactive for transcriptional repression and growth suppression. This has led us to investigate the charged pocket and other features of the BTB domain as therapeutic targets.[105,106] A second repression domain localized more central in the protein bears no similarity to other transcriptional effects domains and binds to ETO (Eight-twenty-one oncoprotein). ETO, the fusion partner of AML1 in t(8;21) associated M2 leukemia binds to mSin3a and the N-Cor and SMRT co-repressors. In contrast the AML1-ETO fusion protein blocks repression by PLZF, in part by inhibiting DNA binding by PLZF and by displacement of PLZF from the nuclear matrix and possibly from the deacetylase structures.[107] This finding suggested that PLZF function might be disrupted in other forms of leukemia. We found that deletion of either the BTB domain or the second repression domain of PLZF severely impaired the ability of PLZF to repress transcription and to cooperate with ETO or other co-repressors.[108] We hence believe that PLZF forms a dimerization/multimerization platform, which may bind to DNA cooperatively and forms a multi-protein complex with co-repressors and histone deacetylase. In accordance with this model histone deacetylase inhibitors block repression by PLZF and the absence of PLZF expression was associated with up-regulation of the endogenous *HoxD11* gene in mice, an effect associated with hyperacetylation of that promoter.[96]

The nature of the endogenous PLZF repressor complex remains incompletely understood although several other proteins have been found in complexes with PLZF. For example the PLZF interacts with Bmi-1 a member of the Polycomb protein family implicated in chromatin remodelling.[96] PLZF which has a predicted molecular weight of 70 kD, binds to DNA in native gels as a high molecular weight complex of nearly 600 kD.[68] Deletion of the BTB domain of PLZF reduces this complex to about 150 kD suggesting that the BTB domain may allow a tetramer of PLZF to form on a DNA binding site. In accordance with these studies atomic force microscopy experiments of PLZF binding to the Hoxd11 promoter concluded that PLZF bound to DNA as a dimer, forming loops when allowed to bind to naked DNA.[96] The BTB multimer may serve as a platform upon which the repressor complex can assemble. DNA looping through the formation of PLZF dimer/multimers might allow repression by PLZF to be propagated at distant sites.

The identity of the other PLZF partners in the 600 kD complex is uncertain. One protein we did identify in the 150 kD complex is the cdc2 protein, which may play a role in the phosphorylation of PLZF. Phosphorylation of PLZF may be critical for its function as treatment of PLZF with a phosphatase leads to inhibition of DNA binding activity. In contrast the RARα-PLZF protein which is devoid of the BTB or other repression domains of the protein binds to DNA as a 40 kD complex, exactly as predicted from its molecular weight, does not bind to cdc2 and is unaffected by phosphatase treatment. These data suggest that RARα-PLZF is not subjected to the same regulation as wild-type PLZF and is consistent with the notion that this leukemia-associated protein is a potent dominant negative inhibitor of PLZF.

Several other proteins were recently found to complex with PLZF. VDUP1 also known as TRX2 (thioredoxin binding protein 2) is a nuclear and cytoplasmic protein mutated in hyperlipidemic syndromes. PLZF was identified in a 2-hybrid screen as a VDUP1 partner[95] and although overexpressed PLZF and VDUP1 can interact it is not yet certain if this is a significant in vivo interaction. However VDUP1 can augment repression by PLZF in reporter gene assays. Since VDUP1 binds thioredoxin is intriguing to speculate that it could affect the redox state of PLZF.

PLZF can also interact with the LIM domain protein DRAL/FHL2, an interaction that required both the BTB domain and the second repression domain of PLZF. Co-expression of DRAL potentiated transcriptional repression by PLZF. DRAL could be localized in both the nucleus and cytoplasm and serum stimulation of cells increased nuclear accumulation of DRAL and further augmented repression by PLZF. This is in accordance with increasing data that PLZF-mediated repression may be subjected to control through growth factors. In a recent paper one group found that PLZF could be modified by SUMO (small ubiquitin like molecule). The modification occurred in the second repression domain of PLZF and curiously appeared to inhibit both DNA binding and transcriptional activity of PLZF.[109] The conditions under which PLZF may be modified by SUMO and the roles of other potential post-translational modifications such as phosphorylation and acetylation is currently under study.

The PLZF Fusion Proteins in APL

The t(11;17)(q23;q21) fusion yields two reciprocal transcripts encoding PLZF-RARα and RARα-PLZF.[61,62] The PLZF-RARα fusion contains the entire N-terminal transcriptional effector region of PLZF and the first two zinc fingers of PLZF. In one case, due to an alternative breakpoint in *PLZF*, three zinc fingers are included in the fusion. The reciprocal RARα-PLZF transcript links the ligand independent activation domain of RARα[110] to the last 7 PLZF zinc fingers αPLZF-RARα is an aberrant retinoic acid receptor which unlike the wild-type retinoic acid receptor can bind to target sites as a homodimer[76,80,111,112] α due to the ability of the PLZF BTB domain to mediate homodimerization. In combination with RXR, PLZF-RARα forms multiple DNA-protein complexes, but the PLZF-RARα/RXR heterodimer binds to retinoic acid response elements (RAREs) with higher affinity than PLZF-RARα homodimers.[76,80] The PLZF-RARα/RXR heterodimer bound to the RARE less efficiently than wild-type RXR/RARα perhaps due to the ability of the BTB domain to multimerize and preclude efficient DNA binding.[71] Hence, PLZF-RARα might sequester RXRα, an essential co-factor for RARα function. Furthermore, PLZF-RAR/RXR heteromers also bound to non-classical RAREs, potentially contributing toleukemogenesis by misdirecting the chimeric protein to novel genes.[111,113] PLZF-RARα is relatively weak trans-activator[76,80,114] and is a dominant negative inhibitor of wild-type RARα. In addition PLZF-RARα may contribute to theleukemic phenotype by blocking vitamin D_3-induced hematopoietic differentiation by sequestering the RXR co-factor.[115] Repression by PLZF-RARα is predominantly due to its high affinity for the HDAC containing co-repressor complexes.[15,16,21,22,103,104,116] This could be due in part to the presence of four docking sites for co-repressors on each oncoprotein homodimer rather than two sites found in normal RXR/RAR heterodimers.[117] However, due to the ability of the PLZF BTB domain to bind the co-repressors in an ATRA insensitive manner, while the PML-RARα protein of typical APL releases these factors in the presence of 10^{-6} M ATRA, PLZF-RARα does not.[15,16,21,22,104] In addition, ETO can bind and augment repression by PLZF-RARα but not PML-RARα and could contribute to the retinoid resistance of t(11;17)-APL.[107] HDAC inhibitors, such as trichostatin A and sodium butyrate, allowed cells harboring PLZF-RARα to respond to ATRA, presumably by inactivating the co-repressor complexes bound to the BTB domain.[21,22,104] However, there remains a paradox in the lack of response of PLZF-RARα associated APL. When APL cells are treated with ATRA the PML-fusion protein degrades which may help account for clinical response. A study with leukemic blasts from a t(11;17) patient showed that ATRA, although unable to induce clinical response did lead to PLZF-RARα degradation.[118] This suggests that the PLZF-RARα may leave a permanent, epigenetic mark on the leukemic cell silencing target genes in a long lasting manner in a way recently described for the PML-RARα fusion protein through the recruitment of DNA methyl transferases to target genes.[119]

Indeed recent evidence indicates that there are significant differences in the genes regulated by PML-RARα and PLZF-RARα. Upon induction of these fusion proteins in the myeloid U937 cell line, 57 genes were down regulated in common by the two proteins but PLZF-RARα down-regulated 31 genes that were unaffected by PML-RAR and PML-RAR exclusively regulated a distinct set of genes.[120] This group identified a subset of genes down-regulated by both fusion proteins, that might explain the similar block in differentiation at the promyelocytic stage observed in t(15:17) and t(11;17)-associated APL. However, some of these commonly regulated genes were only regulated in the presence of ATRA in PML-RARα-expressing cells. For example, the gene encoding a tumor necrosis factor receptor (TNFR1) was repressed by both fusions and only re-activated by PML-RARα. In addition ATRA treatment of PML-RARα cells led to decreased expression of the antiapoptotic Bcl2 gene but similar treatment of PLZF-RARα cells did not regulate Bcl2. These differences in gene expression, potentially due to the differences in the repressive complexes brought to target genes by the two different fusion proteins could help explain the differential sensitivity of the two syndromes to ATRA and other therapies.

Animal models also indicate differences between PLZF-RARα and PML-RARα. Transgenic animals expressing PLZF-RARα developed a CML-like syndrome rather than APL.[21,121] Unlike PML-RARα transgenic mice, the PLZF-RARα mice did not achieve durable remission after ATRA treatment, although leukemic cells from the mice treated with ATRA differentiated ex vivo.[21] The relative insensitivity of the disease correlates with the markedly decreased ability of PLZF-RARα to transactivate due to binding of co-repressors. This idea was confirmed by the result that the combination of a histone deacetylase inhibitor and ATRA synergistically induced differentiation of these leukemic cells.[67] Though explaining the poor response of this disease to ATRA, these animal models do not explain the poor response of some t(11;17)(q23;q21) patients to chemotherapy and the complete resistance of fresh APL cells to high doses of ATRA in vitro.[62,122] These findings implicate the reciprocal RARα-PLZF protein in the aggressive nature of PLZF-RARα associated APL.

The reciprocal transcript encoding RARα-PLZF was expressed in nearly all of the t(11;17)(q23;q21) patients.[62,123] The three zinc fingers of RARα-PLZF can bind to the same site as full-length PLZF.[36,124] While PLZF represses gene transcription, RARα-PLZF activates it.[36] Whereas PLZF is a growth suppressor, RARα-PLZF activates expression of cyclin A2 and enhances cell growth.[38] Hence t(11;17)(q23;q21) may be an ATRA and chemotherapy resistant disease due to the presence of two oncogenes. PLZF-RARα blocks the differentiation inducing action of ATRA, while RARα-PLZF may activate cell cycle regulators, blocking the anti-proliferative effects of ATRA. In accordance with this idea, mice engineered to express the RARα-PLZF protein develop a myeloproliferative syndrome and splenomegaly.[125] When PLZF-RARα mice were crossed with RARα-PLZF mice there was a phenotypic change in the resulting leukemia with the appearance of more primitive, more proliferative cells that were highly resistant to apoptosis and morphologically more similar to human APL.[125] Strikingly, an identical phenotype was seen when PLZF-RARα transgenic mice were bred with PLZF null mice. This information is in accordance with data suggesting that RARα-PLZF is a dominant negative form of PLZF and a growth activator. It is possible that through aberrant expression of PLZF targets such as *HOX* genes, cyclins, c-myc, and other cell cycle regulators, the presence of both fusion genes lead to a more aggressive and undifferentiated form of leukemia.

Though t(11;17)-APL is a very rare disease, the comparison of contrast of this illness with the more typical form of APL solidified the notion of aberrant transcriptional repression as a root cause of leukemia. In addition studies from our group and others have implicated PLZF as an important negative regulator of cell growth with roles in differentiation, development and homeostasis.

The Role of Bcl-6 Protein in Lymphoma

The *Bcl-6* gene was first cloned by virtue of its frequent mutations in B-cell lymphomas, where overexpression or lack of downregulation of Bcl-6 occurs when translocated to a heterologous promoter or mutations introduced by somatic hypermutation damage Bcl-6 regulatory elements.[126-128] Such tumors represent approximately 40% of human B-cell lymphomas and are believed to be biologically dependent on Bcl-6 expression.[129,130] Bcl-6 was subsequently shown to also play a critical role in the development of the immune system.[131-133] The BTB domain of Bcl-6 was shown to interact with the SMRT/N-CoR corepressors and histone deacetylases (HDACs).[15,16,20,134-136] As in the case of PLZF there is a second repression domain in the protein localized to the mid-portion of the protein.

Bcl-6 As a Transcriptional Repressor: Structure-function Studies of the BTB Domain

Along the lines of our studies with PLZF, we performed in depth structure-function analysis of the Bcl-6 BTB domain.[6,40,106,137] The PLZF and Bcl-6 dimer crystals revealed that the two structures are very similar. BTB monomers interact via an extensive interface to form obligate dimers. The most highly conserved surface feature is a charged pocket. The charged pocket is structurally consistent with an interaction site for peptides from putative partner proteins. However, there is a region of conformational variability between the PLZF and Bcl-6 pockets conferred by an α3/β4 loop, where the BCL-6 BTB domain has a unique 4 amino acid insertion relative to PLZF that might obstruct access to this motif in the case of Bcl-6.

We performed extensive mutagenesis of the PLZF and Bcl-6 BTB domain and tested these products in both the full length and isolated BTB context in vitro and in vivo.[6,40,106,137] These results allowed us to assign functions to its various structural motifs. For example, conserved residues in the dimer interface are required for proper folding, indicating that dimerization is an architectural function of BTB domains that is invariant throughout evolution. While the oligomerization surface was required for high molecular weight complex formation by PLZF, which is known to form such oligomers, there was no evidence that Bcl-6 forms similar oligomers through this motif. Therefore, it is possible that the Bcl-6 hydrophobic face is involved in other protein-protein interactions.[6,40,106,137] As in the case of PLZF, mutation of the charged pocket of Bcl6 inhibited both transcriptional repression and interaction with the co-repressors N-Cor and SMRT. Comparing the PLZF, Bcl6 and PLZF-related FAZF BTB domains a single amino acid difference was noted. When this single residue of these three BTB domains was altered to that of one of the other two proteins, the strength of transcriptional repression and interaction with N-Cor or SMRT was modulated. This supports the notion that the charged pocket of the BTB is an important contact site for co-repressors.

Further analysis of the Bcl6 BTB domain indicated that the charged pocket could not completely explain repression by the BTB domain. The Bcl6 BTB domain directly bound N-Cor and SMRT much more tightly than PLZF and a small region of N-Cor or SMRT mediated this direct binding. The co-crystal structure of the Bcl6 BTB domain with a segment of SMRT (aa 1353-1526) showed that this portion of SMRT tightly bound to the lateral face of the BTB domain.[138] Subtle mutations in either the SMRT peptide or in the BCL6 BTB domain abrogated SMRT/Bcl6 interaction and transcriptional repression.

Bcl-6 As a Transcriptional Repressor: Other Functional Domains of the Bcl-6 Protein

The other major structural motif of Bcl-6 is the array of six C_2H_2 zinc fingers at the C-terminal end of the protein, which bind to a specific DNA consensus sequence (TTCCT(A/C)GAA).[139] The zinc fingers also mediate protein-protein interactions. For example, the zinc fingers interact with class II HDACs as well as with the zinc fingers of other BTB proteins such as PLZF and LRF.[140-143] The zinc finger domain also interacted with the c-jun transcription factor, leading to negative regulation of AP1 transcriptional activation by Bcl-6.[144] Recently, we found that the ETO protein which was first cloned as a member of the acute myeloid leukemia (8;21) translocation fusion product, also interacts with Bcl-6 through the zinc fingers.[145] We had previously demonstrated that ETO is a novel corepressor much like N-CoR and SMRT, which interacts with transcriptional repressors and enhances transcriptional repression by amplifying histone deacetylation.[142,146,147] We found that ETO and Bcl-6 are co-expressed in normal and malignant lymphoid tissue, where they interact and colocalize in nuclear speckles. ETO binds to the fourth zinc finger of Bcl-6, enhances Bcl-6 repression of artificial and endogenous genes in an HDAC-dependent manner, and forms a complex with Bcl-6 on the promoters of its endogenous target genes in B-cell lymphoma cells.[145] Therefore, ETO is a bona fide corepressor for Bcl-6 and is the only corepressor conclusively demonstrated to bind outside of the BTB domain in the context of physiological transcriptional repression.

Although the zinc finger domain of Bcl-6 was often used as a dominant negative construct to antagonize Bcl-6 repression, multiple lines of evidence indicate that it may also function as an autonomous repression motif albeit with much less potency than the BTB domain. Although Seyfert et al found that the zinc finger domain did not repress transcription of a Bcl-6 binding site containing reporter,[148] the same group also reported that the zinc finger domain functioned as a weak dominant negative when expressed in Bcl-6 positive Raji cells, indicating that it either does not repress or that it represses less efficiently than full-length Bcl-6.[149] In contrast, a GAL4-zinc finger fusion construct inhibited transcriptional activation of a reporter gene containing LexA and GAL4 sites when it was activated by a LexA-VP16 fusion while in reciprocal experiments, deletion of the zinc fingers caused a loss in potency of Bcl-6 repression.[141] A zinc finger only construct was able to partially mediate Bcl-6 induced growth and cell cycle arrest in U2OS cells although the BTB domain was required to induce apoptosis.[150] Finally, the Bcl-6 zinc fingers alone were able to partially inhibit AP1 transactivation, although full length Bcl-6 as much more powerful.[144] These experiments illustrate the difficulty of determining the functional contribution of the zinc finger domain to transcriptional repression, which is unlikely to be resolved by studying deletion mutants of Bcl-6. Taken together these results seem to indicate that the zinc finger and BTB repression domains both contribute to gene silencing by Bcl-6, although the gene loci they regulate may not fully overlap. Finally, the central region of Bcl-6 may also have autonomous repression activity by virtue of recruiting HDACs and possibly the Sin3A corepressor.[151]

Transcriptional Mechanisms of Action of Bcl-6

Bcl-6 mediates transcriptional repression of a number of endogenous target genes that include surface antigens (for example CD23b, CD44 and CD69), cell cycle mediators (for example Cyclin D2, and $p27^{KIP}$), cytokines, chemokines and their receptors (for example IL2Rα chain, IL5, IL18, MCP-1, MCP-3, MIP1α, and CXCR4), and mediators of B-Cell differentiation such as blimp-1.[132,149,152-156] In most cases Bcl-6 binds directly to its consensus binding site located in its target loci and mediates repression in large part through the BTB domain as discussed above. However, Bcl-6 also represses downstream transcriptional activation by the IL-2, IL-4 and CD40 signal transduction pathways that mediate lymphocyte differentiation.[132,155,157-160] A subset of DNA binding sites for mediators such as STAT6, STAT3 and IRF-4 are similar to the Bcl-6 consensus sequence.[132,155,157,160] Bcl-6 can bind to the same sites as these transactivators and repress expression of their target genes. For example, Bcl-6 inhibits IgE class switching by repressing IL-4 induced activation of the IgE promoter through an overlapping DNA element with STAT6.[160] Consistent with this observation, *Bcl-6-/-* mice demonstrated increased IgE class switching and tissue infiltration with *IgE+* B-lymphocytes, while *Bcl-6-/- STAT6-/-* mice had neither of these features.[160] Bcl-6 represses activation by these transcriptional messengers in a BTB-dependent manner, although it is not known whether the BTB domain is required for an architectural or corepressor recruiting function.

A third mechanism of Bcl-6 transcriptional repression was recently described in which Bcl-6 acts as a corepressor for AP-1 by virtue of binding to c-Jun through the C_2H_2 zinc finger domain. In this way Bcl-6 can behave as a corepressor by repressing AP-1 target genes such as $p27^{kip}$ through c-Jun. Full repression also required the BTB domain in this case, although once again it is unknown whether the BTB contributes architectural or corepressor function.[144]

Bcl-6 was recently shown to induce down regulation of the GATA-3 inducer of Th2 differentiation. This is consistent with the increased levels of Th2 cells in Bcl-6 null animals. Surprisingly, only GATA-3 protein levels were reduced, while the mRNA level didn't change.[161] This implicates Bcl-6 in the regulation of protein stability, an attractive idea in light of emerging evidence of the role of the BTB domain in ubiquitylation. Finally, Bcl-6 was found in the proximity of DNA replicative foci raising the possibility of a role in DNA replication as well.[162] Therefore, Bcl-6 is a multifunctional regulatory factor that represses expression of target genes and proteins through several different mechanisms.

Bcl-6 and the Immune System

Bcl-6 is a critical regulator of specific components of the immune system and its hierarchy in this process is underlined by the severity of the Bcl-6 null phenotype and by the fact that de-regulation of Bcl-6 expression causes cancer.[130] Accordingly, expression of Bcl-6 is tightly regulated and limited to specific subsets of immune system cells.

B-Cells

In the B-cell lineage, Bcl-6 is expressed solely in germinal center (GC) cells.[163,164] Upon T-cell dependent antigen challenge, subsets of activated B-cells (pre-GC cells) enter lymphoid follicles and up-regulate Bcl-6, which induces cells to proliferative and survive and at the same time to, resist further differentiation.[131-133,155,165] Simultaneously, induction of the somatic hypermutation machinery leads to affinity maturation of the immunoglobulin genes. When this process is complete Bcl-6 is downregulated.[131-133,155,165,166] B-cell clones with productive immunoglobulin rearrangements differentiate further into post-GC memory or plasma cells while clones with non-productive rearrangements undergo apoptosis.[155,167] Regulation of Bcl-6 expression at the mRNA level is poorly understood, although Bcl-6 itself can repress transcription of its own message.[168] From the protein standpoint, MAP kinase leads to phosphorylation of Bcl-6 via the ERK1/2 kinases with subsequent ubiquitylation and proteolysis.[166,169] This process is under the control of the B-cell receptor in pre-GC cells and possibly under the control of CD40 in post-GC cells.[166]

The GC specific role of Bcl-6 is underscored by the phenotype of Bcl-6 knockout mice. Such animals are born at the expected frequencies and with all the lymphoid organs fully formed and containing the expected primary lymphoid follicles.[131-133] All lymphoid cell subsets are present in normal numbers, although baseline antibody levels are slightly low.[131,132] However, Bcl-6 null mice exhibited a complete failure to generate GCs in all lymphoid tissues when challenged by T-cell dependent antigens. T-cell dependent antibody generation was deficient, with antibody levels of all classes (Except IgM) lower than control.[131-133] Even more remarkably, these antibodies had not undergone the (GC-dependent) process of affinity maturation as demonstrated by the absence of high affinity binding in antigen binding assays.[131] Not surprisingly, memory B cells in these animals express only low affinity antibodies after booster shots of T-cell dependent antigens, indicating that these cells had not undergone somatic hypermutation as they did not transit through GCs during the course of their terminal differentiation.[167]

The defect in GC formation was entirely B-cell autonomous as demonstrated by transplantation experiments in RAG1 deficient mice in which the immune system was reconstituted with normal T cells plus Bcl-6 deficient B cells or vice versa.[133] This was confirmed in experiments were chimeric mice were generated by injecting *Bcl-6-/-* ES cells into *RAG1, IgM* or *TCRβδ* null blastocysts.[153] In the resulting *RAG1-/-* mice all B and T cells derive from the *Bcl-6-/-* cells, while in *IgM-/-* only the B-cells are exclusively *Bcl-6-/-*, and in the *TCRβδ-/-* cells only T cells are *Bcl-6-/-*.[153] In concordance with the transplantation experiment, the *Bcl-6-/- RAG1* and *IgM* null mice were unable to form GCs, while GCs were able to form in the *Bcl-6-/- TCRβδ* chimeric mice.[153]

Only a few of the Bcl-6 target genes relevant for its biological activity in the GC were identified so far. For example, Bcl-6 directly represses transcription of the Blimp1 transcriptional repressor gene, which in turn represses expression of genes involved in GC processes such as class switching (Ku70), somatic hypermutation (AID), proliferation (myc) and B-cell signaling (STAT6).[149,170] As noted, Bcl-6 binding sites overlap in several loci with binding sties for the STAT3, STAT6 and IRF4 transcriptional activators as shown by EMSA and DNA footprinting.[132,155,157-160] Bcl-6 also physically interacts with these transcriptional activators and represses promoters with overlapping Bcl-6 and STAT sites such as the CD23b promoter.[155,157] In this way GC cells expressing Bcl-6 could resist differentiation signaling from the IL2, IL4 and CD40 pathways.

T-Cells

Bcl-6 plays important regulatory roles in T cells, and is normally expressed in germinal center T cells and in a subset of CD30 positive T cells present in interfollicular regions of lymph nodes.[163,171] Bcl-6 is also expressed in cortical and medullary T cells of the thymus.[172] Bcl-6 null animals appear similar to control littermates when born, but rapidly become sickly and growth retarded.[131-133] At least 50% of these mice die by 5 weeks of age due in large part to an inflammatory infiltrate of monocytes/macrophages and eosinophils, affecting the myocardium, lungs, gut, skin and other mucosas.[131-133] Taken together with observed increases in IgE expressing B-cells, these findings suggested that a powerful Th2-type reaction was occurring in these animals leading to their demise through systemic inflammatory reactions. Accordingly, cells isolated from *Bcl-6-/-* lymph nodes expressed high levels of (Th2 related) IL4, IL5 and IL13 but low levels of (Th1 related) IFNγ at mRNA and protein levels.[131-133] These results implicated Bcl-6 in the skewing of T-cells towards the Th1 lineage or in the inhibition of differentiation into Th2 cells. Since both IL4 and STAT6 null mice lack Th2 cells and given that Bcl-6 is an inhibitor of IL4-STAT6 signaling, it was anticipated that generating mice null for Bcl-6 and deficient of either IL4 or STAT6 might resolve the Th2 skewing.[158] However, both *Bcl-6-/- STAT6-/-* and *Bcl-6-/- IL4-/-* mice exhibited Th2 skewing when mice were challenged with TNP/KHL antigen/hapten. The defect was not due to loss of Th1 differentiation since Bcl-6 single or double knock-out mice were able to generate Th1 cells in response to IL12.[158]

The immunological significance of this phenomenon was illustrated by the following experiment. The 129sv and C57BL/6 strains from which the Bcl-6 null animals were derived are normally able to mount a Th1 immune reaction that heals infections by *Leishmania major* through macrophage stimulation by IFNγ.[159] In contrast, both single and double knock-out Bcl-6/IL4 animals developed progressive infections by *Leishmania major*, since Th2 reactions are inhibitory to macrophages.[159] Therefore, Bcl-6 must normally inhibit Th2 formation through an alternate pathway independent of IL4/STAT6 signaling. One such pathway is mediated through GATA-3 activation of Th2 cytokines in the absence of STAT6.[173] Interestingly, Bcl-6 single and double knock-out with STAT6 mice have high levels of GATA-3.[161] Re-introduction leads to downregulation of GATA-3 and downregulation of Th2 cytokine secretion (IL4 & IL10).[161] As mentioned above, the GATA-3 gene did not have Bcl-6 binding sites in its promoter and Bcl-6 seemed to downregulate only the protein but not the mRNA levels of this protein, indicating a potentially novel regulatory mechanism.[161] Finally, these results indicate that loss of Bcl-6 results in both humoral and cellular immunodeficiency.

Macrophages

A role for Bcl-6 regulation of macrophage function was suggested by the fact that Bcl-6 was shown to be expressed in monocytic cells[174] and the finding that RAG1 null mice transplanted with Bcl-$6^{-/-}$ bone marrow or *RAG-/- Bcl-6-/-* chimeric mice failed to develop eosinophilic infiltrates, indicating that full blown Th2 inflammation requires the participation of additional non-lymphoid cells.[153] A *Bcl-6-/- TCRα-/-* chimeric mouse specifically unable to generate Th2 cells was used to study macrophage function.[153] Interestingly several chemokines involved in inflammatory reactions and eosinophil chemotaxis were upregulated and some of these were direct Bcl-6 target genes.[153] Re-introduction of Bcl-6 suppressed expression of these genes. Therefore, macrophage products such as MCP-1 contribute to Th2 inflammation possibly by promoting antigen presenting cell stimulation of Th2 and though inhibition of the Th1 response by virtue of inhibiting Il-12 production. The triggering event that leads to this inflammatory picture remains obscure, although one of the Bcl-6 null mice models had a high incidence of bacterial infections even though they were housed in a barrier facility.

Other Tissues

Bcl-6 is highly expressed in terminally differentiated tissues such as the heart, skin and skeletal muscle.[175-177] For example, Bcl-6 is upregulated in cardiac myocytes within two weeks of birth. Bcl-6 null mice develop subtle signs of cardiac degeneration within four weeks of birth and apparently prior to infiltration by the Th2 inflammatory disease (as detected by electron microscopy).[175] In these tissues Bcl-6 might be required to maintain a quiescent, differentiated state. In contrast, Bcl-6 is also highly expressed in spermatocytes undergoing meiosis, which like the somatic hypermutation and class switching of GC B-cells, leads to DNA breakage in the setting of rapidly proliferating cells.[178] Interestingly, Bcl-6 protects spermatocytes from apoptosis and Bcl-6 null animals are infertile.[158,178] Therefore, Bcl-6 physiological effects may vary according to cell type and available partner proteins.

Bcl-6 and B-Cell Lymphomas

In spite of the frequent involvement of the Bcl-6 locus in B-cell lymphomas, no definitive gain or loss of function study has demonstrated that deregulated Bcl-6 expression causes lymphoma. However, several lines of evidence are consistent with potential oncogenic effects of Bcl-6.

Most studies of Bcl-6 functions in primary cells and lymphoma cells indicate that Bcl-6 protects cells from apoptosis. For example, Bcl-6 deficiency induces apoptosis in spermatocytes, skeletal muscle cells, and causes degenerative changes in cardiac myocytes.[175,178,179] In B-cell lymphoma cell lines, a dominant negative form of Bcl-6 that releases repression of Bcl-6 target genes induced apoptosis as well.[149] Furthermore, Bcl-6 may prevent apoptosis by directly repressing the PDCD2 pro-apoptotic gene.[180] Most recently, Bcl-6 was discovered in an unbiased screen of cDNA libraries to bypass p53 mediated cellular senescence in mouse embryonal fibroblasts.[181] When Bcl-6 was co-expressed with an activated allele of Ras, these cells became fully transformed and formed colonies in soft agar and tumors in nude mice.[181] Induction of cyclinD1 was required for these effects. Furthermore, primary human tonsilar B cells infected with a Bcl-6 expressing retrovirus were able to grow indefinitely in culture in the presence of CD40L, IL2 and IL4.[181] However, Bcl-6 can also induce apoptosis, although this usually occurs in the context of forced overexpression studies in non-physiologically relevant cell lines such as CV1 and U2OS.[150,182]

The ability of Bcl-6 to block differentiation induced by IL2, IL4 and CD40L pathways explains in part how Bcl-6 mediates the B-cell GC life cycle.[155] Accordingly, expression of Bcl-6 in WIL2 and SUDHL5 cells blocked differentiation while in Raji cells, a dominant negative Bcl-6 zinc finger construct triggered incomplete plasmacytic differentiation.[149]

Since Bcl-6 establishes a proliferative cohort of cell in germinal centers, it would be expected to mediate cell cycle progression. Accordingly, in Raji cells, dominant negative Bcl-6 caused cell cycle arrest in G1 and in BCL-1 cells, expression of Bcl-6 overcomes the G2/M phase cell cycle arrest mediated by IL-2 and IL-5.[149,155] Bcl-6 regulates a number of cell cycle mediators and inhibitors including cyclin D2 gene, cyclin A2, p27 and cyclin D1.[149,154,181] Therefore, it is possible that Bcl-6 may differentially affect cell cycle progression according the milieu of the cell type. Thus, while Bcl-6 may mediate cell cycle progression in GC B cells, it might inhibit cell cycle in terminally differentiated cells such as myocytes and keratinocytes.

Taken together, these data suggest that in germinal center B-cells, de-regulated expression of Bcl-6 causes cells to survive in the face of DNA damage and evade p53 mediated senescence, as well as to transit the cell cycle, block differentiation and undergo continuous mutagenesis through somatic hypermutation characteristic of germinal center B cells.

Other BTB/Zinc Finger Proteins

HIC-1 (Hypermethyalted in Cancer-1)

Recruitment of a histone deacetylase complexes not a universal property of BTB domain containing transcriptional repressors. The HIC-1 gene encodes a protein with five C_2H_2 zinc fingers and an N-terminal BTB domain.[183] Transcription of this ubiquitously expressed gene is activated by p53, but HIC-1 is not expressed in several common types of human cancers due to hypermethylation of its promoter. These facts make HIC-1 a strong candidate for a tumour suppressor gene.[183] In accordance with this notion it was recently shown that a heterozygous disruption of HIC-1 in mice led to a variety of sex-dependent tumors arising after a year. In most tumor tissue, it was found at least one promoter of the wild-type HIC-1 allele was hypermethylated.[184]

Like many BTB domains both the HIC-1 and γFBP-B (an avian homologue) BTB domains can function as autonomous transcriptional repression domains. However, they do not interact with components of the histone deacetylase complex in vitro or in vivo. Furthermore, transcriptional repression by HIC-1 was not alleviated with the HDAC inhibitor trichostatin A, further substantiating the idea of an alternate mechanism of transcriptional repression by these particular BTB domains.[55] This was confirmed when it was determined that HIC-1 could complex with the co-repressor CtBP through a motif more internal within the protein.[185] The BTB domain did not directly interact with CtBP, yet deletion of the BTB domain of Hic-1 severely impaired transcriptional repression by the protein. This and evidence from the study of PLZF and Bcl6 indicates that dimerization/multimerization through the BTB domain is a critical function of the motif in transcriptional repressors, but the BTB domain itself need not be required for direct interaction with co-repressor complexes.

Kaiso

The Kaiso BTB domain protein was identified in a two-hybrid screen as an interactor with $p120^{Ctn}$ a protein similar to catenin with multiple armadillo repeats. Kaiso interaction with $p120^{Ctn}$ was specific and no interaction with E-cadherin, β-catenin or α-catenin could be detected. Kaiso is ubiquitously expressed and contains a N-terminal BTB domain and three C-terminal zinc finger motifs.[102] Kaiso/$p120^{Ctn}$ interaction was mediated by a region of the protein encompoassing the zinc finger motifs and including a region immediately N-terminal to these motifs.[102,186] Kaiso was localized mostly to the nucleus but in some cells cytoplasmic localization was noted suggesting that the Kaiso/$p120^{Ctn}$ interaction may mediate a nuclear/cytoplasmic shuttling. Kaiso was subsequently identified as part of a methyl CpG DNA binding activity known as MeCP1. Kaiso itself bound to DNA specifically in a methylation dependent manner.[186,187] Zinc fingers 2 and 3 of the protein were sufficient for specific DNA binding to two symmetrically methylated CpG dinucleotides in the core sequence. Kaiso could also recognize with high specificity a non-methylated DNA sequence implying a role for the protein in epigenetic silencing as well as broader transcriptional processes.[186] Kaiso repressed transcription of a reporter gene containing a consensus Kaiso binding sequence in a cell line otherwise defective for methylation-dependent silencing due to deletion of another methyl-DNA binding protein. Repression by Kaiso required the BTB domain as well as the zinc fingers of Kaiso.

The function of Kaiso is conserved in evolution as *Xenopus* Kaiso, tethered to the Gal4-DNA binding domain could represses a reporter gene in frog oocytes.[188] This information suggested that Kaiso might play a part in the silencing of methylated genes by recruitment of co-repressors complexes through the BTB and other domains of the protein. Indeed this notion was confirmed when Kaiso was found as a component of a high molecular weight complex containing the co-repressor N-Cor (but not the related SMRT protein) and HDAC3.[189] As in the case of PLZF and Bcl6 the BTB domain mediated interaction of Kaiso with N-Cor and a histone deacetylase inhibitor reversed repression by the Kaiso BTB domain. Chromatin immunoprecipitation demonstrated that N-Cor and Kaiso were present on a methylated promoter and were released from the gene when cells were treated with the de-methylating agent 5-aza-2-deoxycytidine. The presence of Kaiso on an endogenous promoter was associated with de-acetylation histone 3 lysine tail residue 9 and methylation of this residue, a set of changes associated with gene repression.[190] Knockdown of Kaiso by siRNA led to reactivation of a methylated gene and loss of co-repressors from the gene. These studies indicate a major role of Kaiso in epigenetic gene silencing. However the importance of kaiso in regulation of other genes through non-methylated DNA sequences and the role of $p120^{Ctn}$ in Kaiso function in vivo and gene silencing is as yet undetermined. In turn these studies raise the idea that other BTB/zinc finger proteins such as PLZF and Bcl6 may also bind to methylated DNA and play a role in both adjustment of gene expression in a dynamic manner as well as in more permanent inhibition of methylated genes.

Miz-1

The Miz-1 (myc interacting zinc finger-1) BTB protein has come under increasing study, as it appears to be an important mediator of the function of the c-myc oncoprotein. Miz-1 was identified in a two hybrid screen for proteins that interact with c-myc.[191] Miz-1, like other BTB/zinc finger proteins, contains an N-terminal BTB domain and has 12 zinc finger motifs localized in the middle of the protein and a lone zinc finger motif near the C-terminus of the protein. Two regions flanking the 12 zinc fingers mediated interaction with the helix-loop-helix motif of c-myc as well as the related N-myc proteins. Miz-1 bound specifically to a site within the adenovirus major late promoter and unlike the other BTB/zinc finger proteins described activated this promoter. Co-expression of myc impaired transcriptional activation by Miz-1. Curiously, although the BTB domain did not interact with c-myc, the BTB domain was required for the ability of c-myc to inhibit *trans*-activation by Miz-1. Unlike other BTB domain proteins like PLZF, Miz-1 itself was not localized to discrete nuclear foci and was readily extracted from the cell. However, when co-expressed with c-myc, Miz-1 did localize to punctate domains and appears to make the Miz-1 protein insoluble in the cell and extractable from the nucleus only under high-salt conditions. In part this appears to be due to the ability of myc to donate a functional nuclear localization signal to Miz-1. Whether the Miz-1 nuclear domains overlap with those of other BTB/zinc finger proteins or co-localize with co-repressors and histone deacetylases is unknown.

Miz-1 is a growth suppressor and co-expression of c-myc relieves its ability to suppress growth. Deletion of the BTB domain only partially inhibited trans-activation and growth suppression by Miz-1 implying that domains within Miz-1 other than the BTB domain were also critical for its transcriptional function. However the deletion of the BTB domain made growth suppression by Miz-1 insensitive to co-expression of c-myc. Miz-1 can augment transcriptional activation by c-Myc and its obligate partner Max. This correlates with the ability of Miz-1 to stabilize c-Myc and inhibit its proteosome-mediated degradation.[192]

It has become clear that transcriptional repression by c-myc is essential for its ability to transform cells. Miz-1 is a major mediator of myc-mediated repression. Miz-1 binds to the initiator region of both the promoters of the cell cycle regulators $p21^{waf1/Cip1}$ and $p15^{INK4b}$ to activate expression of these genes, accounting for its activity as a growth suppressor. c-myc activates cell growth and inhibits cell senescence least in part due to its ability to bind to Miz-1 on these promoters and lead to the repression of transcription of these cell cycle regulators.[193-196] Myc inhibits Miz-1 mediated activation by preventing p300 binding to Miz-1.[193] A C-terminal region of Miz-1 and not the BTB domain mediates p300 interaction. Myc-Miz-1 interaction may also release a latent repression activity of Miz-1, possibly involving the BTB domain but to date no mechanism of active repression by the Miz-1 has been elucidated. The ability of TGFβ to inhibit cell growth was related to its ability to decrease c-myc levels and inhibit myc-Miz-1-mediated repression of $p15^{INK4b}$.[194] In addition activation and nuclear migration of SMAD3 and SMAD4 proteins allowed these transcription actors to complex with Miz-1 through its zinc finger motifs and further stimulate the $p15^{INK4b}$ promoter. The myc-Miz interaction is also an important regulator of p53 action. In the presence of myc, repression of the p21 promoter through Miz-1 prevents the ability of p53 to inhibit the cell cycle and favors the pro-apoptotic actions of p53.[195]

The Miz-1 BTB domain does not appear to play a direct role in transcriptional activation but does mediate important protein-protein interactions. An interacting partner for this BTB domain, topoisomerase IIβ binding protein (TOBP1), a BRCT domain protein implicated in DNA damage response was recently isolated.[197] Intriguing in light of recent data implicating the BTB domain in protein degradation, the other interacting protein isolated with Miz-1 was an E3 ligase. The TOPBP1 protein did not interact with the Bcl6 BTB domain, which we[20,138] and others showed to be a potent transcriptional repressor domain. The TOPBP1 protein inhibited activation by Miz-1 but was found to be down-regulated in response to UV irradiation, leading to de-repression of Miz-1 target promoters $p21^{waf1/Cip1}$ and $p15^{INK4b}$.[197] Host cell factor (HCF) a protein previously implicated in viral gene regulation also interacts with Miz-1 to inhibit its ability to activate transcription. This interaction can occur independently of the BTB domain but HCF can also interact in vitro with the Miz-1 BTB domain.[198] Like c-myc, HCF blocks activation by Miz-1 by interfering with Miz-1/p300 interaction.

As noted with many of the other BTB/ zinc finger proteins described above, a common theme appears to be emerging in that many of these proteins may shuttle to the cytoplasm and interact with cytoplasmic proteins. Miz-1 was originally found to co-purify with tubulin[191] and this was confirmed in a chemical screen for activators of the gene encoding the low density receptor.[199] Addition of a small chemical to cells released Miz-1 from microtubules and allows Miz-1 to bind and stimulate transcription of this gene. Miz-1 bound to tubulin through its most C-terminal zinc finger motifs, while the BTB domain of Miz-1 was required for the protein to migrate to the nucleus after chemical treatment. The physiological role of Miz-1 sequestration and what natural stimuli encourage its nuclear import are not yet known.

References

1. Koonin EV, Senkevich TG, Chernos VI. A family of DNA virus genes that consists of fused portions of unrelated cellular genes. Trends Biochem Sci 1992; 17:213-214.
2. Godt D, Couderc JL, Cramton SE et al. Pattern formation in the limbs of Drosophila: bric a brac is expressed in both a gradient and a wave-like pattern and is required for specification and proper segmentation of the tarsus. Development 1993; 119:799-812.
3. Numoto M, Niwa O, Kaplan J et al. Transcriptional repressor ZF5 identifies a new conserved domain in zinc finger proteins. Nucleic Acids Res 1993; 21:3767-3775.
4. Zollman S, Godt D, Prive GG et al. The BTB domain, found primarily in zinc finger proteins, defines an evolutionarily conserved family that includes several developmentally regulated genes in Drosophila. Proc Natl Acad Sci USA 1994; 91:10717-10721.

5. Bardwell VJ, Treisman R. The POZ domain: a conserved protein-protein interaction motif. Genes Dev 1994; 8:1664-1677.
6. Ahmad KF, Engel CK, Prive GG. Crystal structure of the BTB domain from PLZF. Proc Natl Acad Sci USA 1998; 95:12123-12128.
7. Li X, Lopez-Guisa JM, Ninan N et al. Overexpression, purification, characterization, and crystallization of the BTB/POZ domain from the PLZF oncoprotein. J Biol Chem 1997; 272:27324-27329.
8. Li X, Peng H, Schultz DC et al. Structure-function studies of the BTB/POZ transcriptional repression domain from the promyelocytic leukemia zinc finger oncoprotein. Cancer Res 1999; 59:5275-5282.
9. Davies JM, Hawe N, Kabarowski J et al. Novel BTB/POZ domain zinc-finger protein, LRF, is a potential target of the LAZ-3/BCL-6 oncogene. Oncogene 1999; 18:365-375.
10. Daniel JM, Reynolds AB. The catenin p120(ctn) interacts with Kaiso, a novel BTB/POZ domain zinc finger transcription factor. Mol Cell Biol 1999; 19:3614-3623.
11. Hoatlin ME, Zhi Y, Ball H et al. A novel BTB/POZ transcriptional repressor protein interacts with the Fanconi anemia group C protein and PLZF. Blood 1999; 94:3737-3747.
12. Okabe S, Fukuda T, Ishibashi K et al. BAZF, a novel Bcl6 homolog, functions as a transcriptional repressor. Mol Cell Biol 1998; 18:4235-4244.
13. Huynh KD, Bardwell VJ. The BCL-6 POZ domain and other POZ domains interact with the co-repressors N-CoR and SMRT. Oncogene 1998; 17:2473-2484.
14. Lin RJ, Nagy L, Inoue S et al. Role of the histone deacetylase complex in acute promyelocytic leukaemia. Nature 1998; 391:811-814.
15. Hong SH, David G, Wong CW et al. SMRT corepressor interacts with PLZF and with the PML-retinoic acid receptor alpha (RARalpha) and PLZF-RARalpha oncoproteins associated with acute promyelocytic leukemia. Proc Natl Acad Sci USA 1997; 94:9028-9033.
16. Guidez F, Ivins S, Zhu J et al. Reduced retinoic acid-sensitivities of nuclear receptor corepressor binding to PML- and PLZF-RARalpha underlie molecular pathogenesis and treatment of acute promyelocytic leukemia. Blood 1998; 91:2634-2642.
17. Wong CW, Privalsky ML. Components of the SMRT corepressor complex exhibit distinctive interactions with the POZ domain oncoproteins PLZF, PLZF-RARalpha, and BCL-6. J Biol Chem 1998; 273:27695-27702.
18. David G, Alland L, Hong SH et al. Histone deacetylase associated with mSin3A mediates repression by the acute promyelocytic leukemia-associated PLZF protein. Oncogene 1998; 16:2549-2556.
19. Dhordain P, Lin RJ, Quief S et al. The LAZ3(BCL-6) oncoprotein recruits a SMRT/mSIN3A/histone deacetylase containing complex to mediate transcriptional repression. Nucleic Acids Res 1998; 26:4645-4651.
20. Dhordain P, Albagli O, Lin RJ et al. Corepressor SMRT binds the BTB/POZ repressing domain of the LAZ3/BCL6 oncoprotein. Proc Natl Acad Sci USA 1997; 94:10762-10767.
21. He LZ, Guidez F, Tribioli C et al. Distinct interactions of PML-RARalpha and PLZF-RARalpha with co- repressors determine differential responses to RA in APL. Nat Genet 1998; 18:126-135.
22. Grignani F, De Matteis S, Nervi C et al. Fusion proteins of the retinoic acid receptor-alpha recruit histone deacetylase in promyelocytic leukaemia. Nature 1998; 391:815-818.
23. Pintard L, Willis JH, Willems A et al. The BTB protein MEL-26 is a substrate-specific adaptor of the CUL-3 ubiquitin-ligase. Nature 2003; 425:311-316.
24. Xu L, Wei Y, Reboul J et al. BTB proteins are substrate-specific adaptors in an SCF-like modular ubiquitin ligase containing CUL-3. Nature 2003; 425:316-321.
25. Krek W. BTB proteins as henchmen of Cul3-based ubiquitin ligases. Nat Cell Biol 2003; 5:950-951.
26. Geyer R, Wee S, Anderson S et al. BTB/POZ domain proteins are putative substrate adaptors for cullin 3 ubiquitin ligases. Mol Cell 2003; 12:783-790.
27. Furukawa M, He YJ, Borchers C et al. Targeting of protein ubiquitination by BTB-Cullin 3-Roc1 ubiquitin ligases. Nat Cell Biol 2003; 5: 1001-1007.
28. Aravind L, Koonin EV. Fold Prediction and Evolutionary Analysis of the POZ Domain: Structural and Evolutionary Relationship with the Potassium Channel Tetramerization Domain. J Mol Biol 1999; 285:1353-1361.
29. Uren AG, Vaux DL. TRAF proteins and meprins share a conserved domain. Trends Biochem Sci 1996; 21:244-245.
30. Xue F, Cooley L. Kelch encodes a component of intercellular bridges in Drosophila egg chambers. Cell 1993; 72:681-693.
31. Kreusch A, Pfaffinger PJ, Stevens CF et al. Crystal structure of the tetramerization domain of the Shaker potassium channel. Nature 1998; 392:945-948.
32. Ramos S, Khademi F, Somesh BP et al. Genomic organization and expression profile of the small GTPases of the RhoBTB family in human and mouse. Gene 2002; 298:147-157.
33. Hamaguchi M, Meth JL, von Klitzing C et al. DBC2, a candidate for a tumor suppressor gene involved in breast cancer. Proc Natl Acad Sci USA 2002; 99:13647-13652.
34. Schulman BA, Carrano AC, Jeffrey PD et al. Insights into SCF ubiquitin ligases from the structure of the Skp1-Skp2 complex. Nature 2000; 408:381-386.
35. Stebbins CE, Kaelin WG, Jr, Pavletich NP. Structure of the VHL-ElonginC-ElonginB complex: implications for VHL tumor suppressor function. Science 1999; 284:455-461.
36. Li JY, English MA, Ball HJ et al. Sequence-specific DNA binding and transcriptional regulation by the promyelocytic leukemia zinc finger protein. J Biol Chem 1997; 272:22447-22455.
37. Ball HJ, Melnick A, Shaknovich R et al. The promyelocytic leukemia zinc finger (PLZF) protein binds DNA in a high molecular weight complex associated with cdc2 kinase. Nucleic Acids Res 1999; 27:4106-4113.
38. Yeyati PL, Shaknovich R, Boterashvili S et al. Leukemia translocation protein PLZF inhibits cell growth and expression of cyclin A. Oncogene 1999; 18:925-934.
39. Dong S, Zhu J, Reid A et al. Amino-terminal protein-protein interaction motif (POZ-domain) is responsible for activities of the promyelocytic leukemia zinc finger- retinoic acid receptor-alpha fusion protein. Proc Natl Acad Sci USA 1996; 93:3624-3629.
40. Melnick A, Ahmad KF, Arai S et al. In-depth mutational analysis of the promyelocytic leukemia zinc finger BTB/POZ domain reveals motifs and residues required for biological and transcriptional functions. Mol Cell Biol 2000; 20:6550-6567.
41. Kim CA, Phillips ML, Kim W et al. Polymerization of the SAM domain of TEL in leukemogenesis and transcriptional repression. EMBO J 2001; 20:4173-4182.
42. Kim CA, Gingery M, Pilpa RM, Bowie JU. The SAM domain of polyhomeotic forms a helical polymer. Nat Struct Biol 2002; 9:453-457.
43. Dhordain P, Albagli O, Ansieau S et al. The BTB/POZ domain targets the LAZ3/BCL6 oncoprotein to nuclear dots and mediates homomerisation in vivo. Oncogene 1995; 11:2689-2697.
44. Espinas ML, Jimenez-Garcia E, Vaquero A et al. The N-terminal POZ domain of GAGA mediates the formation of oligomers that bind DNA with high affinity and specificity. J Biol Chem 1999; 274:16461-16469.
45. Katsani KR, Hajibagheri MA, Verrijzer CP. Co-operative DNA binding by GAGA transcription factor requires the conserved BTB/POZ domain and reorganizes promoter topology. EMBO J 1999; 18:698-708.
46. Farkas G, Gausz J, Galloni M et al. The Trithorax-like gene encodes the Drosophila GAGA factor. Nature 1994; 371:806-808.
47. Pedone PV, Ghirlando R, Clore GM et al. The single Cys2-His2 zinc finger domain of the GAGA protein flanked by basic residues is sufficient for high-affinity specific DNA binding. Proc Natl Acad Sci USA 1996; 93:2822-2826.
48. Granok H, Leibovitch BA, Shaffer CD et al. Chromatin. Ga-ga over GAGA factor. Curr Biol 1995; 5:238-241.
49. Tsukiyama T, Becker PB, Wu C. ATP-dependent nucleosome disruption at a heat-shock promoter mediated by binding of GAGA transcription factor. Nature 1994; 367:525-532.

50. Shopland LS, Hirayoshi K, Fernandes M et al. HSF access to heat shock elements in vivo depends critically on promoter architecture defined by GAGA factor, TFIID, and RNA polymerase II binding sites. Genes Dev 1995; 9:2756-2769.
51. Yoshida C, Tokumasu F, Hohmura KI et al. Long range interaction of cis-DNA elements mediated by architectural transcription factor Bach1. Genes Cells 1999; 4:643-655.
52. Pagans S, Ortiz-Lombardia M, Espinas ML et al. The Drosophila transcription factor tramtrack (TTK) interacts with Trithorax-like (GAGA) and represses GAGA-mediated activation. Nucleic Acids Res 2002; 30:4406-4413.
53. Muto A, Hoshino H, Madisen L et al. Identification of Bach2 as a B-cell-specific partner for small maf proteins that negatively regulate the immunoglobulin heavy chain gene 3' enhancer. EMBO J 1998; 17:5734-5743.
54. Melnick A, Carlile G, Ahmad KF et al. Critical residues within the BTB domain of PLZF and Bcl-6 modulate interaction with corepressors. Mol Cell Biol 2002; 22:1804-1818.
55. Deltour S, Guerardel C, Leprince D. Recruitment of SMRT/N-CoR-mSin3A-HDAC-repressing complexes is not a general mechanism for BTB/POZ transcriptional repressors: the case of HIC-1 and gammaFBP-B. Proc Natl Acad Sci USA 1999; 96:14831-14836.
56. Guerardel C, Deltour S, Leprince D. Evolutionary divergence in the broad complex, tramtrack and bric a brac/poxviruses and zinc finger domain from the candidate tumor suppressor gene hypermethylated in cancer. FEBS Lett 1999; 451:253-256.
57. Pointud JC, Larsson J, Dastugue B et al. The BTB/POZ domain of the regulatory proteins Bric a brac 1 (BAB1) and Bric a brac 2 (BAB2) interacts with the novel Drosophila TAF(II) factor BIP2/dTAF(II)155. Dev Biol 2001; 237:368-380.
58. Zheng N, Schulman BA, Song L et al. Structure of the Cul1-Rbx1-Skp1-F boxSkp2 SCF ubiquitin ligase complex. Nature 2002; 416:703-709.
59. Stebbins CE, Kaelin WG, Jr., Pavletich NP. Structure of the VHL-ElonginC-ElonginB complex: implications for VHL tumor suppressor function. Science 1999; 284:455-461.
60. Chen Z, Brand NJ, Chen A et al. Fusion between a novel Kruppel-like zinc finger gene and the retinoic acid receptor-alpha locus due to a variant t(11;17) translocation associated with acute promyelocytic leukaemia. EMBO J 1993; 12:1161-1167.
61. Chen SJ, Zelent A, Tong JH et al. Rearrangements of the retinoic acid receptor alpha and promyelocytic leukemia zinc finger genes resulting from t(11;17)(q23;q21) in a patient with acute promyelocytic leukemia. J Clin Invest 1993; 91:2260-2267.
62. Licht JD, Chomienne C, Goy A et al. Clinical and molecular characterization of a rare syndrome of acute promyelocytic leukemia associated with translocation (11;17). Blood 1995; 85:1083-1094.
63. Culligan DJ, Stevenson D, Chee Y et al. Acute promyelocytic leukaemia with t(11;17)(q23;q12-21) and a good initial response to prolonged ATRA and combination chemotherapy. Br J Haematol 1998; 100:328-330.
64. Grimwade D, Biondi A, Mozziconacci MJ et al. Characterization of acute promyelocytic leukemia cases lacking the classic t(15;17): results of the European Working Party. Groupe Francais de Cytogenetique Hematologique, Groupe de Francais d'Hematologie Cellulaire, UK Cancer Cytogenetics Group and BIOMED 1 European Community-Concerted Action "Molecular Cytogenetic Diagnosis in Haematological Malignancies". Blood 2000; 96:1297-1308.
65. Jansen JH, de Ridder MC, Geertsma WM et al. Complete remission of t(11;17) positive acute promyelocytic leukemia induced by all-trans retinoic acid and granulocyte colony-stimulating factor. Blood 1999; 94:39-45.
66. Zhang T, Xiong H, Kan LX et al. Genomic sequence, structural organization, molecular evolution, and aberrant rearrangement of promyelocytic leukemia zinc finger gene. Proc Natl Acad Sci USA 1999; 96:11422-11427.
67. Petti MC, Fazi F, Gentile M et al. Complete remission through blast cell differentiation in PLZF/RARalpha-positive acute promyelocytic leukemia: in vitro and in vivo studies. Blood 2002; 100:1065-1067.
68. Ball HJ, Melnick A, Shaknovich R et al. The promyelocytic leukemia zinc finger (PLZF) protein binds DNA in a high molecular weight complex associated with cdc2 kinase. Nucleic Acids Res 1999; 27:4106-4113.
69. Sitterlin D, Tiollais P, Transy C. The RAR alpha-PLZF chimera associated with Acute Promyelocytic Leukemia has retained a sequence-specific DNA-binding domain. Oncogene 1997; 14:1067-1074.
70. Ivins S, Pemberton K, Guidez F et al. Regulation of Hoxb2 by APL-associated PLZF protein. Oncogene 2003; 22:3685-3697.
71. Bardwell VJ, Treisman R. The POZ domain: a conserved protein-protein interaction motif. Genes Dev 1994; 8:1664-1677.
72. Hoatlin ME, Zhi Y, Ball H et al. A Novel BTB/POZ Transcriptional Repressor Protein Interacts With the Fanconi Anemia Group C Protein and PLZF. Blood 1999; 94:3737-3747.
73. Lin W, Lai CH, Tang CJ et al. Identification and gene structure of a novel human PLZF-related transcription factor gene, TZFP. Biochem Biophys Res Commun. 1999; 264:789-795.
74. Miaw SC, Choi A, Yu E et al. ROG, repressor of GATA, regulates the expression of cytokine genes. Immunity 2000; 12:323-333.
75. Tang CJ, Chuang CK, Hu HM et al. The zinc finger domain of Tzfp binds to the tbs motif located at the upstream flanking region of the Aie1 (aurora-C) kinase gene. J Biol Chem 2001; 276:19631-19639.
76. Dong S, Zhu J, Reid A et al. Amino-terminal protein-protein interaction motif (POZ-domain) is responsible for activities of the promyelocytic leukemia zinc finger- retinoic acid receptor-alpha fusion protein. Proc Natl Acad Sci USA 1996; 93:3624-3629.
77. Martin PJ, Delmotte MH, Formstecher P et al. PLZF is a negative regulator of retinoic acid receptor transcriptional activity. Nucl Recept 2003; 1:6.
78. Nanba D, Mammoto A, Hashimoto K et al. Proteolytic release of the carboxy-terminal fragment of proHB-EGF causes nuclear export of PLZF. J Cell Biol 2003; 163:489-502.
79. Reid A, Gould A, Brand N et al. Leukemia translocation gene, PLZF, is expressed with a speckled nuclear pattern in early hematopoietic progenitors. Blood 1995; 86:4544-4552.
80. Licht JD, Shaknovich R, English MA et al. Reduced and altered DNA-binding and transcriptional properties of the PLZF-retinoic acid receptor-alpha chimera generated in t(11;17)-associated acute promyelocytic leukemia. Oncogene 1996; 12:323-336.
81. Koken MH, Reid A, Quignon F et al. Leukemia-associated retinoic acid receptor alpha fusion partners, PML and PLZF, heterodimerize and colocalize to nuclear bodies. Proc Natl Acad Sci USA 1997; 94:10255-10260.
82. Ruthardt M, Orleth A, Tomassoni L et al. The acute promyelocytic leukaemia specific PML and PLZF proteins localize to adjacent and functionally distinct nuclear bodies. Oncogene 1998; 16:1945-1953.
83. Hummel J, Wells R, Dubé I et al. Deregulation of NPM and PLZF in a variant t(5;17) case of acute promyelocytic leukemia. Oncogene 1999; 18:633-641.
84. Melnick A, Licht J. Deconstructing a disease: RARα, its fusion proteins and their roles in the pathogenesis of acute promyelocytic leukemia. Blood 1999; 93:3167-3215.
85. Dai MS, Chevallier N, Stone S et al. The effects of the Fanconi anemia zinc finger (FAZF) on cell cycle, apoptosis, and proliferation are differentiation stage-specific. J Biol Chem 2002; 277:26327-26334.
86. Ward JO, McConnell MJ, Carlile GW et al. The acute promyelocytic leukemia-associated protein, promyelocytic leukemia zinc finger, regulates 1,25-dihydroxyvitamin D(3)-induced monocytic differentiation of U937 cells through a physical interaction with vitamin D(3) receptor. Blood 2001; 98:3290-3300.
87. Labbaye C, Quaranta MT, Pagliuca A et al. PLZF induces megakaryocytic development, activates Tpo receptor expression and interacts with GATA1 protein. Oncogene 2002; 21:6669-6679.
88. Tsuzuki S, Enver T. Interactions of GATA-2 with the promyelocytic leukemia zinc finger (PLZF) protein, its homologue FAZF, and the t(11;17)-generated PLZF-retinoic acid receptor alpha oncoprotein. Blood 2002; 99:3404-3410.

89. Cook M, Gould A, Brand N et al. Expression of the zinc-finger gene PLZF at rhombomere boundaries in the vertebrate hindbrain. Proc Natl Acad Sci USA 1995; 92:2249-2253.
90. Barna M, Hawe N, Niswander L et al. Plzf regulates limb and axial skeletal patterning. Nat Genet 2000; 25:166-172.
91. Fahnenstich J, Nandy A, Milde-Langosch K et al. Promyelocytic leukaemia zinc finger protein (PLZF) is a glucocorticoid- and progesterone-induced transcription factor in human endometrial stromal cells and myometrial smooth muscle cells. Mol Hum Reprod 2003; 9:611-623.
92. Wan Y, Nordeen S. Overlapping but Distinct Gene Regulation Profiles by Glucocorticoids and Progestins in Human Breast Cancer Cells. Mol Endo 2002; 16:1204–1214.
93. Shaknovich R, Yeyati PL, Ivins S et al. The promyelocytic leukemia zinc finger protein affects myeloid cell growth, differentiation, and apoptosis. Mol Cell Biol 1998; 18:5533-5545.
94. McConnell M, Chevallier N, Berkofsky-Fessler W et al. Growth Suppression by Acute Promyelocytic Leukemia-Associated Protein PLZF is Mediated by Repression of c-Myc Expression. Mol Cell Bio. 2003; 23:9375-9388.
95. Han SH, Jeon JH, Ju HR et al. VDUP1 upregulated by TGF-beta1 and 1,25-dihydorxyvitamin D3 inhibits tumor cell growth by blocking cell-cycle progression. Oncogene 2003; 22:4035-4046.
96. Barna M, Merghoub T, Costoya JA et al. Plzf Mediates Transcriptional Repression of HoxD Gene Expression through Chromatin Remodeling. Dev Cell 2002; 3:499-510.
97. Ruthardt M, Testa U, Nervi C et al. Opposite effects of the acute promyelocytic leukemia PML-retinoic acid receptor alpha (RAR alpha) and PLZF-RAR alpha fusion proteins on retinoic acid signalling. Mol Cell Biol 1997; 17:4859-4869.
98. Gilliland DG, Griffin JD. Role of FLT3 in leukemia. Curr Opin Hematol 2002; 9:274-281.
99. Hong SH, Wong CW, Privalsky ML. Signaling by tyrosine kinases negatively regulates the interaction between transcription factors and SMRT (silencing mediator of retinoic acid and thyroid hormone receptor) corepressor. Mol Endocrinol 1998; 12:1161-1171.
100. Zhou Y, Gross W, Hong SH et al. The SMRT corepressor is a target of phosphorylation by protein kinase CK2 (casein kinase II). Mol Cell BioChem 2001; 220:1-13.
101. Hyman J, Chen H, Di Fiore PP et al. Epsin 1 undergoes nucleocytosolic shuttling and its eps15 interactor NH(2)-terminal homology (ENTH) domain, structurally similar to Armadillo and HEAT repeats, interacts with the transcription factor promyelocytic leukemia Zn(2)+ finger protein (PLZF). J Cell Biol 2000; 149:537-546.
102. Daniel JM, Reynolds AB. The catenin p120(ctn) interacts with Kaiso, a novel BTB/POZ domain zinc finger transcription factor. Mol Cell Biol 1999; 19:3614-3623.
103. David G, Alland L, Hong SH et al. Histone deacetylase associated with mSin3A mediates repression by the acute promyelocytic leukemia-associated PLZF protein. Oncogene 1998; 16:2549-2556.
104. Lin RJ, Nagy L, Inoue S et al. Role of the histone deacetylase complex in acute promyelocytic leukaemia. Nature 1998; 391:811-814.
105. Melnick A, Ahmad KF, Arai S et al. In-depth mutational analysis of the promyelocytic leukemia zinc finger BTB/POZ domain reveals motifs and residues required for biological and transcriptional functions. Mol Cell Biol 2000; 20:6550-6567.
106. Melnick A, Carlile G, Ahmad KF et al. Critical residues within the BTB domain of PLZF and Bcl-6 modulate interaction with corepressors. Mol Cell Biol 2002; 22:1804-1818.
107. Melnick A, Carlile GW, McConnell MJ et al. AML-1/ETO fusion protein is a dominant negative inhibitor of transcriptional repression by the promyelocytic leukemia zinc finger protein. Blood 2000; 96:3939-3947.
108. Melnick AM, Westendorf JJ, Polinger A et al. The ETO protein disrupted in t(8;21)-associated acute myeloid leukemia is a corepressor for the promyelocytic leukemia zinc finger protein. Mol Cell Biol 2000; 20:2075-2086.
109. Kang S, Chang W, Cho S et al. Modification of promyelocytic leukemia zinc finger protein (PLZF) by SUMO-1 conjugation regulates its transcriptional repressor activity. J Biol Chem 2003;In Press.
110. Nagpal S, Saunders M, Kastner P et al. Promoter context-and response element-dependent specificity of the transcriptional activation and modulating functions of retinoic acid receptors. Cell 1992; 70:1007-1019.
111. Hauksdottir H, Privalsky ML. DNA recognition by the aberrant retinoic acid receptors implicated in human acute promyelocytic leukemia. Cell Growth Differ 2001; 12:85-98.
112. Perez A, Kastner P, Sethi S et al. PMLRAR homodimers: distinct DNA binding properties and heteromeric interactions with RXR. EMBO J 1993; 12:3171-3182.
113. So CW, Dong S, So CK et al. The impact of differential binding of wild-type RARalpha, PML-, PLZF- and NPM-RARalpha fusion proteins towards transcriptional co-activator, RIP-140, on retinoic acid responses in acute promyelocytic leukemia. Leukemia 2000; 14:77-83.
114. Chen Z, Guidez F, Rousselot P et al. PLZF-RAR alpha fusion proteins generated from the variant t(11;17)(q23;q21) translocation in acute promyelocytic leukemia inhibit ligand-dependent transactivation of wild-type retinoic acid receptors. Proc Natl Acad Sci USA 1994; 91:1178-1182.
115. Puccetti E, Obradovic D, Beissert T et al. AML-associated Translocation Products Block Vitamin D(3)-induced Differentiation by Sequestering the Vitamin D(3) Receptor. Cancer Res 2002; 62:7050-7058.
116. Tomita A, Buchholz DR, Obata K et al. Fusion protein of retinoic acid receptor alpha with promyelocytic leukemia protein or promyelocytic leukemia zinc finger protein recruits N-CoR-TBLR1 corepressor complex to repress transcription in vivo. J Biol Chem 2003; 278:30788-30795.
117. Lin RJ, Evans RM. Acquisition of oncogenic potential by RAR chimeras in acute promyelocytic leukemia through formation of homodimers. Mol Cell 2000; 5:821-830.
118. Koken MH, Daniel MT, Gianni M et al. Retinoic acid, but not arsenic trioxide, degrades the PLZF/RARalpha fusion protein, without inducing terminal differentiation or apoptosis, in a RA-therapy resistant t(11;17)(q23;q21) APL patient. Oncogene 1999; 18:1113-1118.
119. Di Croce L, Raker VA, Corsaro M et al. Methyltransferase recruitment and DNA hypermethylation of target promoters by an oncogenic transcription factor. Science 2002; 295:1079-1082.
120. Park DJ, Vuong PT, De Vos S et al. Comparative analysis of genes regulated by PML/RAR{alpha} and PLZF/RAR{alpha} in response to retinoic acid using oligonucleotide arrays. Blood 2003; 102:3727-3736.
121. Cheng GX, Zhu XH, Men XQ et al. Distinct leukemia phenotypes in transgenic mice and different corepressor interactions generated by promyelocytic leukemia variant fusion genes PLZF-RARalpha and NPM-RARalpha. Proc Natl Acad Sci USA 1999; 96:6318-6323.
122. Guidez F, Huang W, Tong JH et al. Poor response to all-trans retinoic acid therapy in a t(11;17) PLZF/RAR alpha patient. Leukemia 1994; 8:312-317.
123. Grimwade D, Gorman P, Duprez E et al. Characterization of cryptic rearrangements and variant translocations in acute promyelocytic leukemia. Blood 1997; 90:4876-4885.
124. Ball HJ, Melnick A, Shaknovich R et al. The promyelocytic leukemia zinc finger (PLZF) protein binds DNA in a high molecular weight complex associated with cdc2 kinase. Nucleic Acids Res 1999; 27:4106-4113.
125. He L, Bhaumik M, Tribioli C et al. Two critical hits for promyelocytic leukemia. Mol Cell 2000; 6:1131-1141.
126. Kerckaert JP, Deweindt C, Tilly H et al. LAZ3, a novel zinc-finger encoding gene, is disrupted by recurring chromosome 3q27 translocations in human lymphomas. Nat Genet 1993; 5:66-70.
127. Baron BW, Nucifora G, McCabe N et al. Identification of the gene associated with the recurring chromosomal translocations t(3;14)(q27;q32) and t(3;22)(q27;q11) in B-cell lymphomas. Proc Natl Acad Sci USA 1993; 90:5262-5266.
128. Ye BH, Lista F, Lo Coco F et al. Alterations of a zinc finger-encoding gene, BCL-6, in diffuse large-cell lymphoma. Science 1993; 262:747-750.

129. Capello D, Vitolo U, Pasqualucci L et al. Distribution and pattern of BCL-6 mutations throughout the spectrum of B-cell neoplasia. Blood 2000; 95:651-659.
130. Ye BH. BCL-6 in the pathogenesis of non-Hodgkin's lymphoma. Cancer Invest 2000; 18:356-365.
131. Ye BH, Cattoretti G, Shen Q et al. The BCL-6 proto-oncogene controls germinal-centre formation and Th2-type inflammation. Nat Genet 1997; 16:161-170.
132. Dent AL, Shaffer AL, Yu X et al. Control of inflammation, cytokine expression, and germinal center formation by BCL-6. Science 1997; 276:589-592.
133. Fukuda T, Yoshida T, Okada S et al. Disruption of the Bcl6 gene results in an impaired germinal center formation. J Exp Med 1997; 186:439-448.
134. Dhordain P, Lin RJ, Quief S et al. The LAZ3(BCL-6) oncoprotein recruits a SMRT/mSIN3A/histone deacetylase containing complex to mediate transcriptional repression. Nucleic Acids Res 1998; 26:4645-4651.
135. Huynh KD, Bardwell VJ. The BCL-6 POZ domain and other POZ domains interact with the co-repressors N-CoR and SMRT. Oncogene 1998; 17:2473-2484.
136. Wong CW, Privalsky ML. Components of the SMRT corepressor complex exhibit distinctive interactions with the POZ domain oncoproteins PLZF, PLZF-RARalpha, and BCL-6. J Biol Chem 1998; 273:27695-27702.
137. Privé 6. Unpublished results 2003.
138. Ahmad K, Melnick A, Lax S et al. Mechanism of SMRT Corepressor Recruitment by the BCL6 BTB Domain. Molecular Cell 2003; 551-564.
139. Kawamata N, Miki T, Ohashi K et al. Recognition DNA sequence of a novel putative transcription factor, BCL6. Biochem Biophys Res Commun 1994; 204:366-374.
140. Dhordain P, Albagli O, Honore N et al. Colocalization and heteromerization between the two human oncogene POZ/zinc finger proteins, LAZ3 (BCL6) and PLZF. Oncogene 2000; 19:6240-6250.
141. Lemercier C, Brocard MP, Puvion-Dutilleul F et al. Class II histone deacetylases are directly recruited by BCL6 transcriptional repressor. J Biol Chem 2002; 277:22045-22052.
142. Melnick A. Unpublished results. 2003.
143. Davies JM, Hawe N, Kabarowski J et al. Novel BTB/POZ domain zinc-finger protein, LRF, is a potential target of the LAZ-3/BCL-6 oncogene. Oncogene 1999;18:365-375.
144. Vasanwala FH, Kusam S, Toney LM et al. Repression of AP-1 function: a mechanism for the regulation of Blimp-1 expression and B lymphocyte differentiation by the B cell lymphoma-6 protooncogene. J Immunol 2002; 169:1922-1929.
145. Chevallier N, Corcoran C, Lennon C et al. The ETO Protein of t(8;21) AML is a Corepressor for the Bcl-6 B-Cell Lymphoma Oncoprotein. Blood 2004; in press.
146. Melnick A, Carlile GW, McConnell MJ et al. AML-1/ETO fusion protein is a dominant negative inhibitor of transcriptional repression by the promyelocytic leukemia zinc finger protein. Blood 2000; 96:3939-3947.
147. Melnick AM, Westendorf JJ, Polinger A et al. The ETO protein disrupted in t(8;21)-associated acute myeloid leukemia is a corepressor for the promyelocytic leukemia zinc finger protein. Mol Cell Biol 2000; 20:2075-2086.
148. Seyfert VL, Allman D, He Y et al. Transcriptional repression by the proto-oncogene BCL-6. Oncogene 1996; 12:2331-2342.
149. Shaffer AL, Yu X, He Y et al. BCL-6 represses genes that function in lymphocyte differentiation, inflammation, and cell cycle control. Immunity 2000; 13:199-212.
150. Albagli O, Lantoine D, Quief S et al. Overexpressed BCL6 (LAZ3) oncoprotein triggers apoptosis, delays S phase progression and associates with replication foci. Oncogene 1999; 18:5063-5075.
151. Zhang H, Okada S, Hatano M et al. A new functional domain of Bcl6 family that recruits histone deacetylases. Biochim Biophys Acta 2001; 1540:188-200.
152. Takeda N, Arima M, Tsuruoka N et al. Bcl6 is a transcriptional repressor for the IL-18 gene. J Immunol 2003; 171:426-431.
153. Toney LM, Cattoretti G, Graf JA et al. BCL-6 regulates chemokine gene transcription in macrophages. Nat Immunol 2000; 1:214-220.
154. Hosokawa Y, Maeda Y, Seto M. Target genes downregulated by the BCL-6/LAZ3 oncoprotein in mouse Ba/F3 cells. Biochem Biophys Res Commun 2001; 283:563-568.
155. Reljic R, Wagner SD, Peakman LJ et al. Suppression of signal transducer and activator of transcription 3-dependent B lymphocyte terminal differentiation by BCL-6. J Exp Med 2000; 192:1841-1848.
156. Arima M, Toyama H, Ichii H et al. A putative silencer element in the IL-5 gene recognized by Bcl6. J Immunol 2002; 169:829-836.
157. Gupta S, Jiang M, Anthony A et al. Lineage-specific modulation of interleukin 4 signaling by interferon regulatory factor 4. J Exp Med 1999; 190:1837-1848.
158. Dent AL, Hu-Li J, Paul WE et al. T helper type 2 inflammatory disease in the absence of interleukin 4 and transcription factor STAT6. Proc Natl Acad Sci USA 1998; 95:13823-13828.
159. Dent AL, Doherty TM, Paul WE et al. BCL-6-deficient mice reveal an IL-4-independent, STAT6-dependent pathway that controls susceptibility to infection by Leishmania major. J Immunol 1999; 163:2098-2103.
160. Harris MB, Chang CC, Berton MT et al. Transcriptional repression of Stat6-dependent interleukin-4-induced genes by BCL-6: specific regulation of iepsilon transcription and immunoglobulin E switching. Mol Cell Biol 1999; 19:7264-7275.
161. Kusam S, Toney LM, Sato H et al. Inhibition of Th2 differentiation and GATA-3 expression by BCL-6. J Immunol 2003; 170:2435-2441.
162. Albagli O, Lindon C, Lantoine D et al. DNA replication progresses on the periphery of nuclear aggregates formed by the BCL6 transcription factor. Mol Cell Biol 2000; 20:8560-8570.
163. Cattoretti G, Chang CC, Cechova K et al. BCL-6 protein is expressed in germinal-center B cells. Blood 1995; 86:45-53.
164. Onizuka T, Moriyama M, Yamochi T et al. BCL-6 gene product, a 92- to 98-kD nuclear phosphoprotein, is highly expressed in germinal center B cells and their neoplastic counterparts. Blood 1995; 86:28-37.
165. Allman D, Jain A, Dent A et al. BCL-6 expression during B-cell activation. Blood 1996; 87:5257-5268.
166. Niu H, Ye BH, Dalla-Favera R. Antigen receptor signaling induces MAP kinase-mediated phosphorylation and degradation of the BCL-6 transcription factor. Genes Dev 1998; 12:1953-1961.
167. Toyama H, Okada S, Hatano M et al. Memory B cells without somatic hypermutation are generated from Bcl6-deficient B cells. Immunity 2002; 17:329-339.
168. Wang X, Li Z, Naganuma A, Ye BH. Negative autoregulation of BCL-6 is bypassed by genetic alterations in diffuse large B cell lymphomas. Proc Natl Acad Sci USA 2002; 99:15018-15023.
169. Moriyama M, Yamochi T, Semba K et al. BCL-6 is phosphorylated at multiple sites in its serine- and proline-clustered region by mitogen-activated protein kinase (MAPK) in vivo. Oncogene 1997; 14:2465-2474.
170. Shaffer AL, Lin KI, Kuo TC et al. Blimp-1 orchestrates plasma cell differentiation by extinguishing the mature B cell gene expression program. Immunity 2002; 17:51-62.
171. Carbone A, Gloghini A, Gaidano G et al. BCL-6 protein expression in human peripheral T-cell neoplasms is restricted to CD30+ anaplastic large-cell lymphomas. Blood 1997; 90:2445-2450.
172. Hyjek E, Chadburn A, Liu YF. BCL-6 protein is expressed in precursor T-cell lymphoblastic lymphoma and in prenatal and postnatal thymus. Blood 2001; 97:270-276.
173. Ouyang W, Lohning M, Gao Z et al. Stat6-independent GATA-3 autoactivation directs IL-4-independent Th2 development and commitment. Immunity 2000; 12:27-37.
174. Yamochi T, Kitabayashi A, Hirokawa M et al. Regulation of BCL-6 gene expression in human myeloid/monocytoid leukemic cells. Leukemia 1997; 11:694-700.
175. Yoshida T, Fukuda T, Hatano M et al. The role of Bcl6 in mature cardiac myocytes. Cardiovasc Res 1999; 42:670-679.
176. Yoshida T, Fukuda T, Okabe S et al. The BCL6 gene is predominantly expressed in keratinocytes at their terminal differentiation stage. Biochem Biophys Res Commun 1996; 228:216-220.

177. Albagli-Curiel O, Dhordain P, Lantoine D et al. Increased expression of the LAZ3 (BCL6) proto-oncogene accompanies murine skeletal myogenesis. Differentiation 1998; 64:33-44.
178. Kojima S, Hatano M, Okada S et al. Testicular germ cell apoptosis in Bcl6-deficient mice. Development 2001; 128:57-65.
179. Kumagai T, Miki T, Kikuchi M et al. The proto-oncogene Bc16 inhibits apoptotic cell death in differentiation-induced mouse myogenic cells. Oncogene 1999; 18:467-475.
180. Baron BW, Anastasi J, Thirman MJ et al. The human programmed cell death-2 (PDCD2) gene is a target of BCL6 repression: implications for a role of BCL6 in the down-regulation of apoptosis. Proc Natl Acad Sci USA 2002; 99:2860-2865.
181. Shvarts A, Brummelkamp TR, Scheeren F et al. A senescence rescue screen identifies BCL6 as an inhibitor of anti-proliferative p19(ARF)-p53 signaling. Genes Dev 2002; 16:681-686.
182. Yamochi T, Kaneita Y, Akiyama T et al. Adenovirus-mediated high expression of BCL-6 in CV-1 cells induces apoptotic cell death accompanied by down-regulation of BCL-2 and BCL-X(L). Oncogene 1999; 18:487-494.
183. Wales MM, Biel MA, el Deiry W et al. p53 activates expression of HIC-1, a new candidate tumour suppressor gene on 17p13.3. Nat Med 1995; 1:570-577.
184. Chen WY, Zeng X, Carter MG et al. Heterozygous disruption of Hic1 predisposes mice to a gender-dependent spectrum of malignant tumors. Nat Genet 2003; 33:197-202.
185. Deltour S, Pinte S, Guerardel C et al. The human candidate tumor suppressor gene HIC1 recruits CtBP through a degenerate GLDLSKK motif. Mol Cell Biol 2002; 22:4890-4901.
186. Daniel JM, Spring CM, Crawford HC et al. The p120(ctn)-binding partner Kaiso is a bi-modal DNA-binding protein that recognizes both a sequence-specific consensus and methylated CpG dinucleotides. Nucleic Acids Res 2002; 30:2911-2919.
187. Prokhortchouk A, Hendrich B, Jorgensen H et al. The p120 catenin partner Kaiso is a DNA methylation-dependent transcriptional repressor. Genes Dev 2001; 15:1613-1618.
188. Kim SW, Fang X, Ji H et al. Isolation and characterization of XKaiso, a transcriptional repressor that associates with the catenin Xp120(ctn) in Xenopus laevis. J Biol Chem 2002; 277:8202-8208.
189. Yoon HG, Chan DW, Reynolds AB et al. N-CoR mediates DNA methylation-dependent repression through a methyl CpG binding protein Kaiso. Mol Cell 2003; 12:723-734.
190. Jenuwein T, Allis CD. Translating the histone code. Science 2001; 293:1074-1080.
191. Peukert K, Staller P, Schneider A et al. An alternative pathway for gene regulation by Myc. EMBO J 1997; 16:5672-5686.
192. Salghetti SE, Kim SY, Tansey WP. Destruction of Myc by ubiquitin-mediated proteolysis: cancer-associated and transforming mutations stabilize Myc. EMBO J 1999; 18:717-726.
193. Staller P, Peukert K, Kiermaier A et al. Repression of p15INK4b expression by Myc through association with Miz-1. Nat Cell Biol 2001; 3:392-399.
194. Seoane J, Pouponnot C, Staller P et al. TGFbeta influences Myc, Miz-1 and Smad to control the CDK inhibitor p15INK4b. Nat Cell Biol 2001; 3:400-408.
195. Seoane J, Le HV, Massague J. Myc suppression of the p21(Cip1) Cdk inhibitor influences the outcome of the p53 response to DNA damage. Nature 2002; 419:729-734.
196. Wu S, Cetinkaya C, Munoz-Alonso MJ et al. Myc represses differentiation-induced p21CIP1 expression via Miz-1-dependent interaction with the p21 core promoter. Oncogene 2003; 22:351-360.
197. Herold S, Wanzel M, Beuger V et al. Negative regulation of the mammalian UV response by Myc through association with Miz-1. Mol Cell 2002; 10:509-521.
198. Piluso D, Bilan P, Capone JP. Host cell factor-1 interacts with and antagonizes transactivation by the cell cycle regulatory factor Miz-1. J Biol Chem 2002; 277:46799-46808.
199. Ziegelbauer J, Shan B, Yager D et al. Transcription factor MIZ-1 is regulated via microtubule association. Mol Cell 2001; 8:339-349.
200. Mulder NJ, Apweiler R, Attwood TK et al. The InterPro Database, 2003 brings increased coverage and new features. Nucleic Acids Res 2003; 31:315-318.

CHAPTER 21

KRAB Zinc Finger Proteins: A Family of Repressors Mediating Heterochromatin-Associated Gene Silencing

Shiro Iuchi

Abstract

Krüppel-associated box (KRAB) domain, present at the N terminus of mammalian C_2H_2 zinc finger proteins, binds KAP-1 (also named TIF1β or KRIP-1) that binds HP1α and SETDB1 (H3-K9 methyltransferase) as well as NuRD complexes (histone deacetylases). The KRAB domain tethers these binding proteins to a specific gene recognized by zinc fingers, resulting in a heterochromatin-associated, strong, long-term repression of the gene. The importance of such repression is reflected in the large number (two hundred ninety) of the KRAB zinc finger protein genes in the human genome and revealed by early stage-development failure of mouse embryos lacking the common corepressor, KAP-1 (TIF1β). Many reports have described KRAB zinc finger proteins that participate in a variety of cellular processes including development of organs and cell types such as heart, bone, sperm and hematopoietic cells. Some KRAB zinc finger proteins have an additional repression mechanism that is KRAB domain-independent and involves histone deactylases or NSD1 histone lysine methyltrasferase.

Linking of KRAB Zinc Finger Proteins to Gene Silencing

Transcription repressors bind to cis-elements of a given gene, curtailing the gene's expression. Repression mechanisms include (1) exclusion of a transcriptional activator, (2) inactivation of the gene's activator, and (3) recruitment of a corepressor complex (see ref. 51). Mechanisms (1) and (2) result from the repressor's direct contact to the cis-element and activator, respectively, and these repression mechanisms have been long studied since investigation on the *E. coli* lactose operon and the bacteriophage λ regulatory genes.[1,2] Mechanism (3), only recently emerged, requires additional repressive proteins to cause gene silencing (repression) associated with heterochromatin. Heterochromatin was defined and described in 1928 as the chromosome region condensed throughout the cell cycle.[3,4] It is observed in pericentromeric and/or telomeric regions in which repetitive DNA sequences are plentiful and structural genes are rare.[3,5] This chromosomal region exclusively contains heterochromatin protein 1 (HP1), which was first found in *Drosophila* but is now known to be widely distributed from yeast to mammals and the three homologues, HP1α, HP1β and HP1γ, are present in mammals and humans.[4] Although details of the mechanisms have not been revealed, binding of HP1 is thought to render the DNA region into condensed states and result in gene silencing.[4,5] Highly structured DNA and the compartmentalization of the heterochromatin may make promoters within inaccessible to RNA polymerase II and III complexes.[6] This repression is called "gene silencing" or "position effect variegation (PEV)" after its discovery as the gene-shut-off phenomenon associated with chromosome rearrangement in *Drosophila.*[7] Only two years after KRAB was found to be a repressor domain,[8] two groups independently identified a KRAB binding protein and named it KAP-1 or KRIP1,[9,10] a third group found a HP1 binding protein and named it TIF1β,[11] and a fourth group found a KRAB binding protein recognizing it to be TIF1β and KAP-1.[12] These findings firmly established that KRAB domains of zinc finger proteins interact with KAP-1 (TIF1β) which in turn binds HP1 to repress gene expression. This repression turned out to involve several other proteins that affect chromosomal structure (See below).

Structure of KRAB Domains

Krüppel zinc finger protein was defined as a C_2H_2 zinc finger protein with tandem C_2H_2 zinc fingers connected by the well-conserved linker sequence (TGEKP).[13] These proteins often contain highly homologous amino acid sequences in the N terminus outside the zinc finger motif itself, and the sequence was named Krüppel-associated box (KRAB)[14] or finger proceeding box (FPB).[15] Initial screening for the sequence estimated that approximately one third of human Krüppel zinc finger proteins contain KRAB,[14] and a subsequent study confirmed the conclusion and extended the total number of human KRAB-Zinc finger proteins to two hundred ninety.[16] Proteins of this type are also known to be abundant in mammals but not in simpler organisms,[17,22] suggesting that KRAB domains associated with Krüppel zinc finger proteins in a relatively late stage of protein evolution. KRAB domains consist of a few of sub-domains (A, B, b and C), and KRAB zinc finger proteins belong to either AB-, AAB-, Ab-, AbAb-, AC- or A-type (Fig. 1A).[14,18-21] Moreover, KRAB domains coexist with a SCAN domain; and some KRAB zinc finger proteins belong to either (SCAN + A)- or (AB + SCAN + A)-type (see Chapter 22).[19,22] KRAB AB domain, which spans

Zinc Finger Proteins: From Atomic Contact to Cellular Function, edited by Shiro Iuchi and Natalie Kuldell.

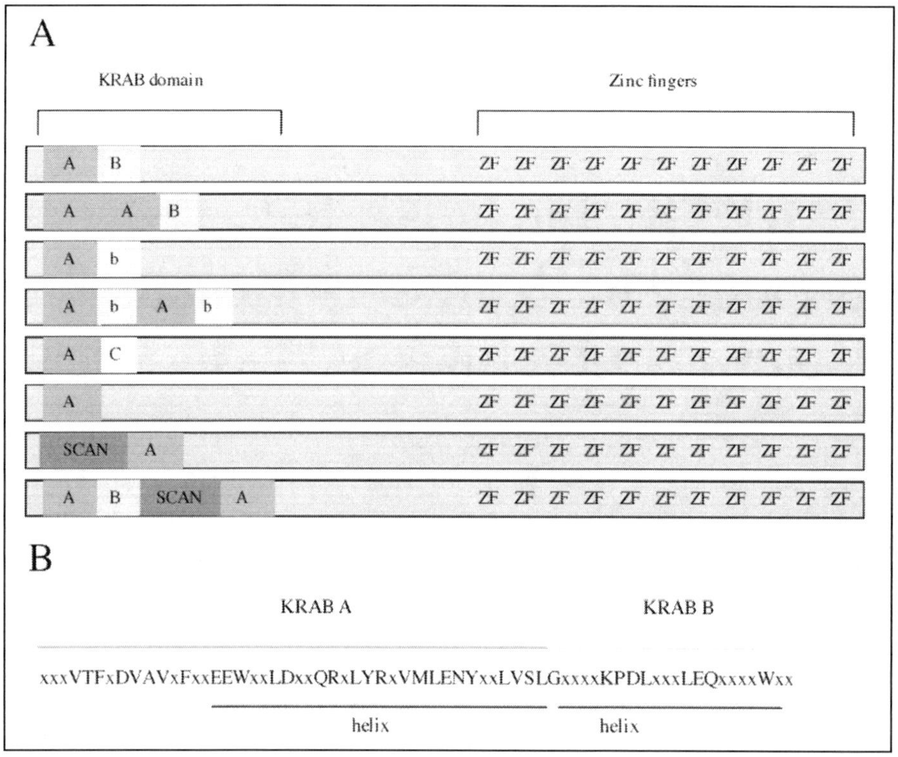

Figure 1. A) A schematic representation of various KRAB domains present in zinc finger proteins. B) A consensus sequence of KRAB AB domain. The sequence was previously described.[26]

about 75-amino acids, is thought to be the origin of all and has diverged to the other types by means of duplication, deletion and mutation. One strong argument for this suggestion has come from the fact that as many as two hundred sixty two genes of zinc finger proteins cluster in the human chromosome 19 and as many as twenty-one genes of KRAB zinc finger proteins cluster in the 700-kb region of human chromosome 19q13.2.[19,23,24] There is no perfect consensus sequence for the KRAB domains among zinc finger proteins, but KRAB A with 42-amino acids is the longest and best conserved of all the sub-domains (Fig. 1B) This sub-domain contains hydrophobic as well as charged amino acids and is supposed to fold into an amphiphilic α-helix.[14] KRAB B with a 32-amino acid long sequence is also well conserved and supposed to fold into an α-helix. KRAB b is, however, highly diverged from KRAB B and less conserved within the sub-domain.[18] KRAB C with a 21-amino acid long sequence is also less conserved.[20]

Functions of KRAB and the Protein Complexes

KRAB and the Interaction with KAP-1(TIF1β)

Because no natural target genes had been identified, studies on the function of KRAB domains began with KRAB AB domain (from KOX1 zinc finger protein) fused to GAL4. GAL4 alone activated transcription of a reporter gene in cells but the expression was abolished by fusing the KRAB domain to the activator, giving the first insight into KRAB domain's function.[8] Since then, the GAL4-KRAB fusion protein and similar constructs have been almost exclusively used for in vivo and in vitro experiments. Analyzing by this method, scientists now know that although KRAB domain structures has several variations (Fig. 1A), sub-domain A is the principal repressor.[8,25] Sub-domain B fortifies the role of sub-domain A to some extent but the sub-domain b barely plays any role.[25] Sub-domain C has been suggested to strengthen the repression by sub-domain A but it has not been demonstrated.[20] Perhaps, it may be concluded that KRAB A sub-domain is necessary and sufficient to repress gene expression as originally stated.[8]

KAP-1 (TIF1β) binds to the KRAB domain through the N-terminal RING, B-boxes, coiled coil (RBCC) (Fig. 2, see Chapter 16).[25-27] The affinity constant is $7x10^6$ M^{-1}.[28] It is roughly comparable to the K_a for zinc fingers-DNA binding (see Chapter 2) and the KRAB-RBCC binding has been well demonstrated by electrophoretic mobility shift assay[27,28] and yeast two hybrid binding assay.[25] The KRAB-RBCC binding is very specific as it has been shown that the KRAB domains of KOX1, MZF22, MTZ1 and MZF13 bind RBCC from KAP-1 (TIF1β) but not any other homologues including RBCC derived from TIF1α, TIF1γ and MID.[25,27] This strict specificity is consistent with the observation that *TIF1*β $^{-/-}$ mouse embryos are affected at an early developmental stage and die by E8.5.[29] Interestingly, one KRAB binds a homotrimer of RBCC in vitro[27] and binding of a KRAB zinc finger protein to KAP-1 (TIF1β) oligomers has been observed in living cells.[30] Thus, a KRAB zinc finger protein is able to amplify the repression through KAP-1 (TIF1β).

The RING finger of KAP-1 (TIF1β) contains many canonical cysteine residues that are supposed to chelate zinc ions and thereby form the stable tertiary structure (Fig. 2 and see Chapter 16). A double mutation of two cysteine residues of the RING finger completely destroys its binding ability to KRAB,[27] and a similar mutational effect occurs in the second B box. The coiled coil domain does not contain conserved cysteine residues but contains many leucine or other hydrophobic residues, which are periodically included in the peptide separated by six hydrophobic

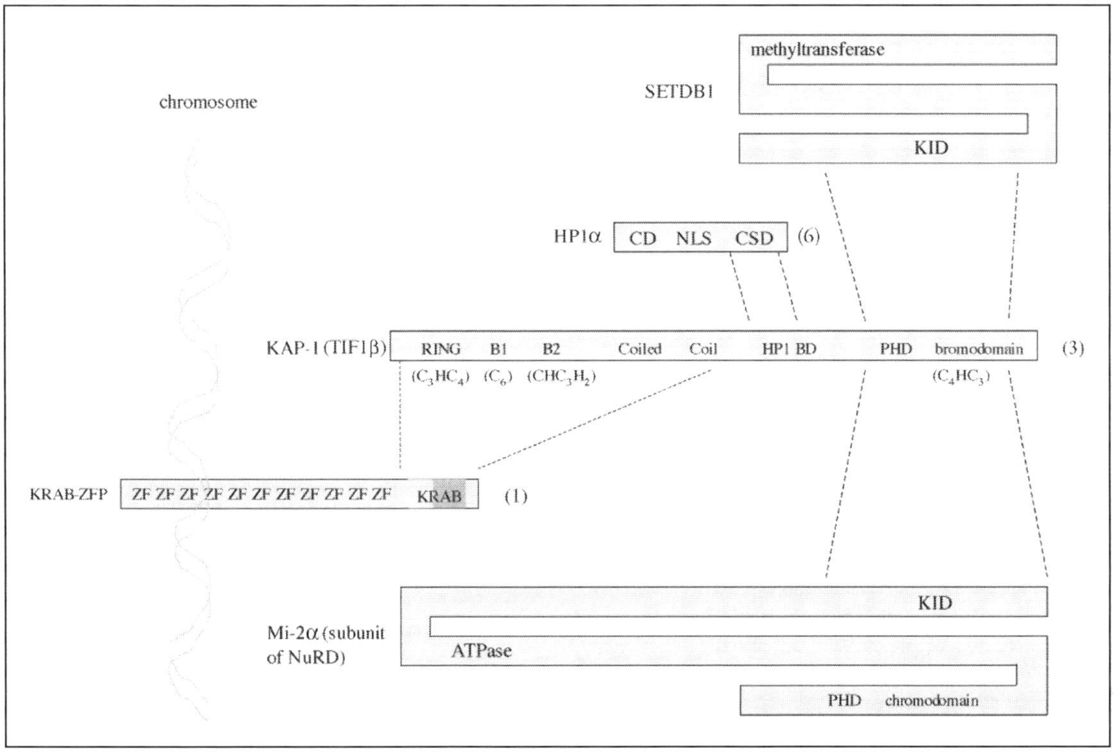

Figure 2. A schematic representation of protein-protein interaction in a KRAB zinc finger protein complex. Domain-domain interactions are shown by dotted lines. The N terminus of proteins is presented at the left and the C terminus at the right, but the KRAB zinc finger protein is drawn in the opposite direction. Numbers (1), (3) and (6) next to the C terminus of proteins indicate molar ratios in the protein complex.

and hydrophilic amino acid residues and thought to compose a primary pole of a leucine zipper-like structure. Mutagenesis of a leucine of each coil domain intermediately or completely abolished the KRAB binding. These results suggest the predicted leucine zipper-like coiled coil structure is present.

KAP-1 (TIF1β) and the Interactions with the Downstream Proteins

From the middle of KAP-1 (TIF1β) protein to the C terminus, there are three conserved domains, HP1 binding box (HP1 BD), PHD finger and bromodomain (Fig. 2).[9-12] These are indispensable for repression. HP1 BD binds HP1α[31,32] with 1: 2 at a K_a value of $1.6 \times 10^7\ M^{-1}$,[32] and PHD finger plus bromodomain bind NuRD complex as well as SETDB1.[33,34] The NuRD complexes remove acetyl groups from the N terminus of histones and allow histones to interact with DNA[33] while SETBD1 methylates the lysine residue at position 9 of histone 3 and stimulates HP1α to interact with DNA.[34] The three different KAP1 (TIF1β) binding proteins together work toward forming heterochromatin. As KAP-1 (TIF1β) is present as a trimer,[27] HP1 can bind six molecules per KRAB zinc finger complex and both NuRD and SETDB1 can bind up to three molecules (Fig. 2). It is likely that the three executers would be associated simultaneously through KAP-1 (TIF1β) to mediate the severe repression.

Gene Silencing by Heterochromatin Formation

A majority of results on KRAB domain that have been collected to date measure either repression or heterochromatin formation, and the relation between the two has not been clear. Recently, however, a series of experiments examined both gene silencing and chromatin structure together by constructing cells that have a chromosome with a reporter gene at one locus and a repressor gene, encoding a KRAB domain fused to a reporter-gene-binding domain whose activity is regulated by a hormone, at another.[35] In these cells the reporter gene was present in euchromatin without the active repressor, but it was located within heterochromatin upon activation of the repressor. Furthermore, the silenced gene was physically bound with KAP-1 (TIF1β), HP1α and SETDB1. This silencing was maintained over fifty cell divisions, suggesting that once heterochromatin is made, the structure persists.[35] Thus, a KRAB zinc finger protein silences expression of the target gene by establishing the heterochromatin through the concerted action of KRAB, KAP-1 (TIF1β), HP1α and SETDB1[35] as well as NuRD complexes.[33] It was also shown that the gene silencing is variegated from one cell to another. The observation is consistent with the idea that the more tightly folding heterochromatin gives rise to the stronger and longer silencing. Since the promoter of the silence reporter gene was hypermethylated at the CpG rich region, DNA methylation may also participate in the gene silencing.[35]

KAP-1 (TIF1β) and HP1α are present in undifferentiated mouse embryonic carcinoma F9 cells, but only HP1α is localized within heterochromatin. When the cells are differentiated by retinoic acid, however, KAP-1 (TIF1β) is compartmented with HP1 within heterochromatin.[36] This suggests that some factor, which is likely to be a specific KRAB zinc finger protein, initiates specific gene's heterochromatin formation resulting in silencing of gene expression. Presence of the KRAB protein with an active form of the zinc fingers would be a critical step to initiate such gene silencing (see Chapter 2). KAP-1 (TIF1β) undergoes autophosphorylation and transphosphorylates HP1.[31] This may have also something to do with regulation of heterochromatin formation and gene silencing.

KRAB-Independent Repression by KRAB Zinc Finger Proteins

The KRAB domain is a repressor domain essential for gene silencing, but there is a rare example in which a corepressor binds somewhere else and mediates repression different from the KAP-1 repression. BRCA1, whose genetic loss in the germ line confers a cumulative lifetime risk of female breast and ovarian cancer, regulates transcription of several DNA damage-inducible genes but does not directly bind to DNA. Recently, a member of the KRAB zinc finger family, designated ZBRK1, has been shown to bind BRCA1.[37] BRCA1 interacts with the finger (5 to 8)-containing C terminus of ZBRK1, and represses the target genes.[38] The repression is almost completely reversed by addition of TSA, suggesting that BRCA1 repression is mediated through histone deacetylases only. Since the role of some zinc fingers belonging to the multiple C_2H_2 zinc finger family is to interact with proteins rather than to direct DNA binding[39] and since KRAB proteins often have multiple zinc fingers, such as more than eleven fingers, it is possible that the example of ZBRK1 predicts that some KRAB zinc finger proteins have a second mode of repression independent of the KRAB-KAP-1 mediated silencing.

Another KRAB zinc finger protein called Nizp1, which belongs to the SCAN-KRAB A type, was recently shown to have a unique interaction with NSD1 histone lysine methyltransferase for Histone 3-K36 and Histone 4-K20.[40] Nizp1 contains a C_2HR structure in the middle of the peptide in addition to two pairs of C_2H_2 zinc fingers at the C terminus, and this C_2HR domain directly binds the metyltransferase without participation of KAP-1 (TIF1β). This would join the KAP-1 (TIF1β)-mediated repression, resulting in stronger and also various kinds of gene silencing. The C_2HR domain is thought to have evolved from one of the tandem C_2H_2 zinc fingers to gain the interaction with the histone methyltransferase.

Identification of the Cellular Function of KRAB Zinc Finger Proteins

Recently, many KRAB zinc finger proteins have been found in developmental stages of organs and different cell types including heart, bone, sperm and hematopoietic cells.[41-46] However, the genes directly controlled by KRAB proteins have not been found. Some target sequences have been determined using a binding assay with synthetic oligonucleotides and the specific sequences were used to deduce the natural cognate genes.[37,44,47,48] The methods work but uncertainty remains if the resulting binding sequences are identical to the natural cognate genes. It would be significantly useful if natural target genes could be found, perhaps in the way the basonuclin 1 target was identified (see Chapter 28). For this zinc finger protein, two different approaches found the same cognate gene. One approach identified the rRNA gene promoter as a target of basonuclin 1 by purifying the sequence from a human genomic library[49] and the other approach identified the same gene by spotting the basonuclin-DNA complex on the chromosome.[50] Either method should work for any KRAB zinc finger proteins.

References

1. Beckwith J, Zipser D, eds. The lactose operon. New York: Cold Spring Harbor Laboratory; 1970.
2. Ptashne M. A genetic switch. Cambridge: Blackwell Scientific Publications; 1986.
3. Wallrath LL. Unfolding the mysteries of heterochromatin. Curr Opin Genet Dev 1998; 8(2):147-153.
4. Eissenberg JC, Elgin SC. The HP1 protein family: getting a grip on chromatin. Curr Opin Genet Dev 2000; 10(2):204-210.
5. Henikoff S. Heterochromatin function in complex genomes. Biochim Biophys Acta 2000; 1470(1):O1-8.
6. Moosmann P, Georgiev O, Thiesen HJ et al. Silencing of RNA polymerases II and III-dependent transcription by the KRAB protein domain of KOX1, a Kruppel-type zinc finger factor. Biol Chem 1997; 378(7):669-677.
7. Howe M, Dimitri P, Berloco M et al. Cis-effects of heterochromatin on heterochromatic and euchromatic gene activity in Drosophila melanogaster. Genetics 1995; 140(3):1033-1045.
8. Margolin JF, Friedman JR, Meyer WK et al. Kruppel-associated boxes are potent transcriptional repression domains. Proc Natl Acad Sci USA 1994; 91(10):4509-4513.
9. Friedman JR, Fredericks WJ, Jensen DE et al. KAP-1, a novel corepressor for the highly conserved KRAB repression domain. Genes Dev 1996; 10(16):2067-2078.
10. Kim SS, Chen YM, O'Leary E et al. A novel member of the RING finger family, KRIP-1, associates with the KRAB-A transcriptional repressor domain of zinc finger proteins. Proc Natl Acad Sci USA 1996; 93(26):15299-15304.
11. Le Douarin B, Nielsen AL, Garnier JM et al. A possible involvement of TIF1 alpha and TIF1 beta in the epigenetic control of transcription by nuclear receptors. Embo J 1996; 15(23):6701-6715.
12. Moosmann P, Georgiev O, Le Douarin B et al. Transcriptional repression by RING finger protein TIF1 beta that interacts with the KRAB repressor domain of KOX1. Nucleic Acids Res 1996; 24(24):4859-4867.
13. Turner J, Crossley M. Mammalian Kruppel-like transcription factors: more than just a pretty finger. Trends Biochem Sci 1999; 24(6):236-240.
14. Bellefroid EJ, Poncelet DA, Lecocq PJ et al. The evolutionarily conserved Kruppel-associated box domain defines a subfamily of eukaryotic multifingered proteins. Proc Natl Acad Sci USA 1991; 88(9):3608-3612.
15. Rosati M, Marino M, Franze A et al. Members of the zinc finger protein gene family sharing a conserved N-terminal module. Nucleic Acids Res 1991; 19(20):5661-5667.
16. Rousseau-Merck MF, Koczan D, Legrand I et al. The KOX zinc finger genes: genome wide mapping of 368 ZNF PAC clones with zinc finger gene clusters predominantly in 23 chromosomal loci are confirmed by human sequences annotated in EnsEMBL. Cytogenet Genome Res 2002; 98(2-3):147-153.
17. Venter JC, Adams MD, Myers EW et al. The sequence of the human genome. Science 2001; 291(5507):1304-1351.
18. Mark C, Abrink M, Hellman L. Comparative analysis of KRAB zinc finger proteins in rodents and man: evidence for several evolutionarily distinct subfamilies of KRAB zinc finger genes. DNA Cell Biol 1999; 18(5):381-396.
19. Looman C, Abrink M, Mark C et al. KRAB zinc finger proteins: an analysis of the molecular mechanisms governing their increase in numbers and complexity during evolution. Mol Biol Evol 2002; 19(12):2118-2130.
20. Looman C, Hellman L, Abrink M. A novel Kruppel-Associated Box identified in a panel of mammalian zinc finger proteins. Mamm Genome 2004; 15(1):35-40.
21. Tian Y, Breedveld GJ, Huang S et al. Characterization of ZNF333, a novel double KRAB domain containing zinc finger gene on human chromosome 19p13.1. Biochim Biophys Acta 2002; 1577(1):121-125.

22. Collins T, Stone JR, Williams AJ. All in the family: the BTB/POZ, KRAB, and SCAN domains. Mol Cell Biol 2001; 21(11):3609-3615.
23. Shannon M, Hamilton AT, Gordon L et al. Differential expansion of zinc-finger transcription factor loci in homologous human and mouse gene clusters. Genome Res 2003; 13(6A):1097-1110.
24. Dehal P, Predki P, Olsen AS et al. Human chromosome 19 and related regions in mouse: conservative and lineage-specific evolution. Science 2001; 293(5527):104-111.
25. Abrink M, Ortiz JA, Mark C et al. Conserved interaction between distinct Kruppel-associated box domains and the transcriptional intermediary factor 1 beta. Proc Natl Acad Sci USA 2001; 98(4):1422-1426.
26. Agata Y, Matsuda E, Shimizu A. Two novel Kruppel-associated box-containing zinc-finger proteins, KRAZ1 and KRAZ2, repress transcription through functional interaction with the corepressor KAP-1 (TIF1beta/KRIP-1). J Biol Chem 1999; 274:16412-16422.
27. Peng H, Begg GE, Schultz DC et al. Reconstitution of the KRAB-KAP-1 repressor complex: a model system for defining the molecular anatomy of RING-B box-coiled-coil domain-mediated protein-protein interactions. J Mol Biol 2000; 295(5):1139-1162.
28. Peng H, Begg GE, Harper SL et al. Biochemical analysis of the Kruppel-associated box (KRAB) transcriptional repression domain. J Biol Chem 2000; 275(24):18000-18010.
29. Cammas F, Mark M, Dolle P et al. Mice lacking the transcriptional corepressor TIF1beta are defective in early postimplantation development. Development 2000; 127(13):2955-2963.
30. Germain-Desprez D, Bazinet M, Bouvier M et al. Oligomerization of transcriptional intermediary factor 1 regulators and interaction with ZNF74 nuclear matrix protein revealed by bioluminescence resonance energy transfer in living cells. J Biol Chem 2003; 278(25):22367-22373.
31. Nielsen AL, Ortiz JA, You J et al. Interaction with members of the heterochromatin protein 1 (HP1) family and histone deacetylation are differentially involved in transcriptional silencing by members of the TIF1 family. Embo J 1999; 18(22):6385-6395.
32. Lechner MS, Begg GE, Speicher DW et al. Molecular determinants for targeting heterochromatin protein 1-mediated gene silencing: direct chromoshadow domain-KAP-1 corepressor interaction is essential. Mol Cell Biol 2000; 20(17):6449-6465.
33. Schultz DC, Friedman JR, Rauscher FJ, 3rd. Targeting histone deacetylase complexes via KRAB-zinc finger proteins: the PHD and bromodomains of KAP-1 form a cooperative unit that recruits a novel isoform of the Mi-2alpha subunit of NuRD. Genes Dev 2001; 15(4):428-443.
34. Schultz DC, Ayyanathan K, Negorev D et al. SETDB1: a novel KAP-1-associated histone H3, lysine 9-specific methyltransferase that contributes to HP1-mediated silencing of euchromatic genes by KRAB zinc-finger proteins. Genes Dev 2002; 16(8):919-932.
35. Ayyanathan K, Lechner MS, Bell P et al. Regulated recruitment of HP1 to a euchromatic gene induces mitotically heritable, epigenetic gene silencing: a mammalian cell culture model of gene variegation. Genes Dev 2003; 17(15):1855-1869.
36. Cammas F, Oulad-Abdelghani M, Vonesch JL et al. Cell differentiation induces TIF1beta association with centromeric heterochromatin via an HP1 interaction. J Cell Sci 2002; 115(Pt 17):3439-3448.
37. Zheng L, Pan H, Li S et al. Sequence-specific transcriptional corepressor function for BRCA1 through a novel zinc finger protein, ZBRK1. Mol Cell 2000; 6(4):757-768.
38. Tan W, Zheng L, Lee WH et al. Functional dissection of transcription factor ZBRK1 reveals zinc fingers with dual roles in DNA-binding and BRCA1-dependent transcriptional repression. J Biol Chem 2004; 279(8):6576-6587.
39. Iuchi S. Three classes of C2H2 zinc finger proteins. Cell Mol Life Sci 2001; 58(4):625-635.
40. Nielsen AL, Jorgensen P, Lerouge T et al. Nizp1, a novel multitype zinc finger protein that interacts with the NSD1 histone lysine methyltransferase through a unique C2HR motif. Mol Cell Biol 2004; 24(12):5184-5196.
41. Yi Z, Li Y, Ma W et al. A novel KRAB zinc-finger protein, ZNF480, expresses in human heart and activates transcriptional activities of AP-1 and SRE. Biochem Biophys Res Commun 2004; 320(2):409-415.
42. Hering TM, Kazmi NH, Huynh TD et al. Characterization and chondrocyte differentiation stage-specific expression of KRAB zinc-finger protein gene ZNF470. Exp Cell Res 2004; 299(1):137-147.
43. Ganss B, Kobayashi H. The zinc finger transcription factor Zfp60 is a negative regulator of cartilage differentiation. J Bone Miner Res 2002; 17(12):2151-2160.
44. Jheon AH, Ganss B, Cheifetz S et al. Characterization of a novel KRAB/C2H2 zinc finger transcription factor involved in bone development. J Biol Chem 2001; 276(21):18282-18289.
45. Looman C, Mark C, Abrink M et al. MZF6D, a novel KRAB zinc-finger gene expressed exclusively in meiotic male germ cells. DNA Cell Biol 2003; 22(8):489-496.
46. Takashima H, Nishio H, Wakao H et al. Molecular cloning and characterization of a KRAB-containing zinc finger protein, ZNF317, and its isoforms. Biochem Biophys Res Commun 2001; 288(4):771-779.
47. Gebelein B, Urrutia R. Sequence-specific transcriptional repression by KS1, a multiple-zinc-finger-Kruppel-associated box protein. Mol Cell Biol 2001; 21(3):928-939.
48. Peng H, Zheng L, Lee WH et al. A common DNA-binding site for SZF1 and the BRCA1-associated zinc finger protein, ZBRK1. Cancer Res 2002; 62(13):3773-3781.
49. Iuchi S, Green H. Basonuclin, a zinc finger protein of keratinocytes and reproductive germ cells, binds to the rRNA gene promoter. Proc Natl Acad Sci USA 1999; 96(17):9628-9632.
50. Tseng H, Biegel JA, Brown RS. Basonuclin is associated with the ribosomal RNA genes on human keratinocyte mitotic chromosomes. J Cell Sci 1999; 112 Pt 18:3039-3047.
51. Roberts SG. Mechanisms of action of transcripton activation and repression domains. Cell Mol Life Sci 2000; 57:1149-1160.

CHAPTER 22

The Superfamily of SCAN Domain Containing Zinc Finger Transcription Factors

Tucker Collins* and Tara L. Sander

Abstract

The SCAN domain is a highly conserved 84 residue motif that is found near the N-terminus of a subfamily of C_2H_2 zinc finger proteins. The SCAN domain, which is also known as the leucine rich region (LeR), functions as a protein interaction domain, mediating self-association or selective association with other proteins. Bioinformatic approaches were used to identify 71 SCAN domains in the human genome and to define the structures of the members in the human SCAN domain family. In addition to a single SCAN domain, the members of the family can have a variable number of zinc fingers (2-22), a KRAB domain, as well as a novel N-terminal motif. The genes encoding SCAN domains are clustered, often in tandem arrays, in both the human and mouse genomes and are capable of generating isoforms that may affect the function of family members. Twenty-three members of the mouse SCAN family appear to be orthologous with human family members, and human-specific cluster expansions were observed. Although the function of most of the family members is unknown, an overview of selected members of this group of transcription factors suggests that the SCAN domain family is involved in the regulation of growth factor gene expression, genes involved in lipid metabolism, as well as other genes involved in cell survival and differentiation. Analysis of the SCAN domain family using phylogenetic and comparative genomics approaches reveals that the SCAN family is vertebrate -specific. Remarkably, the SCAN domains in lower vertebrates are not associated with C_2H_2 zinc finger genes, but are contained in large retrovirus-like polyproteins. Collectively, these studies define a large family of transcriptional regulators that have rapidly expanded during recent evolution.

Introduction

Zinc finger proteins are a large class of regulatory proteins. The zinc finger is a small peptide domain with a secondary structure stabilized by a zinc ion interacting with the cysteine and histidine residues of the finger.[1] Although several types of zinc finger motifs have been identified,[2] the C_2H_2 finger has emerged as the classical zinc finger and is described as $CX_{2-4}CX_{12}HX_{2-6}H$ (reviewed in Chapter 1 of this book).

Accompanying the C_2H_2 zinc finger elements in some of the transcription factors are a series of extended sequence motifs. These structural modules regulate subcellular localization, DNA binding, and gene expression by controlling selective association of the transcription factors with each other, or with other cellular components. In the C_2H_2 class of zinc fingers, these associated modules include the Kruppel-associated box (KRAB),[3] the poxvirus and zinc finger (POZ) domain,[4] which is also known as the BTB domain (Broad-Complex, Tramtrack, and Bric-a-brac),[5] and the SCAN domain.[6] These domains define subgroups within the C_2H_2 family and may provide insights into the functions of the members of this large family of zinc finger transcription factors. Here we will place the SCAN domain family of transcription factors in perspective.

The SCAN Domain

The SCAN domain was originally identified in the C_2H_2 zinc finger transcription factor, *ZNF174*.[6] The name was based on the first letters found in some of the founding members of the family (SREZBP, Ctfin51, AW-1 (ZNF174), and Number 18).

Definition of the SCAN Domain

Based on the analysis of a limited number of SCAN genes, the definition of the SCAN domain has varied, resulting in several forms of the domain in online databases.[7] Following a more complete analysis of the SCAN domains found in the human genome, the definition of the SCAN domain has been refined.[8] As will be discussed below, the human genome has at least 70 genes containing the SCAN domain.[8] An alignment of a select number of these members is illustrated in Figure 1. A more extensive alignment of the highly homologous SCAN domains (as well as other associated domains) can be found online at the Project Website (http://www.scanfamily.org). Based on the sequence alignment, as well as the biochemical properties of the recombinant protein detailed below, the SCAN domain is defined as 84 residues, beginning with E-43 and ending with R-126 of ZNF174. In many sequences, there are proline residues both before and after the domain, helping to delineate the boundaries of the predicted secondary structural elements. Of the 84 residues that comprise the

* *Corresponding author. See list of "Contributors".*

Zinc Finger Proteins: From Atomic Contact to Cellular Function, edited by Shiro Iuchi and Natalie Kuldell.

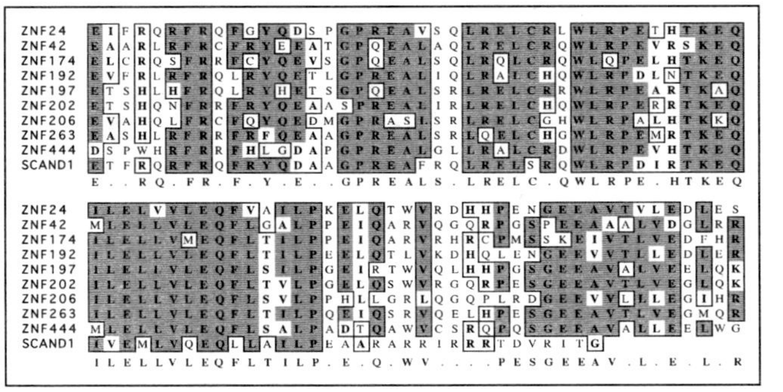

Figure 1. SCAN domains in the human genome. To illustrate the homology of the SCAN domain, the amino acid sequences of a select number of human SCAN domains were aligned by consensus ClustalW sequence alignment using MacVector 7.1.1 (Genetics Computer Group) with an open gap penalty setting of 10.0. Residues that share > 51% identity are outlined with dark shading, conserved amino acid differences are indicated by lighter shading. The consensus sequence was determined for the SCAN sequences of the 64 human family members (excluding possible pseudogenes) and is shown below the alignment.

SCAN domain, a remarkable 62 are conserved in more than 50% of the family members.

The SCAN domain has also been called the "leucine rich region" (LeR).[9] This is an unfortunate designation because the similar term "leucine rich repeat" is used to describe a different domain found in a functionally distinct group of proteins involved in inflammation and cell death. For example, the caspase recruitment domain (CARD) containing protein CARD4 (also called NOD1) is a CED-4/Apaf-1 family member that activates NF-κB signaling and induces apoptosis.[10] CARD4 contains several putative functional domains, including an N-terminal CARD domain and a C-terminal region consisting of leucine-rich repeats that likely function as a site for interactions with upstream signaling components. To complicate matters further, the leucine-rich repeat superfamily of proteins is quite diverse and includes the Toll-like receptors that are involved in transmembrane signaling.[11] Unfortunately, the use of the term "leucine rich region" to describe the SCAN domain, and the related term "leucine rich repeat," found in mediators of immune responses and apoptosis, has resulted in confusion in the literature.

The SCAN Domain Is a Protein Interaction Motif

Unlike the KRAB and BTB/POZ domains, the SCAN domain does not have transcriptional activation or repression capabilities.[12,13] Findings using both mammalian and yeast two-hybrid systems demonstrate that the SCAN domain is an interaction motif.[12-14] For example, a mammalian two-hybrid assay was used to show that the SCAN domain mediates protein-protein interactions.[12] To demonstrate that the SCAN domain can self-associate, the ability of a ZNF174 SCAN-GAL4 construct to activate a reporter was tested in the presence of a ZNF174 SCAN-VP16 fusion construct. Coexpression of both fusion constructs markedly activated transcription of a GAL4-dependent reporter when compared with that of empty GAL4 or VP16 vectors, or with each of the SCAN domain fusion constructs alone. The SCAN domain did not interact with the leucine zipper motifs of either c-FOS or c-JUN, demonstrating that the SCAN domain did not interact nonspecifically with other transcription factors that contain amphipathic α-helices mediating oligomerization.[12] The mammalian two-hybrid system was also used to demonstrate that self-association of the ZNF174 SCAN domain requires the entire SCAN domain, in that smaller regions of the domain failed to interact in the assay.[12] Similarly, forms of the SCAN domain into which mutations were introduced that disrupted the predicted central helix of the domain failed to interact in the two-hybrid assay. Taken together, these findings suggest that the minimum length functional unit is the entire SCAN domain and that structural integrity of this domain is required for self-association with other SCAN domain partners.

The SCAN domain is capable of mediating protein-protein interactions in the context of an intact member of the family. Studies were done in which a tagged full-length form of ZNF174 was cotranslated with another SCAN protein, and associations between the two SCAN protein forms were then examined by immunoprecipitation with an antibody that recognized the tag.[12] These coimmunoprecipitation studies confirmed that the SCAN domain is responsible for self-association in the context of an intact SCAN family member and that the α-helical character of the domain is required for the interaction. Additionally, these in vitro studies demonstrate that intact ZNF174 can selectively bind other members of the SCAN family. Generalizing on these observations, members of the SCAN family can both self-associate and form heterodimers using the SCAN domain as an interaction motif.

The SCAN Domain Forms a Stable Dimer in Solution

In previous studies our group and others demonstrated that the SCAN domain functions as an interaction domain, mediating self-association or association with other proteins bearing SCAN domains. In part of those studies, a fragment of either the N-terminus of ZNF174 or ZNF202 that encompassed the SCAN domain behaved as an oligomeric species.[12,14] However, in those studies the definitive subunit stoichiometry of the oligomeric SCAN domain was not determined. Our group utilized the sequences of the SCAN domains in the human genome to more completely define the limits of the domain. The isolated ZNF174 SCAN domain was then overexpressed in bacteria, purified and characterized. Both size exclusion chromatography and equilibrium sedimentation analysis demonstrate that the isolated ZNF174 SCAN domain forms a homodimer.[7] Additionally, thermal denaturation studies of the ZNF174 domain revealed a high melting temperature (74°C) and demonstrated the stability of the SCAN dimer. These studies resolve the issue of the oligomeric nature of the isolated SCAN domain in that the purified protein forms a stable dimer. However, study of the isolated SCAN domain does not address the important issue of the oligomerization state of the intact SCAN domain family member in cells.

Assembly of a Transcriptional Regulatory System Through SCAN-SCAN Interactions

Based on the ability of the SCAN domain to function as a dimerization domain, it is possible that the domain participates in the assembly of a family of regulatory proteins. To determine which SCAN domains have the ability to interact with one another, pairwise combinations of nine isolated SCAN motifs were tested in the mammalian two-hybrid system.[12] Several general features of SCAN domain interactions could be inferred from the findings in this system. First, not all SCAN domains are able to self-associate. Second, interactions between different SCAN domains are selective. For example, ZNF174 can interact with some, but not all, SCAN domains. Third, there is significant variation in the relative affinities of the SCAN domains. Although these studies with nine SCAN family members represent only a small fraction of the members of the family, they may provide some initial guidelines for the assembly of a SCAN protein network.

In addition to the findings from the mammalian two-hybrid system, various SCAN domains have been used as bait in the yeast two-hybrid system. These studies have identified SCAN containing proteins, revealing new interactions among family members. For example, a region of MZF1 (ZNF42) containing a SCAN domain interacted with RAZ1 (SCAN-related protein associated with MZF1B),[13] which is also known as SDP1 (SCAN-domain containing protein 1) or SCAND1. Similarly, RAZ1 interacted with the SCAN domain of ZNF202 and inhibited its repressive function.[14,15] A summary of the existing interactions between SCAN family members is shown in the accompanying model (Fig. 2). The majority of these observations outlined above were based on studying isolated SCAN-SCAN associations. Whether most of these interactions occur in the context of full-length proteins is uncertain. Additionally, understanding the nature of the partnerships occurring in vivo will be key to defining the family.

Transcription factor dimerization can increase the selectivity of protein-DNA interactions and generate a large amount of diversity from a relatively small number of proteins.[16] With a group of regulators the size of the SCAN family (see below), the range of potential dimers is tremendous. It will be important to

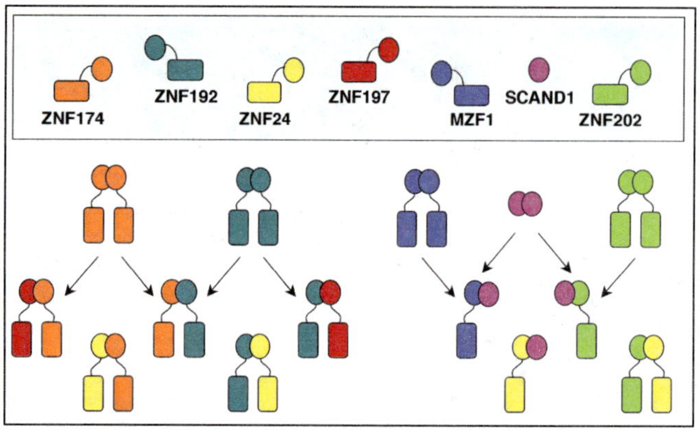

Figure 2. A model network of SCAN family members. A summary of the reported interactions between SCAN family members is shown. Each SCAN member is represented by a different color: ZNF174 (orange), ZNF192 (teal), ZNF24 (yellow), ZNF197 (red), MZF1 (blue), SCAND1 (pink), ZNF202 (green). The SCAN domain of each protein is illustrated with a circle; the C_2H_2 zinc finger motifs are indicated with a box. Note that some of the interactions were not studied within the context of the full-length protein.

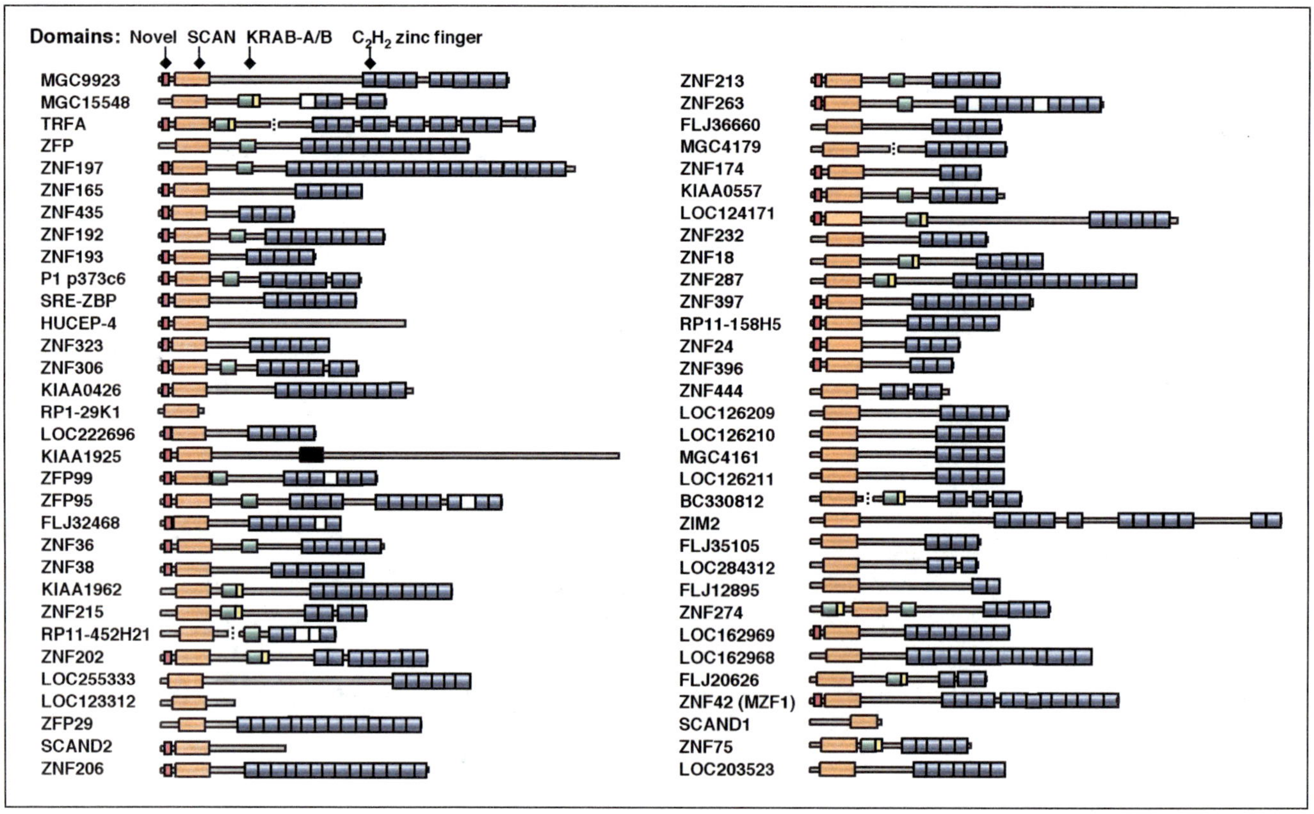

Figure 3. Structural features of the human SCAN domain family. The names and predicted structures of the 64 members of the human SCAN domain family are schematically shown in order of their gene locations, starting with chromosome 1. The conserved domains are indicated with a shaded box: novel region (red), SCAN (orange), KRAB-A (green), KRAB-B (yellow), integrase core domain (black), and C_2H_2 zinc finger domain (blue). Zinc fingers displayed as an open white box are atypical and deviate from the C_2H_2 consensus. A dotted line in some of the structures indicates that the predicted C-terminal domains are questionable because of discrepancies between the databases. The official gene symbols assigned by NCBI (as of November 1, 2003) are shown. In instances where an official (or interim) name was not available, the genomic DNA clone containing the SCAN sequence was given. Note that the SCAN domains of *ZFP29* and *SCAND1* are truncated. Modified from Sander et al, 2003.

determine the structural rules that regulate SCAN dimerization specificity and obtain experimental data both in vitro and in vivo that verifies some of the partners. Based on this information it may be possible to classify many of the SCAN family members based on dimerization properties, as has been done for the B-ZIP proteins.[17]

Superfamily of SCAN-Domain Containing Proteins

The SCAN field was limited by an incomplete description and definition of the members within the SCAN domain family. We reasoned that the release of the human genome could provide the key to defining the entire human SCAN domain family of transcription factors. Therefore, a bioinformatics approach was used to determine the total number of SCAN domains in the human genome and to define the structure and chromosome location of each human SCAN family member.[8]

To determine the total number of genes predicted to encode SCAN domains in the human genome, both public and private human genome databases were screened with a representative human SCAN domain (ZNF174). Once the SCAN domains were identified, DNA sequences adjacent to the motif were annotated to predict the cDNA structures.[18] Analysis of expressed sequence tag (EST) databases and serial analysis of gene expression (SAGE) libraries confirmed that most of the predicted SCAN family members are actively expressed.

This screen revealed the presence of 71 SCAN domain-containing genes in the human genome. Twenty-four of the SCAN domains were known, while 47 were previously unidentified. Thus, the SCAN family constitutes approximately 10% of the estimated 700 C_2H_2 zinc finger genes present in the human genome.[19,20] Sixty-four of the 71 SCAN domains (90%) were found in complete open reading frames, suggesting that they are contained within functional genes. The remaining seven SCAN domains have termination codons that prematurely truncate the open reading frame, suggesting that these domains are contained in pseudogenes. The DNA sequences adjacent to the 64 SCAN domains were computationally analyzed and manually annotated to predict the cDNA structure and protein coding-potential for each member of the human SCAN family. A schematic diagram of the structural features of the human SCAN family, excluding potential pseudogenes, is presented in Figure 3. This approach successfully predicted the structures for all of the 24 previously reported members of the human SCAN family.

Components of SCAN Family Members in Addition to the SCAN Domain

The SCAN family members were predicted to contain conserved modular motifs lying outside of the SCAN domain.

Zinc Fingers

The majority of SCAN domains (91%) are associated with C_2H_2 zinc finger domains. These 58 members contain a variable number (2-22) of C_2H_2 zinc fingers, defined by the consensus sequence $CX_{2\text{-}4}CX_{12}HX_{2\text{-}6}H$ (with X being any amino acid). All of the C_2H_2 zinc finger motifs are located at the C-terminal end of the SCAN proteins (Fig. 3). Most of these C_2H_2 zinc fingers are *Krüppel*-type, as defined by the conserved link (TGEKP(Y/F)X) between the histidine of the preceding finger with the cysteine of the next finger (H-C link). Many members contain multiple-adjacent fingers that are arranged in clusters of 3 or more (e.g., *ZNF197, ZFP95*), while other members contain single or duplicate pairs of fingers (e.g., *FLJ12895, ZNF215*). Since *Krüppel*-type factors are frequently involved in DNA-binding, this variation in zinc finger number and spacing may have an effect on nucleotide base pair recognition.[21] Application of the proposed DNA-zinc finger rules,[22,23] suggests that most of the C-terminal zinc fingers of the family members can participate in DNA binding. Notably, members of the SCAN family with multiple adjacent C_2H_2 zinc fingers may have more than one DNA binding activity, as is seen with other proteins with more than four zinc fingers.[24] Additionally, since C_2H_2 fingers have been implicated in protein-protein interactions,[25] it is possible that some of the zinc fingers in the SCAN family members may interact with other members of the family, or with other target proteins.

KRAB Domain

Twenty-four members of the human SCAN domain family also contain a KRAB domain (Fig. 3). As outlined elsewhere in this book, the KRAB domain is found at the N-terminus of approximately 200 C_2H_2 zinc finger proteins[19,20] and mediates repression of transcription.[26,27] The KRAB domain spans about 75 amino acids and has been divided into two subregions, designated A and B.[3] It has been demonstrated that the KRAB-A domain contains transcriptional repression activity. This domain is predicted to form an amphipathic helix that interacts with KAP1 (KRAB-associated protein-1, also known as TIF-1β, transcription intermediary factor-1β).[28-34] KAP1 can enhance KRAB-A-mediated repression by recruitment of a histone deacetylase complex containing N-CoR (nuclear receptor corepressor),[35] or the Mi-2α subunit of the NuRD complex.[33] KAP1 can also associate with members of the heterochromatin protein 1 (HP1) family, a family of nonhistone heterochromatin-associated proteins with an established gene-silencing function.[31]

A sequence alignment of the KRAB domains from the SCAN family members is found on the project web site. As with other KRAB domain containing genes, the KRAB domain is located at the N-terminus of SCAN family members, usually C-terminal to the SCAN domain. KRAB-containing proteins can be classified into three types:[36] those that contain a KRAB-A domain alone, those with both –A and –B domains, or those that have an –A domain and a divergent –B domain. Twelve members of the human SCAN family contain both KRAB-A and -B domains, while another twelve members have a KRAB-A domain only. It is likely that the SCAN domain will influence the function of the KRAB domain, although this has not been extensively studied.

Novel Domain

Thirty-four members of the SCAN family contain a novel region of homology that was found at the very N-terminus of the predicted protein (Fig. 3). A sequence alignment of a select number of the members is illustrated in Figure 4 and a full alignment of all the members is found on the project web site. This region is 13 residues in length, with the consensus EqEGLLiVKvEEe (where capital letters correspond to amino acids conserved over 51%, and lowercase letters to amino acids conserved over 35%), and may be similar to the previously described N-terminal acidic domain of the murine myeloid-specific zinc finger gene, MZF-2.[37] A search of the human genome with the consensus sequence indicates that this region may be found only in members of the SCAN family.

The core of this conserved N-terminal domain in the SCAN family members contains a sequence for conjugation to the small

ZNF24	P	D	E	E	E	K	I	L	R	V	K	L	E	E	D	P	D
ZNF36	A	Q	E	K	D	G	I	V	I	V	K	V	E	E	E	D	E
ZNF42	P	P	E	D	E	G	P	V	M	V	K	L	E	D	S	E	E
ZNF165	S	P	E	D	E	G	L	L	I	V	K	I	E	E	E	E	F
ZNF174	S	S	K	Q	E	R	H	I	I	A	K	L	E	E	K	R	G
ZNF192	Q	T	P	E	E	D	L	V	I	V	K	V	E	E	D	H	G
ZNF197	A	L	R	Q	E	G	L	V	K	G	K	D	D	T	W	K	W
ZNF202	L	W	E	E	E	G	I	L	M	V	K	L	E	D	D	F	T
ZNF213	P	G	E	G	E	G	L	L	I	V	K	V	E	D	S	S	W
ZNF263	S	Q	E	R	E	G	L	L	I	V	K	L	E	E	D	C	A
			E	q	E	G	L	L	i	V	K	v	E	E	e		

Figure 4. The conserved region N-terminal of the SCAN domain. The amino acid sequences of the N-terminal novel region from selected members of the human SCAN family were aligned by consensus ClustalW sequence alignment using MacVector 7.1.1 (Genetics Computer Group) with a gap penalty setting of 10.0. Residues that share > 51% identity are outlined with dark shading, conserved amino acid differences are indicated by lighter shading. The consensus sequence for the novel region of the 34 human SCAN family members is shown below the alignment, where capital letters correspond to amino acids conserved > 51% and lowercase letters to amino acids conserved > 35%. The putative SUMO acceptor site is underlined.

ubiquitin-related modifier, SUMO. Like ubiquitin, SUMO is a small polypeptide that is covalently attached to lysine residues in substrate proteins.[38] SUMO is attached to substrate proteins by a series of enzymatic reactions similar to those involved in ubiquitination. Following cleavage to expose a C-terminal glycine, SUMO is attached to a heterodimeric E1 SUMO activating enzyme (Aos1/Uba2), then transferred to an E2 SUMO conjugating enzyme (Ubc9), before being covalently attached to a lysine in the substrate protein. A consensus SUMO acceptor site has been identified consisting of the sequence ΨKXE, where Ψ is a large hydrophobic amino acid and K is the site of SUMO conjugation.[39] Contained within the core of the novel N-terminal region (EqEGLLiVKvEEe) is the sequence VKxE, which fits the consensus for a SUMO acceptor site. Additionally, a glycine residue is located five amino acids upstream from the acceptor lysine, consistent with the novel domain containing a sumoylation site (Fig. 4).

Post-translational modification of the SCAN family member by SUMO may have a substantial effect on transcription factor activity. By analogy to other proteins, SUMO modification of the SCAN family member could prevent modification by ubiquitin at that position, potentially preventing degradation of the family member by the proteasome. SUMO may also compete for post-translational modification of lysines by acetylation. Additionally, SUMO modification of the SCAN family member may disrupt protein-protein interactions with some proteins, while promoting interaction with other proteins. In addition to playing a role in regulating transcription factor activity, SUMO modification also affects transcription factor subcellular localization. For example, SUMO-1 modified Sp3 accumulates at the nuclear periphery and in nuclear dots whereas the transcriptionally active form lacking SUMO-1 has a more diffuse nuclear localization.[40] Although experimental verification will be required, the potential sumoylation site N-terminal to the SCAN domain may have an important role in the function of some of the family members.

Overview of Selected SCAN Family Members

The members of the SCAN family of transcription factors can be divided into classes based on the presence of other modular elements.[41] Representatives from the three largest classes of SCAN family members will be outlined in this review. Most SCAN domains are found in transcription factors with a variable number of C-terminal Cys_2-His_2 $(C_2H_2)_x$ type zinc-fingers [SCAN-$(C_2H_2)_x$]. A second group of SCAN family members is composed of proteins predicted to have a SCAN-KRAB-$(C_2H_2)_x$ domain alignment, while a third class of SCAN family members consists of isolated SCAN-domain proteins.

SCAN-$(C_2H_2)_x$

The first group of SCAN domain proteins consists of family members with just a SCAN domain and a variable number of C-terminal C_2H_2 zinc fingers (Fig. 3). There are at least 40 of these genes in the SCAN family, but only a few have been characterized. Some of the better known representatives from this class of genes are indicated in Table 1.

The myeloid zinc finger gene (*MZF1*, also known as *ZNF42*) is expressed preferentially in hematopoietic progenitor cells of the myeloid lineage.[42] Inactivation of the murine *MZF1* gene results in a striking increase in hematopoetic progenitors, with the eventual development of lethal myeloid neoplasias,[43] suggesting that MZF1 is involved in growth, differentiation and tumorigenesis of myeloid progenitors. In both human and mouse cells, several isoforms of MZF proteins are produced from a single gene by the use of alternative promoters and alternative splicing.[37,44] Of the isoforms, MZF1B, and its murine counterpart MZF-2A, seems to represent the major product of the gene. MZF1B shares identity to the carboxy-terminus of MZF1, a C-terminal DNA binding domain consisting of 13 C_2H_2 zinc finger modules, but MZF1B encodes an additional amino terminal extension that contains a novel N-terminal domain, a SCAN domain and a transactivation domain.

To further characterize this member of the SCAN family, proteins that interact with several of the domains of MZF1B (MZF-2A) have been defined. The SCAN domain of MZF1B interacts with RAZ1,[13] a member of the SCAN family that lacks zinc fingers, as discussed below. The transactivation domain of MZF-2A works specifically in myeloid cells and can function as an autonomous transactivation motif when fused to a heterologous DNA binding domain.[45] In studies exploring the regulation of the activity of the transactivation domain, three serine residues were found to be phosphorylated by ERK and p38 MAP kinases.[46] This suggests that the activity of the transactivation region of MZF-2A is negatively regulated through phosphorylation by MAP kinases. To understand the mechanism of MZF-2A dependent transcriptional activation, a yeast two-hybrid Ras recruitment screen was done to look for interacting proteins.[47] A novel SWI2/SNF2-related protein, termed mammalian Domino, was identified as a candidate MZF-2A interacting partner. Notably, mDomino contains a SWI2/SNF2-type ATPase/helicase domain, a SANT domain, and a glutamine rich domain. The C-terminal Q-rich domain physically associated with the transactivation domain of MZF-2A and overexpression of mDomino enhanced MZF-2A mediated activation of a reporter gene.[47] Collectively, these findings suggest that an ATF-dependent chromatin-remodeling complex interacts with MZF-2A to regulate gene expression in myeloid cells.

Table 1. Summary of representative SCAN family members from the three largest classes

SCAN Protein*	Interacting Proteins	Transcriptional Function
SCAN-$(C_2H_2)_X$		
ZNF24 (ZNF191, KOX17, mZF-12)	ZNF174, ZNF192, ZNF202, SCAND1	Binding to TCAT repeat in tyrosine hydroxylase gene
ZNF42 (MZF1, MZF1B, mMZF-2A)	RAZ1, ERK, p38, mDomino	Regulator of myeloid progenitor differentiation/proliferation
ZNF174 (AW-1)	ZNF24, ZNF192, ZNF197	Negative regulator of PDGF-B chain, TGF-β1 expression
ZNF444 (EZF-2)	ND	Positive regulator of SREC gene expression
SCAN-KRAB-$(C_2H_2)_X$		
ZNF192(LD5-1)	ZNF24, ZNF174, ZNF197	ND
ZNF197 (ZnF20)	ZNF174, ZNF192, VHL	Negative regulator of HIF-1α transactivation
ZNF202	ZNF24, SCAND1	Negative regulator of genes involved in lipid metabolism
ZNF263 (FPM315, mNT2)	ND	Negative regulator of α2 (XI) collagen gene (Col11α-2)
SCAN		
SCAND1 (SDP1, RAZ1, mPGC-2)	MZF1 , ZNF24, ZNF202, PPARγ, NF-κB	Co-activator, adipocyte differentiation
SCAND2	ND	ND

The official human gene symbol assigned by NCBI is given; alternate human and mouse (m) gene names are in parentheses. ND= not determined.

To address the key issue of defining genes regulated by MZF1, a DNA binding site was determined for the zinc fingers of MZF1 and consensus elements were found in the promoters of several hematopoietic cell-specific genes, such as CD34, c-myb, lactoferin and myeloperoxidase.[48,49] The MZF1 binding motif is also found in the promoter of the telomerase reverse transcriptase gene,[50] and in an intron in the gene for the high affinity IgE receptor,[51] where MZF1 is likely to function as a negative regulator for gene expression. Interestingly, MZF1 activates gene expression in cells of the hematopoietic lineage, while in nonhematopoietic cells it functions as a transcriptional repressor.[42] It is possible that other SCAN family members may be both positive and negative regulators of target gene expression, depending upon the cell type.

SCAN-KRAB-$(C_2H_2)_x$

A second class of SCAN family members consists of a SCAN-KRAB-$(C_2H_2)_x$ modular architecture (Fig. 3). There are at least 24 genes in this class of SCAN family members. Some of the better known members of this class of SCAN family members are listed in Table 1.

An interesting example of this subgroup of SCAN family members is the candidate hypoalphalipoproteinemia susceptibility gene, ZNF202. A low HDL cholesterol locus on chromosome 11q23 was identified that is distinct from the apoA-I/C-III/A-IV/AV cluster of genes.[52] This new familial susceptibility locus for hypoalphalipoproteinemia contains the SCAN family member, ZNF202. The gene for ZNF202 encodes a protein predicted to contain a SCAN domain, an intact KRAB domain and eight C_2H_2 zinc-finger motifs. Two splice forms have been identified for ZNF202, the m1 form encoding the full-length protein, and the m3 form encoding a smaller version containing only the SCAN domain and lacking the zinc fingers.[53] The ZNF202 gene product is a transcriptional repressor that binds to elements found predominantly in genes that participate in lipid metabolism, such as the group of genes that comprise the apoA-I/C-III/A-IV/A-V gene cluster on chromosome 11, the apoE/C-I/C-IV/C-II gene cluster on chromosome 19, phospholipid transfer protein, and a series of enzymes involved in lipid processing, including lipoprotein lipase, hepatic triglyceride lipase and lecithin cholesteryl ester transferase.[53] Additional targets for repression by ZNF202 include the ATP-binding cassette A1 (ABCA1) and ABCG1.[15] ABCA1 is a key regulator of plasma HDL levels, while ABCG1 supports lipid efflux in human macrophages. ZNF202, which acts as a transcriptional repressor of both genes, is able to reduce phospholipid and cholesterol efflux in transiently transfected macrophages, demonstrating the functional relevance of the SCAN family member.[15] Additionally, ZNF202 expression is inversely regulated with respect to the expression of its target genes ABCA1 and apoE during macrophage differentiation and foam cell formation.[54] Collectively, these findings demonstrate that ZNF202 regulates reverse-cholesterol transport and are consistent with the proposal that ZNF202 controls the balanced expression of genes involved in lipid metabolism.

Isolated SCAN Domain Proteins

For six members of the human SCAN family there is no current evidence for associated C_2H_2 zinc finger motifs (Fig. 3). These SCAN domain-only sequences might represent novel genes without zinc fingers, or splice forms of larger transcripts that contain zinc finger motifs. Indeed, there is cDNA evidence that some SCAN-containing genes are alternatively spliced generating transcripts that lack the SCAN, KRAB, or zinc finger domains, as discussed below.

A murine SCAN domain containing protein was identified as an adipogenic cofactor bound by the differentiation domain of the nuclear receptor peroxisome proliferator-activated receptor γ(PPARγ).[55] This mouse protein, termed PPARγ coactivator-2 (PGC-2), or its human homolog, SCAND1 (also known as SDP1, SCAN-domain containing protein 1; or RAZ1, SCAN-related protein associated with MZF1B), interacts with the ligand-independent activation function region (AF-1) of PPARγ and can potentiate PPARγ-dependent gene expression. This effect is presumably mediated by facilitating the assembly of a coactivation complex and enhancing fat cell formation. PGC-2 does not contain zinc fingers, but does contain a partial SCAN domain that consists of the N-terminal 60 residues of the "authentic" SCAN domain. Interestingly, SCAND1 and PGC-2 may interact with PPARγ through their SCAN domains, even though PPARγ does not contain a SCAN domain.[14,56] Although the SCAN domain was required for interaction with PPARγ, it was not sufficient for coactivator function. Interestingly, SCAND1 has been shown to interact with ZNF202 via its SCAN domain, thereby preventing recruitment of KAP1 and transcriptional repression.[15] Thus by enhancing PPARγ-dependent transcription, and blocking KAP binding to ZNF202, SCAND1 may be an important coregulator of genes controlling the cellular lipid machinery.[54,56]

This isolated SCAN domain protein may play a role in differentiation of stem cells in the bone marrow. Pluripotent mesenchymal stem cells in bone marrow differentiate into adipocytes and other cells. Although balanced cytodifferentiation of stem cells is essential for the formation and maintenance of bone marrow, the mechanisms that control this balance remain largely unknown. Whereas PPARγ is a key inducer of adipogenesis, inflammatory cytokines, such as interleukin-1 and tumor necrosis factor-α, inhibit adipogenesis. Recent results suggest that the ligand induced transactivation function of PPARγ is suppressed by the inflammatory cytokines and that this suppression is mediated through a signaling cascade that results in NF-κB activation. NF-κB blocks PPARγ binding to DNA by forming a complex with PPARγ and its coactivator PGC-2.[57] These findings suggest that expression of the cytokines in bone marrow may alter the fate of stems cells by suppressing PPARγ function, directing cellular differentiation towards osteoblasts rather than adipocytes. Since SCAND1 also interacts with MZF1B,[13,58] a factor that plays a role in the differentiation of myeloid progenitors, it is possible that similar SCAND1-dependent mechanisms may regulate the differentiation of other types of stem cells in the bone marrow.

This brief overview of just one representative gene from each of the three largest classes of SCAN family members demonstrates that the SCAN domain containing transcription factors perform a wide range of functions important in cell development or differentiation.

Isoforms of SCAN Family Members and Functional Diversity

Alternative use of transcriptional start sites, or alternative splicing, could have a significant effect on the function of SCAN family members. Some specific examples illustrate the effects these processes could have on the functional diversity of the SCAN family.

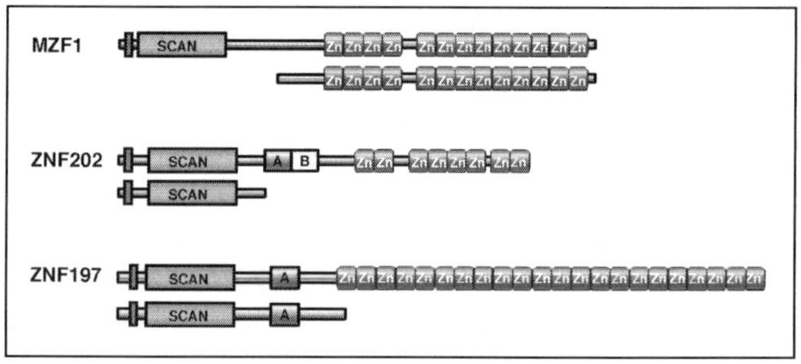

Figure 5. Structure and function of some of SCAN family member isoforms. Representative isoforms from selected SCAN family members discussed in the text. The predicted structure of the isoforms from three representative human SCAN-containing family members is shown. The conserved domains are indicated with a shaded box: novel region (red), SCAN (orange), KRAB-A (green), KRAB-B (yellow), and C_2H_2 zinc finger domain (blue) as shown in Figure 3. A color version of this figure can be viewed at http://www.eurekah.com/chapter.php?chapid=1765&bookid=124& catid=30.

Isoforms Lacking the SCAN Domain

As mentioned above, the *MZF1* gene generates a transcript that encodes a SCAN domain-containing zinc finger protein, MZF1B.[44] MZF1B shares identity to the carboxy-terminus of MZF1, including the 13 C_2H_2 zinc finger modules, but MZF1B encodes an additional amino terminal extension that contains a SCAN domain (Fig. 5). Thus while MZF1 and MZF1B may bind to the same DNA target, it is likely that they exert distinct regulatory effects because of their distinct amino termini. Additionally, it is possible that when MZF1 is bound to a target gene it may act as a dominant-negative inhibitor of MZF1B function.

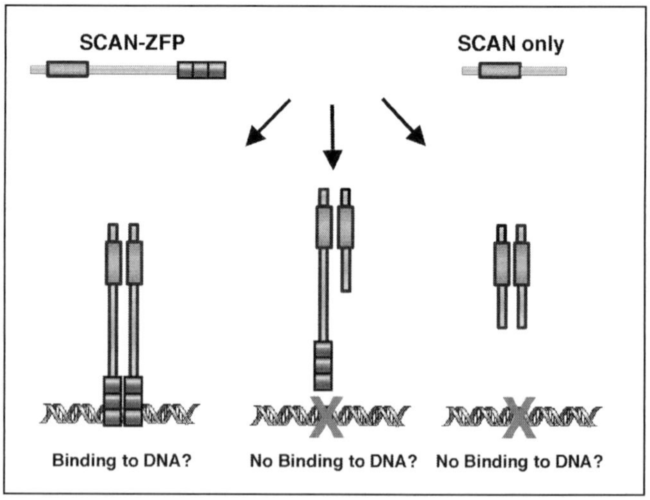

Figure 6. The corresponding model illustrates that selective SCAN dimerization may be a potential mechanism for functional diversity in the SCAN family. In one example, a SCAN-containing gene generates two isoforms; a SCAN domain-containing zinc finger protein (SCAN-ZFP) and a protein that lacks zinc finger motifs (SCAN only). Dimerization of the SCAN-only protein with the SCAN-ZFP might result in altered DNA binding activity, since the heterodimer would lack a pair of zinc finger motifs.

Isoforms Lacking Zinc Fingers

Other SCAN family member genes also generate forms that lack the C-terminal zinc fingers. Thus in a general context, a SCAN domain-only protein may dimerize with another member of the family and alter nucleotide binding, since the resulting heterodimer would lack the pair of DNA binding motifs (Fig. 6). The expression of these alternate transcripts from SCAN containing genes would, therefore, increase the available number of protein combinations within the family. Such forms may help regulate the function of the SCAN family members.

One interesting example of this type of isoform is derived from ZNF197 because it interacts with the von Hipple Lindau tumor suppressor (pVHL). pVHL is a component of an E3 ubiquitin ligase and targets hypoxia-inducible factor-1α (HIF-1α) for ubiquitination and degradation under normoxic conditions.[59] pVHL also directly inhibits HIF-1α transactivation by recruiting histone deacetylases. In a yeast two-hybrid screen for proteins interacting with pVHL, an isoform of ZNF197 was identified which was given the name pVHL-associated KRAB-A domain-containing protein (VHLaK).[60] It contains a SCAN domain and a KRAB A-domain, but lacks the 22 zinc fingers present in ZNF197 (Fig. 5). The KRAB A domain in VHLaK mediates pVHL binding and functions as a transcriptional repression module,[60] consistent with previous reports that the KRAB domain of ZNF197 could repress transcription of a reporter gene.[9] The SCAN domain in VHLaK mediates homo-oligomerization and enhances VHLaK repressive activity. pVHL can recruit VHLaK to repress HIF-1α transcriptional activity and HIF-1α-induced VEGF expression.[60] Additionally, the ZNF197 isoform can recruit both KAP1 and pVHL simultaneously, indicating that KAP1 may participate in pVHL mediated transcriptional repression of HIF-1α.[60]

In summary, SCAN family members can generate isoforms lacking the SCAN domain as well as isoforms lacking the zinc fingers. Expression of these isoforms may be static, but more likely, the production of these isoforms may be biologically relevant. For example, it is possible that in tumors, relative to the corresponding normal tissues, the levels of the various isoforms of the SCAN family members may change.[61]

Identification of Target Genes for SCAN Family Members

The biological function of most of the SCAN family members is not known, although the function of a few of the members has been inferred following targeted mutagenesis of the corresponding mouse gene, or following the identification of interacting protein partners. Another way to determine the function of a SCAN family member is to identify candidate target genes that the member might regulate. Several experimental approaches have been used to identify candidate target genes for SCAN family members

One-Hybrid Approach

Perhaps most compelling are results from groups that have functionally characterized a specific regulatory element in a gene of interest and then used that site in a "one-hybrid" screen to identify the cognate regulatory protein. The yeast one-hybrid system is based on the principle of the yeast two-hybrid system and is used for isolating novel genes encoding proteins that bind to a target, cis-acting regulatory element.[62] Several members of the SCAN family have been identified by this approach. EZF-2 (endothelial zinc finger protein-2 or ZNF444) is a member of the family with four zinc fingers and specifically targets the scavenger receptor expressed by endothelial cells.[63] ZNF263 has nine zinc fingers and a KRAB domain and interacts with an element in the promoter of the α2(XI) collagen gene.[64] ZNF24 (ZNF191) interacts with an intronic polymorphic TCAT repeat in the tyrosine hydroxylase gene, the rate-limiting enzyme in the synthesis of catecholamines.[65]

Binding Site Selection

A second experimental approach to determine the genes regulated by a SCAN family member is to determine its DNA binding site and then utilize this information to identify candidate target genes. As outlined above, based on the binding site determination, ZNF202 has been shown to interact with and control the expression of a series of genes involved in lipid metabolism, including the ATP binding cassette transporter A1.[15] Binding sites for MZF1 (ZNF42) were found in CD34, c-myb and myeloperoxidase, genes that are expressed in hematopoietic cells.

Overexpression

In these studies, members of the SCAN family are overexpressed with promoter-reporter constructs of candidate target genes. In this type of approach, ZNF174 selectively repressed the expression of PDGF-B chain and TGF-β1 promoter-reporters.[6] Establishing the biological relevance of the findings is always the challenge with this type of experimental approach.

To date, chromatin immunoprecipitation (ChIP) has not been used to verify the interaction of SCAN family members with DNA binding sites in an authentic gene or to clone SCAN family member target promoters.[66] Additionally, it should be possible to identify physiologic targets of SCAN family members by looking at altered gene expression profiles using microarrays following overexpression of a SCAN family member.

Genomics of the SCAN Family Members

Human SCAN Family Genes

The genes for the human SCAN domain family have been mapped to specific human chromosome locations.[8] Of the 71 SCAN-containing genes, 14 are isolated single genes. The majority of genes (80%) are found in clusters on human chromosomes 3p21, 6p21.3, 7q22, 15q25, 16p13.3, 17p11.2, 18q12, 19q13.4 and Xq26, and some of these genes are arrayed in tandem. Some of these locations are sites frequently disrupted and associated with cytogenetic abnormalities. For example, ZNF197 is located at 3p21, a region frequently involved in cytogenetic abnormalities associated with epithelial malignancies of the kidney, lung, thyroid and breast, as well as other tumors.[67] The gene for MZF1 is located at the extreme end of the long arm of chromosome 19. Since telomeres are known to shorten as cells age, the location of MZF1 at the telomere may make it vulnerable to dysregulation or disruption during aging.[68]

When the amino acid sequences of human SCAN domains were analyzed using a cladistics program, the majority of SCAN domains within these clustered regions grouped into distinct sets sharing sequence similarities.[8] For example, several SCAN domains from the clustered regions on chromosomes 3, 6, 18 and 19 are grouped into subsets, whereas the isolated SCAN domains on chromosomes 1 and 11 are assorted elsewhere due to lesser sequence similarity. SCAN-containing genes that are tandemly linked present a target for mispairing and unequal crossover, which could result in duplication and divergence of the genes. Over time, these tandemly duplicated SCAN genes could become physically separated through chromosomal rearrangements and translocations. These local duplications may account for the high degree of sequence similarity shared by neighboring genes. Such SCAN gene duplication events taking place in the clustered region on chromosome 19q13.4, for example, would account for the striking similarity of the neighboring genes *MGC4161*, *LOC126209*, *LOC1262110* and *LOC126211*. These four genes encode putative SCAN-containing C_2H_2 zinc finger transcription factors that are 80% identical at the nucleotide level. The SCAN domains are highly conserved (94% identity), whereas differences at the nucleotide level are most predominant in the zinc finger regions (68% identity). Within the five predicted zinc fingers found in each gene, most of the amino acids that differ are found in portions that determine the recognition specificities of the fingers. This suggests that after a possible duplication event, the genes may have acquired changes that permit functional divergence.

This apparent clustered expansion of SCAN genes at chromosome 19q13.4 appears similar to the human-specific expansion of KRAB-containing zinc finger genes clustered at this same locus.[69] In addition to chromosome 19, chromosome 6 contains highly similar SCAN-C_2H_2 genes (*P1 p373c6* and *ZNF306* at 6p21.3) that may represent a duplication event, suggesting that the expansion of SCAN genes at these two clustered sites is similar.

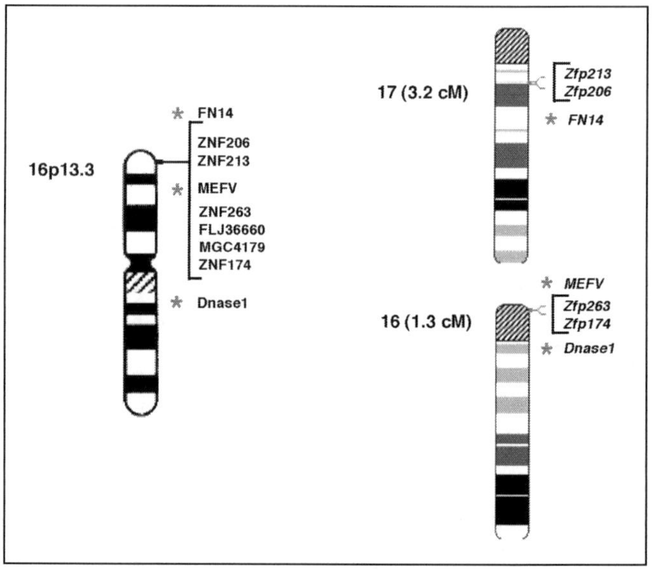

Figure 7. Comparison of selected human and mouse SCAN family members. The SCAN genes on human 16p13.3 and the corresponding regions in the mouse genome are indicated. The cluster of six SCAN genes on human chromosome 16p13.3 is represented by only four SCAN genes in the corresponding conserved segments of mouse chromosomes 16 (3.2 cM) and 17 (1.3 cM). Conserved reference sets of genes (represented by an asterisk) that flank genomic segments containing human SCAN genes on chromosome 16 were located in the mouse genome by MGI, Ensembl, and NCBI. The mouse DNA region between the markers was searched to identify significant matches to DNA sequences corresponding to SCAN domains. A reciprocal comparison of the human and mouse SCAN-containing genes within regions of conserved synteny identified likely orthologs.

Mouse SCAN Family Genes

An analysis of conserved segments between human and mouse chromosomes becomes a useful approach in the identification of likely gene orthologs. To establish potential SCAN-containing gene orthologs, two approaches were taken. First, BLAST searches were performed against available mouse genome databases using representative mouse SCAN domains (Zfp38 or Skz1-pending). Second, conserved reference sets of genes that flank the genomic segment of the human SCAN gene(s) were located in the mouse genome by Mouse Genome Informatics (MGI) and Ensembl. The mouse DNA region between the markers was then searched; significant matches to DNA sequences corresponding to SCAN domains were identified.[8]

A reciprocal comparison of the human and mouse SCAN-containing genes within regions of conserved synteny identified likely orthologs. Based on the SCAN-containing genes represented in the Ensembl database, at least 23 of these members are represented by putative orthologs on conserved segments of the human chromosomes.[8] In several cases, homologous SCAN family members within the human clusters were indistinguishable from each other when compared to the mouse and, as a result, an ortholog assignment was not possible. Interestingly, we found human SCAN clusters that are represented by a smaller number of SCAN genes in the conserved syntenic regions of the mouse. For example, the six clustered SCAN genes on human chromosome 16p13.3 are represented by two clusters of only four SCAN genes in the conserved segment of mouse chromosomes 16 and 17 (Fig. 7). These findings provide evidence for human-specific cluster expansions of SCAN family members. This argues that some genes within the SCAN family are lineage-specific and may have been selected independently since the divergence of primate and rodent lineages.

Similar to the human SCAN family, a large percentage (> 75%) of the mouse SCAN domain-containing genes are predicted to encode one SCAN domain and a variable number of C_2H_2 zinc fingers. Where orthologous relationships could be established, predicted structures of the human and mouse family members were highly similar in that they had comparable amino acid lengths and the same number of conserved domains (KRAB and zinc fingers). For example, the human SCAN gene, *ZNF287*, and its mouse ortholog, *Zfp287/Skat-2*, encode predicted proteins of 754 and 758 amino acids, respectively, containing one SCAN domain, one KRAB domain, and 14 zinc fingers. Interestingly, in some pairs there is substantial conservation of the amino acids in the region of the C_2H_2 zinc fingers that determine the recognition specificity, suggesting that the contact sites in target genes may be conserved.

SCAN Family Members Are Vertebrate Specific

Initial reports of the human genome suggested that SCAN domain-containing C_2H_2 zinc finger proteins (SCAN-ZFP) are unique to vertebrates.[19,20] As expected, the genomes of invertebrate species, such as the fly (*D. melanogaster*), worm (*C. elegans*),

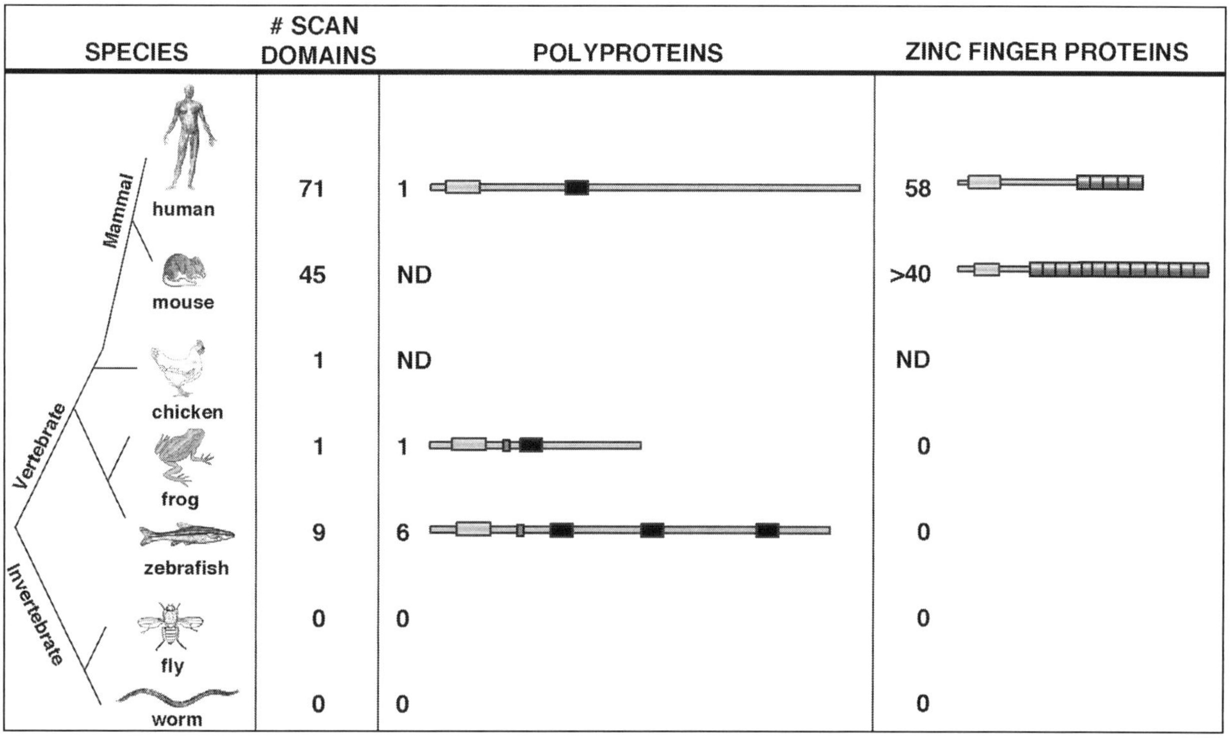

Figure 8. SCAN C_2H_2 proteins are vertebrate specific. The apparent number of SCAN domains found in databases of the worm, fly, zebrafish, frog, chicken, mouse, and human are indicated in the first column. The apparent number of SCAN-containing retrovirus-like polyproteins, compared to the apparent number of SCAN-containing C_2H_2 zinc finger proteins (ZFP), in the represented species is indicated in the second and third columns, respectively (ND, not determined). Searches were limited by the availability of less complete sequence databases for the chicken and frog. The predicted structure of a representative SCAN-containing protein is illustrated to the right of each number. The following conserved regions are indicated: SCAN domain (gray), CCHC zinc knuckle (smallest box), retrovirus-like sequences (black), and C_2H_2 zinc finger (shaded). The representative SCAN proteins are: human polyprotein KIAA1925 (Accession XP_166375), human zinc finger protein MGC4161 (Accession NP_077279), mouse zinc finger protein Zfp29 (Accession NP_033579), frog polyprotein (*Xenopus laevis*, Accession S33978), zebrafish polyprotein (Sanger Centre Accession dZ37K09). A color version of this figure can be viewed at http://www.eurekah.com/chapter.php?chapid=1765&bookid=124&catid=30.

and yeast (*S. cerevisiae*), do not contain SCAN domains. Updated databases were searched for predicted SCAN domains outside of the human and mouse genomes. The search was limited by the content available in large-scale sequence databases that are less complete than their human and mouse counterparts. Nevertheless, apparent SCAN domains were found in the genomes of a number of vertebrate organisms including monkey, cow, pig, mouse, rat, frog and several types of fish. Representative examples of SCAN domains from each species are provided in Sander et al 2003.

Possible Recent Retroviral Origin of SCAN Domains

Remarkably, the SCAN domains found in lower vertebrates are not associated with C_2H_2 zinc finger genes, but are contained in large retrovirus-like polyproteins that are reminiscent of those found in the retrovirus-like polyproteins.[8]

The absence of the SCAN domain in invertebrates, and the nonexistence of SCAN-ZFP in lower vertebrates, suggests that these genes originated and rapidly expanded during recent evolution. The rapid and lineage-specific expansion of the SCAN family may contribute to the diversity that is seen in higher vertebrates (Fig. 8). As a result of the ability of the SCAN domain to mediate dimerization, a diverse network of transcription factor dimers could be generated that may play a key role in the uniqueness of higher vertebrates.

A number of interesting questions remain about the SCAN domain, including a detailed understanding of its structure, mechanism of partner choice and the nature of interacting proteins. Additionally, identifying the target genes regulated by this family will be useful in determining the function of this group of zinc finger transcription factors. Like the POZ/BTB and KRAB domains, the presence of a SCAN domain may provide some mechanistic insights into the large number of C_2H_2 type zinc finger proteins.

References

1. Wolfe SA, Nekludova L, Pago CO. DNA recognition by Cys2His2 zinc finger proteins. Annu Rev Biophys Biomol Struct 2000; 29:183-212.
2. Schwabe JW, Klug A. Zinc mining for protein domains. Nat Struct Biol 1994; 1:345-349.
3. Bellefroid EJ, Poncelet DA, Lecocq PJ et al. The evolutionarily conserved Krüppel-associated box domain defines a subfamily of eukaryotic multifingered proteins. Proc Natl Acad Sci USA 1991; 88:3608-3612.
4. Bardwell VJ, Treisman R. The POZ domain: A conserved protein-protein interaction motif. Genes Dev 1994; 8:1664-1677.
5. Albagli O, Dhordain P, Deweindt C et al. The BTB/POZ domain: A new protein-protein interaction motif common to DNA- and actin-binding proteins. Cell Growth Differ 1995; 6:1193-1198.
6. Williams AJ, Khachigian LM, Shows T et al. Isolation and characterization of a novel zinc-finger protein with transcriptional repressor activity. J Biol Chem 1995; 270:22143-22152.

7. Stone JR, Maki JL, Blacklow SC et al. The SCAN domain of ZNF174 is a dimer. J Biol Chem 2002; 277:5448-5452.
8. Sander TL, Stringer KL, Maki JL et al. The SCAN domain defines a large family of zinc finger transcription factors. Gene 2003; 310:29-38.
9. Pengue G, Calabrò V, Bartoli PC et al. Repression of transcriptional activity at a distance by the evolutionary conserved KRAB domain present in a subfamily of zinc finger proteins. Nucleic Acids Res 1994; 22:2908-2914.
10. Bertin J, Nir WJ, Fischer CM et al. Human CARD4 protein is a novel CED-4/Apaf-1 cell death family member that activates NF-kappaB. J Biol Chem 1999; 274:12955-12958.
11. Schuster JM, Nelson PS. 2000. Toll receptors: An expanding role in our understanding of human disease. J Leukoc Biol 2000; 67:767-73.
12. Williams AJ, Blacklow SC, Collins T. The zinc finger-associated SCAN box is a conserved oligomerization domain. Mol Cell Biol 1999; 19:8526-8535.
13. Sander TL, Haas AL, Peterson MJ et al. Identification of a novel SCAN box-related protein that interacts with MZF1B: The leucine-rich SCAN box mediates hetero- and homoprotein associations. J Biol Chem 2000; 275:12857-12867.
14. Schumacher C, Wang H, Honer C et al. The SCAN-domain mediates selective oligomerization. J Biol Chem 2000; 275:17173-17179.
15. Porsch-Özcürümez M, Langmann T, Heimer S et al. The zinc finger protein 202 (ZNF202) is a transcriptional repressor of ATP binding cassette transporter A1 (ABCA1) and ABCG1 gene expression and a modulator of cellular lipid efflux. J Biol Chem 2001; 276:12427-12433.
16. Lamb P, McKnight SL. Diversity and specificity in transcriptional regulation: The benefits of heterotypic dimerization. Trends Biochem Sci 1991; 11:417-422.
17. Vinson C, Myakishev M, Acharya A et al. Classification of human B-ZIP proteins based on dimerization properties. Mol Cell Biol 2002; 22:6321-6335.
18. Zhang MQ. Computational predication of eukaryotic protein-coding genes. Nature Rev Genet 2002; 3:698-702.
19. Venter JC, Adams MD, Myers EW et al. The sequence of the human genome. Science 2001; 291:1305-1351.
20. Lander ES, Linton LM, Birren B et al. Initial sequencing and analysis of the human genome. Nature 2001; 409:860-921.
21. Nagaoka M, Nomura W, Shiraishi Y et al. Significant effect of linker sequence on DNA recognition by multi-zinc finger protein. Biochem Biophys Res Comm 2001; 282:1001-1007.
22. Desjarlais JR, Berg JM. Toward rules relating zinc finger protein sequences and DNA binding site preferences. Proc Natl Acad Sci USA 1992; 89:7345-7349.
23. Suzuki M, Gerstein M, Yagi N. Stereochemical basis of DNA recognition by Zn fingers. Nucleic Acids Res 1994; 22:3397-3405.
24. Iuchi S. Three classes of C2H2 zinc finger proteins. Cell Mol Life Sci 2001; 58:625-635.
25. Makay JP, Crossley M. Zinc fingers are sticking together. Trends Biochem Sci 1998; 23:1-4.
26. Margolin JF, Friedman JR, Meyer WK et al. Krüppel-associated boxes are potent transcriptional repression domains. Proc Natl Acad Sci USA 1994; 91:4509-4513.
27. Witzgall R, O'Leary E, Leaf A et al. The Kruppel-associated box-A (KRAB-A) domain of zinc finger proteins mediates transcriptional repression. Proc Natl Acad Sci USA 1994; 91:4514-4518.
28. Friedman JR, Fredericks WJ, Jensen DE et al. KAP-1, a novel corepressor for the highly conserved KRAB repression domain. Genes Dev 1996; 10:2067-2078.
29. Kim SS, Chen YM, O'Leary E et al. A novel member of the RING finger family, KRIP-1, associates with the KRAB-A transcriptional repressor domain of zinc finger proteins. Proc Natl Acad Sci USA 1996; 93:15299-15304.
30. Moosmann P, Georgiev O, Le Douarin B et al. Transcriptional repression by RING finger protein TIF1β that interacts with the KRAB repressor domain of KOX1. Nucleic Acids Res 1996; 24:4859-4867.
31. Ryan RF, Schultz DC, Ayyanathan K et al. KAP-1 corepressor protein interacts and colocalizes with heterochromatic and euchromatic HP1 proteins: A potential role for Krüppel-associated box-zinc finger proteins in heterochromatin-mediated gene silencing. Mol Cell Biol 1999; 19:4366-4378.
32. Lechner MS, Begg GE, Speicher DW et al. Molecular determinants for targeting heterochromatin protein 1-mediated gene silencing: Direct chromoshadow domain-KAP-1 corepressor interaction is essential. Mol Cell Biol 2000; 20:6449-6465.
33. Schultz DC, Friedman JR, Rauscher III FJ. Targeting histone deacetylase complexes via KRAB-zinc finger proteins: The PHD and bromodomains of KAP-1 form a cooperative unit that recruits a novel isoform of the Mi-2a subunit of NuRD. Genes Dev 2001; 15:428-443.
34. Schultz DC, Ayyanathan K, Negorev D et al. SETDB1: A novel KAP-1-associated histone H3, lysine 9-specific methyltransferase that contributes to HP1-mediated silencing of euchromatic genes by KRAB zinc-finger proteins. Genes Dev 2002; 16:919-932.
35. Underhill C, Qutob MS, Yee SP et al. A novel nuclear receptor corepressor complex, N-CoR, contains components of the mammalian SWI/SNF complex and the corepressor KAP-1. J Biol Chem 2000; 275:40463-40470.
36. Mark C, Åbrink M, Hellman L. Comparative analysis of KRAB zinc finger proteins in rodents and man: Evidence for several evolutionarily distinct subfamilies of KRAB zinc finger genes. DNA Cell Biol 1999; 18:381-396.
37. Murai K, Murakami H, Nagata S. A novel form of the myeloid-specific zinc finger protein (MZF-2). Genes to Cells 1997; 2:581-591.
38. Gill G. Post-translational modification by the small ubiquitin-related modifier SUJO has big effects on transcription factor activity. Curr Opin Genetics Dev 2003; 12:108-113.
39. Rodriguez MS, Dargemont C, Hay RT. SUMO-1 conjugation in vivo requires both a consensus modification motif and nuclear targeting. J Biol Chem 2001; 276:12654-12659.
40. Ross S, Best JL, Zon LI et al. SUMO-1 modification represses Sp3 transcriptional activation and modulates its subnuclear localization. Mol Cell 2002; 10:831-842.
41. Collins T, Stone JR, Williams AJ. All in the family: The BTB/POZ, KRAB, and SCAN domains. Mol Cell Biol 2001; 21:3609-3615.
42. Hromas R, Davis B, Rauscher 3rd FJ et al. Hematopoietic transcriptional regulation by the myeloid zinc finger gene, MZF-1. Curr Top Microbiol Immunol 1996; 211:159-64.
43. Gaboli M, Kotsi PA, Gurrieri C et al. Mzf1 controls cell proliferation and tumorigenesis. Genes Dev 2001; 15:1625-1630.
44. Peterson MJ, Morris JL. Human myeloid zinc finger gene MZF produces multiple transcripts and encodes a SCAN box protein. Gene 2000; 254:105-18.
45. Murai K, Murakami H, Nagata S. Myeloid-specific transcriptional activation by murine myeloid zinc-finger protein 2. Proc Natl Acad Sci USA 1998; 95:3461-3466.
46. Ogawa H, Murayama A, Nagata S et al. Regulation of myeloid zinc finger protein 2A (MZF-2A) transactivation activity through phosphorylation by MAP kinases. J Biol Chem 2003a; 278:2921-2927.
47. Ogawa H, Ueda T, Aoyama T et al. A SW12/SNF2-type ATPase/helicase protein, mDomino, interacts with myeloid zinc finger protein 2A (MZF-2A) to regulate its transcriptional activity. Genes to Cells 2003b; 8:235-329.
48. Morris JF, Hromas R, Rauscher FJ. Characterization of the DNA-binding properties of the myeloid zinc finger protein MZF1: Two independent DNA-binding domains recognize two DNA consensus sequences with a common G-rich core. Mol Cell Biol 1994; 14:1786-1795.
49. Morris JF, Rauscher FJ, Davis B et al. The myeloid zinc finger gene, MZF-1, regulates the CD34 promoter in vitro. Blood 1995; 86:3640-3647.
50. Fujimoto K, Kyo S, Takakura M et al. Identification and characterization of negative regulatory elements of the human telomerase catalytic subunit (hTERT) gene promoter: Possible role of MZF-2 in transcriptional repression of hTERT. Nucleic Acids Res 2000; 28:2557-2562.

51. Takahashi K, Nishiyama C, Hasegawa M et al. Regulation of the high affinity IgE receptor β-chain gene expression via an intronic element. J Immunol 2003; 171:2478-2484.
52. Kort EN, Ballinger DG, Ding W et al. Evidence of linkage of familial hypoalphalipoproteinemia to a novel locus on chromosome 11q23. Am J Hum Genet 2000; 66:1845-1856.
53. Wagner S, Hess MA, Ormonde-Hanson P et al. A broad role for the zinc finger protein ZNF202 in human lipid metabolism. J Biol Chem 2000; 275:15685-15690.
54. Langmann T, Schumacher C, Morham SG et al. ZNF202 is inversely regulated with its target genes ABCA1 and apoE during macrophage differentiation and foam cell formation. J Lipid Res 2003; 44:968-977.
55. Castillo G, Brun RP, Rosenfield JK et al. An adipogenic cofactor bound by the differentiation domain of PPARgamma. EMBO J 1999; 18:3676-3687.
56. Babb R, Bowen BR. SDP1 is a peroxisome-proliferator-activated receptor gamma 2 coactivator that binds through its SCAN domain. Biochem J 2003; 270:719-727.
57. Suzawa M, Takada I, Yanagisawa J et al. Cytokines suppress adipogenesis and PPAR-gamma function through the TAK1/TAB1/NIK cascade. Nat Cell Biol 2003; 5:224-230.
58. Sander TL, Morris JF. Characterization of the SCAN box encoding RAZ1 gene: Analysis of cDNA transcripts, expression, and cellular localization. Gene 2002; 293:53-64.
59. Safran M, Kaelin WG. HIF hydroxylation and the mammalian oxygen-sensing pathway. J Clin Invest 2003; 111:779-783.
60. Li Z, Wang D, Na X et al. The VHL protein recruits a novel KRAB-A domain protein to repress HIF-1alpha transcriptional activity. EMBO J 2003; 22:1857-1867.
61. Gonsky R, Knauf JA, Elisei R et al. Identification of rapid turnover transcripts overexpressed in thyroid tumors and thyroid cancer cell lines: Use of a targeted differential RNA display method to select for mRNA subsets. Nucleic Acids Res 1997; 25:3823-3831.
62. Wang MM, Reed RR. Molecular cloning of the olfactory neuronal transcription factor Olf-1by genetic selection in yeast. Nature 1993; 364:121-126.
63. Adachi H, Tsujimoto M. Characterization of the human gene encoding the scavenger receptor expressed by endothelial cell and its regulation by a novel transcription factor, endothelial zinc finger protein-2. J Biol Chem 2002; 277:24014-240212.
64. Tanaka K, Tsumaki N, Kozak C et al. A kruppel-associated box-zinc finger protein, NT2, represses cell-type-specific promoter activity of the α2(XI) collagen gene. Mol Cell Biol 2002; 22:4256-4267.
65. Albanese V, Biguet NF, Kiefer H et al. Quantitative effects of gene silencing by allelic variation at a tetranucleotide microsatellite. Hum Mol Genet 2001; 10:1785-1792.
66. Weinmann AS, Farnham PJ. Identification of unknown target genes of human transcription factors using chromatin immunoprecipitation. Methods 2002; 26:37-47.
67. Imreh S, Klein G, Zabarovsky ER. Search for unknown tumor-antagonizing genes. Genes Chromosomes Cancer 2003; 38:307-321.
68. Hoffman SM, Hromas R, Amemiya C et al. The location of MZF-1 at the telomere of human chromosome 19q makes it vulnerable to degeneration in aging cells. Leuk Res 1996; 20:281-283.
69. Dehal P, Predki P, Olsen AS et al. Human chromosome 19 and related regions in mouse: Conservation and lineage-specific evolution. Science 2001; 293:104-111.

CHAPTER 23

Sp1 and Huntington's Disease

Dimitri Krainc

Abstract

Sp1, a triple C_2H_2 zinc finger protein, has been studied for more than two decades and biochemical features of the protein have been well documented. However, its biological roles in humans are still ambiguous. Recent studies on Huntington's disease have shown that the mutant huntingtin proteins with expanded polyglutamine tracts interact with transcription factors, such as TFIID and Sp1, and have provided new insights into mechanisms of gene transcription as well as Huntington's disease. Here, I describe aspects of Sp1 function in the context of Huntington's disease research.

Huntington's Disease

Huntington's disease (HD) is an autosomal, dominantly inherited disorder characterized phenotypically by chorea, involuntary movements, dystonia, intellectual impairment and emotional disturbances.[1] The gene, huntingtin, which is mutated in HD patients, was mapped to chromosome 4p16.3 in 1983 and cloned a decade later.[2] The mutation is an expanded polyglutamine repeat, $(CAG)_n$, within exon 1 of the gene. The number of diseases identified to be caused by $(CAG)_n$ (in general n=35 repeats) continues to grow and a common mechanism could underlie these diseases. Most of the diseases caused by expanded CAG repeats, $(CAG)_n$, share common features, which include neurodegeneration, a dominant pattern of inheritance and genetic anticipation.[3-6] The CAG trinucleotide repeats in all these genes are found in the coding region and are translated into a string of glutamines. Despite the widespread tissue distribution of the protein for each of these genes, the affected region is primarily the brain, and the regions of neuronal loss are somewhat selective and specific for each given disease.[7]

In HD, the number of CAG repeats ranges from 6-35 whereas in individuals affected by HD the repeat length ranges from 40-121. The age of onset of HD is inversely correlated with CAG-repeat length.[1] Although the gene encoding huntingtin is expressed ubiquitously,[8] selective cell loss and fibrillary astrocytosis is observed in the brain, particularly in the caudate and putamen of the striatum, and to a lesser extent in hippocampus and subthalamus. In addition, cerebellar degeneration was also observed in post-mortem examination of juvenile-onset cases.[9-13]

Mutant Huntingtin As "a Transcription Factor"

The functional significance of the *expanded* polyglutamine tract is not well understood. Many other proteins containing polyglutamine-rich regions are transcription factors. Therefore, one possibility is that this tract plays a role in gene transcription. Through the use of two-hybrid screens, a number of proteins have been identified that interact with proteins containing polyglutamines; some of these interactions are sensitive to polyglutamine-tract length. A recent survey of all *S. cerevisiae* proteins containing polyglutamine tracts showed that a significant majority of these tracts are in proteins involved in transcription. Thus, when the mutant form of the HD gene expresses an abnormally long polyglutamine moiety, it might resemble a transcription factor.[14-16]

Nuclear Localization of Mutant Huntingtin

Normal huntingtin is localized in the cytoplasm,[17,18] but mutant huntingtin in addition to being found in the cytoplasm is also found localized in the nucleus. The processes governing nuclear translocation are unclear, although higher numbers of polyglutamine residues produce more nuclear localization in a variety of experimental systems.[19-25] Although polyglutamine repeat length seems to be one feature governing nuclear entry of huntingtin, another is the length of the whole molecule. Shorter fragments of huntingtin are translocated to the nucleus more efficiently, with longer fragments tending to form aggregates in the cytoplasm.[19-24] In transgenic mouse models, lines that synthesize truncated forms of huntingtin[26,27] demonstrate more nuclear inclusion formation and more abnormal phenotypic behavior than transgenic lines that synthesize full-length huntingtin.[25,28,29] Similar results have been obtained in cellular models and in *Drosophila* models.[29,30] It is not clear if huntingtin possesses active nuclear localization signals (NLS). However, in cellular experimental models, addition of nuclear export signals greatly reduces nuclear accumulation of huntingtin. Nuclear localization of huntingtin corresponded to increased cellular toxicity, although the mechanism by which this causes cellular damage is not yet clear.[31-33] It has been demonstrated that the proteolytic cleavage of a glutathione-S-transferase-huntingtin fusion protein leads to the formation of insoluble high molecular weight protein aggregates only when the polyglutamine expansion is in the pathogenic range.[34] Furthermore, electron microscopic studies of polyglutamine aggregates revealed a fibrillary structure that is reminiscent of scrapie prions and -amyloid fibrils in Alzheimer's disease. So far, similar aggregates have been found in human brains affected with CAG-repeat disorders, in animal models for CAG-repeat-associated diseases and in cell-culture models for these disorders. It is becoming evident that the formation of intranuclear

Zinc Finger Proteins: From Atomic Contact to Cellular Function, edited by Shiro Iuchi and Natalie Kuldell.

and perinuclear aggregates is a characteristic feature of CAG-repeat disorders, and might act as a marker of the disease process. However, the direct link to the disease process is still elusive. On the basis of HD mouse-model studies,[31] SCA1 mouse-model studies[35] and the above in vitro culture studies,[36,37] evidence is accumulating to suggest that the formation of aggregates is not required for the initiation of cell death. On the contrary, it has been proposed that aggregate formation might actually have a protective role.[37] However, it remains to be determined what the true role of inclusions is in all CAG-repeat-associated disorders, particularly HD.

Interactions of Huntingtin with Transcription Factors

The combination of aberrant nuclear localization and altered protein-protein interactions of mutant huntingtin suggests that mutant huntingtin might interact with nuclear proteins. Mutant huntingtin can recruit transcription factors into aggregates. For example, recruitment of TATA-box binding protein (TBP) into aggregates has been shown in vitro as well as in human HD post-mortem brains.[38] Mutant huntingtin has also been shown to recruit CREB-binding protein (CBP) into aggregates in vitro[39] and CBP is recruited into aggregates with the androgen receptor in cell culture, in a mouse model of spinobulbar muscular atrophy (SBMA), and in SBMA samples from humans.[40] Similarly, atrophin, the abnormal protein in the polyglutamine disease, dentatorubropallidoluysian atrophy (DRPLA), can recruit Eto. Eto is a transcriptionally active molecule, which is part of the histone deacetylase complex involved in transcriptional repression.[41] In addition, an interaction between the N-terminal pathogenic region of huntingtin and p53 has been shown in vitro by coimmunoprecipitation and observation of coaggregation within inclusions generated in cell culture.[42] Interactions of mutant huntingtin with transcriptional repressors have also been shown. A yeast two-hybrid system was used to identify an interaction of huntingtin with the transcription factor, nuclear corepressor (N-CoR)[43] More recently, huntingtin has been shown to inhibit acetylation of histones via its association with CBP,[44] and interfere with CBP- and BDNF-mediated transcription.[45-47]

These interactions support the possibility that huntingtin and other CAG-repeat disease proteins might affect cellular functioning via alteration of the normal expression pattern of genes, through abnormal association, intracellular redistribution and/or sequestration of crucial transcription factors. If huntingtin is involved in regulating gene transcription it is important to determine which may be target genes affected by normal and/or mutant huntingtin. Some obvious choices are genes altered in patients or in animal models of HD.

Changes in Gene Expression in HD

Neurotransmitter alterations have been described in early-stage human HD autopsy material; many of these changes have been confirmed in transgenic mouse models of HD.[48,49] In transgenic HD mice, neurotransmitter receptors are affected selectively; downregulation of specific receptors argues against a generalized problem with receptor production. This receptor downregulation occurs only in transgenic lines with abnormal CAG-repeat numbers[49] Decreases in receptor binding are corroborated by decreases in receptor mRNA levels, indicating that affected receptors decrease in number, rather than changing their binding characteristics. Because receptor downregulation occurs before the onset of symptoms or observable cell loss, receptor decreases are not merely epiphenomena of disease progression; supporting the notion that neuronal dysfunction is just as relevant for the etiology of symptoms as neuronal death. Especially prominent are early decreases in the levels of dopamine D_1 and D_2 receptors, both of which regulate gene expression, suggesting that certain alterations in mRNA levels observed in HD are secondary to the loss of neurotransmitter receptors.[50,51] Although decreased mRNA levels suggest that receptor alteration occurs at the level of gene transcription, these observations cannot exclude the possibility that these changes occur at the level of mRNA degradation. However, the observation that many mRNA transcripts are unchanged argues against a general defect with mRNA processing. In HD mice crossed with caspase-1 dominant negative mice, delayed symptom onset and prolonged survival correspond not only to delayed nuclear inclusion formation, but also to delayed loss of neurotransmitter receptor binding and a delayed decrease of the corresponding receptor mRNAs.[52]

Gene expression arrays on DNA microchips have recently been used to examine the scope of transcriptional changes in transgenic HD mice.[53] Using the murine Mu6500 chip (Affymetrix, Santa Clara, CA, USA) mRNA transcripts from striata of 6- and 12-week-old R6/2 mice were compared with striatal transcripts from age-matched controls. The results showed that the genes whose expression levels were altered were limited to several key molecular systems. The affected systems included neurotransmitter receptors, intracellular signaling mechanisms, retinoic acid receptor machinery and calcium homeostasis systems. Overall, only 1.2% of the genes examined had significantly altered levels of expression. The pattern of affected genes shows that a single mutant protein affects the expression of many different groups of genes.

Interestingly, when we examined the known regulatory sequences of the genes identified by the DNA microchips it became apparent that all of them contained binding sites for transcription factor Sp1 in their TATA-less promoters. Huntingtin physically interacts with Sp1 and its cognate coactivator TAF130 and inhibits Sp1-mediated transcription of the D2 dopamine receptor gene (Fig. 1).[54,55]

Transcription Factors Associated with Huntingtin and Sp1

A major target of Sp1 is the general transcription factor TFIID that is recruited to DNA by Sp1[56-63] Once bound to DNA, TFIID mediates initiation and elongation of transcription. TFIID is a multisubunit complex made up of the TATA box-binding protein and multiple[64] TATA box-binding protein-associated factors (TAFs). Previous in vitro studies have pointed to the role of TAFs as coactivators that interact with and mediate activation by selected transcriptional regulators[65-68] In addition, TAF_{II}s interact with core promoter elements and play a role in promoter recognition. The recent discovery of a subset of TAFs shared by TFIID and chromatin remodeling complexes further points to multiple functions of TAFs in regulating eukaryotic gene expression[69-71] To determine contribution that TAFs make to transcriptional activation, several studies have examined the interplay between transcriptional regulators and TAFs. A number of TAFs from *Drosophila*, yeast, and humans have been isolated and shown to interact with distinct activators in vitro[72] In vivo studies in yeast have indicated that TAFs are not required for the transcription of all genes[73] but that certain TAF's are required for the transcription of distinct groups of genes[76-79] Genome-wide expression profiling confirmed these findings; 16% of the genes in the yeast genome were affected by the inactivation of yTAFII145 (yeast homologue of hTAF(II)250)[80] Thus, TAFs serve important functions for a subset of genes.

TAF(II)130 and TAF(II)250

The involvement of one of the human TAFs (hTAF130) in activator-TAF interactions has been examined in detail. hTAFII130 has been shown to interact with cellular transcriptional activators such as Sp1, the cAMP-response element binding protein (CREB), CCAAT-binding factor CBF, as well as nuclear receptors, viral transcriptional regulators, and a transcriptional repressor. Sp1 and CREB interact with one or more of the four central glutamine-rich domains Q1-Q4 of hTAFII130 as determined by in vitro binding and yeast two-hybrid assays.[81-85]

Since both CREB/CBP[40,42,45] and Sp1[54,55] have been shown to interact with huntingtin, it was postulated that TAF(II)130 might play an important role in huntingtin-mediated transcriptional regulation[54] It was found that both Sp1 and TAFII130 are required for maximal protection from the effects of mutant huntingtin on gene transcription and cell toxicity.

We examined whether other TAFs, focusing on TAF(II)250 may also be involved in huntingtin-mediated gene transcription. Recent excitement about TAF250 stems from its involvement in regulating the association of TFIID with the core promoter initiator element and its intrinsic enzymatic activities that post-translationally modify general transcription factors and histones.[86] It has recently been shown that huntingtin inhibits histone acetyltransferase (HAT) activity of CBP.[45] Therefore, it was of interest to determine if huntingtin interferes with HAT activity of TAFII250 and directly modulates the activity of TFIID complex.

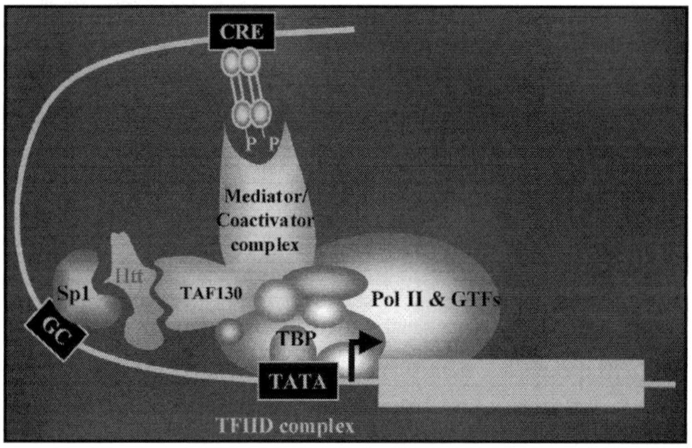

Figure 1. Mutant huntingtin disrupts transcriptional activation by Sp1 and TAFII130 in HD. The transcription factor Sp1 binds to DNA elements called GC boxes in cellular promoters. A specific protein-protein interaction between the glutamine-rich regions of Sp1 and the TAFII130 subunit of TFIID is required for recruitment of the general transcriptional machinery, which includes transcription factors TFIIA, B, D, E, F, and H. This glutamine interface serves to bridge Sp1 to the machinery required to recruit RNA pol II. Once correctly targeted to the target gene, RNA pol II initiates transcription of an mRNA copy of the target gene. In HD, the glutamine expansion in huntingtin (Htt) disrupts transcriptional activation by Sp1 and TAFII130. An amino-terminal fragment of mutant huntingtin, containing an expanded polyglutamine tract, accumulates in the nucleus. Here, this mutant protein associates with Sp1 and TAFII130, preventing Sp1 from binding to the GC box, and ultimately disrupting the ability of Sp1 and TAFII130 to interact. Without proper targeting by the general transcriptional machinery, RNA pol II cannot properly locate the target promoter region and the gene cannot be transcribed. The deregulated gene expression was noted in HD brain samples isolated from presymptomatic HD patients suggesting that it may represent an early step in the neurodegenerative process.

The essential nature of TAFII250 can be attributed to its broad requirement during RNA-pol-II-dependent transcription. The signal instructing TFIID where to bind the genome comes from activators, such as Sp1, that bind both cis-regulatory regions of target genes and TFIID to recruit TFIID to particular genes. TAFII250 plays a fundamental role in this recruitment process, serving as a binding partner for activators and as a scaffold for other TAF subunits of TFIID. The arsenal of TAFII250 enzymatic functions, such as histone acetylation (HAT) and phosphorylation, may be employed at this point to provide accessibility to nucleosome-embedded core promoters. Requirements for individual TAFII250 activities could be gene specific, since impairment of phosphorylation or HAT activity affects the expression of different genes.[87] In yeast, genome-wide microarray analysis indicates that expression of 14-27% of RNA-pol-II-transcribed genes is downregulated more than twofold upon TAFII250 inactivation.[88] In ts13 hamster cells that harbor a temperaturesensitive allele of TAFII250 (mutated HAT domain of TAFII250), similar analysis indicates that the transcription of 18% of genes transcribed by RNA pol II is affected more than twofold at the nonpermissive temperature.[87] Mizzen et al using in-gel activity assays showed that TAFII250 has HAT activity.[89] The HAT domain maps to the central, most conserved portion of metazoan $TAF_{II}250$· Many studies have established a correlation between acetylation of highly conserved lysine residues in the N-terminal tails of histones and transcriptional activation.[90] One model to explain how histone acetylation affects gene expression proposes that the extent of chromatin condensation is directly related to the level of histone acetylation. Accordingly, hyperacetylation reduces the affinity of histone tails for DNA, resulting in less compact chromatin and increased accessibility of transcription factors to DNA. TAFII250-mediated structural changes within chromatin could allow initiation and elongation of transcription of a subset of target genes.[86,91]

Mutant Huntingtin Inhibits Sp1 Acetylation by TAF250

These detailed in vitro and in vivo analyses have uncovered a plethora of TAFII250 activities that together underscore the importance of TAFII250 as a regulator of RNA pol II transcription. In our most recent work, we showed that mutant huntingtin-mediated repression of gene transcription is rescued by an overexpression of normal TAF250 but not mutant forms of human TAF250 with mutated histone acetyltransferase domain (HAT), suggesting that the effect of TAF250 on huntingtin-mediated repression depends on the HAT activity of TAF250. Furthermore, we found that mutant huntingtin inhibits acetylation of Sp1 by TAF250, which in turn leads to decreased binding of Sp1 to DNA and decreased expression of Sp1-driven genes. These findings are particulary intriguing in light of recent reports that Sp1 can be acetylated in vivo and that HDAC inhibitors can augment Sp1 acetylation and interfere with Sp1/DNA binding.[92,93]

These data suggests that huntingtin interacts with transcriptional coactivators in the context of TFIID complex and that Sp1 transcription plays a central role in the process. Therefore, we hypothesize that mutant huntingtin may alter the composition and DNA contacts of transcriptional complexes in HD. It will of interest to identify individual components in multiprotein complexes in HD and examine the complexes in different brain regions and brain samples from presymptomatic patients as well as transgenic animals. For example, striatum may be compared with hippocampus, cerebellum and cortex in presymptomatic and symptomatic mice or postmortem human brain samples. In addition, other Sp1 gene targets, such as Sp1-genes that were altered in HD (i.e., D2) and genes that were not altered (i.e., NMDAR1) should be analyzed. Such regional and gene-specific approach may identify specific molecular alterations in transcriptional complexes in presymptomatic and symptomatic HD.

In addition to Sp1, neuronal cells also express transcription factors Sp3 and Sp4 that are structurely and functionally related to Sp1.[94] Sp1, Sp3 and Sp4 comprise most of the GC box binding activity in post-mitotic neurons, suggesting that the Sp family of transcription factors is of importance for the normal function and activity of post-mitotic neurons. Sp1, Sp3 and Sp4 recognize and bind GC box with similar affinities dependent on three highly conserved zinc finger motifs in the carboxyl terminus. In addition, each of these Sp family proteins contains two glutamine-rich regions in the amino terminus that contribute to transcriptional activation.

Genetic Approaches for Understanding of Function of Sp1 and the Sp Protein Family

Research on Sp1 and other Sp proteins has made rapid advances in the last few years, with more and more family members being identified. Most of the knowledge is derived from biochemical analyses, in vitro protein-DNA interaction studies and transfection experiments. The knockouts published to date show that individual family members do have specific functions in the mouse. However, even knockout mice may not reflect the whole truth because so little is known about the extent of overlapping functions. For instance, inactivation of Sp1 was expected to have dramatic effects at the cellular level since it was generally believed that Sp1 is the essential transcription factor for expression of housekeeping genes. Possibly, Sp1 is indeed engaged in the expression of these genes in vivo, but functionally replaced by Sp3 under knockout conditions. In addition, since deletion of the Sp1 gene leads to lethality early in development, the role of Sp1 in fully developed tissues remains an open question. In this regard, it is interesting to note that mutations of Sp1-binding sites in the LDL receptor and collagen type alpha genes have been associated with hypercholesterolaemia and osteoporosis in humans.

Although Sp1, Sp3 and Sp4 are highly related proteins, they may have distinct biological roles. Sp3, unlike Sp1 and Sp4 contains an inhibitory domain that suppresses transcriptional activation and may contribute to active repression. Whether Sp3 functions as activator or a repressor depends on promoter architecture and cell type. Although both Sp1 and Sp3 are ubiquitously expressed, Sp4 mRNA and protein are highly expressed in the central nervous system and the brain. Gene disruption of Sp1, Sp3 or Sp4 in mice produces different phenotypes. Interestingly, knockout of Sp4 genes in mice produces a complex phenotype that includes behavioral defects, supporting a role for Sp4 in regulating transcription of genes in the nervous system. In addition, the relative contribution of Sp family of transcription factors to GC box binding activity was observed to change during neuronal differentiation, with a increase in Sp3 and Sp4 binding and a decrease in Sp1 binding, suggesting the ratio change of the Sp family of transcription factors may also play an important role in regulating transcription in post mitotic neurons. Since the zinc finger motif in Sp1 that has been shown to interact with the huntingtin protein are conserved in Sp3 and Sp4, the zinger finger motifs in Sp3 and Sp4 may also interact with huntingtin, leading to interruption of Sp3 and Sp4 transcriptional function as well. Because of the distinct role for each Sp family of transcription factors in neurons, the contribution of these interaction between huntingtin and different factors to the pathogenesis may vary.

It is worth noting that in those cases that have key residues differing between members of the family, the DNA-binding specificity is frequently altered. For example, Sp2, which has a leucine residue within the first zinc-finger motif in place of the histidine found in the corresponding region of Sp1, preferentially recognizes the GT box (5'-GGTGTGGGG-3'), found in many different promoters, rather than the GC box. Previous studies also suggested that the linkers between the zinc-finger motifs contribute to high-affinity binding of certain zinc-finger proteins. The linker regions of members of the Sp1-like family have several potentially relevant differences, but it is currently not known if all of these contribute to differences in DNA-binding activity.

By regulating the expression of a large number of genes that have GC-rich promoters, Sp1 and other zinc-finger transcription regulators may take part in virtually all facets of cellular function, including cell proliferation, apoptosis, differentiation, and neoplastic transformation. Individual members of the Sp1-like family can function as activators or repressors depending on which promoter they bind and the coregulators with which they interact. A long-standing research aim has been to define the mechanisms by which Sp1-like factors regulate gene expression and cellular function in a cell- and promoter-specific manner. Gene arrays studies in HD have shown that transcription of many Sp1 target genes is altered by mutant huntingtin. Therefore, it will be of interest to determine which are the specific cellular functions that are altered by the dysfunction of Sp1-like factors in the context of HD. Moreover, it will be of interest to examine whether Sp1-like zinc finger proteins play a more general role in the pathogenesis of other polyglutamine disorders. For example, androgen receptor and Sp1 have been shown to interact with one another and cooperatively regulate gene transcription suggesting that Sp1 transcription may be deregulated in spinobulbar muscular atrophy (SBMA).

Conditional disruption of the Sp1 gene in specific tissues at any given stage of development will be an important step to further unravel the physiological role of Sp1 in vivo. The identification of target genes presents another formidable challenge. Recent advances in the detection of differentially expressed genes and the development of DNA microchip arrays provide very useful tools for high throughput analysis of target genes. This should be of great help in understanding the downstream effects that result in the knockout phenotypes.

To fully understand the physiological function of the Sp proteins, it will be equally important to understand the mechanisms of their mode of action. Detailed characterization of the individual transcription factors, identification of specific interaction partners, careful analysis of protein modifications and the identification of signals and transduction pathways by which these proteins are regulated will be essential for a mechanistic understanding of transcriptional control by this growing family of transcription factors. Another important area of research pertains to the transcriptional regulatory function of the amino-terminal domains of these proteins. What are the molecular mechanisms that regulate how these domains interact with coregulatory complexes and thereby repress or activate gene expression? Further insights into the functions of different Sp1-like proteins have the potential to change the partially informative classification of these proteins, which is currently based on primary structure. Lastly, because these proteins are important in morphogenesis, it is likely that they may play a significant role in the mechanisms underlying human diseases that are characterized by aberrant growth and differentiation. Future studies of the Sp1-like proteins have a large potential for defining the machinery that not only regulates physiological processes but may also modulate human diseases such as HD.

Acknowledgement

I thank Naoko Tanese, Anne Young, Anthone Dunah, Hyun Jeong, Libin Cui, Edith Wang, Maral Mouradian, Steven Hersch and Robert Tjian for their contributions to the experiments described in this chapter.

References

1. Vonsattel JP, DiFiglia M. Huntington's disease. J Neuropathol Exp Neurol 1998; 57:369-384.
2. Huntington's disease collaborative research group. A novel gene containing a trinucleotide repeat that is unstable in Huntington's disease chromosomes. Cell 1993; 72:971-983.
3. Albin RL, Tagle DA. Genetics and molecular biology of Huntington's disease. Trends Neurosci 1995; 18:11-14.
4. Warren ST. The expanding world of trinucleotide repeats. Science 1996; 271:1374-1375.
5. Zoghbi HY, Orr HT. Glutamine repeats and neurodegeneration. Annu Rev Neurosci 2000; 23:217-47.
6. Reddy PS, Housman DE. The complex pathology of trinucleotide repeats. Curr Opin Cell Biol 1997; 9:364-372.
7. MacDonald ME et al. Huntington's disease. Neuromolecular Med 2003; 4(1-2):7-20.
8. Strong TV et al. Widespread expression of the human and rat Huntington's disease gene in brain and nonneural tissues. Nat Genet 1995; 5:259-263.
9. Spargo E et al. Neuronal loss in the hippocampus in Huntington's disease: Acomparison with HIV infection. J Neurol Neurosurg Psychiatry 1993; 56:487-491.
10. Byers RK et al. Huntington's disease in children. Neuropathologic study of four cases. Neurology 1973; 23:561-569.
11. Lange HG et al. Morphometric studies of the neuropathological changes in choreatic diseases. J Neurol Sci 1976; 28:401-425.
12. Landwehrmeyer GB et al. Huntington's disease gene: Regional and cellular expression in brain of normal and affected individuals. Ann Neurol 1995; 37:218-230.
13. Ferrante RJ et al. Heterogeneous topographic and cellular distribution of huntingtin expression in the normal human neostriatum. J Neurosci 1997; 17:3052-3063.
14. Jou YS, Myers RM. Evidence from antibody studies that the CAG repeat in the Huntington disease gene is expressed in the protein. Hum Mol Genet 1995; 4:465-469.
15. Trottier Y et al. Cellular localization of the Huntington's disease protein and discrimination of the normal and mutated form. Nat Genet 1995; 10:104-110.
16. Cha J. Transcriptional dysregulation in Huntington's disease. Trends Neurosci 2000; 23:87-392.
17. DiFiglia M et al. Huntington is a cytoplasmic protein associated with vesicles in human and rat brain neurons. Neuron 1995; 14:1075-1081.
18. Wood JD et al. Partial characterization of the murine huntingtin and apparent variations in the subcellular localization of huntingtin in human, mouse and rat brain. Hum Mol Genet 1996; 5:481-487.
19. Yang W et al. Aggregated polyglutamine peptides delivered to nuclei are toxic to mammalian cells. Hum Mol Genet 2002; 11(23):2905-2917.
20. Chen S et al. Huntington's disease age-of-onset linked to polyglutamine aggregation nucleation. Proc Natl Acad Sci USA 2002; 99(18):11884-11889.
21. Kalchman MA et al. Huntingtin is ubiquitinated and interacts with a specific ubiquitin-conjugating enzyme. J Biol Chem 1996; 271:19385-19394.
22. Burke JR et al. Huntingtin and DRPLA proteins selectively interact with the enzyme GAPDH. Nat Med 1996; 2:347-350.
23. Sittler A et al. SH3GL3 associates with the Huntingtin exon 1 protein and promotes the formation of polygln-containing protein aggregates. Mol Cell 1998; 2:427-436.
24. Faber PW et al. Huntingtin interacts with the WW domain proteins. Hum Mol Genet 1998; 7:1463-1474.
25. Nasir J et al. Targeted disruption of the Huntington's disease gene results in embryonic lethality and behavioral and morphological changes in heterozygotes. Cell 1995; 81:811-823.
26. Zeitlin SZ et al. Increased apoptosis and early embryonic lethality in mice nullizygous for the Huntington's disease gene homologue. Nat Genet 1995; 11:155-163.
27. Goldberg YP et al. Absence of the disease phenotype and intergenerational stability of the CAG repeat in transgenic mice expressing the human Huntington disease transcript. Hum Mol Genet 1996; 5:177-185.
28. Mangiarini L et al. Exon 1 of the HD gene with an expanded CAG repeat is sufficient to cause a progressive neurological phenotype in transgenic mice. Cell 1996; 87:493-506.
29. Burright EN et al. SCA1 transgenic mice: A model for neurodegeneration caused by an expanded CAG trinucleotide repeat. Cell 1996; 82:937-948.
30. Kawaguchi Y et al. CAG expansions in a novel gene for Machado-Joseph disease at chromosome 14q32.1. Nat Genet 1996; 8:221-228.
31. Reddy PH et al. Bahavioural abnormalities and selective neuronal loss in HD transgenic mice expressing mutated full-length HD cDNA. Nat Genet 1998; 20:198-202.
32. Hodgson JG et al. Human huntingtin derived from YAC transgenes compensates for loss of murine huntingtin by rescue of the embryonic lethal phenotype. Hum Mol Genet 1996; 5:1875-1885.
33. Davies SW et al. Formation of neuronal intranuclear inclusions underlies the neurological dysfunction in mice transgenic for the HD mutation. Cell 1997; 90:537-548.
34. Scherzinger E et al. Huntingtin-encoded polyglutamine expansions form amyloid-like protein aggregates in vitro and in vivo. Cell 1998; 90:549-558.
35. Klement S et al. Ataxin-1 nuclear localization and aggregation: Roll in polyglutamine-induced disease in SCA1 transgenic mice. Cell 1998; 95:41-53.
36. Kim M et al. Mutant huntingtin expression in clonal striatal cells: Dissociation of inclusion formation and neuronal survival by caspase inhibition. J Neurosci 1998; 19:964-973.
37. Saudou F et al. Huntingtin acts in the nucleus to induce apoptosis but death does not correlate with the formation of intranuclear inclusions. Cell 1998; 95:55-66.
38. Huang CC et al. Amyloid formation by mutant huntingtin: Threshold, progressivity and recruitment of normal polyglutamine proteins. Somat Cell Mol Genet 1998; 24:217-233.
39. Kazantsev A et al. Insoluble detergent-resistant aggregates form between pathological and nonpathological lengths of polyglutamine in mammalian cells. Proc Natl Acad Sci USA 1999; 96:11404-11409.
40. McCampbell ARV et al. Colocalization of CBP with expanded polyglutamine containing androgen receptor. Am J Hum Genet 1999; 65 (Suppl:A):106.
41. Ross CA et al. Polyglutamine pathogenesis. Philos Trans R Soc London B Biol Sci 1999; 354:1005-1011.
42. Steffan JS et al. The Huntington's disease protein interacts with p53 and CBP and represses transcription. Proc Natl Acad Sci USA 2000; 97:6763-6768.

43. Boutell JM et al. Aberrant interactions of transcriptional repressor proteins with the Huntington's disease gene product, huntingtin. Hum Mol Genet 1999; 8:1647-1655.
44. Jones AL. The localization and interactions of huntingtin. Philos Trans R Soc Lond B Biol Sci 1999; 354:1021-1027.
45. Steffan JS et al. Histone deacetylase inhibitors arrest polyglutamine-dependent neurodegeneration in Drosophila. Nature 2001; 413:739-743.
46. Nucifora FC et al. Interference by huntingtin and atrophin-1 with CBP-mediated transcription leading to cellular toxicity. Science 2001; 291:2423-2428.
47. Zuccato C et al. Loss of Huntingtin-mediated BDNF gene transcription in Huntington's disease. Science 2001; 293:493-498.
48. Augood SJ et al. Dopamine D1 and D2 receptor gene expression in the striatum in Huntington's disease. Ann Neurol 1997; 42:215-221.
49. Cha J-HJ et al. Altered brain neurotransmitter receptors in transgenic mice expressing a portion of an abnormal human Huntington disease gene. Proc Natl Acad Sci USA 1998; 95:6480-6485.
50. Berke JD et al. A complex program of striatal gene expression induced by dopaminergic stimulation. J Neurosci 1998; 18:5301-5310.
51. Gerfen CR et al. D_1 and D_2 dopamine receptor-regulated gene expression of striatonigral and striatopallidal neurons. Science 1990; 250:1429-1432.
52. Ona VO et al. Inhibition of caspase-1 slows disease progression in a mouse model of Huntington's disease. Nature 1999; 399:263-267.
53. Luthi-Carter R et al. Decreased expression of striatal signalling genes in a mouse model of Huntington's disease. Hum Mol Genet in press.
54. Dunah AW, Jeong H, Griffin A et al. Sp1 and TAF130 transciptional activity disrupted in early Huntington's Disease. Science 2002; 296(5576):2238-2243.
55. Li SH, Cheng AL, Zhou H et al. Interaction of Huntington disease protein with transcriptional activator Sp1. Mol Cell Biol 2002; 22(5):1277-1287.
56. Kadonaga JT et al. Isolation of cDNA encoding transcription factor Sp1 and functional analysis of the DNA binding domain. Cell 1987; 51:1079-1090.
57. Dynan WS, Tjian R. Control of eukaryotic messanger RNA synthesis by sequence-specific DNA binding proteins. Nature 1985; 316:774-778.
58. Black AR et al. Sp1 and Kruppel-like factor family of transcription factors in cell growth regulation and cancer. J Cell Physiol 2001; 188:143-160.
59. Jackson SP, Tjian R. O-Glycosylation of eukaryotic transcription factors: Implications for mechanisms of transcriptional regulation. Cell 1988; 55:125-133.
60. Hagen G et al. Cloning by recognition site screening of two novel GT box binding proteins: A family of Sp1 related genes. Nucleic Acids Res 1992; 20:5519-5525.
61. Suske G. The Sp-family of transcription factors. Gene 1999; 238:291-300.
62. Sjottem E et al. The promoter activity of long terminal repeats of the HERV-H family of human retrovirus-like elements is critically dependent on Sp1 family proteins interacting with a GC/GT box located immediately 3' to the TATA box. J Virol 1996; 70:188-198.
63. Dennig J et al. An inhibitor domain in Sp3 regulates its glutamine-rich activation domains. EMBO J 1996; 15:5659-5667.
64. Burley SK et al. Biochemistry and structural biology of transcription factor IID (TFIID). Annu Rev Biochem 1996; 65:769-799.
65. Chen JL et al. Assembly of recombinant TFIID reveals differential coactivator requirements for distinct transcriptional activators. Cell 1994; 79:93-105.
66. Hahn S. The role of TAFs in RNA polymerase II transcription. Cell 1998; 95:579-582.
67. Lee TI, Young, RA. Regulation of gene expression by TBP-associated proteins. Genes Dev 1998; 12:1398-1408.
68. Verrijzer CP, Tjian R. TAFs mediate transcriptional activation and promoter selectivity. Trends Biochem Sci 1996; 21:338-342.
69. Struhl K, Moqtaderi Z. The TAFs in the HAT. Cell 1998; 94:1-4.
70. Bjorklund S et al. Global transcription regulators of eukaryotes. Cell 1999; 96:759-767.
71. Hampsey M. Molecular genetics of the RNA polymerase II general transcriptional machinery. Microbiol Mol Biol Rev 1998; 62:465-503.
72. Sauer F, Tjian R. Mechanisms of transcriptional activation: Differences and similarities between yeast, Drosophila, and man. Curr Opin Genet Dev 1997; 7:176-181.
73. Moqtaderi Z et al. TBP-associated factors are not generally required for transcriptional activation in yeast. Nature 1996; 383:188-191.
76. Walker SS et al. Yeast $TAF_{II}145$ required for transcription of G1/S cyclin genes and regulated by the cellular growth state. Cell 1997; 90:607-614.
77. Wang, EH, Tjian R. Promoter-selective transcriptional defect in cell cycle mutant ts13 rescued by hTAFII250. Science 1994; 263:811-814.
78. Wang, EH et al. TAFII250-dependent transcription of cyclin A is directed by ATF activator proteins. Genes Dev 1997; 11:2658-2669.
79. Suzuki-Yagawa Y et al. The ts13 mutation in the TAF(II)250 subunit (CCG1) of TFIID directly affects transcription of D-type cyclin genes in cells arrested in G1 at the nonpermissive temperature. Mol Cell Biol 1997; 17:3284-3294.
80. Holstege FC et al. Dissecting the regulatory circuitry of a eukaryotic genome. Cell 1998; 95:717-728.
81. Tanese N et al. Molecular cloning and analysis of two subunits of the human TFIID complex: HTAFII130 and hTAFII100. Proc Natl Acad Sci USA 1996; 93:13611-13616.
82. Hoey T et al. Molecular cloning and functional analysis of Drosophila TAF110 reveal properties expected of coactivators. Cell 1993; 72:247-260.
83. Rojo-Niersbach E et al. Genetic dissection of hTAF130 defines a hydrophobic surface required for interaction with glutamine-rich activators. J Biol Chem 1999; 274:33778-33784.
84. Saluja D et al. Distinct subdomains of human TAFII130 are required for interactions with glutamine-rich transcriptional activators. Mol Cell Biol 1998; 18:5734-5743.
85. Ferreri K et al. The cAMP-regulated transcription factor CREB interacts with a component of the TFIID complex. Proc Natl Acad Sci USA 1994; 91:1210-1213.
86. Wassarman DA et al. $TAF_{II}250$: A transcription toolbox. J Cell Sci 2001; 114:2895-2902.
87. O'Brien T et al. Different functional domains of $TAF_{II}250$ modulate expression of distinct subsets of mammalian genes. Proc Natl Acad Sci USA 2000; 97:2456-2461.
88. Lee TI et al. Redundant roles for the TFIID and SAGA complexes in global transcription. Nature 2000; 405:701-704.
89. Mizzen CA et al. Transcription. New insights into an old modification. Science 2000; 289:2290-2291.
90. Strahl BD et al. The language of covalent modification. Nature 2000; 403:41-45.
91. Dunphy EL et al. Requirement for $TAF_{II}250$ acetyltransferase activity in cell cycle progression. Mol Cell Biol 2000; 20:1134-1139.
92. Ryu H et.al. Histone deacetylase inhibitors prevent oxidative neuronal death independent of expanded polyglutamine repeats via an Sp1-dependent pathway. Proc Natl Acad Sci USA 2003; 100(7):4281-6.
93. Ferrante R et al. Histone deacetylase inhibition by sodium butyrate chemotherapy ameliorates the neurodegenerative phenotype in Huntington's disease mice. J Neurosci 2003; 23(28):9418-9427.
94. Kaczynski J et al. Sp1- and Kruppel-like transcription factors. Genome Biol 2003; 4(2):206-210.

CHAPTER 24

The Role of WT1 in Development and Disease

Sean Bong Lee,* Hongjie Li and Ho-Shik Kim

Abstract

Wilms tumor suppressor gene product WT1 is a Cys_2His_2 (C_2H_2) zinc finger protein, one of the largest protein families in higher eukaryotes. Since the cloning of *WT1* in 1990, we have learned a great deal about the functions of WT1. Now, even greater molecular details of WT1 are beginning to emerge with a possibility of the unexpected role of WT1 in the post-transcriptional processes. This chapter does not include all aspects of WT1, but is intended to focus on and highlight the biological function of WT1 (for detailed reviews see refs. 1-4). The chapter consists largely of two parts, the biochemical characterization of WT1 and the genetics of WT1 in both the human and the mouse, and ends with future perspectives on WT1 research.

Biochemical Characterization of WT1

Alternative Splicing and Generation of Multiple WT1 Isoforms

WT1 was cloned more than a decade ago,[5,6] and the sequence revealed a *Drosophila* Krüppel-like zinc finger protein with four C_2H_2 zinc fingers at its carboxyl terminus as the DNA binding domain (Fig. 1). The amino-terminal domain contains a proline-glutamine rich region, which is frequently found in many transcriptional activators. *WT1* transcripts undergo numerous modifications such as two alternative splicing events, RNA editing at codon 280 and alternative usage of either the upstream CUG or the internal in-frame AUG initiations that result in 24 possible isoforms.[3] The exact in vivo functions of the RNA-edited, upstream CUG- and internal-AUG initiated gene products are not clear and represent minor species relative to the 4 major isoforms, which arise as the result of two alternative splicing events (Fig. 1). The first alternative splice event includes or excludes 17 amino acids (aa) encoded by exon 5 while the second alternative splice involves differential usage of two splice donor sites located 9 nucleotides apart at the end of exon 9;[7] this alternative splice II results in the exclusion or insertion of three amino acids, termed KTS (lysine, threonine, and serine), between the zinc fingers three and four. The KTS-containing (+KTS) transcripts are the predominant form of *WT1*, at least two to three times more abundant than the –KTS, throughout development.[7] Maintaining this ratio is critical to the normal function of WT1 as demonstrated by Frasier syndrome,[8] where decreased production of the +KTS leads to severe developmental defects (see below).

Evolutionary Conservation of WT1

The high degree of evolutionary conservation of WT1 is remarkable throughout vertebrate species.[9] The overall sequence similarity between the fish (*Fugu*) and human WT1 is 68% (57% identical) but exclusive comparison of all vertebrate WT1 in the zinc finger domain alone indicates the greatest conservation in the DNA binding domain (more than 90-95% similar across species). Furthermore, the alternative KTS splicing (KPS in *Fugu*) has been conserved in all WT1 orthologs, whereas the alternative splice I (17 aa) is absent in chicken, reptile, and fish.[9] This suggests that the first alternative splicing occurred rather recently with the appearance of placental mammals, while the KTS splicing arose with the earliest forms of WT1 and thus likely to serve a more fundamental role in WT1 function. *WT1* transcripts of lower vertebrates also do not utilize multiple translation initiations and give rise to only two isoforms:[10] WT1-KTS and the +KTS. Thus, the function of non-KTS isoforms may be mammal specific or serve minor or redundant functions of WT1. Indeed, the *wt1* knockout mice, which disrupt only the first alternative splice (exon 5)[11] or the upstream CUG-initiation,[12] do not display any gross abnormalities (see below).

The remarkable conservation of the KTS sequence throughout vertebrates may indicate a special role of the KTS residues (such as a phosphorylation site). However, a recent study suggests that the KTS can be altered in both its sequence and length without affecting its apparent function.[13] Then, why is the KTS so well conserved from fish to man? The most likely answer comes from the studies of Davies et al[13] who showed that the nucleotides that code for the KTS as well as the immediate surrounding sequences are extremely well conserved throughout evolution. Because maintaining the KTS alternative splicing is central to the WT1 function as exemplified by the Frasier syndrome[8] (see below), any changes in the nucleotides within or surrounding the KTS during evolution would have led to defects in KTS splicing and subsequent demise of the organism.[13] Thus, KTS conservation exemplifies a situation where the nucleotides rather than amino acid sequences served as the determining factor behind strict evolutionary conservation.

* *Corresponding author. See list of "Contributors".*

Zinc Finger Proteins: From Atomic Contact to Cellular Function, edited by Shiro Iuchi and Natalie Kuldell.

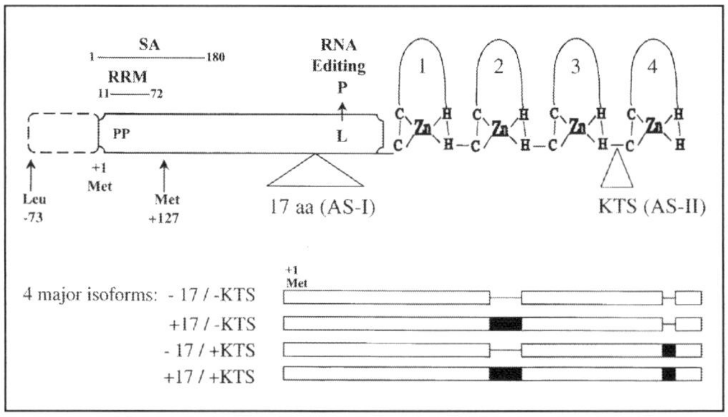

Figure 1. Schematic representation of various WT1 isoforms. N-terminal domain contains the transcriptional regulatory domain rich in proline (PP) residues and the C-terminal domain contains four C_2H_2 zinc fingers that bind to DNA or RNA. Two alternative splicing events (AS-I and AS-II) are marked by 17 aa and KTS respectively. RNA editing at codon 280 highlights leucine (L) to proline (P) change. Two alternative initiation sites (Leu at -73 and Met at +127) are marked by arrows. A self-association (SA) domain (residues 1-180) and a potential RNA recognition motif (RRM) within residues 11-72 are noted. Four major isoforms of WT1 (starting at +1 Met) as the result of two alternative splicing events are indicated at the bottom. Black boxes indicate two alternative spliced exons.

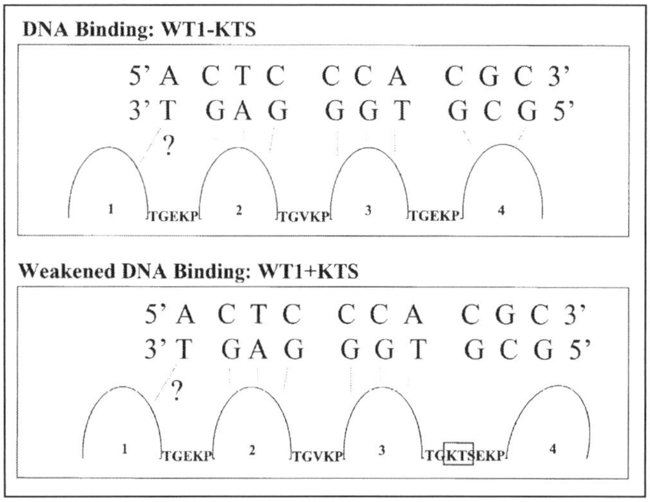

Figure 2. Strong and weak DNA binding by the –KTS and the +KTS isoforms of WT1. Each zinc finger (fingers 2-4) recognizes a set of three nucleotides (marked by dotted lines) with the exception of zinc finger 1, which is thought to recognize only one nucleotide. Conserved TGEKP linker sequences are shown. Insertion of the KTS between the zinc fingers 3 and 4 is marked by a box. The KTS-containing linker remains its flexible nature, not allowing the zinc finger 4 to make specific contacts with the DNA (schematically shown by a slanted zinc finger 4 and the absence of dotted lines). The nucleotide sequence is the optimal 10 base pair WT1 binding element identified by Nakagama et al,[20] and the DNA-protein alignment is shown in an anti-parallel manner as predicted by the EGR1-DNA crystal structure.[22]

Structure and Function of WT1

The amino-terminal domain of WT1, which contains both the repression and the activation domains, has been extensively studied for its dichotomous property in transcriptional regulation (for review see ref. 1). A number of different studies provide conflicting results on the exact location of the activation and the repression domains[1] and the discrepancies may reflect the variability in cellular and promoter context in different studies. A more recent study has mapped the activation domain to residues 245-297, which include the alternatively spliced 17 aa encoded by exon 5.[14] The 17 aa domain was shown to interact with Par4, prostate apoptosis response factor 4, which is a previously identified WT1-interacting protein.[15] The same 245-297 aa domain of WT1 was shown to interact with the basic transcription factor TFIID, though whether this occurs directly or indirectly through other proteins is not clear.[14] In contrast, a previous study has mapped the Par4-binding domain to the zinc finger portion of WT1 and suggested that this interaction mediates transcriptional repression.[15] Despite these discrepancies, mice lacking the 17 aa of WT1 develop normally[11] suggesting that WT1 isoforms lacking the 17 aa are capable of serving all the necessary functions of WT1. Thus, the exact location of the transcriptional regulatory domains within the amino terminus of WT1 awaits further clarification. Other notable features of the amino terminus include the homodimerization domain in the first 180 residues of WT1,[16] and a potential RNA recognition motif (RRM) as predicted by computer modeling (Fig. 1).[17]

The carboxyl-terminal domain contains four zinc fingers of C_2H_2 type that mediate DNA binding as well as protein-protein interaction (see below). The nuclear localization signals also reside in the zinc finger region.[18] Sequence analysis of the WT1 zinc fingers provided clues to its DNA binding sequences as the last three fingers of WT1 matched closely (67% identical) to the three zinc fingers of the early growth response gene 1, EGR1.[19] Indeed, WT1 bound to the GC-rich EGR-1 binding site in vitro albeit with less affinity than EGR1. Many studies have since examined the exact binding site of WT1,[3] and it is now generally regarded that WT1-KTS binds to at least two different DNA elements: (1) GC-rich, EGR-like sequences to which WT1 binds with higher affinity than EGR1[20] and (2) TCC-repeats.[21] It is also assumed based on the crystal structure of EGR1 bound to its cognate DNA[22] that each of the four zinc fingers of WT1 binds to a set of three nucleotides, resulting in 12 nucleotides as the potential WT1 binding element. However, because the first zinc finger of WT1 contributes only weakly to DNA binding at least in vitro,[20] a 9 or 10 base pair binding site for WT1-KTS was proposed where the zinc finger 1 of WT1 either does not participate in specific DNA binding or would only specifically interact with one out of three nucleotides (Fig. 2). The +KTS form on the other hand does not bind well to the -KTS DNA binding sites as the insertion of KTS disrupts the conserved linker TGEKP sequence between the zinc fingers 3 and 4.[23]

The structure of WT1 either in the free or DNA-bound state was recently resolved by NMR spectroscopy.[23] Remarkably, both the –KTS and the +KTS zinc fingers adopt very similar structures in free solution with the linkers between the fingers displaying great flexibility. Upon binding to DNA, the linkers become more rigid and adopt a highly conserved structure, which limits the movement of adjacent fingers in the -KTS to allow stable interaction with the major groove of DNA.[23] The +KTS containing linker TGKTSEKP, however, maintains its flexible nature even in the presence of DNA indicating that the KTS-containing linker cannot form structures similar to the other linkers. This is probably due to disruption of the highly conserved TGEKP sequence by the insertion of KTS. As a result, the fourth zinc finger makes very little contact with the major groove leading to a weakened or complete loss of DNA binding (Fig. 2).

WT1 in Transcriptional Regulation

Much of the early work on WT1 was focused on transcriptional repression as many GC-rich containing promoters were repressed by WT1 in co-transfection studies.[1] Thus, WT1 was thought to function solely as a transcriptional repressor. This notion prevailed until it was shown that WT1 could activate transcription depending on the promoter-context.[24,25] Since then, a growing number of studies have shown WT1 to be an activator of transcription.[3] Dichotomous nature of WT1 in transcription regulation is likely to be cellular and promoter dependent with proteins that interact with WT1 determining the outcome of the transcriptional regulation of WT1 (Fig. 3). This is demonstrated, as aforementioned, by the WT1-Par4 interaction where WT1 either activates or represses transcription upon interacting with Par4.[14,15] Other proteins shown to interact and modulate WT1 function are the prototypical tumor suppressor TP53 (p53)[26] and its homolog, p73.[27] Interaction with p53 can convert WT1 from an activator to a repressor on a given WT1-regulated promoter, and conversely, WT1 can cooperate and enhance the transcriptional activity of p53 on p53-target promoters.[26] On the other hand, association of WT1 with p73 results in inhibition of both WT1- and p73-mediated transcriptional activation.[27]

Perhaps transcriptional activation of WT1 is best understood by its interaction with the transcriptional coactivator CREB-binding protein (CBP). Interaction of WT1 with CBP requires the zinc fingers 1 and 2 and the two proteins act synergistically to activate the *Amphiregulin* promoter.[28] Although the exact mechanism of WT1-mediated transcriptional activation is unclear, it may be envisioned that WT1 influences the basal transcription machinery through interacting with proteins such as CBP. As mentioned, the basic transcription factor TFIID, comprised of many subunits, has been shown to interact either directly or indirectly with WT1[14] and it is well established that CBP interacts with TFIID.[29] Thus, CBP may serve as a bridging molecule and allow WT1 to exert its influence on the basal transcription machinery. Synergistic activation of transcription is also observed with other WT1-interacting proteins such as Steroidogenic Factor 1 (SF1) on the *Müllerian inhibiting substance (MIS)* promoter,[30] and four-and-half LIM domain protein (FHL2) on *MIS* and *Dax1* promoters.[31] An alternative mechanism of transcriptional activation by WT1 may also involve chromatin remodeling (Fig. 3). In this scenario, CBP, a histone acetyltransferase,[32] may be recruited to the sites of WT1-bound promoters and induce chromatin remodeling through histone acetylation. The relaxed or 'opened' chromatin will then allow access to other activators and basic transcription factors to activate transcription. Although similar mechanisms can be imagined on the transcriptional repression by WT1 through interactions with the corepressors, histone deacetylases and methyltransferases, no such interaction with WT1 has been documented thus far. Another surprising consequence of the WT1-CBP interaction is that the zinc finger domain is involved in both DNA binding and interaction with CBP, presumably simultaneously. How WT1 zinc fingers can interact simultaneously with both DNA and protein is currently not clear and may be revealed by the co-crystal structure of WT1, CBP and DNA.

Subnuclear Localization of WT1

Even though the +KTS isoforms represent the most abundant form of WT1 in the cell,[7] the specific function of the WT1+KTS has remained elusive since it did not bind to DNA as well as the –KTS and was inactive in many promoter-reporter assays. The

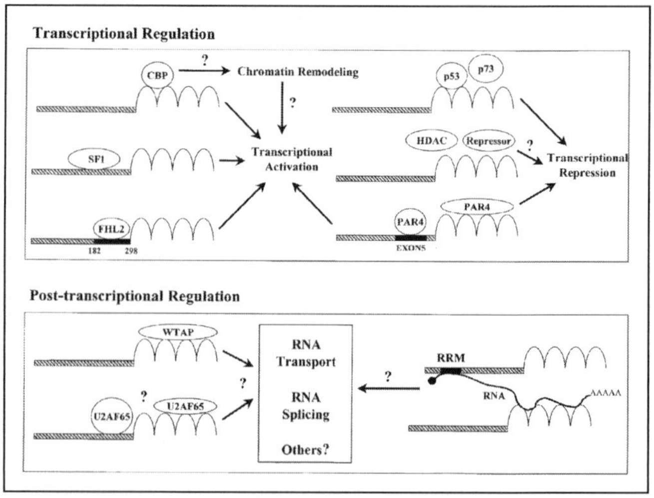

Figure 3. WT1 activity is determined by various WT1-interacting proteins. Transcriptional regulation. Transcriptional activation can be mediated by WT1 binding to SF1, FHL2, PAR4 and CBP, while interactions with p53 and PAR4 result in transcriptional repression. Potential mechanisms of WT1-induced chromatin remodeling leading to either transcription activation (WT1-CBP) or repression (WT1-HDAC or -repressor) are shown. Notice PAR4 interacts with at least two separate domains of WT1, Exon 5 and the zinc finger domain. Post-transcriptional regulation. WT1 may participate in various RNA-related processes by interacting with splicing factors (U2AF65 and WTAP) and/or by binding to RNA via the RRM and/or the zinc finger domain of WT1. Interaction with U2AF65 may require both the N terminus and the zinc fingers of WT1. The dark circle in the RNA represents the 5' m(7)GpppN cap structure.

first functional clue of WT1+KTS was provided by examining the subcellular localization of both WT1-KTS and the +KTS.[33] It was shown that the +KTS form of WT1 displayed a "speckled" pattern of nuclear distribution which was clearly different from more diffuse nuclear distribution of the –KTS. This was subsequently confirmed by another report[34] and perhaps best demonstrated by the –KTS-specific knockout mouse study where the endogenous +KTS was observed to form the "speckles" in vivo in the absence of the WT1–KTS.[35] Distinct speckled nuclear distribution is a feature of splicing factors, and subsequently, a number of studies have shown that both WT1±KTS can interact with the ubiquitous splicing factor U2AF65[36] and a putative splicing factor hFL(2)D/WTAP (WT1-associating protein),[37] a human homolog of *Drosophila female-lethal(2)-d (fl(2)-d)* which is required for splicing regulation of pre-mRNAs in flies.[38] Recently, hFL(2)D/WTAP was also identified by mass spectroscopy as a component of spliceosome, a splicing machinery comprised of small nuclear RNAs and proteins.[39] These observations have led to an intriguing possibility of WT1 participating in the post-transcriptional processes.

WT1 in Post-Transcriptional Regulation

Soon after the initial observation of speckled nuclear distribution of WT1+KTS, studies describing RNA binding by WT1[40,41] and a report of RNA-recognition motif (RRM) in the N terminus of WT1 as predicted by computer modeling[17] emerged. Both WT1-KTS and the +KTS was demonstrated to bind to the RNA sequences in exon 2 of the *Insulin-like Growth Factor 2*,[40] although the +KTS bound with significantly higher affinity. Deletion of the zinc finger 1 abolished the RNA binding, which is in direct contrast to WT1-DNA interaction where the zinc finger 1 appears to contribute minimally.[20] A recent study has also demonstrated the importance of zinc finger 1 in RNA binding in vivo.[42] Another study using systematic evolution of ligands by exponential enrichment (SELEX) has identified a group of RNA sequences bound by the WT1-KTS;[41] however, this approach did not yield a consensus RNA binding sequences nor any common secondary structures were evident in the identified RNA. WT1 protein has also been isolated from the nuclear poly(A)$^+$ ribonucleoprotein fraction using oligo(dT) chromatography.[43] Together, a body of evidence is starting to emerge in favor of WT1-RNA interaction and a possible role in RNA processing in vivo. The major challenge now is to identify the in vivo RNA targets and the exact role of WT1 in the post-transcriptional processes. The dual property of WT1 in DNA and RNA binding is not unique to WT1 as another C_2H_2 zinc finger protein TFIIIA in *Xenopus laevis* also binds specifically to the intragenic region of the *5S rRNA* gene[44] and the 5S rRNA.[45]

Expression of WT1 in Development

Expression of human WT1 has been documented in fetal kidneys and gonads,[46] but more detailed knowledge about WT1 expression is available from the mouse studies.[47-49] Expression of WT1 in developing mouse embryos is restricted to specific tissues, most of which undergo mesenchyme to epithelial transition during organogenesis.[48] These tissues include kidneys, gonads, mesothelial lining of the abdominal organs including the epicardium, spleen, spinal cord, brain stem, ganglion cells of developing retina and digits. In adults, Wt1 (mouse WT1) expression becomes even more restricted and is maintained only in very few cells such as the podocytes of glomeruli, Sertoli cells in testes, granulosa cells of the ovary, uterine wall,[47] and the immature CD34$^+$/CD38$^-$ hematopoietic precursor cells in the bone marrow.[50] Thus, *wt1 (*mouse *WT1 gene)* encodes a transcription factor (and also a RNA-binding protein) whose expression is developmentally regulated and likely has a limited but specialized functions in only a few organs after organogenesis is complete.

Genetic Characterization of WT1

Human Genetics of WT1

Identification of diseases affected by the mutations of the *WT1* and their symptoms has provided a great deal of knowledge about the possible functions of *WT1* (Table 1). A rare human disease WAGR syndrome (Wilms tumor, aniridia, genitourinary malformations, and mental retardation) provided the first clue about the location of *WT1*. WAGR patients are distinguished by constitutional hemizygous deletion of the chromosome 11p13 band,[51,52] and these children are at high risk for Wilms tumor, a childhood kidney cancer and one of the most common solid tumors among young children. Thus, a tumor suppressor gene, as well as genes responsible for other defects associated with WAGR, was thought to reside in 11p13. This has led to the positional cloning of a zinc finger-containing gene *WT1*,[5,6] which was subsequently found to harbor mutations in about 10-15% of all sporadic Wilms tumors.[53] In most cases, Wilms tumor samples from WAGR patients lose the remaining wild type *WT1* allele,[54,55] suggesting that complete loss of *WT1* is required to develop Wilms tumor. *WT1* hemizygosity is also responsible for the gonadal defects and renal failures associated with WAGR patients.[53] In addition to WAGR, two other rare syndromes are also affected by germ line heterozygous *WT1* mutations: Denys-Drash syndrome (DDS) and Frasier syndrome (FS). Symptoms of DDS patients include defects in genito-urinary development (such as diffuse mesangial sclerosis and male-female sex reversal phenotype) and high incidents of Wilms tumor.[56] The majority of *WT1* mutations found in DDS patients are point mutations within the critical residues of the zinc fingers 2 and 3, which abolish DNA binding in vitro.[1] Wilms tumors arising in DDS, like WAGR, show a loss of the remaining *WT1* allele.[56] In contrast, most FS mutations are found in the non-coding, splice junction of intron 9, effectively disrupting the production of the alternatively spliced WT1+KTS isoform.[57] This imbalance of the +/-KTS isoforms in FS leads to severe genito-urinary defects as many of FS patients suffer from male-female sex reversal phenotype (as seen in some DDS patients) and end-stage renal failure resulting from focal segmental glomerular sclerosis.[8] FS patients, unlike DDS patients, however, are not predisposed for development of Wilms tumor, but they suffer from a greater risk of gonadoblastoma.[57] The role of *WT1* in gonadoblastoma is not known. Thus, these human studies highlight the importance of WT1 function in renal and gonadal development and the involvement of *WT1* inactivation in Wilms tumor. Of note, a homozygous point mutation of *WT1* is also found in non-asbestos-related mesothelioma,[58] a tumor derived from the peritoneal lining, although further reports of *WT1* mutations in mesothelioma are warranted.

WT1 mutations are also found in about 10-15% of leukemias,[59-61] mostly in adult acute myeloid leukemia. But unlike Wilms tumors, many but not all of the *WT1* mutations are heterozygous in leukemias,[61] raising the possibility that the mutations act in a dominant or dominant-negative fashion, or perhaps these mutations represent secondary effects of leukemia. There are also conflicting results regarding the role of WT1 in leukemia as some data argue for an oncogenic role of WT1 (reviewed in refs 62, 63). In fact, based on the clinical correlation of high WT1 expression and poor prognosis in some leukemias, immunocancer therapy using anti-WT1-specific cytotoxic T-cells is being explored for treatment of leukemia.[64,65] The paradoxical effects of WT1 in the hematopoietic system may be partially reconciled by the observation of Ellisen et al.[66] In this study, the ectopic expression of WT1 in primary CD34$^+$cells isolated from human umbilical cord blood, which include very rare multipotent stem cells and lineage-committed but undifferentiated progenitor cells, resulted in two different populations of cells in culture: (1) undifferentiated, but lineage-committed progenitor cells rapidly underwent spontaneous differentiation with ectopic expression of WT1-KTS; (2) WT1-KTS expression in a subset of CD34$^+$/CD38$^-$ cells which represent the most primitive hematopoietic precursor cells led to persistent quiescence of these cells in culture. This may be a reflection of the biphasic nature of WT1 expression during normal hematopoiesis, as low level expression of WT1 is detected in a subset of very primitive CD34$^+$/CD38$^-$ cells[50] and is dramatically downregulated as these cells begin to differentiate.[67] Expression of WT1 reappears as the lineage-committed cells begin to differentiate into granulocyte, monocyte and B-cell lineages.[66] Thus, it can be postulated that WT1 is involved in both maintaining the cells in multipotent progenitor state as well as promoting differentiation in later stages, partially explaining the anti- and pro-differentiation properties of WT1, respectively.

WT1 is also involved in a reciprocal chromosomal translocation t(11;22)(p13;q12) in a rare tumor called Desmoplastic small

Table 1. Human diseases associated with* WT1 *mutation

Syndrome	Clinical Manifestations	Types of Mutations
Wilms tumor	Childhood kidney cancer	Point mutations, frameshift (small insertions or deletions), homozygous, somatic
WAGR	Wilms tumor Genitourinary defects	Hemizygous interstitial deletion of 11p13 (includes *WT1*) in the germ line, somatic in the tumor (homozygous)
Denys-Drash	Wilms tumor Genitourinary defects Diffuse mesangial sclerosis Male-female sex reversal phenotype	Point mutations usually in the zinc finger (i.e., Arg 394 to Trp), heterozygous in the germ line, somatic in the tumor (homozygous)
Frasier	Genitourinary defects Focal segmental glomerular sclerosis Male-female sex reversal phenotype Gonadoblastoma[a]	KTS splice donor site point mutation (i.e., position +2 in intron 9: **TG** to **T**<u>**C**</u>) heterozygous in the germ line (reduction of +KTS form)
Leukemia	Acute myeloid leukemia	Frameshifts, usually heterozygous, (few homozygous), somatic
Desmoplastic small round cell tumor	Small round cells in the abdomen (propensity for young adolescent males)	Reciprocal chromosome translocation t(11;22) (p13;q12), heterozygous, somatic

[a] Status of *WT1* mutation is not known

round cell tumor,[68] which fuses the N terminus of *Ewings sarcoma gene (EWS)* with the last three zinc fingers of *WT1*. The *EWS/WT1* fusion transcript retains the KTS splicing and thus produces two isoforms: EWS/WT1 -/+KTS. EWS/WT1–KTS is thought to function as a novel transcription factor and some of its target genes have been identified (ref. 69 and the references therein). The function of the +KTS version of the fusion protein, however, remains unclear. EWS/WT1-KTS but not the +KTS can transform NIH3T3 cells in vitro demonstrating the oncogenic potential of the fusion gene product.[70]

Mouse Genetics of wt1

The initial description of *wt1* heterozygotes reported no abnormalities,[71] but upon prolonged examination of the $wt1^{+/-}$ strain showed that 80% of the mice die of terminal renal failure.[72] Autopsy results revealed that these mice suffered from severe glomerular sclerosis. Renal dysfunction and premature death of the $wt1^{+/-}$ strain were also reported by another study.[73] These observations are consistent with the human studies of WAGR and DDS syndromes where *WT1* hemizygosity leads to end-stage renal failure. Unlike WAGR and DDS patients, however, $wt1^{+/-}$ strain do not develop Wilms tumor.[72,73]

Mice with homozygous inactivation of *wt1* are embryonic lethal at around E13.5.[71] Examination of the $wt1^{-/-}$ embryos showed that these animals lack kidneys and gonads in addition to defects in diaphram, epicardium and mesothelium lining the abdominal organs. The cause of death was likely due to malformation of epicardium, a tissue that surrounds the heart, leading to excessive bleeding and heart failure. Renal agenesis was due to a developmental arrest during the early stage of kidney development as metanephric mesenchymes, where Wt1 is expressed, die of massive apoptosis. In vitro kidney organ culture assay demonstrated that $wt1^{-/-}$ mesenchymal cells also fail to differentiate in response to signals from strong inducers of kidney differentiation such as a spinal cord. Thus, during kidney development, Wt1 is required for the survival of the mesenchymal cells, as well as promoting outgrowth of ureteric bud and responding to the signals generated by ureteric bud. In addition to the early stages of renal development, continuous expression of Wt1 from both alleles is required for the homeostasis of mature kidneys as demonstrated by frequent renal failures observed in $wt1^{+/-}$ mice and in individuals with only one functional *WT1* allele (Table 1). The hematopoietic system of *wt1*-null embryos appeared to develop normally, although early lethality of the mutant animals prevented a more detailed study.[71]

Out-breeding the mixed background of the original $wt1^{+/-}$ strain to a pure M1 strain background led to prolonged survival of the *wt1*-null embryos, but the animals still lacked kidneys and gonads, and died just before or soon after birth.[74] However, this allowed the investigators to uncover additional defects within *wt1*-null embryos as these animals lacked spleen and adrenal gland.[74] The absence of spleen was shown to be due to 4 to 5-fold increase in apoptosis within the spleen primordium. Another defect was uncovered more recently in the eyes of $wt1^{-/-}$ embryos which were smaller than the wild type littermates.[75] Specifically, $wt1^{-/-}$ retina showed less retinal cell proliferation and increased apoptosis, resulting in about 40% less cell number in the ganglion cell layer of the mutant retina. This was correlated with the absence of the POU-domain transcription factor, pou4f2, which is required for the survival of the retinal ganglion cells. Recurrent observations of decreased proliferation and differentiation along with increased apoptosis in various $wt1^{-/-}$ *organs* suggest that one of the primary functions of wt1 is in regulating genes that are critical for cell proliferation, differentiation and survival, either transcriptionally or post-transcriptionally.

A more detailed insight into specific functions of the WT1 isoforms was provided by the generation of either the –KTS or +KTS-specific knockout mice by introducing mutations in the splice donor sites at the end of exon 9.[35] The first surprise came when these mice were born with incomplete kidneys and gonads. As expected, both—and +KTS knockout mice died soon after birth due to renal dysfunction. Because the *wt1*-null animals completely lack kidneys and gonads,[71] this suggests that both the –KTS and the +KTS isoforms have limited redundant functions at least during the initial steps of renal and gonadal development. This is both remarkable and unexpected given the dramatic differences in the DNA binding and the distinct nuclear distribution displayed by the two isoforms. However, there are also clear functional differences between the two isoforms as neither knockout animals fully develop in the absence of the other isoform.[35] The –KTS-null kidneys and gonads are much more reduced in size and undifferentiated compared to the +KTS-null counterparts, suggesting that the –KTS has more fundamental role in proliferation, differentiation and survival. The +KTS-null animals, on the other hand, display complete male to female sex reversal phenotype. This is reminiscent of the Frasier syndrome in humans although in Frasier syndrome, the symptoms are manifested by a heterozygous mutation[8] while in mouse it takes homozygous KTS splice mutations to reveal the phenotype.[35] The lack of Wt1+KTS expression in the +KTS-null gonads resulted in a drastically decreased expression of two important male-specific genes, the Y-chromosome associated *mSry* (sex-determining region Y gene) and *mSox9* (*Sry*-related HMG box). Both of these genes encode transcription factors that are critical for the male sex determination pathway.[4] Another feature of Frasier syndrome that is recapitulated by the +KTS mice is that 70% of the *wt1+KTS* heterozygous mice die of renal dysfunction due to glomerular nephropathy. No abnormalities are reported in the kidneys and gonads of *wt1-KTS* heterozygous mice. These observations demonstrate the importance of Wt1+KTS during a male-specific differentiation of gonads, and the severe consequences of reducing the gene dosage of the +KTS but not the –KTS in maintaining proper glomerular function. Interestingly, homozygous inactivation of two mammal-specific *wt1* isoforms, the alternative, upstream CUG-initiated isoform[12] and the exon 5-containing isoform,[11] had no effect on the mouse development, survival and fertility. Thus, these two isoforms of Wt1 probably have functions that are redundant with the major Wt1 isoforms.

Conclusion and Future Perspectives

Over the years, WT1 research has contributed greatly to our general concept of how zinc finger proteins work as well as to our understanding of development and cancer. This chapter briefly discussed the structure and molecular details of WT1 zinc fingers and aimed to focus on the biology of WT1 in the context of development and disease. Biochemical studies of WT1 illuminate its role in both transcriptional and post-transcriptional processes while the genetic studies underlie the importance of WT1 in development and in different types of human disease. It is evident that future studies on WT1 will provide even greater insights to our knowledge of zinc finger proteins, and continue to serve as a paradigm for studying the link between development and cancer. Current advances in WT1 have raised a new set of questions and opened new fields of research, and the latest adventures of WT1 research will take us into the world of post-transcriptional mechanisms. There is also a growing interest in the role of WT1 in hematopoiesis and leukemia. In this regard, a provocative and intriguing research in the development of WT1-based immunocancer therapy is worth noting. At the same time, questions that remain unanswered from the past efforts, such as identifying the physiological transcriptional target genes, delineating the transcriptional regulation mechanisms and the mechanism of tissue-specific regulation of WT1, will continue to fuel and reshape our efforts on WT1. Combining the biochemical, cellular and genetical approaches with new tools in genomics and proteomics will continue to unravel the fascinating stories of WT1 in development and cancer.

Acknowledgments

Due to space limitation, many important findings on WT1 could not be discussed here. Thus, we would like to acknowledge all the investigators in the WT1 field for their many contributions and apologize to many of our colleagues whose important work could not be mentioned. We thank Drs. Rick Proia, Connie Noguchi and Barbara Christensen for critical reading and comments on the manuscript. SBL is indebted to Dr. Dan Haber for his encouragement and unwavering support.

References

1. Reddy JC, Licht JD. The WT1 Wilms' tumor suppressor gene: how much do we really know? Biochim Biophys Acta 1996; 1287(1):1-28.
2. Little M, Holmes G, Walsh P. WT1: what has the last decade told us? Bioessays 1999; 21(3):191-202.
3. Lee SB, Haber DA. Wilms tumor and the WT1 gene. Exp Cell Res 2001; 264(1):74-99.
4. Wagner KD, Wagner N, Schedl A. The complex life of WT1. J Cell Sci 2003; 116(Pt 9):1653-1658.
5. Call KM, Glaser T, Ito CY et al. Isolation and characterization of a zinc finger polypeptide gene at the human chromosome 11 Wilms' tumor locus. Cell 1990; 60(3):509-520.
6. Gessler M, Poustka A, Cavenee W et al. Homozygous deletion in Wilms tumours of a zinc-finger gene identified by chromosome jumping. Nature 1990; 343(6260):774-778.
7. Haber DA, Sohn RL, Buckler AJ et al. Alternative splicing and genomic structure of the Wilms tumor gene WT1. Proc Natl Acad Sci U S A 1991; 88(21):9618-9622.
8. Barbaux S, Niaudet P, Gubler MC et al. Donor splice-site mutations in WT1 are responsible for Frasier syndrome. Nat Genet 1997; 17(4):467-470.
9. Kent J, Coriat AM, Sharpe PT et al. The evolution of WT1 sequence and expression pattern in the vertebrates. Oncogene 1995; 11(9):1781-1792.
10. Schedl A, Hastie N. Multiple roles for the Wilms' tumour suppressor gene, WT1 in genitourinary development. Mol Cell Endocrinol 1998; 140(1-2):65-69.
11. Natoli TA, McDonald A, Alberta JA et al. A mammal-specific exon of WT1 is not required for development or fertility. Mol Cell Biol 2002; 22(12):4433-4438.
12. Miles CG, Slight J, Spraggon L et al. Mice lacking the 68-amino-acid, mammal-specific N-terminal extension of WT1 develop normally and are fertile. Mol Cell Biol 2003; 23(7):2608-2613.
13. Davies RC, Bratt E, Hastie ND. Did nucleotides or amino acids drive evolutionary conservation of the WT1 +/-KTS alternative splice? Hum Mol Genet 2000; 9(8):1177-1183.
14. Richard DJ, Schumacher V, Royer-Pokora B et al. Par4 is a coactivator for a splice isoform-specific transcriptional activation domain in WT1. Genes Dev 2001; 15(3):328-339.
15. Johnstone RW, See RH, Sells SF et al. A novel repressor, par-4, modulates transcription and growth suppression functions of the Wilms' tumor suppressor WT1. Mol Cell Biol 1996; 16(12):6945-6956.
16. Moffett P, Bruening W, Nakagama H et al. Antagonism of WT1 activity by protein self-association. Proc Natl Acad Sci U S A 1995; 92(24):11105-11109.
17. Kennedy D, Ramsdale T, Mattick J et al. An RNA recognition motif in Wilms' tumour protein (WT1) revealed by structural modelling. Nat Genet 1996; 12(3):329-331.

18. Bruening W, Moffett P, Chia S et al. Identification of nuclear localization signals within the zinc fingers of the WT1 tumor suppressor gene product. FEBS Lett 1996; 393(1):41-47.
19. Rauscher FJ, 3rd, Morris JF, Tournay OE et al. Binding of the Wilms' tumor locus zinc finger protein to the EGR-1 consensus sequence. Science 1990; 250(4985):1259-1262.
20. Nakagama H, Heinrich G, Pelletier J et al. Sequence and structural requirements for high-affinity DNA binding by the WT1 gene product. Mol Cell Biol 1995; 15(3):1489-1498.
21. Wang ZY, Qiu QQ, Enger KT et al. A second transcriptionally active DNA-binding site for the Wilms tumor gene product, WT1. Proc Natl Acad Sci U S A 1993; 90(19):8896-8900.
22. Pavletich NP, Pabo CO. Zinc finger-DNA recognition: crystal structure of a Zif268-DNA complex at 2.1 A. Science 1991; 252(5007):809-817.
23. Laity JH, Dyson HJ, Wright PE. Molecular basis for modulation of biological function by alternate splicing of the Wilms' tumor suppressor protein. Proc Natl Acad Sci USA 2000; 97(22):11932-11935.
24. Wang ZY, Qiu QQ, Deuel TF. The Wilms' tumor gene product WT1 activates or suppresses transcription through separate functional domains. J Biol Chem 1993; 268(13):9172-9175.
25. Reddy JC, Hosono S, Licht JD. The transcriptional effect of WT1 is modulated by choice of expression vector. J Biol Chem 1995; 270(50):29976-29982.
26. Maheswaran S, Park S, Bernard A et al. Physical and functional interaction between WT1 and p53 proteins. Proc Natl Acad Sci U S A 1993; 90(11):5100-5104.
27. Scharnhorst V, Dekker P, van der Eb AJ et al. Physical interaction between Wilms tumor 1 and p73 proteins modulates their functions. J Biol Chem 2000; 275(14):10202-10211.
28. Wang W, Lee SB, Palmer R et al. DA. A functional interaction with CBP contributes to transcriptional activation by the Wilms tumor suppressor WT1. J Biol Chem 2001; 276(20):16810-16816.
29. Dallas PB, Yaciuk P, Moran E. Characterization of monoclonal antibodies raised against p300: both p300 and CBP are present in intracellular TBP complexes. J Virol 1997; 71(2):1726-1731.
30. Nachtigal MW, Hirokawa Y, Enyeart-VanHouten DL et al. Wilms' tumor 1 and Dax-1 modulate the orphan nuclear receptor SF-1 in sex-specific gene expression. Cell 1998; 93(3):445-454.
31. Du X, Hublitz P, Gunther T et al. The LIM-only coactivator FHL2 modulates WT1 transcriptional activity during gonadal differentiation. Biochim Biophys Acta 2002; 1577(1):93-101.
32. Bannister AJ, Kouzarides T. The CBP co-activator is a histone acetyltransferase. Nature 1996; 384(6610):641-643.
33. Larsson SH, Charlieu JP, Miyagawa K et al. Subnuclear localization of WT1 in splicing or transcription factor domains is regulated by alternative splicing. Cell 1995; 81(3):391-401.
34. Englert C, Vidal M, Maheswaran S et al. Truncated WT1 mutants alter the subnuclear localization of the wild-type protein. Proc Natl Acad Sci U S A 1995; 92(26):11960-11964.
35. Hammes A, Guo JK, Lutsch G et al. Two splice variants of the Wilms' tumor 1 gene have distinct functions during sex determination and nephron formation. Cell 10 2001; 106(3):319-329.
36. Davies RC, Calvio C, Bratt E et al. WT1 interacts with the splicing factor U2AF65 in an isoform-dependent manner and can be incorporated into spliceosomes Oct 15 1998; 12(20):3217-3225.
37. Little NA, Hastie ND, Davies RC. Identification of WTAP, a novel Wilms' tumour 1-associating protein. Hum Mol Genet 2000; 9(15):2231-2239.
38. Granadino B, Campuzano S, Sanchez L. The Drosophila melanogaster fl(2)d gene is needed for the female-specific splicing of Sex-lethal RNA. Embo J 1990; 9(8):2597-2602.
39. Zhou Z, Licklider LJ, Gygi SP et al. Comprehensive proteomic analysis of the human spliceosome. Nature 2002; 419(6903):182-185.
40. Caricasole A, Duarte A, Larsson SH et al. RNA binding by the Wilms tumor suppressor zinc finger proteins. Proc Natl Acad Sci U S A 1996; 93(15):7562-7566.
41. Bardeesy N, Pelletier J. Overlapping RNA and DNA binding domains of the wt1 tumor suppressor gene product. Nucleic Acids Res 1998; 26(7):1784-1792.
42. Ladomery M, Sommerville J, Woolner S et al. Expression in Xenopus oocytes shows that WT1 binds transcripts in vivo, with a central role for zinc finger one. J Cell Sci 2003; 116(Pt 8):1539-1549.
43. Ladomery MR, Slight J, Mc Ghee S et al. Presence of WT1, the Wilm's tumor suppressor gene product, in nuclear poly(A)(+) ribonucleoprotein. J Biol Chem 1999; 274(51):36520-36526.
44. Engelke DR, Ng SY, Shastry BS et al. Specific interaction of a purified transcription factor with an internal control region of 5S RNA genes. Cell 1980; 19(3):717-728.
45. Honda BM, Roeder RG. Association of a 5S gene transcription factor with 5S RNA and altered levels of the factor during cell differentiation. Cell 1980; 22(1 Pt 1):119-126.
46. Pritchard-Jones K, Fleming S, Davidson D et al. The candidate Wilms' tumour gene is involved in genitourinary development. Nature 1990; 346(6280):194-197.
47. Pelletier J, Schalling M, Buckler AJ et al. Expression of the Wilms' tumor gene WT1 in the murine urogenital system. Genes Dev 1991; 5(8):1345-1356.
48. Armstrong JF, Pritchard-Jones K, Bickmore WA et al. The expression of the Wilms' tumour gene, WT1, in the developing mammalian embryo. Mech Dev 1993; 40(1-2):85-97.
49. Moore AW, Schedl A, McInnes L et al. YAC transgenic analysis reveals Wilms' tumour 1 gene activity in the proliferating coelomic epithelium, developing diaphragm and limb. Mech 1998; 79(1-2): 169-184.
50. Baird PN, Simmons PJ. Expression of the Wilms' tumor gene (WT1) in normal hemopoiesis. Exp Hematol 1997; 25(4):312-320.
51. Riccardi VM, Sujansky E, Smith AC et al. Chromosomal imbalance in the Aniridia-Wilms' tumor association: 11p interstitial deletion. Pediatrics 1978; 61(4): 604-610.
52. Francke U, Holmes LB, Atkins L et al. Aniridia-Wilms' tumor association: evidence for specific deletion of 11p13. Cytogenet Cell Genet 1979; 24(3):185-192.
53. Hastie ND. The genetics of Wilms' tumor-a case of disrupted development. Annu Rev Genet 1994; 28:523-558.
54. Baird PN, Groves N, Haber DA et al. Identification of mutations in the WT1 gene in tumours from patients with the WAGR syndrome. Oncogene 1992; 7(11):2141-2149.
55. Brown KW, Watson JE, Poirier V et al. Inactivation of the remaining allele of the WT1 gene in a Wilms' tumour from a WAGR patient. Oncogene 1992; 7(4):763-768.
56. Pelletier J, Bruening W, Kashtan CE et al. Germline mutations in the Wilms' tumor suppressor gene are associated with abnormal urogenital development in Denys-Drash syndrome. Cell 1991; 67(2): 437-447.
57. Koziell A, Charmandari E, Hindmarsh PC et al. Frasier syndrome, part of the Denys Drash continuum or simply a WT1 gene associated disorder of intersex and nephropathy? Clin Endocrinol (Oxf) 2000; 52(4):519-524.
58. Park S, Schalling M, Bernard A et al. The Wilms tumour gene WT1 is expressed in murine mesoderm-derived tissues and mutated in a human mesothelioma. Nat Genet 1993; 4(4):415-420.
59. King-Underwood L, Renshaw J, Pritchard-Jones K. Mutations in the Wilms' tumor gene WT1 in leukemias. Blood 1996; 87(6):2171-2179.
60. Miyagawa K, Hayashi Y, Fukuda T et al. Mutations of the WT1 gene in childhood nonlymphoid hematological malignancies. Genes Chromosomes Cancer 1999; 25(2):176-183.
61. King-Underwood L, Pritchard-Jones K. Wilms' tumor (WT1) gene mutations occur mainly in acute myeloid leukemia and may confer drug resistance. Blood 1998; 91(8):2961-2968.
62. Pritchard-Jones K, King-Underwood L. The Wilms tumour gene WT1 in leukaemia. Leuk Lymphoma 1997; 27(3-4):207-220.
63. Algar E. A review of the Wilms' tumor 1 gene (WT1) and its role in hematopoiesis and leukemia. J Hematother Stem Cell Res 2002; 11(4):589-599.
64. Bellantuono I, Gao L, Parry S et al. Two distinct HLA-A0201-presented epitopes of the Wilms tumor antigen 1 can function as targets for leukemia-reactive CTL. Blood 2002; 100(10):3835-3837.
65. Scheibenbogen C, Letsch A, Thiel E et al. CD8 T-cell responses to Wilms tumor gene product WT1 and proteinase 3 in patients with acute myeloid leukemia. Blood 2002; 100(6):2132-2137.

66. Ellisen LW, Carlesso N, Cheng T et al. The Wilms tumor suppressor WT1 directs stage-specific quiescence and differentiation of human hematopoietic progenitor cells. Embo J 2001; 20(8):1897-1909.
67. Maurer U, Brieger J, Weidmann E et al. The Wilms' tumor gene is expressed in a subset of CD34+ progenitors and downregulated early in the course of differentiation in vitro. Exp Hematol 1997; 25(9):945-950.
68. Ladanyi M, Gerald W. Fusion of the EWS and WT1 genes in the desmoplastic small round cell tumor. Cancer Res 1994; 54(11):2837-2840.
69. Palmer RE, Lee SB, Wong JC et al. Induction of BAIAP3 by the EWS-WT1 chimeric fusion implicates regulated exocytosis in tumorigenesis. Cancer Cell 2002; 2(6):497-505.
70. Kim J, Lee K, Pelletier J. The desmoplastic small round cell tumor t(11;22) translocation produces EWS/WT1 isoforms with differing oncogenic properties. Oncogene 1998; 16(15):1973-1979.
71. Kreidberg JA, Sariola H, Loring JM et al. WT-1 is required for early kidney development. Cell 1993; 74(4):679-691.
72. Menke AL, Clarke AR, Leitch A et al. Genetic interactions between the Wilms' tumor 1 gene and the p53 gene. Cancer Res 2002; 62(22):6615-6620.
73. Guo JK, Menke AL, Gubler MC et al. WT1 is a key regulator of podocyte function: reduced expression levels cause crescentic glomerulonephritis and mesangial sclerosis. Hum Mol Genet 2002; 11(6):651-659.
74. Herzer U, Crocoll A, Barton D et al. The Wilms tumor suppressor gene wt1 is required for development of the spleen. Curr Biol 1999; 9(15):837-840.
75. Wagner KD, Wagner N, Vidal VP et al. The Wilms' tumor gene Wt1 is required for normal development of the retina. Embo J 2002; 21(6):1398-1405.

CHAPTER 25

Yin Yang 1

Huifei Liu and Yang Shi*

Abstract

The transcription factor Yin Yang 1 (YY1) (also known as NF-E1, δ, CP-1 and UCRBP) is a GLI-Kruppel zinc finger protein.[1-4] YY1 is highly conserved from *Xenopus* to mammalian YY1 and has been demonstrated to play an essential role in mouse embryonic development and other physiological and pathological conditions.[5,6] As one of the first mammalian transcription factors that have been identified to possess both repression and activation activities, YY1 has served as a model for understanding how a single molecule can conduct opposite transcriptional functions, that is, repression and activation. Recent studies suggest that YY1 is a mammalian counterpart of the *Drosophila* Polycomb Group (PcG) protein *Pleiohomeotic* (Pho),[7,8] providing a new framework in which to further explore YY1's biological functions and mechanism of action in vivo. In this chapter, we discuss progress on YY1 since its discovery.

Introduction

YY1 Family of Proteins

YY1 is evolutionarily conserved from human to *Xenopus* with significant sequence homology spanning the entire protein (Table 1). The four zinc fingers that are responsible for sequence-specific DNA-binding are identical among *Xenopus*, rat, mouse and human YY1, while the N-terminal regions are more divergent.[2,5,9,10] Recently, a YY1-related protein, which we term YY2, has been predicted based on the genome sequence.[11] YY2 shares 55% homology with YY1 and 86% homology in the zinc finger regions. Preliminary data suggest that it is likely to be an expressed gene (Shi and Seto labs, unpublished results). However, the function and mechanism of action of YY2 remain unclear. A third mammalian protein related to YY1 is Rex-1 (Reduced Expression 1).[1,12] Sequence homology between YY1 and Rex-1 is limited to the four zinc fingers (75% homology).[1] Rex-1 was isolated as a protein whose expression decreases during F9 cell differentiation induced by retinoic acid.[12] Recently Rex-1 has been classified as one of the definitive markers for embryonic stem cells.[13-15]

Two *Drosophila* PcG proteins, Pho and Pho-like share significant homology with YY1 within the four zinc fingers.[8,19] The overall zinc finger sequence similarity between Pho and YY1 is as high as 96%, with zinc fingers 2 and 3 being completely identical.[8] Interestingly, outside the zinc finger region, a stretch of 22 amino acids (aa) (human YY1 aa205-226, Pho aa148-169) is conserved as well with 80% sequence similarity, suggesting that this region may be important for yet to be identified, evolutionarily conserved functions.[8]

Table 1. YY1 Homologues

Name	Species	# aa	Chromosome Localization	Refs
YY1	Human	414	14q32	16
YY2	Human	372	Xq22. 1-22.2	11
Delta	Mouse	414	12F1	17
YY1	Rat	411	6q32	10,18
cYY1	Chicken	415		
FIII/xYY1	*Xenopus*	373		9
YY1	*Danio rerio* (Zebrafish)	357		
Pho	*Drosophila*	520	4	8
Pho-like	*Drosophila*	669	3	19
Rex-1	Mouse	288	8	1,12,20

YY1 Structure

YY1 protein contains modular features characteristic of many transcription factors: a DNA binding motif and domains important for transcriptional repression and activation (Fig. 1). The four-zinc-finger DNA-binding domain is located at its C-terminus (aa 298-320, 327-347, 355-377 and 385-407), which overlaps with one of the two repression domains, and is also the most conserved motif among YY1 across species (from *Drosophila* to human). The second repression domain is located in the central region (aa 170-200). Both repression domains have been shown to recruit histone deacetylases (HDACs) for repression.[21] The activation domain of YY1 lies at the N-terminus and is generally believed to be bipartite. It contains an acidic activation domain (aa16-29 and 43-53) and a glutamine-proline rich region (aa 80-100).[22] The acidic, glutamine- and proline-rich regions are common features to many transcription activators.[1,23-26] The fact that both repression and activation domains are present in YY1 provides the structural basis for the dual repression and activation activities of YY1. Between the two activation domains, there

* *Corresponding author. See list of "Contributors".*

Zinc Finger Proteins: From Atomic Contact to Cellular Function, edited by Shiro Iuchi and Natalie Kuldell.

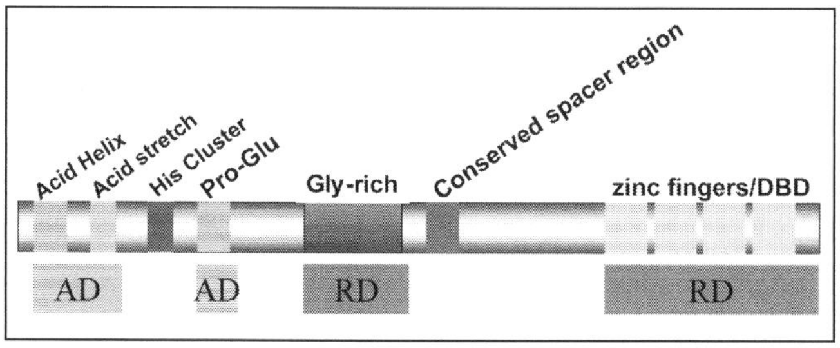

Figure 1. Structure of YY1. AD= activation domain; RD= repression domain; His= histidine; Pro= proline; Glu= glutamine; Gly= glycine; DBD= DNA binding domain.

are a glycine-rich region (aa44-69) and a cluster of 11 consecutive histidines (aa70-80).[5,22,27] The functions of these domains, if any, are not clear. Interestingly, these glycine- and histidine-rich regions are not conserved in *Xenopus* YY1 (xYY1), suggesting that they are either not important or they may engage in some mammalian-specific functions of YY1. However, deletion of the histidines did not affect the ability of human YY1 to repress transcription in the GAL4 fusion protein-based assays.[28] YY1 also contains sequences predicted to bind mono- and di-nucleotides (aa176-182 FFKKSGK and aa189-195 GAGAAG). However, bacterially purified YY1 does not appear to be capable of binding ATP or NAD (Shi lab, unpublished). It remains to be determined whether endogenous YY1 may bind NTPs or NAD/NADH and whether binding may affect YY1 activities.

YY1 Consensus Binding Site

The concensus binding site of YY1 contains a conserved 5'-CAT-3' core flanked by variable nucleotides. The following sequences probably represent the majority of YY1 binding sites, GACATNTT, VDCCATNWY (V= not T, D= not C, W=A or T, M=A or C, Y=C or T, N= any base) and CGGCATNTT.[29,30] The amino acids of human YY1 that contact DNA, identified by X-ray crystallography, are conserved in YY1 across species from human to *Drosophila*.[8,19,31] Consistent with this, homologues of YY1 from *Drosophila* to human bind to the same DNA consensus sequence.[8,19] Although the core nucleotides CAT normally determine the binding specificity of YY1, the flanking nucleotides contribute to the binding affinity of YY1 as illustrated by both binding site mutational studies and the observation of differential binding affinity of YY1 binding sites from different gene promoters.[32-36]

Subcellular Localization of YY1

With the exception of mouse and *Xenopus* early embryongenesis, YY1 is found to be a nuclear protein in both the soluble nuclear fraction and the insoluble nuclear matrix with a significant amount of YY1 concentrating in the nucleoli.[37-39] Although YY1 does not have a conventional nuclear localization signal, residues 332-398 have been shown to mediate nuclear entry and residues 256-340 target YY1 to the nuclear matrix.[40] Sub-nuclear locations of YY1 have been suggested to be important for regulating YY1's transcriptional activity.[40] For example, in vitro studies showed that B23, a nucleolar protein, could release the repression activity of YY1.[39,41] Thus, it would be interesting to identify signals that regulate YY1 nuclear distribution, proteins interacting with YY1 at these different locations and the potential influence on YY1 function.

The sub-cellular localization of YY1 in mouse early embryonic stages is unique in that YY1 is present in the cytoplasm of unfertilized oocytes and fertilized zygotes.[6] It begins to accumulate in the nucleus at the two-cell stage, coinciding with the onset of zygotic transcription. *Xenopus* YY1 remains as a cytoplasmic protein from oocytes to at least the neurula stage embryos which are 12 hr past the onset of zygotic transcription.[42] Further characterization suggests that xYY1 is part of a 480 KDa ribonucleoprotein complex which includes some maternally expressed mRNAs. However, it seems that this association is not responsible for preventing YY1 from entering the nucleus since intracellular injection of RNase released YY1 from this large complex but did not induce its nuclear localization.[42,43] These findings suggest possible novel regulation of YY1 during early development. Localization of YY1 in the cytoplasm may prevent it from participating in the transcriptional regulation of the zygotic genome during early embryogenesis. It is also possible that cytoplasmic YY1 may perform some yet to be identified functions (nontranscriptional) that are important for the early development of the zygotes such as the storage, metabolism and expression of the maternally derived mRNA. In this respect, it is interesting to note that the YY1 corebinding site is CAT/ATG, which is the signal for translational initiation. However, whether YY1 binds to mRNA directly needs further investigation.

Mechanisms of YY1-Mediated Transcription

Transcription is composed of multiple distinct steps with transcriptional initiation being the most extensively investigated process. Transcriptional initiation can be regulated at two levels: the assembly of the RNA polymerase-containing transcriptional machinery at the promoters and the local chromatin structure, which affects the accessibility of promoter DNA to transcription factors. Recent studies have provided convincing evidence supporting a critical role for histone modifications in transcription regulation by remodeling chromatin structure. As discussed below, YY1 appears to regulate RNA polymerase II-mediated transcription initiation by affecting both processes. YY1 may also regulate transcription mediated by RNA polymerase I and III.[5]

Chromatin Remodeling Is an Important Mechanism of YY1-Mediated Transcriptional Regulation

The most recent advance on the understanding of transcriptional control is the regulation of chromatin structure by post-translational modifications of histones and other chromatin packing proteins.[44-47] It is generally believed that histone hyper-acetylation and hypo-acetylation are associated with transcriptional activation and repression, respectively.[44] The histone acetylation status is regulated by HATs and HDACs which can interact with YY1 directly.[21,44,48,49] Histone methylation can either activate or repress transcription. Accumulating data support the idea that YY1 may activate or repress transcription through recruiting histone modifying enzymes such as histone acetytransferases (HATs), deacetylases (HDACs) and methytransferases (HMTs). YY1 has been shown to recruit the histone H4 (Arginine 3)-specific methytransferase, PRMT1, through an intermediate protein to YY1-activated promoters.[50] In addition, several HMTases protein complexes contain YY1-interacting proteins,[51-53] suggesting a potential indirect interaction between HMTases and YY1. However, it remains an outstanding issue as to the mechanisms that control the differential recruitment of these histone-modifying enzymes by YY1.

Histone Acetylases Interact with YY1 Directly and Regulate the Transcriptional Activities of YY1

The YY1 interacting proteins p300 and CBP are two closely related proteins that have HAT activity and function as transcriptional coactivators by acetylating histone tails (Fig. 2a).[21,44] However, their role in YY1-mediated transcription is not completely understood. An earlier study showed that adenovirus E1A could relieve YY1-mediated repression via interaction with p300, suggesting that p300 might play a role in YY1-mediated repression.[54] Consistent with this idea, a recent study showed that upon acetylation by p300/PCAF (p300/CBP associated factor), the acetylated YY1 has decreased DNA binding ability and increased affinity to HDACs.[21] Taken together, these findings suggest that interactions of YY1 with p300 may in some cases help maintain YY1 as a repressor. On the other hand, a positive role of CBP has been shown in the 25-hydroxyvitamin D (3)-24-hydroxylase promoter where it relieves YY1-mediated repression by preventing YY1 from contacting and inhibiting TFIIB function.[34] However, this action of CBP does not appear to require its HAT activity since a HAT defective CBP mutant is still able to relieve YY1-mediated repression. In contrast, a CBP mutant defective for YY1 interaction failed to relieve YY1 repression, highlighting the importance of protein-protein interactions.[34] It will be interesting to use a combination of ChIP and RNA interference (RNAi) technology to determine the histone acetylation status at these YY1-repressed promoters, which may provide insight into whether p300/CBP HAT activities play an active role in transcriptional regulation at these promoters.

Histone Deacetylases Are Required for Repression by YY1 in Several Promoters

YY1 can repress transcription through direct interactions with HDACs or indirect recruitment of HDACs (Fig. 2b). YY1 binds HDAC1 and 2 directly and this interaction is critical for YY1-mediated transcription repression at several promoters such as the HIV-1 long terminal repeat (LTR) and MHC class II gene.[49,55,56] YY1 may recruit HDACs indirectly through bridge proteins as well. Several YY1 binding proteins, including SAP30, FKBP25 (FK506-binding protein) and MTA2 (metastasis-associated protein 2), have been demonstrated to bind HDACs.[57-59] The interaction between YY1 and HDACs is regulated by post-translational modification of both YY1 and HDACs. Acetylated YY1 has increased affinity for HDACs,[21] while phosphorylation of HDAC1 disrupts its interaction with YY1.[60]

YY1 May Recruit Histone Methyltransferases to Regulate Transcription

Histone methylation has been demonstrated to play a role in both transcriptional repression and activation.[50,61-64] YY1 has recently been shown to activate transcription by recruiting the histone H4 (Arg 3)-specific methyltransferase, PRMT1, through DRBP76, an alternative splicing form of ILF3 (interleukin enhancer binding factor 3) which binds both YY1 and PRMT1 directly (Fig. 2c).[50] PRMT1 has also been demonstrated to facilitate transcriptional activation by nuclear hormone receptor.[64]

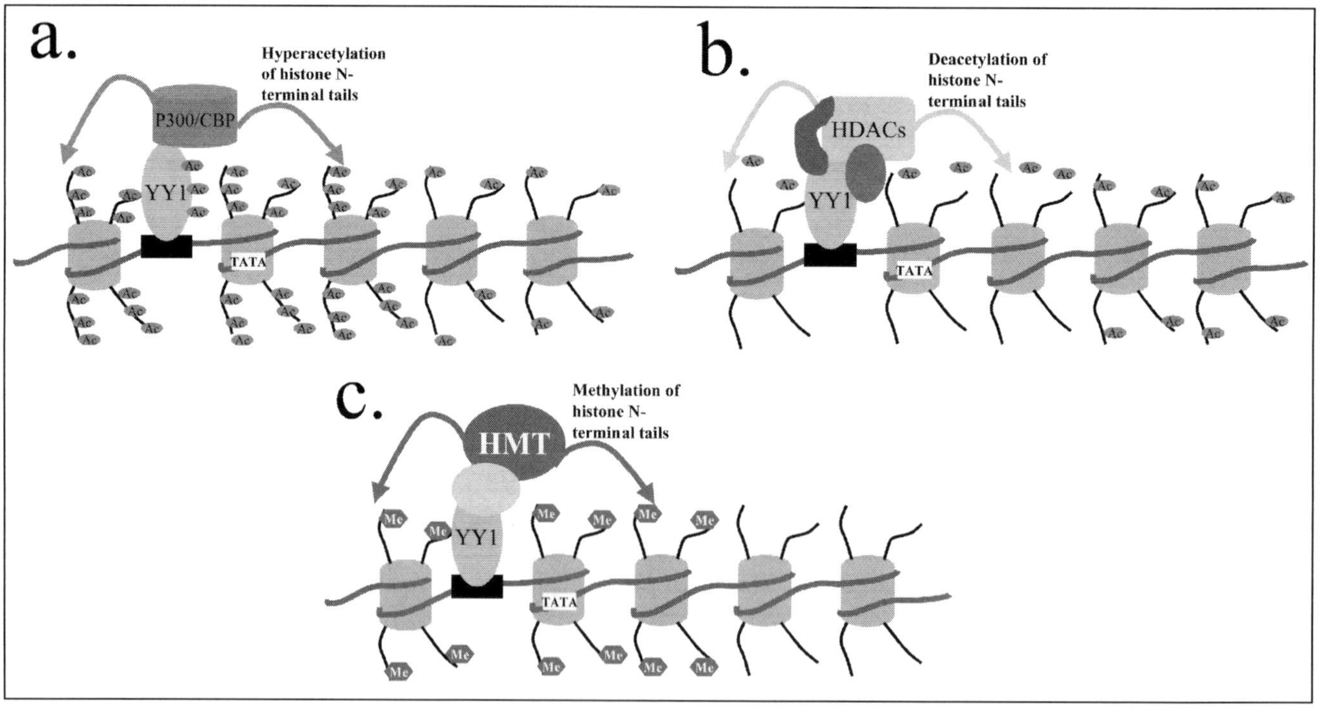

Figure 2. a) YY1 regulates transcription through DNA bending or chromatin remodeling. YY1 recruits coactivators such as p300 and CBP to acetylate histone tails or YY1 itself; b) YY1 interacts with HDACs either directly or indirectly through bridge proteins. HDACs then deacetylate histone tails and result in transcriptional repression; c) YY1 tethers HMT through intermediate molecules, HMT then regulates transcription by methylating histone tails. CBP= cAMP response element binding protein; HDACs= histone deacetylases; HMT= histone methyltransferase; Ac= acetyl; Me= methyl.

Recent studies suggest that YY1 may also recruit HMT-containing protein complexes to regulate transcription. Elegant genetic studies in Drosophila showed that YY1 repression requires the transcriptional corepressor CtBP.[7] CtBP was initially identified as an E1A-binding protein and has been shown to repress transcription by recruiting both HDACs and HMTs.[51,65] Thus, YY1-mediated repression may occur through the recruitment of the HMT-containing CtBP complex.

In summary, recruitment of histone modifiers appears to be an important mechanism for YY1-mediated transcriptional regulation. The role of HATs in YY1-mediated transcription appears more complicated and will require further investigation.

YY1 Regulates Transcription by Affecting the Basal Transcription Machinery

In addition to the chromatin-remodeling model, YY1 has been shown to regulate transcription by influencing the basal transcription machinery. YY1 may substitute the function of TBP (TATA-box binding protein) and directly recruit TFIIB (transcription factor IIB) and RNA polymerase II to the promoter region to initiate transcription. YY1 has also been shown to directly interact with several components of the basal transcription machinery. YY1 may also regulate the basal transcription machinery indirectly by interacting with other transcription factors, coactivators and corepressors.

YY1 May Function As an Initiator Element Binding Protein to Initiate Transcription in Vitro

Several studies suggest a potential role of YY1 to initiate transcription. YY1 binds to the transcription initiation sites of several promoters such as human PCNA, mouse Surf-1 and the AAV P5 promoter.[5,66] YY1 directly interacts with TFIIB and RNA polymerase II.[67] It was demonstrated that the combination of YY1, TFIIB and RNA polymerase II is sufficient to initiate transcription from the AAV P5 promoter in vitro.[67,68] The ability of YY1 to initiate transcription is augmented by the presence of a TATA motif or binding sites for the transcription factor Sp1, which interacts with YY1 directly.[69,70] Finally, the cocrystal structure of YY1 zinc fingers with the AAV P5 initiator element provided structural basis for understanding the YY1-dependent, TBP-independent transcription initiation.[31] The zinc fingers of YY1 bind to both strands of DNA upstream of the transcription initiation site while only the template strand downstream of the initiation site binds to YY1, allowing access of RNA polymerase II to the transcription initiation sites. This YY1-DNA binding permits transcription to occur only from the downstream of the YY1 binding sites. Transcription started upstream or in the opposite direction will be blocked by YY1-DNA binding which prevents the double strand separation. Since all these studies were conducted in vitro with the AAV P5 initiator, their in vivo significance remains to be investigated. Nevertheless, YY1 has been identified to exist in a large RNA polymerase II complex isolated from HeLa nuclear extracts that contains several general transcription factors but not TFIIB, TBP and TBP-associated factors,[71] suggesting a potential function for YY1 to initiate transcription in vivo (Fig. 3a) and providing a new angle to consider transcription initiation from TATA-less promoters.

YY1 May Regulate Transcription by Directly Contacting the General Transcription Factors (GTFs)

Several YY1-interacting proteins are members of the RNA polymerase II associated basal transcription machinery including RNA polymerase II, TBP, TFIIB and TAFII55 (TBP-associated factor II) (Fig. 3b).[22,67,72] The interaction between YY1 and these GTFs and the subsequent impact on transcriptional regulation remain largely unclear. It was shown that binding to TFIIB significantly increases the binding affinity of YY1 to its cognate sites while YY1 may repress transcription by sequestration of TFIIB to prevent it from interacting with transcription activators.[34,67]

Competition for Promoter Occupancy and Physical Interactions with Other Transcription Factors

An earlier model for YY1-mediated repression is competition with other activators for overlapping DNA binding sites in a promoter (Fig. 3c).[5] This model is supported by the convergence of recognition sites between YY1 and several transcription factors such as serum response factor (SRF) on the skeletal α-actin and c-fos promoter, NF-κB on the serum amyloid A 1 (SAA1) promoter. The competitive binding between YY1 and these transcription factors to these cognate sequences are regulated in different cellular conditions, contributing to differential gene expression.[73-75]

A second model is a combinatorial regulation of gene transcription by multiple transcription factors binding to their respective consensus sites in the same promoter region. In this case, the YY1 recognition sequence is close to but not overlapping with the DNA binding sites of other transcription factors. The close proximity of YY1 with other transcription factors facilitates their physical interactions at the promoter, which result in new patterns of gene expression. For example, binding of YY1 or E2F2/3 to their consensus sites in the cdc6 promoter had little effect on gene transcription by themselves.[76] However, the coordination of YY1 and E2F2/3 through the cofactor RYBP allows the synergistic activation of cdc6 transcription (Fig. 3d). On the other hand, it was shown that binding of LexA-YY1 fusion protein to the LexA-binding sites is important to the YY1-mediated repression on several activators binding to the nearby GAL4 binding sites in a reporter plasmid.[77] Although direct protein-protein interaction between YY1 and these transcription factors may occur, it is not necessary for at least some, if not all, combinatorial regulation.[77] The exact mechanism for the combinatorial regulation is also not clear. It is possible that this may allow differential recruitment of histone modifiers and/or interactions with the basal transcription machinery. It is often the case that multiple nonoverlapping binding sites are present in a single promoter, making this model a potentially common mechanism for transcriptional control involving YY1.

YY1 may also activate or repress transcription through direct interactions with other transcription factors without binding to DNA itself (Fig. 3e).[78] For example, YY1 inhibits the serum response element binding protein (SREBP)-mediated activation of the low-density lipoprotein (LDL) receptor gene in a dose-dependent manner.[78] This inhibition does not require DNA binding of YY1 but relies on the direct interaction between YY1

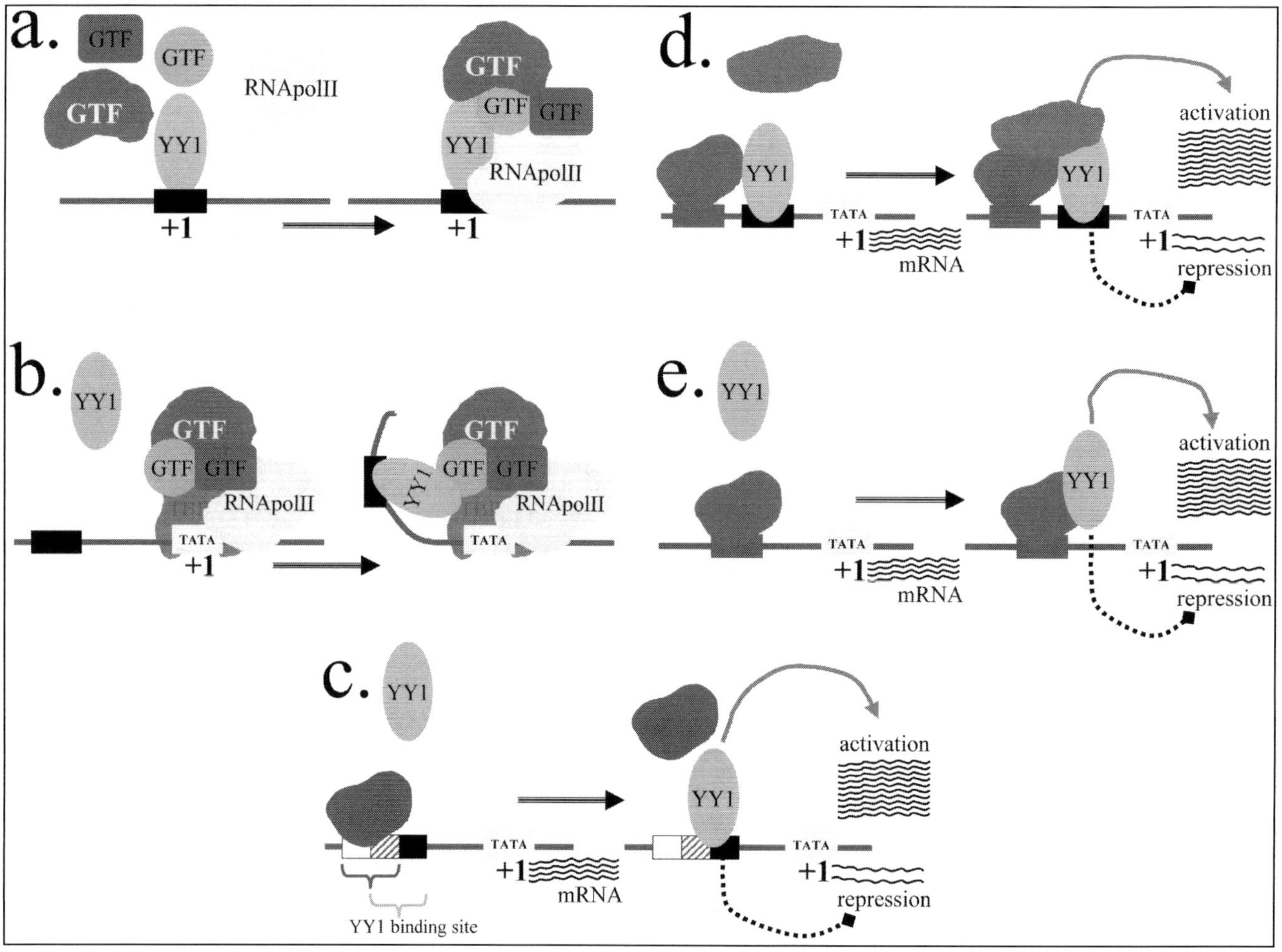

Figure 3. a) YY1 regulates transcription through protein-protein interaction. YY1 functions as an initiator element binding protein to assemble the basal transcription machinery. b) YY1 regulates transcription by contacting general transcription factors. c) YY1 regulates transcription by competing DNA binding with other transcription factors. d) Combinatorial regulation of transcription by YY1 and other transcription factors binding to the same promoter. e) YY1 either activates or represses transcription by interacting with other transcription factors without binding to DNA. GTF= general transcription factor; RNApolII= RNA polymerase II.

and Sp1, which may subsequently inhibit the Sp1-SREBP interaction necessary for transcription activation.[78] This model is further supported by another study showing YY1 repression of several activator- mediated transcription of a reporter gene containing no YY1 binding sites.[77] This DNA-binding independent activity of YY1 is expected to increase the repertoire of genes that can be regulated by YY1.

DNA Bending Plays a Role in Some YY1-Regulated Promoters

DNA bending induced by DNA-binding transcription factors such as TBP has been suggested to be an important transcription regulatory mechanism. Previous studies suggested that YY1 might regulate promoters such as c-fos by inducing DNA bending.[79,80] However, the absence of DNA bending in the cocrystal structure of the YY1 zinc fingers and the AAV P5 initiator element argue against DNA bending as a general feature of YY1-mediated transcription.[31]

In summary, YY1 mediates transcriptional regulation through multiple nonexclusive mechanisms. Current studies of YY1 transcription mechanisms have mostly been carried out in vitro. The in vivo significance of these models under different physiological and pathologic conditions awaits further studies.

Regulation of YY1's Transcriptional Activity

Previous studies suggest that YY1 can be regulated at multiple levels. Activity of YY1 can be modified by its DNA binding sites and its-interacting proteins. In addition, YY1 function also depends on its cellular localization and post-translational modifications. These different factors may help achieve cell and developmental specific control of YY1-mediated transcription.

DNA Binding Sites May Regulate the Transcriptional Activity of YY1

YY1 Binding to its consensus sites is likely to be influenced by several features of the DNA binding sites including the flanking nucleotides, CpG methylation and local chromatin structure. The differential binding of YY1 to a selected set of promoters in a specific cellular environment is critical for its tissue and developmental specific control of gene expression.

The local chromatin structure is important in determining whether a putative binding site is available for YY1 binding. For example, the developmentally-regulated Ig light chain κ gene expression by YY1 is probably due to a structural alteration of the chromatin which exposes the YY1 binding site and allows subsequent occupancy by YY1 in the plasma cell but not in early stages of B lymphocytes.[81]

DNA binding of YY1 has been shown to be sensitive to CpG methylation.[35,82,83] For example, methylation of the CpG sequences in the YY1 cognate sequences of the Peg3 promoter (paternally expressed gene 3) blocks YY1 binding completely.[82] It was further demonstrated that methylation of YY1 binding sites occurs only at the repressed maternal allele and YY1-binding is limited to the active paternal allele of the Peg3 in vivo.[82]

The nucleotide sequence of YY1 binding sites may affect YY1 binding affinity. YY1 sites from several promoters have been shown to display differential YY1 binding affinity. These sites, arranged in the order of decreasing affinity, are as follows: upstream enhancer of murine intracisternal A particle, human Ig κ light-chain 3' enhancer, upstream conserved region from the Moloney murine leukemia virus (MMLV) promoter.[35] Both mutational and polymorphism studies illustrate that even a single base change can cause elimination or generation of a YY1 binding site, and/or significant binding-affinity change.[32,33,84-89] Furthermore, the nucleotide sequence of a binding motif may contribute to determining whether YY1 functions as a repressor or an activator. This is probably because DNA can function as an allosteric regulator to change the three-dimensional conformation of transcription factors upon DNA binding as demonstrated by cocrystal structure of several DNA binding domains such as POU domain and their DNA binding sites from different gene promoters.[90,91] Minor changes of nucleotide sequence in a DNA motif may therefore affect the conformation of the bound transcription factor and subsequently influence the function of the transcription factors and their ability to recruit cofactors.

Regulation YY1 Activity by Its Protein Binding Partners

Most YY1 binding proteins are either transcription factors such as Sp1, c-myc, c-myb, CREB, or transcription coactivators and corepressors such as HDACs, p300, CBP, YAF2 (YY1-associated factor 2) and RYBP (ring1- and YY1- binding protein).[5,21,22,44,54,69,76,77,92-96] Binding of different proteins may exert dramatic functional consequence on the promoter context-dependent YY1 transcription activity. The C/EBP β-YY1 interaction in the upstream regulatory region (URR) of the human papillomavirus (HPV) type 18 P105 promoter is one of the best examples of how an interacting protein can influence YY1's transcriptional activity.[97-99] YY1 represses transcription of the HPV p105 promoter in HepG2 cells but activates its transcription in HeLa cells. This is because HeLa but not HepG2 cells express C/EBP β, which binds the p105 promoter and interacts with the adjacently bound YY1. This interaction is believed to cause a conformational change of YY1 and allows YY1 to function as an activator of the p105 promoter. In the absence of C/EBP β, YY1 functions as a repressor of the promoter as is the case in HepG2 cells. Consistently, ectopic expression of C/EBP β in HepG2 cells can convert YY1 from a repressor to an activator of the HPV P105 promoter, similar to what has been observed in HeLa cells.

Regulation of YY1 Transcription

YY1 is expressed in a biallelic manner that is further confirmed by the observation that cells isolated from heterozygous YY1 knockout mice display 50% reduction of YY1 protein level.[6,100] Thus far little is known about transcriptional regulation of the yy1 gene. Sequence analysis identified several Sp1 binding sites in both mouse and human YY1 promoter regions.[16,101] However, the specific role of Sp1 in yy1 gene transcription is not clear.

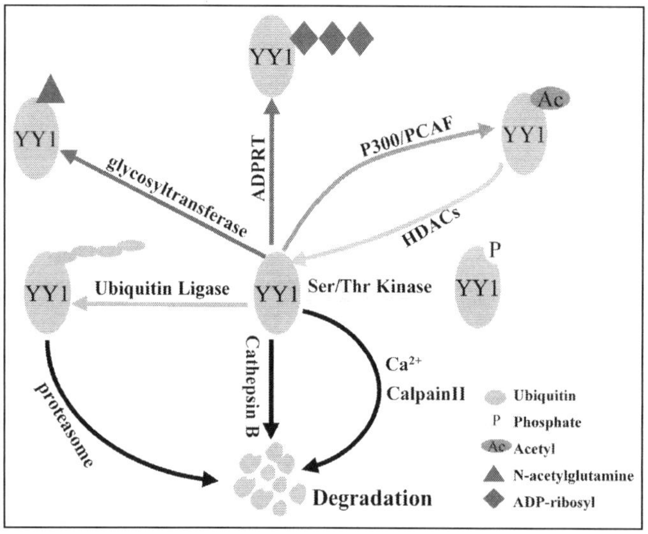

Figure 4. Modification and degradation pathways of YY1 protein. ADPRT= poly(ADP-ribosyl) transferase; PCAF= p300/CBP associated factor; HDACs= histone deacetylases; ser= serine; thr= threonine.

Regulation of YY1 Protein Levels by Protein Degradation

YY1 protein is relatively stable at different developmental stages and under different conditions. However, down-regulation of YY1 protein level, especially through increased protein degradation, may be critical for myogenic and myeloid differentiation in culture.[102-104] It has been demonstrated that YY1 can be a substrate for the calcium-activated neutral protease calpain II, the cathepsin B-like nuclear protease and the Ubiquitin-mediated 26 S proteasome (Fig. 4).[104,105] In addition, the proteolytic process of YY1 may be regulated by cellular factors such as YAF2 and extracellular signals such as interleukin 2 (IL-2).[96,106]

Post-Translational Modifications of YY1

YY1 has been shown to be subjected to various post-translational modifications, including phosphorylation, acetylation, glycosylation, ADP-ribosylation and ubiquitination (Fig. 4). These modifications appear to exert different effects on the activities and stability of YY1.

YY1 has been shown to be a relatively stable phospho-protein.[22] There are a total of eight potential serine/threonine (ser/thr) phosphorylation sites in human YY1 with two of them localized within the zinc finger domain.[107] Previous studies suggested that phosphorylation might influence YY1's DNA-binding ability. While phosphatase treatment abolished YY1 binding to the murine leukemia virus long terminal repeat (LTR) in Jurkat T cells and to the skeletal actin gene in neonatal rat cardiac myocytes, it increased DNA-binding of YY1 in stage VI *Xenopus* oocytes and to the α-element of mouse histone H3.2 gene promoter.[42,107-109] Unlike many other phospho-proteins, phosphatase treatment did not change the size of *Xenopus* YY1 on SDS gels.[42] What complicates the interpretation is that the phosphatase treatments were conducted with nuclear extracts. Therefore the observed effects on YY1 could also be due to dephosphorylation of YY1-interacting proteins. For instance, the dephosphorylated retinoblastoma protein (Rb) has been shown to bind YY1 and prevents it from binding to DNA while phosphorylated Rb releases YY1, permitting subsequent YY1 binding to DNA.[110]

YY1 can also be modified by glycosylation through enzymatic addition of N-acetylglucosamine (GlcNAc) to serine and threonine residues by nucleocytoplasmic glycosyltransferase in HeLa cells.[111] Glycosylation appears to disrupt the interaction between YY1 and Rb, adding a potential mechanism by which YY1 regulates cell cycle.[111]

Transient poly(ADP-ribosyl)ation of YY1 has been detected in HeLa cells immediately after genotoxic treatment.[112,113] Poly(ADP-ribosylation) is catalyzed by the enzyme poly(ADP-ribosyl) transferase (ADPRT) and is believed to be a cellular protective response to DNA damage.[114] Poly(ADP-ribosylation) decreased the binding affinity of YY1 to DNA and thus may prevent transcription from damaged DNA.[112,115] Furthermore, ADP-ribosylated YY1 may expedite the process of DNA repair mediated by ADPRT by increasing its enzymatic activity through protein-protein interaction.[112-116] Thus, YY1 may play a role in DNA repair.

YY1 can also be acetylated by p300 and PCAF (p300/CBP associated factor) and deacetylated by HDAC1 and HDAC2.[21] Sequence analysis reveals that there are 6 lysine residues that are potential substrates for acetylation within one of the two YY1's HDAC interaction domains (residues 170-200). Acetylation of YY1 residues 170-200 might be important for YY1-mediated transcription repression, perhaps due to an increased recruitment of the repressive HDACs.[21] Acetylation of YY1 at residues 261-333 overlaps with the DNA binding domains (residues 298-414) and decreases YY1's DNA-binding ability.[21] In addition, HDACs do not deacetylate YY1 at residues 261-333 in vitro.[21] The question is how these two sets of enzymes with antagonistic functions regulate YY1 in vivo on different promoters.

Several studies suggest that YY1 can be ubiquitinated and ubiquitin-mediated proteasomal degradation allows the cells to rapidly reduce YY1 protein level in response to various signals. It was demonstrated that purified YY1 could be readily polyubiquitinated in reticulocyte lysates.[105] Treatment with the proteasome inhibitors MG132 and lactacystin stabilizes YY1 during in vitro muscle cell differentiation.[105] In addition, down-regulation of YY1 protein level during serum deprivation induced apoptosis in cultured mouse muscle satellite cells and is accompanied by the accumulation of ubiquitinated YY1.[117]

In summary, it is possible that all the above-discussed mechanisms contribute to the regulation of YY1 activities, allowing this ubiquitously expressed stable protein to achieve tissue- and developmental-specific control of gene expression. Future studies will provide us with a better understanding on how these different mechanisms function and cooperate at the molecular levels.

The Role of YY1 During Development

YY1 probably plays important roles in multiple stages of development. The embryonic death of the constitutive YY1 knockout mice provided the first evidence of a crucial function for YY1 during embryogenesis.[6] However, the molecular basis for this early lethality is currently unclear. The neurulation defects observed in a subset of YY1 heterozygous mice and the studies of YY1 on *Xenopus* development suggest a critical role for YY1 during embryonic neural development.[6] Cell culture studies support a role for YY1 in cell proliferation and differentiation, implicating a potential role for YY1 during development of other systems and at a later developmental stage. Consistent with the above hypothesis, YY1 is likely to be the mammalian homolog of the *Drosophila* PcG protein Pho. The multiple phenotypic defects associated with the Pho mutants suggest a similar role for YY1 during mammalian development through its PcG function.[118] Below we discuss the role of YY1 in cell proliferation, cell differentiation, neural development and its PcG function.

Cell Proliferation and Differentiation

Cell culture studies suggest that YY1 promote cell growth and proliferation by activating transcription of genes whose products are required for cell cycle transition and progression. For instance, previous studies showed that YY1 activates transcription of the proto-oncogene, c-myc.[119] Increased c-myc protein stimulates G1 to S transition by activating transcription of E2F2 and E2F3 which subsequently activate several other molecules required for S phase entry.[76] YY1 also collaborates with E2F2 and E2F3 to activate transcription of genes essential for DNA replication such as cdc6.[76] In addition, YY1 has been shown to activate transcription of such genes as ribonucleotide reductase, proliferating cell nuclear antigen (PCNA) and the replication-dependent histones.[109,120-122] The protein products of these genes are either essential for S phase DNA synthesis or important for packaging newly synthesized DNA. Consistent with a role for YY1 in cell proliferation, YY1 heterozygous cells have been shown to display significantly impaired cell proliferation, supporting a dosage-dependent role for YY1 in cell growth control (Shi lab, unpublished result). It should be noted that although the above-mentioned genes are mostly engaged in G1/S transition and S phase progression, it has not been well characterized as to how YY1 regulates different phases of cell cycle.

Numerous studies show that YY1 is important in the transcriptional control of many differentiation-specific genes in cell culture, suggesting a potential role for YY1 in cell differentiation. However, the exact role for YY1 in differentiation remains largely unclear, and sometimes controversial. It appears that YY1 may promote cell differentiation in some systems by activating the transcription of differentiation associated gene expression. These include peripheral B lymphocyte differentiation, erythroid differentiation and myeloid differentiation.[2,81,110,123-128] In contrary, it has been shown that YY1 inhibits cell differentiation in several cultured systems including myeloid differentiation, myogenic differentiation, neuronal differentiation and mesenchymal differentiation.[73,75,102-104,129-133] This idea is supported by the observation that transient down-regulation of YY1 protein level and/or activity is correlated with cell differentiation,[73,75,102-104,129-132] while over-expression of YY1 promotes cell proliferation and inhibits cell differentiation.

It must be pointed out that these cell culture studies may or may not reflect the actual in vivo function of YY1. Direct genetic evidence is necessary and crucial to establishing a role for YY1 in cell proliferation and differentiation.

Neural Development

Two lines of evidence strongly support a role for YY1 in the induction and patterning of embryonic neural development. First, some YY1 heterozygous mouse embryos exhibit neurulation defects resembling exencephaly.[6] Second, the role for YY1 in *Xenopus* neural development is shown by both over-expression and knockdown studies.[52,134] Knocking down xYY1 by antisense oligonucleotide resulted in abnormal antero-posterior axial pattern and reduction of head structures in *Xenopus* embryo after the tailbud stage. This is accompanied by reduced expression of the pan-neural marker NCAM, the midbrain/hindbrain junction marker En2, the hindbrain marker Krox20 and the forebrain marker Otx2 but not the spinal cord marker HoxB9.[134] Consistent with this, over-expression of exogenous XYY1 induced ectopic neural axis at the injection site in *Xenopus* embryos with the induction of NCAM and two other neural markers NPR-1 and ANF.[52] Furthermore, over-expression of xEED or *Drosophila* Pho results in similar phenotype and over-expression of xYY1 and xEED together resulted in synergistic effects, suggesting that the interplay between YY1 and EED may be important for the ectopic induction of the neural structure.[52] Thus, characterization of the direct YY1 targets will help us understand the critical events during neural development.

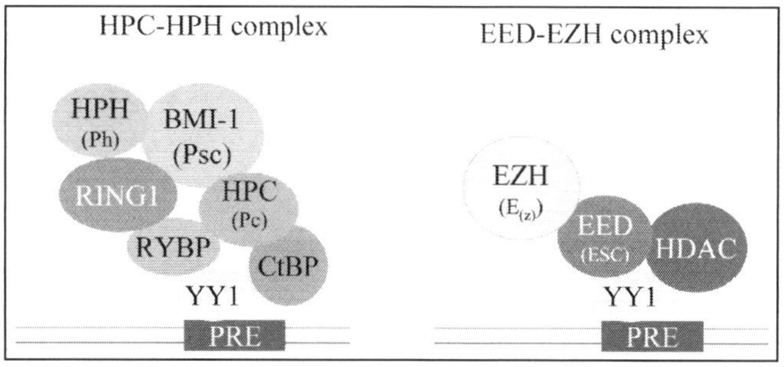

Figure 5. Model for YY1 to recruit PcG complexes to PREs. PRE= polycomb response element; HPH= human homolog of Polyhomeotic; HPC= human homolog of Polycomb; RYBP= ring 1- and YY1- binding protein; BMI-1= B-cell-specific Moloney murine leukemia virus integration site 1; CtBP= c-terminal binding protein; EZH= human homolog of enhancer of Zeste; EED= embryonic ectoderm development.

YY1 As a PcG Protein

Both sequence analysis and functional studies suggest that YY1 is a mammalian counterpart of the *Drosophila* PcG protein Pho. The DNA binding domain of YY1 and Pho share 96% sequence homology and binds to the same consensus DNA sites.[8] Over-expression of exogenous *Xenopus* YY1 or Pho in *Xenopus* embryos caused similar phenotypes that are dependent on another PcG protein EED.[52] YY1 also partially rescue the Pho mutants.[7] Furthermore, in vivo YY1-mediated repression seems to depend on other PcG proteins including Pc (polycomb), Pcl (polycomb-like), Scm (Sex combs on midleg), E(z) and Esc.[7]

PcG proteins function as high molecular weight complexes and regulate transcriptional repression of target genes, mainly homeotic genes, by binding to the DNA sequence called PcG response element (PRE).[135-137] There are two multimeric PcG complexes identified thus far, the human HPC-HPH complex and the EED-EZH2 complex, corresponding to the *Drosophila* PRC 1-complex and E(z)/Esc complex, respectively.[135] It is largely unclear how PcG complexes are recruited to their target genes since most of the PcG proteins do not have sequence-specific DNA binding ability except for the *Drosophila* Pho and the mammalian YY1 and Mel-18.[7,19,138] Previous studies demonstrated that Pho can link both PcG complexes to PREs by a direct interaction with Esc from E(z)/Esc complex and Pc, Ph from PRC 1-complex.[52,139] Pho-binding sites are found in most PREs such as those from the *engrailed*, the *Ultrabithorax* and the *Abdominal-B* promoter regions and are necessary for the PRE function in PcG repression.[8,19,136,139,140] Furthermore, immunolocalization studies show that pho and pho-like proteins collaborate to anchor Pc, Psc, and Scm to some polytene chromosomal sites.[19] YY1 also interacts with EED directly and may also interact with the HPC-HPH complexes indirectly through bridging proteins such as CtBP or RYBP.[7,52,76,93,141] Thus, one attractive hypothesis is that YY1, the mammalian homologue of Pho, may also tether PcG proteins to target genes as illustrated in Figure 5. It is reported that YY1 binds the promoters of several homeobox genes such as gp91, Msx2, Hoxb4 and Hoxb7.[128,142-144] However, YY1 seems to activate, rather than repress, transcription of gp91 and Msx2 and therefore contributes to myeloid differentiation and early mouse craniofacial and limb morphogenesis.[21,142] Future experiments that examine the function of YY1 in mammalian PcG repression will provide more direct evidence to resolve this issue.

YY1 may regulate development through its PcG mediated activities. Knockout mice of YY1, EED and EZH2 all die at early embryonic stages, suggesting related developmental functions.[6,145,146] As discussed earlier, YY1 is probably essential for embryonic neural development and its function relies on the interplay with EED.[52,134] Mutant animal studies have also demonstrated a critical role for several PcG proteins such as EED, EZH2 and Bim1 in mouse lymphopoiesis.[135,147-149] It is possible that YY1 may play a role in lymphocyte development through its interactions with these proteins.

YY1 and Diseases

YY1 may play a role in a myriad of pathological conditions including viral infection and cancer. The YY1-related pathology can be the result of loss of function as demonstrated by decreased YY1 protein level or elimination of YY1 binding sites, or a gain of function due to YY1 over-expression or a newly created YY1 binding site. Although it is currently unclear whether the disease-associated YY1 changes are the cause of the pathogenesis, abnormal YY1 activities are likely to contribute to the development, progress and complications of diseases.

Viral Infection

YY1 has been suggested to influence infection by viruses such as HPV, Epstein Bar (EB) virus, human Immunodeficiency virus type I (HIV-1).[48,150,151] It appears that YY1 may play a positive role in limiting acute viral infection and inhibit viral oncogenesis by activating multiple protective mechanisms in the host against viral infection and by inhibiting viral gene transcription. For instance, YY1 induces transcription of β-interferon, one of the first responses of cell to viral infection.[152] YY1 may help decrease viral infection by down-regulating cell surface expression of viral receptors as demonstrated in the HIV infection.[153-155] YY1 may reduce virion production by repressing viral immediate early gene expression and by interfering with viral protein function through protein-protein interaction.[48,55,150,156-159] All these different host-responses mediated by YY1 may contribute to viral latency control. YY1 may also inhibit virus-induced malignant transformation by repressing viral oncogene expression such as HPV E6 and E7.[151,160,161] An effective mechanism developed by the virus is to create mutations at YY1 binding sites in the viral promoter region as identified in many HPV 16-positive cervical tumors.[160,162-164] These mutations decrease or eliminate YY1 binding and are expected to release the YY1-mediated repression, resulting in over-expression of viral oncoproteins, which promote oncogenesis.[165-167] Thus, understanding the function of YY1 will facilitate our understanding in both short-term and long-term host-viral interaction, and provide information on how to fight against viral infection and viral-related cancer.

Cancer

Both loss of YY1 function and over-expression of YY1 have been associated with malignancy. Abnormal YY1 protein level is detected in the retroviral-induced mouse leukemia/lymphoma, some cases of human acute myeloid leukemia, carcinogen-induced hepatocarcinoma and breast tumors.[102,168,169] Mutation of YY1 binding sites has been associated with the occurrence or recurrence of malignancy such as cervical cancer and cutaneous basal cell carcinoma.[162-165,170-172] Abnormal YY1 function may promote malignant transformation by interfering with both oncogenes and tumor suppressor genes. For example, YY1 regulates transcription of both cellular and viral oncogenes such as c-myc and the HPV E6 and E7.[161,167,173-175] YY1 may also regulate expression of tumor suppressor genes such as p53 and CTCF.[169,176-178] YY1 also directly binds oncoproteins and tumor suppressors and these interactions may in turn affect cell proliferation and transformation.[1,110,173] For example, modulation of YY1 activity is essential for adenovirus E1A to induce oncogenic transformation and to inhibit differentiation.[5] Thus, modulating YY1 function and protein level may represent a potential strategy in the prevention and treatment of cancer.

Others

YY1 may play a role in other disease processes as well. A recent study has implicated YY1 in the autosomal dominant inherited myopathy, facioscapulo-humeral muscular dystrophy (FSHD).[179] A YY1 containing protein complex is required to repress human chromosome 4q35 gene transcription by binding to an internal tandem repeats termed D4Z4. Loss of D4Z4 results in over-expression of 4q35 genes including ANT1, FRG1 and FRG2 with subsequent myopathy. YY1 may also be associated with the development and severity of allergies/asthma and cystic fibrosis by activating transcription of TGF-β and cystic fibrosis trans-membrane conductance regulator (CFTR) gene, respectively.[85,180]

Summary

Significant advances have been made in the past decade towards understanding of YY1. These studies have identified a crucial role for YY1 in both normal development and some pathological situations. Using conditional gene targeting technology coupled with global gene expression profiling and molecular analysis, the next decade will witness tremendous strides in understanding the role of YY1 under physiological and pathological conditions, and the molecular mechanisms that underlie the in vivo functions of YY1. Importantly, insights into YY1 may help future prevention and treatment of several diseases such as viral infection and cancer.

Acknowledgements

We thank Nathan Wall for critical reading of this manuscript and Guangchao Sui for information on YY1. YS is supported by NIH GM53874.

References

1. Shi Y, Seto E, Chang LS et al. Transcriptional repression by YY1, a human GLI-Kruppel-related protein, and relief of repression by adenovirus E1A protein. Cell 1991; 67:377-88.
2. Park K, Atchison ML. Isolation of a candidate repressor/activator, NF-E1 (YY-1, delta), that binds to the immunoglobulin kappa 3' enhancer and the immunoglobulin heavy-chain mu E1 site. Proc Natl Acad Sci USA 1991; 88:9804-8.
3. Hariharan N, Kelley DE, Perry RP. Delta, a transcription factor that binds to downstream elements in several polymerase II promoters, is a functionally versatile zinc finger protein. Proceedings of the National Academy of Sciences of the United States of America 1991; 88:9799-803.
4. Flanagan JR, Becker KG, Ennist DL et al. Cloning of a negative transcription factor that binds to the upstream conserved region of Moloney murine leukemia virus. Molecular & Cellular Biology 1992; 12:38-44.
5. Shi Y, Lee JS, Galvin KM. Everything you have ever wanted to know about Yin Yang 1. Biochim Biophys Acta 1997; 1332:F49-66.
6. Donohoe ME, Zhang X, McGinnis L et al. Targeted disruption of mouse Yin Yang 1 transcription factor results in peri-implantation lethality. Mol Cell Biol 1999; 19:7237-44.
7. Atchison L, Ghias A, Wilkinson F et al. Transcription factor YY1 functions as a PcG protein in vivo. Embo J 2003; 22:1347-58.
8. Brown JL, Mucci D, Whiteley M et al. The Drosophila Polycomb group gene pleiohomeotic encodes a DNA binding protein with homology to the transcription factor YY1.[comment]. Molecular Cell 1998; 1:1057-64.
9. Pisaneschi G, Ceccotti S, Falchetti ML et al. Characterization of FIII/YY1, a Xenopus laevis conserved zinc-finger protein binding to the first exon of L1 and L14 ribosomal protein genes. Biochem Biophys Res Commun 1994; 205:1236-42.
10. Chen KS, Paladugu A, Aldaz CM et al. Cloning and chromosomal localization of the rat Stat5 and Yy1 genes. Cytogenet Cell Genet 1996; 74:277-80.
11. Sulston JE, Waterston R. Toward a complete human genome sequence. Genome Res 1998; 8:1097-108.
12. Hosler BA, LaRosa GJ, Grippo JF et al. Expression of REX-1, a gene containing zinc finger motifs, is rapidly reduced by retinoic acid in F9 teratocarcinoma cells. Molecular & Cellular Biology 1989; 9:5623-9.
13. Nishiguchi S, Joh T, Horie K et al. A survey of genes expressed in undifferentiated mouse embryonal carcinoma F9 cells: Characterization of low-abundance mRNAs. Journal of Biochemistry 1994; 116:128-39.
14. Scherer CA, Chen J, Nachabeh A et al. Transcriptional specificity of the pluripotent embryonic stem cell. Cell Growth & Differentiation 1996; 7:1393-401.
15. Ramalho-Santos M, Yoon S, Matsuzaki Y et al. "Stemness": Transcriptional profiling of embryonic and adult stem cells. Science 2002; 298:597-600.
16. Yao YL, Dupont BR, Ghosh S et al. Cloning, chromosomal localization and promoter analysis of the human transcription factor YY1. Nucleic Acids Res 1998; 26:3776-83.
17. Zhu W, Lossie AC, Camper SA et al. Chromosomal localization of the transcription factor YY1 in the mouse and human. Mamm Genome 1994; 5:234-6.
18. Nishiyama C, Yokota T, Nishiyama M et al. Molecular cloning of rat transcription factor YY1. Biosci Biotechnol Biochem 2003; 67:654-8.
19. Brown JL, Fritsch C, Mueller J et al. The Drosophila pho-like gene encodes a YY1-related DNA binding protein that is redundant with pleiohomeotic in homeotic gene silencing. Development 2003; 130:285-94.
20. Hosler BA, Rogers MB, Kozak CA et al. An octamer motif contributes to the expression of the retinoic acid-regulated zinc finger gene Rex-1 (Zfp-42) in F9 teratocarcinoma cells. Molecular & Cellular Biology 1993; 13:2919-28.
21. Yao YL, Yang WM, Seto E. Regulation of transcription factor YY1 by acetylation and deacetylation. Mol Cell Biol 2001; 21:5979-91.

22. Austen M, Luscher B, Luscher-Firzlaff JM. Characterization of the transcriptional regulator YY1. The bipartite transactivation domain is independent of interaction with the TATA box-binding protein, transcription factor IIB, TAFII55, or cAMP-responsive element-binding protein (CPB)-binding protein. J Biol Chem 1997; 272:1709-17.
23. Xiao H, Friesen JD, Lis JT. A highly conserved domain of RNA polymerase II shares a functional element with acidic activation domains of upstream transcription factors. Molecular & Cellular Biology 1994; 14:7507-16.
24. Brou C, Wu J, Ali S et al. Different TBP-associated factors are required for mediating the stimulation of transcription in vitro by the acidic transactivator GAL-VP16 and the two nonacidic activation functions of the estrogen receptor. Nucleic Acids Research 1993; 21:5-12.
25. Defossez PA, Baert JL, Monnot M et al. The ETS family member ERM contains an alpha-helical acidic activation domain that contacts TAFII60. Nucleic Acids Research 1997; 25:4455-63.
26. Kim TK, Roeder RG. Proline-rich activator CTF1 targets the TFIIB assembly step during transcriptional activation. Proceedings of the National Academy of Sciences of the United States of America 1994; 91:4170-4.
27. Bushmeyer S, Park K, Atchison ML. Characterization of functional domains within the multifunctional transcription factor, YY1. J Biol Chem 1995; 270:30213-20.
28. Lee JS, See RH, Galvin KM et al. Functional interactions between YY1 and adenovirus E1A. Nucleic Acids Res 1995; 23:925-31.
29. Yant SR, Zhu W, Millinoff D et al. High affinity YY1 binding motifs: Identification of two core types (ACAT and CCAT) and distribution of potential binding sites within the human beta globin cluster. Nucleic Acids Res 1995; 23:4353-62.
30. Hyde-DeRuyscher RP, Jennings E, Shenk T. DNA binding sites for the transcriptional activator/repressor YY1. Nucleic Acids Res 1995; 23:4457-65.
31. Houbaviy HB, Usheva A, Shenk T et al. Cocrystal structure of YY1 bound to the adeno-associated virus P5 initiator. Proc Natl Acad Sci USA 1996; 93:13577-82.
32. Ma SL, Lovmand J, Sorensen AB et al. Triple basepair changes within and adjacent to the conserved YY1 motif upstream of the U3 enhancer repeats of SL3-3 murine leukemia virus cause a small but significant shortening of latency of T-lymphoma induction. Virology 2003; 313:638-44.
33. Sitwala KV, Adams K, Markovitz DM. YY1 and NF-Y binding sites regulate the transcriptional activity of the dek and dek-can promoter. Oncogene 2002; 21:8862-70.
34. Raval-Pandya M, Dhawan P, Barletta F et al. YY1 represses vitamin D receptor-mediated 25-hydroxyvitamin D(3)24-hydroxylase transcription: Relief of repression by CREB-binding protein. Mol Endocrinol 2001; 15:1035-46.
35. Satyamoorthy K, Park K, Atchison ML et al. The intracisternal A-particle upstream element interacts with transcription factor YY1 to activate transcription: Pleiotropic effects of YY1 on distinct DNA promoter elements. Mol Cell Biol 1993; 13:6621-8.
36. L'Honore A, Lamb NJ, Vandromme M et al. MyoD distal regulatory region contains an SRF binding CArG element required for MyoD expression in skeletal myoblasts and during muscle regeneration. Mol Biol Cell 2003; 14:2151-62.
37. McNeil S, Guo B, Stein JL et al. Targeting of the YY1 transcription factor to the nucleolus and the nuclear matrix in situ: The C-terminus is a principal determinant for nuclear trafficking. J Cell Biochem 1998; 68:500-10.
38. Guo B, Odgren PR, van Wijnen AJ et al. The nuclear matrix protein NMP-1 is the transcription factor YY1. Proc Natl Acad Sci USA 1995; 92:10526-30.
39. Inouye CJ, Seto E. Relief of YY1-induced transcriptional repression by protein-protein interaction with the nucleolar phosphoprotein B23. J Biol Chem 1994; 269:6506-10.
40. Bushmeyer SM, Atchison ML. Identification of YY1 sequences necessary for association with the nuclear matrix and for transcriptional repression functions. J Cell Biochem 1998; 68:484-99.
41. Chan PK, Chan FY, Morris SW et al. Isolation and characterization of the human nucleophosmin/B23 (NPM) gene: Identification of the YY1 binding site at the 5' enhancer region. Nucleic Acids Res 1997; 25:1225-32.
42. Ficzycz A, Eskiw C, Meyer D et al. Expression, activity, and subcellular localization of the Yin Yang 1 transcription factor in *Xenopus oocytes* and embryos. J Biol Chem 2001; 276:22819-25.
43. Ficzycz A, Ovsenek N. The Yin Yang 1 transcription factor associates with ribonucleoprotein (mRNP) complexes in the cytoplasm of *Xenopus oocytes*. J Biol Chem 2002; 277:8382-7.
44. Thomas MJ, Seto E. Unlocking the mechanisms of transcription factor YY1: Are chromatin modifying enzymes the key? Gene 1999; 236:197-208.
45. Peterson CL. Chromatin remodeling enzymes: Taming the machines. Third in review series on chromatin dynamics. EMBO Reports 2002; 3:319-22.
46. Nielsen AL, Oulad-Abdelghani M et al. Heterochromatin formation in mammalian cells: Interaction between histones and HP1 proteins. Molecular Cell 2001; 7:729-39.
47. Galarneau L, Nourani A, Boudreault AA et al. Multiple links between the NuA4 histone acetyltransferase complex and epigenetic control of transcription. Molecular Cell 2000; 5:927-37.
48. He G, Margolis DM. Counterregulation of chromatin deacetylation and histone deacetylase occupancy at the integrated promoter of human immunodeficiency virus type 1 (HIV-1) by the HIV-1 repressor YY1 and HIV-1 activator Tat. Mol Cell Biol 2002; 22:2965-73.
49. Yang WM, Inouye C, Zeng Y et al. Transcriptional repression by YY1 is mediated by interaction with a mammalian homolog of the yeast global regulator RPD3. Proc Natl Acad Sci USA 1996; 93:12845-50.
50. Rezai-Zadeh N, Zhang X, Namour F et al. Targeted recruitment of a histone H4-specific methyltransferase by the transcription factor YY1. Genes Dev 2003; 17:1019-29.
51. Shi Y, Sawada J, Sui G et al. Coordinated histone modifications mediated by a CtBP co-repressor complex. Nature 2003; 422:735-8.
52. Satijn DP, Hamer KM, den Blaauwen J et al. The polycomb group protein EED interacts with YY1, and both proteins induce neural tissue in Xenopus embryos. Mol Cell Biol 2001; 21:1360-9.
53. Ogawa H, Ishiguro K, Gaubatz S et al. A complex with chromatin modifiers that occupies E2F- and Myc-responsive genes in G0 cells. Science 2002; 296:1132-6.
54. Lee JS, Galvin KM, See RH et al. Relief of YY1 transcriptional repression by adenovirus E1A is mediated by E1A-associated protein p300. Genes Dev 1995; 9:1188-98.
55. Coull JJ, Romerio F, Sun JM et al. The human factors YY1 and LSF repress the human immunodeficiency virus type 1 long terminal repeat via recruitment of histone deacetylase 1. J Virol 2000; 74:6790-9.
56. Osborne A, Zhang H, Yang WM et al. Histone deacetylase activity represses gamma interferon-inducible HLA-DR gene expression following the establishment of a DNase I-hypersensitive chromatin conformation. Mol Cell Biol 2001; 21:6495-506.
57. Huang NE, Lin CH, Lin YS et al. Modulation of YY1 activity by SAP30. Biochemical & Biophysical Research Communications 2003; 306:267-75.
58. Yao YL, Yang WM. The metastasis-associated proteins 1 and 2 form distinct protein complexes with histone deacetylase activity. J Biol Chem 2003; 278:42560-8.
59. Yang WM, Yao YL, Seto E. The FK506-binding protein 25 functionally associates with histone deacetylases and with transcription factor YY1. Embo J 2001; 20:4814-25.
60. Galasinski SC, Resing KA, Goodrich JA et al. Phosphatase inhibition leads to histone deacetylases 1 and 2 phosphorylation and disruption of corepressor interactions. J Biol Chem 2002; 277:19618-26.
61. Sewalt RG, Lachner M, Vargas M et al. Selective interactions between vertebrate polycomb homologs and the SUV39H1 histone lysine methyltransferase suggest that histone H3-K9 methylation contributes to chromosomal targeting of Polycomb group proteins. Molecular & Cellular Biology 2002; 22:5539-53.

62. Kuzmichev A, Nishioka K, Erdjument-Bromage H et al. Histone methyltransferase activity associated with a human multiprotein complex containing the Enhancer of Zeste protein. Genes & Development 2002; 16:2893-905.
63. Cao R, Wang L, Wang H et al. Role of histone H3 lysine 27 methylation in Polycomb-group silencing. Science 2002; 298:1039-43.
64. Wang H, Huang ZQ, Xia L et al. Methylation of histone H4 at arginine 3 facilitating transcriptional activation by nuclear hormone receptor. Science 2001; 293:853-7.
65. Sewalt RG, Gunster MJ, van der Vlag J et al. C-Terminal binding protein is a transcriptional repressor that interacts with a specific class of vertebrate Polycomb proteins. Molecular & Cellular Biology 1999; 19:777-87.
66. Seto E, Shi Y, Shenk T. YY1 is an initiator sequence-binding protein that directs and activates transcription in vitro. Nature 1991; 354:241-5.
67. Usheva A, Shenk T. YY1 transcriptional initiator: Protein interactions and association with a DNA site containing unpaired strands. Proc Natl Acad Sci USA 1996; 93:13571-6.
68. Usheva A, Shenk T. TATA-binding protein-independent initiation: YY1, TFIIB, and RNA polymerase II direct basal transcription on supercoiled template DNA. Cell 1994; 76:1115-21.
69. Lee JS, Galvin KM, Shi Y. Evidence for physical interaction between the zinc-finger transcription factors YY1 and Sp1. Proc Natl Acad Sci USA 1993; 90:6145-9.
70. Seto E, Lewis B, Shenk T. Interaction between transcription factors Sp1 and YY1. Nature 1993; 365:462-4.
71. Maldonado E, Shiekhattar R, Sheldon M et al. A human RNA polymerase II complex associated with SRB and DNA-repair proteins.[erratum appears in Nature 1996 Nov 28;384(6607):384]. Nature 1996; 381:86-9.
72. Chiang CM, Roeder RG. Cloning of an intrinsic human TFIID subunit that interacts with multiple transcriptional activators. Science 1995; 267:531-6.
73. Lee TC, Shi Y, Schwartz RJ. Displacement of BrdUrd-induced YY1 by serum response factor activates skeletal alpha-actin transcription in embryonic myoblasts. Proc Natl Acad Sci USA 1992; 89:9814-8.
74. Lu SY, Rodriguez M, Liao WS. YY1 represses rat serum amyloid A1 gene transcription and is antagonized by NF-kappa B during acute-phase response. Mol Cell Biol 1994; 14:6253-63.
75. Gualberto A, LePage D, Pons G et al. Functional antagonism between YY1 and the serum response factor. Mol Cell Biol 1992; 12:4209-14.
76. Schlisio S, Halperin T, Vidal M et al. Interaction of YY1 with E2Fs, mediated by RYBP, provides a mechanism for specificity of E2F function. Embo J 2002; 21:5775-86.
77. Galvin KM, Shi Y. Multiple mechanisms of transcriptional repression by YY1. Mol Cell Biol 1997; 17:3723-32.
78. Bennett MK, Ngo TT, Athanikar JN et al. Co-stimulation of promoter for low density lipoprotein receptor gene by sterol regulatory element-binding protein and Sp1 is specifically disrupted by the yin yang 1 protein. J Biol Chem 1999; 274:13025-32.
79. Natesan S, Gilman MZ. DNA bending and orientation-dependent function of YY1 in the c-fos promoter. Genes Dev 1993; 7:2497-509.
80. Kim J, Shapiro DJ. In simple synthetic promoters YY1-induced DNA bending is important in transcription activation and repression. Nucleic Acids Res 1996; 24:4341-8.
81. Roque MC, Smith PA, Blasquez VC. A developmentally modulated chromatin structure at the mouse immunoglobulin kappa 3' enhancer. Mol Cell Biol 1996; 16:3138-55.
82. Kim J, Kollhoff A, Bergmann A et al. Methylation-sensitive binding of transcription factor YY1 to an insulator sequence within the paternally expressed imprinted gene, Peg3. Hum Mol Genet 2003; 12:233-45.
83. Gaston K, Fried M. CpG methylation has differential effects on the binding of YY1 and ETS proteins to the bi-directional promoter of the Surf-1 and Surf-2 genes. Nucleic Acids Res 1995; 23:901-9.
84. Hobbs K, Negri J, Klinnert M et al. Interleukin-10 and transforming growth factor-beta promoter polymorphisms in allergies and asthma. Am J Respir Crit Care Med 1998; 158:1958-62.
85. Pulleyn LJ, Newton R, Adcock IM et al. TGFbeta1 allele association with asthma severity. Hum Genet 2001; 109:623-7.
86. Costa M, Grant PJ, Rice GI et al. Human endothelial cell-derived nuclear proteins that recognise polymorphic DNA elements in the von Willebrand factor gene promoter include YY1. Thromb Haemost 2001; 86:672-9.
87. Field JM, Tate LA, Chipman JK et al. Identification of functional regulatory regions of the connexin32 gene promoter. Biochim Biophys Acta 2003; 1628:22-9.
88. Hines RN, Luo Z, Hopp KA et al. Genetic variability at the human FMO1 locus: Significance of a basal promoter yin yang 1 element polymorphism (FMO1*6). J Pharmacol Exp Ther 2003; 306:1210-8.
89. de Souza AP, Trevilatto PC, Scarel-Caminaga RM et al. Analysis of the TGF-beta1 promoter polymorphism (C-509T) in patients with chronic periodontitis. J Clin Periodontol 2003; 30:519-23.
90. Rao A. New functions for DNA binding domains. Sci STKE 2001; 2001:E1.
91. Lefstin JA, Yamamoto KR. Allosteric effects of DNA on transcriptional regulators. Nature 1998; 392:885-8.
92. Austen M, Cerni C, Luscher-Firzlaff JM et al. YY1 can inhibit c-Myc function through a mechanism requiring DNA binding of YY1 but neither its transactivation domain nor direct interaction with c-Myc. Oncogene 1998; 17:511-20.
93. Garcia E, Marcos-Gutierrez C, del Mar Lorente M et al. RYBP, a new repressor protein that interacts with components of the mammalian Polycomb complex, and with the transcription factor YY1. Embo J 1999; 18:3404-18.
94. Lee JS, Zhang X, Shi Y. Differential interactions of the CREB/ATF family of transcription factors with p300 and adenovirus E1A. J Biol Chem 1996; 271:17666-74.
95. Sawa C, Yoshikawa T, Matsuda-Suzuki F et al. YEAF1/RYBP and YAF-2 are functionally distinct members of a cofactor family for the YY1 and E4TF1/hGABP transcription factors. J Biol Chem 2002; 277:22484-90.
96. Kalenik JL, Chen D, Bradley ME et al. Yeast two-hybrid cloning of a novel zinc finger protein that interacts with the multifunctional transcription factor YY1. Nucleic Acids Res 1997; 25:843-9.
97. Bauknecht T, Angel P, Royer HD et al. Identification of a negative regulatory domain in the human papillomavirus type 18 promoter: Interaction with the transcriptional repressor YY1. Embo J 1992; 11:4607-17.
98. Bauknecht T, Jundt F, Herr I et al. A switch region determines the cell type-specific positive or negative action of YY1 on the activity of the human papillomavirus type 18 promoter. J Virol 1995; 69:1-12.
99. Bauknecht T, See RH, Shi Y. A novel C/EBP beta-YY1 complex controls the cell-type-specific activity of the human papillomavirus type 18 upstream regulatory region. J Virol 1996; 70:7695-705.
100. Yevtodiyenko A, Carr MS, Patel N et al. Analysis of candidate imprinted genes linked to Dlk1-Gtl2 using a congenic mouse line. Mamm Genome 2002; 13:633-8.
101. Safrany G, Perry RP. Characterization of the mouse gene that encodes the delta/YY1/NF-E1/UCRBP transcription factor. Proc Natl Acad Sci USA 1993; 90:5559-63.
102. Erkeland SJ, Valkhof M, Heijmans-Antonissen C et al. The gene encoding the transcriptional regulator Yin Yang 1 (YY1) is a myeloid transforming gene interfering with neutrophilic differentiation. Blood 2003; 101:1111-7.
103. Lee TC, Zhang Y, Schwartz RJ. Bifunctional transcriptional properties of YY1 in regulating muscle actin and c-myc gene expression during myogenesis. Oncogene 1994; 9:1047-52.
104. Pizzorno MC. Nuclear cathepsin B-like protease cleaves transcription factor YY1 in differentiated cells. Biochim Biophys Acta 2001; 1536:31-42.
105. Walowitz JL, Bradley ME, Chen S et al. Proteolytic regulation of the zinc finger transcription factor YY1, a repressor of muscle-restricted gene expression. J Biol Chem 1998; 273:6656-61.
106. Bovolenta C, Camorali L, Lorini AL et al. In vivo administration of recombinant IL-2 to individuals infected by HIV down-modulates the binding and expression of the transcription factors ying-yang-1 and leader binding protein-1/late simian virus 40 factor. J Immunol 1999; 163:6892-7.

107. Becker KG, Jedlicka P, Templeton NS et al. Characterization of hUCRBP (YY1, NF-E1, delta): A transcription factor that binds the regulatory regions of many viral and cellular genes. Gene 1994; 150:259-66.
108. Patten M, Wang W, Aminololama-Shakeri S et al. IL-1 beta increases abundance and activity of the negative transcriptional regulator yin yang-1 (YY1) in neonatal rat cardiac myocytes. J Mol Cell Cardiol 2000; 32:1341-52.
109. Eliassen KA, Baldwin A, Sikorski EM et al. Role for a YY1-binding element in replication-dependent mouse histone gene expression. Mol Cell Biol 1998; 18:7106-18.
110. Gordon SJ, Saleque S, Birshtein BK. Yin Yang 1 is a lipopolysaccharide-inducible activator of the murine 3' Igh enhancer, hs3. J Immunol 2003; 170:5549-57.
111. Hiromura M, Choi CH, Sabourin NA et al. YY1 is regulated by O-linked N-acetylglucosaminylation (O-glcNAcylation). J Biol Chem 2003; 278:14046-52.
112. Oei SL, Griesenbeck J, Schweiger M et al. Regulation of RNA Polymerase II-dependent Transcription by Poly(ADP-ribosyl)ation of Transcription Factors. J Biol Chem 1998; 273:31644-31647.
113. Oei SL, Shi Y. Poly(ADP-ribosyl)ation of transcription factor Yin Yang 1 under conditions of DNA damage. Biochem Biophys Res Commun 2001; 285:27-31.
114. Oei SL, Griesenbeck J, Schweiger M et al. Interaction of the transcription factor YY1 with human poly(ADP-ribosyl) transferase. Biochem Biophys Res Commun 1997; 240:108-11.
115. Griesenbeck J, Ziegler M, Tomilin N et al. Stimulation of the catalytic activity of poly(ADP-ribosyl) transferase by transcription factor Yin Yang 1. FEBS Lett 1999; 443:20-4.
116. Oei SL, Shi Y. Transcription factor Yin Yang 1 stimulates poly(ADP-ribosyl)ation and DNA repair. Biochem Biophys Res Commun 2001; 284:450-4.
117. Mampuru LJ, Chen SJ, Kalenik JL et al. Analysis of events associated with serum deprivation-induced apoptosis in C3H/Sol8 muscle satellite cells. Exp Cell Res 1996; 226:372-80.
118. Girton JR, Jeon SH. Novel embryonic and adult homeotic phenotypes are produced by pleiohomeotic mutations in Drosophila. Dev Biol 1994; 161:393-407.
119. Riggs KJ, Saleque S, Wong KK et al. Yin-yang 1 activates the c-myc promoter. Mol Cell Biol 1993; 13:7487-95.
120. Wu F, Lee AS. YY1 as a regulator of replication-dependent hamster histone H3.2 promoter and an interactive partner of AP-2. J Biol Chem 2001; 276:28-34.
121. Johansson E, Hjortsberg K, Thelander L. Two YY-1-binding proximal elements regulate the promoter strength of the TATA-less mouse ribonucleotide reductase R1 gene. J Biol Chem 1998; 273:29816-21.
122. Labrie C, Lee BH, Mathews MB. Transcription factors RFX1/EF-C and ATF-1 associate with the adenovirus E1A-responsive element of the human proliferating cell nuclear antigen promoter. Nucleic Acids Res 1995; 23:3732-41.
123. Hehlgans T, Strominger JL. Activation of transcription by binding of NF-E1 (YY1) to a newly identified element in the first exon of the human DR alpha gene. J Immunol 1995; 154:5181-7.
124. Zabel MD, Wheeler W, Weis JJ et al. Yin Yang 1, Oct1, and NFAT-4 form repeating, cyclosporin-sensitive regulatory modules within the murine CD21 intronic control region. J Immunol 2002; 168:3341-50.
125. Raich N, Clegg CH, Grofti J et al. GATA1 and YY1 are developmental repressors of the human epsilon-globin gene. Embo J 1995; 14:801-9.
126. Peters B, Merezhinskaya N, Diffley JF et al. Protein-DNA interactions in the epsilon-globin gene silencer. J Biol Chem 1993; 268:3430-7.
127. Wandersee NJ, Ferris RC, Ginder GD. Intronic and flanking sequences are required to silence enhancement of an embryonic beta-type globin gene. Mol Cell Biol 1996; 16:236-46.
128. Jacobsen BM, Skalnik DG. YY1 binds five cis-elements and trans-activates the myeloid cell-restricted gp91(phox) promoter. J Biol Chem 1999; 274:29984-93.
129. Galvagni F, Cartocci E, Oliviero S. The dystrophin promoter is negatively regulated by YY1 in undifferentiated muscle cells. J Biol Chem 1998; 273:33708-13.
130. Patten M, Hartogensis WE, Long CS. Interleukin-1beta is a negative transcriptional regulator of alpha1-adrenergic induced gene expression in cultured cardiac myocytes. J Biol Chem 1996; 271:21134-41.
131. Liu SH, Peng BH, Ma JT et al. Serum response element associated transcription factors in mouse embryos: Serum response factor, YY1, and PEA3 factor. Dev Genet 1995; 16:229-40.
132. Chen CY, Schwartz RJ. Competition between negative acting YY1 versus positive acting serum response factor and tinman homologue Nkx-2.5 regulates cardiac alpha-actin promoter activity. Mol Endocrinol 1997; 11:812-22.
133. Kurisaki K, Kurisaki A, Valcourt U et al. Nuclear factor YY1 inhibits transforming growth factor beta- and bone morphogenetic protein-induced cell differentiation. Molecular & Cellular Biology 2003; 23:4494-510.
134. Kwon HJ, Chung HM. Yin Yang 1, a vertebrate polycomb group gene, regulates antero-posterior neural patterning. Biochem Biophys Res Commun 2003; 306:1008-13.
135. Raaphorst FM, Otte AP, Meijer CJ. Polycomb-group genes as regulators of mammalian lymphopoiesis. Trends Immunol 2001; 22:682-90.
136. Brock HW, van Lohuizen M. The Polycomb group—no longer an exclusive club? Current Opinion in Genetics & Development 2001; 11:175-81.
137. Gould A. Functions of mammalian Polycomb group and trithorax group related genes. Current Opinion in Genetics & Development 1997; 7:488-94.
138. Kanno M, Hasegawa M, Ishida A et al. mel-18, a Polycomb group-related mammalian gene, encodes a transcriptional negative regulator with tumor suppressive activity. EMBO Journal 1995; 14:5672-8.
139. Mohd-Sarip A, Venturini F, Chalkley GE et al. Pleiohomeotic can link polycomb to DNA and mediate transcriptional repression. Molecular & Cellular Biology 2002; 22:7473-83.
140. Mishra RK, Mihaly J, Barges S et al. The iab-7 polycomb response element maps to a nucleosome-free region of chromatin and requires both GAGA and pleiohomeotic for silencing activity. Mol Cell Biol 2001; 21:1311-8.
141. Trimarchi JM, Fairchild B, Wen J et al. The E2F6 transcription factor is a component of the mammalian Bmi1-containing polycomb complex. Proceedings of the National Academy of Sciences of the United States of America 2001; 98:1519-24.
142. Tan DP, Nonaka K, Nuckolls GH et al. YY1 activates Msx2 gene independent of bone morphogenetic protein signaling. Nucleic Acids Res 2002; 30:1213-23.
143. Meccia E, Bottero L, Felicetti F et al. HOXB7 expression is regulated by the transcription factors NF-Y, YY1, Sp1 and USF-1. Biochim Biophys Acta 2003; 1626:1-9.
144. Gilthorpe J, Vandromme M, Brend T et al. Spatially specific expression of Hoxb4 is dependent on the ubiquitous transcription factor NFY. Development 2002; 129:3887-99.
145. O'Carroll D, Erhardt S, Pagani M et al. The polycomb-group gene Ezh2 is required for early mouse development. Mol Cell Biol 2001; 21:4330-6.
146. Faust C, Lawson KA, Schork NJ et al. The Polycomb-group gene eed is required for normal morphogenetic movements during gastrulation in the mouse embryo. Development 1998; 125:4495-506.
147. Lessard J, Schumacher A, Thorsteinsdottir U et al. Functional antagonism of the Polycomb-Group genes eed and Bmi1 in hemopoietic cell proliferation. Genes Dev 1999; 13:2691-703.
148. Su IH, Basavaraj A, Krutchinsky AN et al. Ezh2 controls B cell development through histone H3 methylation and Igh rearrangement. Nat Immunol 2003; 4:124-31.
149. Lessard J, Sauvageau G, Kwon HJ et al. Polycomb group genes as epigenetic regulators of normal and leukemic hemopoiesis. Exp Hematol 2003; 31:567-85.
150. Zalani S, Coppage A, Holley-Guthrie E et al. The cellular YY1 transcription factor binds a cis-acting, negatively regulating element in the Epstein-Barr virus BRLF1 promoter. J Virol 1997; 71:3268-74.

151. Dong X, Liu H, Pfister H. YY1 and its repressive effect on human papillomavirus 16 early promoter P97 existed widely among human epithelial cell lines. Chinese Journal of Experimental & Clinical Virology 1998; 12:217-22.
152. Weill L, Shestakova E, Bonnefoy E. Transcription factor YY1 binds to the murine beta interferon promoter and regulates its transcriptional capacity with a dual activator/repressor role. J Virol 2003; 77:2903-14.
153. Moriuchi M, Moriuchi H, Margolis DM et al. USF/c-Myc enhances, while Yin-Yang 1 suppresses, the promoter activity of CXCR4, a coreceptor for HIV-1 entry. J Immunol 1999; 162:5986-92.
154. Moriuchi M, Moriuchi H. YY1 transcription factor down-regulates expression of CCR5, a major co-receptor for HIV-1. J Biol Chem 2003; 5:5.
155. Cristillo AD, Bierer BE. Regulation of CXCR4 expression in human T lymphocytes by calcium and calcineurin. Mol Immunol 2003; 40:539-53.
156. Margolis DM, Somasundaran M, Green MR. Human transcription factor YY1 represses human immunodeficiency virus type 1 transcription and virion production. J Virol 1994; 68:905-10.
157. Romerio F, Gabriel MN, Margolis DM. Repression of human immunodeficiency virus type 1 through the novel cooperation of human factors YY1 and LSF. J Virol 1997; 71:9375-82.
158. Lee D, Kim H, Lee Y et al. Identification of sequence requirement for the origin of DNA replication in human papillomavirus type 18. Virus Res 1997; 52:97-108.
159. Lee KY, Broker TR, Chow LT. Transcription factor YY1 represses cell-free replication from human papillomavirus origins. J Virol 1998; 72:4911-7.
160. Park JS, Hwang ES, Lee CJ et al. Mutational and functional analysis of HPV-16 URR derived from Korean cervical neoplasia. Gynecol Oncol 1999; 74:23-9.
161. O'Connor MJ, Tan SH, Tan CH et al. YY1 represses human papillomavirus type 16 transcription by quenching AP-1 activity. J Virol 1996; 70:6529-39.
162. Schmidt M, Kedzia W, Gozdzicka-Jozefiak A. Intratype HPV16 sequence variation within LCR of isolates from asymptomatic carriers and cervical cancers. J Clin Virol 2001; 23:65-77.
163. Veress G, Murvai M, Szarka K et al. Transcriptional activity of human papillomavirus type 16 variants having deletions in the long control region. Eur J Cancer 2001; 37:1946-52.
164. Stephen AL, Thompson CH, Tattersall MH et al. Analysis of mutations in the URR and E6/E7 oncogenes of HPV 16 cervical cancer isolates from central China. Int J Cancer 2000; 86:695-701.
165. Kozuka T, Aoki Y, Nakagawa K et al. Enhancer-promoter activity of human papillomavirus type 16 long control regions isolated from cell lines SiHa and CaSki and cervical cancer biopsies. Jpn J Cancer Res 2000; 91:271-9.
166. Dong XP, Stubenrauch F, Beyer-Finkler E et al. Prevalence of deletions of YY1-binding sites in episomal HPV 16 DNA from cervical cancers. Int J Cancer 1994; 58:803-8.
167. May M, Dong XP, Beyer-Finkler E et al. The E6/E7 promoter of extrachromosomal HPV16 DNA in cervical cancers escapes from cellular repression by mutation of target sequences for YY1. Embo J 1994; 13:1460-6.
168. Parija T, Das BR. Involvement of YY1 and its correlation with c-myc in NDEA induced hepatocarcinogenesis, its prevention by d-limonene. Mol Biol Rep 2003; 30:41-6.
169. Nayak BK, Das BR. Differential binding of NF1 transcription factor to P53 gene promoter and its depletion in human breast tumours. Mol Biol Rep 1999; 26:223-30.
170. Kalantari M, Blennow E, Hagmar B et al. Physical state of HPV16 and chromosomal mapping of the integrated form in cervical carcinomas. Diagn Mol Pathol 2001; 10:46-54.
171. Yengi L, Inskip A, Gilford J et al. Polymorphism at the glutathione S-transferase locus GSTM3: Interactions with cytochrome P450 and glutathione S-transferase genotypes as risk factors for multiple cutaneous basal cell carcinoma. Cancer Res 1996; 56:1974-7.
172. Rose BR, Thompson CH, Zhang J et al. Sequence variation in the upstream regulatory region of HPV 18 isolates from cervical cancers. Gynecol Oncol 1997; 66:282-9.
173. Shrivastava A, Saleque S, Kalpana GV et al. Inhibition of transcriptional regulator Yin-Yang-1 by association with c-Myc. Science 1993; 262:1889-92.
174. Austen M, Cerni C, Henriksson M et al. Regulation of cell growth by the Myc-Max-Mad network: Role of Mad proteins and YY1. Curr Top Microbiol Immunol 1997; 224:123-30.
175. Shrivastava A, Yu J, Artandi S, Calame K. YY1 and c-Myc associate in vivo in a manner that depends on c-Myc levels. Proc Natl Acad Sci USA 1996; 93:10638-41.
176. Wang J, Mager J, Chen Y et al. Imprinted X inactivation maintained by a mouse Polycomb group gene. Nat Genet 2001; 28:371-5.
177. Klenova EM, Fagerlie S, Filippova GN et al. Characterization of the chicken CTCF genomic locus, and initial study of the cell cycle-regulated promoter of the gene. J Biol Chem 1998; 273:26571-9.
178. Furlong EE, Rein T, Martin F. YY1 and NF1 both activate the human p53 promoter by alternatively binding to a composite element, and YY1 and E1A cooperate to amplify p53 promoter activity. Mol Cell Biol 1996; 16:5933-45.
179. Gabellini D, Green MR, Tupler R. Inappropriate gene activation in FSHD: A repressor complex binds a chromosomal repeat deleted in dystrophic muscle. Cell 2002; 110:339-48.
180. Romey MC, Pallares-Ruiz N, Mange A et al. A naturally occurring sequence variation that creates a YY1 element is associated with increased cystic fibrosis transmembrane conductance regulator gene expression. J Biol Chem 2000; 275:3561-7.

CHAPTER 26

The Multiple Cellular Functions of TFIIIA

Natalie Kuldell

Abstract

Transcription factor IIIA (TFIIIA) is a single polypeptide with several distinct functions in the cell. In this chapter I will review the role of TFIIIA in transcription initiation of the 5S rRNA gene. I will also describe the model by which TFIIIA associates with the 5S rRNA itself to regulate ribosome biosynthesis and assembly. Finally, I will compare these functions of TFIIIA in various eukaryotic cells, from yeast to vertebrate. In all cases, the part played by the protein's zinc fingers will be emphasized.

Introduction

Although it is notoriously difficult to identify genes from sequencing data,[1,2] whole-genome analyses have led to the unexpected conclusion that an organism's absolute number of genes is not directly related to its morphological, developmental or behavioral complexity. Humans, whose genes are estimated at 30,000 or fewer[3,4] seem more than twice as complex as flies or worms, whose genes are estimated at fewer than 14,000 for *D. melanogaster*[5] and 20,000 for *C. elegans*.[6] Not only are there unexpectedly few protein-encoding genes in metazoan genomes but also their sequences represent only a small fraction of the total. A mere 2% of the human genome is thought to encode proteins.[7]

If an organism's complexity does not arise from the absolute number of proteins in its genome or the amount of genomic "real estate" that is dedicated to them, then perhaps it is their regulation that accounts for interspecies diversity. Perhaps increasing complexity arises from a limited number of proteins with increasingly complex expression patterns. Indeed, the more complex an organism, the more transcription regulatory proteins are encoded by its genome,[7] and since regulatory proteins often multimerize with other regulatory proteins, a small increase in the number of regulatory proteins could give rise to a significant number of alternative regulatory molecules. Also consistent with this "few proteins, greater regulation" model is the observation that, in complex organisms, the regulatory sequences upstream of protein-encoding genes are more dense with binding sites for regulatory factors and extend further upstream from the gene's start site. For example the regulatory sequence for a yeast gene is typically within 200 basepairs of the transcription start site while human genes can have multiple regulatory factors binding sequences 50 or even 100 kb upstream or downstream from the gene.[7]

An alternative explanation for the gene counting conundrum is the "few proteins, multiple functions" model. Perhaps a protein with a single function in a simple organism can have multiple functions in the cells of more complex organisms. This model runs counter to the classic "one gene, one protein function" idea from the Beadle and Tatum experiments in *Neurospora*, but there are several examples of "multitasking" proteins,[8,9] including TFIIIA.

TFIIIA was the first eukaryotic transcription factor to be purified and it is required for the in vitro transcription of 5S rRNA gene.[10] A second, distinct function was subsequently assigned to TFIIIA, namely rRNA localization and storage.[11] Both functions of TFIIIA modulate 5S rRNA availability. TFIIIA, in its role as a DNA-binding transcription factor, modulates the synthesis of 5S rRNA and then, in its role as an RNA-binding nuclear/cytoplasmic shuttle, controls the assembly of 5S rRNA into ribosomes. The zinc fingers of TFIIIA (Fig. 1) are critical for both its DNA and RNA binding activities.

DNA Binding by TFIIIA and Transcription of the 5S rRNA Gene

The work of transcription in a eukaryotic cell is divided among three RNA polymerases. In general, RNA polymerase I is used for transcription of rRNAs, while RNA polymerase II is the enzyme that transcribes protein-encoding genes, and RNA polymerase III primarily transcribes tRNAs and small nuclear RNAs. Each multisubunit polymerase requires additional proteins for accurate initiation of transcription. These proteins, collectively called "basal" or "general" transcription factors, tether and position the polymerase at a gene's promoter for accurate and efficient transcription initiation. TFIIIA is one such basal transcription factor.

Unlike most rRNAs, the 5S rRNA is transcribed by RNA polymerase III in conjunction with the transcription factors TFIIIA, TFIIIB, and TFIIIC (Fig. 2). Transcription begins with the recognition of a 5S DNA sequence by TFIIIA. This recognition event is unusual in several ways. First, TFIIIA is used exclusively for transcription of the 5S rRNA gene. No other genes transcribed by RNA polymerase III are known to require TFIIIA for expression, thus the term "general transcription factor" may not strictly apply. The designation may be more related to the history of TFIIIA since one of the first eukaryotic genes to be transcribed in vitro required purified TFIIIA.[10] Second, the sequence recognized by TFIIIA is within the 5S gene itself as opposed to upstream of the gene where transcription initiation events traditionally occur. The structural details of this protein-DNA interaction have been characterized[12-14] and

Zinc Finger Proteins: From Atomic Contact to Cellular Function, edited by Shiro Iuchi and Natalie Kuldell.

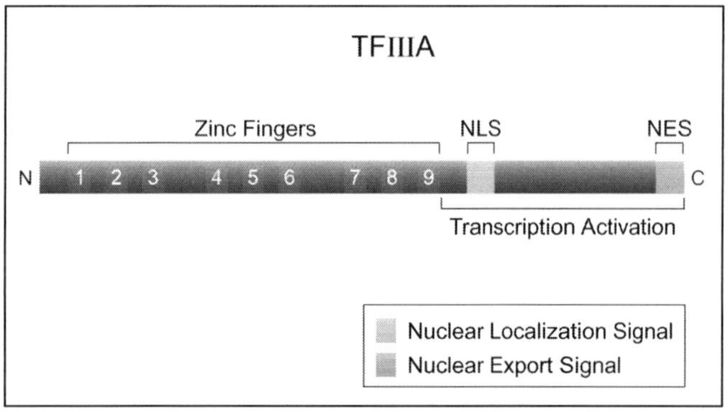

Figure 1. A schematic diagram of TFIIIA. Relevant sequence and structural features of the protein are shown. The nine zinc fingers are numbered and their arrangement within the protein is indicated. The areas of the protein critical for nuclear import (NLS) and export (NES) are shown. Transcription activation is mediated by the C-terminal portion of the protein, as indicated by the bracket. This figure is a composite of information known about TFIIIA from different species and, as described in the text, many variations of this consensus diagram are known.

molecular contacts responsible for TFIIIA-DNA recognition are described in Chapter 3 of this book.[15] Finally, the sequence that is recognized by TFIIIA is comparatively long. The TFIIIA recognition sequence, called the internal control region (ICR), is approximately 55 base pairs long. This can be compared with the consensus TATA-binding protein (TBP) recognition sequence, TATAAA, the recognition of which initiates RNA polymerase II transcription.

The zinc fingers of TFIIIA mediate binding to the ICR of the 5S rRNA gene as well as the protein's interaction with the 5S rRNA.[16] The region of the TFIIIA sequence that is downstream of the zinc fingers is critical for transcription activation of the 5S rRNA gene[17] as well as for TFIIIA to shuttle 5S rRNA out of the nucleus, a function discussed in detail below. A yeast two-hybrid experiment identified proteins in rat and human cells that interact with the carboxyl-terminus of frog but not yeast TFIIIA.[18] The protein found by this interaction trap is leucine and glutamate rich, contains numerous potential leucine zippers as well as a C_2HC zinc-binding motif, and has no apparent homolog in the *S. cerevisiae* or *C. elegans* genomes. Analysis of TFIIIA from *S. pombe* revealed no DNA-binding role for the tenth zinc finger found at the protein's C-terminus. Deletion of the sequences that separate the ninth and tenth fingers also had no effect on the affinity of TFIIIA for the 5S rRNA gene, but deletion the protein's C-terminus severely reduced transcription activation.[19]

A transcription initiation complex can assemble on the 5S rRNA gene, beginning with the association of TFIIIA and the ICR (Fig. 2). The ICR has three sequence elements within it.[20] These sequences are called the A-block, the intermediate element, and the C-block. In the *Xenopus* 5S rRNA gene, these sequences are found from +50 to +64, from +67 to +72 and from +80 to +97, respectively. Mutational studies, such as references 21 and 22, and structural data[13,14] indicate that the first three zinc fingers of TFIIIA make base-specific contacts with the C-block of the 5S rRNA gene and that these contacts contribute the majority of the binding energy to this protein-DNA interaction. Additional contacts are seen between zinc finger five and the major groove of the intermediate element.[14] Given the number of contacts between TFIIIA and the DNA, it is perhaps surprising that their dissociation constant is relatively high ($K_d = \sim 10^{-9}$ M).[23]

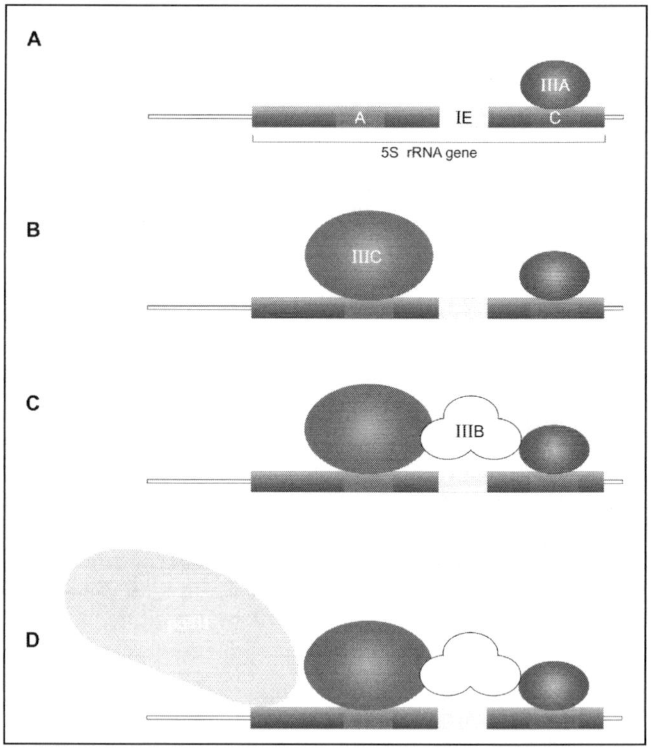

Figure 2. Transcription Initiation at the 5S rRNA gene. The ordered assembly of transcription factors that leads to synthesis of 5S rRNA is depicted. Critical contacts made by each factor with the A-block, the C-block and the intermediate element (IE) of the internal control region are emphasized. Contacts made by TFIIIB with upstream DNA sequences are not shown. To see the image in color please visit http://www.eurekah.com/chapter.php?chapid=1548&bookid=124&catid=30.

Once TFIIIA is bound, it recruits TFIIIC, a multisubunit factor, resulting in a more stable complex at the ICR. In contrast with TFIIIA, which binds principally to the C-block, TFIIIC contacts the A-block of the ICR.[24] Once TFIIIA and TFIIIC are bound, TFIIIB is recruited to the complex.[25] This transcription factor contains the TATA-binding protein, and does not directly contact sequences in the ICR. Rather it is recruited and maintained through protein-protein interactions with the ICR-bound transcription factors and through protein-DNA interactions upstream of the transcription start site.[25,26] Finally, transcription of 5S rRNA begins after RNA polymerase III associates with TFIIIB. It is uncertain if TFIIIA and TFIIIC remain associated with the gene during transcription.[27-29] Assembly of the human transcription initiation complex may differ slightly from that of *Xenopus* in that some of the components of TFIIIC may enter the complex after TFIIIB to recruit RNA polymerase III to the 5S rRNA gene.[30]

The synthesis of 5S rRNA is developmentally regulated in *Xenopus* oocytes and nucleosomal structure has long been recognized for its importance in this regulation.[31,32] Nucleosomes wrap eukaryotic genomes and condense the DNA, allowing it to fit into the nucleus. They also limit access of sequence-specific DNA binding proteins. The nucleosomes that are positioned over the 5S rRNA gene inhibit the interaction of TFIIIA with the ICR in vitro.[33,34] Conversely, the RNA polymerase III transcription complex bound to the 5S rRNA gene prevents the association of nucleosomes with this sequence in vitro.[35] H2A, H2B, H3 and H4 are the four core histones that comprise the nucleosome octomer.

Tetramers containing only H3 and H4 can be assembled on the 5S rRNA gene but these do not inhibit TFIIIA binding or transcription initiation.[34,36] Repression of 5S transcription by nucleosomes has been explored further by examining the amino-terminal tails that extend from each of the globular histones. These tails can be removed by limited trypsin digestion of each histone. Removal of the tails from H2A and H2B does not relieve nucleosomal repression of the 5S rRNA gene, whereas proteolytic removal of the H3 and H4 tails allows TFIIIA to bind the ICR and increases the in vitro transcription efficiency from the nucleosomal-DNA complex.[37] This data suggests it is the histone tails of H3 and H4 that are critical for nucleosomal repression of 5S rRNA transcription. Interestingly, hydroxyl radical footprinting indicates that deletion of the histone tails does not remove or reposition nucleosomes on the template. Thus it is not simply the presence or absence of nucleosomes over the ICR that determines its interaction with TFIIIA. The in vitro transcription of *Xenopus* 5S rRNA has allowed for extensive characterization of TFIIIA-nucleosome interactions and the effects of histone modification on the reaction.[38]

The zinc fingers of TFIIIA cannot simultaneously interact with the 5S rRNA gene and the rRNA product[39] and competition for TFIIIA binding creates a feedback loop that may help regulate the cellular concentration of 5S rRNA (ref. 11, reviewed in ref. 9). In this model (Fig. 3), free TFIIIA binds the ICR C-block to activate transcription of the 5S rRNA gene. Once the 5S rRNA is synthesized, it titrates TFIIIA from the ICR causing transcription of the gene to decrease. Less transcription lowers the concentration of the 5S product, freeing TFIIIA to bind the ICR so transcription of the gene increases again. In vitro analysis supports the proposal that 5S rRNA inhibits its own synthesis by directly competing with the DNA for TFIIIA.[11] The behavior of TFIIIA mutants also supports the feedback model for regulating 5S rRNA concentration. For example mutations that decrease the affinity of TFIIIA for 5S rRNA increase transcription of the 5S rRNA gene, presumably because more TFIIIA is available to bind the ICR.[40,41]

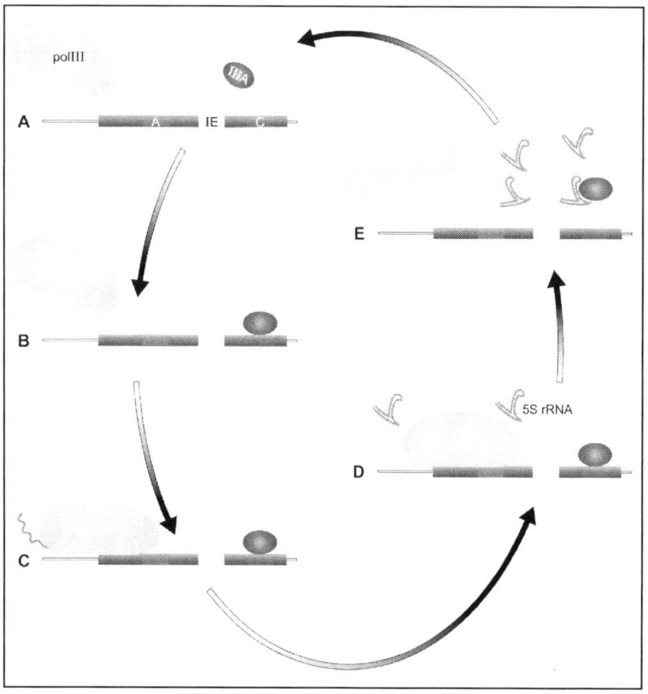

Figure 3. A feedback loop regulating 5S rRNA synthesis. Panels A and B: the binding of TFIIIA to the 5S rRNA gene recruits RNA polymerase III and leads to transcription. Other factors required for this process are not shown. Panels C and D: Synthesis of the 5S rRNA titrates TFIIIA away from the DNA, eventually leading to less transcription. Panel E: transcription of the 5S rRNA gene stops, eventually causing the concentration of 5S rRNA to drop, thus freeing TFIIIA to bind the DNA once again, as shown in panel A.

RNA Binding by TFIIIA and Regulation of Ribosome Biosynthesis

TFIIIA spends time in both the cytoplasm and the nucleus. After translation of the mRNA for TFIIIA, the protein is imported into the nucleus to initiate 5S rRNA synthesis. TFIIIA is later exported from the nucleus with 5S rRNA as part of the 7S RNP. Nuclear import signals have been mapped just C-terminal to last zinc finger,[42] and a leucine-rich, nuclear export signal (NES), important for localization of 7S RNP, is at the extreme C-terminus of TFIIIA[43] (Fig. 1). Interestingly, the NES of TFIIIA was identified based on its homology to the NES found in Rev, a protein that shuttles mRNAs from the nucleus of cells infected by human immunodeficiency virus. Indeed, the sequence identified as the nuclear export signal in TFIIIA can functionally replace the nuclear export signal in Rev.[43]

The remarkable ability of TFIIIA to bind both the 5S gene and its RNA product led many researchers to probe the molecular basis of these interactions. The resulting wealth of mutational, structural and biochemical data makes clear that the interaction of TFIIIA with DNA is fundamentally different from its interaction with encoded RNA. Its interaction with RNA is based on particular secondary and tertiary structures rather than primary sequence of the 5S rRNA (some refs. are 46-48 and the TFIIIA-5S RNA cocrystal structure in ref. 47), while the TFIIIA-DNA interaction is characterized by base-specific protein contacts (some refs. are 21, 22, 48, 49). Furthermore, a different subset of the protein's zinc fingers mediate its interaction with DNA and with RNA (see references 8 and 47). Further insight into the binding of DNA and RNA by TFIIIA has been gained by study of p43.[50] p43 is a nine zinc finger protein with affinity for 5S RNA, but, in contrast with TFIIIA, p43 does not bind DNA. A detailed comparison of these proteins and their interactions with 5S rRNA can be found in reference 39.

Like TFIIIA, the 5S rRNA also spends time in both the nucleus and the cytoplasm. Newly synthesized 5S rRNA can bind TFIIIA to form the 7S RNP. It can also bind the L5 ribosomal protein (called YL5, Y1 or L1 in yeast) to form the 5S RNP.[51] The binding sites for TFIIIA and L5 overlap but the proteins recognize different structural elements.[52,53] 7S RNP and 5S RNP accumulate in the cytoplasm until they are needed for ribosome assembly.[54] At that time, only 5S rRNA that is bound to L5 is shuttled back into the nucleoli to be assembled into the large ribosomal subunit. Thus any 5S rRNA that is bound to TFIIIA must transfer to L5 if it is to be incorporated into the ribosome.

The idea that TFIIIA and L5 coordinate to shuttle 5S rRNA between the nucleus and the cytoplasm is upheld by mutational analysis of the shuttling components. For example, 5S rRNA mutations that inhibit TFIIIA and L5 binding cause retention of the mutant rRNA in the nucleus.[55] Overexpression of TFIIIA enhances transcription of the 5S rRNA gene[56] as do mutations that reduce its affinity for 5S rRNA.[41] Point mutations in L1, the yeast homolog of L5, that affect its in vitro and in vivo binding to 5S rRNA also cause a polyribosome phenotype in *S. cerevisiae*, suggesting defects in ribosome assembly.[53] In *Xenopus* embryos, overexpression of L5 results in increased expression of 5S rRNA, presumably because L5 displaces TFIIIA for the existing pool of 5S rRNA, allowing TFIIIA to re-enter the nucleus and initiate new transcription of the 5S rRNA gene.[54] These studies provide compelling evidence for a TFIIIA/L5/5S rRNA network. Further study is needed to understand the mechanisms that regulate this network and coordinate it with the synthesis and assembly of the other ribosome components.

Recent experiments show TFIIIA can be phosphorylated by casein kinase II,[57] a tantalizing observation given that this modification is so often associated with regulation of protein activity. The site of phosphorylation appears to be the first zinc finger. This serine/threonine kinase is associated with RNA-binding proteins in *Xenopus* oocytes[58] and can phosphorylate TFIIIA in vitro.[59] Indeed, early purifications of TFIIIA were "contaminated" with kinase activity.[60] Interestingly, human L5 protein was isolated in a yeast two-hybrid screen for proteins that interact with the regulatory subunit of casein kinase II.[61] The phosphorylation of L5 decreases its affinity for 5S rRNA in vitro,[62] an observation consistent with the idea that phosphorylation regulates the nuclear/cytoplasmic shuttle of 5S rRNA.

The physiological significance of TFIIIA phosphorylation is still unclear. Most TFIIIA that is bound to 5S rRNA is not phosphorylated, but it is certainly possible that the modification is transient or occurs on a small but active fraction of the 7S RNP. Alternatively, phosphorylation of TFIIIA could influence its DNA binding activity or transcription activation function. Another potentially relevant observation is the ATP-ase activity detected for TFIIIA,[63] an activity that is dependent on DNA. How these observations connect and fit into existing models for TFIIIA function and regulation awaits further investigation.

TFIIIA in Yeast, Plants and Complex Eukaryotes

The sequences of TFIIIA from a wide range of organisms has been determined and are widely divergent (reviewed in reference 64). The presence of nine zinc fingers is perhaps the only common feature of the TFIIIA sequences described so far, and even this commonality is violated by TFIIIA from the yeast *S. pombe* which bears a tenth zinc finger. The yeast *S. cerevisiae* also has some noteworthy differences in the arrangement of its TFIIIA zinc fingers. The transcription activation domain and protein's localization signals are found between the eighth and ninth zinc fingers in this protein (compare to Fig. 1). TFIIIA from vertebrate organisms is similarly varied. Among frog species, *X. laevis* is only 84% and 63% identical to *X. borealis* and *R. catesbiana*, respectively.[20]

The recent cloning of TFIIIA from the plant *A. thaliana* further emphasizes the flexibility of this protein.[65] The authors identified the *A. thaliana* TFIIIA homolog despite its limited sequence identity with the *S. cerevisiae* and *X. laevis* proteins, reported as 17 and 26% respectively. The most significant alignment of these proteins was found in the sequences of the zinc fingers themselves. Outside of these motifs, however, there was little to indicate that the cloned cDNA was, in fact, TFIIIA. The clone has two unusual spacer sequences separating zinc fingers 1 and 2, and zinc fingers 4 and 5. In addition, there are no sequences reminiscent of the nuclear import or export signals or the transcription activation domains found in TFIIIA from other organisms. Nevertheless, the identified cDNA was used to make a recombinant protein and that protein sequence-specifically bound the 5S rRNA gene as well as its rRNA product in vitro. Furthermore, a TFIIIA-GFP fusion protein localized to the nucleolus, suggesting the presence of a nuclear localization signal though none could be identified by sequence gazing.

Given the variety of TFIIIA arrangements it is interesting that the sequence of the 5S rRNA gene, and consequently the rRNA itself, is highly conserved. The gene is relatively short, only 120 nucleotides in most organisms, and the minor sequence variations that occur are primarily in the ICR.

Summary

Studies of TFIIIA have revealed multiple cellular functions for this protein. Its ability to bind both DNA and RNA has made this protein a wonderful tool for exploration of distinct cellular activities, providing insight into important regulatory mechanisms. The zinc fingers of TFIIIA play critical roles in the protein's functions. Recent evidence indicates the zinc fingers are essential for yet another cellular function, namely recognition of damaged DNA and targeting of DNA repair.[66] Studies of TFIIIA continue to generate much interest and excitement.

References

1. Snyder M, Gerstein M. Defining Genes in the Genomics Era. Science 2003; 300:258-260.
2. Pennisi E. Gene Counters Struggle to Get the Answer Right. Science 2003; 301:1040-1041.
3. Lander ES, Linton LM, Birren B et al. Initial sequencing and analysis of the human genome. Nature 2001; 409:860-921.
4. Venter JC, Adams MD, Myers EW et al. The sequence of the human genome. Science 2001; 291:1304-1351.
5. Adams MD, Celniker SE, Holt RA et al. The genome sequence of Drosophila melanogaster. Science 2000; 287:2185-2195.
6. Ruvkun G, Hobert O. The taxonomy of developmental control in Caenorhabditis elegans. Science 1998; 282:2033-2041
7. Levine M, Tjian R. Transcription regulation and animal diversity. Nature 2003; 424:147-151.
8. Pieler T, Theunissen O. TFIIIA: nine fingers-three hands? Trends Biochem Sci 1993; 18:226-230.
9. Cassiday LA, Maher LJ 3rd. Having it both ways: transcription factors that bind DNA and RNA. Nucleic Acids Res 2002; 30:4118-4126.
10. Engelke DR, Ng S-Y, Shastry BS et al. Specific Interaction of a Purified Transcription Factor with an Internal Control Region of 5S RNA Genes. Cell 1980; 19:717-728.
11. Pelham HRB, Brown DD. A specific transcription factor that can bind either the 5S rRNA gene or 5S RNA. Proc Natl Acad Sci USA 1980; 77:4170-4174.
12. Wuttke DS, Foster MP, Case DA et al. Solution structure of the first three zinc fingers of TFIIIA bound to the cognate DNA sequence: determinants of affinity and sequence specificity. J Mol Biol 1997; 273:183-206.
13. Foster MP, Wuttke DS, Radhakrishnan I et al. Domain packing and dynamics in the DNA complex of the N-terminal zinc fingers of TFIIIA. Nat Struct Biol 1997; 4:605-608.
14. Nolte RT, Conlin RM, Harrison SC et al. Differing roles for zinc fingers in DNA recognition: structure of a six-finger transcription factor IIIA complex. Proc Natl Acad Sci USA 1998; 95:2938-2943.
15. Brown RS. TFIIIA: A Sophisticated Zinc Finger Protein. In Iuchi, S and Kuldell N, eds. Zinc Finger Proteins: from atomic contacts to ... Landes Bioscience, 2004: pgs XXX.
16. Theunissen O, Rudt F, Guddat U et al. RNA and DNA binding zinc fingers in Xenopus TFIIIA. Cell 1992; 71:679-690
17. Mao X and Darby MK. A position-dependent transcription-activating domain in TFIIIA. Mol Cell Biol 1993; 13: 7496-7506.
18. Moreland RJ, Dresser ME, Rodgers JS et al. Identification of a transcription factor IIIA-interacting protein. Nucleic Acids Res 2000; 28:1986-1993.
19. Schulman DB, Setzer DR. Functional analysis of the novel C-terminal domains of S pombe transcription factor IIIA. J Mol Biol 2003; 331:321-330.

20. Paule MR, White RJ. Transcription by RNA polymerases I and III. Nucleic Acids Res 2000; 28:1283-1298.
21. Del Rio S, Menezes SR, Setzer DR. The function of individual zinc fingers in sequence-specific DNA recognition by transcription factor IIIA. J Mol Biol 1993; 233:567-579.
22. Veldhoen N, You Q, Setzer DR et al. Contribution of individual base pairs to the interaction of TFIIIA with the Xenopus 5S RNA gene. Biochemistry 1994; 33:7568-7575.
23. Hanas JS, Bogenhagen DF, Wu CW. Cooperative model for the binding of Xenopus transcription factor A to the 5S RNA gene. Proc Natl Acad Sci USA 1983; 80:2142-2145.
24. Braun BR, Bartholomew B, Kassavetis GA et al. Topography of transcription factor complexes on the Saccharomyces cerevisiae 5 S RNA gene. J Mol Biol 1992; 228:1063-1077.
25. Kassavetis GA, Braun BR, Nguyen LH et al. TFIIIB is the transcription initiation factor proper of RNA polymerase III, while TFIIIA and TFIIIC are assembly factors. Cell 1990; 60:235-245.
26. Kassavetis GA, Joazeiro CA, Pisano M et al. The role of the TATA-binding protein in the assembly and function of the multisubunit yeast RNA polymerase III transcription factor, TFIIIB. Cell 1992; 71:1055-1064.
27. Bogenhagen DF, Wormington WM, Brown DD. Stable transcription complexes of Xenopus 5S RNA genes: a means to maintain the differentiated state. Cell 1982; 28:413-421.
28. Engelke DR, Gottesfeld JM. Chromosomal footprinting of transcriptionally active and inactive oocyte-type 5S RNA genes of Xenopus laevis. Nucleic Acids Res 1990; 18:6031-6037.
29. Bardeleben C, Kassavetis GA, Geiduschek EP. Encounters of Saccharomyces cerevisiae RNA polymerase III with its transcription factors during RNA chain elongation. J Mol Biol 1994; 235:1193-1205.
30. Weser S, Riemann J, Seifart KH et al. Assembly and isolation of intermediate steps of transcription complexes formed on the human 5S rRNA gene. Nucleic Acids Res 2003; 31:2408-2416.
31. Wolffe AP, Brown DD. Developmental regulation of two 5S ribosomal RNA genes. Science 1988; 241:1626-1632.
32. Hansen JC, Wolffe AP. A role for histones H2A/H2B in chromatin folding and transcriptional repression. Proc Natl Acad Sci U S A 1994; 91:2339-2343.
33. Hansen JC, Wolffe AP. Influence of chromatin folding on transcription initiation and elongation by RNA polymerase III. Biochemistry 1992; 31: 7977-7988.
34. Hayes JJ, Wolffe AP. Histone H2A/H2B inhibit the interaction of TFIIIA with a nucleosome including the Xenopus borealis somatic 5S RNA gene. Proc Natl Acad Sci USA 1992; 89:1229-1233.
35. Tremethick D, Zucker K, Worcel A. The transcription complex of the 5 S RNA gene, but not transcription factor IIIA alone, prevents nucleosomal repression of transcription. J Biol Chem 1990; 265:5014-5023.
36. Tse C, Fletcher TM, Hansen JC. Enhanced transcription factor access to arrays of histone H3/H4 tetramer:DNA complexes in vitro: implications for replication and transcription. Proc Natl Acad Sci USA 1998; 95:12169-12173.
37. Vitolo JM, Thiriet C, Hayes JJ. The H3-H4 N-terminal tail domains are the primary mediators of transcription factor IIIA access to 5S DNA within a nucleosome. Mol Cell Biol 2000; 20:2167-2175.
38. Yang Z, Hayes JJ. Xenopus transcription factor IIIA and the 5S nucleosome: development of a useful in vitro system. Biochem Cell Biol 2003; 81:177-184.
39. Romaniuk PJ. TFIIIA and p43: Binding to 5S Ribosomal RNA. In: Iuchi S, Kuldell N, eds. Zinc Finger Proteins: From Atomic Contacts to Cellular Function. Landes Bioscience and Kluwer Academic Press, 2004.
40. Setzer DR, Menezes SR, Del Rio S et al. Functional interactions between the zinc fingers of Xenopus transcription factor IIIA during 5S rRNA binding. RNA 1996; 2:1254-1269.
41. Rollins MB, Del Rio S, Galey AL et al. Role of TFIIIA zinc fingers in vivo: analysis of single-finger function in developing Xenopus embryos. Mol Cell Biol 1993; 13:4776-4783.
42. Hanas JS, Hocker JR, Cheng Y-G et al. cDNA cloning, DNA binding, and evolution of mammalian transcription factor IIIA. Gene 2002; 282:43-52.
43. Fridell RA, Fischer U, Luhrmann R et al. Amphibian transcription factor IIIA proteins contain a sequence element functionally equivalent to the nuclear export signal of human immunodeficiency virus type 1 Rev. Proc Natl Acad Sci USA 1996; 93:2936-2940.
44. You QM, Romaniuk PJ. The effects of disrupting 5S RNA helical structures on the binding of Xenopus transcription factor IIIA. Nucleic Acids Res 1990; 18:5055-5062.
45. Theunissen O, Rudt F, Pieler T. Structural determinants in 5S RNA and TFIIIA for 7S RNP formation. Eur J Biochem. 1998;258:758-767.
46. Searles MA, Lu D, Klug A. The role of the central zinc fingers of transcription factor IIIA in binding to 5 S RNA. J Mol Biol 2000; 301:47-60.
47. Lu D. Searles MA, Klug A. Crystal structure of a zinc-finger-RNA complex reveals two modes of molecular recognition. Nature 2003; 426:96-100.
48. You QM, Veldhoen N, Baudin F et al. Mutations in 5S DNA and 5S RNA have different effects on the binding of Xenopus transcription factor IIIA. Biochemistry 1991; 30:2495-2500.
49. Bumbulis MJ, Wroblewski G, McKean D et al. Genetic analysis of Xenopus transcription factor IIIA. J Mol Biol 1998; 284: 1307-1322.
50. Darby MK and Joho KE. Differential binding of zinc fingers from Xenopus IIIA and p43 to 5S RNA and the 5S RNA gene. Mol Cell Biol 1992; 12:3155-3164.
51. Guddat U, Bakken AH, Pieler T. Protein-mediated nuclear export of RNA: 5S rRNA containing small RNPs in xenopus oocytes. Cell 1990; 60:619-628.
52. Scripture JB, Huber PW. Analysis of the binding of Xenopus ribosomal protein L5 to oocyte 5 S rRNA. The major determinants of recognition are located in helix III-loop C. J Biol Chem 1995; 270:27358-27365.
53. Deshmukh M, Stark J, Yeh L-C C et al. Multiple Regions of Yeast Ribosomal Protein L1 are Important for its Interaction with 5S rRNA and Assembly into Ribosomes. J Biol Chem 1995; 270:30148-30156.
54. Pittman RH, Andrews MT, Setzer DR. A feedback loop coupling 5 S rRNA synthesis to accumulation of a ribosomal protein. J Biol Chem 1999; 274:33198-33201.
55. Rudt F, Pieler T. Cytoplasmic retention and nuclear import of 5S ribosomal RNA containing RNPs. EMBO J 1996; 15:1383-1391.
56. Andrews MT, Brown DD. Transient activation of oocyte 5S RNA genes in Xenopus embryos by raising the level of the trans-acting factor TFIIIA. Cell 1987; 51:445-453.
57. Westmark CJ, Ghose R, Huber PW. Phosphorylation of Xenopus transcription factor IIIA by an oocyte protein kinase CK2. Biochem J 2002; 362: 375-382.
58. Kandror KV, Stepanov AS. RNA-binding protein kinase from amphibian oocytes is a casein kinase II. FEBS Lett 1984; 170: 33-37.
59. Stepanov AS, Kandror KV, Elizarov SM. Protein kinase activity in RNA-binding proteins of Amphibia oocytes. FEBS Lett 1982; 141: 157-160.
60. Bieker JJ, Roeder RG. Physical properties and DNA-binding stoichiometry of a 5 S gene-specific transcription factor. J Biol Chem 1984; 259:6158-6164.
61. Kim JM, Cha JY, Marshak DR et al. Interaction of the beta subunit of casein kinase II with the ribosomal protein L5. Biochem Biophys Res Commun 1996; 226: 180-186.
62. Park JW, Bae YS. Phosphorylation of ribosomal protein L5 by protein kinase CKII decreases its 5S rRNA binding activity. Biochem Biophys Res Commun 1999; 263: 475-481.
63. Hazuda DJ, Wu CW. DNA-activated ATPase activity associated with Xenopus transcription factor A. J Biol Chem 1986; 261:12202-12208.
64. Huang Y, Maraia RJ. Comparison of the RNA polymerase III transcription machinery in Schizosaccharomyces pombe, Saccharomyces cerevisiae and human. Nucleic Acids Res 2001; 29:2675-2690.
65. Mathieu O, Yukawa Y, Prieto JL et al. Identification and characterization of transcription factor IIIA and ribosomal protein L5 from Arabidopsis thaliana. Nucleic Acids Res 2003; 31:2424-2433.
66. Kwon Y and Smerdon MJ Binding of Zinc Finger Protein TFIIIA to its Cognate DNA Sequence with Single UV Photoproducts at Specific Sites and its Effect on DNA Repair J Biol Chem 2003; 278:45451-45459.

CHAPTER 27

The Role of the *Ikaros* Gene Family in Lymphocyte Development

Pablo Gómez-del Arco, Taku Naito, John Seavitt, Toshimi Yoshida, Christine Williams and Katia Georgopoulos*

Abstract

In many developmental systems, nuclear regulators have been implicated in coupling key events in gene expression with specific cell fate and lineage decisions. In the hemo-lymphoid system, the Ikaros gene family of zinc finger DNA binding factors controls lymphocyte specification and homeostasis from the hemopoietic stem cell (HSC) throughout development. The dependence of hemo-lymphoid differentiation on Ikaros DNA binding activity together with the presence of Ikaros proteins within higher order chromatin remodeling complexes supports the hypothesis that Ikaros plays a key role in the lineage-specific remodeling of chromatin. Association of Ikaros and its remodeling partners with the chromatin of key lineage-specific genes, and the dependence of these genes on Ikaros complexes for their expression supports this hypothesis and provides unique paradigms to study chromatin regulation of differentiation.

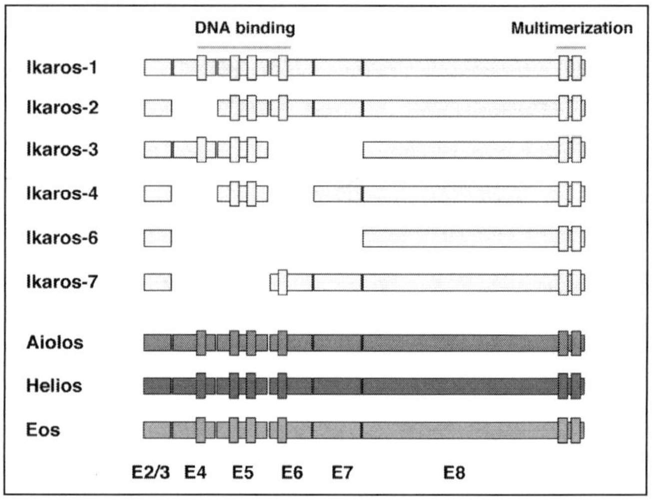

Figure 1. Ikaros family proteins. Schematic representation of functional domains in Ikaros family member proteins and various splicing isoforms. Exons are illustrated individually with zinc fingers shown as boxes. Regions required for DNA-binding and multimerization are indicated.

The *Ikaros*-Gene Family

Functional Domains of Ikaros *Gene Family*

The *Ikaros* gene family encodes a unique group of DNA-binding proteins characterized by the presence of two functionally distinct domains containing Krüppel-type zinc fingers.[1-3] Initially identified as a regulator of lymphoid-specific genes, Ikaros was subsequently determined to be the archetype for a family of proteins including Aiolos, Helios, and Eos, expression of which extends from hemo-lymphoid to neuro-epithelial tissues (Fig. 1).[4-7] cDNAs for the Ikaros family members have been isolated from human, mouse, chicken, amphibian, fish, and protochordate species and comparison of the protein sequences indicates a strong conservation of domain function.[8,9]

The full length Ikaros protein is characterized by an amino-terminal zinc finger-dependent DNA binding domain and a carboxy-terminal zinc finger domain required for multimer assembly (Fig. 1).[2,3] *Ikaros* is comprised of two alternatively used noncoding exons followed by seven coding exons.[10] Alternative splicing of exons four through seven gives rise to ten recognized isoforms. These isoforms vary in the inclusion of the four amino-terminal zinc fingers required for DNA binding. Mutagenesis studies indicate that the function and presence of the core pair of zinc fingers, F2 and F3, are critical for high affinity binding to the core motif 5'-c/TGGGAAT/c-3'.[2,11] The presence of at least one additional flanking zinc finger, F1 or F4, further influences binding to this motif.[2] The amino-terminal zinc finger domain is also required for nuclear localization.[2,11] All recognized splicing isoforms retain the carboxy-terminal pair of zinc fingers. This region may represent an evolutionarily conserved interaction module, as it shares homology with the *Drosophila* hunchback and mammalian Pegasus and tricho-rhino-phalangeal TRPS1 proteins.[1,12-14] This multimerization domain allows assembly between Ikaros proteins and other family members. Multimerization between Ikaros isoforms that can and cannot bind DNA compromises the ability of the resulting complex to bind DNA. This suggests that the nonbinding isoforms may represent a naturally occurring dominant negative mechanism to regulate the activity of this family of proteins.[15] Nevertheless, the

* *Corresponding author. See list of "Contributors".*

Zinc Finger Proteins: From Atomic Contact to Cellular Function, edited by Shiro Iuchi and Natalie Kuldell.

DNA-binding isoforms Ik1 and Ik2 are most abundant in normal lymphocytes, although increased expression of the nonDNA binding isoforms is observed in some human hematologic malignancies.[2,4]

Expression

Ikaros is highly expressed in the earliest sites of hemopoiesis in the developing embryo including the blood islands of the yolk sac at E8, the Aorta Gonad Mesonephros (AGM) and the fetal liver at E9.5. *Ikaros* mRNA is subsequently detected in the fetal thymus from E10.5, concomitant with the onset of its population with fetal lymphoid precursors.[1] In the adult, *Ikaros* is expressed in hemopoietic progenitor populations enriched for the pluripotent and self-renewing HSC, and in myelo-erythroid precursors.[4,5,16] Upon differentiation of these precursors to monocytes, macrophages and erythrocytes, *Ikaros* message level is downregulated, but maintained at significant levels in granulocytes.[6] In contrast, *Ikaros* is upregulated in T and B cell precursors as they expand and differentiate and remains expressed in mature lymphocytes in the fetus and adult. Among the lymphoid populations, *Ikaros* expression is highest in double positive thymocytes and mature T cells, populations that display strong haplo-insufficiency phenotypes in mice heterozygous for Ikaros mutations.

In contrast to *Ikaros*, expression of the closely related family member *Aiolos* is restricted to the lymphoid system.[4] It is first detected at low levels in the common lymphoid progenitor (CLP) and is progressively upregulated during B and T cell differentiation.[4,17] *Aiolos* expression peaks in mature and recirculating peripheral B cells, populations that are most affected by Aiolos deficiency. The Helios pattern of expression in early hemopoietic progenitors is similar to that of Ikaros but is detected at much lower levels.[5] Unlike Ikaros, Helios is additionally expressed in a variety of epithelial tissues. Finally, Eos is expressed in a variety of tissues outside the hemopoietic system in very low amounts.[12]

Mechanisms of Action

To gain insight into the mechanisms by which Ikaros exerts its effects on gene expression and to identify associated proteins, biochemical purification of Ikaros was carried out. Purification of Ikaros from the thymus of a Flag-tagged Ikaros-expressing transgenic line revealed that it is associated with two different classes of chromatin remodeling complexes.[18] A major fraction of Ikaros associates with components of a previously described NuRD (nucleosome remodeling and deacetylating) complex that contains the ATPase Mi-2 and HDAC1/2. A smaller fraction of Ikaros was found associated with Swi/Snf, another chromatin remodeling complex containing the ATPase Brg1 and BAFs. Ikaros complexes were also purified from an erythroleukemia cell line with both classes of chromatin remodelers.[19] This group suggested that Ikaros may form a mega-complex containing both NuRD and Swi/Snf components. Outside the NuRD complex, a small fraction of Ikaros can also associate with another HDAC-interacting factor, Sin3.[20] Ikaros can therefore recruit HDAC activity using two independent pathways. Ikaros has also been shown to repress gene expression in an HDAC-independent manner through CtBP and CtIP.[21]

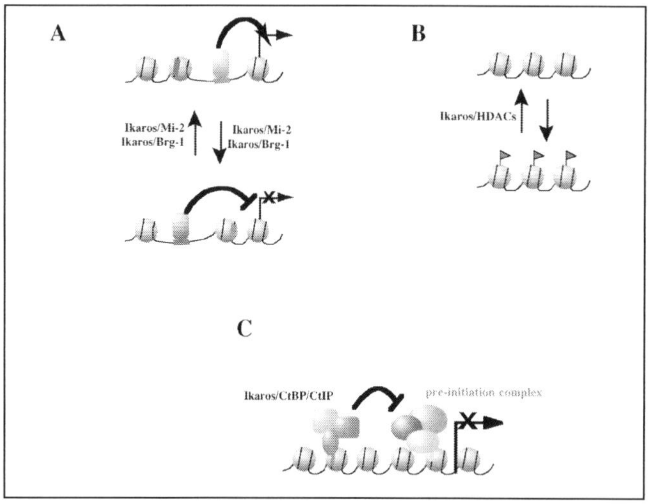

Figure 2. Modes of transcriptional regulation by Ikaros. A) Through chromatin remodeling. Chromatin remodeling can result in both activation and repression by facilitating the binding of a transcriptional activator or a repressor. B) Through histone modifications. The Ikaros/NuRD and Ikaros/Sin3 complexes contain histone deacetylases (HDACs). Thus Ikaros can affect transcription by changing the histone acetylation status in the vicinity of its target sites. C) Through direct perturbation of basal transcription factors. Both CtBP and CtIP can interact with basal transcription factors. It has been proposed that these interactions can directly inhibit the activity or affect the stability of the preinitiation complex, thus contributing an HDAC-independent repression mechanism.

The Ikaros interactors described above reveal that there are at least three different mechanisms that are potentially utilized by Ikaros to alter gene expression (Fig. 2).

a. Targeting of ATP-dependent chromatin remodeling to catalyze chromatin fluidity, conceptualized as a dynamic equilibrium between closed and open chromatin states. As many transcription factors cannot bind to their recognition sequences in the context of a restrictive nucleosomal structure, nucleosome remodeling targeted by an Ikaros complex can facilitate binding of downstream transcription factors. This may promote or inhibit gene expression, depending on the nature of the factors recruited to the region (panel A). Alternatively, Ikaros mediated chromatin remodeling may prohibit binding of transcription factors and limit the potential for regulation.

b. The authoring of a "histone code" by Ikaros complexes. Core histones are biochemically modified at specific residues in a variety of ways, including acetylation, methylation, phosphorylation and ubiquitination. It is widely theorized that a combination of specific histone modifications dictates a code establishing a downstream event. Indeed, several conserved protein motifs are known to recognize specific histone modifications. The Ikaros/NuRD and Ikaros/Sin3 complexes contain HDAC1/2 that can remove acetyl groups from histone tails and establish a repressive histone code (panel B).

c. Ikaros may repress gene expression in an HDAC-independent mechanism, which involves CtBP and CtIP. Repression by these molecules is thought to be achieved through a direct inhibition of basal transcription factors[21] (panel C).

Ikaros mouse model systems

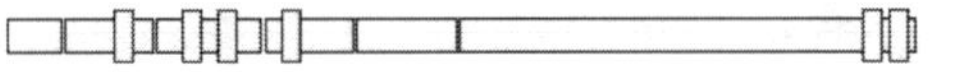

Mutation	Phenotype
Ikaros$^{null/null}$	Reduction in HSC activity 30-40X Lack of all B, fetal T cell precursors and NK cells Diminished thymic DC and γδ T cells Relative increase in myeloid precursors and their progeny Deregulated CD8 expression Clonal T cell expansions
Ikaros$^{dn/dn}$ (ΔDNA BD)	Reduction in HSC activity >100X Lack of all B and T cell precursors and downstream progeny Lack of NK and DC
Ikaros$^{null/wt}$	B and T cell populations are normal in number and cell surface phenotype Augmented TCR-mediated proliferation Development of T cell malignancies
Ikaros$^{dn/wt}$	Normal lymphocyte development Augmented TCR-mediated proliferation Rapid development of T cell malignancies
Aiolos$^{null/null}$	Reduction in recirculating, marginal zone, peritoneal, and plasma B cells Production of autoantibodies Augmented BCR-mediated proliferation Development of B cell malignacies
Aiolos$^{null/null}$ Ikaros$^{null/wt}$	Exaggerated Aiolos -/- phenotypes in periphery Accelerated development of T/B cell malignacies

Figure 3. Mouse model systems. Summary of phenotypes for Ikaros null, dominant negative (dn) and Aiolos null mice.

Ikaros Mutant Mice

In order to understand how Ikaros expression is critical for hemo-lymphopoiesis, two strains of Ikaros-deficient mice were produced (Fig. 3). Deletion of the last translated exon of the Ikaros protein (exon 8) eliminates the C-terminal half of the protein including the C-terminal zinc fingers, which are involved in protein-protein interactions. No wild-type Ikaros protein is produced after deletion, and instead a virtually undetectable truncated protein is translated then rapidly degraded. These mice are therefore true Ikaros null mutants.[22] Alternatively, deletion of exons 5 and 6, encoding zinc fingers 2, 3 and 4 including the DNA binding domain core, results in the production of a stable mutant protein. As the dimerization and transcriptional activation domains remain intact, this Ikaros mutant form functions as a dominant negative (dn) when produced in excess.[23] The Ikaros dn mutation displays a similar trend of phenotypes to the null mutation, but the severity is far greater, indicating negative interactions with other compensating family members.

Effects of Ikaros in HSC and Early Hemopoietic Progenitors

Studies on Ikaros null mice revealed that Ikaros is required at several different steps of the hemopoietic hierarchy. Ikaros null mice show a significant reduction in HSC activity (> 30 fold reduction in long-term hemopoietic lineage repopulating activity) both in the fetus and in the adult.[24] No HSC activity is detected in mice homozygous for the Ikaros dn mutation. This suggests that Ikaros and family members are required for normal production and perhaps maintenance of HSC. HSC that are produced in the Ikaros null mice fail to generate B cells, fetal T cells and Natural Killer cells.[22] The block in lymphoid differentiation is manifested very early, as the earliest described precursors for both the B and T cell pathways, including the common lymphoid progenitor (CLP), are absent.[22] The only exception to the early and severe block in lymphoid differentiation imposed by the Ikaros mutation is a postnatal wave of αβ-T cells. In addition to demonstrating that production of lymphoid progenitors is dependent on Ikaros, these studies also reveal the existence of an alternative pathway for T cell differentiation that is not dependent on the common lymphoid progenitor and is less affected by Ikaros deficiency. No fetal or adult waves of lymphoid differentiation (including all T cells) are detected in the Ikaros$^{dn/wt}$ mice.

Ikaros null HSC, although reduced in activity and possibly in number, make a near-normal number of myeloid precursors, indicating an increase in their production.[24] The number of mature myelocytes is also increased in Ikaros deficient mice. Taken together, these studies on the lymphoid versus myeloid populations in Ikaros deficient mice have provided us with the hypothesis that Ikaros plays a pivotal role in the differentiation of the HSC or of its immediate multipotent progeny. At the level of a multipotent hemopoietic progenitor, Ikaros provides positive regulation for the lymphoid and negative regulation for the myeloid differentiation programs.

Effects of Ikaros on T Cell Differentiation

T Cell Development in the Absence of Ikaros

The normal fetal waves of T cell development are completely abrogated in Ikaros null mice, and no T cells are detectable in the underdeveloped thymus of newborn animals.[22] At 3-5 days after birth, small numbers of T cell precursors become detectable in the thymus, and by one week of age a thymic cortex becomes histologically distinguishable. Although at one week after birth the Ikaros null thymus contains 100-300 fold fewer thymocytes than age-matched controls, the adult thymus is of near normal cellularity. Immature $CD25^+$ double negative (DN2 and DN3) thymocytes are 10-20 fold reduced in both young and adult Ikaros null mice, however, suggesting that Ikaros deficiency either reduces the input of hematopoietic precursors from the bone marrow or that early thymocyte precursors fail to expand normally in the absence of Ikaros.[25]

An apparent increase in the number of $CD4^+$ SP thymocytes is detected in the Ikaros null thymus, with a concomitant decrease in DP thymocytes.[22] As will be discussed below, this is in part due to a failure of a significant fraction of DP cells to express CD8[26]. Lymph node development is severely impaired in the absence of Ikaros.[22] This defect in lymph node and Peyer's patch formation is not corrected by transplantation of wild-type bone marrow into these mice, suggesting that Ikaros is also crucial for the development of normal lymphatic structures.[24]

Young Ikaros dominant negative heterozygous (dn/wt) mice have a largely normal lymphoid compartment.[27] At 2-3 months of age, $CD4^+CD8^{int}$ and $CD4^{int}CD8^+$ transitional thymocytes begin to accumulate abnormally, and T cells expressing autoreactive T cell receptor specificities can be detected. Together, these observations suggest that normal selection of thymocytes with nonself receptor specificities may be impaired in the Ikaros$^{dn/wt}$ background. Ikaros$^{dn/dn}$ animals display a more severe defect in T cell development when compared to Ikaros null mice; in these animals there is a complete block in all fetal and adult T cell lineages and only a rudimentary thymus structure develops.[23]

Effects of Ikaros-Deficiency on T Cell Activation

Mature CD4 and CD8 αβT cells isolated from Ikaros null mice hyperproliferate compared to wild-type in response to in vitro TCR stimulation.[25] The lower threshold required for activation correlates with decreasing Ikaros levels, as Ikaros$^{dn/wt}$ cells show a greater proliferative response to TCR stimulation than Ikaros$^{null/wt}$ cells. In addition, Ikaros mutant T cells exhibit an

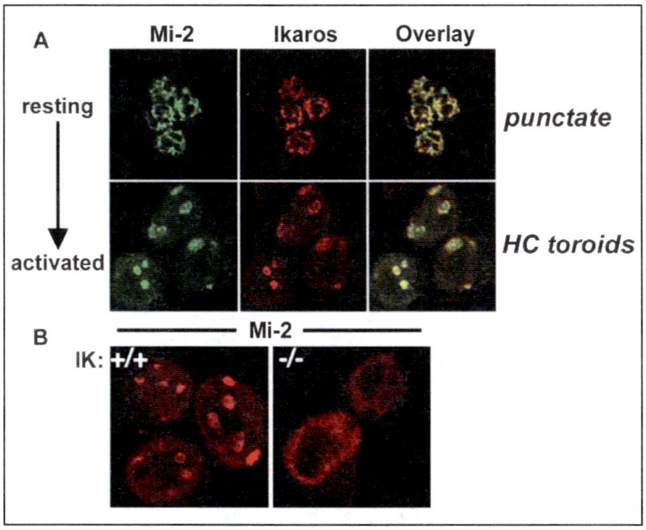

Figure 4. Changes in Ikaros-Mi-2 localization during the cell cycle. A) Dynamic redistribution of Ikaros and Mi-2β into heterochromatin (HC) upon T cell activation. B) Mi-2β localization to heterochromatin in cycling cells is Ikaros dependent.

accelerated entry into S phase following activation (16 hours post-stimulation versus 26-28 hours for wild-type cells), indicating that a reduction in Ikaros results in the early initiation of DNA replication.[25]

Nuclear staining reveals that in resting T cells, Ikaros is distributed into a fine speckled pattern.[25] Upon activation, Ikaros proteins coalesce into foci and form distinct toroidal structures when the cells are in G1/S phases of the cell cycle.[25,28] These toroids are associated with DNA replication clusters in heterochromatin. Upon T cell activation, the Mi-2 chromatin remodeler (of the NuRD complex) colocalizes with Ikaros into toroidal structures[18] (Fig. 4). In the absence of Ikaros, the ability of Mi-2 to redistribute into heterochromatin is impaired (Fig. 4), suggesting that the NuRD complex may be targeted to heterochromatic areas by Ikaros-DNA binding in activated T cells.

Ikaros Deficiency Leads to T Cell Lymphomagenesis

Ikaros-deficient primary T cells require a lower activation threshold to enter the cell cycle and accumulate chromosomal abnormalities.[25] Ikaros mutant mice develop T cell leukemias and lymphomas with a high penetrance and short latency.[27] At 2-3 months of age, a clonal expansion of T cells can be observed in Ikaros$^{dn/wt}$ mice, and by 3-6 months of age, there is a 100% penetrance of T cell leukemias and lymphomas. The development of these tumors is always associated with the mutation being at both *Ikaros* loci.[27] Ikaros$^{null/wt}$ mice are also tumor-prone, but develop T cell malignancies at a lower frequency than the dominant negative.[22] Interestingly, when the Ikaros$^{dn/wt}$ mutation is bred onto recombinase activating gene deficient mice (RAG$^{-/-}$), which cannot form functional TCR's, no lymphomagenesis is observed.[25] This suggests that preTCR or TCR signaling is required for transformation of Ikaros-deficient T cells.

Interestingly, deletion of the Ikaros-associated NuRD subunit Mi-2β from peripheral T cells causes a dramatic (5-7 fold) reduction in TCR-mediated in vitro proliferative responses (C. Williams et al, in press). This finding implicates Mi-2β as a positive regulator in TCR-mediated proliferation and suggests that Ikaros and Mi-2 may play opposing roles in lymphocyte homeostasis. During T cell activation, the loss of Ikaros may increase Mi-2 availability (or activity), which may be responsible for increased proliferative responses and the oncogenic transformation of lymphocytes.

The Role of Aiolos in B Cell Differentiation and Function

Mice homozygous for a null mutation in *Aiolos* (Aio$^{-/-}$) show that, as predicted from its pattern of expression, *Aiolos* is not required for B or T lineage commitment from the early HSC or CLP. However, Aio$^{-/-}$ mice show defects in B cell maturation and B cell effector function.[29] Aio$^{-/-}$ mice exhibit a severe reduction in the numbers of long-lived recirculating B220^{++}IgM^{+} B cells in the bone marrow, in the number of marginal zone B cells in the spleen and of peritoneal B1-a cells.[29,30] Splenic B cells in these mice, although normal in numbers, display an activated cell surface phenotype. Additionally, Aio$^{-/-}$ B cells form germinal centers in the absence of immunization, suggesting that they are responding to self-antigens or to low amounts of antigen that would not normally elicit a response in wild type B cells. Consistent with these observations, Aio$^{-/-}$ B cells display hyperproliferative responses to B cell receptor, LPS and CD40 stimulation in vitro. This result is strikingly similar to the hyperproliferative responses observed in Ikaros deficient T cells. As the Aio$^{-/-}$ mice age, they develop nuclear autoantibodies and symptoms of systemic lupus erythematosus, as well as B cell lymphomas.[29,31]

A further role of Aiolos in mature B cell effector function has been revealed by the study of T-cell dependent immune responses. Aiolos is necessary for the generation of long-lived high-affinity plasma cells (or antibody-forming cell, AFC) resident in the bone marrow (BM).[32] These BM AFC are responsible for immunity to antigens by sustaining long-term antibody production. Chimera reconstitutions demonstrated that the BM plasma cell defect is intrinsic to the Aiolos-deficient B cells. Since lack of Aiolos does not alter expression of any of the previously described factors required for general plasma cell differentiation, it was proposed that the differentiation program leading to the formation of the high-affinity plasma cell population in the BM is distinct from that of other plasma cells and is uniquely dependent on Aiolos.

In summary, Aiolos, in contrast to Ikaros, is not required for differentiation or commitment to the lymphoid lineages. Instead, Aiolos is critical at later stages of B cell differentiation: for the production or survival of long-lived B cell populations like the marginal zone B cells, the peritoneal B1-a B cells and the long-lived high-affinity plasma cells in the BM.

Ikaros-Family Gene Targets

Gene expression studies on the hemopoietic system of mice with Ikaros-family deficiencies have begun to reveal biologically important targets for this regulatory network. DP thymocytes from Ikaros and Ikaros/Aiolos deficient mice fail to express the CD8 gene appropriately, suggesting that Ikaros regulates CD8 expression. Chromatin immunoprecipitation revealed that Ikaros directly associates with the CD8 enhancer DNaseI hypersensitive sites (DHS). Studies of transgenes driven by the CD8 enhancer elements showed a dependency for their expression on Ikaros. Genetic and molecular evidence therefore indicates that Ikaros and its family members are required for normal CD8 activation during T cell differentiation.[26] A recent report on SWI/SNF (Brg-1) inactivation in T cells showed a similar deregulation of CD8 expression.[33] It is likely that the Ikaros-SWI/SNF complex described in T cells[18] is responsible for activating CD8 expression during development. Activation of *Rag-1* and *Rag-2* gene expression is also dependent on Ikaros. *Rag* gene expression is abolished in the HSC compartment (S. Ng, unpublished data) of Ikaros null mice and *Rag* BAC transgenes are downregulated in these cell types (T. Naito, unpublished data).

TdT is another potential Ikaros gene target during development. An Ikaros binding site in the *TdT* promoter overlaps that of Elf-1, a putative activator of *TdT*. In vitro binding of Ikaros and Elf-1 to the site is mutually exclusive, suggesting that Ikaros may down-regulate *TdT* by competing away Elf-1.[34] Ikaros has also been implicated in the repression of the lambda5 gene.[35] Here again, competition with E2A at a common binding site has been suggested as a repression mechanism.

Regulation of Ikaros

Post-Translational

A progressive reduction of Ikaros protein levels in T cells facilitates their activation and causes the rapid development of leukemias and lymphomas, suggesting Ikaros functions as a potent tumor suppressor. Indeed, Ikaros provides for critical control over cell cycle progression by negatively regulating the G1-S transition through its DNA binding activity.[36]

In search of post-translational modifications that regulate Ikaros activity during the cell cycle, three regions of phosphorylation were identified in Ikaros proteins.[36,37] Phosphorylation impacts Ikaros' DNA binding ability at different stages of the cell cycle (Fig. 5). In the G1 phase, Ikaros is expressed in a de-phosphorylated state, becomes phosphorylated during the replicative phase (S), and remains so through M phase. Ikaros phosphorylation is lost in the G1 phase of the next cycle and is reestablished during S phase (Fig. 5B).

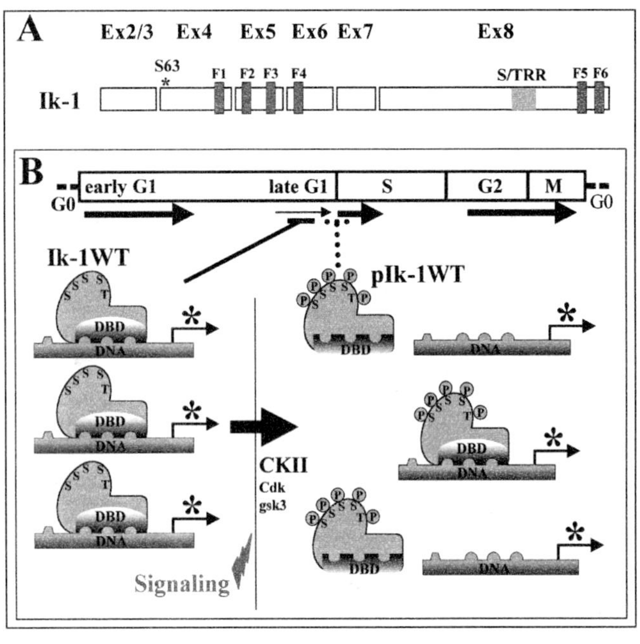

Figure 5. Phosphorylation controls Ikaros' ability to negatively regulate the G1 to S phase transition. A) Schematic representation of Ik-1 showing translated exons (horizontal rectangles) and its 6 Zinc Fingers (F). The asterisk and grey rectangle represent residues where Ikaros is phosphorylated during the G1-S transition. B) Model for Ikaros' control of the G1 to S transition and its regulation by Ikaros phosphorylation.

Of the three major areas of Ikaros phosphorylation detected during the G1-S transition, phosphorylation of the S/TRR residues in exon 8 negatively regulates the ability of the protein to arrest the cell cycle. Ikaros phosphorylation mutants (Serine to Alanine) at the S/TRR residues act as stronger cell cycle inhibitors than the wild type protein and increase Ikaros' DNA binding activity. Casein Kinase II (CKII) was found to be responsible for the Ikaros phosphorylation events during the G1-S transition. During mitosis, Ikaros becomes further phosphorylated in the linker region of the four DNA-binding zinc fingers which further reduces Ikaros DNA binding.[37] Thus, there appears to be a gradual inactivation of Ikaros DNA binding activity during the cell cycle that is mediated by phosphorylation. Further studies are required to determine how these post-transcriptional modifications modulate Ikaros' activity as a transcriptional regulator of cell cycle control genes.

Transcriptional Regulation of Ikaros

Ikaros expression in the hemopoietic system is controlled by a large number of regulatory regions that are scattered throughout its 120kb locus on mouse Chromosome 11.[10] The *Ikaros* gene consists of two untranslated exons (1a and 1b) and seven translated exons (exon 2 to 8). The two untranslated exons are independently spliced to exon 2. Several hemopoietic-specific transcription start sites were detected in their vicinity that are utilized at different frequencies indicating the existence of two promoter regions upstream of exon 1a and 1b (Fig. 6). These regions do not contain any TATA or CAAT box canonical promoter elements and are GC rich in composition, like other hemopoietic and lymphoid-specific promoters.

Ten clusters of hemopoietic tissue-specific DNase I hypersensitivity sites (DHS) in the *Ikaros* locus were identified which demarcate regions of chromatin accessibility found in the vicinity of active regulatory regions (Fig. 6). Some of these DHS are detected in both spleen and thymus and others are more restricted to the thymus. Studies on cross species sequence conservation have identified eight conserved regions, most of which are found in the proximity of the DHS clusters described for the mouse *Ikaros* genomic locus (Fig. 6).

The activity of the two genomic fragments that contain the *Ikaros* promoter regions has been reported in mouse transgenic studies. The DHS-C2 (A) region containing the promoter upstream of exon 1a drives expression of the enhanced green fluorescent protein (EGFP) reporter in granulocytes (Mac-1$^+$/Gr-1$^+$) (Fig. 6, A-p-GFP). The DHS-C3 (B) region containing the promoter upstream of exon 1b provides reporter activity predominantly in B cells and in some myeloid cells (Mac-1$^+$/Gr-1$^-$) (Fig. 6, B-p-GFP). In both cases, the reporter is subject to position effect variegation (PEV) indicating that critical regulatory elements that counteract the restrictive effects of heterochromatin in the Ikaros locus are missing. The DHS-C3 region is also active in the hemopoietic stem cell compartment (HSCs). DHS-C3 activity is maintained throughout B cell development and in mature B cells; however, during T cell development, it starts to decline at the DN3 stage (CD44$^-$CD25$^+$) and minimal activity is detected at the DP stage. Thus additional elements are required to maintain DHS-C3 activity past the DN3 stage of T cell differentiation.

The DHS-C3 regulatory region, in combination with the intronic DHS-C6 (C) region, provides for almost 100% reporter activity throughout T cell development in the thymus (Fig. 6, B-p-GFP-C). This combination of regulatory elements in part

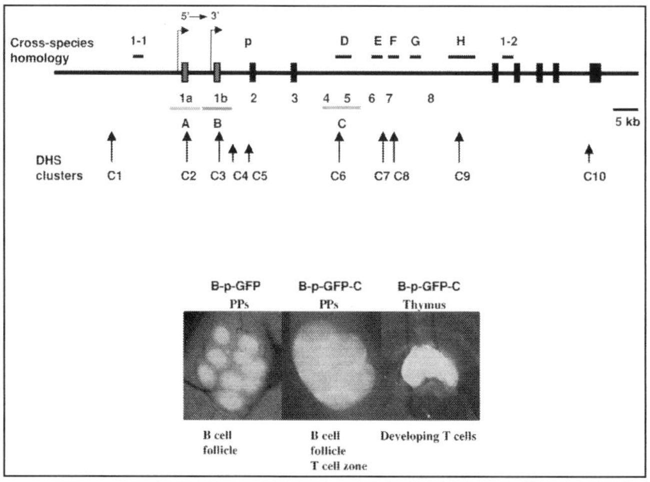

Figure 6. The Murine *Ikaros* gene locus and its regulatory elements. (Upper) Gray and black boxes indicate untranslated and translated exons, respectively. Cross-species homology and DNaseI hypersensitive sites (DHS) clusters are shown above and below the genomic locus, respectively. The shaded bars, (P), (A), (B) and (C) represent regulatory regions used in transgenic reporters. (Lower) The activity of Ikaros expression cassettes is shown in the Peyer's patches (PPs) and the thymus. The bright areas in the left and middle panels indicate B cell follicles. The areas between the B cell follicles represent T cell zones and exhibit the brightest reporter (GFP) expression (middle panel).

recapitulates the endogenous *Ikaros* expression in adult hemopoietic tissues but is not active in the fetal liver, the fetal thymus, and in the neonatal brain (T. Yoshida, unpublished data), indicating that additional regulatory element(s) are required to recapitulate endogenous *Ikaros* expression in these tissues.

Conclusions

The Krüppel-type zinc finger-containing Ikaros family of proteins play a critical role in both the development and function of cells in the lymphoid system, possibly through their association with a variety of chromatin remodeling and histone code-editing nuclear complexes. It is exciting to note that post-translational modifications such as phosphorylation appear to regulate the function of Ikaros proteins in a cell cycle and possibly differentiation-dependent manner. Moreover, the dynamic transcriptional regulation of *Ikaros* during the earliest stages of lymphoid development suggests that an investigation of the mechanisms directing its expression will be illuminating for understanding this process. Ikaros family proteins provide a lineage-restricted DNA-binding specificity to otherwise ubiquitously expressed protein complexes, and may represent a critical regulatory addition that contributed to the evolution of novel developmental and functional features within the hemo-lymphoid system. Studies into the recruitment of Ikaros into different complexes with various roles in chromatin remodeling and authoring of histone codes will permit insight into the fundamental processes that these activities play in cellular and epigenetic regulation.

References

1. Georgopoulos K, Moore D, Derfler B. Ikaros an early lymphoid restricted transcription factor, a putative mediator for T cell commitment. Science 1992; 258:808-812.
2. Molnár Á, Georgopoulos K. The Ikaros gene encodes a family of functionally diverse zinc finger DNA binding proteins. Mol Cell Biol 1994; 14:785-794.
3. Hahm K, Ernst P, Lo K et al. The lymphoid transcription factor LyF-1 is encoded by specific, alternatively spliced mRNAs derived from the Ikaros gene. Mol Cell Biol 1994; 14(11):7111-7123.
4. Morgan B, Sun L, Avitahl N et al. Aiolos, a lymphoid restricted transcription factor that interacts with Ikaros to regulate lymphocyte differentiation. EMBO J 1997; 16:2004-2013.
5. Kelley CM, Ikeda T, Koipally J et al. Helios, a novel dimerization partner of Ikaros expressed in the earliest hematopoietic progenitors. Curr Biol 1998; 8:508-515.
6. Hahm K, Cobb BS, McCarty AS et al. Helios, a T-cell restricted Ikaros family member that quantitatively associates with Ikaros at centromeric heterochromatin. Genes Dev 1998; 12:782-796.
7. Honma Y, Kiyosawa H, Mori T et al. Eos: A novel member of the Ikaros gene family expressed predominantly in the developing nervous system. FEBS Lett 1999; 447(1):76-80.
8. Molnár Á, Wu P, Largespada D et al. The Ikaros gene encodes a family of lymphocyte restricted zinc finger DNA binding proteins, highly conserved in human and mouse. J Immunol 1996; 156:585-592.
9. Haire RN, Miracle AL, Rast JP et al. Members of the Ikaros gene family are present in early representative vertebrates. J Immunol 2000; 165(1):306-312.
10. Kaufmann C, Yoshida T, Perotti EA et al. A complex network of regulatory elements in Ikaros and their activity during hemo-lymphopoiesis. EMBO J 2003; 22(9):2211-2223.
11. Cobb BS, Morales-Alcelay S, Kleiger G et al. Targeting of Ikaros to pericentromeric heterochromatin by direct DNA binding. Genes Dev 2000; 14(17):2146-2160.
12. Perdomo J, Holmes M, Chong B et al. Eos and pegasus, two members of the Ikaros family of proteins with distinct DNA binding activities. J Biol Chem 2000; 275(49):38347-38354.
13. Momeni P, Glockner G, Schmidt O et al. Mutations in a new gene, encoding a zinc-finger protein, cause tricho- rhino-phalangeal syndrome type I. Nat Genet 2000; 24(1):71-74.
14. Malik TH, Shoichet SA, Latham P et al. Transcriptional repression and developmental functions of the atypical vertebrate GATA protein TRPS1. EMBO J 2001; 20(7):1715-1725.
15. Sun L, Liu A, Georgopoulos K. Zinc finger -mediated protein interactions modulate Ikaros activity, a molecular control of lymphocyte development. EMBO J 1996; 15:5358-5369.
16. Klug CA, Morrison SJ, Masek M et al. Hematopoietic stem cells and lymphoid progenitors express different Ikaros isoforms, and Ikaros is localized to heterochromatin in immature lymphocytes. Proc Natl Acad Sci USA 1998; 95:657-662.
17. Akashi K, Traver D, Miyamoto T et al. A clonogenic common myeloid progenitor that gives rise to all myeloid lineages. Nature 2000; 404(6774):193-197.
18. Kim J, Sif S, Jones B et al. Ikaros DNA binding proteins direct formation of chromatin remodeling complexes in lymphocytes. Immunity 1999; 10:345-355.
19. O'Neill DW, Schoetz SS, Lopez RA et al. An ikaros-containing chromatin-remodeling complex in adult-type erythroid cells. Mol Cell Biol 2000; 20(20):7572-7582.
20. Koipally J, Renold A, Kim J et al. Repression by Ikaros and Aiolos is mediated through histone deacetylase complexes. EMBO J 1999; 18(11):3090-3100.
21. Koipally J, Georgopoulos K. Ikaros interactions with CtBP reveal a repression mechanism that is independent of histone deacetylase activity. J Biol Chem 2000; 275(26):19594-19602.
22. Wang J, Nichogiannopoulou A, Wu L et al. Selective defects in the development of the fetal and adult lymphoid system in mice with an Ikaros null mutation. Immunity 1996; 5:537-549.
23. Georgopoulos K, Bigby M, Wang J-H et al. The Ikaros gene is required for the development of all lymphoid lineages. Cell 1994; 79:143-156.
24. Nichogiannopoulou N, Trevisan M, Naben S et al. Defects in hemopoietic stem cell activity in Ikaros mutant mice. J Exp Med 1999; 190(9):1201-1214.

25. Avitahl N, Winandy S, Friedrich C et al. Ikaros sets thresholds for T cell activation and regulates chromosome propagation. Immunity 1999; 10:333-343.
26. Harker N, Naito T, Cortes M et al. The CD8alpha Gene Locus Is Regulated by the Ikaros Family of Proteins. Mol Cell 2002; 10(6):1403-1415.
27. Winandy S, Wu P, Georgopoulos K. A dominant mutation in the Ikaros gene leads to rapid development of leukemia and lymphoma. Cell 1995; 83:289-299.
28. Brown KE, Guest SS, Smale ST et al. Association of transcriptionally silent genes with Ikaros complexes at centromeric heterochromatin. Cell 1997; 91:845-854.
29. Wang J-H, Avitahl N, Cariappa A et al. Aiolos regulates B cell activation and maturation to effector state. Immunity 1998; 9:543-553.
30. Cariappa A, Tang M, Parng C et al. The follicular versus marginal zone B lymphocyte cell fate decision is regulated by Aiolos, Btk, and CD21. Immunity 2001; 14(5):603-615.
31. Sun J, Matthias G, Mihatsch MJ et al. Lack of the transcriptional coactivator OBF-1 prevents the development of systemic lupus erythematosus-like phenotypes in Aiolos mutant mice. J Immunol 2003; 170:1699-1706.
32. Cortes M, Georgopoulos K. Aiolos is required for the generation of high affinity bone marrow plasma cells responsible for long-term immunity. J Exp Med 2004; 199(2):209-219.
33. Chi TH, Wan M, Zhao K et al. Reciprocal regulation of CD4/CD8 expression by SWI/SNF-like BAF complexes. Nature 2002; 418(6894):195-199.
34. Trinh LA, Ferrini R, Cobb BS et al. Down-regulation of TDT transcription in CD4(+)CD8(+) thymocytes by Ikaros proteins in direct competition with an Ets activator. Genes Dev 2001; 15(14):1817-1832.
35. Sabbattini P, Lundgren M, Georgiou A et al. Binding of Ikaros to the lambda5 promoter silences transcription through a mechanism that does not require heterochromatin formation. EMBO J 2001; 20(11):2812-2822.
36. Gomez-del Arco P, Maki K, Georgopoulos K. Phosphorylation controls Ikaros' ability to negatively regulate the G1-S transition. MCB 2004; 24(7):2797-2807.
37. Dovat S, Ronni T, Russell D et al. A common mechanism for mitotic inactivation of C2H2 zinc finger DNA-binding domains. Genes Dev 2002; 16(23):2985-2990.

CHAPTER 28

Basonuclin: A Zinc Finger Protein of Epithelial Cells and Reproductive Germ Cells

Howard Green* and Hung Tseng

Abstract

Basonuclin is a C_2H_2 zinc finger protein first discovered in the keratinocytes of stratified squamous epithelia and in certain keratinocytes of the hair follicles. It was later detected in the reproductive germ cells of the testis and ovary and in the cells of the ocular lens epithelium. The only known function of basonuclin is that of a transcription factor in the synthesis of ribosomal RNA, but the interesting question remains why this function of basonuclin should be required by a small but diverse group of epithelial cell types. The recent discovery of a second basonuclin with strong resemblances in structure and function to the first, but with some important differences in cellular distribution and in evolutionary behavior, adds to the interest of basonuclin as a regulatory protein of epithelial cell types.

The Discovery of Basonuclin

Basonuclin was discovered during a search for keratinocyte transcription regulators containing a helix-loop-helix (HLH) domain. This domain is found in the myogenic family of proteins such as MyoD and Myf-5, which cooperate to initiate myogenesis in myoblasts by inducing transcription of genes specifically expressed in muscle.[1] On the possibility that the control of keratinocyte differentiation might rely on similar proteins, we screened a cDNA library prepared from cultured human keratinocytes with a cDNA encoding the HLH domain of Myf-5.[2] Keratinocyte cDNAs selected by the screening contained a sequence of only 26 nucleotides similar to those of the Myf-5 coding region; these nucleotides encoded the so-called Ω loop[3,4] conserved in the myogenic proteins as NPNQRL.[5] The selected keratinocyte cDNAs did not encode an HLH protein but rather a zinc finger protein containing the loop sequence NPNPRL between the two histidines of the second zinc finger. The cDNA was sequenced and found to contain 4611 nucleotides encoding an unknown protein with six typical C_2H_2 zinc fingers, arranged in three separate pairs.[2] Aside from the short loop sequence, there was no similarity between basonuclin and the HLH proteins.

Structural Features of Basonuclin

The Zinc Fingers

The C-terminal two-thirds of basonuclin contain the three pairs of zinc fingers, each pair separated from its neighbors by 150-320 amino acids (Fig. 1). The first finger of each pair (Fig. 2) has a short spacer between the two cysteines ($C\text{-}X_2\text{-}C$) and between the two histidines ($H\text{-}X_4\text{-}H$), whereas the second finger has a longer spacer in each location ($C\text{-}X_4\text{-}C$ and $H\text{-}X_{6\text{-}8}\text{-}H$). The

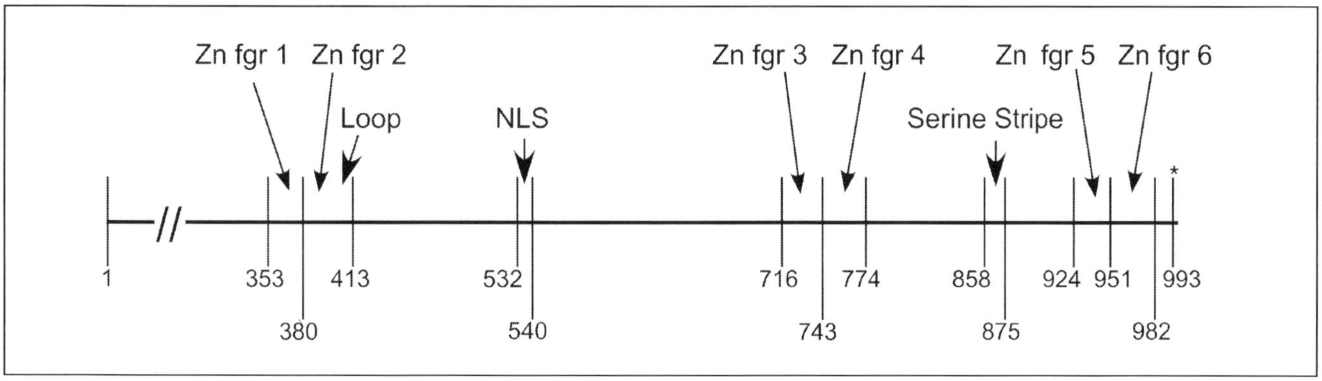

Figure 1. Principal features of basonuclin 1. Below the line are the numbers of the amino acid sequence beginning or ending each structural element. Each zinc finger ends on the second histidine. * indicates C-terminus of the protein.

* *Corresponding author. See list of "Contributors".*

Zinc Finger Proteins: From Atomic Contact to Cellular Function, edited by Shiro Iuchi and Natalie Kuldell.

```
Zn fgr      Amino
 no.        Acid
            no.
1 hu bn      353   KGRVFCTACEKTFYDKGTLKIHYNAVH
                                    +    +
1 disco-r    472   ·K··Q·NV·L···C···A····FS···

2 hu bn      380   LKIKHKCTIEGCNMVFSSLRTRNRHSANPNPRLH
                    +      ++    +     +          +
2 disco-r    502   ·REM····VD··S·M···R·S··········K··

3 hu bn      716   ENRFQCDICKKTFKNACSVKIHHKNMH
                     +
3 disco-r   1023   ··PLK·TA·GEI·Q·FHLKTH·QSVHL

4 hu bn      743   VKEMHTCTVEGCNATFPSRRSRDRHSSNLNLH
                           ++    +   +
4 disco-r   1053   KL-H·K·NID····A···K·············

5 hu bn      924   GLPITCHLCQKTYSNKGTFRAHYKTVH

6 hu bn      951   LRQLHKCKVPGCNTMFSSVRSRNRHSQNPNLH
```

Figure 2. Comparison of the four zinc fingers of *Drosophila* discorelated with the first four zinc fingers of human basonuclin. Dots indicate identity. + indicates conservative substitution. Sequence of discorelated from Mahaffey et al[10] Sequence of human basonuclin from Tseng and Green.[2] Included are the sequences of zinc fingers 5 and 6 of basonuclin, which have no counterpart in discorelated.

distance between the second cysteine and the first histidine is a constant 12 residues typical of C_2H_2 fingers.[6]

The Nuclear Localization Signal

The nuclear localization signal (NLS) of basonuclin is located near the middle of the protein between the second and third zinc fingers (Fig. 1). Its sequence, PKKKSR-K539, matches well the consensus sequence of the class A NLS, in which a helix breaker (a proline or a glycine) is followed by at least three basic amino acids. Residues 540 and 541 are both serines. Deletion of this region severely reduced nuclear localization of the protein.[7]

Basonuclin of intact keratinocytes was found to incorporate ^{32}Pi. The only phosphorylated amino acid detected was serine. When phosphorylation of a 33 amino acid sequence containing the NLS was studied in keratinocytes, it was found that substitution of aspartic acid or alanine for serine 541 eliminated phosphorylation completely or nearly completely, respectively. Substitutions at serine 537 had weak effects and at serine 540 had no effect on phosphorylation. Serine 541 was therefore the principal phosphorylated residue. Alanine replacement experiments showed that phosphorylation of serine 541 was the principal determinant of cytoplasmic localization of basonuclin. Inhibition of phosphatase action on wild type basonuclin in intact keratinocytes led to cytoplasmic localization.[7]

The state of phosphorylation of this serine is likely to be regulated by extra-cellular signals. In samples of epidermis of human and rat the basonuclin was predominantly cytoplasmic. When these cells were put into cultivation with 3T3 feeders (conditions supporting more rapid keratinocyte proliferation) the basonuclin became predominantly nuclear. This translocation of basonuclin was reversible, since when the 3T3 feeders were removed, the basonuclin returned to the cytoplasm, and the rate of keratinocyte proliferation was reduced.[8] Conditions that promote keratinocyte proliferation are evidently associated with nuclear localization of basonuclin.

Relatedness of Zinc Fingers of Basonuclin to Those of a Drosophila *Protein*

In our original paper describing the sequence of human basonuclin,[2] we pointed out the similarity of its first and second zinc fingers to those of the *Drosophila* protein disco,[9] whose function in the larval optic nerve seems to have no relation to that of basonuclin and whose amino acid sequence, apart from the pair of zinc fingers, also has no similarity to that of basonuclin.

The recently discovered *Drosophila* gene discorelated[10] also has a function in the development of the larval head, and this function overlaps that of disco. Discorelated has two pairs of zinc fingers resembling the first two pairs of fingers of human basonuclin (Fig. 2). In sequences extending from the residue preceding the second cysteine to the second histidine, zinc fingers 4 are the most similar (91% identity); the only two mismatches are due to conservative substitutions. The zinc fingers 2 are 79% identical, and this finger of discorelated even possesses the loop sequence NPNPKL in the same position as the NPNPRL of basonuclin. Zinc fingers 1 are 73% identical, and only zinc finger 3 is poorly conserved. The strong similarity of the shared zinc fingers 1, 2 and 4 show that these fingers are of ancient origin. Since the rest of the two proteins and their functions are completely different, each protein must either alter the nucleotide recognition of the fingers or adapt a similar recognition pattern to achieve completely different results.

Conservation of Basonuclin during Mammalian Evolution: Comparison of Its Amino Acid Sequences in Human and Mouse

Overall, murine and human basonuclin are 88% identical.[11] This value is intermediate between values for highly conserved proteins and highly divergent proteins of mouse and human.[12] But the amino acid sequence of certain regions of basonuclin is highly conserved (Table 1). Although fingers 3 and 4 have diverged considerably, zinc fingers 1 and 6 are identical in the two species, and 2 and 5 differ in a single replacement. The regions flanking the NLS and the serine stripe have also been highly conserved. There are numerous silent nucleotide substitutions in all of these conserved regions, supporting the idea that the function of the conserved regions does not tolerate evolutionary amino acid replacement.

Structure of the *Basonuclin* Gene

The human *basonuclin* gene contains five exons spanning nearly 29 kb of genomic DNA.[13] The first three exons are relatively small and all the zinc fingers are encoded in the fourth and fifth exons. A notable feature of the gene is its large second intron of 17 kb. The gene was initially mapped to human chromosome 15 by screening a panel of human-rodent somatic cell hybrids[13] and was later mapped more precisely to 15q25 by fluorescence in-situ hybridization.[14] This assignment was confirmed by genomic sequencing (Celera Discovery System, Gene number hCG27264).

Initial Northern analysis showed *basonuclin* mRNA as a single band of 4.6 kb. But analysis of RNase resistance, using genomic DNA to protect the RNA, showed mRNAs with 5' untranslated

Table 1. Comparison of murine with human basonuclin (first cysteine to second histidine of each finger)

Zinc Finger	Non-Synonymous Nucleotide Substitutions	Synonymous Nucleotide Substitutions	Ratio Synonymous Mutations/ Total Mutations (%)
1	0/66 (0.0%)	8/66 (12.1%)	100
2	1/84 (1.2%)	6/84 (7.1%)	86
3	4/66 (6.1%)	6/66 (9.1%)	60
4	3/78 (3.8%)	3/78 (3.8%)	50
5	1/66 (1.5%)	3/66 (4.5%)	75
6	0/78 (0.0%)	5/78 (6.4%)	100
28 amino acid sequence containing the NLS	0/84 (0.0%)	4/84 (8.3%)	100
34 amino acid sequence containing the Serine Stripe	0/102 (0.0%)	7/102 (6.9%)	100
The Loop Sequence of zinc finger 2	0/18 (0.0%)	1/18 (5.6%)	100

Adapted from Matsuzaki et al;[11] human zinc finger 2 corrected

regions of different lengths. The 5' end of the longest mRNA was located approximately 350 nucleotides upstream of the 5' end of the original cDNA clone.[15] This discrepancy might be explained by the fact that the sequence near the 5' end of the cDNA clone (-250 to +187) has a GC content of over 80%,[13] in contrast to the average of 40-50% in the flanking sequences. RNA with high GC content usually contains stable secondary structure that may hinder reverse transcription. Indeed, evidence from computer analysis showed that a large and stable (ΔG = 210 kcal/mol) stem-loop structure resembling a cruciform is likely formed by the sequence between –212 to –59 and may explain interference with reverse transcription by the avian enzyme.[15] Study of a coupled transcription-translation system in vitro, from DNA increasingly deleted into the stem-loop region, showed that full deletion of the stem-loop sequence was required to produce the full-length basonuclin protein. It was suggested that the stem-loop region might be the basis for regulation of basonuclin synthesis at the level of translation.[15]

It has been reported[16] that in both human and mouse there is an alternative exon (*BSN1b*) located downstream from the *basonuclin* exon I reported earlier.[13] Although the human sequence *BSN1b* is present in a bacterial artificial chromosome derived from human chromosome 15, we do not know of any closely related murine sequence present in GenBank.

The basonuclin promoter was found to be TATA-less.[13,15] Transcription from such a promoter usually requires Sp1 as an activator protein. Many Sp1 binding sites were found within the GC rich region, but all were located downstream of the start sites of the longer mRNAs.[13,15]

Despite the heterogeneity at their 5' ends, all human *basonuclin* mRNAs shared the same two neighboring AUG codons at +89 and +179 of which the first is the initiator codon.[15] As *basonuclin* mRNA of the mouse lacked the AUG orthologous to the initiator AUG of human *basonuclin*, the mouse protein had a slightly greater mobility in SDS-PAGE electrophoresis than the human protein.[11] It is likely that the nucleotide sequence orthologous to that located between the first two human AUGs is not translated.

The Cell Types Containing Basonuclin

Keratinocytes of Stratified Squamous Epithelia

Basonuclin was first detected in the nucleus of basal keratinocytes of plantar epidermis.[17] At the tip of the rete ridges, where the epidermis is thickest, some cells in the first supra-basal layer also contained basonuclin. Beyond this layer, basonuclin was usually absent from the terminally differentiating cells; a typical protein of terminal differentiation, involucrin, was never found in the same cells as basonuclin.[17] Basonuclin of the basal cells may be either nuclear or cytoplasmic; the latter location seemed to be more common in the epidermis of elderly humans.[8]

The expression pattern of basonuclin in the corneal epithelium has been studied in detail in the mouse. In adults, basonuclin was present and nuclear in virtually all cells of the basal layer of the epithelium.[18] The basonuclin was sometimes present in small intranuclear aggregates or dots, suggesting association of the protein with a sub-nuclear structure, possibly the nucleolus. This was also true of esophageal and vaginal keratinocytes.

During corneal development in the mouse, basonuclin first appeared in a few cells of the epithelium at post-natal day 4 and was present in all basal cells by day 20. Curiously, basonuclin was consistently absent from the limbal region, where the stem cells of the corneal epithelium reside.

Hair Follicle Cells

In the anagen or growth phase of the hair cycle in adult humans and mice, basonuclin-containing cells were found within the outer root sheath throughout the length of the follicle including the bulge region, in the adjacent sebaceous gland, and in those cells of the bulb in contact with the mesenchymal cells of the follicular papilla. In these regions, the basonuclin appeared strictly nuclear, whereas in the basal layer of the epidermis and adjacent outer region of the follicle, basonuclin was located in the cytoplasm.[8,19] The protein also was cytoplasmic in regions associated with terminal differentiation such as the supramatrical region of the bulb.[17,19]

In the mouse, strong immunofluorescent staining of basonuclin was first detected in the nuclei of precursor keratinocytes of the hair follicles on embryonic day 14.[19] This shows that the synthesis of basonuclin is an early part of the program of follicular genesis. As the newly formed hair peg grew into the dermis, basonuclin-containing cells were concentrated in the nuclei of the leading tip of the peg, the outer root sheath and the bulb close to the follicular papilla. In catagen, basonuclin remained nuclear in the retreating epithelial strand and adjacent outer root sheath. In telogen, the resting phase of the cycle, basonuclin persisted only in the peg and adjacent outer root sheath of the club. This is the region from which the new follicle regenerates during the next anagen. Basonuclin therefore seems to remain in cells with proliferative potential for a long period during which they are not yet proliferating.[17,19]

Germ Line Cells of Testis and Ovary

Surveys of mouse and human RNA of different tissues examined by Northern analysis and RT-PCR revealed that *basonuclin* mRNA was present not only in epidermis and related stratified squamous epithelia but also in the testis,[16] where it was even more abundant.[20] When the sensitivity of detection was increased by hybridizing the RT-PCR products with labeled *basonuclin* cDNA, bands of *basonuclin* mRNA were detected in mammary gland and thymus at perhaps one order of magnitude lower abundance.[20] This may be explained by the presence in these tissues of some keratinocytes or related cell types. Placenta and spleen gave signals of intensity possibly two orders of magnitude lower than those of epidermis and testis, but the responsible cell types remain unknown.

As basonuclin of mouse testis was absent from the Sertoli cells, the protein was restricted to the germ cell lineage which forms the seminiferous tubules. At first, *basonuclin* mRNA was found in the spermatogonia, spermatocytes and spermatids, but for some reason the protein was not detected.[16] Later, use of a different antiserum revealed the presence of basonuclin protein at all stages of spermatogenesis, including the earliest precursors, the spermatogonia, where it was located in the nuclei.[20] By late stages of differentiation in the spermatozoa, the basonuclin had moved to an extra-nuclear location and was found associated with the midpiece, the centrioles and the acrosomes.[16,20]

Basonuclin was shown to be present not only in the male germ line but also in developing oocytes, the only ovarian cell type that stained intensively for basonuclin.[20] In tissue sections made through growing primary oocytes and secondary oocytes that had undergone the first meiotic division, basonuclin was found distributed evenly through the oocyte nucleoplasm and nucleolus.[21]

Epithelium of the Ocular Lens

Basonuclin was shown to be present in the lens epithelium at birth, considerably earlier than in the corneal epithelium. The protein was confined to the preequatorial epithelium and was absent from equatorial cells that contained $p57^{KIP2}$, an early differentiation marker for these cells.[18] Although the first appearance of basonuclin in the lens epithelium coincided with a period of proliferative activity, the protein remained in the cells during a long subsequent period of mitotic inactivity. In adults, only 1% of the cells in the preequatorial region of the lens epithelium could be labeled by BrdU, but basonuclin remained in all the cells. This observation suggests that the presence of basonuclin in the lens epithelial cells is a reflection of their potential for future proliferation, for these cells are known to be capable of greatly increasing their rate of multiplication in response to injury.

An unusual aspect of basonuclin in the anterior lens epithelium is that the protein was always located only in the cytoplasm, concentrated in its apical part.[18] Even when the cells were proliferating, the protein was never detected in the nuclei.

Location of Intranuclear Basonuclin

Knowledge of the structure of the gene for ribosomal RNA and the role of the transcription factors UBF and SL owes much to the work of R. Tjian[22-25] and R. H. Reeder.[23,26] This form of regulation of ribosomal RNA synthesis is thought to be present in all cell types.

Basonuclin is associated with human mitotic chromosomes at eight highly specific sites corresponding to the short arms of the acrocentric chromosomes.[14] These sites are known to contain the genes for rRNA. This was confirmed by demonstration of colocalization on those chromosomes of basonuclin and UBF1.[14]

Cultured keratinocytes, like other cell types, contain UBF1 (S. Iuchi and H. Green, unpublished experiments). Unlike UBF1, which is entirely nucleolus-associated, basonuclin is present in the nucleoplasm as well. In tissue sections through the mouse ovary, basonuclin of the oocyte, like that of the keratinocyte, was not confined to the nucleolus and instead appeared uniformly distributed in the nucleoplasm and nucleolus.[21] But when growing oocytes were isolated and permeabilized with Triton X100 under conditions that favored Pol I but not Pol II transcription, nucleolar basonuclin became predominant. Under these conditions, active Pol I transcription could be simultaneously detected.[21]

Basonuclin and The Synthesis of Ribosomal RNA

The presence of zinc fingers in basonuclin suggested that it was a transcriptional regulator with ability to interact with a specific DNA sequence. This led to a search for the target gene. Two groups, independently and by different routes, reached the conclusion that the zinc fingers of basonuclin interact with sites within the promoter region of human rDNA, where, like UBF1, the protein acts as a transcription factor for the synthesis of rRNA.[14,27]

In order to isolate human genomic sequences that interact with basonuclin, a human genomic library was incubated with bacterial inclusions containing human basonuclin and selected for genomic sequences with high affinity for the protein.[27] After five rounds of selection, a 2 kb DNA fragment became predominant. By electrophoresis mobility shift assay, a 258-bp sequence located within this fragment was found to interact with the first pair of human basonuclin zinc fingers but not with the second. This DNA fragment contained the rDNA promoter region and all the cis-elements: the up-stream control element or UCE, the CORE, and the transcription start site. Within the rDNA promoter, basonuclin inclusion bodies protected two regions, BBS-1 and BBS-2, from DNase I digestion. BBS-1 was largely located within a palindrome containing short repeats and both BBS-1 and 2 were also part of a larger dyad sequence (5'-CGCGTCGCCX_{18}GGCGACGCG).

Independent studies by electrophoretic mobility shift assay also revealed that the DNA-binding sites of basonuclin were located within the promoter sequence of rDNA.[14] A highly purified preparation containing the recombinant N-terminal pair of basonuclin zinc fingers (designated B10B) protected against DNase I digestion and generated footprints on each strand of the DNA, corresponding to two binding sites, A and B.[14] The location of binding site A agrees well with that of BBS-1 (nucleotides -95 to -135 vs. -88 to 138) and that of B with BBS-2 (-64 to -79 vs -61 to -74). A third binding site C was located within the transcribed sequence between nucleotides +16 and +30.

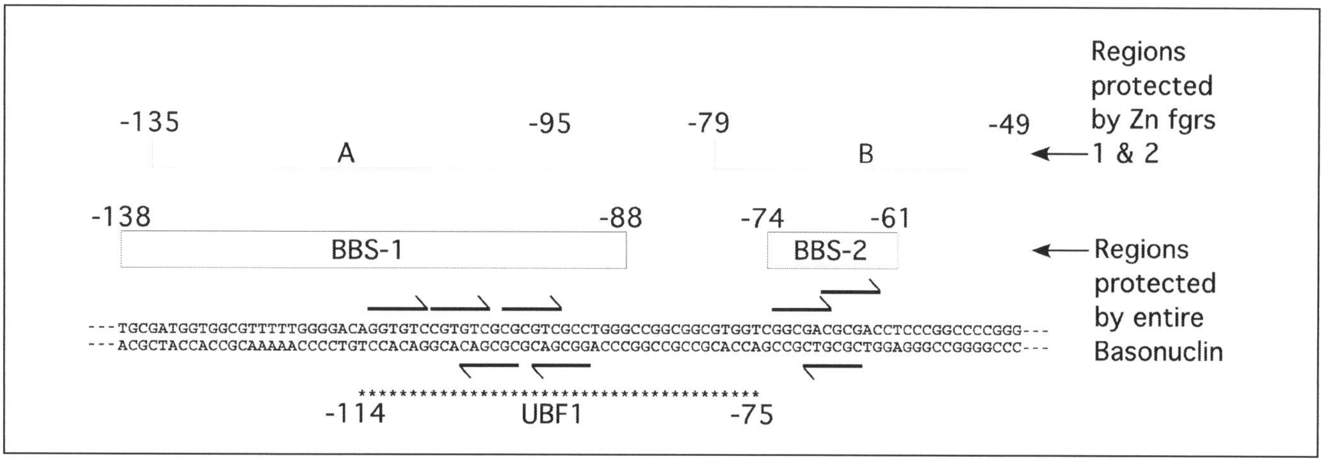

Figure 3. Affinity of basonuclin for the promoter region of the ribosomal RNA gene. Shown above is a summary of DNase I footprint experiments on the promoter of ribosomal DNA by intact basonuclin[27] and by zinc fingers 1 and 2.[14] By both methods, two regions of protection (boxes) were found overlapping and adjacent to the region protected by UBF1.[25] Both protected regions contain sequence with dyad symmetry.[27] In these regions are numerous copies of a hexanucleotide conforming to the consensus sequence 5' (G/C) G (C/T) G (A/T) C 3' (heavy black arrows). There are five such sequences in the RNA-like strand and three in the antisense strand.

The binding site A/BBS1 (Fig. 3) is of particular interest because it has the highest affinity for the N-terminal zinc finger pair and overlaps with the binding site of UBF1. Both A/BBS1 and B/BBS2 contain numerous repeats of a short sequence (Fig. 3) within the large dyad.[27] A consensus sequence for these short repeats was proposed: 5'-(G/C)G(C/T)G(A/T)C, in which the G at the second and the fourth positions and the C at the sixth position are invariable.[14]

The N-terminal pairs of zinc fingers of human and mouse basonuclin are nearly identical, but the rDNA promoter sequences of human and mouse are quite different (only 30% identical). Nevertheless, there are similar binding sites for this pair of zinc fingers on the rDNA promoter of the mouse.[21] This region contains four copies of a hexanucleotide corresponding to a more relaxed consensus 5' (A/G/C) G (C/T) G (G/A/T) C 3'. Furthermore, the first pair of zinc fingers of basonuclin can inhibit Pol I transcription by a dominant-negative effect on the mouse gene, as detected by a run-on assay.[21] This demonstrates that there is functional interaction of basonuclin and the rDNA promoter in the mouse as well as the human.

The Discovery of Basonuclin 2

It has recently been discovered independently in the laboratory of S. Sinha, by Romano et al[28] and by Vanhoutteghem and Djian[29] that there exists a gene encoding a second basonuclin (Bn2). The relation between Bn2 and the original basonuclin (now Bn1) is of great interest. The following is a brief summary of their findings, some of which were discovered by both laboratories, others by one laboratory and not yet confirmed by the other. There were no serious inconsistencies between the results of the two laboratories. The reader is referred to the original articles for more precise attribution.

1. The amino acid sequences of Bn1 and Bn2 of the mouse are only 44% identical; but the gene for Bn2 encodes all the principal features of Bn1 described above. The sequence identity is higher in the zinc fingers, the region of the NLS and the region of the serine stripe than in the rest of the molecule.
2. The gene encoding Bn2 is considerably larger than that of Bn1 (300 - 400 kb compared with 29 kb for Bn1).
3. There are six exons in the gene for Bn2, compared with five for Bn1, but in both cases most of the protein is encoded by two exons: 4 and 5 of Bn1, 5 and 6 of Bn2. The splice point between these exons is identical, interrupting zinc finger 4 at the same point. It is therefore clear that the genes encoding Bn1 and Bn2 have a common evolutionary origin. Nevertheless, the gene for Bn2 is located on human chromosome 9, not chromosome 15, which contains the gene for Bn1.
4. The evolutionary origins of Bn1 and Bn2 are ancient, since both genes are present not only in mammals, but also in birds and fishes. The amino acid sequences of Bn1 of human and mouse are 88% identical, but the sequences of Bn2 of those species are 97.5% identical. Moreover, the Bn2 of the mouse, the chicken and the zebrafish are over 99% identical. This indicates highly unusual evolutionary stability of the Bn2 sequence, comparable to that of p63, a transcription factor essential for the keratinocyte lineage.[30]
5. Conservation of zinc fingers, the NLS and the serine stripe by Bn1 and Bn2 and divergence of the rest of their amino acid sequence suggests that the recognition system is similar or identical, but that resulting functions may be different. Since the sequence of Bn2 is conserved over a broad evolutionary range, it is possible that its gene is the older of the two and after its duplication to produce Bn1, the latter was free to evolve in other directions, while Bn2 remained practically invariant.
6. Like Bn1, Bn2 binds to the promoter region of the gene for rRNA.
7. *Bn2* mRNA is expressed in the same cell types that possess abundant *Bn1* mRNA: keratinocytes and the germ cells of ovary and testis. But there is evidence supporting wider expression of *Bn2* mRNA in epithelial cell types: mouse kidney and lung, human embryonic kidney cells and HeLa cells, small intestine, uterus and even (to a small extent) in brain.
8. In most of these tissues or cell types, *Bn2* mRNA is more abundant than that of *Bn1*. But this relation may depend on the circumstances. In mouse epidermis, whose Bn1 is cytoplasmic, there is rather more mRNA encoding Bn2 than Bn1. But in

cultured human epidermal cells, in which Bn1 is strongly nuclear, the amount of *Bn1* mRNA greatly exceeds the amount of *Bn2.*

9. Previous work on immunocytological detection of Bn1 was carried out with polyclonal antisera prepared against antigen containing most or all of the Bn1 protein. In view of the similarities between Bn1 and Bn2, we believe that these antisera are not likely to distinguish between the two basonuclins. In order to detect the protein of each independently, it will be necessary to obtain antibodies to specific determinants not shared by Bn1 and Bn2.

Possible Functions of Basonuclin

Although the evidence is by no means conclusive, it seems to us that the function of basonuclin is concerned with the potential for cell proliferation.[31] Since all studies on which this hypothesis is based were performed prior to the discovery of basonuclin 2, it will be necessary to understand to what extent the functions of the two proteins overlap and to what extent they may differ.

While some cell types containing basonuclin remain to be identified, the cell types that have already been identified are all epithelial cells or reproductive germ cells. Basonuclin has not been found in any cells of mesenchymal origin.

The one identified function of basonuclin is that of a transcription factor in the synthesis of ribosomal RNA. It is not clear why epithelial cell types and reproductive germ cells should require a specific transcription factor for this purpose, while other cell types appear to use only the widely distributed UBF1-SL1 system. It seems possible that basonuclin also has a nucleoplasmic function in the regulation of transcription of genes transcribed by Pol II. It is unlikely that cytoplasmic basonuclin has any function except to constitute a pool that can be quickly mobilized for transport into the nucleoplasm, but in this case it would be unexplained why basonuclin in cells of the ocular lens epithelium has so far been found only in the cytoplasm.[18] Perhaps basonuclin might have RNA- or protein-binding properties, such as have been found for some other zinc finger proteins.[6]

Finally, although basonuclin is a very early marker in the reproductive germ line, it is not a very early marker in the development of the keratinocyte lineage from embryonic stem cells, for the transcription factor ΔNp63 appears in this lineage considerably earlier.[32]

References

1. Lassar AB, Weintraub H. The Myogenic Helix-Loop-Helix Family: Regulators of Skeletal Muscle Determination and Differentiation. In: McKnight S, Yamamoto K, eds. Transcriptional Regulation. Cold Spring Harbor Laboratory Press, 1992:2:1037-1061.
2. Tseng H, Green H. Basonuclin: A keratinocyte protein with multiple paired zinc fingers. Proc Natl Acad Sci USA 1992; 89(21):10311-10315.
3. Leszczynski JF, Rose GD. Loops in globular proteins: A novel category of secondary structure. Science 1986; 234(4778):849-855.
4. Murre C, McCaw PS, Baltimore D. A new DNA binding and dimerization motif in immunoglobulin enhancer binding, daughterless, MyoD, and myc proteins. Cell 1989; 56(5):777-783.
5. Murre C, Baltimore D. The Helix-Loop-Helix Motif: Structure and Function. In: McKnight S, Yamamoto K, eds. Transcriptional Regulation. Cold Spring Harbor Laboratory Press, 1992:2:861-879.
6. Iuchi S. Three classes of C2H2 zinc finger proteins. Cell Mol Life Sci 2001; 58(4):625-635.
7. Iuchi S, Green H. Nuclear localization of basonuclin in human keratinocytes and the role of phosphorylation. Proc Natl Acad Sci USA 1997; 94(15):7948-7953.
8. Iuchi S, Easley K, Matsuzaki K et al. Alternative subcellular locations of keratinocyte basonuclin. Exp Dermatol 2000; 9(3):178-184.
9. Heilig JS, Freeman M, Laverty T et al. Isolation and characterization of the disconnected gene of Drosophila melanogaster. Embo J 1991; 10(4):809-815.
10. Mahaffey JW, Griswold CM, Cao QM. The Drosophila genes disconnected and discorelated are redundant with respect to larval head development and accumulation of mRNAs from deformed target genes. Genetics 2001; 157(1):225-236.
11. Matsuzaki K, Iuchi S, Green H. Conservation of human and mouse basonuclins as a guide to important features of the protein. Gene 1997; 195(1):87-92.
12. Makalowski W, Zhang J, Boguski MS. Comparative analysis of 1196 orthologous mouse and human full-length mRNA and protein sequences. Genome Res 1996; 6(9):846-857.
13. Teumer J, Tseng H, Green H. The human basonuclin gene. Gene 1997; 188(1):1-7.
14. Tseng H, Biegel JA, Brown RS. Basonuclin is associated with the ribosomal RNA genes on human keratinocyte mitotic chromosomes. J Cell Sci 1999; 112 Pt 18:3039-3047.
15. Tang W, Tseng H. A GC-rich sequence within the 5' untranslated region of human basonuclin mRNA inhibits its translation. Gene 1999; 237(1):35-44.
16. Yang Z, Gallicano GI, Yu QC et al. An unexpected localization of basonuclin in the centrosome, mitochondria, and acrosome of developing spermatids. J Cell Biol 1997; 137(3):657-669.
17. Tseng H, Green H. Association of basonuclin with ability of keratinocytes to multiply and with absence of terminal differentiation. J Cell Biol 1994; 126(2):495-506.
18. Tseng H, Matsuzaki K, Lavker RM. Basonuclin in murine corneal and lens epithelia correlates with cellular maturation and proliferative ability. Differentiation 1999; 65(4):221-227.
19. Weiner L, Green H. Basonuclin as a cell marker in the formation and cycling of the murine hair follicle. Differentiation 1998; 63(5):263-272.
20. Mahoney MG, Tang W, Xiang MM et al. Translocation of the zinc finger protein basonuclin from the mouse germ cell nucleus to the midpiece of the spermatozoon during spermiogenesis. Biol Reprod 1998; 59(2):388-394.
21. Tian Q, Kopf GS, Brown RS et al. Function of basonuclin in increasing transcription of the ribosomal RNA genes during mouse oogenesis. Development 2001; 128(3):407-416.
22. Learned RM, Cordes S, Tjian R. Purification and characterization of a transcription factor that confers promoter specificity to human RNA polymerase I. Mol Cell Biol 1985; 5(6):1358-1369.
23. Bell SP, Pikaard CS, Reeder RH et al. Molecular mechanisms governing species-specific transcription of ribosomal RNA. Cell 1989; 59(3):489-497.
24. Haltiner MM, Smale ST, Tjian R. Two distinct promoter elements in the human rRNA gene identified by linker scanning mutagenesis. Mol Cell Biol 1986; 6(1):227-235.
25. Learned RM, Learned TK, Haltiner MM et al. Human rRNA transcription is modulated by the coordinate binding of two factors to an upstream control element. Cell 1986; 45(6):847-857.
26. Reeder RH. Regulation of transcription by RNA polymerase I. Transcription Regulation. Cold Spring Harbor: Cold Spring Harbor Press, 1992.
27. Iuchi S, Green H. Basonuclin, a zinc finger protein of keratinocytes and reproductive germ cells, binds to the rRNA gene promoter. Proc Natl Acad Sci USA 1999; 96(17):9628-9632.
28. Romano R, Li H, Ramakumar T et al. Molecular cloning and characterization of Basonuclin2, a DNA-binding Zinc Finger protein expressed in germ tissues and skin keratinocytes. Genomics 2004; 83(5):821-833.
29. Vanhoutteghem A, Djian P. Basonuclin 2: An extremely conserved homologue of the zinc finger protein basonuclin. Proc Natl Acad Sci USA 2004; 101(10):3468-3473.
30. Yang A, Kaghad M, Wang Y et al. p63, a p53 homolog at 3q27-29, encodes multiple products with transactivating, death-inducing, and dominant-negative activities. Mol Cell 1998; 2(3):305-316.
31. Tseng H. Basonuclin, a zinc finger protein associated with epithelial expansion and proliferation. Front Biosci 1998; 3:D985-988.
32. Green H, Easley K, Iuchi S. Marker succession during the development of keratinocytes from cultured human embryonic stem cells. Proc Natl Acad Sci USA 2003; 100(26):15625-15630.

CHAPTER 29

ZAS Zinc Finger Proteins: The Other κB-Binding Protein Family

Carl E. Allen and Lai-Chu Wu*

Abstract

The *ZAS* gene family includes *ZAS1*, *ZAS2*, and *ZAS3* in mammals as well as *Schnurri* in *Drosophila melanogaster*. The *ZAS* genes encode large proteins with two separate C_2H_2 zinc finger pairs that independently bind to specific DNA sequences, including the κB motif. In vitro experiments have shown that ZAS proteins bind to target sequences and regulate transcription of a variety of genes involved in cell growth, development, and metastasis. ZAS proteins also function in signal transduction: Schnurri interacts with Smad homologs in TGF-β-related pathways and ZAS3 interacts with TNF receptor-associated factors in TNF-α pathways. In vivo experiments have shown that disruption of ZAS expression alters normal cell growth as well as proliferation and results in spontaneous tumor formation and impaired lymphocyte development.

Introduction

Three *ZAS* genes, *ZAS1, ZAS2* and *ZAS3*, have each been characterized from human, mouse and rat species. A related gene, *Schnurri*, has been identified in *Drosophila melanogaster*. Typically, a ZAS protein is enormous (ranging from 2348 to 2717 amino acid residues) and contains four to five zinc fingers. Two sets of the C_2H_2 zinc fingers pairs form the conserved proximal and distal DNA binding domains, ZAS-N and ZAS-C, respectively. Functionally, these proteins regulate transcription through binding of specific DNA targets as well as by influencing signal transduction pathways. They have emerged as a protein family that may play significant roles in the regulation of immunity as well as normal cell growth, differentiation and proliferation.

Nomenclature

We propose naming this gene family *ZAS*. The *ZAS* genes have been characterized by many independent laboratories and have been named according to target genes or motifs, which has resulted in several names for the same protein. For example, human ZAS1 has been called major histocompatability complex enhancer binding protein, MBP1,[1] human immunodeficiency virus type I enhancer binding protein, HIV-EP1,[2] positive regulatory domain II binding factor, PRDII-BF1,[3] and zinc finger 40.[4] Mouse ZAS1 has been named αA-crystallin binding protein, αACRYBP1.[5,6] Rat ZAS1 has been named α1-antitrypsin promoter binding protein, AT-BP2.[7]

The ZAS nomenclature recognizes the unique protein domain shared by this family. A ZAS domain is a modular protein structure consisting of a pair of C_2H_2 zinc fingers with an acidic-rich region and a serine/threonine-rich sequence.[8] Currently, the human and mouse *ZAS* genes are named *HIVEP-1, -2,* and *-3* in the Unigene database.[9] We feel that the ZAS name more accurately conveys a distinctive family moniker as the HIV enhancer is only one of the many ZAS protein targets. Over the past 15 years, more than 20 cDNA sequences corresponding to nine *ZAS* genes have been cloned from humans, mice and rats (Table 1). An invertebrate protein, Schnurri in *Drosophila*, with zinc fingers similar to those of vertebrate ZAS proteins has also been identified.[10-12]

Structure

Gene Organization

Human *ZAS1, ZAS2,* and *ZAS3* (*hZAS1, hZAS2,* and *hZAS3*) have been mapped to chromosomal loci on 6p24, 6q24, and 1p34, respectively.[13-15] All three *ZAS* genes are linked to *endothelin (EDN)* and/or transcription factor *AP2 (TFAP2)* genes, suggesting that a single *-ZAS-EDN-TFAP2* ancestral gene cluster was expanded by tandem duplications.[16]

The gene structures of the *hZAS1, hZAS2,* and *hZAS3* are largely similar (Fig. 1). Each gene spans over 150 kb and contains at least 10 exons. The 5' ends of the *ZAS* genes have not yet been reported, though the 5'-untranslated region (UTR) for *hZAS3* is at least 888 nucleotides and the 5'-UTR for *rZAS2 (rat ZAZ2)* is at least 766 nucleotides. Both 5'-UTRs are much larger than the average, 210 nucleotides for human and 186 nucleotides for rodent.[17] The most notable feature of the *ZAS* genes is an exceptionally large exon, ranging from 5582 to 5978 nucleotides. For comparison, the average vertebrate exon size is only 137 bp and few are greater than 600 bp. The putative translation initiation codon for *hZAS1* is located at exon 2. Therefore, its largest exon, exon 4, contains exclusively protein-coding sequences. However, the corresponding exons of *ZAS2* and *ZAS3* contain UTRs at the

* *Corresponding author. See list of "Contributors".*

Zinc Finger Proteins: From Atomic Contact to Cellular Function, edited by Shiro Iuchi and Natalie Kuldell.

Table 1. The ZAS protein family

Protein	Species	Name	CDNA Origin	Binding Site	Binding Sequence	References
A.						
ZAS1	Human	MBP-1	B cell	MHC class I enhancer	TGGGGA-TTCCCCA	1,44
				Igκ enhancer	AGGGGACTTTCCC	
				SV40 enhancer	TGGGGACTTTCCA	
				HIV LTR-1 enhancer	AAGGGACTTTCCG	
				HIV LTR-2 enhancer	TGGGGACTTTCCA	
				β_2-microglobulin promoter	AAGGGACTTTCCC	
		HIV-EP1	B cell	HIV LTR-1 enhancer	TAGGGACTTTCCG	2
				HIV LTR-2 enhancer	TGGGGACTTTCCA	
		PRDII-BF1	osteosarcoma	IFN-β promoter	GGGAAATTCC	3
		ZnF-40				4
	Mouse	αACRYBP1	lens epithelium	αA crystallin promoter	GGGAAATCCC	6
	Rat	AT-BP2	thyroid	α1-antitrypsin promoter	TGGGGATTCACCA	7
ZAS2	Human	MBP-2	retinal cells, preB, osteosarcoma	MHC Class I enhancer	TGGGG(N_{14})ATTCCCCA	78,79
				Igκ enhancer	AGGGGACTTTCCC	
				β_2-microglobulin promoter	AAGGGACTTTCCC	
		HIV-EP2	T cell	HIV LTR-1 enhancer	TAGGGACTTTCCG	54
				HIV LTR-2 enhancer	TGGGGACTTTCCA	
	Mouse	Schnurri-2				77
	Rat	MIBP-1	liver, brain	c-myc intron I	GGGTAGGCC	80
		AT-BP1	thyroid, liver	α1-antitrypsin promoter	TGGGG(N_{14})ATTCACCA	7
ZAS3	Human	KBP-1	B cell lymphoma	MHC Class I enhancer	TGGGGATTCCCCA	79
				Igκ enhancer	AGGGGACTTTCCC	
				β_2-microglobulin promoter	AAGGGACTTTCCC	
		HIV-EP3	brain-stem, fetal brain			15
	Mouse	Rc	thymocyte, brain	Igκ enhancer	GGGACTTTCCC	31
				MHC Class I enhancer		
		KRC	adenocarcinomas, thymus, brain, B cells	S100A4/mts-1 enhancer	GGGGTTTTTCCAC	8,47
Schnurri	Fly	Schnurri		Ultrabithorax enhancer B	GGGGGGAGCCA	10,11,50
					CGGGTGCACCC	
CONSENSUS					**GGGGNNNNCC**	
B.						
ZAS1	Mouse		limb bud	α1(II) Collagen intron I	TTGAGAAAAGCC	33
ZAS2	Mouse		Brain	Somatostatin receptor type II promoter	TTCCTCTTTTCC	34
ZAS3	Mouse		thymocyte	Recombination signal sequences	CACAGTG GGTTTTTTG	31,45,72

5' region. In addition, the largest exons of *ZAS1* and *ZAS3* have been shown to be involved in alternative splicing, which results in truncated proteins without ZAS-N.[18,19] The *Schnurri* gene structure is different from the mammalian *ZAS* genes, suggesting divergence of exons and introns during evolution from invertebrates to vertebrates (Fig. 1).

The three exons following the largest exon are similar among the human *ZAS* genes with respect to size, sequence, and exon-intron boundaries. One caveat to the faithful conservation of the 3' region of the *ZAS* genes is that *hZAS3* has acquired an additional exon, exon 5b, between exon 5 and exon 6.[15] The divergence of the upstream exons and the acquisition of a novel exon in *hZAS3* reflect the evolution dynamics of the *ZAS* gene family.

Protein Domains and Motifs

The ZAS proteins are enormous and contain four C_2H_2 zinc fingers, which divided into two pairs. A solitary medial C_2HC zinc finger is present in ZAS1, ZAS3 and Schnurri, but not in ZAS2. Schnurri has an additional C-terminal zinc fingers cluster, giving it a total of eight zinc fingers (Fig. 2).

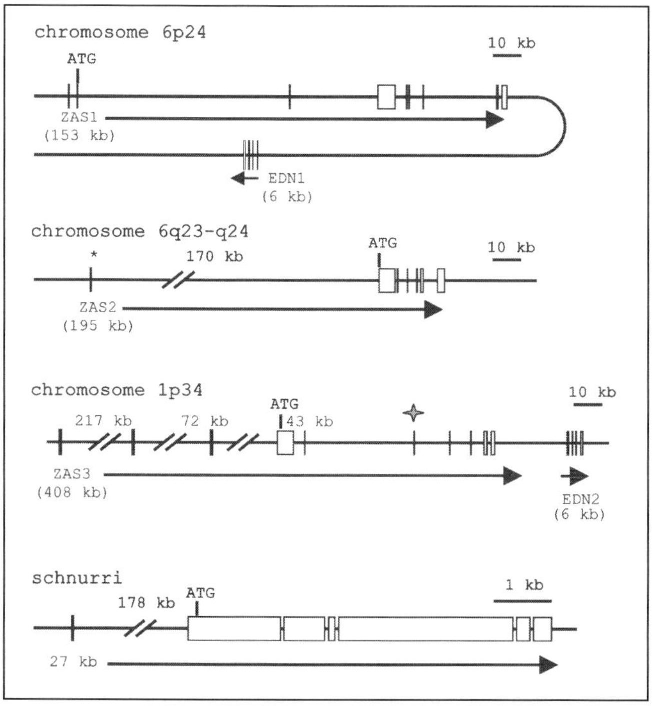

Figure 1. Exon intron structures of human *ZAS* and *Schnurri* genes. Arrows indicate the 5' to 3' transcriptional orientations. An asterisk indicates a putative *ZAS2* exon in which corresponding human cDNA has not been reported. However, that region is 90% identical to the first exon of rat *ZAS2* (GenBank accession number d37951). A large star represents an additional transcriptional termination exon, exon 5b, in *ZAS3*.

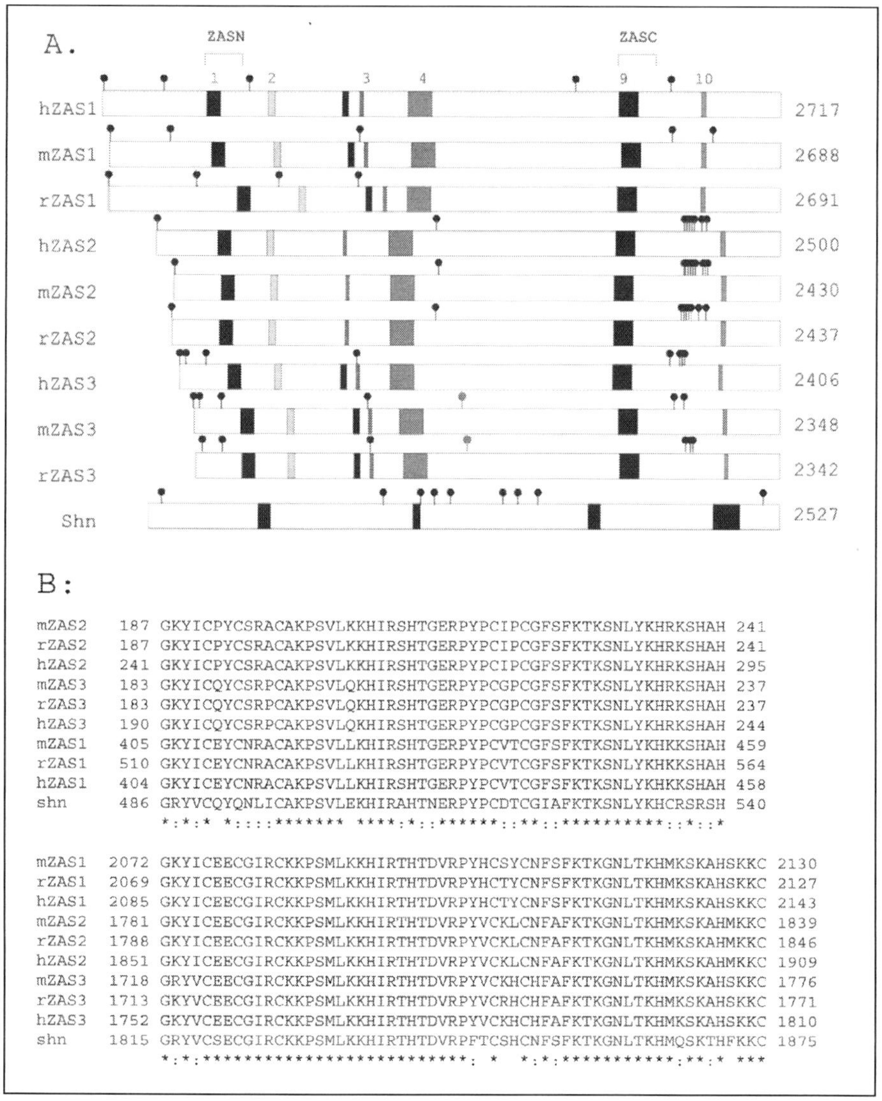

Figure 2. Protein domains in the ZAS proteins. A) Structural domains and motifs. Black boxes, one to three zinc fingers; black lollipops, Ser/Thr-Pro-X-Arg/Lys sequences; and red lollipops, P-region of GTPase motif. The numbering on top of hZAS1 indicates the location of some conserved regions. B) Multiple sequence alignments showing conservation of the proximal and distal zinc finger pairs. Completely conserved residues are marked with (*), and conserved residues are marked with (:).

The C_2H_2 zinc fingers are common nucleic acid binding motifs, but their sequence and organization is unique in the ZAS proteins. There are three classes of zinc fingers: multiple-adjacent, triple, and separate-paired. The ZAS, basonuclin, and tramtrack proteins constitute the separate-paired class.[20] Sequence comparison of zinc finger families across species found the ZAS family zinc fingers to be distinct.[21] The overall structure of the ZAS domain, with a tandem zinc finger pair, acidic domain, and serine/threonine-rich region to form a composite structure, is also unique. Sequence identities among the corresponding paralogous and orthologous ZAS zinc fingers are very high, including the backbone residues and the linker regions, suggesting conserved DNA-binding function (Fig. 2B).

Excluding the zinc finger pairs, there are eight conserved regions scattered within the vertebrate ZAS proteins that are not present in Schnurri. The largest conserved region is a stretch of ~100 amino acid residues between the ZAS domains (region 4 in Fig. 2A). It begins with a glutamate- and proline-rich region, followed by a putative nuclear localization signal, and ends with a serine stripe, which is also found in basonuclin.[15] As aforementioned, basonuclin and the ZAS proteins belong to the separate-paired zinc finger class. Basonuclin is expressed in proliferating cells and appears to be involved in regulation of cell growth.[22,23] Three other conserved regions have notable features. Region 2 contains 27 residues, 40% of which are serine or proline. Region 3 contains 14 residues and is bipolar with are three lysines and three arginines at the N-terminus, and two glutamic acids and two aspartic acids at the other end. Finally, region 10 contains 27 residues, and is named the linker region. Those conserved region do not have obvious resemblance to known protein domains.[7,15] However, their high degree of conservation in the ZAS proteins suggests they may have functional importance.

Throughout the ZAS proteins there are many potential sites for protein modification. There are three or four tyrosine kinase sites clustered within or near the ZAS-C domain. Phosphorylation of these sites has been shown to affect DNA binding affinity of ZAS proteins.[24,25] There are 7 to 12 copies of the serine/threonine-proline-X-arginine/lysine (SPXK) motif in each ZAS protein.[8] The SPXK motif, frequently found in gene regulatory proteins,[26] can bind AT-rich DNA sequences,[27] and may be phosphorylated.[28]

There are protein motifs found only in some ZAS proteins. Leucine zipper motifs, which may confer protein-protein interactions, are found in mZAS2 at residues 751-772 and in hZAS3 at residues 2024-2052. Three conserved GTPase motifs are found in ZAS3, suggesting the use of GTP as an energy source or allosteric effector. These and other unique domains may confer specific functions to individual ZAS proteins.

Protein Isoforms and Alternative RNA Processing

The largest open reading frame of each ZAS is >250 kD (Fig. 2A). Smaller proteins have been identified by immunoblot analysis using ZAS1[29] or ZAS3[25,30] antibodies. These proteins are likely ZAS isoforms generated by alternative splicing and/or protein processing. Multiple alternative splicing events have been characterized in *ZAS1* and *ZAS3* transcripts.[5-7,18,19,31] Significantly, alternative splicing events involve sequences encoding the ZAS-N or ZAS-C regions[8,18,19,31] and could lead to the production of at least three truncated protein isoforms: One contains only the ZAS-N domain, one contains only the ZAS-C domain, and one has no ZAS domain at all. Some alternative splicing events of *ZAS3* transcripts occur in a tissue-specific manner. For example, whereas the 176 bp exon is present in most *ZAS3* transcripts in the thymus, it is spliced out in brain transcripts.[19] The latter transcripts encode ZAS3 proteins that lack the carboxyl one third of the protein including the ZAS-C domain. Therefore, ZAS3 may have distinct function between lymphoid and neuronal tissues. Variations in the number of the DNA binding domains should render truncated proteins with functional characteristics distinct from full-length proteins. Functional control via RNA processing is supported by the finding that full-length ZAS1 activates the HIV-LTR enhancer,[32] whereas two truncated isoforms lacking either ZAS-N or ZAS-C domains repress transcription.[18]

Alternative splicing of *mZAS3* transcripts has been observed in the largest exon of 5487 bp and a 176 bp exon. These exons encode sequences including the ZAS-N domain and the ZAS-C domain, respectively. Additionally, a unique alternative splicing event within that enormous exon has been identified.[19] Some *mZAS3* transcripts lack a sequence of 459 nucleotides within this exon, which encodes the ZAS-N domain. This intra-exon deleted sequence starts with "GC" and ends with "AG", which matches splice donor and acceptor dinucleotides. This 459 nucleotide sequence may be regarded as an unusual intron which, when retained in mRNA, encodes a DNA binding domain. Such an alternative splicing event has not been demonstrated for *ZAS1*, *ZAS2*, or *Schnurri*.

Human *ZAS3* produces a shorter 7 kb transcript by transcription termination at an upstream exon. As discussed above, *hZAS3* has an additional 310 nucleotide exon, exon 5b.[15] Exon 5b encodes 69 amino acid residues, a 3' UTR of 103 nucleotides, and contains a hexanucleotide AAUAAA located 39 nucleotides from the 3' end of the exon. This AAUAAA sequence is the canonical polyadenlyation signal. Therefore, *hZAS3* is unique among the *ZAS* genes in that its transcription may termination at either exon 5b, resulting in a 7 kb truncated transcript, or at exon 9, resulting in the full-length 9.5 kb transcript.

Tissue Distribution of ZAS Transcripts

In adult mouse tissues, *ZAS1* and *ZAS2* are ubiquitously expressed,[5,6] where *ZAS3* is expressed only in lymphoid and neural tissues.[8,19,31] *ZAS* genes are all expressed in developing mouse embryos, though the embryonic stage and patterns of expression differ: *ZAS1* is first detected at E8.5,[33] *ZAS2* is detected at E10.5,[34] and *ZAS3* is detected at E13.5.[35] *ZAS1* and *ZAS2* are expressed relatively uniformly in developing tissues, where *ZAS3* expression is limited to developing trigeminal ganglion, dorsal root ganglia, cerebral cortex, and thymus.[35] The distinct spatial and temporal expression patterns of *ZAS* genes suggest that they may perform general functions as well as specific functions during lymphoid and neural development.

Function

DNA Binding

The Rel/NF-κB and ZAS both bind to the κB motif. Sen and Baltimore first described NF-κB as a constitutively active protein in the nuclei of B cells which binds a specific site in the immunoglobulin enhancer.[36] Since then, the NF-κB/Rel family has been extensively studied.[37-39] The NF-κB/Rel proteins are thought to function as a stress-response mechanism whose function is activated by exposure to bacteria, viruses, cytokines, mitogens, and environmental chemicals.[40] Activated NF-κB/Rel proteins in turn mediate transcription of genes encoding for cytokines, immunoreceptors, cell adhesion molecules, acute phase proteins, stress response molecules, apoptotic regulators, growth factors, and other transcription factors.[40] Mutation studies support the physiologic importance of the Rel/NF-κB proteins in regulation of immunity, inflammation, development, apoptosis, cell proliferation, and oncogenesis.[40-43] The initial studies leading to the identification of the ZAS genes sought to characterize transcription factors that share binding affinity with the Rel proteins for the NF-κB binding sites.

ZAS1 was the original clone to be isolated from a human B cell cDNA expression library by screening with radio-labeled MHC Class I gene enhancer, a sequence similar to the κB motif, 5'-TGGGGATTCCCA-3'.[44] *ZAS2* and *ZAS3* were subsequently cloned using similar strategies using specific DNA fragments as ligands. Specific affinities of the ZAS proteins for promoter and enhancer gene targets have been studied extensively. The known DNA binding targets of ZAS proteins are listed in Table 1. All of the ZAS proteins, including Schnurri, bind *cis*-acting gene regulatory elements with a consensus sequence similar to the κB motif: 5'-GGGG($N_{4\text{-}5}$)CC-3' (Table 1A). In addition to the κB-like targets, each protein has also been shown to bind additional target sequence: mZAS1 binds to a sequence (5'-GAGAAAAGCC-3') at the core enhancer of the *type II collagen gene* (*Col2α1*);[33] mZAS2 binds to a TC-rich sequence in the *somatostatin receptor type II (SSTR2)* promoter;[34] and mZAS3 binds to the *cis*-acting *RSS* (recombination signal sequences) essential for somatic V(D)J recombination of the antigen receptor genes (Table 1B).[31,45]

Regulation of Transcription

From *Drosophila* to mammals, the ZAS zinc fingers are highly conserved (Fig. 2B) and retain the ability to bind to the κB motif (Table 1A). While κB-binding by the Rel family proteins activates transcription, the consequences of DNA-binding by the ZAS proteins are not so straightforward: ZAS-DNA binding may result in transcriptional activation or inhibition. ZAS1 activates transcription of *HIV-1 LTR* and *αA-crystallin*.[6,32] However, ZAS1 binding to the type II collagen gene enhancer inhibits transcription due to blocking the binding of Sox9, a transcriptional activator.[33] Similarly, ZAS2 has a dual function in regulating gene transcription. Murine ZAS2 binds a TC-box in the *somatostatin type II receptor* gene promoter, and activates transcription by recruiting a basic-helix-loop-helix transcription factor, SEF-2.[34] However, rat ZAS2 inhibits transcription at the major *c-myc* promoter.[46] In addition, rat ZAS2 interacts with the Ski-interacting protein (SKIP), although the consequence of this interaction is unknown.[46] ZAS3 activates transcription of *S100A4/mts1*.[47] It also has the unique characteristic of specifically binding the conserved signal sequences of V(D)J recombination (RSS), which are not typically thought of as gene regulatory elements. The ZAS transcriptional functions are summarized in Figure 3A. These isolated mammalian ZAS-DNA interactions described above provide glimpses of the range of potential complexity of ZAS protein function, from independent transcriptional activators or inhibitors to adapter proteins that recruit or sterically hinder other transcriptional proteins.

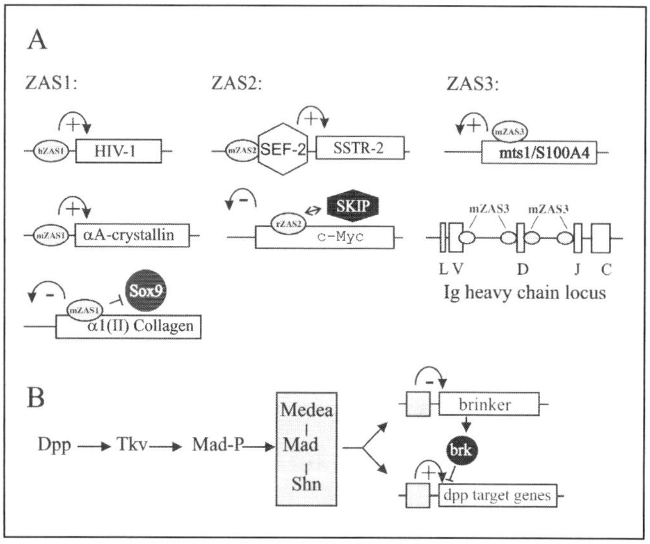

Figure 3. Transcriptional regulation by ZAS proteins. A) ZAS-mediated gene transcription. Probable interaction of ZAS3 with *RSS* flanking the variable (*V*), diversity (*D*), and joining (*J*) gene segments of the immunoglobulin heavy chain gene is also shown. B) Schnurri is essential for dpp-mediated transcription. Signal from Dpp induces phosphorylation of Mad by Tkv. Phosphorylated Mad forms a complex with Medea and Schnurri. This complex regulates Dpp target genes by direct activation, or by inhibition of the repressor brinker.

Signal Transduction

Signal Transduction by Schnurri and TGF-β

Of the *ZAS* genes, the physiological function of Schnurri is the best studied. Genetic studies initially identified Schnurri as a transcription factor that transduces signal from decapentaplegic (dpp), a TGF-β-related ligand homologous to mammalian bone morphology protein, BMP-4. Mutation of *Schnurri* result in dorsal patterning and wing formation defects.[10-12] BMPs signal through a heteromeric complex of type I and type II receptor serine-threonine kinases in which a ligand-receptor complex results in phosphoryation of the type I receptor by the type II receptor. The activated type I receptor then phosphorylates a receptor-specific Smad protein. In vertebrates, the phosphorylated Smad binds Smad4, and the resulting complex is translocated to the nucleus. The Smads have weak DNA-binding ability and recognize sequence of low complexity, and are chaperoned to target genes by association with specific transcription factors.[48] In *Drosophila*, a model has emerged in which dpp signal is transduced through the type I receptor thick veins (tkv), which phosphorylates the Smad1 ortholog, Mad, which is translocated to the nucleus with the Smad4 ortholog, Medea.[49] A Shnurri-Mad-Medea complex is essential to transduce Dpp-mediated transcription: Specific target genes, such as *ultrabithorax*, are directly activated, whereas the gene encoding a general transcription repressor, *brinker*, is repressed. Repression of *brinker* subsequently activates multiple additional Dpp-target genes (Fig. 3B).[50-52] As a transcription factor that regulates expression of global regulators, such as brinker, Schnurri illustrates the potential for the ZAS proteins as master regulators of gene transcription.

Signal Transduction by ZAS3 and TRAF Proteins

A role for the ZAS proteins in mammalian TGF-β pathways has not yet been established. However, biochemical properties of ZAS3 suggest the ZAS proteins to be well-suited to mediate signal transduction as DNA binding is regulated by posttranslational modification: ZAS3 is a substrate for epidermal growth factor kinase and p34cdc2 in vitro, and phosphorylation on serine and tyrosine residues increases DNA binding affinity of a ZAS3 fusion protein.[24] Recent evidence suggests that ZAS3 may function in signal transduction. The result of a yeast two-hybrid screen suggests that ZAS3 may interact with TRAF proteins which transduces TFN receptor signaling, resulting in down-regulation of TRAF-dependent JNK activation.[53] The ZAS proteins, therefore, may have transcriptional as well as nontranscriptional functions.

Proliferation and Carcinogenesis

While molecular studies continue to identify ZAS DNA targets and explore the role of ZAS proteins in signal transduction pathways, functional studies demonstrate that the ZAS proteins have important roles in the regulation of cellular proliferation and growth. Expression of all three *ZAS* genes can be induced by mitogens, such as fetal bovine serum, PHA/PMA and lipopolysaccharide.[1,35,53,54] In *Drosophila*, expression of Schnurri in cyst cells is required to restrict proliferation and mitosis of neighboring germ cells.[55] In the case of *ZAS3*, expression is normally restricted to lymphoid and neuronal tissues. However, *ZAS3* transcripts are detected in every cultured cell line examined, regardless of origin (unpublished observation). When ZAS3 expression was decreased by an antisense construct in a preB cell line, 38B9, cell proliferation accelerated.[30] Conversely, when ZAS3 protein expression was increased, cell proliferation slowed. Furthermore, decreased ZAS3 protein expression in HeLa cells conferred anchorage independence and induced formation of terminal multinucleated giant cells.[30] A similar phenotype was observed with p53 anti-sense expression in HeLa cells.[56] These findings support a role for ZAS3 in regulation of normal cell growth, proliferation, and cell death.

The generation of embryonic stem cells in which endogenous *ZAS3* is disrupted has provided clues the physiological function of ZAS3 protein. In a $RAG2^{-/-}$;$ZAS3^{-/-}$ complementation assay, a tumor spontaneously formed in the axilla of one chimera which was composed of > 90% $ZAS3^{-/-}$ cells. The tumor was first noted at 3-weeks as a 5 mm mass which grew to more than 20% of the total body mass by 5 weeks. DNA analysis showed that the tumor was derived entirely from $ZAS3^{-/-}$ cells. It was diagnosed as a malignant teratoma due to the presence of poorly-differentiated carcinoma cells coupled with multiple tissues of diverse origin, including hyaline cartilage, skeletal muscle, fat, lymphocytes, and primitive neural elements. Permanent cell lines were subsequently derived from teratoma cells, which continue to grow without viral transformation for over 2 years.[57] The spontaneous development of an extragonadal teratoma is very rare. In the past 25 years, only one case of such an event has been reported in mice.[58] Teratomas, however, have been induced in animals with *p53* mutations,[59] double *p53/atm* mutations,[60] and double *mdm2/p53* mutations.[61] The changes in cell growth induced by disruption of *ZAS3* in cell and animal models suggest that ZAS3 may have clinical relevance as a tumor suppressor. Chromosomal analysis of multiple tumors identify mutations around the *ZAS3* locus, 1p34: Allelic loss at 1p34 has been reported in multiple tumors including breast cancer,[62-65] liver cancers,[66-68] pheochromocytomas,[69] and B cell lymphomas.[70] Furthermore, changes in expression of ZAS1 and ZAS2 proteins have been associated with poor prognosis in chronic lymphocytic leukemia patients.[71]

Immunity

Review of known ZAS target genes (Table 1A) implicates the ZAS family in transcriptional regulation of genes involved in immune response as well as cell proliferation. In addition to the κB motif, ZAS3 has been shown to specifically bind the recombination signal sequences (RSS) which flank gene segments eligible for somatic recombination in the variable regions of immunoglobulin and T cell receptor genes.[31,45,72] These conserved motifs are *cis*-acting gene elements essential for recombination activating gene (*RAG-1* and *RAG-2*) mediated recombination.[73,74] In fact, ZAS3 is the most abundant RSS-binding species in the thymus.[35] What a transcription factor might be doing binding the RSS is not yet clear, however the DNA-binding ability of ZAS3 does correlate with activation of V(D)J recombination.[25] In the context of the accessibility model, ZAS3 may potentially activate transcription at specific target loci, rendering them accessible to the V(D)J recombinase.[75,76] Additionally, ZAS3 could theoretically recruit other transcriptional proteins to RSS to activate recombination or block the access of RAGs. Ultimately, ZAS3 does not appear to be essential for V(D)J recombination as $RAG2^{-/-}$;$ZAS3^{-/-}$ chimeric mice do develop mature thymocytes. However, there is a progressive depletion of the double-positive ($CD4^+/CD8^+$) T cell compartment over time.[57] Similarly, targeted disruption of *ZAS2* also results in T cell maturation defects. Thymi of $ZAS2^{-/-}$ mice contain few single positive ($CD4^+/CD8^-$ and $CD4^-/CD8^+$) thymocytes but with normal levels of double positive ($CD4^+/CD8^+$) thymocytes, reflecting defects in positive selection.[77] These data suggest that individual ZAS proteins may be important at different stages of thymocyte development.

Future Directions

From *Drosophila* to mammals, the ZAS C_2H_2 zinc finger pairs specifically bind the κB motif. The snapshots of ZAS-mediated transcription collected to date suggest a model in which ZAS proteins may function as adaptors that recruit or inhibit other transcription factors to gene regulatory sites, resulting in activation or repression of transcription of target genes. Binding of the κB motif by the Rel/NF-κB proteins has been extensively studied over the past decade (reviewed in 38-41). While ZAS and Rel/NF-κB share the same DNA target, little is known about their competition or interaction. It is reasonable to hypothesize that ZAS proteins may stimulate or inhibit NF-κB-mediated transcription.

The ZAS proteins are subject to post-translational regulation, a key feature of proteins involved in signal transduction. In *Drosophila*, Schnurri functions as a master-regulator of transcription by repressing transcription of a super-repressor gene, *brinker*. While no brinker homolog has yet been identified in vertebrates, it will be interesting to investigate a role for ZAS proteins in Smad interactions in TGF-β signal transduction.

Disruption of *ZAS* expression in cell models suggests a fundamental role for this protein family in regulation of cell growth and proliferation. Targeted disruption of *ZAS* genes in animal models resulted in spontaneous tumor formation as well as aberrant thymocyte differentiation. Clearly, ZAS proteins individually influence fundamental cellular processes. However, it is not yet clear to what extent individual ZAS functions are redundant. In *Drosophila*, disruption of *Schnurri* is lethal, while disruption of mammalian *ZAS* is not. It will be intriguing to compare the phenotypes of individual *ZAS* knock-out animals with double-knockouts and, ultimately, a triple knock-out. Ultimately, as ZAS functions become further defined, post-translational modifications of ZAS proteins or pharmacology strategies based on ZAS zinc finger structure may become useful in modifying ZAS protein function in human disease.

References

1. Baldwin AS, LeClair KP, Singh H et al. A large protein containing zinc finger domains binds to related sequence elements in the enhancers of the class I major histocompatibility complex and kappa immunoglobulin genes. Mol Cell Biol 1990; 10(4):1406-1414.
2. Maekawa T, Sakura H, Sudo T et al. Putative metal finger structure of the human immunodeficiency virus type 1 enhancer binding protein HIV-EP1. J Biol Chem 1989; 264(25):14591-14593.
3. Fan CM, Maniatis T. A DNA-binding protein containing two widely separated zinc finger motifs that recognize the same DNA sequence. Genes Dev 1990; 4(1):29-42.
4. Fountain JW, Dracopoli NC, Housman DE et al. MspI RFLP detected by a ZNF-40 cDNA sequence. Nucleic Acids Res 1991; 19(9):2514.
5. Nakamura T, Donovan DM, Hamada K et al. Regulation of the mouse alpha A-crystallin gene: Isolation of a cDNA encoding a protein that binds to a cis sequence motif shared with the major histocompatibility complex class I gene and other genes. Mol Cell Biol 1990; 10(7):3700-3708.
6. Brady JP, Kantorow M, Sax CM et al. Murine transcription factor alpha A-crystallin binding protein I. Complete sequence, gene structure, expression, and functional inhibition via antisense RNA. J Biol Chem 1995; 270(3):1221-1229.
7. Mitchelmore C, Traboni C, Cortese R. Isolation of two cDNAs encoding zinc finger proteins which bind to the alpha 1-antitrypsin promoter and to the major histocompatibility complex class I enhancer. Nucleic Acids Res 1991; 19(1):141-147.
8. Wu LC, Liu Y, Strandtmann J et al. The mouse DNA binding protein Rc for the kappa B motif of transcription and for the V(D)J recombination signal sequences contains composite DNA-protein interaction domains and belongs to a new family of large transcriptional proteins. Genomics 1996; 35(3):415-424.
9. Boguski MS, Schuler GD. ESTablishing a human transcript map. Nat Genet 1995; 10(4):369-371.
10. Arora K, Dai H, Kazuko SG et al. The Drosophila schnurri gene acts in the Dpp/TGF beta signaling pathway and encodes a transcription factor homologous to the human MBP family. Cell 1995; 81(5):781-790.
11. Grieder NC, Nellen D, Burke R et al. Schnurri is required for Drosophila Dpp signaling and encodes a zinc finger protein similar to the mammalian transcription factor PRDII-BF1. Cell 1995; 81(5):791-800.
12. Staehling-Hampton K, Laughon AS, Hoffmann FM. A Drosophila protein related to the human zinc finger transcription factor PRDII/MBPI/HIV-EP1 is required for dpp signaling. Development 1995; 121(10):3393-3403.
13. Sudo T, Ozawa K, Soeda EI et al. Mapping of the human gene for the human immunodeficiency virus type 1 enhancer binding protein HIV-EP2 to chromosome 6q23-q24. Genomics 1992; 12(1):167-170.
14. Olavesen MG, Bentley E, Mason RV et al. Fine mapping of 39 ESTs on human chromosome 6p23-p25. Genomics 1997; 46(2):303-306.
15. Hicar MD, Liu Y, Allen CE et al. Structure of the human zinc finger protein HIVEP3: Molecular cloning, expression, exon-intron structure, and comparison with paralogous genes HIVEP1 and HIVEP2. Genomics 2001; 71(1):89-100.
16. Wu LC. ZAS: C_2H_2 Zinc finger proteins involved in growth and development. Gene Expression 2002; 10(4):137-152.
17. Mignone F, Gissi C, Liuni S et al. Untranslated regions of mRNAs. Genome Biol 2002; 3(3):REVIEWS0004.
18. Muchardt C, Seeler JS, Nirula A et al. Regulation of human immunodeficiency virus enhancer function by PRDII-BF1 and c-rel gene products. J Virol 1992; 66(1):244-250.
19. Mak CH, Li Z, Allen CE et al. KRC transcripts: Identification of an unusual alternative splicing event. Immunogenetics 1998; 48(1):32-39.
20. Iuchi S. Three classes of C2H2 zinc finger proteins. Cell Mol Life Sci 2001; 58(4):625-635.
21. Knight RD, Shimeld SM. Identification of conserved C2H2 zinc-finger gene families in the Bilateria. Genome Biol 2001; 2(5):RESEARCH0016.
22. Tian Q, Kopf GS, Brown RS et al. Function of basonuclin in increasing transcription of the ribosomal RNA genes during mouse oogenesis. Development 2001; 128(3):407-416.
23. Tseng H, Matsuzaki K, Lavker RM. Basonuclin in murine corneal and lens epithelia correlates with cellular maturation and proliferative ability. Differentiation 1999; 65(4):221-227.
24. Bachmeyer C, Mak CH, Yu CY et al. Regulation by phosphorylation of the zinc finger protein KRC that binds the kappaB motif and V(D)J recombination signal sequences. Nucleic Acids Res 1999; 27(2):643-648.
25. Wu LC, Hicar MD, Hong J et al. The DNA-binding ability of HIVEP3/KRC decreases upon activation of V(D)J recombination. Immunogenetics 2001; 53(7):564-571.
26. Suzuki M. SPXX, a frequent sequence motif in gene regulatory proteins. J Mol Biol 1989; 207(1):61-84.
27. Brzeski J, Grycuk T, Lipkowski AW et al. Binding of SPXK- and APXK-peptide motifs to AT-rich DNA. Experimental and theoretical studies. Acta Biochim Pol 1998; 45(1):221-231.
28. Kennelly PJ, Krebs EG. Consensus sequences as substrate specificity determinants for protein kinases and protein phosphatases. J Biol Chem 1991; 266(24):15555-15558.
29. Kantorow M, Becker K, Sax CM et al. Binding of tissue-specific forms of alpha A-CRYBP1 to their regulatory sequence in the mouse alpha A-crystallin-encoding gene: Double-label immunoblotting of UV-crosslinked complexes. Gene 1993; 131(2):159-165.
30. Allen CE, Wu LC. Downregulation of KRC induces proliferation, anchorage independence, and mitotic cell death in HeLa cells. Exp Cell Res 2000; 260(2):346-356.
31. Wu LC, Mak CH, Dear N et al. Molecular cloning of a zinc finger protein which binds to the heptamer of the signal sequence for V(D)J recombination. Nucleic Acids Res 1993; 21(22):5067-5073.
32. Seeler JS, Muchardt C, Suessle A et al. Transcription factor PRDII-BF1 activates human immunodeficiency virus type 1 gene expression. J Virol 1994; 68(2):1002-1009.
33. Tanaka K, Matsumoto Y, Nakatani F et al. A zinc finger transcription factor, alpha A-crystallin binding protein 1, is a negative regulator of the chondrocyte-specific enhancer of the alpha1(II) collagen gene. Mol Cell Biol 2000; 20(12):4428-4435.
34. Dorflinger U, Pscherer A, Moser M et al. Activation of somatostatin receptor II expression by transcription factors MIBP1 and SEF-2 in the murine brain. Mol Cell Biol 1999; 19(5):3736-3747.
35. Hicar MD, Robinson ML, Wu LC. Embryonic expression and regulation of the large zinc finger protein KRC. Genesis 2002; 33(1):8-20.
36. Sen R, Baltimore D. Multiple nuclear factors interact with the immunoglobulin enhancer sequences. Cell 1986; 46(5):705-716.
37. Baldwin AS. The NF-kappa B and I kappa B proteins: New discoveries and insights. Annu Rev Immunol 1996; 14:649-683.
38. Ghosh S, May MJ, Kopp EB. NF-kappa B and Rel proteins: Evolutionarily conserved mediators of immune responses. Annu Rev Immunol 1998; 16:225-260.

39. Perkins ND. The Rel/NF-kappa B family: Friend and foe. Trends Biochem Sci 2000; 25(9):434-440.
40. Pahl HL. Activators and target genes of Rel/NF-kappaB transcription factors. Oncogene 1999; 18(49):6853-6866.
41. Chen F, Castranova V, Shi X et al. New insights into the role of nuclear factor-kappaB, a ubiquitous transcription factor in the initiation of diseases. Clin Chem 1999; 45(1):7-17.
42. Karin M, Cao Y, Greten FR et al. NF-kappaB in cancer: From innocent bystander to major culprit. Nat Rev Cancer 2002; 2(4):301-310.
43. Gerondakis S, Grossmann M, Nakamura Y et al. Genetic approaches in mice to understand Rel/NF-kappaB and IkappaB function: Transgenics and knockouts. Oncogene 1999; 18(49):6888-6895.
44. Singh H, LeBowitz JH, Baldwin AS et al. Molecular cloning of an enhancer binding protein: Isolation by screening of an expression library with a recognition site DNA. Cell 1988;52(3):415-423.
45. Allen CE, Mak CH, Wu LC. The kappaB transcriptional enhancer motif and signal sequences of V(D)J recombination are targets for the zinc finger protein HIVEP3/KRC: A site selection amplification binding study. BMC Immunology 2002; 3(1):10.
46. Fukuda S, Yamasaki Y, Iwaki T et al. Characterization of the biological functions of a transcription factor, c-myc intron binding protein 1 (MIBP1). J Biochem (Tokyo) 2002; 131(3):349-357.
47. Hjelmsoe I, Allen CE, Cohn MA et al. The kappaB and V(D)J recombination signal sequence binding protein KRC regulates transcription of the mouse metastasis-associated gene S100A4/mts1. J Biol Chem 2000; 275(2):913-920.
48. Massague J. TGF-beta signal transduction. Annu Rev Biochem 1998; 67:753-791.
49. Raftery LA, Sutherland DJ. TGF-beta family signal transduction in Drosophila development: From Mad to Smads. Dev Biol 1999; 2 10(2):251-268.
50. Dai H, Hogan C, Gopalakrishnan B et al. The zinc finger protein schnurri acts as a Smad partner in mediating the transcriptional response to decapentaplegic. Dev Biol 2000; 227(2):373-387.
51. Marty T, Muller B, Basler K et al. Schnurri mediates Dpp-dependent repression of brinker transcription. Nat Cell Biol 2000; 2(10):745-749.
52. Torres-Vazquez J, Park S, Warrior R et al. The transcription factor Schnurri plays a dual role in mediating Dpp signaling during embryogenesis. Development 2001; 128(9):1657-1670.
53. Oukka M, Kim ST, Lugo G et al. A mammalian homolog of Drosophila schnurri, KRC, regulates TNF receptor-driven responses and interacts with TRAF2. Mol Cell 2002; 9(1):121-131.
54. Nomura N, Zhao MJ, Nagase T et al. HIV-EP2, a new member of the gene family encoding the human immunodeficiency virus type 1 enhancer-binding protein. Comparison with HIV-EP1/PRDII-BF1/MBP-1. J Biol Chem 1991; 266(13):8590-8594.
55. Matunis E, Tran J, Gonczy P et al. punt and schnurri regulate a somatically derived signal that restricts proliferation of committed progenitors in the germline. Development 1997; 124(21):4383-4391.
56. Iotsova V, Stehelin D. Antisense p53 provokes changes in HeLa cell growth and morphology. Eur J Cell Biol 1995; 68(2):122-132.
57. Allen CE, Muthusamy N, Weisbrode SE et al. Developmental anomalies and neoplasia in animals and cells deficient in the large zinc finger protein KRC. Genes Chromosomes Cancer 2002; 35(4):287-298.
58. Artzt K, Damjanov I. Spontaneous extragonadal teratocarcinoma in a mouse. Lab Anim Sci 1978; 28(5):584-586.
59. Jacks T, Remington L, Williams BO et al. Tumor spectrum analysis in p53-mutant mice. Curr Biol 1994; 4(1):1-7.
60. Westphal CH, Rowan S, Schmaltz C et al. atm and p53 cooperate in apoptosis and suppression of tumorigenesis, but not in resistance to acute radiation toxicity. Nat Genet 1997; 16(4):397-401.
61. McDonnell TJ, Montes de Oca Luna R, Cho S et al. Loss of one but not two mdm2 null alleles alters the tumour spectrum in p53 null mice. J Pathol 1999; 188(3):322-328.
62. Emi M, Yoshimoto M, Sato T et al. Allelic loss at 1p34, 13q12, 17p13.3, and 17q21.1 correlates with poor postoperative prognosis in breast cancer. Genes Chromosomes Cancer 1999; 26(2):134-141.
63. Tsukamoto K, Ito N, Yoshimoto M et al. Allelic loss on chromosome 1p is associated with progression and lymph node metastasis of primary breast carcinoma. Cancer 1998; 82(2):317-322.
64. Utada Y, Emi M, Yoshimoto M et al. Allelic loss at 1p34-36 predicts poor prognosis in node-negative breast cancer. Clin Cancer Res 2000; 6(8):3193-3198.
65. Hirano A, Utada Y, Haga S et al. Allelic losses as prognostic markers for breast cancers. Int J Clin Oncol 2001; 6(1):6-12.
66. Li SP, Wang HY, Li JQ et al. Genome-wide analyses on loss of heterozygosity in hepatocellular carcinoma in Southern China. J Hepatol 2001; 34(6):840-849.
67. Parada LA, Hallen M, Tranberg KG et al. Frequent rearrangements of chromosomes 1, 7, and 8 in primary liver cancer. Genes Chromosomes Cancer 1998; 23(1):26-35.
68. Sun M, Eshleman JR, Ferrell LD et al. An early lesion in hepatic carcinogenesis: Loss of heterozygosity in human cirrhotic livers and dysplastic nodules at the 1p36-p34 region. Hepatology 2001; 33(6):1415-1424.
69. Benn DE, Dwight T, Richardson AL et al. Sporadic and familial pheochromocytomas are associated with loss of at least two discrete intervals on chromosome 1p. Cancer Res 2000; 60(24):7048-7051.
70. Liu J, Sen R, Rothstein TL. Abnormal kappa B-binding protein in the cytoplasm of a plasmacytoma cell line that lacks nuclear expression of NF-kappa B. Mol Immunol 1993; 30(5):479-489.
71. Aalto Y, El Rifa W, Vilpo L et al. Distinct gene expression profiling in chronic lymphocytic leukemia with 11q23 deletion. Leukemia 2001; 15(11):1721-1728.
72. Mak CH, Strandtmann J, Wu LC. The V(D)J recombination signal sequence and kappa B binding protein Rc binds DNA as dimers and forms multimeric structures with its DNA ligands. Nucleic Acids Res 1994; 22(3):383-390.
73. Oettinger MA, Schatz DG, Gorka C et al. RAG-1 and RAG-2, adjacent genes that synergistically activate V(D)J recombination. Science 1990; 248(4962):1517-1523.
74. Schatz DG, Oettinger MA, Baltimore D. The V(D)J recombination activating gene, RAG-1. Cell 1989; 59(6):1035-1048.
75. Yancopoulos GD, Alt FW. Developmentally controlled and tissue-specific expression of unrearranged VH gene segments. Cell 1985; 40(2):271-281.
76. Schlissel MS, Stanhope-Baker P. Accessibility and the developmental regulation of V(D)J recombination. Semin Immunol 1997; 9(3):161-170.
77. Takagi T, Harada J, Ishii S. Murine Schnurri-2 is required for positive selection of thymocytes. Nat Immunol 2001; 2(11):1048-1053.
78. van 't Veer LJ, Lutz PM, Isselbacher KJ et al. Structure and expression of major histocompatibility complex-binding protein 2, a 275-kDa zinc finger protein that binds to an enhancer of major histocompatibility complex class I genes. Proc Natl Acad Sci USA 1992; 89(19):8971-5.
79. Rustgi AK, Van 't Veer LJ, Bernards R. Two genes encode factors with NF-kappa B- and H2TF1-like DNA-binding properties. Proc Natl Acad Sci USA 1990; 87(22):8707-10.
80. Makino R, Akiyama K, Yasuda J et al. Cloning and characterization of a c-myc intron binding protein (MIBP1). Nucleic Acids Res 1994; 22(25):5679-5685.

CHAPTER 30

Role of GATA Factors in Development

Marc Haenlin* and Lucas Waltzer

Abstract

Members of the GATA transcription factor family are found throughout eukaryotes, including plants, fungi, invertebrates and vertebrates. In this review, we discuss some of the roles of GATA factors during vertebrate and invertebrate development. We place particular emphasis on their function in hematopoiesis, in heart and in endoderm development, phenomena in which several different organisms have shown striking molecular and developmental similarities.

Introduction

Members of the GATA transcription factor family are so called because they bind to the DNA consensus site WGATAR.[1,2] They all contain one or two conserved zinc finger DNA-binding domains, with the characteristic consensus sequence Cys-X_2-Cys-X_{17-18}-Cys-X_2-Cys followed by a basic region. GATA factors are found throughout eukaryotes, from yeast to vertebrates,[3] where they regulate a variety of developmental processes. In yeast, fungi and plants, they participate in nitrogen metabolism, blue-light-regulated morphogenesis and circadian rhythm.[4] In animals, GATA factors have been implicated in endoderm formation, hematopoiesis and heart development in numerous species,[5] in adipogenesis,[6] urogenital development[7] and neurogenesis[8,9] in vertebrates, and in patterning and ectoderm differentiation in invertebrates.[10-14] Here we shall concentrate on the conserved functions of GATA factors in the animal kingdom during hematopoiesis, heart formation and endoderm development and we shall discuss the implication of GATA genes in human diseases.

Hematopoieisis

The founding member of the GATA family, *GATA1*, was first identified as an erythroid-specific transcription factor that binds to functionally important *cis*-regulatory sequences in the *globin* genes.[15,16] Subsequent cloning of the other vertebrate GATA factors revealed that three of them, *GATA1*, *2* and *3*, were prominently expressed in blood cells. Targeted disruptions of each of these three genes in mouse demonstrate that they play critical roles in hematopoiesis. Furthermore, a growing body of evidence shows that GATA factor function during hematopoiesis has been conserved through evolution.

In Vertebrates

During vertebrate development, hematopoiesis occurs in two successive waves (reviewed in ref. 17). In mammals, the blood islands of the yolk sac produce transient embryonic red blood cells (primitive hematopoiesis). At midgestation, definitive hematopoietic stem cells (HSC), specified in the AGM (aorta-gonad-mesonephros) region, colonise foetal and definitive hematopoietic sites, such as the embryonic liver and the bone marrow. Subsequent hematopoietic differentiation is usually assumed to proceed in a hierarchical fashion: HSC give rise to multipotent progenitors and then to committed progenitors that will differentiate into one or several specialised blood cell types (Fig. 1). "Hematopoietic" GATA factors display distinct, but overlapping, expression profiles during hematopoietic development, in which they participate from the emergence of the HSC to terminal differentiation of blood cells.

GATA2 is expressed in the primitive and definitive HSC as well as in mast cells and megakaryocytes and at lower levels in erythroid cells.[18-21] Homozygous *GATA2* mutant mice die by embryonic day 10 (E10) after having undergone severely reduced primitive hematopoiesis.[22] Furthermore, *GATA2* null embryonic stem cells (ES) fail to contribute to definitive hematopoiesis in chimeric animals.[23] Thus *GATA2* is essential for the development of the blood cell progenitors. One key target of GATA2 in the HSC is probably *SCL/Tal1*, a basic-Helix-Loop-Helix (bHLH)-encoding gene, required for all blood cell development, that contains functionally important GATA2 binding sites in its enhancer.[24] The expression pattern of *GATA2* in zebrafish and *Xenopus* suggests that the role of this factor in hematopoietic progenitor development has been conserved.[25-27] Interestingly, GATA2 can either promote proliferation at the expense of differentiation or induce differentiation, depending on the hematopoietic cell type in which it is expressed and on its expression level. Indeed, GATA2 is required at high levels to promote proliferation of early hematopoietic progenitors and/or survival,[23,28] but it has to be downregulated to allow erythroid differentiation to occur.[29,30] In addition, GATA2 is required for mast cell differentiation.[23]

GATA1 is expressed at a high level in erythroid cells, mast cells, megakaryocytes, and eosinophils, and at a low level in multipotent progenitors.[15,16] Several lines of evidence show that *GATA1* plays a central role in erythroid differentiation. In chimeric mice, *GATA1-/-* ES cells contribute to white blood cells

* *Corresponding author. See list of "Contributors".*

Zinc Finger Proteins: From Atomic Contact to Cellular Function, edited by Shiro Iuchi and Natalie Kuldell.

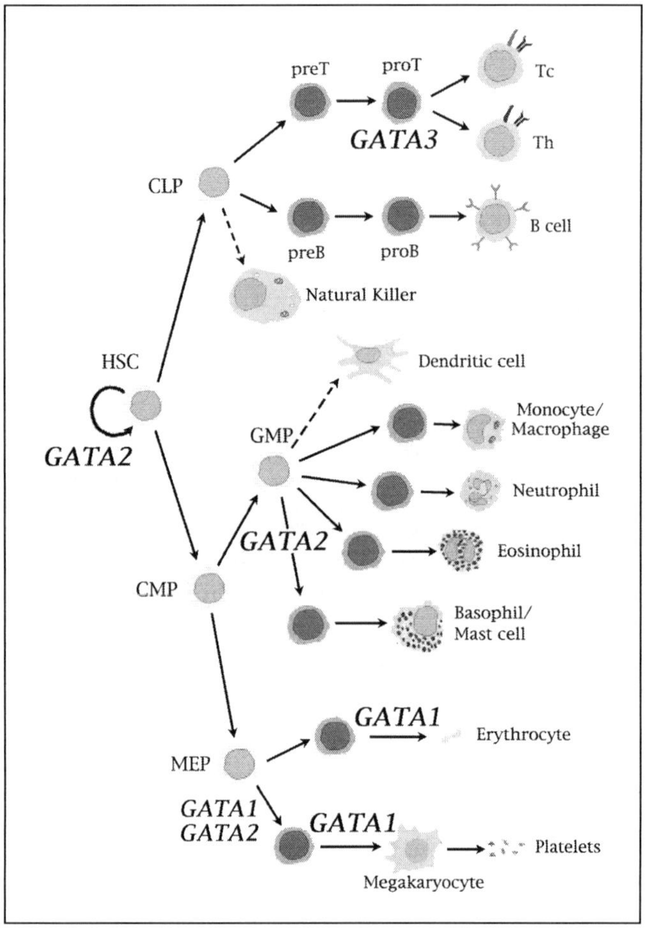

Figure 1. GATA factors during hematopoiesis in mouse. Schematic representation of the hematopoietic lineage differentiation pathway from pluripotential stem cells to mature blood cells. The different GATA genes are indicated at the stage where differentiation is blocked in their absence. See text for further details. Dotted arrows indicate uncertain lineage relationships. HSC: hematopoietic stem cells, CLP: common lymphoid precursor, CMP: common myeloid precursor, GMP: granulocyte/monocyte precursor, MEP: myeloid/erythroid precursor, Tc: T-cytolytic lymphocyte, Th: T-helper lymphocyte.

but not to mature red blood cells.[31] Knock-out and knock-down of *GATA1* ablate primitive erythropoiesis, whereas definitive erythropoiesis stops at the proerythroblast stage.[32,33] Similarly, in zebrafish, the mutant *vlad tepes*, which carries a null mutation in *GATA1*, shows specific defects in the erythroid pathway.[34] Furthermore, lineage-specific ablation of mouse *GATA1* showed that it is required for terminal differentiation of megakaryocytes.[35,36] An essential partner of GATA1 during erythroid and megakaryocyte differentiation is FOG1 (Friend of GATA-1). Members of the FOG family are multiple-zinc-finger proteins that interact with the N-terminal zinc finger of GATA factors and act as co-activators or co-repressors.[37-43] *FOG-1-/-* murine embryos die because of arrested erythroid cell maturation and complete failure of megakaryopoiesis.[37] It was shown that the interaction between FOG1 and GATA1 is critical for erythroid cell differentiation in vitro and in vivo[44,45] and that FOG1 functions with GATA1 and GATA2 to control an early stage of megakaryopoiesis in vivo.[45]

The rescue of *GATA1*-deficient mice with transgenes expressing various GATA1domains under the control of its own enhancer have shown that primitive and definitive erythropoiesis require different domains of GATA1.[46] While the C-terminal zinc finger that mediates DNA-binding is required both for primitive and definitive hematopoiesis, the N-terminal zinc finger or the N-terminal transactivation domains are necessary only for the latter. These two domains may provide additional regulatory functions (notably the interaction with FOG1 and autoregulation of GATA1 expression) required for the more complex lineage specification process occurring during definitive hematopoiesis. Interestingly, in a similar assay, *GATA2* and *GATA3* also rescued *GATA1-/-* embryos from lethality; however rescued mice displayed anaemia and thrombocytopenia, suggesting that GATA1 has unique functional properties that are distinct from those of other GATA factors.[47]

During definitive erythropoiesis, concomitant upregulation of *GATA1* and downregulation of *GATA2* expression is essential to commit proliferating precursors to become mature erythroid cells. In mouse, chicken and zebrafish, several GATA sites in the *GATA1* promoter are required for its erythroid-specific expression. Thus, GATA2 may trigger *GATA1* expression, which may then be maintained by autoregulation.[48-52] Conversely, GATA1 has been shown to inhibit *GATA2* expression by establishing a repressive chromatin structure and by displacing GATA2 from its own promoter, thereby disrupting its positive autoregulation.[53] Consistent with this analysis, *GATA1*-deficient progenitors express high levels of GATA2 and are blocked at the proerythroblast stage.[54]

GATA1 is not only required for erythroid and megakaryocytic differentiation but it is also sufficient to promote the emergence of these lineages. For instance, in *Xenopus*, overexpression of GATA1 suffices to induce *globin* gene expression in animal cap assays [55] and, in combination with the bHLH SCL and the LIM domain protein LMO2, with which its forms a transcriptionally active complex,[56] GATA1 can induce ectopic erythropoiesis in the whole embryo.[55] Moreover, in vitro gain-of-function analysis, particularly with retrovirally transformed chicken hematopoiteic progenitors, have shown that GATA1 can reprogram them toward erythroid, eosinophil and thromboblast/megakaryocyte lineages.[57-63] These assays also brought to light a cross-regulation between a combination of transcription factors whose equilibrium in the progenitors is critical for cell fate choice. Notably, the lineage choice appeared to be highly dependent on the level of expression of GATA1 and of its partners such as FOG1, PU.1 and C/EBPβ (reviewed in ref. 64).

GATA3 was originally identified as a T-cell specific transcription factor that bound the enhancer of the T cell receptor (TCR),[65] and it is the only GATA factor to be expressed in T lymphocytes. *GATA3* is also expressed in other tissues, including the placenta, the central nervous system, the kidney, the adrenal gland, the primitive thymus and the skin.[65-68] *GATA3*-deficient mice display anaemia and central nervous system defects and die on embryonic day 12 thus precluding a detailed assessment of its role in T cell development.[69] Generation of chimeric animals showed that differentiation of *GATA3-/-* T cells is blocked at an early stage, that is, at or before the $CD4^-/CD8^-$ stage of thymocyte development.[70,71] In addition to this very early requirement, GATA3 promotes the development of $CD4^+/CD8^+$ thymocytes into T helper (TCR high, $CD4^+$) to the detriment of cytolytic T cells ($CD8^+$) in response to TCR signalling.[72] Finally, GATA3 induces the differentiation of $CD4^+$ T cells into class 2 T helper (Th2) versus class 1 T helper cells (Th1).[73-75] It is proposed that interleukin 4 (IL4) signalling induces Th2 differentiation by increasing GATA3 expression in naive T helper cells. GATA3 then represses Th1 development by inhibiting Th1-specific genes and induces Th2 cytokine expression by direct transactivation of the IL5 and IL13 gene, as well as through remodelling/demethylation of the IL4 locus.[76-79] Conversely, IL12 signalling inhibits *GATA3* expression/function and induces Th1 development by activating STAT4 and T-bet expression.[78] Thus, the precise regulation of *GATA3* expression at several stages is essential for proper T cell differentiation.

In Drosophila

Despite the physiological differences between *Drosophila* and vertebrate blood cells, various aspects of hematopoietic development have been evolutionarily conserved.[80,81] As in vertebrates, *Drosophila* hematopoiesis (from mesodermally-derived progenitors) occurs in two waves: one during embryogenesis and the second during the larval stages.[82,83] Blood cells (hemocytes) differentiate into two major classes: plasmatocytes and crystal cells.[82] Plasmatocytes act as macrophages, whereas crystal cells participate in melanisation, an insect-specific process involved in the encapsulation of foreign bodies and in wound healing. Most strikingly, it appeared recently that transcription factors of the GATA, FOG and RUNX families, which play critical roles during hematopoiesis in vertebrates, are also involved in controlling blood cell development in *Drosophila*. Like GATA2, the *Drosophila* GATA factor Serpent (Srp) is expressed in blood cell precursors and is required for the determination, proliferation and maintenance of this population.[84,85] Furthermore, Srp is still detected in mature plasmatocytes and, at lower levels, in crystal cells. Reminiscent of *GATA2* regulation during erythropoiesis, down-regulation of *srp* expression in crystal cells may be important, as suggested by the fact that overexpression of Srp in these cells inhibits their differentiation.[86] Nonetheless, Srp is required for crystal cell differentiation in combination with the RUNX factor Lozenge (Lz), with which it synergistically interacts to trigger the crystal cell-specific genetic program (Waltzer et al, in press). Interestingly, the interaction and cooperation between GATA and RUNX factors appear to be conserved in vertebrates and it has been suggested that GATA1 cooperates with RUNX1 during megakaryopoiesis.[87] Furthermore, the *Drosophila* FOG factor U-shaped (Ush) represses crystal cell fate by interacting with Srp,[86,88] suggesting that a competition between Ush and Lz may modulate Srp activity and thereby influence the subsequent development into plasmatocytes rather than crystal cells or vise versa. Such cross-interactions between GATA factors and their partners may constitute a conserved means of regulating lineage choice and differentiation during hematopoiesis.

Heart Formation

Recent studies have shown that heart formation is controlled by a developmental program in which many molecular and developmental aspects are conserved between fly and man. Whereas heart morphology in *Drosophila melanogaster* is simpler than the complex multichambered heart of mammals, both are constituted of spontaneous contractile myocardial cells. In both vertebrates and *Drosophila*, the heart develops from bilaterally symmetrical regions of the mesoderm, where heart progenitors migrate to the midline and fuse to form a linear, beating heart tube (Fig. 2). In vertebrates, this tube undergoes further specific morphogenetic movements that lead to the definitive looped and multichambered heart. In parallel with these developmental similarities, there is a remarkable evolutionary conservation of transcription factors involved in the cardiogenic program (reviewed in ref. 89). Of these, the GATA factors appear to play a critical role.

In Drosophila

During *Drosophila* embryogenesis, the GATA factor-encoding gene *pannier* (*pnr*) is expressed in the dorsal mesoderm from which the cardiac cells develop.[90] Its expression overlaps that of the Nk2 type homeodomain protein-encoding gene *tinman* (*tin*), a key regulator of heart development.[91] The expression of both genes in the dorsal mesoderm depends on the signalling molecule Dpp, a member of the bone morphogenetic protein (BMP) family, which is expressed in the overlaying dorsal ectoderm. In embryos deficient for *pnr* function, there results a dramatic reduction in the number of cardiac progenitors. Moreover, in *pnr* mutant embryos, *tin* expression is greatly diminished along the dorsal edge of mesoderm, where the heart precursors form. In *tin* loss-of-function mutants, no heart cells develop at all, which suggests that the extensive loss of heart cells in *pnr* mutant is probably a consequence of this strong reduction of *tin* expression. The few cells that still express *tin* develop normally.[92]

Interestingly, whereas *pnr* expression requires *tin*, *pnr* may also regulate *tin* expression, since GATA binding sites have been identified in its promotor region. Besides these transcriptional cross-regulations, Tin and Pnr have also been shown to interact to activate the MADs box transcription factor dMef2, another gene essential for cardiac myogenesis that is conserved in vertebrates.[93,94] Finally, the evolutionary conservation of the structure and function of GATA factors is strikingly demonstrated by the fact that mouse *GATA4* can substitute for *pnr* during Drosophila cardiogenesis.[90]

In Vertebrates

The vertebrate heart develops from a bilaterally symmetrical region of the anterior lateral plate mesoderm, the cardiac crescent, which adopts a cardiac fate in response to BMP signalling from the adjacent endoderm.[95,96] In mouse, three GATA factors, *GATA4*, *5* and *6*, are expressed at various stages during heart development and can regulate cardiac-specific gene expression in cell culture.[97-100] At E7.5, *GATA4* and *GATA6* are both expressed in the mesoderm that gives rise to the cardiogenic plate whereas *GATA5* is expressed in a restricted manner within the precardiac mesoderm. Subsequently, *GATA4*, *5* and *6* are expressed throughout the primitive heart tube, but only *GATA4* and *6* continue to be expressed in the heart after E17 and during postnatal development. Expression of GATA5 becomes restricted to the atrial endocardium and is extinguished at E16.5. These three GATA factors are among the earliest markers of the cardiac myocyte lineage and are expressed as early as the vertebrate ortholog of *tin*, *Nkx2-5*.[101] In the mesoderm, *Nkx2-5* is expressed in the cardiogenic crescent on each side of the embryo in a pattern very similar to that of *GATA4*.

The role of GATA factors in cardiogenesis has been analysed by gene inactivation studies. In *GATA4* knockout mice, the heart tube fails to form, but this does not appear to be a consequence of defective cardiomyogenesis. Instead, it appears to result from anomalous rostral-to-caudal and lateral-to-ventral folding of the embryo, which reflects a generalized disruption of the ventral body pattern.[102,103] Notably, in homozygous mutant embryos, differentiated cardiomyocytes form and markers of terminal cardiac differentiation are expressed normally. This could well be due to functional redundancy among the GATA factors.

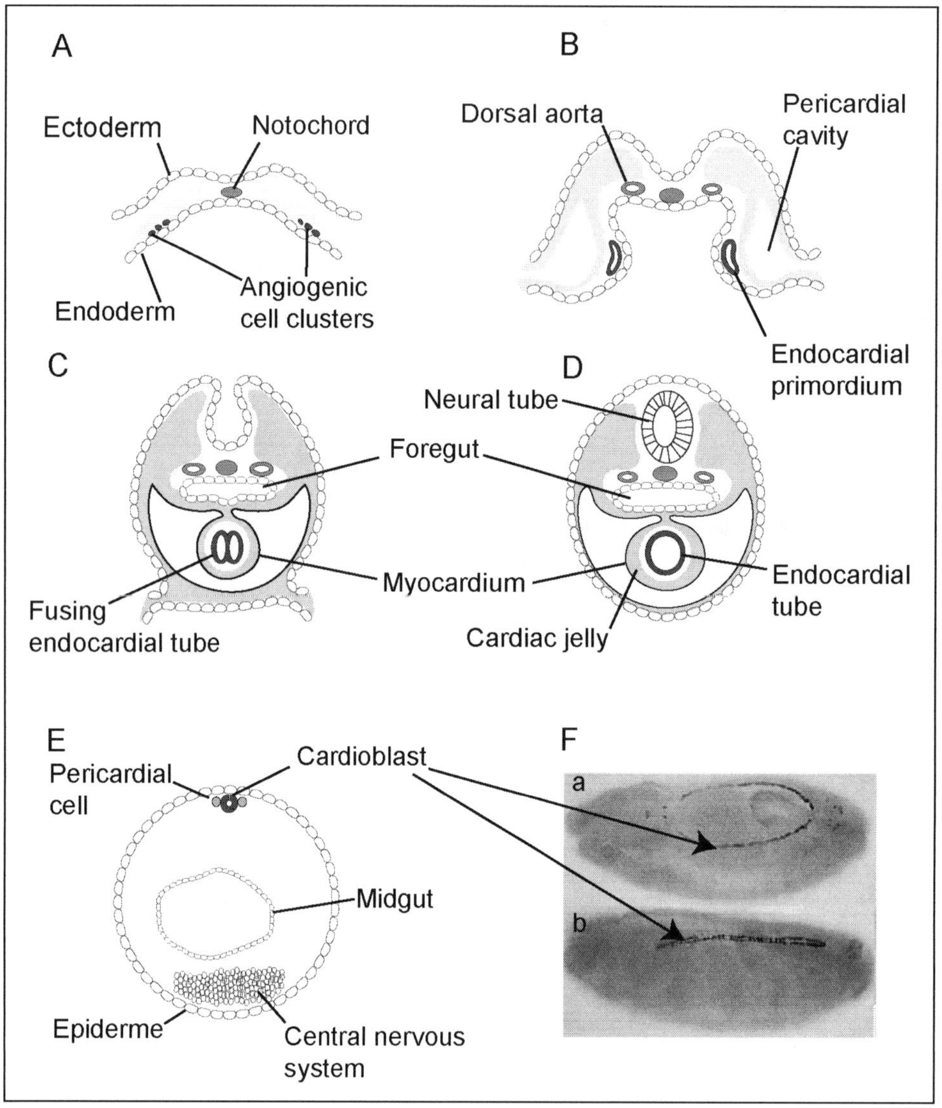

Figure 2. Formation of the heart in vertebrates and Drosophila. Schematic cross sections of human embryos aged 17 days (A), 18 days (B), 21 days (C), 22 days (D), and a Drosophila embryo aged 20 hrs (E). F) Drosophila embryos aged 11 hrs (a) and 20 hrs (b) respectively. a) cardioblast precursors on both sides of the dorsal mesoderm, are stained. b) they meet at the dorsal midline and form a tube that includes the heart.

Consistent with this suggestion, the level of *GATA6* transcripts is increased in *GATA4* knockout embryos. *GATA5* knockout mice are viable, but females display an abnormal urogenital tract.[104] *GATA6* null mice die during early development (stage E5.5-E7.5) as a result of extraembryonic defects (see below),[105,106] which precludes analysis of cardiogenesis.

Interestingly, in zebrafish, the mutation *faust* (*fau*), which maps to the *GATA5* gene, is embryonic lethal and leads to a cardiac phenotype similar to that oberserved in *GATA4* null mice.[107] Analysis of *fau* mutants indicates that *GATA5* is required early in embryogenesis for the production of normal numbers of myocardial precursors and the expression of several myocardial genes, including *Nkx2.5*. Furthermore, overexpression of zebrafish *GATA5* activates ectopic expression of *Nkx2.5* and induces the formation of ectopic regions of rhythmically contracting tissue.[107] These observations suggest a role for zebrafish *GATA5* in the cardiogenic program, including control of growth, morphogenesis and differentiation.

Further experiments in mouse using the Cre-lox technique to generate tissue-specific mutants of *GATA4* and *6*, both as single and double knockouts, should shed light on the precise role of these factors during cardiogenesis.

Contrary to what is observed in zebrafish, it has been shown in mice that neither *GATA4* nor *Nkx2-5* alone can initiate cardiogenesis, although either protein can potentiate it in committed cells. Studies on the atrial natriuretic factor (ANF) promoter have revealed a physical interaction and functional cooperation between *GATA4* and *Nkx2-5*.[108-110] Thus, as in *Drosophila*, these factors may cooperate in regulating cardiogenesis in vertebrates.

Control of Endoderm Formation

All metazoan phyla contain an endoderm and as this germ layer was probably invented only once during evolution, at least some elements of the gene regulatory network controlling its formation may have been conserved during evolution. It has become clear over the last decade that GATA factors are among the most pervasive members of this regulatory network. Indeed, GATA factors play a crucial role in endoderm development in *C. elegans*, *Drosophila*, sea urchin, *Xenopus*, zebrafish and mouse.

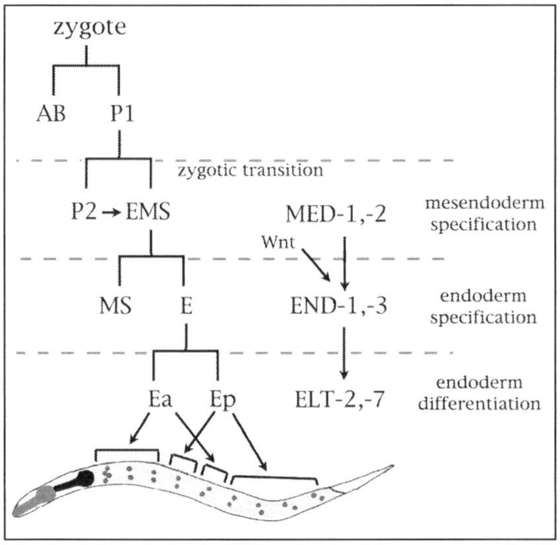

Figure 3. GATA factors during *Caenorhabditis elegans* endoderm formation. The early cell lineage of *C. elegans* is represented, together with the location in the adult worm of the endodermal cells derived from the E blastomere. The three successive tiers of GATA factor pairs controlling endoderm formation are shown to the right of the panel.

In Invertebrates

The most compelling evidence concerning the involvement of GATA factors in endoderm development has come from studies in the worm *Caenorhabditis elegans* (for review see ref. 111). In *C. elegans*, the endoderm appears at the 7-cell stage from a single progenitor, the E cell, which divides 5 times to give the 20 cells that form the juvenile intestine (Fig. 3). The parental cell EMS expresses *med-1* and *med-2*, two functionally redundant genes that code for GATA factors that display 98% identity,[112] and divides asymmetrically to produce an endoderm founder cell (E) and a mesodermal founder cell (MS). In the MS cell, MED-1 and- 2 activate the mesodermal program, whereas in the E cell, which receives a maternally contributed Wnt signal, they induce the endodermal program by directly activating the expression of another pair of GATA factors: END-1 and END-3.[112] These two GATA factors are the first to be specifically expressed in the endoderm and act redundantly in its determination.[113] Furthermore, ectopic expression of END-1 in early embryonic blastomeres is sufficient to trigger endodermal differentiation and to repress other cell fates.[114] However, *end-1/3* expression in the E cell lineage is transient and other factors act downstream of them during differentiation of endoderm cells. Remarkably, a third pair of GATA factors, encoded by *elt-2*, and *elt–7*, are employed during this process.[111] Although the precise role of *elt-7* has not yet been ascertained, it was shown that *elt-2* is expressed in the descendants of the E cell and is required continuously to maintain endodermal differentiation.[115] The *elt-2* promoter contains GATA binding sites and END-1/3 may be direct transactivators. Once initiated, *elt-2* expression is probably maintained by an autoregulatory loop, since ELT-2 binds and activates its own promoter in vivo.[116] It is worth noting that several genes that function in the differentiated intestine of the worm contain functional GATA binding sites in their promoter and are genuine targets of ELT-2 and -7.[116-119]

In summary, a network of GATA factors that act in redundant pairs controls endoderm formation in *C. elegans*. The 3 different pairs of GATA factors are activated in cascade at 3 successive steps of endoderm formation: specification of the mesendoderm progenitor (MED-1/2), establishment of the endoderm (END-1/3) and differentiation of the intestine (ELT-2/7).

GATA factors have been implicated in endoderm formation in other invertebrates. For instance, in *Drosophila*, the GATA factor Srp is essential.[85,120] At the blastoderm stage, Srp is expressed in the anterior and posterior midgut anlage (the precursors of the endodermal tissues) and in a *srp* mutant, ectodermally-derived foregut and hindgut replace the endodermally-derived midgut.[120] This transformation of endoderm into ectoderm is reminiscent of that observed in *C elegans med-1/2* or *end-1/3* mutants. It is striking that Srp combines functions of both the 'blood-specific' (GATA1/2/3) and the 'endoderm-specific' (GATA4/5/6) vertebrate GATAs. Srp is alternatively spliced to give rise to different isoforms, containing either only a C-finger or both a C- and an N-finger, that have different functional properties.[86] The expression of these isoforms of Srp extends the regulatory potentials of *srp* and may account for the broad range of functions assumed by this gene during development. It is noteworthy that two other GATA factors, *Grain*[121] and *dGATAD* (our unpublished results), are expressed in the *Drosophila* midgut primordia after the onset of *srp* expression. However, their role in endoderm formation is unknown. Finally, it was recently shown, in an extensive genomic regulatory network study, that the GATA factor GATA-E plays a central role in endoderm specification in the sea urchin embryo.[122]

In Vertebrates

In vertebrates, the implication of GATA factors in endoderm development has been invoked for a long time. Three of them, GATA4, 5 and 6, are expressed in the extraembryonic and embryonic endoderm during development as well as in some endoderm-derived organs in the adult.[107,123-131] Moreover, several endodermal genes (expressed in the gut epithelium, the lung epithelium or in the liver) contain GATA binding sites in their promoter and can be activated by expression of GATA factors in non-endodermal cell lines.[105,132-136] Finally, in vivo footprinting analysis and in vitro assays have demonstrated that GATA factors (notably GATA4) bind to and open chromatin at the liver-specific transcriptional enhancer of the *serum albumin* gene in liver cell precursors.[137,138]

Gain-of-function analysis also suggests that GATA factors control endoderm development in vertebrates. For instance, overexpression of *Xenopus GATA5* (or *GATA4*) in the whole embryo or in animal caps is sufficient to induce the expression of early endodermal genes and to redirect prospective ectodermal and mesodermal cells toward an endodermal fate.[139] Consistent with a role for GATA factors in *Xenopus* development, it was shown that *Xenopus GATA5* is expressed early in response to two key inducers of the endoderm, VegT or Nodal signalling.[139,140] In the mouse, the forced expression of *GATA6* or *GATA4* in embryonic stem cells induces their differentiation into extraembryonic endoderm and represses mesodermal and ectodermal cell fates.[141] Thus, GATA factors probably constitute master regulators of extraembryonic endoderm development in mammals.

So far, gene inactivations in mouse have failed to demonstrate a clear role for *GATA* genes in the development of definitive endoderm. Chimera analysis and in vitro differentiation of *GATA4-/-* embryoid bodies suggest that *GATA4-/-* cells are unable to generate visceral endoderm.[142,143] However, *GATA4* null embryos form morphologically normal visceral endoderm that expresses higher levels of GATA6.[102,103] Thus *GATA6* may compensate for *GATA4* loss of function. Furthermore, although the definitive endoderm forms normally in *GATA4* null embryos,[102,103] analysis of chimeras indicates that *GATA4*-deficient cells cannot differentiate correctly into endodermally-derived gastric epithelial cells.[144] *GATA6* knockout mice die in the early postimplantation period and exhibit a specific block in visceral endoderm differentiation.[105] However, *GATA6*-deficient cells contributed to gut-derived organs, indicating that GATA6 is not required for definitive endoderm specification.[105,145] Interestingly though, GATA6 is required for the differentiation of endodermal cells into a specialised lung cell type as well as for lung branching morphogenesis.[145,146] Finally, *GATA5* knockout mice are viable, as mentioned earlier, and do not show defects in the endoderm.[104] Since *GATA4*, *5* and *6* are coexpressed in the prospective primitive endoderm in the mouse, the fact that endoderm specification is normal after knockout of each of the individual genes may reflect functional redundancy among them.

Interestingly, it is in zebrafish that the role for vertebrate GATA factors in endoderm formation has been most clearly demonstrated. The *fau* mutation specifically affects the zebrafish *GATA5* gene, which is expressed in all endodermal and in some mesodermal progenitors. *fau* mutants present major defects in myocardial differentiation and gut morphogenesis.[107] Notably, there is a substantial decrease in the number of endodermal cells, which do not differentiate normally into hepatic, pancreatic and pharyngeal derivatives.[147] Since the *fau* mutation is not a null mutation in *GATA5*, the extent of the endoderm defect due to total loss of zebrafish *GATA5* may be even stronger (see for example the phenotype of *GATA4* $^{-/-}$ in mice). It has also been shown that *GATA5* expression depends on the endoderm-inducing signalling of Nodal,[148] and overexpression of *GATA5* in zebrafish embryos induced endodermal genes even when Nodal signalling was reduced.[147] Thus *GATA5* is required for endoderm formation and differentiation in zebrafish.

GATA factors thus seem to be part of a cascade of genes controlling endoderm specification that has been be conserved in *Xenopus*, zebrafish and mouse (for review see ref. 149). In these different species, Nodal-related signalling contributes to the expression of GATA factors as well as of Mix-type homeodomain proteins in endoderm precursors and both of these factors regulate the expression of Sox17-related factors. Finally, members of the Forkhead family of transcription factors, like GATA factors, have been implicated in endoderm formation, not only in vertebrates but also in *Drosophila* and *C. elegans*. Thus, several aspects of endodermal development have been conserved during evolution.

GATA Factors and Diseases

Although GATA factors play crucial roles during normal development, direct evidence for their involvement in human diseases has been relatively scarce until recently. The first category of human inherited disorders linked to GATA factor dysfunction consists of point mutations of GATA binding sites located in the *cis*-regulatory regions of GATA1 target genes. For example, three independent single nucleotide transitions affecting GATA1 binding were found in the ***∂-globin*** gene of patients with delta-thalassemia, a hereditary disease characterised by reduced synthesis of ∂-globin.[150-152] Mutations in GATA binding sites potentially affecting the expression of a relevant gene have been observed in other congenital diseases.[153-155] Whether these mutations do indeed affect GATA-mediated regulation of these genes in vivo remains to be shown. Of note, disruption of a GATA motif in the *Duffy* gene promoter was observed in individuals with Duffy-negative blood group antigen.[156] As this gene code for the erythrocyte receptor for the *Plasmodium vivax* malaria parasite, this mutation may be linked to the resistance to infection by this parasite in West African populations.[157]

The second category of human genetic diseases linked to GATA factor dysfunction comprises mutations directly affecting *GATA* genes. The first evidence for the implication of a *GATA* gene mutation in human disease was obtained after a study of a congenital heart defect (CHD), found to be linked to a deletion in human chromosome 8p23.1, which contains *GATA4*.[158] Recently, two different heterozygous point mutations in *GATA4* were found in two families with CHD.[159] The missense mutation G296S results in diminished GATA4 DNA binding and transcriptional activity, whereas the frameshift mutation E359del gives rise to a shorter, transcriptionally inactive GATA4 protein. Strikingly, the interaction bewteen GATA4 and the T box factor TBX5, which is also mutated in a subset of cardiac septal defects, was disrupted by missense mutations in *GATA4* or in *TBX5* that cause heart defects. Thus, consistent with genetic analysis in model organisms, GATA4 plays an essential role, possibly in association with TBX5, during human heart formation.

Haploinsufficient mutations in other GATA factors have been associated with human congenital disease. Several point mutations affecting GATA1 have been identified in patients with X-linked thrombocytopenia.[160-164] These missense mutations are located in the N-terminal zinc finger and several of them disrupt the interaction with the cofactor FOG1. The phenotype obserevd are consistent with a role for the GATA1/FOG complex during megakaryocytic differentiation in human. Furthermore, *GATA3* haploinsufficiency causes human HDR (Hypoparathyroidism, sensorineural Deafness, Renal anomaly) syndrome.[165] Surprisingly though, whereas human and mouse *GATA3* are expressed in the developing kidney, otic vesicle and parathyroid, no HDR-related phenotype was observed in *GATA3* knockout or heterozygous mice. Finally, it was shown that the tricho-rhino-phalangeal syndrome type I (TRPS1), characterized by craniofacial and skeletal abnormalities, was due to monoallelic nonsense mutations in the *TRPS1* gene.[166] *TRPS1* codes for an atypical GATA protein containing only one C-terminal GATA zinc finger and other unrelated zinc fingers. Interestingly, in a severe form of TRPS (type III), mutations in *TRPS1* are confined to the GATA domain.[167] Consistent with these phenotypes, targeted disruption of the GATA-finger encoding exon in mice mimicked several features of human TRPS as well as more profound and generalized skeletal dysplasia.[168]

Finally, somatically acquired mutations leading to GATA gene dysregulation may participate in cancer development. Indeed, *GATA* gene expression has been shown to be upregulated in a variety of primary tumors and cancer cell lines. Moreover, GATA2 may be implicated in leukemia, since it is located near the 3q21 breakpoint found in some myeloid leukemias and is overexpressed in these leukemias.[169] Recently, acquired somatic mutations in *GATA1* have been specifically associated with megakaryocytic leukemia and transient myeloproliferative disorder in children with Down's syndrome.[170,171] These mutations result in the expression of a truncated GATA1 protein that lacks the N-terminal transactivation domain and they are probably initiating events in these leukemia. Since RUNX1, the gene coding for which is located on chromosome 21, interacts and cooperates with GATA1 to induce megakaryocytic differentiation in vitro, it is tempting to speculate that the interaction between GATA1 and RUNX1 participates in the development of these megakaryocytoses in human.

Conclusions

During the last decade, much has been learned concerning the functions of GATA factors in animal development. In vertebrates, based on protein sequence, exon/intron structure, expression pattern and functions, GATA factors can be assigned to two different families (GATA1/2/3 and GATA4/5/6) that arose by duplications of two ancestral deuterostome GATAs.[3] The precise evolutionary relationship between GATA factors from invertebrates and vertebrates remains unclear. However, it is striking that members of the GATA family play evolutionarily conserved roles in heart, blood and endoderm development from invertebrates to mammals. Moreover, some of their partners and of their target genes have also been functionally conserved in these processes. Altogether, GATA factors appear to be critical regulators of proliferation, differentiation and survival of various cell lineages in the three germ layers. They achieve these diverse functions either through the successive use of different sets of GATA factors harbouring intrinsic specificity or through the reiterative use of the same factor but at different thresholds or in combination with various cofactors.

In the future, the use of cell-type-specific targeted gene disruption in mice and the analysis of human inherited disease caused by GATA dysfunction should help understand the physiological roles and mechanisms of action of *GATA* genes. Furthermore, genetic screens in zebrafish, *Drosophila*, *C. elegans* and other model organisms may also provide unexpected insights into GATA functions and regulations during development.

Acknowledgements

We are grateful to J. Smith for critically reading the manuscript. We thank B. Savelli for art work. This work was supported by the Centre National de Recherche Scientifique (CNRS) and grants from the Association pour la Recherche sur le Cancer (ARC).

References

1. Martin DI, Orkin SH. Transcriptional activation and DNA binding by the erythroid factor GF-1/NF-E1/Eryf 1. Genes Dev 1990; 4(11):1886-1898.
2. Merika M, Orkin SH. DNA-binding specificity of GATA family transcription factors. Mol Cell Biol 1993; 13(7):3999-4010.
3. Lowry JA, Atchley WR. Molecular evolution of the GATA family of transcription factors: conservation within the DNA-binding domain. J Mol Evol 2000; 50(2):103-115.
4. Scazzocchio C. The fungal GATA factors. Curr Opin Microbiol 2000; 3(2):126-131.
5. Patient RK, McGhee JD. The GATA family (vertebrates and invertebrates). Curr Opin Genet Dev 2002; 12(4):416-422.
6. Tong Q, Dalgin G, Xu H et al. Function of GATA transcription factors in preadipocyte-adipocyte transition. Science 2000; 290(5489):134-138.
7. Tevosian SG, Albrecht KH, Crispino JD et al. Gonadal differentiation, sex determination and normal Sry expression in mice require direct interaction between transcription partners GATA4 and FOG2. Development 2002;129(19):4627-4634.
8. Xu RH, Kim J, Taira M et al. Differential regulation of neurogenesis by the two Xenopus GATA-1 genes. Mol Cell Biol 1997; 17(1):436-443.
9. Nardelli J, Thiesson D, Fujiwara Y et al. Expression and genetic interaction of transcription factors GATA-2 and GATA-3 during development of the mouse central nervous system. Dev Biol 1999; 210(2):305-321.
10. Ramain P, Heitzler P, Haenlin M et al. pannier, a negative regulator of achaete and scute in Drosophila, encodes a zinc finger protein with homology to the vertebrate transcription factor GATA-1. Development 1993; 119(4):1277-1291.
11. Calleja M, Herranz H, Estella C et al. Generation of medial and lateral dorsal body domains by the pannier gene of Drosophila. Development 2000;127(18): 3971-3980.
12. Herranz H, Morata G. The functions of pannier during Drosophila embryogenesis. Development 2001; 128(23):4837-4846.
13. Koh K, Peyrot SM, Wood CG et al. Cell fates and fusion in the C. elegans vulval primordium are regulated by the EGL-18 and ELT-6 GATA factors-apparent direct targets of the LIN-39 Hox protein. Development 2002; 129(22):5171-5180.
14. Page BD, Zhang W, Steward K et al. ELT-1, a GATA-like transcription factor, is required for epidermal cell fates in Caenorhabditis elegans embryos. Genes Dev 1997; 11(13):1651-1661.
15. Tsai SF, Martin DI, Zon LI et al. Cloning of cDNA for the major DNA-binding protein of the erythroid lineage through expression in mammalian cells. Nature 1989; 339(6224):446-451.
16. Evans T, Felsenfeld G. The erythroid-specific transcription factor Eryf1: a new finger protein. Cell 1989; 58(5):877-885.
17. Orkin SH, Zon LI. Hematopoiesis and stem cells: plasticity versus developmental heterogeneity. Nat Immunol 2002; 3(4):323-328.
18. Leonard M, Brice M, Engel JD et al. Dynamics of GATA transcription factor expression during erythroid differentiation. Blood 1993; 82(4):1071-1079.
19. Jippo T, Mizuno H, Xu Z et al. Abundant expression of transcription factor GATA-2 in proliferating but not in differentiated mast cells in tissues of mice: demonstration by in situ hybridization. Blood 1996; 87(3):993-998.
20. Harigae H, Takahashi S, Suwabe N et al. Differential roles of GATA-1 and GATA-2 in growth and differentiation of mast cells. Genes Cells 1998; 3(1):39-50.
21. Minegishi N, Ohta J, Yamagiwa H et al. The mouse GATA-2 gene is expressed in the para-aortic splanchnopleura and aorta-gonads and mesonephros region. Blood 1999; 93(12):4196-4207.
22. Tsai FY, Keller G, Kuo FC et al. An early haematopoietic defect in mice lacking the transcription factor GATA-2. Nature 1994; 371(6494):221-226.
23. Tsai FY, Orkin SH. Transcription factor GATA-2 is required for proliferation/survival of early hematopoietic cells and mast cell formation, but not for erythroid and myeloid terminal differentiation. Blood 1997; 89(10):3636-3643.
24. Gottgens B, Nastos A, Kinston S et al. Establishing the transcriptional programme for blood: the SCL stem cell enhancer is regulated by a multiprotein complex containing Ets and GATA factors. EMBO J 2002; 21(12):3039-3050.
25. Kelley C, Yee K, Harland R et al. Ventral expression of GATA-1 and GATA-2 in the Xenopus embryo defines induction of hematopoietic mesoderm. Dev Biol 1994; 165(1):193-205.
26. Walmsley ME, Guille MJ, Bertwistle D et al. Negative control of Xenopus GATA-2 by activin and noggin with eventual expression in precursors of the ventral blood islands. Development 1994; 120(9):2519-2529.
27. Detrich HW, 3rd, Kieran MW, Chan FY et al. Intraembryonic hematopoietic cell migration during vertebrate development. Proc Natl Acad Sci USA 1995;92(23):10713-10717.
28. Kitajima K, Masuhara M, Era T et al. GATA-2 and GATA-2/ER display opposing activities in the development and differentiation of blood progenitors. EMBO J 2002; 21(12):3060-3069.
29. Persons DA, Allay JA, Allay ER et al. Enforced expression of the GATA-2 transcription factor blocks normal hematopoiesis. Blood 1999; 93(2):488-499.
30. Kumano K, Chiba S, Shimizu K et al. Notch1 inhibits differentiation of hematopoietic cells by sustaining GATA-2 expression. Blood 2001; 98(12):3283-3289.

31. Pevny L, Simon MC, Robertson E et al. Erythroid differentiation in chimaeric mice blocked by a targeted mutation in the gene for transcription factor GATA-1. Nature 1991; 349(6306):257-260.
32. Fujiwara Y, Browne CP, Cunniff K et al. Arrested development of embryonic red cell precursors in mouse embryos lacking transcription factor GATA-1. Proc Natl Acad Sci USA 1996; 93(22):12355-12358.
33. Takahashi S, Komeno T, Suwabe N et al. Role of GATA-1 in proliferation and differentiation of definitive erythroid and megakaryocytic cells in vivo. Blood 1998; 92(2):434-442.
34. Lyons SE, Lawson ND, Lei L et al. A nonsense mutation in zebrafish gata1 causes the bloodless phenotype in vlad tepes. Proc Natl Acad Sci USA 2002; 99(8):5454-5459.
35. Shivdasani RA, Fujiwara Y, McDevitt MA et al. A lineage-selective knockout establishes the critical role of transcription factor GATA-1 in megakaryocyte growth and platelet development. EMBO J 1997; 16(13):3965-3973.
36. Vyas P, Ault K, Jackson CW et al. Consequences of GATA-1 deficiency in megakaryocytes and platelets. Blood 1999; 93(9):2867-2875.
37. Tsang AP, Visvader JE, Turner CA et al. FOG, a multitype zinc finger protein, acts as a cofactor for transcription factor GATA-1 in erythroid and megakaryocytic differentiation. Cell 1997; 90(1):109-119.
38. Haenlin M, Cubadda Y, Blondeau F et al. Transcriptional activity of pannier is regulated negatively by heterodimerization of the GATA DNA-binding domain with a cofactor encoded by the u-shaped gene of Drosophila. Genes Dev 1997; 11(22):3096-3108.
39. Cubadda Y, Heitzler P, Ray RP et al. u-shaped encodes a zinc finger protein that regulates the proneural genes achaete and scute during the formation of bristles in Drosophila. Genes Dev 1997; 11(22):3083-3095.
40. Lu JR, McKinsey TA, Xu H et al. FOG-2, a heart- and brain-enriched cofactor for GATA transcription factors. Mol Cell Biol 1999; 19(6):4495-4502.
41. Svensson EC, Tufts RL, Polk CE et al. Molecular cloning of FOG-2: a modulator of transcription factor GATA-4 in cardiomyocytes. Proc Natl Acad Sci USA 1999; 96(3):956-961.
42. Tevosian SG, Deconinck AE, Cantor AB et al. FOG-2: A novel GATA-family cofactor related to multitype zinc-finger proteins Friend of GATA-1 and U-shaped. Proc Natl Acad Sci USA 1999; 96(3):950-955.
43. Deconinck AE, Mead PE, Tevosian SG et al. FOG acts as a repressor of red blood cell development in Xenopus. Development 2000; 127(10):2031-2040.
44. Crispino JD, Lodish MB, MacKay JP et al. Use of altered specificity mutants to probe a specific protein-protein interaction in differentiation: the GATA-1:FOG complex. Mol Cell Feb 1999; 3(2):219-228.
45. Chang AN, Cantor AB, Fujiwara Y et al. GATA-factor dependence of the multitype zinc-finger protein FOG-1 for its essential role in megakaryopoiesis. Proc Natl Acad Sci USA 2002; 99(14):9237-9242.
46. Shimizu R, Takahashi S, Ohneda K et al. In vivo requirements for GATA-1 functional domains during primitive and definitive erythropoiesis. EMBO J 2001; 20(18):5250-5260.
47. Takahashi S, Shimizu R, Suwabe N et al. GATA factor transgenes under GATA-1 locus control rescue germline GATA-1 mutant deficiencies. Blood 2000;96(3):910-916.
48. Vyas P, McDevitt MA, Cantor AB et al. Different sequence requirements for expression in erythroid and megakaryocytic cells within a regulatory element upstream of the GATA-1 gene. Development 1999; 126(12):2799-2811.
49. Onodera K, Takahashi S, Nishimura S et al. GATA-1 transcription is controlled by distinct regulatory mechanisms during primitive and definitive erythropoiesis. Proc Natl Acad Sci USA 1997; 94(9):4487-4492.
50. Nishimura S, Takahashi S, Kuroha T et al. A GATA box in the GATA-1 gene hematopoietic enhancer is a critical element in the network of GATA factors and sites that regulate this gene. Mol Cell Biol 2000; 20(2):713-723.
51. McDevitt MA, Fujiwara Y, Shivdasani RA, Orkin SH. An upstream, DNase I hypersensitive region of the hematopoietic-expressed transcription factor GATA-1 gene confers developmental specificity in transgenic mice. Proc Natl Acad Sci USA 1997; 94(15):7976-7981.
52. Kobayashi M, Nishikawa K, Yamamoto M. Hematopoietic regulatory domain of gata1 gene is positively regulated by GATA1 protein in zebrafish embryos. Development 2001; 128(12):2341-2350.
53. Grass JA, Boyer ME, Pal S et al. GATA-1-dependent transcriptional repression of GATA-2 via disruption of positive autoregulation and domain-wide chromatin remodeling. Proc Natl Acad Sci USA 2003; 100(15):8811-8816.
54. Weiss MJ, Keller G, Orkin SH. Novel insights into erythroid development revealed through in vitro differentiation of GATA-1 embryonic stem cells. Genes Dev 1994; 8(10):1184-1197.
55. Mead PE, Deconinck AE, Huber TL et al. Primitive erythropoiesis in the Xenopus embryo: the synergistic role of LMO-2, SCL and GATA-binding proteins. Development 2001; 128(12):2301-2308.
56. Wadman IA, Osada H, Grutz GG et al. The LIM-only protein Lmo2 is a bridging molecule assembling an erythroid, DNA-binding complex which includes the TAL1, E47, GATA-1 and Ldb1/NLI proteins. EMBO J 1997; 16(11):3145-3157.
57. Visvader JE, Elefanty AG, Strasser A et al. GATA-1 but not SCL induces megakaryocytic differentiation in an early myeloid line. EMBO J 1992; 11(12):4557-4564.
58. Zhang P, Behre G, Pan J et al. Negative cross-talk between hematopoietic regulators: GATA proteins repress PU.1. Proc Natl Acad Sci USA 1999; 96(15):8705-8710.
59. Zhang P, Zhang X, Iwama A et al. PU.1 inhibits GATA-1 function and erythroid differentiation by blocking GATA-1 DNA binding. Blood 2000; 96(8):2641-2648.
60. Rekhtman N, Radparvar F, Evans T et al. Direct interaction of hematopoietic transcription factors PU.1 and GATA-1: functional antagonism in erythroid cells. Genes Dev 1999; 13(11):1398-1411.
61. Nerlov C, Querfurth E, Kulessa H et al. GATA-1 interacts with the myeloid PU.1 transcription factor and represses PU.1-dependent transcription. Blood 2000; 95(8):2543-2551.
62. Querfurth E, Schuster M, Kulessa H et al. Antagonism between C/EBPbeta and FOG in eosinophil lineage commitment of multipotent hematopoietic progenitors. Genes Dev 2000; 14(19):2515-2525.
63. Kulessa H, Frampton J, Graf T. GATA-1 reprograms avian myelomonocytic cell lines into eosinophils, thromboblasts, and erythroblasts. Genes Dev 1995; 9(10):1250-1262.
64. Cantor AB, Orkin SH. Transcriptional regulation of erythropoiesis: an affair involving multiple partners. Oncogene 2002; 21(21):3368-3376.
65. Ho IC, Vorhees P, Marin N et al. Human GATA-3: a lineage-restricted transcription factor that regulates the expression of the T cell receptor alpha gene. EMBO J 1991; 10(5):1187-1192.
66. George KM, Leonard MW, Roth ME et al. Embryonic expression and cloning of the murine GATA-3 gene. Development 1994; 120(9):2673-2686.
67. Lieuw KH, Li G, Zhou Y et al. Temporal and spatial control of murine GATA-3 transcription by promoter-proximal regulatory elements. Dev Biol 1997; 188(1):1-16.
68. Oosterwegel M, Timmerman J, Leiden J et al. Expression of GATA-3 during lymphocyte differentiation and mouse embryogenesis. Dev Immunol 1992; 3(1):1-11.
69. Pandolfi PP, Sonati F, Rivi R et al. Targeted disruption of the housekeeping gene encoding glucose 6-phosphate dehydrogenase (G6PD): G6PD is dispensable for pentose synthesis but essential for defense against oxidative stress. EMBO J 1995; 14(21):5209-5215.
70. Ting CN, Olson MC, Barton KP et al. Transcription factor GATA-3 is required for development of the T-cell lineage. Nature 1996; 384(6608):474-478.
71. Hendriks RW, Nawijn MC, Engel JD et al. Expression of the transcription factor GATA-3 is required for the development of the earliest T cell progenitors and correlates with stages of cellular proliferation in the thymus. Eur J Immunol 1999; 29(6):1912-1918.

72. Hernandez-Hoyos G, Anderson MK, Wang C et al. GATA-3 expression is controlled by TCR signals and regulates CD4/CD8 differentiation. Immunity 2003; 19(1):83-94.
73. Zheng W, Flavell RA. The transcription factor GATA-3 is necessary and sufficient for Th2 cytokine gene expression in CD4 T cells. Cell 1997; 89(4):587-596.
74. Zhang DH, Cohn L, Ray P et al. Transcription factor GATA-3 is differentially expressed in murine Th1 and Th2 cells and controls Th2-specific expression of the interleukin-5 gene. J Biol Chem 1997; 272(34):21597-21603.
75. Murphy KM, Reiner SL. The lineage decisions of helper T cells. Nat Rev Immunol 2002; 2(12):933-944.
76. Lee GR, Fields PE, Flavell RA. Regulation of IL-4 gene expression by distal regulatory elements and GATA-3 at the chromatin level. Immunity 2001; 14(4):447-459.
77. Ouyang W, Ranganath SH, Weindel K et al. Inhibition of Th1 development mediated by GATA-3 through an IL-4-independent mechanism. Immunity 1998; 9(5):745-755.
78. Usui T, Nishikomori R, Kitani A et al. GATA-3 suppresses Th1 development by downregulation of Stat4 and not through effects on IL-12Rbeta2 chain or T-bet. Immunity 2003; 18(3):415-428.
79. Lee HJ, Takemoto N, Kurata H et al. GATA-3 induces T helper cell type 2 (Th2) cytokine expression and chromatin remodeling in committed Th1 cells. J Exp Med 2000; 192(1):105-115.
80. Evans CJ, Banerjee U. Transcriptional regulation of hematopoiesis in Drosophila. Blood Cells Mol Dis 2003; 30(2):223-228.
81. Fossett N, Schulz RA. Functional conservation of hematopoietic factors in Drosophila and vertebrates. Differentiation 2001; 69(2-3):83-90.
82. Lebestky T, Chang T, Hartenstein V et al. Specification of Drosophila hematopoietic lineage by conserved transcription factors. Science 2000; 288(5463):146-149.
83. Tepass U, Fessler LI, Aziz A et al. Embryonic origin of hemocytes and their relationship to cell death in Drosophila. Development 1994; 120(7):1829-1837.
84. Sam S, Leise W, Hoshizaki DK. The serpent gene is necessary for progression through the early stages of fat-body development. Mech Dev 1996; 60(2):197-205.
85. Rehorn KP, Thelen H, Michelson AM et al. A molecular aspect of hematopoiesis and endoderm development common to vertebrates and Drosophila. Development 1996; 122(12):4023-4031.
86. Waltzer L, Bataillé L, Peyrefitte S et al. Two isoforms of Serpent containing either one or two GATA zinc fingers have different roles in Drosophila haematopoiesis. EMBO J 2002; 21(20):5477-5486.
87. Elagib KE, Racke FK, Mogass M et al. RUNX1 and GATA-1 coexpression and cooperation in megakaryocytic differentiation. Blood 2003; 101(11):4333-4341.
88. Fossett N, Tevosian SG, Gajewski K et al. The Friend of GATA proteins U-shaped, FOG-1, and FOG-2 function as negative regulators of blood, heart, and eye development in Drosophila. Proc Natl Acad Sci USA 2001; 98(13):7342-7347.
89. Cripps RM, Olson EN. Control of cardiac development by an evolutionarily conserved transcriptional network. Dev Biol 2002; 246(1):14-28.
90. Gajewski K, Fossett N, Molkentin JD et al. The zinc finger proteins Pannier and GATA4 function as cardiogenic factors in Drosophila. Development 1999; 126(24):5679-5688.
91. Azpiazu N, Frasch M. tinman and bagpipe: two homeo box genes that determine cell fates in the dorsal mesoderm of Drosophila. Genes Dev 1993; 7(7B):1325-1340.
92. Klinedinst SL, Bodmer R. Gata factor Pannier is required to establish competence for heart progenitor formation. Development 2003; 130(13):3027-3038.
93. Lilly B, Zhao B, Ranganayakulu G et al. Requirement of MADS domain transcription factor D-MEF2 for muscle formation in Drosophila. Science 1995; 267(5198):688-693.
94. Ranganayakulu G, Zhao B, Dokidis A et al. A series of mutations in the D-MEF2 transcription factor reveal multiple functions in larval and adult myogenesis in Drosophila. Dev Biol 1995; 171(1):169-181.
95. Schultheiss TM, Xydas S, Lassar AB. Induction of avian cardiac myogenesis by anterior endoderm. Development 1995; 121(12):4203-4214.
96. Schultheiss TM, Burch JB, Lassar AB. A role for bone morphogenetic proteins in the induction of cardiac myogenesis. Genes Dev 1997; 11(4):451-462.
97. Grepin C, Dagnino L, Robitaille L et al. A hormone-encoding gene identifies a pathway for cardiac but not skeletal muscle gene transcription. Mol Cell Biol 1994; 14(5):3115-3129.
98. Ip HS, Wilson DB, Heikinheimo M et al. The GATA-4 transcription factor transactivates the cardiac muscle-specific troponin C promoter-enhancer in nonmuscle cells. Mol Cell Biol 1994; 14(11):7517-7526.
99. Lyons I, Parsons LM, Hartley L et al. Myogenic and morphogenetic defects in the heart tubes of murine embryos lacking the homeo box gene Nkx2-5. Genes Dev 1995; 9(13):1654-1666.
100. Molkentin JD, Kalvakolanu DV, Markham BE. Transcription factor GATA-4 regulates cardiac muscle-specific expression of the alpha-myosin heavy-chain gene. Mol Cell Biol 1994; 14(7):4947-4957.
101. Harvey RP. NK-2 homeobox genes and heart development. Dev Biol 1996; 178(2):203-216.
102. Kuo CT, Morrisey EE, Anandappa R et al. GATA4 transcription factor is required for ventral morphogenesis and heart tube formation. Genes Dev 1997; 11(8):1048-1060.
103. Molkentin JD, Lin Q, Duncan SA et al. Requirement of the transcription factor GATA4 for heart tube formation and ventral morphogenesis. Genes Dev 1997; 11(8):1061-1072.
104. Molkentin JD, Tymitz KM, Richardson JA et al. Abnormalities of the genitourinary tract in female mice lacking GATA5. Mol Cell Biol 2000; 20(14):5256-5260.
105. Morrisey EE, Tang Z, Sigrist K et al. GATA6 regulates HNF4 and is required for differentiation of visceral endoderm in the mouse embryo. Genes Dev 1998; 12(22):3579-3590.
106. Koutsourakis M, Langeveld A, Patient R et al. The transcription factor GATA6 is essential for early extraembryonic development. Development 1999; 126(4):723-732.
107. Reiter JF, Alexander J, Rodaway A et al. Gata5 is required for the development of the heart and endoderm in zebrafish. Genes Dev 1999; 13(22):2983-2995.
108. Lee Y, Shioi T, Kasahara H et al. The cardiac tissue-restricted homeobox protein Csx/Nkx2.5 physically associates with the zinc finger protein GATA4 and cooperatively activates atrial natriuretic factor gene expression. Mol Cell Biol 1998; 18(6):3120-3129.
109. Durocher D, Charron F, Warren R et al. The cardiac transcription factors Nkx2-5 and GATA-4 are mutual cofactors. EMBO J 1997; 16(18):5687-5696.
110. Charron F, Nemer M. GATA transcription factors and cardiac development. Semin Cell Dev Biol 1999; 10(1):85-91.
111. Maduro MF, Rothman JH. Making worm guts: the gene regulatory network of the Caenorhabditis elegans endoderm. Dev Biol 2002; 246(1):68-85.
112. Maduro MF, Meneghini MD, Bowerman B et al. Restriction of mesendoderm to a single blastomere by the combined action of SKN-1 and a GSK-3beta homolog is mediated by MED-1 and -2 in C. elegans. Mol Cell 2001; 7(3):475-485.
113. Zhu J, Hill RJ, Heid PJ et al. end-1 encodes an apparent GATA factor that specifies the endoderm precursor in Caenorhabditis elegans embryos. Genes Dev 1997; 11(21):2883-2896.
114. Zhu J, Fukushige T, McGhee JD et al. Reprogramming of early embryonic blastomeres into endodermal progenitors by a Caenorhabditis elegans GATA factor. Genes Dev 1998; 12(24):3809-3814.
115. Fukushige T, Hawkins MG, McGhee JD. The GATA-factor elt-2 is essential for formation of the Caenorhabditis elegans intestine. Dev Biol 1998; 198(2):286-302.
116. Fukushige T, Hendzel MJ, Bazett-Jones DP et al. Direct visualization of the elt-2 gut-specific GATA factor binding to a target promoter inside the living Caenorhabditis elegans embryo. Proc Natl Acad Sci USA 1999; 96(21):11883-11888.
117. Britton C, McKerrow JH, Johnstone IL. Regulation of the Caenorhabditis elegans gut cysteine protease gene cpr-1: requirement for GATA motifs. J Mol Biol 1998; 283(1):15-27.

118. Hawkins MG, McGhee JD. elt-2, a second GATA factor from the nematode Caenorhabditis elegans. J Biol Chem 1995; 270(24):14666-14671.
119. Moilanen LH, Fukushige T, Freedman JH. Regulation of metallothionein gene transcription. Identification of upstream regulatory elements and transcription factors responsible for cell-specific expression of the metallothionein genes from Caenorhabditis elegans. J Biol Chem 1999; 274(42):29655-29665.
120. Reuter R. The gene serpent has homeotic properties and specifies endoderm versus ectoderm within the Drosophila gut. Development 1994; 120(5):1123-1135.
121. Lin WH, Huang LH, Yeh JY et al. Expression of a Drosophila GATA transcription factor in multiple tissues in the developing embryos. Identification of homozygous lethal mutants with P-element insertion at the promoter region. J Biol Chem 1995; 270(42):25150-25158.
122. Davidson EH, Rast JP, Oliveri P et al. A genomic regulatory network for development. Science 2002; 295(5560):1669-1678.
123. Arceci RJ, King AA, Simon MC et al. Mouse GATA-4: a retinoic acid-inducible GATA-binding transcription factor expressed in endodermally derived tissues and heart. Mol Cell Biol 1993; 13(4):2235-2246.
124. Laverriere AC, MacNeill C, Mueller C et al. GATA-4/5/6, a subfamily of three transcription factors transcribed in developing heart and gut. J Biol Chem 1994; 269(37):23177-23184.
125. Morrisey EE, Ip HS, Lu MM et al. GATA-6: a zinc finger transcription factor that is expressed in multiple cell lineages derived from lateral mesoderm. Dev Biol 1996; 177(1):309-322.
126. Morrisey EE, Ip HS, Tang Z et al. GATA-5: a transcriptional activator expressed in a novel temporally and spatially-restricted pattern during embryonic development. Dev Biol 1997; 183(1):21-36.
127. Jiang Y, Evans T. The Xenopus GATA-4/5/6 genes are associated with cardiac specification and can regulate cardiac-specific transcription during embryogenesis. Dev Biol 1996; 174(2):258-270.
128. Gove C, Walmsley M, Nijjar S et al. Over-expression of GATA-6 in Xenopus embryos blocks differentiation of heart precursors. EMBO J 1997; 16(2):355-368.
129. Huggon IC, Davies A, Gove C et al. Molecular cloning of human GATA-6 DNA binding protein: high levels of expression in heart and gut. Biochim Biophys Acta 1997; 1353(2):98-102.
130. Suzuki E, Evans T, Lowry J et al. The human GATA-6 gene: structure, chromosomal location, and regulation of expression by tissue-specific and mitogen-responsive signals. Genomics 1996; 38(3):283-290.
131. Tamura S, Wang XH, Maeda M et al. Gastric DNA-binding proteins recognize upstream sequence motifs of parietal cell-specific genes. Proc Natl Acad Sci USA 1993; 90(22):10876-10880.
132. Liu C, Glasser SW, Wan H et al. GATA-6 and thyroid transcription factor-1 directly interact and regulate surfactant protein-C gene expression. J Biol Chem 2002; 277(6):4519-4525.
133. Shaw-White JR, Bruno MD, Whitsett JA. GATA-6 activates transcription of thyroid transcription factor-1. J Biol Chem 1999; 274(5):2658-2664.
134. Bruno MD, Korfhagen TR, Liu C et al. GATA-6 activates transcription of surfactant protein A. J Biol Chem 2000; 275(2):1043-1049.
135. Gao X, Sedgwick T, Shi YB et al. Distinct functions are implicated for the GATA-4, -5, and -6 transcription factors in the regulation of intestine epithelial cell differentiation. Mol Cell Biol 1998; 18(5):2901-2911.
136. Maeda M, Kubo K, Nishi T et al. Roles of gastric GATA DNA-binding proteins. J Exp Biol 1996; 199 (Pt 3):513-520.
137. Cirillo LA, Lin FR, Cuesta I et al. Opening of compacted chromatin by early developmental transcription factors HNF3 (FoxA) and GATA-4. Mol Cell 2002; 9(2):279-289.
138. Bossard P, Zaret KS. GATA transcription factors as potentiators of gut endoderm differentiation. Development 1998; 125(24):4909-4917.
139. Weber H, Symes CE, Walmsley ME et al. A role for GATA5 in Xenopus endoderm specification. Development 2000; 127(20):4345-4360.
140. Kelley C, Blumberg H, Zon LI et al. GATA-4 is a novel transcription factor expressed in endocardium of the developing heart. Development 1993; 118(3):817-827.
141. Fujikura J, Yamato E, Yonemura S et al. Differentiation of embryonic stem cells is induced by GATA factors. Genes Dev 2002; 16(7):784-789.
142. Narita N, Bielinska M, Wilson DB. Wild-type endoderm abrogates the ventral developmental defects associated with GATA-4 deficiency in the mouse. Dev Biol 1997; 189(2):270-274.
143. Soudais C, Bielinska M, Heikinheimo M et al. Targeted mutagenesis of the transcription factor GATA-4 gene in mouse embryonic stem cells disrupts visceral endoderm differentiation in vitro. Development 1995; 121(11):3877-3888.
144. Jacobsen CM, Narita N, Bielinska M et al. Genetic mosaic analysis reveals that GATA-4 is required for proper differentiation of mouse gastric epithelium. Dev Biol 2002; 241(1):34-46.
145. Keijzer R, van Tuyl M, Meijers C et al. The transcription factor GATA6 is essential for branching morphogenesis and epithelial cell differentiation during fetal pulmonary development. Development 2001; 128(4):503-511.
146. Yang H, Lu MM, Zhang L et al. GATA6 regulates differentiation of distal lung epithelium. Development 2002; 129(9):2233-2246.
147. Reiter JF, Kikuchi Y, Stainier DY. Multiple roles for Gata5 in zebrafish endoderm formation. Development 2001; 128(1):125-135.
148. Rodaway A, Takeda H, Koshida S et al. Induction of the mesendoderm in the zebrafish germ ring by yolk cell-derived TGF-beta family signals and discrimination of mesoderm and endoderm by FGF. Development 1999; 126(14):3067-3078.
149. Stainier DY. A glimpse into the molecular entrails of endoderm formation. Genes Dev 2002; 16(8):893-907.
150. Angelotti T, Krishna G, Scott J et al. Nodular invasive tracheobronchitis due to Aspergillus in a patient with systemic lupus erythematosus. Lupus 2002; 11(5):325-328.
151. Moi P, Loudianos G, Lavinha J et al. Delta-thalassemia due to a mutation in an erythroid-specific binding protein sequence 3' to the delta-globin gene. Blood 1992; 79(2):512-516.
152. Matsuda M, Sakamoto N, Fukumaki Y. Delta-thalassemia caused by disruption of the site for an erythroid-specific transcription factor, GATA-1, in the delta-globin gene promoter. Blood 1992; 80(5):1347-1351.
153. Ludlow LB, Schick BP, Budarf ML et al. Identification of a mutation in a GATA binding site of the platelet glycoprotein Ibbeta promoter resulting in the Bernard-Soulier syndrome. J Biol Chem 1996; 271(36):22076-22080.
154. Manco L, Ribeiro ML, Maximo V et al. A new PKLR gene mutation in the R-type promoter region affects the gene transcription causing pyruvate kinase deficiency. Br J Haematol 2000; 110(4):993-997.
155. Solis C, Aizencang GI, Astrin KH et al. Uroporphyrinogen III synthase erythroid promoter mutations in adjacent GATA1 and CP2 elements cause congenital erythropoietic porphyria. J Clin Invest 2001; 107(6):753-762.
156. Tournamille C, Colin Y, Cartron JP et al. Disruption of a GATA motif in the Duffy gene promoter abolishes erythroid gene expression in Duffy-negative individuals. Nat Genet 1995; 10(2):224-228.
157. Parasol N, Reid M, Rios M et al. A novel mutation in the coding sequence of the FY*B allele of the Duffy chemokine receptor gene is associated with an altered erythrocyte phenotype. Blood 1998; 92(7):2237-2243.
158. Pehlivan T, Pober BR, Brueckner M et al. GATA4 haploinsufficiency in patients with interstitial deletion of chromosome region 8p23.1 and congenital heart disease. Am J Med Genet 1999; 83(3):201-206.
159. Garg V, Kathiriya IS, Barnes R et al. GATA4 mutations cause human congenital heart defects and reveal an interaction with TBX5. Nature 2003; 424(6947):443-447.
160. Mehaffey MG, Newton AL, Gandhi MJ et al. X-linked thrombocytopenia caused by a novel mutation of GATA-1. Blood 2001; 98(9):2681-2688.
161. Nichols KE, Crispino JD, Poncz M et al. Familial dyserythropoietic anaemia and thrombocytopenia due to an inherited mutation in GATA1. Nat Genet 2000; 24(3):266-270.

162. Freson K, Devriendt K, Matthijs G et al. Platelet characteristics in patients with X-linked macrothrombocytopenia because of a novel GATA1 mutation. Blood 2001; 98(1):85-92.
163. Freson K, Matthijs G, Thys C et al. Different substitutions at residue D218 of the X-linked transcription factor GATA1 lead to altered clinical severity of macrothrombocytopenia and anemia and are associated with variable skewed X inactivation. Hum Mol Genet 2002; 11(2):147-152.
164. Yu C, Niakan KK, Matsushita M et al. X-linked thrombocytopenia with thalassemia from a mutation in the amino finger of GATA-1 affecting DNA binding rather than FOG-1 interaction. Blood 2002;100(6):2040-2045.
165. Van Esch H, Devriendt K. Transcription factor GATA3 and the human HDR syndrome. Cell Mol Life Sci 2001; 58(9):1296-1300.
166. Momeni P, Glockner G, Schmidt O et al. Mutations in a new gene, encoding a zinc-finger protein, cause tricho-rhino-phalangeal syndrome type I. Nat Genet 2000; 24(1):71-74.
167. Ludecke HJ, Schaper J, Meinecke P et al. Genotypic and phenotypic spectrum in tricho-rhino-phalangeal syndrome types I and III. Am J Hum Genet 2001; 68(1):81-91.
168. Malik TH, Von Stechow D, Bronson RT et al. Deletion of the GATA domain of TRPS1 causes an absence of facial hair and provides new insights into the bone disorder in inherited tricho-rhino-phalangeal syndromes. Mol Cell Biol 2002; 22(24):8592-8600.
169. Wieser R, Volz A, Vinatzer U et al. Transcription factor GATA-2 gene is located near 3q21 breakpoints in myeloid leukemia. Biochem Biophys Res Commun. 2000; 273(1):239-245.
170. Wechsler J, Greene M, McDevitt M et al. Acquired mutations in GATA1 in the megakaryoblastic leukemia of Down syndrome. Nat Genet 2002; 32(1):148-152.
171. Groet J, McElwaine S, Spinelli M et al. Acquired mutations in GATA1 in neonates with Down's syndrome with transient myeloid disorder. Lancet 2003; 361(9369):1617-1620.

CHAPTER 31

The Androgen Receptor and Spinal and Bulbar Muscular Atrophy

Federica Piccioni, Charlotte J. Sumner and Kenneth H. Fischbeck*

Abstract

The androgen receptor belongs to the superfamily of nuclear receptors, which are ligand-dependent transcription factors. In the absence of the androgen, the receptor is localized to the cytoplasm where it is associated with heat shock proteins. Upon ligand binding, the receptor translocates into the nucleus and interacts with specific DNA sequences, called androgen response elements. The DNA-bound receptor interacts with the transcription initiation complex to regulate transcription.

The structural organization of the androgen receptor is very similar to the other members of the steroid receptor family, with an N-terminal transcriptional regulatory domain, a centrally positioned C_2C_2 zinc finger DNA binding domain, and a C-terminal ligand binding domain. The N-terminal domain contains a polymorphic polyglutamine tract encoded by a trinucleotide (CAG) repeat; the polyglutamine tract normally consists of 9-36 glutamines. Pathological expansion of the androgen receptor polyglutamine tract to 40-62 glutamines causes spinal and bulbar muscular atrophy, a slowly progressive, X-linked motor neuron disease.

Androgen Receptor

The androgen receptor (AR) belongs to the superfamily of nuclear receptors, which are ligand dependent transcription factors that specifically regulate the expression of target genes involved in metabolism, development and reproduction. Androgen receptor mediates the transcriptional response to androgens such as testosterone and 5α-dihydrotestosterone (DHT) (for review see Zhou et al[1]). The AR has a higher affinity for DHT than testosterone.[2,3] Testosterone dissociates from AR three times faster than DHT, and degradation of AR is faster in the presence of testosterone compared to DHT, indicating that ligand occupancy of the receptor is a major factor in receptor stabilization.[4] It is possible that androgens slow AR degradation by prolonging nuclear retention and limiting recycling from the nucleus to the cytoplasm, preventing proteolysis of AR.[1] Although testosterone and DHT bind the same receptor, they may have different effects in sexual differentiation as a result of activation of different target genes due to differences in activity.[5-7]

Molecular Mechanism of Action

AR is localized to the cytoplasm in the absence of androgen,[1] where it is associated with heat shock proteins (HSPs), including HSP90, HSP70 and HSP56.[8-10] Upon testosterone or DHT binding, the AR undergoes several sequential processes to interact with specific DNA sequences, called androgen response elements (AREs), located in the regulatory regions of androgen-dependent genes.

Regulation of steroid hormone receptor action occurs, in part, by post-translational modification, including phosphorylation, acetylation, and sumoylation (addition of the small ubiquitin-related modifier SUMO).[11] AR is rapidly phosphorylated after synthesis, suggesting that this modification is important for the hormone binding properties of the receptor.[12] AR undergoes further phosphorylation in the presence of ligand.[13] AR is acetylated in a ligand-dependent manner, and acetylation increases its transcriptional activity, implicating this post-translational modification as a regulatory mechanism.[14] Aberrant acetylation and phosphorylation state lead to ligand-independent activation of mutant AR.[15] AR is covalently modified by SUMO-1 in an androgen-enhanced fashion, and sumoylation has been shown to correlate with downregulation of transcriptional activity.[16]

Binding of the inactive AR, complexed with HSP, by ligand induces transformation of the receptor: the receptor changes to a more compact conformation, a second conformation change likely results in dissociation of heat shock proteins and precedes nuclear uptake and the formation of a transcriptionally active complex; the receptor then dimerizes and binds to DNA. The DNA-bound receptor interacts with the transcription initiation complex to regulate transcription (Fig. 1).[17] AR can interact with the basal transcription machinery by direct interaction with general transcription factors[7,18] or indirectly through coactivators such as steroid receptor coactivator-1 (SRC-1) and CBP, which are proteins that interact with nuclear receptors and enhance their transactivation.[19] Transcriptional activation by AR requires the recruitment of RNA polymerase II. Transcription by RNA polymerase II is a complex processes which requires assembly of a set of basal transcription factors (TF) to initiate transcription at the promoters. First, TATA binding protein (TBP) (a component of the transcription factor TFIID) recognizes the TATA element of

* *Corresponding author. See list of "Contributors".*

Zinc Finger Proteins: From Atomic Contact to Cellular Function, edited by Shiro Iuchi and Natalie Kuldell.

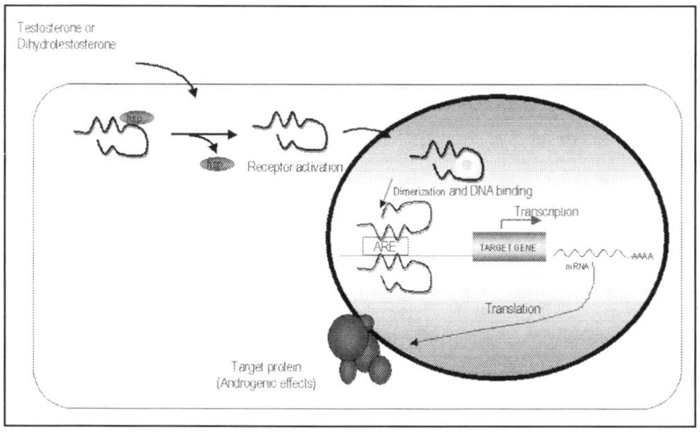

Figure 1. Molecular mechanism of AR function. Testosterone and dihydrotestosterone diffuse through the plasma membrane and bind to the AR. Binding of the inactive AR, complexed with HSP, by ligand induces release of heat shock proteins and activation of the receptor. Within the cell nucleus, the AR dimerizes and binds to androgen response element (ARE), in the regulatory regions of target genes. This interaction affects gene transcription.

a promoter. Not all genes contain a TATA box; in these cases transcription initiation is mediated through other factors and elements. After the recognition, TFIIB binds the preformed TBP-DNA complex, which is in turn recognized by a complex of polymerase II and TFIIF, followed by TFIIE and TFIIH. Once the preinitiation complex is complete, in presence of nucleoside triphosphates strand separation occurs and polymerase II initiates transcription. Following formation, the preinitiation complex is activated, resulting in initiation, promoter clearance, elongation, and termination.[20]

Structure of the Androgen Receptor

The AR gene is located on the X chromosome at Xq11-12 and contains eight exons, all of which contribute to the protein-coding sequence.[21-23] The structural organization is very similar to the genes encoding the other members of the steroid receptor superfamily, with an N-terminal transcriptional regulatory domain (AF-1), a centrally positioned DNA binding domain (DBD), a hinge region with nuclear localization signal (NLS), and a C-terminal ligand binding domain (LBD).[7] The entire amino-terminal domain is encoded by the first exon, the two zinc-fingers of the DNA binding domain by exons 2 and 3, and the hormone binding domain by exons 4-8. There is a small region between the hormone and DNA-binding domain (the hinge region) that contains the nuclear targeting signal and is encoded by part of exon 4 (Fig. 2).

Ligand Binding Domain

The LBD is located in the C-terminal region of the AR. Recently its crystal structure was determined.[24,25] The LBD has a large hydrophobic cavity which forms a ligand-binding pocket. Crystal structure of the human AR LBD complexed with the synthetic ligand R1881 shows that 18 amino acids within the LBD interact with the ligand, mostly through hydrophobic interactions involving the steroid scaffold. LBDs of the nuclear receptors share a similar structural motif consisting of 10-12 α-helices (numbered H1-H12) in an antiparallel sandwich motif. After association of the LBD with an agonist, the H12 amphipathic alpha-helix is repositioned in such a way that it covers the

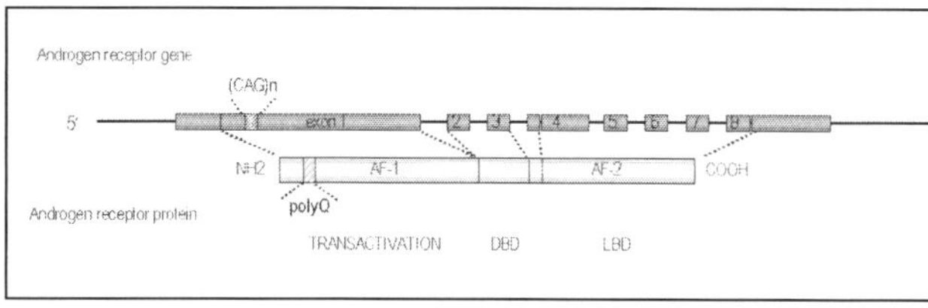

Figure 2. Structure of the AR gene and protein. Top) The gene contains eight exons. Bottom) The protein structure shows how the exon organization translates into discrete functional regions of the receptor: the N-terminal transactivation region, the DNA binding domain (DBD) and the ligand binding domain (LBD).

ligand-binding cavity as a lid and retards the dissociation of the bound ligand, stabilizing the ligand-protein interactions.[26] The unoccupied androgen receptor is associated with several heat shock proteins; androgen binding induces conformational change of the receptor that destabilizes the interaction between the receptor LBD and HSPs.[27] The chaperone complex interacts with unliganded steroid receptors to facilitate the folding of the receptor and stabilize it in a conformation receptive to ligand binding;[27] overexpression of some members of the chaperone complex has been found to alter the affinity of steroids receptor to their ligand.[28,29] In the LBD there is an autonomous transactivating function (AF-2) that appears to be weak, since deletion of LBD results in a molecule that is able to mediate transcription of a reporter gene to the same extent as the full-length receptor in the presence of androgen.[30]

DNA Binding Domain

The androgen receptor has a highly-conserved DBD, encoded by exons 2 and 3. The DBD includes 8 cysteine residues that form two zinc fingers encoded by separate exons.[31] The zinc fingers specifically recognize DNA consensus sequences, usually located in the promoter regions of responsive genes. The AR-ligand complex binds as a dimer to 15-base pair palindromic sequence referred to as androgen-responsive element (ARE) GGTACAnnnTGTTCT-3', as well as to more complex response elements.[32-34] The highly conserved cysteine-rich region folds into two zinc-coordinated loop structures, each with a zinc atom tetrahedrically coordinated to four cysteines.[35] In the nuclear receptors, the first zinc finger contains several hydrophobic residues and the second is more basic. The first zinc finger contains the so-called P-box and other residues involved in the direct binding of the specific recognition sequence of the ARE. During DNA binding, this helix is inserted in the major groove of the response element.[32] It determines the sequence specificity of the interaction;[36] amino acid substitutions of cysteine residues in the first zinc finger alter receptor function due to effects on the coordination of metal ions to form an essential component of the DNA-binding structure.[37] Three amino acids of the first zinc finger (Gly577, Ser528, Val581) appear critical and interact with hormone response elements.[38] Mutation of Val581 has been reported in patients with androgen sensitivity syndrome.[39] The second zinc finger contains the so-called D-box and is involved in receptor dimerization and stabilizes the binding of the receptor to the DNA by contacting the sugar-phosphate backbone at the adjacent sequences.[35] It has been observed that androgen resistance also results from the mutation at Arg615 in the second zinc finger, which is not conserved in all steroid receptors. This mutation led to a defective DNA binding by changing the conformation of the tertiary structure of this portion of the DBD, which likely exists in an alpha-helical conformation.[37] Mutations in the DBD impair the capacity of the receptor to activate transcription, but have little or no effect on androgen binding. In the absence of hormone, steroid receptors have low affinity for DNA. The mechanism by which ligand induces steroid receptors to bind DNA is of interest; the LBD or associated HSPs could interfere with DNA binding in the absence of ligand; ligand binding changes the structure and exposes the DNA binding domain or releases HSPs, relieving the inhibition.[40] Sequences outside the zinc finger DNA binding domain can also contribute to DNA binding. Deletions in the N-terminal region reduce the affinity for DNA, suggesting that the N-terminal domain contributes to the stability of the receptor-DNA complex.[41] Steroid receptors contain a nuclear localization signal (NLS) located in the carboxyl-terminal of the DBD.[42] This region is encoded by the 5'-portion of exon 4 and contains a cluster of basic residues which mediates the transfer of androgen receptor from cytoplasm to the nucleus.[43] Deletions of the NLS still permit nuclear localization in the presence of androgen,[30] suggesting the presence of a second, hormone-dependent NLS located in the C-terminal half of the protein, overlapping the LBD.[44]

N-Terminal Domain

The N-terminal domain is the transcriptional regulatory region of the protein and contains two separate domains important for transactivation. Almost the entire N-terminal domain (residues 1-485) is involved in transcription activation. However, in the absence of the LBD a different region (residues 360 to 528), mediates the activation. The capability of the AR to use different and unique regions of its N-terminal domain for transcriptional activity raises the possibility that different activation functions are responsible for the regulation of different genes, resulting in cell-specific and AR-specific gene expression.[45] The AR N-terminal modulates transcription by directly interacting with the general transcriptional machinery through TFIIF[18,46] and with coregulators such as CBP.[47] The N-terminal also contributes to the ligand-induced three dimensional structure of AR by interaction with the LBD.[48] The AR N and C-terminal regions interact, and this interaction is facilitated by coactivators and is important in stabilizing ligand binding.[47,49] The N-terminal domain of the AR contains three homopolymeric repeated regions, polyglutamine, polyproline and polyglycine, which can influence transactivation. Deletion of the polyglycine tract (normally ranging from 4 to 24 glycines), results in a 30% reduction in AR transcriptional activity in vitro,[50] suggesting that this repeat could be important for the proper functioning of the nearby AF-1 domain.

The N-terminal contains a polymorphic polyglutamine (polygln) tract encoded by a (CAG)n trinucleotide repeat; the polyglutamine tract normally consists of 9-36 glutamines. Glutamine rich regions are found in a number of transcription factors and coregulatory proteins, including CBP and Sp1, and may regulate protein-protein interactions.[51,52]

The polyglutamine tract can act as a repressor of transcriptional activation; longer polyglutamine tract length is associated with decreased AR transcriptional activity in vitro[53,54] and clinically with impaired spermatogenesis and infertility indicative of reduced androgen sensitivity.[55] Deletion of the polyglutamine tract enhances rather than diminishes target gene transactivation. Decreased tract length may be associated with predisposition to prostate cancer.[56-58] Polymorphism of the tract length in humans and the relative lack of cross species conservation indicate that the tract length is not critical to normal AR function. Pathological expansion of the androgen receptor polyglutamine tract to 40-62 glutamines causes spinal and bulbar muscular atrophy (SBMA, or Kennedy's disease).[59]

Table 1. Clinical features of SBMA

X-linked inheritance (only men affected)
Onset 4th- 5th decade
Slowly progressive weakness (over 2-3 decades)
No upper motor neuron signs
Prominent perioral fasciculations ("quivering" of chin)
Mild sensory involvement
Signs of androgen insensitivity (gynecomastia)

Spinal and Bulbar Muscular Atrophy

In 1991, La Spada and colleagues described an expansion of the trinucleotide (CAG) repeat in the AR gene as the disease-causing mutation in SBMA, an inherited neurodegenerative disease.[59] Although first described in the Japanese literature, this disease often goes by the name Kennedy's disease after William Kennedy, who described the clinical features of the disease in eleven male members of two families over thirty years ago.[60]

Clinical Features of SBMA

SBMA affects approximately 1 in 40,000 people. It is a slowly progressive, X-linked motor neuron disease, affecting males beginning at age 30-60 years (see Table 1).[60-64] Early manifestations of muscle cramps and fasciculations are followed by progressive weakness and atrophy of brainstem-innervated (bulbar) and limb muscles. Proximal limb muscles are generally more affected than distal muscles, with progressive impairment of gait and wheelchair dependence, usually within 2 to 3 decades of disease onset. Frequently, involvement of lower facial and tongue muscles causes characteristic perioral muscle twitching (fasciculation) and slurred speech. Patients may also experience pharyngeal and respiratory muscle weakness, with consequent swallowing difficulty and shortness of breath. Occasional patients develop severe jaw closure weakness causing the jaw to hang open.[65,66] The extraocular muscles are always spared. Tendon reflexes are reduced or absent, and no upper neuron features such as spasticity are present. Patients may have a postural tremor of the hands, resembling essential tremor, that occurs early or late in the course of the disease. Patients rarely complain of sensory symptoms. However, clinical and electrophysiological testing may reveal mild sensory deficits.[61-64,67,68] Pathologic studies have confirmed dorsal root ganglia cell loss in addition to the motor neuron loss in the brainstem and spinal cord of patients with SBMA.[63]

Many patients with SBMA also show signs of mild androgen insensitivity. Findings of gynecomastia, testicular atrophy, oligospermia, and erectile dysfunction may have onset as early as adolescence.[69-71] Usually female carriers of SBMA are asymptomatic, however a minority experience muscle cramps or tremor. Some female carriers also show mild electromyography (EMG) abnormalities.[72,73] It had been postulated that women were protected from the full expression of SBMA by mutant X-chromosome inactivation.[73] Recently, however, Schmidt and colleagues described two women who were homozygous for mutant AR.[74] In these patients, symptoms were limited to occasional muscle cramps and neurological examinations were essentially normal. The mildness of symptoms and signs in these homozygous females suggests that females do not develop the full manifestations of SBMA because they have reduced levels of the AR ligand, testosterone, rather than because of X-chromosome inactivation.

Confirmation of SBMA diagnosis can be made with genetic testing. CAG repeat number is determined by polymerase chain reaction amplification of the CAG repeat region within the AR gene (chromosomal locus Xq11-q12).[75] Accurate diagnosis is important in order to provide information about prognosis and genetic counseling for the patient and his family members as well as to avoid inappropriate treatment. Most patients with SBMA have a normal lifespan and do not die of direct consequences of their disease, although late in the course of illness some patients are at risk for aspiration pneumonia or respiratory failure. Currently, treatment is supportive.

Molecular Pathophysiology of SBMA

SBMA is caused by expansion of the trinucleotide CAG repeat in the AR gene from a normal length of 9-36 to a disease-associated length of 40-62.[59] SBMA is one of nine disorders that share this type of mutation. These disorders, known as the polyglutamine expansion diseases, include Huntington's disease (HD), dentatorubropallidoluysian atrophy (DRPLA), and six forms of spinocerebellar ataxia (SCA).[76] They are caused by CAG expansions in unrelated genes and are characterized by selective neuronal degeneration in different nervous system regions. Except for SBMA, they all exhibit autosomal dominant inheritance with a similar age of onset and rate of progression. All occur when the CAG tract achieves a critical threshold length, usually of approximately 40 repeats. Above this threshold, further increased repeat length correlates with earlier disease onset and increased disease severity.[77-79] These shared features suggest that there is a common pathogenic mechanism underlying these disorders.

Toxic Gain of Function

The CAG repeat expansion within the 5' end of the AR gene codes for an elongated tract of polyglutamines near the N-terminal transactivation domain of the AR protein. This expanded polyglutamine tract does not interfere with AR ligand binding or intracellular localization.[80] It does result, however, in some reduction of normal target gene transactivation.[81,82] This defect may account for signs of mild androgen insensitivity in SBMA, however loss of function of the androgen receptor does not appear to be the principal cause of motor neuron cell death. After all, complete loss of function of the androgen receptor is associated with the androgen insensitivity (testicular feminization) syndrome, a disorder that is characterized by feminization, but no abnormality of motor neurons. Mutant AR and other expanded polyglutamine-containing proteins likely cause neuronal degeneration not strictly due to a loss of function, but rather by a toxic gain of function. That is, the structure or function of the protein is altered in such a manner that it becomes deleterious to the cell. Experimental models affirm that polyglutamine-mediated neurodegeneration results from a toxic gain of function. In the many cell culture and transgenic animal models that now exist, polyglutamine disease is recapitulated only with expression of the mutant gene, and not with knock-out of expression of the gene. In some models, an expanded polyglutamine protein fragment alone has been shown to be sufficient to confer cellular toxicity.[83,84] Nonetheless, the particular host protein context for the polyglutamine expansion may play a critical role in the specific mechanisms of neuronal degeneration characteristic of each of the polyglutamine diseases. In the case of SBMA, recent work in transgenic fly[85] and mouse models,[86] has demonstrated that full-length AR requires the presence of ligand in order to cause toxicity. Depletion of the ligand in male mice by castration prevents motor neuron degeneration and administration of androgens to female mice induces motor neuronal toxicity.[86] This ligand-dependency explains the gender specificity of SBMA.

Inclusions and Aggregates

A common pathologic feature of the polyglutamine diseases is the presence of inclusion bodies in susceptible neurons.[87] In SBMA, inclusions have been identified in spinal cord motor neurons from patients[88] and in animal and cell culture models. These inclusions contain N-terminal epitopes of the AR protein that may arise from cleavage of mutant AR by caspases.[89] The inclusions also contain ubiquitin, components of the proteasome complex, and several molecular chaperones, all of which are cellular components involved in normal protein folding and degradation. Their presence in inclusions suggests that there has been accumulation of nondegraded mutant AR protein within the cell. Expanded polyglutamine tracts have been shown to cause protein misfolding and aggregation in a repeat length dependent manner.[90,91] It is postulated that aggregates result from the ability of long polyglutamine stretches to form self-associating β-sheets.[92,93] Aggregation of mutant AR protein into inclusions may be a central toxic event in polyglutamine-mediated neurodegeneration, however in some models, toxicity has been dissociated from the presence of microscopically visible inclusions. Recent work suggests that microscopically visible inclusions represent structures formed by the cell in order to accelerate degradation of excess misfolded protein, while monomeric or micro-aggregated intermediates of polyglutamine protein are the toxic species.[94]

Possible Mechanisms of Toxicity

Mutant AR and other polyglutamine containing proteins may cause neuronal cell death through one or more mechanisms (Fig. 3). In the cytoplasm, mutant AR protein may interrupt the vesicular transport machinery,[95] cause mitochondrial toxicity, or trigger aberrant apoptosis. Excess aggregated polyglutamine proteins may disrupt normal cellular function by overwhelming the ubiquitin-proteasome system (which degrades proteins) or sequestering molecular chaperones (which function in protein repair as well as in ubiquitin-dependent proteasomal degradation). In several models, over-expression of chaperone proteins has been shown to decrease aggregation frequency and mitigate polyglutamine-induced toxicity, perhaps by enhancing proper protein folding and degradation.[96-98]

Other experimental evidence indicates that mutant AR translocation to the nucleus is a key step in toxicity.[85,86] This has led to the hypothesis that toxicity results from a modification of the normal transcriptional interactions of the AR protein. This idea is supported by evidence that altered gene transcription appears to be an early event in polyglutamine disease pathogenesis.[82] Many transcription factors contain a polyglutamine tract that could be bound and sequestered by a polyglutamine expansion protein. Several transcription factors colocalize with polyglutamine-containing nuclear inclusions, including CREB-binding protein (CBP).[99-101] CBP is a ubiquitous transcription factor that normally functions as a histone acetyltransferase. Histone acetyltransferases work in opposition to histone deacetylases. The

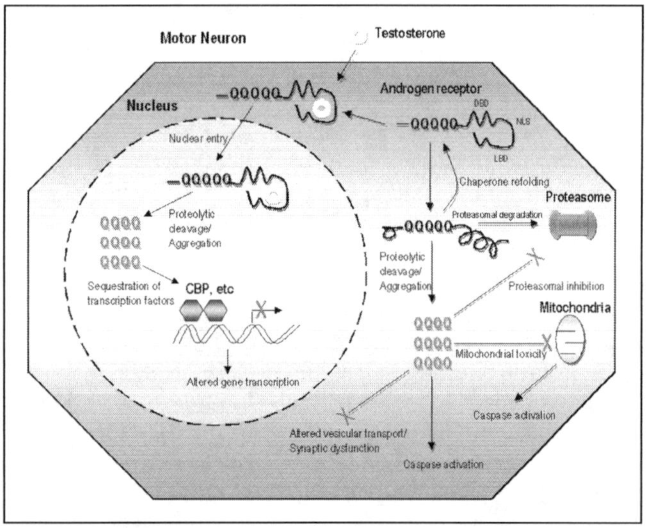

Figure 3. Mutant AR containing a polyglutamine expansion (QQQQ) may cause neuronal degeneration through one or more mechanisms. After proteolytic cleavage (in the nucleus or cytoplasm), the polyglutamine tract assumes an altered conformation, resulting in aggregation and the formation of inclusions. This abnormally folded protein may cause proteasome inhibition, mitochondrial toxicity, altered transport, or transcriptional dysregulation by sequestration of transcription factors. DBD=DNA binding domain, NLS=nuclear localization signal, LBD=ligand binding domain. (Adapted from Ross CA.)[108]

acetylation state of histones is an important determinant of transcriptionally active regions of chromatin; hyper-acetylation of histones marks transcriptionally active regions and hypo-acetylation marks transcriptionally silent regions. Sequestration of CBP by mutant polyglutamine could have effects on gene expression that result in cellular toxicity. In a fly model, polyglutamine-induced neurodegeneration has been shown to be associated with a histone acetylation defect and an alteration in the transcriptional profile. These abnormalities can be reversed with over-expression of CBP.[94] Restoration of the balance of histone acetylation might also be achieved by administering inhibitors of the histone deacetylase enzymes. Indeed, in cell, fly, and mouse models of polyglutamine disease, histone deacetylase inhibitors have been shown to reverse the histone acetylation defect and reduce polyglutamine–induced cell death.[102-104]

With recent advances in the understanding of the basic pathophysiologic mechanisms of SBMA, potential therapeutic strategies can now be envisaged. One strategy might be to deplete the AR ligand, testosterone, thereby preventing mutant AR translocation to the nucleus. It has been recently shown that leuprolide, a lutenizing hormone-releasing hormone agonist that reduces testosterone release from the testis, rescues the behavioral and pathologic phenotypes in a transgenic mouse model of SBMA.[105] Plans are underway to evaluate the efficacy of this compound in patients. Another strategy might be to up-regulate the expression or enhance the function of molecular chaperones. This might promote mutant AR refolding and/or degradation. A third strategy might be to restore the normal transcriptional state within cells. If polyglutamine containing proteins alter histone acetylation levels by sequestering histone acetyltransferases, normal histone acetylation might be restored with administration of histone deacetylase inhibitors. Screening of chemical compound libraries has been done recently to identify agents that mitigate toxicity of mutant AR and other expanded polyglutamine containing proteins.[106,107] Further work is needed to determine whether these strategies will be effective in patients with SBMA. Nonetheless, recent progress in research makes prospects for developing effective therapy very promising.

References

1. Zhou ZX, Wong CI, Sar M et al. The androgen receptor: An overview. Recent Prog Horm Res 1994; 49:249-274.
2. Wilson EM, French FS. Binding properties of androgen receptors. Evidence for identical receptors in rat testis, epididymis, and prostate. J Biol Chem 1976; 251:5620-5629.
3. Wilbert DM, Griffin JE, Wilson JD. Characterization of the cytosol androgen receptor of the human prostate. J Clin Endocrinol Metab 1983; 56:113-120.
4. Zhou ZX, Lane MV, Kemppainen JA et al. Specificity of ligand-dependent androgen receptor stabilization: Receptor domain interactions influence ligand dissociation and receptor stability. Mol Endocrinol 1995; 9:208-218.
5. Brinkmann AO. Molecular basis of androgen insensitivity. Mol Cell Endocrinol 2001; 179:105-109.
6. Brinkmann AO, Trapman J. Androgen receptor mutants that affect normal growth and development. Cancer Surv 1992; 14:95-111.
7. MacLean HE, Warne GL, Zajac JD. Localization of functional domains in the androgen receptor. J Steroid Biochem Mol Biol 1997; 62:233-242.
8. Pratt WB, Welsh MJ. Chaperone functions of the heat shock proteins associated with steroid receptors. Semin Cell Biol 1994; 5:83-93.
9. Smith DF, Toft DO. Steroid receptors and their associated proteins. Mol Endocrinol 1993; 7:4-11.
10. Veldscholte J, Berrevoets CA, Zegers ND et al. Hormone-induced dissociation of the androgen receptor-heat-shock protein complex: Use of a new monoclonal antibody to distinguish transformed from nontransformed receptors. Biochemistry 1992; 31:7422-7430.
11. Moudgil VK. Phosphorylation of steroid hormone receptors. Biochim Biophys Acta 1990; 1055:243-258.
12. Brinkmann AO, Blok LJ, de Ruiter PE et al. Mechanisms of androgen receptor activation and function. J Steroid Biochem Mol Biol 1999; 69:307-313.
13. Jenster G, de Ruiter PE, van der Korput HA et al. Changes in the abundance of androgen receptor isotypes: Effects of ligand treatment, glutamine-stretch variation, and mutation of putative phosphorylation sites. Biochemistry 1994; 33:14064-14072.
14. Fu M, Wang C, Reutens AT et al. p300 and p300/cAMP-response element-binding protein-associated factor acetylate the androgen receptor at sites governing hormone-dependent transactivation. J Biol Chem 2000; 275:20853-20860.
15. Lieberman AP, Harmison G, Strand AD et al. Altered transcriptional regulation in cells expressing the expanded polyglutamine androgen receptor. Hum Mol Genet 2002; 11:1967-1976.
16. Poukka H, Karvonen U, Janne OA et al. Covalent modification of the androgen receptor by small ubiquitin-like modifier 1 (SUMO-1). Proc Natl Acad Sci USA 2000; 97:14145-14150.
17. Kuil CW, Berrevoets CA, Mulder E. Ligand-induced conformational alterations of the androgen receptor analyzed by limited trypsinization. Studies on the mechanism of antiandrogen action. J Biol Chem 1995; 270:27569-27576.
18. Lee DK, Duan HO, Chang C. From androgen receptor to the general transcription factor TFIIH. Identification of cdk activating kinase (CAK) as an androgen receptor NH(2)-terminal associated coactivator. J Biol Chem 2000; 275:9308-9313.
19. McKenna NJ, Lanz RB, O'Malley BW. Nuclear receptor coregulators: Cellular and molecular biology. Endocr Rev 1999; 20:321-344.
20. Nikolov DB, Burley SK. RNA polymerase II transcription initiation: A structural view. Proc Natl Acad Sci USA 1997; 94:15-22.
21. Chang CS, Kokontis J, Liao ST. Molecular cloning of human and rat complementary DNA encoding androgen receptors. Science 1988; 240:324-326.
22. Lubahn DB, Joseph DR, Sullivan PM et al. Cloning of human androgen receptor complementary DNA and localization to the X chromosome. Science 1988; 240:327-330.

23. Tilley WD, Marcelli M, Wilson JD et al. Characterization and expression of a cDNA encoding the human androgen receptor. Proc Natl Acad Sci USA 1989; 86:327-331.
24. Matias PM, Donner P, Coelho R et al. Structural evidence for ligand specificity in the binding domain of the human androgen receptor. Implications for pathogenic gene mutations. J Biol Chem 2000; 275:26164-26171.
25. Sack JS, Kish KF, Wang C et al. Crystallographic structures of the ligand-binding domains of the androgen receptor and its T877A mutant complexed with the natural agonist dihydrotestosterone. Proc Natl Acad Sci USA 2001; 98:4904-4909.
26. Moras D, Gronemeyer H. The nuclear receptor ligand-binding domain: Structure and function. Curr Opin Cell Biol 1998; 10:384-391.
27. Pratt WB, Toft DO. Steroid receptor interactions with heat shock protein and immunophilin chaperones. Endocr Rev 1997; 18:306-360.
28. Reynolds PD, Ruan Y, Smith DF et al. Glucocorticoid resistance in the squirrel monkey is associated with overexpression of the immunophilin FKBP51. J Clin Endocrinol Metab 1999; 84:663-669.
29. Smith DF. Chaperones in progesterone receptor complexes. Semin Cell Dev Biol 2000; 11:45-52.
30. Jenster G, van der Korput HA, van Vroonhoven C et al. Domains of the human androgen receptor involved in steroid binding, transcriptional activation, and subcellular localization. Mol Endocrinol 1991; 5:1396-1404.
31. Evans RM. The steroid and thyroid hormone receptor superfamily. Science 1988; 240:889-895.
32. Claessens F, Verrijdt G, Schoenmakers E et al. Selective DNA binding by the androgen receptor as a mechanism for hormone-specific gene regulation. J Steroid Biochem Mol Biol 2001; 76:23-30.
33. Kasper S, Rennie PS, Bruchovsky N et al. Cooperative binding of androgen receptors to two DNA sequences is required for androgen induction of the probasin gene. J Biol Chem 1994; 269:31763-769.
34. Schoenmakers E, Alen P, Verrijdt G et al. Differential DNA binding by the androgen and glucocorticoid receptors involves the second Zn-finger and a C-terminal extension of the DNA-binding domains. Biochem J 1999; 341(Pt 3):515-521.
35. Beato M. Gene regulation by steroid hormones. Cell 1989; 56:335-344.
36. Umesono K, Evans RM. Determinants of target gene specificity for steroid/thyroid hormone receptors. Cell 1989; 57:1139-1146.
37. Zoppi S, Marcelli M, Deslypere JP et al. Amino acid substitutions in the DNA-binding domain of the human androgen receptor are a frequent cause of receptor-binding positive androgen resistance. Mol Endocrinol 1992; 6:409-415.
38. Quigley CA, De Bellis A, Marschke KB et al. Androgen receptor defects: Historical, clinical, and molecular perspectives. Endocr Rev 1995; 16:271-321.
39. Lumbroso S, Lobaccaro JM, Belon C et al. A new mutation within the deoxyribonucleic acid-binding domain of the androgen receptor gene in a family with complete androgen insensitivity syndrome. Fertil Steril 1993; 60:814-819.
40. Tsai MJ, O'Malley BW. Molecular mechanisms of action of steroid/ thyroid receptor superfamily members. Annu Rev Biochem 1994; 63:451-486.
41. Kallio PJ, Palvimo JJ, Mehto M et al. Analysis of androgen receptor-DNA interactions with receptor proteins produced in insect cells. J Biol Chem 1994; 269:11514-11522.
42. Jenster G, Trapman J, Brinkmann AO. Nuclear import of the human androgen receptor. Biochem J 1993; 293(Pt 3):761-768.
43. Simental JA, Sar M, Lane MV et al. Transcriptional activation and nuclear targeting signals of the human androgen receptor. J Biol Chem 1991; 266:510-518.
44. Picard D, Yamamoto KR. Two signals mediate hormone-dependent nuclear localization of the glucocorticoid receptor. EMBO J 1987; 6:3333-3340.
45. Jenster G, van der Korput HA, Trapman J et al. Identification of two transcription activation units in the N-terminal domain of the human androgen receptor. J Biol Chem 1995; 270:7341-7346.
46. McEwan IJ, Gustafsson J. Interaction of the human androgen receptor transactivation function with the general transcription factor TFIIF. Proc Natl Acad Sci USA1997; 94:8485-8490.
47. Ikonen T, Palvimo JJ, Janne OA. Interaction between the amino- and carboxyl-terminal regions of the rat androgen receptor modulates transcriptional activity and is influenced by nuclear receptor coactivators. J Biol Chem 1997; 272:29821-29828.
48. He B, Kemppainen JA, Voegel JJ et al. Activation function 2 in the human androgen receptor ligand binding domain mediates interdomain communication with the NH(2)-terminal domain. J Biol Chem 1999; 274:37219-37225.
49. He B, Wilson EM. The NH(2)-terminal and carboxyl-terminal interaction in the human androgen receptor. Mol Genet Metab 2002; 75:293-298.
50. Gao T, Marcelli M, McPhaul MJ. Transcriptional activation and transient expression of the human androgen receptor. J Steroid Biochem Mol Biol 1996; 59:9-20.
51. Liu YZ, Chrivia JC, Latchman DS. Nerve growth factor up-regulates the transcriptional activity of CBP through activation of the p42/ p44(MAPK) cascade. J Biol Chem 1998; 273:32400-32407.
52. Tanese N, Saluja D, Vassallo MF et al. Molecular cloning and analysis of two subunits of the human TFIID complex: hTAFII130 and hTAFII100. Proc Natl Acad Sci USA 1996; 93:13611-13616.
53. Chamberlain NL, Driver ED, Miesfeld RL. The length and location of CAG trinucleotide repeats in the androgen receptor N-terminal domain affect transactivation function. Nucleic Acids Res 1994; 22:3181-3186.
54. Kazemi-Esfarjani P, Trifiro MA, Pinsky L. Evidence for a repressive function of the long polyglutamine tract in the human androgen receptor: Possible pathogenetic relevance for the (CAG)n-expanded neuronopathies. Hum Mol Genet 1995; 4:523-527.
55. Tut TG, Ghadessy FJ, Trifiro MA et al. Long polyglutamine tracts in the androgen receptor are associated with reduced trans-activation, impaired sperm production, and male infertility. J Clin Endocrinol Metab 1997; 82:3777-3782.
56. Schoenberg MP, Hakimi JM, Wang SP et al. Microsatellite Mutation (Cag24->18) in the Androgen Receptor Gene in Human Prostate Cancer. Biochemical and Biophysical Research Communications 1994; 198:74-80.
57. Hsing AW, Gao Y-T, Wu G et al. Polymorphic CAG and GGN Repeat Lengths in the Androgen Receptor Gene and Prostate Cancer Risk: A Population-based Case-Control Study in China. Cancer Res 2000; 60:5111-5116.
58. Giovannucci E, Stampfer MJ, Krithivas K et al. The CAG repeat within the androgen receptor gene and its relationship to prostate cancer. Proc Natl Acad Sci USA 1997; 94:3320-3323.
59. La Spada AR, Wilson EM, Lubahn DB et al. Androgen receptor gene mutations in X-linked spinal and bulbar muscular atrophy. Nature 1991; 352:77-79.
60. Kennedy WR, Alter M, Sung JH. Progressive proximal spinal and bulbar muscular atrophy of late onset. A sex-linked recessive trait. Neurology 1968; 18:671-680.
61. Harding AE, Thomas PK, Baraitser M et al. X-linked recessive bulbospinal neuronopathy: A report of ten cases. J Neurol Neurosurg Psychiatry 1982; 45:1012-1019.
62. Hausmanowa-Petrusewicz I, Borkowska J, Janczewski Z. X-linked adult form of spinal muscular atrophy. J Neurol 1983; 229:175-188.
63. Sobue G, Hashizume Y, Mukai E et al. X-linked recessive bulbospinal neuronopathy. A clinicopathological study. Brain 1989; 112:209-232.
64. Olney RK, Aminoff MJ, So YT. Clinical and electrodiagnostic features of X-linked recessive bulbospinal neuronopathy. Neurology 1991; 41:823-828.
65. Suzuki T, Endo K, Igarashi S et al. Isolated bilateral masseter atrophy in X-linked recessive bulbospinal neuronopathy. Neurology 1997; 48:539-540.
66. Sumner CJ, Fischbeck KH. Jaw drop in Kennedy's disease. Neurology 2002; 59:1471-1472.
67. Antonini G, Gragnani F, Romaniello A et al. Sensory involvement in spinal-bulbar muscular atrophy (Kennedy's disease). Muscle Nerve 2000; 23:252-258.
68. Polo A, Teatini F, D'Anna S et al. Sensory involvement in X-linked spino-bulbar muscular atrophy (Kennedy's syndrome): An electrophysiological study. J Neurol 1996; 243:388-392.
69. Arbizu T, Santamaria J, Gomez JM et al. A family with adult spinal and bulbar muscular atrophy, X-linked inheritance and associated testicular failure. J Neurol Sci 1983; 59:371-382.

70. Nagashima T, Seko K, Hirose K et al. Familial bulbo-spinal muscular atrophy associated with testicular atrophy and sensory neuropathy (Kennedy-Alter-Sung syndrome). Autopsy case report of two brothers. J Neurol Sci 1988; 87:141-152.
71. Warner CL, Griffin JE, Wilson JD et al. X-linked spinomuscular atrophy: A kindred with associated abnormal androgen receptor binding. Neurology 1992; 42:2181-2184.
72. Sobue G, Doyu M, Kachi T et al. Subclinical phenotypic expressions in heterozygous females of X-linked recessive bulbospinal neuronopathy. J Neurol Sci 1993; 117:74-78.
73. Mariotti C, Castellotti B, Pareyson D et al. Phenotypic manifestations associated with CAG-repeat expansion in the androgen receptor gene in male patients and heterozygous females: A clinical and molecular study of 30 families. Neuromuscul Disord 2000; 10:391-397.
74. Schmidt BJ, Greenberg CR, Allingham-Hawkins DJ et al. Expression of X-linked bulbospinal muscular atrophy (Kennedy disease) in two homozygous women. Neurology 2002; 59:770-772.
75. Wang Z, Thibodeau SN. A polymerase chain reaction-based test for spinal and bulbar muscular atrophy. Mayo Clin Proc 1996; 71:397-398.
76. Taylor J, Lieberman A, Fischbeck K. Repeat expansion and neurological diseases. In: Asbury AM, GM McDonald, WI Goadsby, PJ McArthur JC, eds. Diseases of the Nervous System. 3rd ed. Cambridge: Cambridge University Press, 2002.
77. La Spada AR, Roling DB, Harding AE et al. Meiotic stability and genotype-phenotype correlation of the trinucleotide repeat in X-linked spinal and bulbar muscular atrophy. Nat Genet 1992; 2:301-304.
78. Doyu M, Sobue G, Mukai E et al. Severity of X-linked recessive bulbospinal neuronopathy correlates with size of the tandem CAG repeat in androgen receptor gene. Ann Neurol 1992; 32:707-710.
79. Igarashi S, Tanno Y, Onodera O et al. Strong correlation between the number of CAG repeats in androgen receptor genes and the clinical onset of features of spinal and bulbar muscular atrophy. Neurology 1992; 42:2300-2302.
80. Brooks BP, Paulson HL, Merry DE et al. Characterization of an expanded glutamine repeat androgen receptor in a neuronal cell culture system. Neurobiol Dis 1997; 3:313-323.
81. Mhatre AN, Trifiro MA, Kaufman M et al. Reduced transcriptional regulatory competence of the androgen receptor in X-linked spinal and bulbar muscular atrophy. Nat Genet 1993; 5:184-188.
82. Lieberman AP, Harmison G, Strand AD et al. Altered transcriptional regulation in cells expressing the expanded polyglutamine androgen receptor. Hum Mol Genet 2002; 11:1967-1976.
83. Kazemi-Esfarjani P, Benzer S. Genetic Suppression of Polyglutamine Toxicity in Drosophila. Science 2000; 287:1837-1840.
84. Ordway JM, Tallaksen-Greene S, Gutekunst CA et al. Ectopically expressed CAG repeats cause intranuclear inclusions and a progressive late onset neurological phenotype in the mouse. Cell 1997; 91:753-763.
85. Takeyama K, Ito S, Yamamoto A et al. Androgen-dependent neurodegeneration by polyglutamine-expanded human androgen receptor in Drosophila. Neuron 2002; 35:855-864.
86. Katsuno M, Adachi H, Kume A et al. Testosterone reduction prevents phenotypic expression in a transgenic mouse model of spinal and bulbar muscular atrophy. Neuron 2002; 35:843-854.
87. Davies SW, Beardsall K, Turmaine M et al. Are neuronal intranuclear inclusions the common neuropathology of triplet-repeat disorders with polyglutamine-repeat expansions? Lancet 1998; 351:131-133.
88. Li M, Miwa S, Kobayashi Y et al. Nuclear inclusions of the androgen receptor protein in spinal and bulbar muscular atrophy. Ann Neurol 1998; 44:249-254.
89. Ellerby LM, Hackam AS, Propp SS et al. Kennedy's disease: Caspase cleavage of the androgen receptor is a crucial event in cytotoxicity. J Neurochem 1999; 72:185-195.
90. Scherzinger E, Lurz R, Turmaine M et al. Huntingtin-encoded polyglutamine expansions form amyloid-like protein aggregates in vitro and in vivo. Cell 1997; 90:549-558.
91. Scherzinger E, Sittler A, Schweiger K et al. Self-assembly of polyglutamine-containing huntingtin fragments into amyloid-like fibrils: Implications for Huntington's disease pathology. Proc Natl Acad Sci USA 1999; 96:4604-4609.
92. Perutz MF, Johnson T, Suzuki M et al. Glutamine repeats as polar zippers: Their possible role in inherited neurodegenerative diseases. Proc Natl Acad Sci USA 1994; 91:5355-5358.
93. Perutz MF, Pope BJ, Owen D et al. Aggregation of proteins with expanded glutamine and alanine repeats of the glutamine-rich and asparagine-rich domains of Sup35 and of the amyloid beta-peptide of amyloid plaques. Proc Natl Acad Sci USA 2002; 99:5596-5600.
94. Taylor JP, Tanaka F, Robitschek J et al. Aggresomes protect cells by enhancing the degradation of toxic polyglutamine-containing protein. Hum Mol Genet 2003; 12:749-757.
95. Piccioni F, Pinton P, Simeoni S et al. Androgen receptor with elongated polyglutamine tract forms aggregates that alter axonal trafficking and mitochondrial distribution in motor neuronal processes. Faseb J 2002; 16:1418-1420.
96. Kobayashi Y, Kume A, Li M et al. Chaperones Hsp70 and Hsp40 suppress aggregate formation and apoptosis in cultured neuronal cells expressing truncated androgen receptor protein with expanded polyglutamine tract. J Biol Chem 2000; 275:8772-8778.
97. Bailey CK, Andriola IF, Kampinga HH et al. Molecular chaperones enhance the degradation of expanded polyglutamine repeat androgen receptor in a cellular model of spinal and bulbar muscular atrophy. Hum Mol Genet 2002; 11:515-523.
98. Adachi H, Katsuno M, Minamiyama M et al. Heat shock protein 70 chaperone overexpression ameliorates phenotypes of the spinal and bulbar muscular atrophy transgenic mouse model by reducing nuclear-localized mutant androgen receptor protein. J Neurosci 2003; 23:2203-2211.
99. McCampbell A, Taylor JP, Taye AA et al. CREB-binding protein sequestration by expanded polyglutamine. Hum Mol Genet 2000; 9:2197-2202.
100. Steffan JS, Kazantsev A, Spasic-Boskovic O et al. The Huntington's disease protein interacts with p53 and CREB-binding protein and represses transcription. Proc Natl Acad Sci USA 2000; 97:6763-6768.
101. Nucifora Jr FC, Sasaki M, Peters MF et al. Interference by huntingtin and atrophin-1 with cbp-mediated transcription leading to cellular toxicity. Science 2001; 291:2423-2428.
102. McCampbell A, Taye AA, Whitty L et al. Histone deacetylase inhibitors reduce polyglutamine toxicity. Proc Natl Acad Sci USA 2001; 98:15179-15184.
103. Steffan JS, Bodai L, Pallos J et al. Histone deacetylase inhibitors arrest polyglutamine-dependent neurodegeneration in Drosophila. Nature 2001; 413:739-743.
104. Hockly E, Richon VM, Woodman B et al. Suberoylanilide hydroxamic acid, a histone deacetylase inhibitor, ameliorates motor deficits in a mouse model of Huntington's disease. Proc Natl Acad Sci USA 2003; 100:2041-2046.
105. Katsuno M, Adachi H, Doyu M, et al. Leuprorelin rescues polyglutamine-dependent phenotypes in a transgenic mouse model of spinal and bulbar muscular atrophy. Nat Med 2003; 9:768-73.
106. Piccioni F, Taylor J, Fischbeck K. Screen for compounds that inhibit polyglutamine-induced Caspase-3 activation. (abstract) Neurology 2003; 60 (Suppl 1):A529.
107. Heemskerk J, Tobin AJ, Ravina B. From chemical to drug: Neurodegeneration drug screening and the ethics of clinical trials. Nat Neurosci 2002; 5 (Suppl):1027-1029.
108. Ross CA. Polyglutamine pathogenesis: Emergence of unifying mechanisms for Huntington's disease and related disorders. Neuron 2002; 35:819-822.

CHAPTER 32

The Role of XPA in DNA Repair

Takahisa Ikegami and Masahiro Shirakawa*

Abstract

XPA, a 273 amino acid protein, is involved in the early stage of the nucleotide excision repair process, by which a variety of DNA lesions are removed from the genome. NMR was used to analyze the structure of the central domain of XPA, which encompasses residues 98 to 219, and contains a zinc coordinating motif. Following chemical shift assignments of the backbone and side-chain ^{1}H, ^{15}N, and ^{13}C nuclei, the tertiary structure was determined by multi-dimensional and multi-resonance NMR methods. The structure shows that the central domain consists of two subdomains, a zinc-containing subdomain and a C-terminal subdomain, which are connected by a short linker sequence. The fold adopted by the zinc-containing subdomain is similar to those of zinc fingers in transcription factors, which bind to DNA in a sequence specific manner. In contrast to these zinc fingers, the zinc-containing subdomain of XPA is dominantly negatively charged, and thus unlikely to directly bind to DNA. The interaction of the central domain of XPA with a damaged DNA was investigated by a chemical shift perturbation experiment, which suggests that the DNA interact with the positively charged cleft in the C-terminal subdomain, and not with the zinc-containing subdomain. The backbone dynamics were analyzed with ^{15}N longitudinal (T_1), transverse (T_2), and *NOE* (*nuclear Overhauser effect*) relaxation data. The results show that the proposed DNA-binding surface exhibits a highly dynamical feature, which may be necessary for interacting structurally with various DNA damages.

Introduction

Ultraviolet (UV) irradiation and certain chemical agents can damage DNA, which is potentially mutagenic or lethal to cells. Usually in living cells, DNA repair machinery efficiently recognizes and eliminates the legions. Nucleotide excision repair (NER) is an important pathway, which removes various kinds of structurally unlike DNA damages caused by UV and chemical carcinogens from the genome.[1,2] The NER pathway first recognizes lesions, and then excises the oligomer carrying the damaged bases, and finally synthesizes a repair patch using the opposite strand as the template.[3] Defects in NER cause a human inherited (autosomal recessive) disease, *xeroderma pigmentosum* (*XP*), which is characterized by an extreme hyper-sensitivity to sunlight, a high frequency of skin cancers on sun-exposed areas as well as by many cancers in other organs, and progressive neurological abnormalities.[4] Cells from *XP* patients have defects in NER, and therefore, are hypersensitive to UV irradiation. Complementation analyses have identified eight genetic complementation groups in XP: (XP-A through XP-G) and a variant (XP-V).[5]

The gene that complements XP group A cells encodes a zinc-finger protein, XPA, which is composed of 273 amino acids. XPA binds preferentially in vitro to a variety of DNA damages including (6-4) photoproducts and crosslinks caused by UV and chemicals. Therefore, it has been suggested that XPA is involved in the damage recognition step of NER.[6-10] XPA also binds directly to many other repair factors: RPA (replication protein A), ERCC1 (excision repair cross-complementing rodent repair deficiency 1)/XPF(ERCC4) heterodimer, and TFIIH (transcription factor IIH), which contains XPB and XPD as its subunits.[11-18] All of these factors are essential for the early steps of NER.[1-19] These observations suggest that XPA also plays a role in loading the incision protein complex onto the damaged site as a multifunctional protein that coordinates the early steps of NER. Of these proteins that function in the early stage of NER, the TFIIH components, XPB and XPD, have ATP-dependent helicase activities to the 3' to 5' and 5' to 3' directions, respectively, and form an open DNA complex in the damaged site. XPG may be recruited to the lesion by its interaction with RPA. Likewise, the ERCC1-XPF complex may be recruited to the lesion by the interactions between ERCC1 and XPA[18] and between ERCC1-XPF and RPA. An oligonucleotide of about 28 nucleotides including the damage is excised by XPG and XPF, which are responsible for nicking the 3'- and 5'-ends, respectively. As the final step of the NER process, the DNA polymerase δε synthesizes a new DNA strand. PCNA (proliferating cell nuclear antigen) and RFC (replication factor C) may be required by RPA to displace the incision proteins from the post-incision complex, followed by ligation of the DNA strand by DNA ligase I.[3]

Recently, Tanaka et al have succeeded in making XPA gene-knockout mice (*XPA* $^{-/-}$ mice).[20] The mice were defective in NER and highly susceptible to UVB- or DMBA-induced skin tumorigenesis, and are therefore expected to be a good animal model for further study of XP-A patients.

Separate Functional Domains of XPA

XPA consists of several distinct functional domains. The amino-terminal part contains a region (residues 4 to 74) that binds

* *Corresponding author. See list of "Contributors".*

Zinc Finger Proteins: From Atomic Contact to Cellular Function, edited by Shiro Iuchi and Natalie Kuldell.

to a subunit of RPA, RPA34, and a region (residues 59 to 97) that binds to ERCC1.[13,14,16-18] The carboxyl-terminal part of XPA (residues 226 to 273) binds to TFIIH and recruits it to the damaged site.[14,15] The central domain (residues M98 to F219), designated as MF122 by Kuraoka et al has been identified as the minimal polypeptide essential for a preferential binding to damaged DNA, by means of a combination of limited proteolysis and deletion analysis.[10] This domain also includes a region essential for binding to another subunit of RPA, RPA70.[12,13]

Structure Determination by NMR

Tanaka et al and the authors of this chapter have determined the solution structure of the central domain of XPA by nuclear magnetic resonance (NMR). The proteins were labeled uniformly with stable isotopes, ^{15}N and ^{13}C, by growing the bacteria, *Escherichia coli*, carrying a plasmid for expression of the central domain in M9 minimum media containing $^{15}NH_4Cl$ and ^{13}C-glucose as the sole nitrogen and carbon sources. Deuterium, ^{2}H, was further introduced in some samples using M9 media composed of 80% D_2O/20% H_2O to enhance the intensities of NMR peaks derived from the ^{13}C and amide ^{1}H spins in the proteins. Typical samples comprised 1.2 mM of the XPA central domain in 50 mM deuterated Tris-HCl (pH 7.3 at 30°C), 150 mM KCl, 10 mM dithiothreitol (DTT), and 20 μM $Zn(CH_3COO)_2$. Two-dimensional (2D), 3D, and 4D NMR spectra were recorded at the proton resonance frequency of 500 or 800 MHz at 30°C, and were used to assign the chemical shifts of the ^{15}N, ^{13}C, and ^{1}H nuclei, and to obtain structural information.

The structure of the central domain of XPA was determined by simulated annealing calculations with 1389 distance and 83 dihedral angle restraints.[21] The quality of the structures obtained was sufficient for elucidation of the function-structure relationships, as indicated by the root-mean-squaredeviation (RMSD) of 0.76 Å in superposing all the backbone heavy atoms of the best 30 calculated structures, excluding parts of the flexible two loops and both terminals. Figure 1a depicts the structure of the central domain determined by NMR, showing that it consists of a zinc-containing subdomain (residues 102 to 129) and a C-terminal subdomain (residues 138 to 209), which are connected by an eight amino acid linker sequence. Buchko and Kennedy et al also determined the structure of the same domain by NMR, and recently refined it using restraints derived from pseudo-contact shifts measured with the cobalt-substituted domain in addition to the normal kinds of restraints based on *NOE*- and *J*-coupling constants (Fig. 1b).[22]

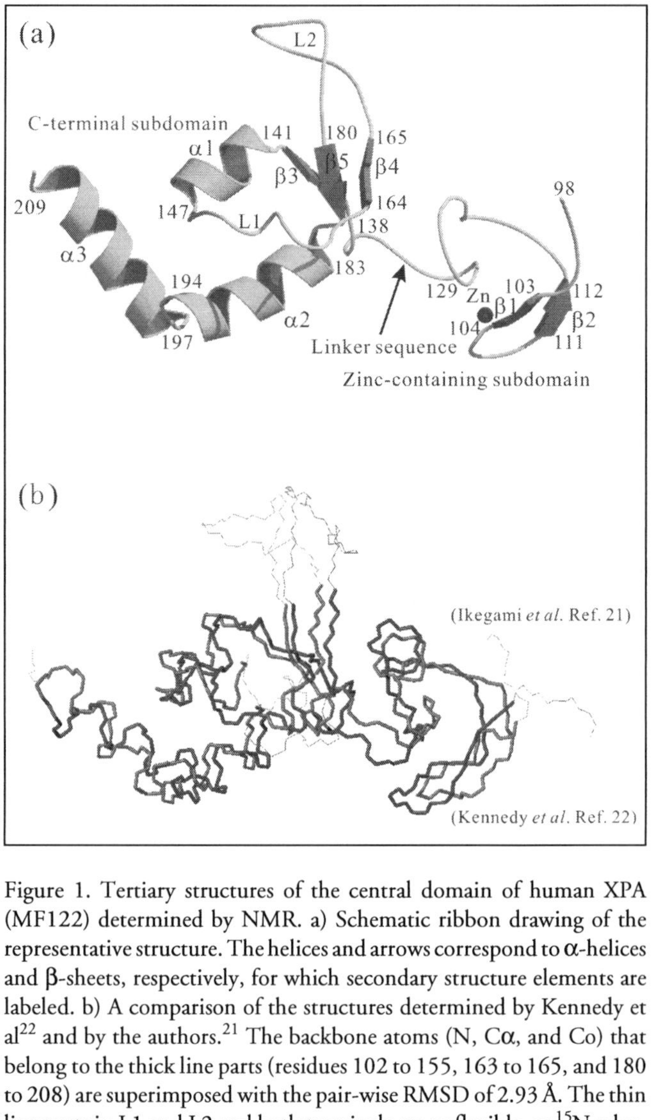

Figure 1. Tertiary structures of the central domain of human XPA (MF122) determined by NMR. a) Schematic ribbon drawing of the representative structure. The helices and arrows correspond to α-helices and β-sheets, respectively, for which secondary structure elements are labeled. b) A comparison of the structures determined by Kennedy et al[22] and by the authors.[21] The backbone atoms (N, Cα, and Co) that belong to the thick line parts (residues 102 to 155, 163 to 165, and 180 to 208) are superimposed with the pair-wise RMSD of 2.93 Å. The thin line parts in L1 and L2 and both terminals are so flexible, as ^{15}N relaxation rates have confirmed, that these parts are not overlaid. (Reproduced with permission from Ikegami et al. Nature Struct Biol 1998; 5:701-706. Nature Publishing Group.)

The Structure of the Zinc Finger in XPA

The central domain of XPA contains five cysteine residues, Cys105, Cys108, Cys126, Cys129 and Cys153. Miyamoto et al showed that a mutation at each of the first four cysteine residues resulted in a drastic reduction in the UV resistance of the cells, as compared to the wild-type XPA protein,[23] and that these mutations caused protein aggregation when expressed as a recombinant protein in *E. coli*, probably due to misfolding, suggesting that these cysteine residues are indispensable for the structural integrity of the protein. In contrast, a mutation at the remaining cysteine at position 153 to serine caused no apparent change in UV resistance, and solubility of the recombinant protein was as high as that of the wild type one. These results suggested that Cys105, Cys108, Cys126 and Cys129 were involved in a C_2C_2 type zinc binding motif, $C_{(i)}$-X_2-$C_{(i+3)}$-X_{17}-$C_{(i+21)}$-X_2-$C_{(i+24)}$, where X can be any amino acid, and Cys153 does not function as a metal-coordinating site.[24,25] A ^{113}Cd-NMR analysis combined with atomic absorption spectra has shown that the four cysteine residues in the zinc binding sequence tetrahedrally coordinate one zinc ion.[26] For the NMR measurement, the proteins binding $^{113}Cd^{2+}$ instead of Zn^{2+} ions were used. In the NMR spectrum, single ^{113}Cd resonance was observed at the chemical shift value of 709 ppm, which is typical for ^{113}Cd coordinated tetrahedrally by four cysteine residues.

Structure calculation using NMR-derived constraints without any information for the zinc coordination provided relatively well-converged protein structures around the coordinated zinc, due to the obtained results of a condensed network of *NOE*-based inter-proton distance restraints. Thus, in the early phase of the structure determination process, we could easily judge which type of zinc-fingers XPA is classified into. The C_2C_2 type fingers can be divided into two pairs of cysteine residues along their primary sequences, and the N-terminal (Cys105 and Cys108) and C-terminal (Cys126 and Cys129) pairs together associate with a

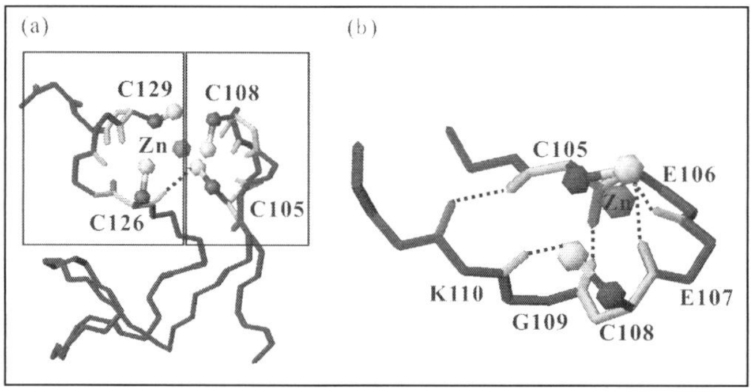

Figure 2. Local conformation of the XPA zinc finger. The main chain, the side chains of the cysteine residues that coordinate a zinc ion, and the amide NH and carbonyl CO bonds that participate in the hydrogen bond network often found in the Rd knuckle-type zinc fingers are indicated. a) The right box in the figure containing the N-terminal two cysteine residues (Cys105 and Cys108) corresponds to one array and the left box containing the C-terminal two cysteine residues (Cys126 and Cys129) to the other array. The inter-array hydrogen bond between the amide NH of $C_{(i+21)}$ in the second array and the sulfur of $C_{(i)}$ in the first array is indicated by the dashed line. b) The first array alone is shown with the hydrogen bonds that exhibit the similarity of the local conformation of the array to that of the Rd knuckle.

zinc ion (Fig. 2a). The local conformation of the segment containing the N-terminal pair resembles that of the iron-binding domain of rubredoxin (referred to as "Rd knuckle" in ref. 27) as displayed in Figure 2b. The pattern of the hydrogen bond network between amide NH and carbonyl CO atoms in the main chain, and between amide NH and the sulfur S atoms of the cysteine residues, is nearly the same as that observed in the Rd knuckle. The glycine residue (Gly109 in XPA) is often found next to the second cysteine in the sequences of many C_2C_2 type fingers. The segment containing the C-terminal pair of the cysteines of the XPA Zn-finger displays a pattern of hydrogen bonds similar to that of the Rd knuckle, except for the $C_{(i+24)}$-X_2 part, which the main chain directs away from the zinc position and takes a helical conformation. The residue following the fourth cysteine ($C_{(i+24)}$) of XPA is not a glycine, but Arg130, which is also found in the C-terminal finger of the transcription factor GATA-1. The hydrogen bond between the amide of $Cys_{(i+21)}$ and the sulfur of $C_{(i)}$ is found in both XPA and GATA-1.[28] Thus, the local structure around the coordinated zinc of XPA most resembles that of GATA-1. Introduction of the distance restrains based on the above mentioned hydrogen bond network into the structure calculation further refined the structure of the zinc-containing subdomain. These newly added restraints did not conflict with those derived from the experimental *J*-coupling and *NOE* data.

The structure of the zinc-containing subdomain is composed of an antiparallel β-sheet and α helical turn (Fig. 3a). Residues 103 to 112 form a β-hairpin structure that contains Cys105 and Cys108, residues 116 to 121 form a helical turn, and residues 122 to 129 form a loop that contains the remaining two cysteine residues, Cys126 and Cys129. The structure of the zinc-containing subdomain is stabilized by the coordination of a zinc ion by the Sγ atoms of the four cysteine residues, and by the hydrophobic core formed by Val103, Phe112 and Met118.

Function of the Zinc Finger

The function of the zinc finger motif in XPA is still unclear. It was previously thought to be involved in DNA binding by analogy to the zinc fingers of transcription factors, most of which bind to particular DNA sequences and possess highly positively charged surfaces. However, the structure of the central domain of XPA shows that it contains many acidic residues, a total of eight glutamic and aspartic acids, which are exposed on the zinc-containing subdomain comprised of 36 amino acids (Fig. 3b). This highly acidic feature makes the zinc-containing subdomain of XPA distinct from those of the transcription factors. This feature is common in XPAs from other species as shown by the well conserved acidic resides. Therefore, in spite of its high

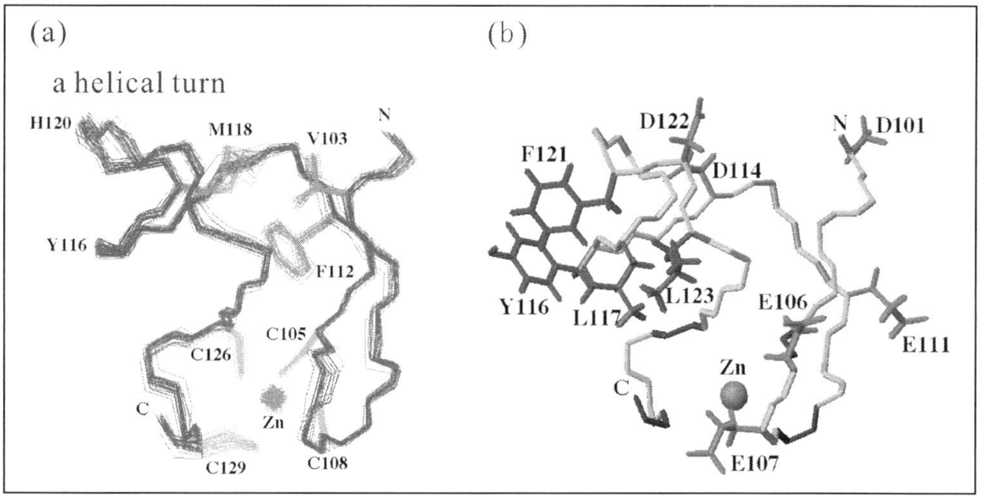

Figure 3. Structures of the XPA zinc-containing subdomain. a) A superposition of the backbone atoms of the 30 best structures of the zinc-containing subdomain (residues 102 to 129). The side chains of the residues in the hydrophobic core (Val103, Phe112, and Met118) and those of the zinc-binding cysteine residues (Cys105, Cys108, Cys126 and Cys129) are also shown. b) Stick representation of the zinc-containing subdomain. Negatively charged residues (Asp101, Glu106, Glu107, Glu111, Asp114 and Asp122) in the vicinity of the β-hairpin, which form an acidic patch, are indicated. Hydrophobic residues (Tyr116, Leu117, Phe121 and Leu123) on the helical turn, which interact with those from the C-terminal subdomain, are also shown. (Reproduced with permission from: Ikegami et al. Nature Struct Biol 1998; 5:701-706. Nature Publishing Group.)

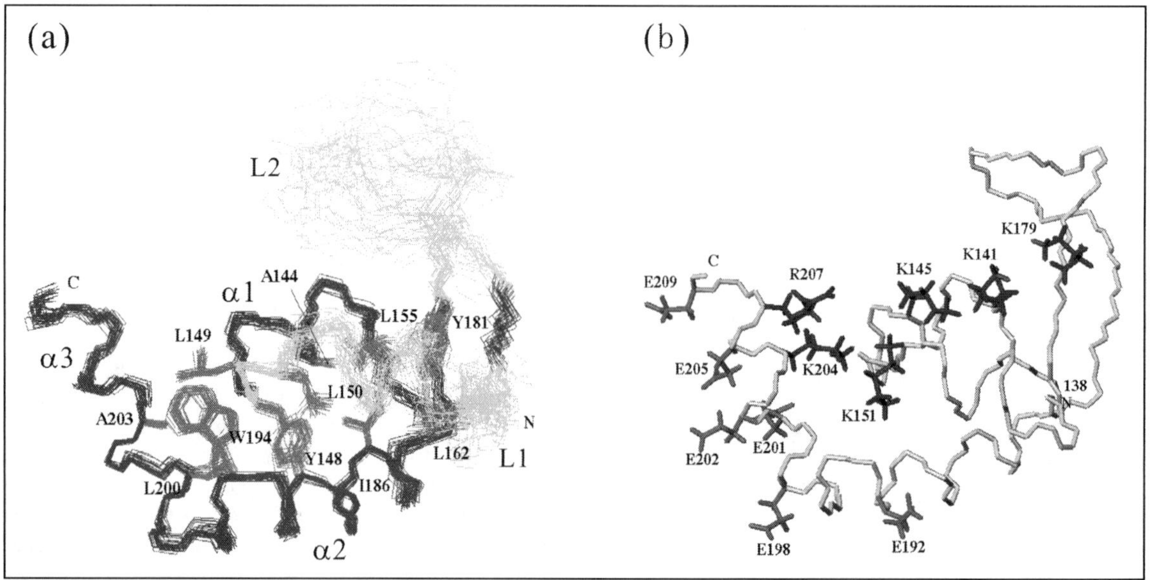

Figure 4. Structures of the XPA C-terminal subdomain. a) A superposition of the backbone atoms of the 30 best structures of the C-terminal subdomain (residues 138 to 209). The backbone atoms of residues 138 to 155, 163 to 165, and 180 to 209 are superimposed. The side chains of the residues in the hydrophobic core are also shown. They serve to fix the relative positions among the α-helices and β-sheet in the subdomain. b) Stick representation of the C-terminal subdomain. The well-conserved arginine and lysine residues that contribute to the positive charges in the basic cleft are shown. The six glutamic acids form an acidic patch on the outer surface of helix α3. The function of this acidic patch is still unknown. It may be used for an electrostatic interaction with other parts of XPA, or with other NER factors. (Reproduced with permission from: Ikegami et al. Nature Struct Biol 1998; 5:701-706. Nature Publishing Group.)

structural similarity with the DNA-binding zinc finger of GATA-1, it is unlikely that the zinc-containing subdomain of XPA directly contacts with DNA, due to electrostatic repulsive forces between negative charges in the subdomain and DNA.

The Structure of the C-Terminal Subdomain

The C-terminal subdomain consists of a sheet-helix-loop region (residues 138 to 182), and a helix-turn-helix region (residues 183 to 209) (Fig. 4). The sheet-helix-loop region is composed of an antiparallel β-sheet (strands β3-5), helix α1, and two long loops, L1 and L2. Loop L1 (residues 148 to 163) connects helix α1 and strand β4, and loop L2 (residues 166 to 179) connects strands β4 and β5. A heteronuclear $\{^1H\}$-^{15}N steady-state *NOE* experiment showed that a part of L1 (residues 157 to 162), L2, and the C-terminal sequence (residues 211 to 219) are highly mobile in solution. L2 apparently formed no definite structure since few long-range *NOE*s were observed. The signals for residues 171 to 174 in L2 were not observed in the ^{15}N-^1H-HSQC spectra, probably due to local exchange broadening and/or rapid exchange of the amide protons with solvent. The helix-turn-helix region consists of two long helices, α2 (residues 183 to 194) and α3 (residues 197 to 209), connected by the conserved Gly195-Ser196 sequence forming a short turn. This region is packed tightly against helix α1, loop L1, and the β-sheet of the sheet-helix-loop region, stabilized by a hydrophobic core containing Ala144, Tyr148, Leu149, Leu150, Leu155, Leu162, Tyr181 and Ile186 (Fig. 4a). The C-terminal subdomain has a large cleft between the sheet-helix-loop region and the helix-turn-helix region. The surface electrostatic potentials show that many positively charged side chains, which are contributed by well-conserved Lys141, Lys145, Lys151, Lys179, Lys204 and Arg207, are present in the cleft (Fig. 4b).

Hydrophobic Contacts between the Two Subdomains

The zinc-containing subdomain offers some hydrophobic residues, Tyr116, Leu117, Phe121 and Leu123, on the outer surface of the helical turn (Fig. 3b), which make contacts with hydrophobic residues of the C-terminal subdomain, Leu138, Ile165, Leu182 and Leu184. These contacts fix the three-stranded β-sheet in the C-terminal subdomain with respect to the zinc-containing subdomain (Fig. 5), indicating that the zinc-containing subdomain is not a structurally independent domain. This feature is in contrast to the zinc fingers of transcriptional factors, most of which are thought to exist as independent domains, separated from other domains by flexible linkers in modular proteins. This character probably makes finding a biochemical function of the isolated zinc-containing subdomain difficult.[10,13]

Exon/Intron Boundaries and Tertiary Structure Elements

It has been shown that the human *xpa* gene contains six exons.[29] Interestingly, the genomic structure of the *xpa* gene has a good correlation with the tertiary structure elements. Exon 3 encodes residues 96 to 130 of XPA, which correspond to the entire zinc-containing subdomain, and exon 4 (residues 131 to 185) encodes the region containing the linker (residues 130 to 137) and the sheet-helix-loop region (residues 138 to 182). Finally, the region encoded by exon 5, residues 186 to 224, covers most of the helix-turn-helix region (residues 183 to 209) and the C-terminal flanking sequence (residues 211 to 219) (Fig. 6). This indicates that the exon/intron junctions are located at the sites corresponding to the inter-module junctions of the central domain of XPA. Such a correlation between the functional domains of XPA and its genomic structure does not seem to be limited to

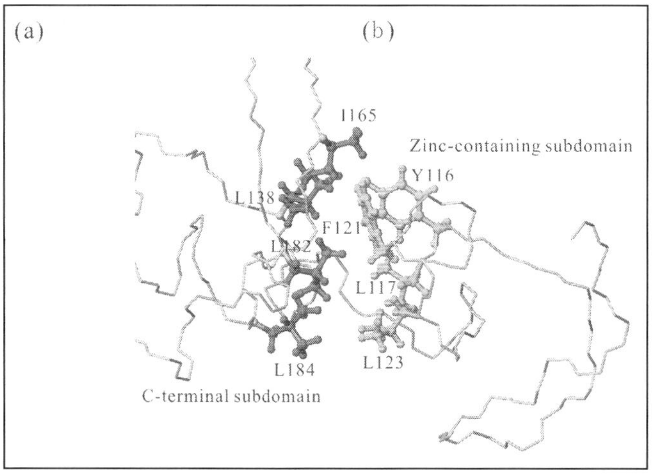

Figure 5. The hydrophobic core formed between the zinc-containing subdomain on the right side and the C-terminal subdomain on the left side. Reproduced with permission from Ikegami et al. Nature Struct Biol 1998; 5:701-706. Nature Publishing Group.

the central domain. Previous study has shown that the ERCC1 binding domain (residues 59 to 97) and the TFIIH binding site (residues 226 to 273) nearly correspond to regions encoded by exons 2 (residues 58 to 95) and 6 (residues 225 to 273), respectively.[13,14,16-18] Future structural studies will reveal the relationship between the exons/introns and the tertiary structure elements of these regions.[30,31]

Interaction with DNA

Chemical shift perturbation experiments are often employed to estimate the surface regions of proteins responsible for interactions with ligands. In the experiments, two-dimensional spectra, for example ^{1}H-^{15}N HSQC spectra, acquired in the presence and absence of the target ligand are compared, and cross peaks that show resonance shifts and/or broadening upon the ligand binding are regarded as being assigned to nuclei in the interaction region. Although there is a disadvantage that resonance shifts can be caused by conformational changes upon binding, which may occur in regions distant from the direct interaction sites, and are difficult to distinguish from those induced by direct interactions, this method provides a convenient way for prediction of interaction surfaces.

For the identification of the DNA binding surface of the central domain of XPA, chemical shift perturbation experiments were performed using the ^{15}N-labeled central domain. Selective chemical shift perturbation and/or peak broadening were observed for signals in a ^{15}N-^{1}H-HSQC spectrum upon mixing with an equimolar amount of a 24-mer oligonucleotide treated with cisplatin. This chemotherapeutic agent reacts with the oligonucleotide to form a single 1,3-intrastrand d(GpTpG)-cisplatin crosslink, and to distort the DNA helix.[32] Almost all of the signals that exhibited large chemical shift perturbations or broadening were attributed to the amide residues in the positively charged cleft in the C-terminal subdomain (Fig. 7), suggesting that the cleft and the surrounding regions are involved in the interaction with DNA, but that the zinc-containing subdomain is not. A similar spectral change, however, was observed upon the addition of the nondamaged oligonucleotide to the central domain.

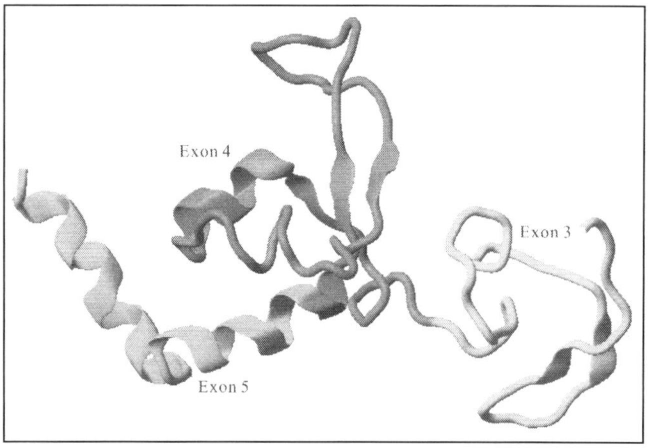

Figure 6. Relationship between the genomic structure and tertiary structure elements of the central domain of XPA.

Residues in loops L1 and L2 exhibited large chemical shift changes upon DNA binding. The conformation of parts of L1 and L2 was not well converged in the structure calculations of the free central domain of XPA, due to the smaller number of *NOE*s observed in the regions. To investigate the flexibility of the central domain, we employed ^{15}N-relaxation experiments under two different magnetic fields, 11.7 and 18.8 T, where the longitudinal- and transverse-relaxation times and static ^{15}N-{^{1}H} *NOE* value were measured for each amide ^{15}N nucleus, and local flexibility was estimated from these data.[33] The results indicated flexibilities in the parts of L1 and L2, for which the local structures were not well determined. The internal motions of these regions may be important for the interaction of the central domain of XPA with various kinds of damaged DNA by altering the conformation of the interaction surface to fit the structures of damaged DNAs.

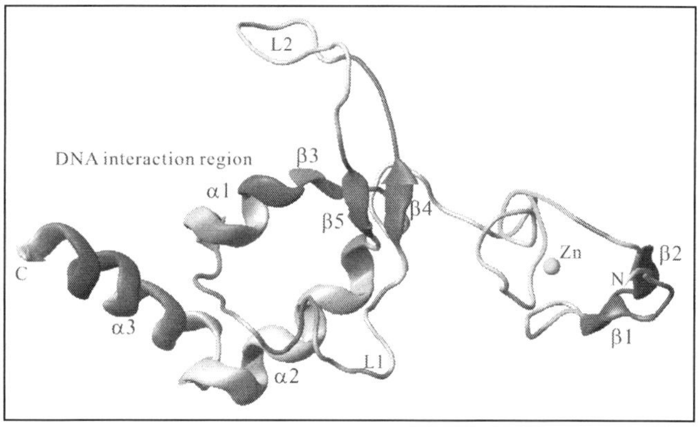

Figure 7. Identification of the binding surfaces for DNA and RPA70. Residues for which the amide signals showed chemical shift perturbation and/or broadening upon binding to the oligonucleotide treated with cisplatin or to $RPA70_{181\text{-}422}$ are drawn. The signals perturbed by the interaction with DNA and RPA70 are restricted within the C-terminal subdomain and the zinc-containing subdomain, respectively. Both bindings are expected to be through electrostatic interactions rather than hydrophobic ones. However, whether $RPA70_{181\text{-}422}$ specifically interacts with the zinc-containing subdomain is not yet certain, as described in the text. (Reproduced with permission from: Ikegami et al. Nature Struct Biol 1998; 5:701-706. Nature Publishing Group.)

Buchko and Kennedy et al have recently proposed a model of the complex between the XPA central domain and a single-stranded DNA nonamer, dCCAATAACC.[34] They monitored chemical shift changes or peak broadenings in ^{31}P resonances of the nonamer DNA as well as in amide ^{15}N-^{1}H resonances of the XPA central domain in titration experiments, and calculated the model structure using these chemical shift perturbation data with the assumption that electrostatic interactions drive the complex formation. The model shows that the DNA covers a surface of the C-terminal subdomain, in particular $\alpha 1$ and the triple stranded β-sheets in the subdomain, and is thus consistent with the result of our perturbation experiment.

Interaction with RPA70

RPA is a heterotrimeric single-stranded-DNA-binding protein, which is involved in DNA replication, homologous recombination, and NER. For NER, the interaction between RPA and XPA has been shown to be essential for the opening of damaged DNA regions by TFIIH.[11,12,35,36] Although XPA binds to the two largest subunits, RPA70 and RPA34, a deletion analysis of XPA showed that RPA70 contributes more predominately to the XPA-binding than RPA34 does, and the central domain of XPA contains the XPA70 binding site.[11-13] By means of limited proteolysis, mutational analysis, and crystal structure determination, Pfuetzner et al have shown that residues 181 to 422 of RPA70, $RPA70_{181-422}$, form a compact structural domain that maintains single-stranded DNA binding activity.[37,38]

In order to examine the interaction between $RPA70_{181-422}$ and the central domain of XPA, we performed chemical shift perturbation experiments using the ^{15}N-labeled central domain. Selective signal losses were observed in a ^{15}N-^{1}H-HSQC spectrum upon mixing with an equimolar amount of $RPA70_{181-422}$ when the salt concentration was sufficiently low (see below). In general, signal losses occur when the chemical shifts are perturbed considerably by interaction, when the exchange rate between the free and bound states is comparable to the chemical shift difference between the two states, or when the ligand binding enhances the amide proton exchange rate with a solvent. Almost all of the signals for which the peak intensities decreased to < 65% of the original values were attributed to the residues in the zinc-containing subdomain (Fig. 7). Buchko and coworkers reported that not only the zinc-containing subdomain but also parts of the C-terminal subdomain, the β-sheet and loop L2, exhibited perturbed chemical shifts upon addition of RPA70ΔC327, which ranges from M1 to Y326 and contains the N-terminal 146 residues (60%) out of $RPA70_{181-422}$.[22] They suggested that RPA may interact only with the zinc-containing subdomain, but that the interaction may generate a conformational change in the C-terminal subdomain to enhance XPA binding to damaged DNA.

The interaction between the XPA central domain and $RPA70_{181-422}$ is highly sensitive to the salt concentration. When the KCl concentration was increased from 28 to 77 mM, the specific broadening observed for the zinc-containing subdomain of XPA became significantly smaller. Since the surface of the zinc-containing subdomain is highly negatively charged, we cannot exclude the possibility that a positively charged patch in $RPA70_{181-422}$ nonspecifically interacts with the zinc-containing subdomain, due primarily to the electrostatic force.

Through deletion analyses, Saijo et al have shown that residues 98 to 187 of XPA, which include the zinc-containing subdomain and the sheet-helix-loop region (residues 138 to 182) of the C-terminal subdomain, are sufficient for binding to RPA70, but that further deletion of the sheet-helix-loop region causes the loss of RPA70 binding.[13] This observation raises the possibility that the sheet-helix-loop region of XPA serves as another and indispensable binding surface for the N- or C-terminal region of RPA70, outside of $RPA70_{181-422}$. Apparently, future experiments need to be directed at identifying the major RPA-binding site of XPA. Chazin and coworkers have shown that the C-terminal domain of RPA32 interacts with the N-terminal 98 residues of XPA.[39]

Role of XPA in NER

Although various experiments conducted so far, including the ones described here, have shown that XPA interacts with damaged DNA, the role of XPA in damage recognition is still unclear. Wakasugi et al have observed that the rate of damage removal was five times faster when the combination of XPA and RPA first bound to a single (6-4) photoproduct in a 136-bp duplex than when XPC first bound to the damaged DNA.[40] They proposed a model for assembly of the NER-related proteins, in which XPA+RPA first forms a complex with damaged DNA with a low specificity and then becomes more discriminatory upon recruiting XPC and TFIIH. In contrast, recent findings have suggested a two-step mechanism for damage recognition during global genome repair, in which various types of damages are detected by XPC-HR23B first,[41] and then verified by XPA.[42] You et al have suggested that damage recognition occurs in a multi-step process, whereby XPC-HR23B initiates damage recognition and then XPC-HR23B is replaced by a combined action of XPA and RPA.[43] As for the identification of cyclobutane pyrimidine dimers (CPDs), XPE may be involved.[44]

Relationships between XPA Mutations and Disease

The patients suffering from XP have mutations in the genes encoding the proteins involved in nucleotide excision repair processes, XPA to XPG. The mutations lead the deficiency in effective repairs of DNA lesions caused by UV and chemicals. Of these, mutations in the *xpa* gene (shown in Fig. 8) tend to cause more severe symptoms than others, suggesting the functional significance of XPA in NER. The missense mutation of Cys108 (TGT) to Phe108 (TTT) most likely has a global structural consequence to the zinc-containing subdomain because this cysteine coordinates the zinc-ion. The nonsense mutation of Tyr116 (TAT) results in deletion of a large part of the C-terminal subdomain of the central domain together with the following region of XPA, and thus, probably results in a total loss of function of XPA. The patient bearing the nonsence mutation of Arg207 (CGA) was reported to exhibit a severe symptom, whereas the mutation of Arg228 (CGA) leads only to a mild symptom. This observation suggests that the C-terminal part after the central domain, which contains the TFIIH binding site, is not essential for NER, but that the further deletion of a large part of the cleft region of the C-terminal subdomain of the central domain of XPA leads to a severe deficiency in NER.

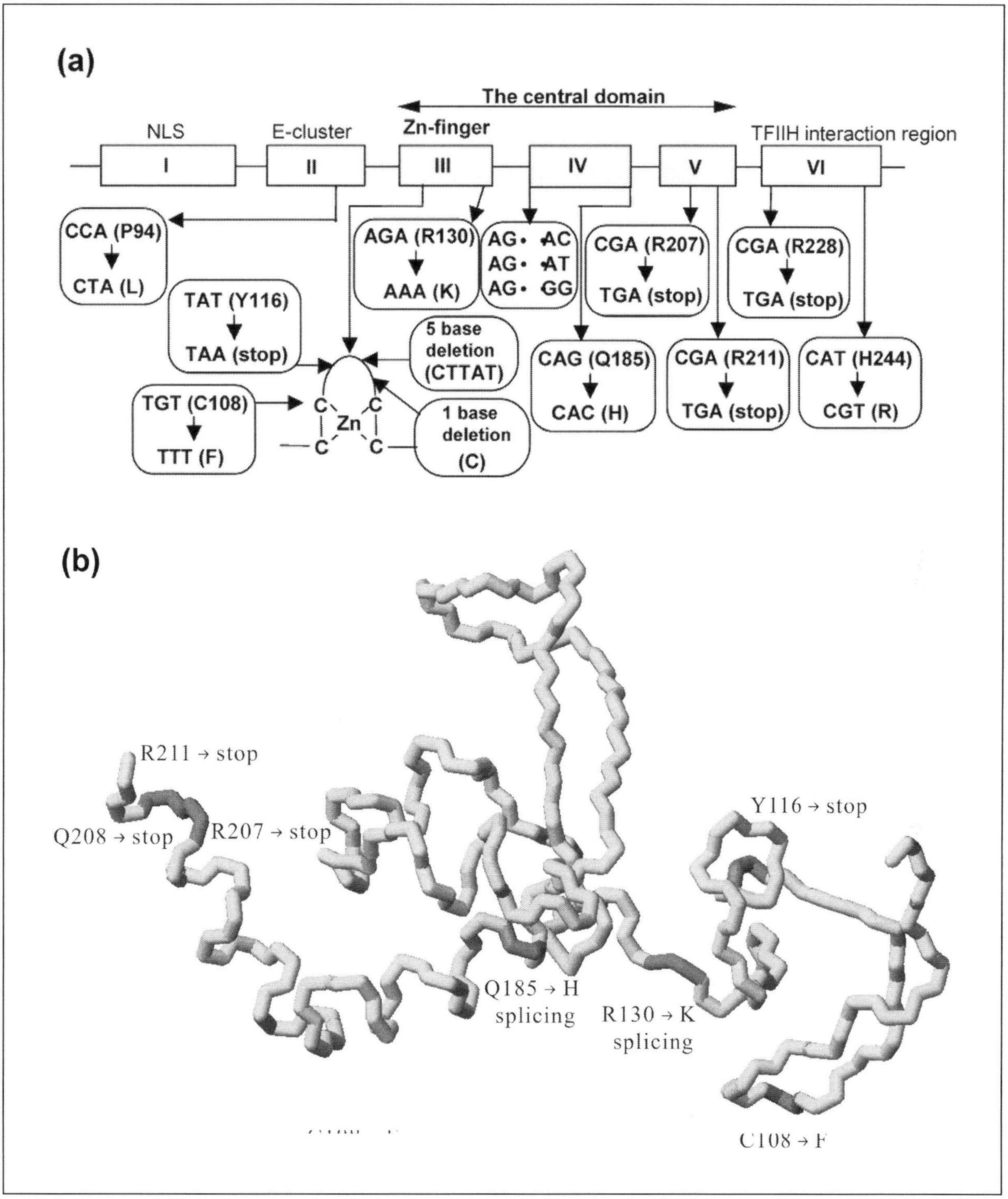

Figure 8. Mutations in the *xpa* gene found in XP patients. a) The mutations are shown on the map of the xpa gene, which contains six exons (I to VI). The central domain corresponds to the region composed of exons III, IV, and V. b) Some mutations in the central domain are mapped on the tertiary structure. The terms "stop" and "splicing" represent nonsense and splicing mutations, respectively.

References

1. Aboussekhra A, Wood RD. Repair of UV-damaged DNA by mammalian cells and Saccharomyces cerevisiae. Curr Opin Genet Dev 1994; 4:212-220.
2. Cline SD, Hanawalt PC. Who's on first in the cellular response to DNA damage? Nat Rev Mol Cell Biol 2003; 4:361-372.
3. Thompson LH. Nucleotide excision repair in DNA damage and repair. In: Nickoloff JA, Hoekstra MF, eds. DNA Repair in Higher Eukaryotes. New Jersey: Human Press, 1998:II:335-393.
4. Cleaver JE, Kraemer KH. Xeroderma pigmentosum. In: Scriver CR, ed. The Metabolic Basis of Inherited Disease, 7th ed. New York: McGraw-Hill, 1995:4393-4419.
5. Hoeijmakers JH, Bootsma D. Molecular genetics of eukaryotic DNA excision repair. Cancer Cells Mon Rev 1990; 2:311-320.
6. Tanaka K, Miura N, Satokata I et al. Analysis of a human DNA excision repair gene involved in group A xeroderma pigmentosum and containing a zinc-finger domain. Nature 1990; 348:3-76.
7. Robins P, Jones CJ, Biggerstaff M et al. Complementation of DNA repair in xeroderma pigmentosum group A cell extracts by a protein with affinity for damaged DNA. EMBO J 1991; 10:3913-3921.
8. Jones CJ, Wood RD. Preferential binding of the xeroderma pigmentosum group A complementing protein to damaged DNA. Biochemistry 1993; 32:12096-12104.
9. Asahina H, Kuraoka I, Shirakawa M et al. The XPA protein is a zinc metalloprotein with an ability to recognize various kinds of DNA damage. Mutat Res 1994; 315:229-237.
10. Kuraoka I, Morita EH, Saijo M et al. Identification of a damaged-DNA binding domain of the XPA protein. Mutat Res 1996; 362:87-95.
11. He Z, Henricksen LA, Wold MS et al. RPA involvement in the damage-recognition and incision steps of nucleotide excision repair. Nature 1995; 374:566-569.
12. Li L, Lu X, Peterson CA et al. An interaction between the DNA repair factor XPA and replication protein A appears essential for nucleotide excision repair. Mol Cell Biol 1995; 15:5396-5402.
13. Saijo M, Kuraoka I, Masutani C et al. Sequential binding of DNA repair proteins RPA and ERCC1 to XPA in vitro. Nucleic Acids Res 1996; 24:4719-4724.
14. Park CH, Mu D, Reardon JT et al. The general transcription-repair factor TFIIH is recruited to the excision repair complex by the XPA protein independent of the TFIIE transcription factor. J Biol Chem 1995; 270:4896-4902.
15. Nocentini S, Coin F, Saijo M et al. DNA damage recognition by XPA protein promotes efficient recruitment of transcription factor II H. J Biol Chem 1997; 272:22991-22994.
16. Li L, Elledge SJ, Peterson CA et al. Specific association between the human DNA repair proteins XPA and ERCC1. Proc Natl Acad Sci USA 1994; 91:5012-5016.
17. Li L, Peterson CA, Lu X et al. Mutations in XPA that prevent association with ERCC1 are defective in nucleotide excision repair. Mol Cell Biol 1995; 15:1993-1998.
18. Nagai A, Saijo M, Kuraoka I et al. Enhancement of damage-specific DNA binding of XPA by interaction with the ERCC1 DNA repair protein. Biochem Biophys Res Commun 1995; 211:960-966.
19. Mu D, Hsu DS, Sancar A. Reaction mechanism of human DNA repair excision nuclease. J Biol Chem 1996; 271:8285-8294.
20. Tanaka K, Kamiuchi S, Ren Y et al. UV-induced skin carcinogenesis in xeroderma pigmentosum group A (XPA) gene-knockout mice with nucleotide excision repair-deficiency. Mutat Res 2001; 477:31-40.
21. Ikegami T, Kuraoka I, Saijo M et al. Solution structure of the DNA- and RPA-binding domain of the human repair factor XPA. Nature Struct Biol 1998; 5:701-706.
22. Buchko GW, Daughdrill GW, Lorimier R et al. Interactions of human nucleotide excision repair protein XPA with DNA and RPA70ΔC327: Chemical shift mapping and ^{15}N NMR relaxation studies. Biochemistry 1999; 38:15116-15128.
23. Miyamoto I, Miura N, Niwa H et al. Mutational analysis of the structure and function of the xeroderma pigmentosum group A complementing protein. Identification of essential domains for nuclear localization and DNA excision repair. J Biol Chem 1992; 267:12182-12187.
24. Lipton AS, Buchko GW, Sears JA et al. ^{67}Zn solid-state NMR spectroscopy of the minimal DNA binding domain of human nucleotide excision repair protein XPA. J Am Chem Soc 2001; 123:992-993.
25. Hess NJ, Buchko GW, Conradson SD et al. Human nucleotide excision repair protein XPA: Extended X-ray absorption fine-structure evidence for a metal-binding domain. Protein Sci 1998; 7:1970-1975.
26. Morita EH, Ohkubo T, Kuraoka I et al. Implications of the zinc-finger motif found in the DNA-binding domain of the human XPA protein. Genes Cells 1996; 1:437-442.
27. Summers MF. Zinc fingers. In: Grant DM, Harris RK, eds. Encyclopedia of Nuclear Magnetic Resonance. West Sussex, England: John Wiley and Sons Ltd., 1996:8:5063-5071.
28. Omichinski JG, Clore GM, Schaad O et al. NMR structure of a specific DNA complex of Zn-containing DNA binding domain of GATA-1. Science 1993; 261:438-446.
29. Satokata I, Iwai K, Matsuda T et al. Genomic characterization of the human DNA excision repair-controlling gene XPAC. Gene 1993; 136:345-348.
30. Go M. Protein structures and split genes. Adv Biophys 1985; 19:91-131.
31. Nakashima T, Sekiguchi T, Sunamoto H et al. Structure of the human CCG1 gene: Relationship between the exons/introns and functional domain/modules of the protein. Gene 1994; 141:193-200.
32. Moggs JG, Yarema KJ, Essigmann JM et al. Analysis of incision sites produced by human cell extracts and purified proteins during nucleotide excision repair of a 1,3-intrastrand d(GpTpG)-cisplatin adduct. J Biol Chem 1996; 271:7177-7186.
33. Ikegami T, Kuraoka I, Saijo M et al. Resonance assignments, solution structure, and backbone dynamics of the DNA- and RPA-binding domain of human repair factor XPA. J Biochemistry 1999; 125:495-506.
34. Buchko GW, Tung CS, McAteer K et al. DNA-XPA interactions: A ^{31}P NMR and molecular modeling study of dCCAATAACC association with the minimal DNA-binding domain (M98-F219) of the nucleotide excision repair protein XPA. Nucleic Acids Res 2001; 29:2635-2643.
35. Guzder SN, Habraken Y, Sung P et al. Reconstitution of yeast nucleotide excision repair with purified Rad proteins, replication protein A, and transcription factor TFIIH. J Biol Chem 1995; 270:12973-12976.
36. Aboussekhra A, Biggerstaff M, Shivji MK et al. Mammalian DNA nucleotide excision repair reconstituted with purified protein components. Cell 1995; 80:859-868.
37. Bochkarev A, Pfuetzner RA, Edwards AM et al. Structure of the single-stranded-DNA-binding domain of replication protein A bound to DNA. Nature 1997; 385:176-181.
38. Pfuetzner RA, Bochkarev A, Frappier L et al. Replication protein A: Characterization and crystallization of the DNA binding domain. J Biol Chem 1997; 272:430-434.
39. Mer G, Bochkarev A, Gupta R et al. Structural basis for the recognition of DNA repair proteins UNG2, XPA, and RAD52 by replication factor RPA. Cell 2000; 103:449-456.
40. Wakasugi M, Sancar A. Order of assembly of human DNA repair excision nuclease. J Biol Chem 1999; 274:18759-18768.
41. Hey T, Lipps G, Sugasawa K et al. The XPC-HR23B complex displays high affinity and specificity for damaged DNA in a true-equilibrium fluorescence assay. Biochemistry 2002; 41:6583-6587.
42. Sugasawa K, Ng JM, Masutani C et al. Xeroderma pigmentosum group C protein complex is the initiator of global genome nucleotide excision repair. Mol Cell 1998; 2:223-232.
43. You JS, Wang M, Lee SH. Biochemical analysis of the damage recognition process in nucleotide excision repair. J Biol Chem 2003; 278:7476-7485.
44. Tang JY, Hwang BJ, Ford JM et al. Xeroderma pigmentosum p48 gene enhances global genomic repair and suppresses UV-induced mutagenesis. Mol Cell 2000; 5:737-744.

CHAPTER 33

MOF, an Acetyl Transferase Involved in Dosage Compensation in *Drosophila*, Uses a CCHC Finger for Substrate Recognition

Asifa Akhtar and Peter B. Becker*

Abstract

The histone acetyltransferase MOF is central to the process of dosage compensation complex in *Drosophila*, which assures that the transcriptional activity of large parts of the male X chromosome is enhanced by about two-fold. Inactivation of the acetyl transferase activity of MOF is lethal to affected male flies. MOF belongs to the MYST family of acetylases that are involved in a wide variety of cellular processes. One characteristic structural feature of this family is the presence of a CCHC zinc finger adjacent to their acetyl-CoA binding motif. Structure-function analyses of several family members suggest that this region is important for enzymatic activity. Our detailed characterization of MOF shows that the CCHC-type zinc finger is required for interaction with the histone substrate and hence catalytic activity.

Introduction

The organization of eukaryotic genomes as chromatin represents a considerable barrier for a variety of nuclear processes with DNA substrate, such as transcription of genes, the replication of the chromosome and the repair of damaged DNA. DNA is coiled around histone octamers to form successions of nucleosomes with beads-on-a-string appearance at low salt.[1-3] The folding of the nucleosomal fiber into compact structures under physiological conditions can render the wrapped DNA inaccessible to interacting regulators. The degree and type of folding and hence the functional state of a locus is defined by interactions between nucleosomes and interactions between nucleosomes and nonhistone proteins. Main targets for these interactions are the histone N-terminal domains, their 'tails' which protrude from the compact globular part of the nucleosome and are hence available for interactions. Their potential for interactions can be modified by post-translational modifications. The histone N termini contain a number of conserved residues that can be acetylated, methylated, phosphorylated, ubiquitinated and ribosylated. Distinct histone modification patterns serve as semaphore 'flags' which are decoded by nonhistone regulators of chromatin folding.[4] The reversible acetylation of the ε-amino group of lysines within the histone N-terminal tails has emerged as a major principle underlying dynamic transitions between permissive and repressed chromatin structure.

The acetylation status of chromosomal loci is negotiated through a dynamic interplay between histone acetyl transferases (HATs) and corresponding deacetylases (HDACs), which may vary in concentration and activity, and may be targeted to their sites of action by sequence-specific DNA binding factors.[5] In general, acetylation of core histones is associated with transcriptionally active genes, whereas deacetylation of histones often correlates with repression,[6] but exceptions to these rules exist.[7]

HDACs and HATs can be categorized according to their domain structure, function and substrate specificity.[3,8,9] The cytoplasmic HATs (of B-type) are involved in histone storage, transport and assembly. The A-type HATs are nuclear regulators and can be grouped into the GNAT family, the p300/CBP family and the MYST family.[8,10] We will be focusing on MYST acetylases, which feature a CCHC zinc finger as an integral part of their acetyl transferase domain.

The MYST Family of Acetyl Transferases

The 'MYST-signature' emerged when similarities in the domain structure of the founding members hMOZ, ySas2, yYbf2/Sas3 and hTip60 was noted. Later, additional members such as yeast Esa1, Drosophila MOF, human MORF and HBO1 were added to the list.[8,11,12] MYST proteins are mainly related by similarity around their acetyl transferase domains, which includes a $CxxCx_{12}HxxxC$ zinc finger (all except Esa1, see below) (Fig. 1). Apart from the 'MYST signature' the proteins differ in size and structure, although some recurring domains, such as chromodomains and PHD fingers can be noted (Fig. 1). MYST histone acetyl transferases are evolutionarily conserved proteins involved in a wide variety of regulatory functions. Although they were originally identified as histone acetyl transferases (HATs) it is quite likely that they may modify other, nonhistone substrates as well, as was recently shown for MOF (see below).

* *Corresponding author. See list of "Contributors".*

Zinc Finger Proteins: From Atomic Contact to Cellular Function, edited by Shiro Iuchi and Natalie Kuldell.

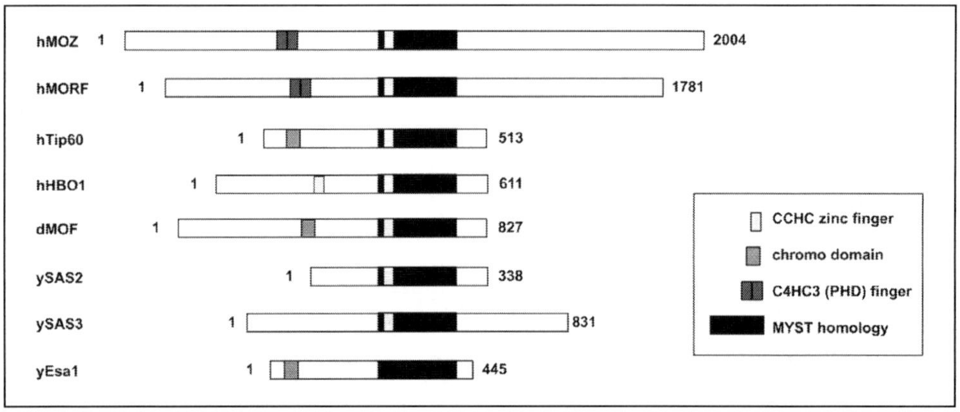

Figure 1. Schematic representation of the domain structure of various MYST family members. The total length of the proteins is indicated.

Since MYST acetylases were largely identified through sequence similarity screens, our knowledge of their function is rudimentary in many cases. However, they all are found in large multisubunit complexes involved in gene regulation. Tip60, for example, the first human MYST protein member, was identified in a yeast 2-hybrid screen in a search for Tat (HIV-1 transcactivator) interacting proteins.[13] It associates with at least 11 other proteins in a complex that may have other functions besides the nucleosomal acetylase and that can be recruited to chromatin through interactions with the transcription factor MYC.[14] It also interacts with the TEL/ETV6 repressor and has been suggested to have corepressor function.[15] Deregulation of MOZ (monocytic leukemia zinc finger protein) leads to oncogenic transformations.[16,17] In acute myeloid leukemia, the MOZ gene is translocated to fuse with the acetylase CBP gene at various breakpoints.[18] MOZ is itself an acetyl transferase able to interact with transcription factors, but additionally containing transcription repression and activation domains.[19,20] The yeast Sas2 and Sas3/Ybf2 proteins are involved in silencing of mating type loci (Sas: 'something about silencing'). Sas2 was discovered in a screen for epigenetic silencing, while Sas3 was discovered because of its homology with Sas2. Sas3 is part of the NuA3 complex, which acetylates histone H3 in nucleosomes.[21,22] Sas2, unlike Sas3, does have not HAT activity as a recombinant protein but does have HAT activity as part of a complex with Sas4 and Sas5.[23] Most is known about the regulatory role of MOF, a *Drosophila* HAT which is central to the process of dosage compensation, whereby transcription of many genes on the single male X chromosome is increased by two-fold as compared the transcription from either of the two female X genes.[24] As will be detailed, the role of the CCHC zinc finger in MOF function is particularly well documented. Thus, MOF serves as a paradigm for the activity of MYST acetyl transferases.

The Role of MOF in Dosage Compensation

Dosage compensation mechanisms ensure that, despite of the fact that male and female cells contain different numbers of X chromosomes (one in males and two in females in humans and Drosophila) the expression of X-linked genes is equal in both sexes. Dosage compensation is achieved differently in different organisms. In mammals, one of the two female X chromosomes is inactivated by epigenetic means. [25] In *Caenorhabditis elegans*, the activity of each of the two hermaphrodite X chromosomes is halved.[26] By contrast, in *Drosophila*, both female X chromosomes are transcribed, but transcription from the majority of genes on the single male X chromosome is increased two-fold to make up for the reduced chromosome dosage.[27] Drosophila MOF is essential to this process: inactivation of MOF leads to male-specific lethality, a striking example of the drastic effect a two-fold difference in gene expression may have. MOF is part of a multisubunit dosage compensation complex (DCC), which contains at least four other proteins, the precise function of which is not clear yet.[28,29]

A hallmark of the hypertranscribed male X chromosome is the acetylation of histone H4 at lysine 16 (H4K16ac).[28-30] This epigenetic mark is set by MOF and failure to do so leads to lack of dosage compensation and male-specific lethality. Whether embedded in the DCC or as a recombinant protein, MOF specifically acetylates chromatin.[31,32] This single modification suffices to relieve chromatin-mediated repression of transcription in vitro and in vivo.[31] MOF binds chromatin nonspecifically in vitro[31] and hence the selective association with the male X chromosome observed in vivo depends on other parts of the DCC. MOF interacts with other DCC subunits (Morales, Mengus, Akhtar and Becker, submitted),[33-35] and most interestingly also requires a noncoding RNA, roX, for proper association with the X chromosome.[36-38] Besides histones, MOF is also able to acetylate other associated subunits within DCC, such as MSL1 and MSL3, suggesting a coordinated modification of the histone infrastructure and the nonhistone regulators themselves (Morales, Mengus, Akhtar and Becker, submitted).[35]

Drosophila MOF is highly related to human MOF (hMOF) (52% identity and 69% similarity), which acetylates histones with a specificity similar to the other members of the MYST family.[39] The functional role of hMOF protein is currently unknown. The closest relative of MOF in yeast is Esa1, which is part of the multisubunit NuA4 complex involved in transcriptional activation.[40,41] Intriguingly, the NuA4 complex has other similarities to the DCC in flies: the NuA4 subunit Eaf3 is homologous to Drosophila MSL3.[40,42]

The MOF CCHC Zinc Finger Is Involved in Substrate Recognition

Part of the sequence similarity that characterises MYST acetyl transferases is a $CxxCx_{12}HxxxC$ zinc finger close to their HAT domains. Zinc fingers of this type are frequently involved in protein-protein interactions.[43] Although they fold as independent units,[44,45] they frequently perform their functions in combination with other elements, such as CCHH fingers in LIM domains,[46] where they are again involved in protein-protein contacts (see reference 47 and references therein). The CCHC fingers of the HIV1 nucleocapsid proteins bind single-stranded DNA during the replication of the viral genome.[48]

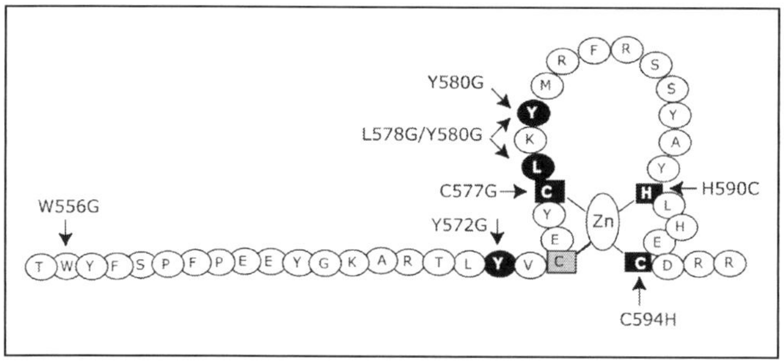

Figure 2. Schematic drawing of the MOF CCHC zinc finger region and position of the different point mutations. Each point mutation is shown by the original residue, the position and the mutant residue. Residues which, if mutated, affect chromatin binding and acetylase function, are marked in black.

Since MOF is known to interact with proteins as well as RNA within the DCC we tested for a function of the MYST zinc finger in various assays. While the RNA binding activity of MOF was unaffected by mutation of the finger structure, we found that it was essential for the HAT activity of MOF, in addition to the well-established HAT domain (Fig. 2). Mutation of single, conserved, hydrophic residues close to or within the predicted beta sheet/loop regions abolish HAT activity. These mutations apparently do not affect the catalytic activity of the enzyme, but rather the recognition of the substrate, since zinc finger mutations also abolish the interaction of the enzyme with the N terminus of histone H4.[49] We also wondered, whether the CCHC coordination of the zinc atom could be replaced by more typical CCCC or CCHH coordination. Strikingly, converting the CCHC pattern (Fig. 2) to CCHH (C594H) or CCCC (H590C) also abolishes HAT activity and chromatin binding (Fig. 3). These results point to a specific involvement of the coordinating amino acids in histone binding, either directly or through subtle changes of the geometry of the zinc finger's secondary structures. In the case of the GATA-FOG interaction, a replacement of a cysteine in the FOG1 CCHC finger by a histidine (yielding a CCHH structure) leads to a repositioning of two aromatic residues in the loop, and hence abolishes interaction with GATA.[43]

Interestingly, the HIV nucleocapsid protein (NC) requires two CCHC-type zinc fingers for the DNA chaperonin activity of NC during virus replication. Converting the fingers into canonical CCCC or CCHH coordination significantly impairs their activity.[48] In light of these results we tested for the interactions of wild type MOF and its C594H and H590C derivatives with nucleic acids. The previously documented RNA binding activity of MOF through the chromodomain[36] was unaffected by zinc finger mutations (data not shown). However, we found that the poor interaction of wild type MOF with single stranded DNA was improved by the (H590C) substitution (Fig. 4A). This binding was sensitive to the zinc chelator 1,10-phenanthroline (Fig. 4B), pointing to an involvement of the finger structure in ssDNA binding and thus confirming the formation of the finger. Collectively, our data thus suggest a specific role of the MOF CCHC zinc finger in substrate recognition. They also illustrate the functional integration of the finger into the specialized HAT domain of MYST-type acetylases, which was already suggested by the extent of sequence conservation. Recently, the CCHC importance of the CCHC zinc finger for the function of other MYST family members emerged: the CCHC finger of Tip60 was shown to be

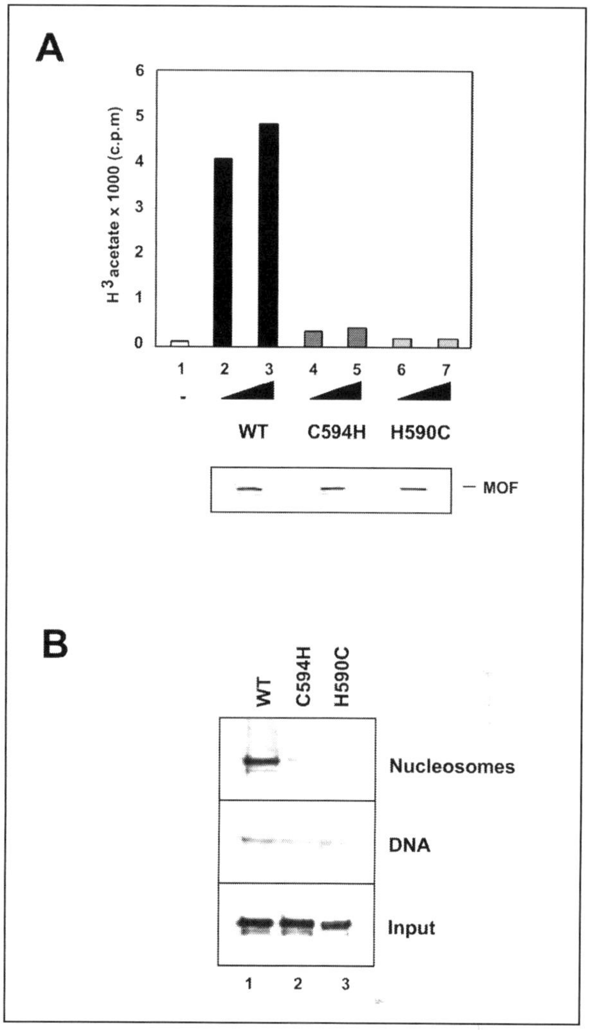

Figure 3. A) Altered specificity zinc finger mutants are inactive as HAT enzymes. Top panel: 1 µg of histones were used as a substrate for HAT assay using 50 ng (lanes 2, 4, 6) or 100 ng (lanes, 3, 5, 7) of either wild type, or the C594H, H590C MOF derivatives. Lanes 1 shows a reaction without enzyme. Lower panel: MOF amounts were analysed by Western blot analysis using a specific antibody. B) Altered specificity zinc finger mutants fail to interact with chromatin. 300 ng of either nucleosomal (N) or free DNA (D) was used for each binding reaction (according to ref. 49) using 100 ng of MOF derivatives. 'Input' shows 20% of input protein. The Western blots were probed with affinity-purified antibody against MOF protein.

required for interaction with the TEL protein.[15] It will be interesting to see, whether TEL is an acetylation substrate for Tip60. A fusion of MOZ with the nuclear receptor coactivator TIF2 during a chromosomal translocation is associated with acute myeloid leukemia. The transforming potential of the fusion protein requires the CCHC nucleosome recognition motif of MOZ, but interestingly not the acetylase activity. This indicates that the nucleosome targeting function of MOZ may be functionally uncoupled from acetylation. Since TIF2 interacts with the acetylase CBP, histone acetylation may be provided 'in trans'.[16]

There appeared to be one exception to the rule that the zinc finger was important: the MYST member Esa1 does not contain an obvious zinc finger motif, since three out of four coordinating residues are lacking. The apparent problem was solved recently

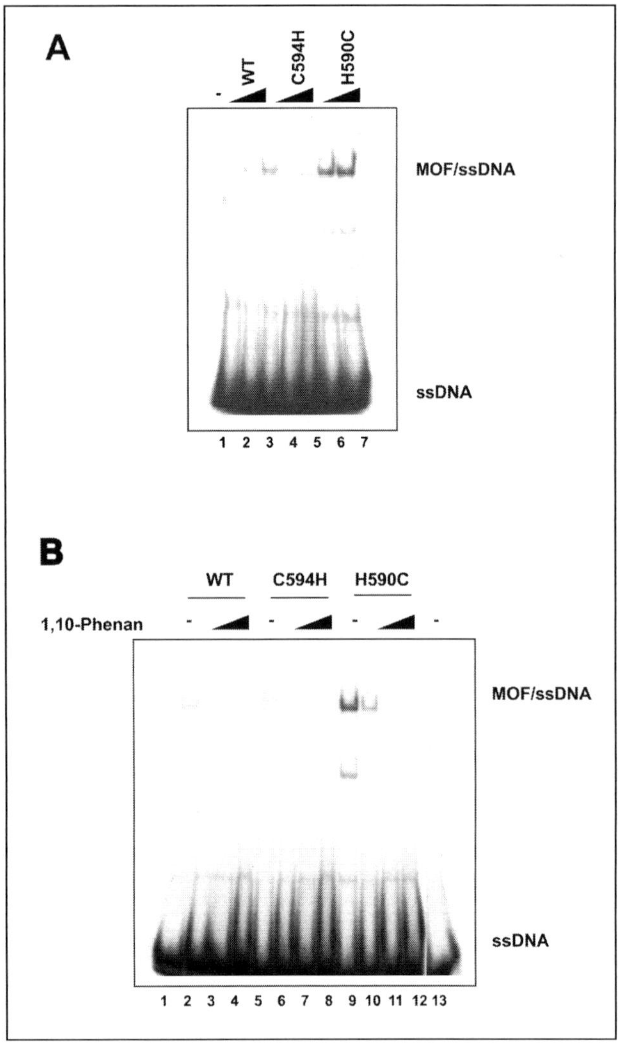

Figure 4. A) Nucleic acid binding of zinc finger mutants. DNA electrophoretic mobility shift assay (EMSA) were performed as described in reference 36, except 75 bp ss DNA oligo was used as a probe. 150 ng (lanes 2, 4, 6) or 300 ng (lanes 3, 5, 7) of WT MOF, C594H or H590C were used respectively. B) DNA EMSA as in (A), but including 5, 10 or 15 mM 1,10-phenanthroline (lanes 2-4, 6-8, 10-12, respectively). Lane 13, ss DNA probe alone.

when Marmorstein and colleagues determined the structure of the HAT domain of Esa1 in complex with coenzyme A and found that the corresponding region forms a classical TFIIIA-type zinc finger fold entirely in the absence of zinc.[50] This finding provides unexpected support to the idea of the importance of the zinc finger fold in MYST acetylase function.

Concluding Remarks

The detailed analysis of Esa1 structure revealed fundamentally different catalytic principles between MYST family members and, for example, acetylases of the GCN5/PCAF family.[51] Although both classes of enzymes target similar histone substrates, their mode of substrate recognition may also be very different. However, while MYST acetylases are specialized in using a $CxxCx_{12}HxxxC$ zinc finger for substrate recognition, other histone modification enzymes require different types of zinc fingers for full function. For example, the acetyl transferases P300/CBP contain PHD fingers adjacent to their catalytic domains.[52] Interestingly, although p300 and CBP perform very similar functions in many respects, they differ when it comes to the functional importance of these structures: whereas the PHD finger of P300 was dispensable for HAT activity, the corresponding structure of CBP are required for HAT function.[52,53] Mutations in the PHD finger of CBP give rise to the developmental disorder Rubinstein-Taybi syndrome.[54] Furthermore, cysteine-rich regions are not restricted to HATs, but the catalytic SET domain of histone methyl transferases are frequently flanked by cysteine-rich regions, which are important for enzymatic activity.[55-58] The phenomenology of diverse zinc finger structures in chromatin regulators is colorful, and we are looking forward to the emergence of common principles and differences, which may explain the exquisite substrate specificities that many of these enzymes show, as more structures are being solved.

Acknowledgements

We thank Gerit Vriend for fruitful discussions and Mitzi Kuroda for anti-MOF antibodies. Research on MOF in the lab of P.B. is supported by the Deutsche Forschungsgemeinschaft.

References

1. Hayes JJ, Hansen JC. Nucleosomes and the chromatin fiber. Curr Opin Genet Dev 2001; 11(2):124-129.
2. Woodcock CL, Dimitrov S. Higher-order structure of chromatin and chromosomes. Curr Opin Genet Dev 2001; 11(2):130-135.
3. Gregory PD, Wagner K, Horz W. Histone acetylation and chromatin remodeling. Exp Cell Res 2001; 265(2):195-202.
4. Jenuwein T, Allis CD. Translating the histone code. Science 2001; 293(5532):1074-1080.
5. Eberharter A, Becker PB. Histone acetylation: A switch between repressive and permissive chromatin. Second in review series on chromatin dynamics. EMBO Rep 2002; 3(3):224-229.
6. Allfrey VG, Pogo BG, Littau VC et al. Histone acetylation in insect chromosomes. Science 1968; 159(812):314-316.
7. Suka N, Luo K, Grunstein M. Sir2p and Sas2p opposingly regulate acetylation of yeast histone H4 lysine16 and spreading of heterochromatin. Nat Genet 2002; 32(3):378-383.
8. Sterner DE, Berger SL. Acetylation of histones and transcription-related factors. Microbiol Mol Biol Rev 2000; 64(2):435-459.
9. de Ruijter AJ, van Gennip AH, Caron HN et al. Histone deacetylases (HDACs): Characterization of the classical HDAC family. Biochem J 2003; 370(Pt 3):737-749.
10. Struhl K. Histone acetylation and transcriptional regulatory mechanisms. Genes Dev 1998; 12(5):599-606.
11. Timmermann S, Lehrmann H, Polesskaya A et al. Histone acetylation and disease Cell Mol Life Sci. May 2001; 58(5-6):728-736.
12. Utley RT, Cote J. The MYST family of histone acetyltransferases. Curr Top Microbiol Immunol 2003; 274:203-236.
13. Kamine J, Elangovan B, Subramanian T et al. Identification of a cellular protein that specifically interacts with the essential cysteine region of the HIV-1 Tat transactivator. Virology 1996; 216(2):357-366.
14. Frank SR, Parisi T, Taubert S et al. MYC recruits the TIP60 histone acetyltransferase complex to chromatin. EMBO Rep 2003; 4(6):575-580.
15. Nordentoft I, Jorgensen P. The acetyltransferase Tip60 interacts with TEL/ETV6 and functions as a transcriptional corepressor. Biochem J 2003; Pt.
16. Deguchi K, Ayton PM, Carapeti M et al. MOZ-TIF2-induced acute myeloid leukemia requires the MOZ nucleosome binding motif and TIF2-mediated recruitment of CBP. Cancer Cell 2003; 3(3):259-271.
17. Borrow J, Stanton Jr VP, Andresen JM et al. The translocation t(8;16)(p11;p13) of acute myeloid leukaemia fuses a putative acetyltransferase to the CREB-binding protein. Nat Genet 1996; 14(1):33-41.

18. Panagopoulos I, Isaksson M, Lindvall C et al. Genomic characterization of MOZ/CBP and CBP/MOZ chimeras in acute myeloid leukemia suggests the involvement of a damage-repair mechanism in the origin of the t(8;16)(p11;p13). Genes Chromosomes Cancer 2003; 36(1):90-98.
19. Champagne N, Pelletier N, Yang XJ. The monocytic leukemia zinc finger protein MOZ is a histone acetyltransferase. Oncogene 2001; 20(3):404-409.
20. Pelletier N, Champagne N, Stifani S et al. MOZ and MORF histone acetyltransferases interact with the Runt-domain transcription factor Runx2. Oncogene 2002; 21(17):2729-2740.
21. Reifsnyder C, Lowell J, Clarke A et al. Yeast SAS silencing genes and human genes associated with AML and HIV-1 Tat interactions are homologous with acetyltransferases. Nat Genet 1996; 14(1):42-49.
22. John S, Howe L, Tafrov ST. The something about silencing protein, Sas3, is the catalytic subunit of NuA3, a yTAF(II)30-containing HAT complex that interacts with the Spt16 subunit of the yeast CP (Cdc68/Pob3)-FACT complex. Genes Dev 2000; 14(10):1196-1208.
23. Sutton A, Shia WJ, Band D et al. Sas4 and Sas5 are required for the histone acetyltransferase activity of Sas2 in the SAS complex. J Biol Chem 2003; 278(19):16887-16892.
24. Hilfiker A, Hilfiker-Kleiner D, Pannuti A et al. mof, a putative acetyl transferase gene related to the Tip60 and MOZ human genes and to the SAS genes of yeast, is required for dosage compensation in Drosophila. EMBO J 1997; 16(8):2054-2060.
25. Lyon M. Gene action in the X-chromosome of the mouse (Mus musculus). Nature 1961; 190:372-373.
26. Meyer BJ, Casson LP. Caenorhabditis elegans compensates for the difference in X chromosome dosage between the sexes by regulating transcript levels. Cell 1986; 47(6):871-881.
27. Mukherjee AS, Beermann W. Synthesis of ribonucleic acid by the X-chromosomes of Drosophila melanogaster and the problem of dosage compensation. Nature 1965; 207(998):785-786.
28. Lucchesi JC. Dosage compensation in flies and worms: The ups and downs of X-chromosome regulation. Curr Opin Genet Dev 1998;8(2):179-184.
29. Akhtar A. Dosage compensation: An intertwined world of RNA and chromatin remodelling. Curr Opin Genet Dev 2003; 13(2):161-169.
30. Lucchesi JC. Dosage compensation in Drosophila and the "complex' world of transcriptional regulation. Bioessays 1996; 18(7):541-547.
31. Akhtar A, Becker PB. Activation of transcription through histone H4 acetylation by MOF, an acetyltransferase essential for dosage compensation in Drosophila. Mol Cell 2000; 5(2):367-375.
32. Smith ER, Pannuti A, Gu W et al. The drosophila MSL complex acetylates histone H4 at lysine 16, a chromatin modification linked to dosage compensation. Mol Cell Biol 2000; 20(1):312-318.
33. Palmer MJ, Richman R, Richter L et al. Sex-specific regulation of the male-specific lethal-1 dosage compensation gene in Drosophila. Genes Dev 1994;8(6):698-706.
34. Scott MJ, Pan LL, Cleland SB et al. MSL1 plays a central role in assembly of the MSL complex, essential for dosage compensation in Drosophila. EMBO J 2000; 19(1):144-155.
35. Buscaino A, Kocher T, Kind JH et al. MOF-regulated acetylation of MSL-3 in the Drosophila dosage compensation complex. Mol Cell 2003; 11(5):1265-1277.
36. Akhtar A, Zink D, Becker PB. Chromodomains are protein-RNA interaction modules. Nature 2000; 407(6802):405-409.
37. Meller VH, Gordadze PR, Park Y et al. Ordered assembly of roX RNAs into MSL complexes on the dosage-compensated X chromosome in Drosophila. Curr Biol 2000; 10(3):136-143.
38. Meller VH, Rattner BP. The roX genes encode redundant male-specific lethal transcripts required for targeting of the MSL complex. EMBO J 2002; 21(5):1084-1091.
39. Neal KC, Pannuti A, Smith ER et al. A new human member of the MYST family of histone acetyl transferases with high sequence similarity to Drosophila MOF. Biochim Biophys Acta 2000; 1490(1-2):170-174.
40. Eisen A, Utley RT, Nourani A et al. The yeast NuA4 and Drosophila MSL complexes contain homologous subunits important for transcription regulation. J Biol Chem 2001; 276(5):3484-3491.
41. Clarke AS, Lowell JE, Jacobson SJ et al. Esa1p is an essential histone acetyltransferase required for cell cycle progression. Mol Cell Biol 1999; 19(4):2515-2526.
42. Gavin AC, Bosche M, Krause R et al. Functional organization of the yeast proteome by systematic analysis of protein complexes. Nature 2002; 415(6868):141-147.
43. Kowalski K, Liew CK, Matthews JM et al. Characterization of the conserved interaction between GATA and FOG family proteins. J Biol Chem 2002; 277(38):35720-35729.
44. Hammarstrom A, Berndt KD, Sillard R et al. Solution structure of a naturally-occurring zinc-peptide complex demonstrates that the N-terminal zinc-binding module of the Lasp-1 LIM domain is an independent folding unit. Biochemistry 1996; 35(39):12723-12732.
45. Yao X, Perez-Alvarado GC, Louis HA et al. Solution structure of the chicken cysteine-rich protein, CRP1, a double-LIM protein implicated in muscle differentiation. Biochemistry 1999; 38(18):5701-5713.
46. Schwabe JW, Klug A. Zinc mining for protein domains. Nat Struct Biol 1994; 1(6):345-349.
47. Schmeichel KL, Beckerle MC. Molecular dissection of a LIM domain. Mol Biol Cell 1997; 8(2):219-230.
48. Guo J, Wu T, Kane BF et al. Subtle alterations of the native zinc finger structures have dramatic effects on the nucleic acid chaperone activity of human immunodeficiency virus type 1 nucleocapsid protein. J Virol 2002; 76(9):4370-4378.
49. Akhtar A, Becker PB. The histone H4 acetyltransferase MOF uses a C2HC zinc finger for substrate recognition. EMBO Rep 2001; 2(2):113-118.
50. Yan Y, Barlev NA, Haley RH et al. Crystal structure of yeast Esa1 suggests a unified mechanism for catalysis and substrate binding by histone acetyltransferases. Mol Cell 2000; 6(5):1195-1205.
51. Yan Y, Harper S, Speicher DW et al. The catalytic mechanism of the ESA1 histone acetyltransferase involves a self-acetylated intermediate. Nat Struct Biol 2002; 9(11):862-869.
52. Bordoli L, Husser S, Luthi U et al. Functional analysis of the p300 acetyltransferase domain: The PHD finger of p300 but not of CBP is dispensable for enzymatic activity. Nucleic Acids Res 2001; 29(21):4462-4471.
53. Kalkhoven E, Teunissen H, Houweling A et al. The PHD type zinc finger is an integral part of the CBP acetyltransferase domain. Mol Cell Biol 2002; 22(7):1961-1970.
54. Kalkhoven E, Roelfsema JH, Teunissen H et al. Loss of CBP acetyltransferase activity by PHD finger mutations in Rubinstein-Taybi syndrome. Hum Mol Genet 2003; 12(4):441-450.
55. Rea S, Eisenhaber F, O'Carroll D et al. Regulation of chromatin structure by site-specific histone H3 methyltransferases. Nature 2000; 406(6796):593-599.
56. Zhang X, Tamaru H, Khan SI et al. Structure of the Neurospora SET domain protein DIM-5, a histone H3 lysine methyltransferase. Cell 2002; 111(1):117-127.
57. Xiao B, Jing C, Wilson JR et al. Structure and catalytic mechanism of the human histone methyltransferase SET7/9. Nature 2003; 421(6923):652-656.
58. Jacobs SA, Harp JM, Devarakonda S et al. The active site of the SET domain is constructed on a knot. Nat Struct Biol 2002; 9(11):833-838.

CHAPTER 34

MDM2: RING Finger Protein and Regulator of p53

Liqing Wu and Carl G. Maki*

Abstract

MDM2 is an oncogene that is frequently over-expressed in various human cancers, including sarcomas, gliomas, melanomas, and breast cancers.[1] The primary function of MDM2 is to inhibit the activity of the p53 tumor suppressor protein. P53 inhibits cell proliferation in response to DNA damage and other stresses by activating the transcription of genes that mediate either cell-cycle arrest or apoptosis. MDM2 binds the transactivation domain of p53 and inhibits the ability of p53 to activate transcription. Importantly, MDM2 binding also promotes the ubiquitin-mediated degradation of p53, as well as the export of p53 from the nucleus to the cytoplasm. The *MDM2* gene itself is transcriptionally activated by p53, thus forming an autoregulatory feedback loop in which p53 promotes the expression of its own inhibitory factor. The ability of MDM2 to promote the degradation and nuclear export of p53 depends on an intact RING finger domain located in the MDM2 C terminus. This chapter will discuss the mechanisms by which MDM2 inhibits p53 function, with an emphasis on the requirement of the MDM2 RING finger domain in p53 inhibition.

Introduction

The *MDM2* gene was first discovered in 1987 as a gene that was amplified in spontaneously transformed murine 3T3 fibroblasts.[2] Subsequent studies revealed that the MDM2 protein interacts strongly with the tumor suppressor protein p53 and inhibits p53 activity.[3,4] It is now well-established that the primary function of MDM2 is to inhibit p53.[5,6] The importance of p53 inhibition by MDM2 is best illustrated through the attempted generation of mice that lack *MDM2*. These *MDM2* "knockout" mice are not viable, dying at an early stage of embryonic development.[7,8] Importantly, however, the concomitant knockout of *p53* allows survival of the mice. These results demonstrate that MDM2 normally inhibits p53 function, and that this inhibition is critically important during embryonic development.

Human MDM2 is a 491 amino acid protein that localizes mostly in the nucleus, though it can also localize in the cytoplasm.[6] The *MDM2* gene has been mapped to chromosome 12q13-14 and includes several exons that encode the MDM2 protein. Alternative splicing of the primary mRNA transcript can result in the generation of different MDM2 isoforms, some of which inhibit the activity of the wild-type protein in a dominant-negative fashion.[9,10] Analysis of the *MDM2* coding sequence has revealed several structural and functional domains (Fig. 1). These include the p53-binding domain located in the N terminus between residues 19-102, the p300-binding domain between residues 217-246, an acidic domain located between residues 222-272, a zinc finger domain located between residues 289-331, and a RING finger domain located in the MDM2 C terminus between residues 438-478. RING fingers are zinc binding domains with a defined octet of cysteine (C) and histidine (H) residues spaced at varying intervals in the coding sequence.[11-14] In the RING H2 subtype, the cysteine and histidines are arranged in the order $C_3H_2C_3$, while in the RING HC subtype the order is C_3HC_4. The MDM2 RING finger is of the C_3HC_4 subtype and, like other RING fingers, is predicted to bind two zinc atoms. RING finger domains are believed to function in the assembly and architecture of large protein complexes, including enzyme complexes of the ubiquitin system of protein degradation.[15,16] As we will see below, the MDM2 protein can function as an ubiquitin-system E3 ligase that promotes the specific degradation of p53, and this activity requires the MDM2 RING finger domain.

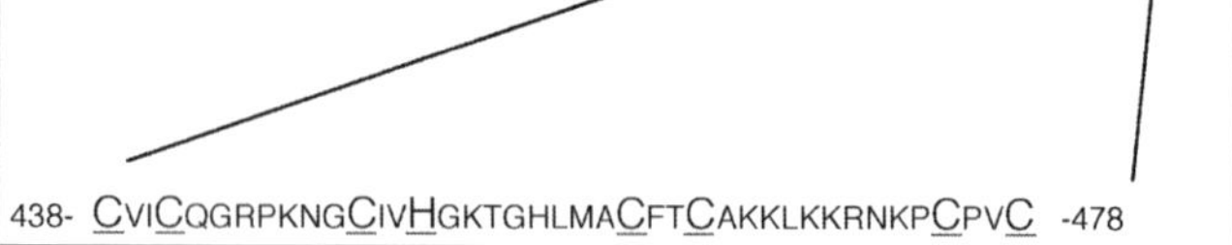

Figure 1. The MDM2 protein. Full-length human MDM2 (also called HDM2) is a 491 amino acid protein. Positions of the nuclear localization signal (NLS), nuclear export signal (NES), and nucleolar localization signal (NoLS) are indicated. Also indicated are the various functional and structural domains of MDM2, including the RING-finger located in the C terminus.

* *Corresponding author. See list of "Contributors".*

Zinc Finger Proteins: From Atomic Contact to Cellular Function, edited by Shiro Iuchi and Natalie Kuldell.

MDM2 Promotes p53 Degradation

In 1997, research groups led by Moshe Oren in Israel and Karen Vousden in Frederick, Maryland were investigating the binding interactions between MDM2 and p53.[17,18] These studies were complicated by the fact that p53 levels were consistently decreased with coexpression of MDM2 (see Fig. 2) as an example), which made a quantitative analysis of the p53:MDM2 interaction difficult. The researchers decided to investigate the molecular basis for the decreased levels of p53. Small-molecule inhibitors of the proteasome prevented the decrease in p53 levels observed with MDM2 expression, suggesting that MDM2 was somehow promoting p53 degradation by the proteasome. Moreover, a high molecular weight ladder of p53 species was generated when p53 was coexpressed with MDM2 (see Fig. 2) as an example). The size of these p53 species was consistent with the addition of multiple ubiquitin-moieties to p53, and analysis with anti-ubiquitin antibodies identified them as p53:ubiquitin conjugates. These seminal studies provided the first evidence that MDM2 can promote the degradation of p53 through the ubiquitin-proteasome pathway.[17,18]

MDM2 Is an E3 Ubiquitin-Protein Ligase

The ubiquitin-proteasome system is the major pathway in the cell for selective protein degradation.[19,20] The covalent attachment of multiple ubiquitin molecules to lysine residues of a target protein serves to signal its recognition and rapid degradation by the 26S proteasome. Ubiquitination of a protein substrate requires the concerted action of three classes of enzymes designated E1, E2, and E3. The ubiquitin activating enzyme E1 initially activates ubiquitin in an ATP-dependent reaction through the formation of a thiol-ester bond between the carboxyl terminus of ubiquitin and the thiol group of a specific cysteine residue of E1. Ubiquitin is then transferred to a specific cysteine residue on one of several ubiquitin-conjugating enzymes (Ubcs or E2s). E2 enzymes in turn may transfer the ubiquitin either directly to a substrate or to the final class of enzymes known as ubiquitin protein ligases (or E3s). The E3 enzymes catalyze the formation of an isopeptide bond between the carboxyl terminus of ubiquitin and the ε-amino group of lysine residues on the target protein. A substrate may then be multiply ubiquitinated through the attachment of additional ubiquitin molecules to specific lysine residues of ubiquitin itself. In many cases the E1, E2, and E3 enzymes involved form large, multi-protein complexes. This increases the efficiency of the process by allowing the rapid thiol-ester transfer of ubiquitin molecules between proteins.

In vitro studies with purified E1 enzyme, E2 enzyme, MDM2, and p53 showed that MDM2 could form a direct thiol-ester linkage with ubiquitin at Cys464 in the MDM2 RING finger domain, and could transfer the ubiquitin to p53.[21] Removal of either the E1 or E2 enzymes prevented formation of the thiol-ester bond between ubiquitin and MDM2, and also prevented the transfer of ubiquitin to p53. Moreover, mutations that converted Cys464 to Ala464 (C464A) also inhibited the transfer of ubiquitin to p53. Taken together, these data support a model in which MDM2 functions as an E3 ubiquitin ligase that can accept activated ubiquitin from an E2 enzyme, and then transfer the ubiquitin to p53 (Fig. 3).

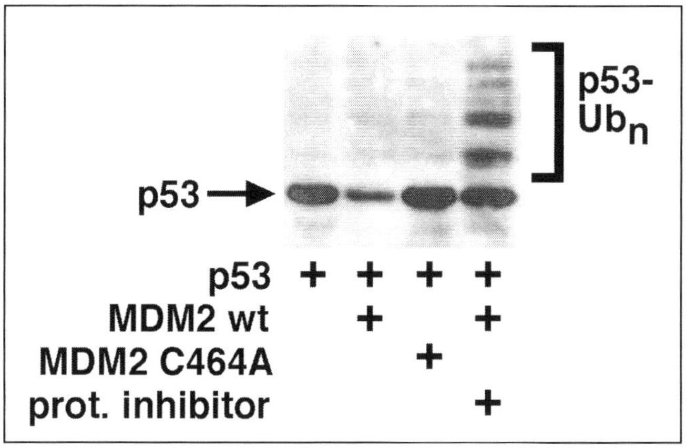

Figure 2. MDM2 promotes p53 ubiquitination and degradation. Saos-2 cells (p53 null) were transfected with DNAs encoding p53, wild-type MDM2, and a RING-finger mutant of MDM2 (C464A). P53 levels were then assessed by immunoblotting with an anti-p53 antibody. In some cases, cells were treated with a proteasome inhibitor for 6 hrs. prior to analysis. Wild-type MDM2 promotes p53 degradation while MDM2 C464A does not, indicating the RING-finger of MDM2 is required for p53 degradation. A ladder of high molecular weight p53 species is revealed when cells transfected with p53 and wild-type MDM2 are treated with the proteasome inhibitor. These bands are ubiquitinated forms of p53.

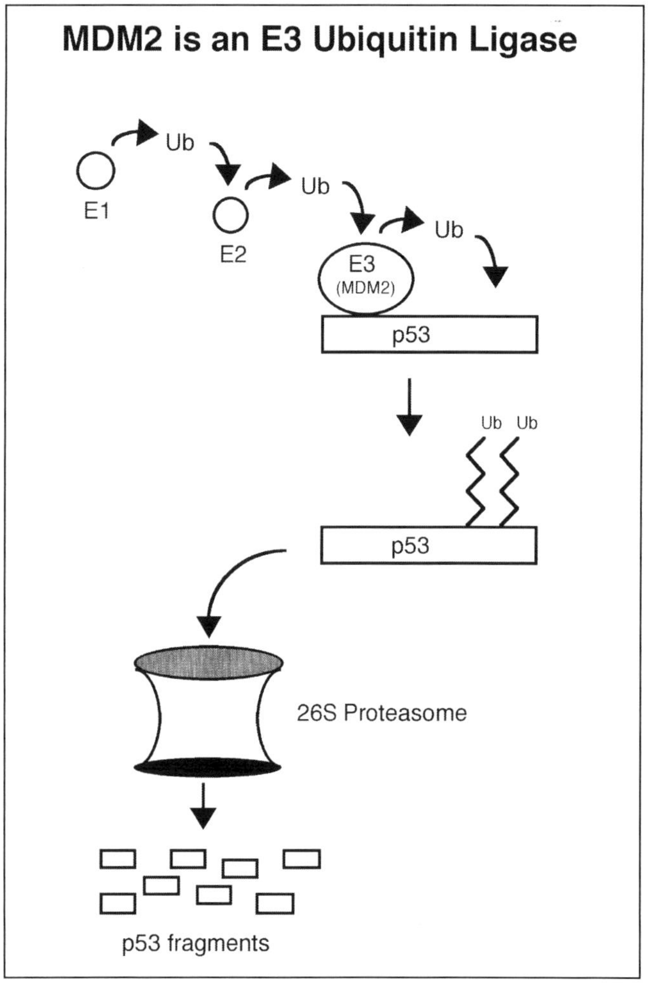

Figure 3. MDM2 is an E3 ubiquitin-ligase for p53. Current models suggest that MDM2 accepts activated ubiquitin from an E2 enzyme and transfers the ubiquitin to multiple lysine residues in the p53 C terminus.

The nature of p53 ubiquitination by MDM2 has not been fully clarified. MDM2 can stimulate the formation of a "ladder" of p53:ubiquitin conjugates in cells transiently expressing both proteins (see Fig. 2 as an example). This ladder could result from the addition of single ubiquitin molecules to multiple lysine residues in p53, or from the formation of poly-ubiquitin chains of varying lengths on one or only a few lysines. Mutational analysis of p53, as well as in vitro studies using ubiquitin-aldehyde, support the notion that MDM2 mono-ubiquitinates p53, transferring single ubiquitin moieties to multiple lysine residues located in the extreme C terminus of p53.[22,23] However, the fact that formation of poly-ubiquitin chains is necessary to target a protein to the proteasome[24] indicates that a so called "E4" enzyme must exist that can promote p53 poly-ubiquitination. Recent studies suggest p300 may be such a factor. P300 is a large (approx. 300 kilodalton) protein that can bind the central region of MDM2 (Fig. 1), as well as two separate regions of p53. While purified MDM2 can promote p53 mono-ubiquitination in an in vitro system, addition of p300 stimulates the poly-ubiquitination of p53,[25] probably through the formation of a p53:MDM2:p300 trimeric complex. These and other studies indicate that p300 can participate with MDM2 to promote p53 degradation by promoting p53 poly-ubiquitination[25-27] (Fig. 4, left). However, it is important to note that p300 can also function as an activator of p53 by promoting p53 acetylation at one or more lysine residues in the p53 C terminus (Fig 4, right). This acetylation increases the DNA binding activity of p53, thus leading to an activation of p53-responsive genes, and will be discussed in more detail below. Regulatory interactions between p53, MDM2, and p300 remain both interesting and complex.

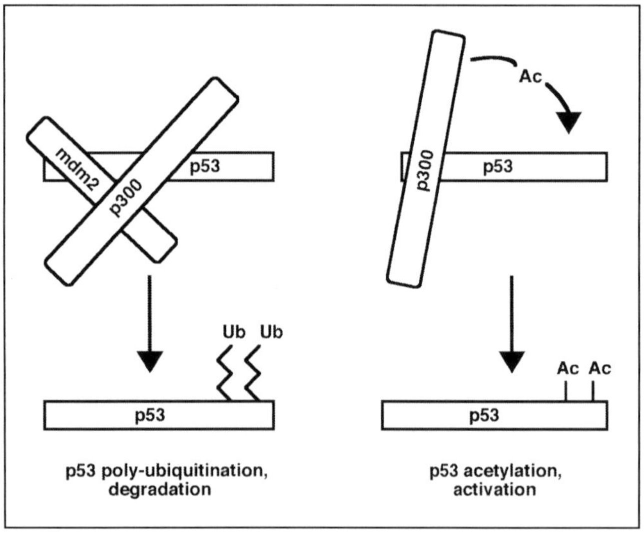

Figure 4. Dual roles for p300 in regulating p53 levels and activity. Left) MDM2 promotes the addition of single ubiquitin moieties to multiple lysine residues in p53. p300 is believed to act in conjunction with MDM2 to promote the poly-ubiquitination of p53, and the subsequent degradation of p53 by the proteasome. This process likely involves formation of a trimeric complex between p53, MDM2 and p300 in which p300 binds simultaneously to MDM2 and a central region of p53. Right) p300 can bind the N-teminus of p53 and promote the acetylation of lysines in the p53 C terminus. This acetylation increases p53 DNA binding activity, leading to an activation of p53 responsive genes.

P53 Is Stabilized and Its Levels Increase in Response to Stress

P53 levels increase in response to multiple genotoxic and nongenotoxic stresses, including DNA damage, hypoxia, ribonucleotide depletion, and the abnormal activation of oncogene signaling pathways, among others.[28-30] The effect of increasing p53 is to inhibit cell proliferation through either cell cycle arrest or apoptosis (Fig. 5). These effects are mediated by p53-responsive genes, such as *p21* and *bax*. The importance of signaling either a cell cycle arrest or apoptosis in response to these stresses is abundantly clear. DNA damaging stresses have the potential to damage chromosomes, alter genomic integrity, and cause mutations that may activate oncogenes and/or inhibit tumor suppressors. By triggering a cell cycle arrest or apoptosis in response to DNA damaging stress, p53 prevents the propagation of cells that may have an altered genome or acquired potentially cancer-promoting mutations. Similarly, the mutational activation of oncogenes is an early step in the development of most or all human cancers. By triggering an arrest or apoptosis in response to the mutational activation of oncogenes, p53 prevents the growth of cells that would otherwise be prone to becoming cancerous. It is worth noting that many chemotherapeutic drugs used in cancer treatment in some way modify DNA, resulting in the generation of single or double-stranded breaks. In cancers that retain wild-type p53, these DNA breaks trigger an increase in p53 levels, and a resulting activation of p53-induced cell death pathways.[31,32] In contrast, the same thing does not occur in cancers that express mutant p53 or in which the wild-type p53 gene has been deleted.[31,32] This explains the increased sensitivity to chemotherapeutics observed in cancers with wild-type p53 compared with cancers that lack wild-type p53 expression.

The increase in p53 levels following stress results from a stabilization of the p53 protein.[33,34] This is illustrated in Figure 6. RKO cells (human colon cancer cells that express wild-type p53) were untreated, or exposed to UV or ionizing (IR) radiation. Six hours after radiation exposure, the cells were treated with cyclohexamide (CHX) to inhibit de novo protein synthesis, and p53 steady state levels were monitored at various time points after CHX addition. The rate at which p53 levels decrease under these conditions is a measure of protein stability. In untreated cells, p53 levels decrease rapidly following CHX treatment, with a half-life ($T_{1/2}$) of approximately 30 min. In contrast, p53 levels

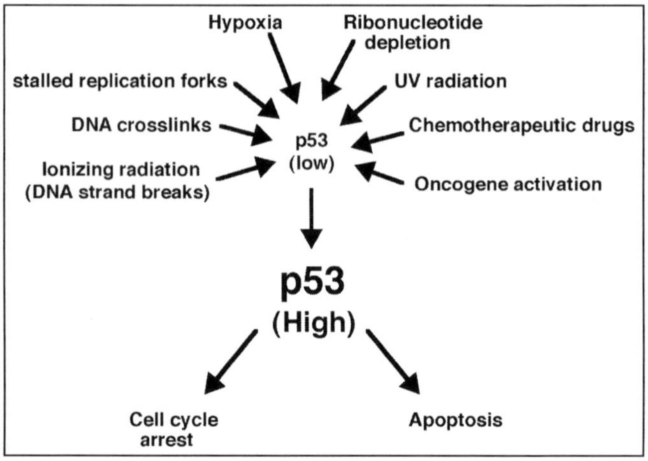

Figure 5. p53 levels increase in response to diverse stresses, resulting in cell cycle arrest or apoptosis.

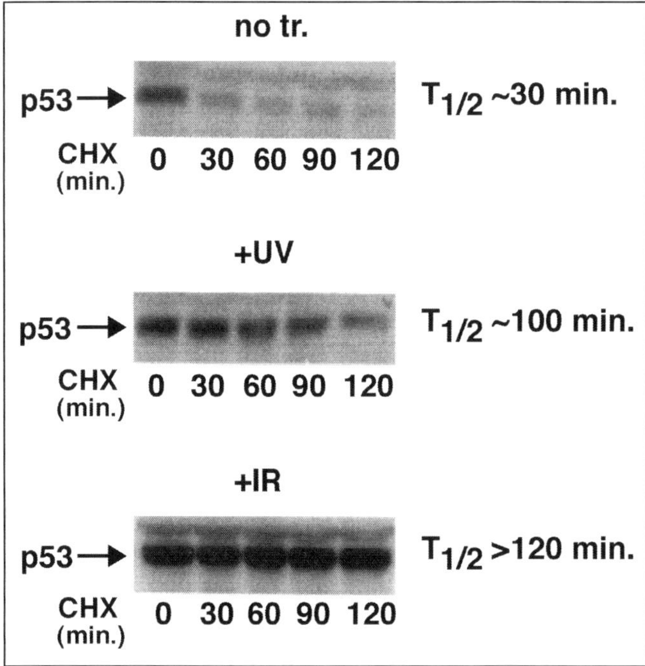

Figure 6. p53 is stabilized in response to DNA damaging stress. RKO cells (wild-type p53) were untreated or exposed to either ionizing radiation (IR) or UV radiation. 6 hrs after treatment, cells were treated with cyclohexamide (CHX) for the indicated amount of time, and p53 levels were monitored by immunoblotting. The estimated half-life (T1/2) of p53 is indicated. P53 levels increase and the p53 protein is stabilized following IR or UV treatment.

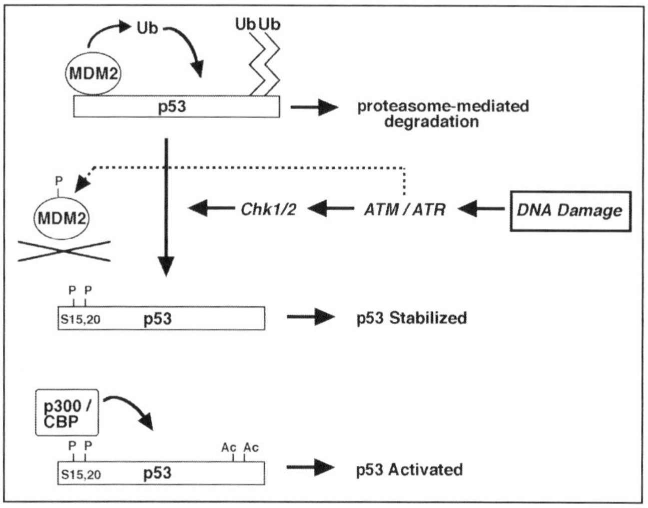

Figure 7. Mechanism for p53 stabilization in response to DNA damage. DNA damage activates two related kinases, ATM and ATR. These kinases can phosphorylate p53 directly at serines-15 and -37. ATM has been reported to also phosphorylate MDM2. ATM activates the kinase Chk2, and ATR activates the kinase Chk1. Activated Chk1 and Chk2 phosphorylate p53 at S20. Current models suggest that S20 phosphorylation stabilizes p53 by blocking the interaction between p53 and MDM2, whereas S15 phosphorylation activates p53 by recruiting p300 (or CBP), which acetylate the p53 C terminus.

are elevated in the UV and IR-treated cells and remain elevated for up to 2 hours. following CHX treatment. The half-life ($T_{1/2}$) of p53 is increased following UV or IR-treatment to about 100 minutes or greater than 2 hours, respectively. These results demonstrate that the stress-induced increase in p53 levels is due, at least in part, to an increase in p53 protein stability.

Stress Inhibits MDM2-Dependent p53 Degradation

Given that MDM2 promotes p53 ubiquitination and degradation, it is perhaps not surprising that the stress-induced stabilization of p53 results from an inhibition of these MDM2 activities.[34] Agents that damage or modify DNA, such as IR and UV radiation, DNA cross-linkers, and chemotherapeutic drugs, activate a signal transduction cascade that culminates in the phosphorylation of p53 at multiple sites, including sites within or near the N-terminal MDM2-binding domain (Fig. 7).[28-30,35] Phosphorylation of Ser15 and Ser20 in p53 is critical for the stabilization and activation of p53 in response to DNA damage,[36-38] and kinases that can phosphorylate p53 at these sites have been identified. One of these kinases, referred to as the ataxia telangiectasia mutated (ATM) kinase, is activated in response to DNA strand breaks, such as those that might occur in response to IR, and can directly phosphorylate p53 at Ser15 and other residues[28-30,35] (Fig. 7). A second kinase that is related to ATM and referred to as the ATM and Rad3-related (ATR) kinase, is activated in response to stalled replication forks, such as might occur in response to UV radiation. ATR can also phosphorylate p53 at Ser15. P53 phosphorylation at Ser20 requires ATM or ATR, but is not phosphorylated by either of these kinases directly. Instead, ATM and ATR activate other kinases referred to as Chk1 (activated by ATR) and Chk2 (activated by ATM) that can directly phosphorylate p53 at Ser20.[39] Most current models propose that Ser20 phosphorylation stabilizes p53 by blocking the interaction between p53 and MDM2, whereas the phosphorylated Ser15 residue is believed to activate p53 transcriptional activity by recruiting p300 or CBP.[40,41] As mentioned earlier, p300 activates p53 promoting its acetylation at C-terminal lysine residues. This acetylation is believed to increase the DNA binding affinity of p53, thus leading to an activation of p53 responsive genes. It is important to note that ATM can also phosphorylate MDM2 (Fig. 7), and this phosphorylation has been reported to diminish MDM2-mediated degradation and nuclear export of p53.[42,43]

The stabilization of p53 in response aberrant oncogene signaling occurs through a pathway different from that which stabilizes p53 following DNA damage.[44,45] In this case, activated oncogenes signal an increase in the transcription of a tumor suppressor protein referred to as either p14 or Arf (referred to as p19/Arf in the mouse). P14 forms a direct complex with MDM2 and inhibits the MDM2-mediated ubiquitination of p53 (Fig. 8). Most current models suggest that p14 inactivates MDM2 by sequestering it in the nucleolus,[44-47] though there was some suggestion that p14 binding with MDM2 in the nucleoplasm is enough to inhibit MDM2 activity. Regardless, the effect of inhibiting MDM2 is to stabilize p53 and thus increase p53 levels. The tumor suppressor PML has also been implicated in the oncogene-mediated activation of p53. PML is the major component in multi-protein sub-nuclear complexes termed PML-nuclear bodies (PML-NBs).[48-49] PML can interact with several different proteins and recruit them to PML-NBs, including p53[50] and p300/CBP.[51] Further, PML can stimulate p53 transcriptional activity in transiently transfected cells, and this requires recruitment of p53 to PML-NBs.[50] Studies from the laboratories of Pier Pandolfi at Memorial Sloan Kettering in New York and Scott

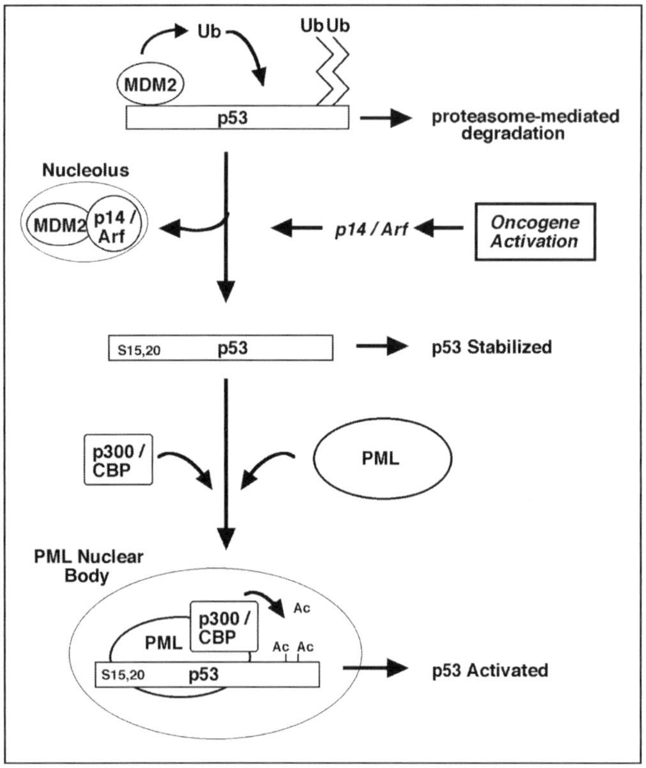

Figure 8. Oncogene-mediated activation of p53. Activated oncogenes stimulate expression of the tumor suppressor protein p14/Arf, which stabilizes p53 by sequestering MDM2 in the nuclelous. P53 may then be recruited to PML-nuclear bodies (PML-NBs) through its association with PML. P300/CBP in the nuclear body can activate p53 by promoting its acetylation.

Lowe at the Cold Spring Harbor Laboratories in New York have provided compelling evidence that the oncogene-mediated activation of p53 requires PML. In these studies, p53 activity was measured in $PML^{+/+}$ and $PML^{-/-}$ mouse embryo fibroblasts (MEFs) infected with retroviruses expressing an activated ras oncogene (Ras V12). Ras V12 expression in $PML^{+/+}$ cells resulted in the activation of p53, recruitment of p53 into PML-NBs, and the induction of premature senescence. In contrast, p53 was neither activated nor recruited into PML-NBs in $PML^{-/-}$ cells, and the cells were resistant to ras-induced senescence.[52,53] From these studies emerge a model in which p14/Arf stabilizes p53 in response to aberrant oncogene signaling by inhibiting the activity of MDM2, whereas the activation of p53 involves recruitment of p53 to PML-NBs, and its subsequent acetylation by p300/CBP[52-54] (Fig. 8).

MDM2 Promotes p53 Nuclear Export

In addition to promoting p53 degradation, MDM2 can also promote the export of p53 from the nucleus to the cytoplasm, and this is expected to further inhibit the transcriptional activity of p53. P53 contains two leucine-rich nuclear export signals (NESs), one located in the N-terminal MDM2-binding domain of p53, and the second located in the C terminus within the tetramerization domain.[55,56] Both of these NESs can support nuclear export when fused to a heterologous, nuclear protein. Disruption of the C-terminal NES causes p53 to display a more pronounced nuclear localization;[55] however, a similar finding has not been demonstrated for the N-terminal NES. Despite the identification of these NESs, the molecular mechanisms that control p53 nuclear export and the role of MDM2 in this process have remained a matter of debate. Initial studies found that endogenous wild-type p53 accumulated in the nucleus and was stabilized in cells exposed to leptomycin B (LMB), a specific inhibitor of nuclear export.[55,57] This suggested that nuclear export is required for p53 degradation. Further studies reported that MDM2 contains an NES within its sequence, and that blocking nuclear export could inhibit the ability of MDM2 to promote p53 degradation.[57-59] Based on these findings a model was proposed in which MDM2 transports p53 from the nucleus to the cytoplasm for p53 to be degraded, and this transport depends on a functional NES within MDM2. In conflict with this model were reports that p53 contains its own NES, and that this NES is sufficient to promote p53 nuclear export even in cells that lack MDM2 expression.[55] The NES identified in these studies was the C-terminal NES located within the p53 tetramerization domain, and structural analyses indicated that this NES would be inaccessible to the export machinery when p53 is a tetramer, but would be exposed and accessible when p53 is in either a dimeric or monomeric form. Taken together, these latter findings supported a second model in which p53 is exported from the nucleus via its own NES and independent of MDM2, and that this export may be regulated by changes in p53 tetramerization.

In an attempt to clarify the role of MDM2 in p53 nuclear export, cultured cells were transiently transfected with DNAs

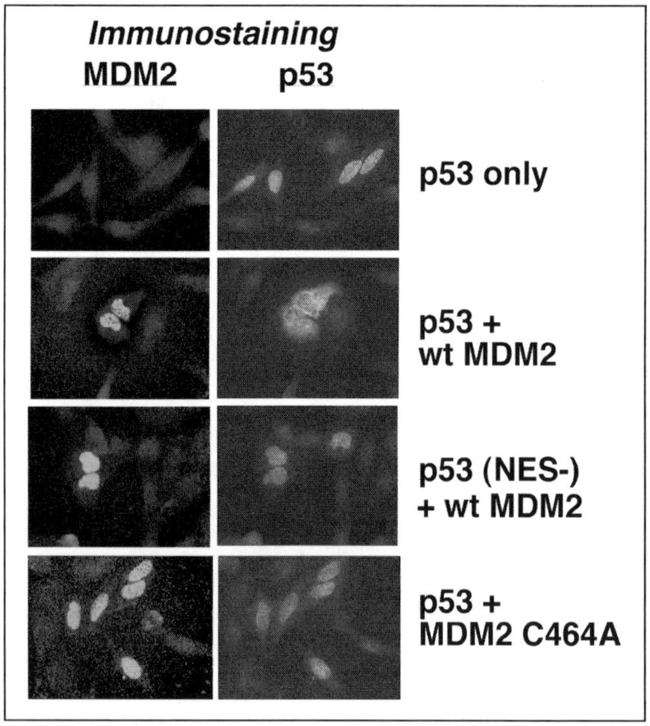

Figure 9. MDM2 promotes p53 nuclear export. Cells were transfected with DNAs encoding the indicated forms of p53 and MDM2. Localization of p53 and MDM2 was then monitored by immunofluorescence staining. Representative pictures are shown. Wild-type p53 localizes to the nucleus when expressed alone, but is exported to the cytoplasm when expressed with MDM2. This nuclear export of p53 requires the NES located in the p53 C terminus, and the RING finger domain of MDM2.

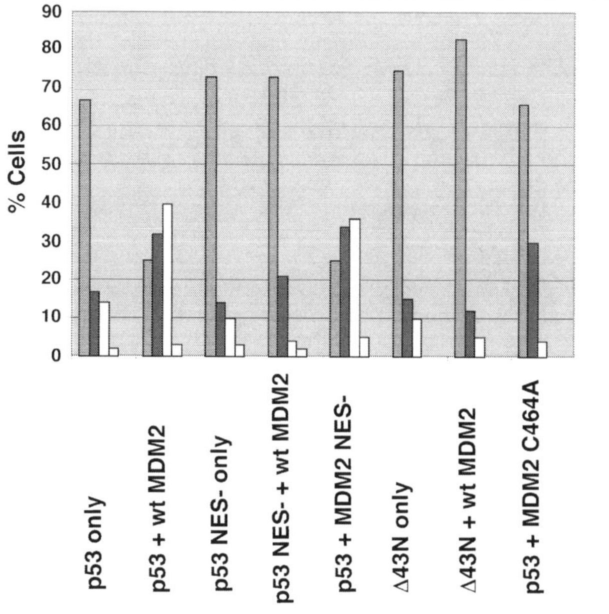

Figure 10. Quantification of p53 localization in transfected cells. The localization pattern for p53 in cells expressing the indicated p53 and MDM2 DNAs was scored for 150 cells in 3 separate experiments. Graph shows the percentage of cells with the indicated p53 localization patterns. p53 Δ42N lacks the MDM2-binding domain in the p53 N terminus. MDM2 exhibited complete nuclear localization in greater than 90% of transfected cells when expressed alone or with p53. Cells in which MDM2 did not exhibit complete nuclear localization were excluded from the analysis.

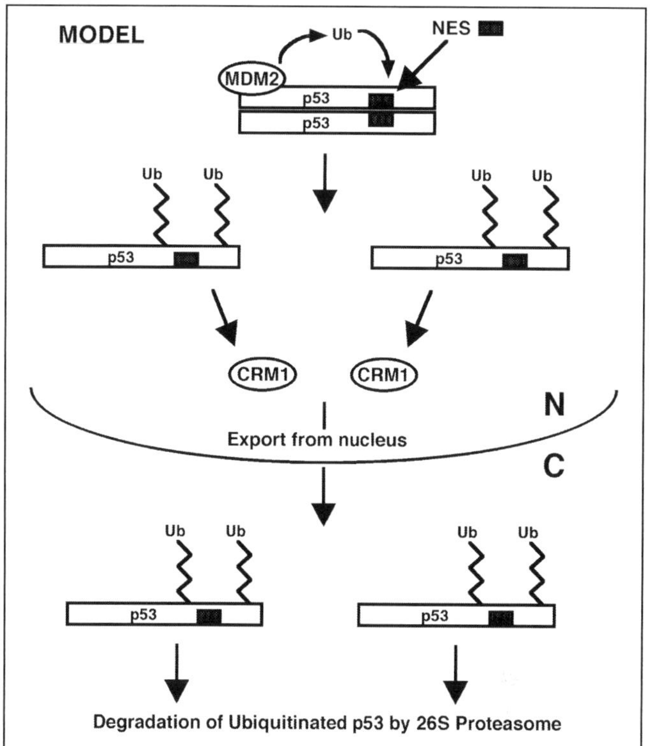

Figure 11. A model for MDM2-dependent nuclear export of p53. According to this model, MDM2 binds p53 in the nucleus and promotes p53 ubiquitination. Addition of ubiquitin moieties to the p53 C terminus are predicted to cause dissolution of p53 tetramers into either dimers or monomers, thus exposing the C-terminal NES. P53 is exported via binding of the nuclear export factor CRM1 to the NES.

encoding p53 and MDM2, and localization of p53 and MDM2 was then monitored by immunofluorescence staining.[60,61] As shown in Figures 9 and 10, p53 displayed a predominantly nuclear localization when expressed alone, but was relocalized to the cytoplasm in cells coexpressing MDM2. These results are consistent with the hypothesis that MDM2 can promote p53 nuclear export. Importantly, MDM2 remained in the nucleus despite the fact that p53 was relocalized to the cytoplasm, suggesting that nuclear MDM2 can promote p53 nuclear export. Mutants of p53 and MDM2 deficient in their nuclear export signals (NES^- mutants) were also tested in these experiments. The MDM2 NES^- mutant promoted the nuclear export of p53 to an extent similar to wild-type MDM2 (Fig. 10), indicating that the MDM2 NES was not required for this effect. In contrast, the p53 NES^- mutant was nuclear when expressed alone, and remained nuclear when coexpressed with wild-type MDM2 (Figs. 9 and 10). These results indicate that the MDM2-mediated nuclear export of p53 requires the NES of p53, but not the NES of MDM2. Finally, to assess whether the RING-finger domain of MDM2 has a function in p53 nuclear export, p53 localization was monitored in cells cotransfected with wild-type p53 and the MDM2 C464A mutant. As mentioned earlier, the C464A mutation in the MDM2 RING-finger domain inhibits the ability of MDM2 to promote p53 ubiquitination. As shown in Figures 9 and 10, MDM2 C464A was unable to promote the nuclear export of p53, indicating that an intact MDM2 RING finger domain is required for p53 nuclear export. It is important to note that the C464A mutation did not affect the ability of MDM2 to bind p53.[60,61] Taken together, these results support the current model for p53 nuclear export (Fig. 11). According to this model, MDM2 can bind p53 and promote its ubiqutination in the nucleus, and this leads to the export of p53 from the nucleus to the cytoplasm. This export requires the p53 NES located in p53s C terminus. Based on the structural studies mentioned earlier, it is proposed that the addition of ubiquitin exposes the C-terminal NES, probably by causing dissolution of p53 tetramers into p53 monomers or dimers. This allows recognition of the C-terminal NES by the export factor CRM1. This model was further supported by reports demonstrating that mutations in specific ubiquitination sites in p53's C terminus can inhibit the nuclear export of p53 that is mediated by MDM2.[62,63]

Several studies have supported the notion that nuclear export is an important component of p53 regulation. For example, various tumor types have been described in which p53 is functionally inactivated due to its abnormal sequestration in the cytoplasm, including inflammatory breast carcinoma, undifferentiated neuroblastoma, colorectal carcinoma, and retinoblastoma.[64-68] Sequestration of p53 in the cytoplasm is believed to contribute to the generation of these cancers by inhibiting p53 activity. In some

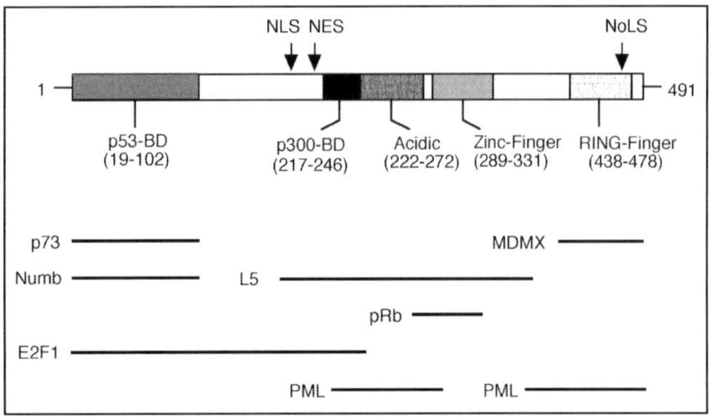

Figure 12. MDM2 interacts with multiple different proteins other than p53. The region of MDM2 involved in binding the indicated protein is shown.

cases, treatments that either block nuclear export (exposure to the nuclear export inhibitor leptomycin B), or inhibit endogenous MDM2 expression, result in the redistribution of p53 into the nucleus.[55,69] These findings suggest that the cytoplasmic localization of p53 in these cancers results, at least in part, from excessive nuclear export that is mediated by MDM2.

Interactions between MDM2 and Other Proteins

Notwithstanding its effects on p53, there is abundant evidence that MDM2 has p53-independent activities that may contribute to its oncogenic properties. This evidence includes the following: First, rare cancers bearing both p53 mutations and MDM2 gene amplifications have been described and are more aggressive than those with alterations of either gene alone.[70] This suggests that MDM2 over-expression may accelerate tumor growth even in the absence of functional p53. Second, MDM2 over-expression has been reported to transform cells in culture in the absence of a functioning p53.[71] Finally, targeted MDM2 over-expression in the mammary glands of mice leads to abnormal cell proliferation and mammary hypertrophy to a similar extent in both a *p53* $^{+/+}$ and *p53* $^{-/-}$ background.[72] Accordingly, considerable effort has been aimed at identifying cell-cycle regulatory proteins other than p53 with which MDM2 can interact. A number of such factors have been identified, including the p53-related protein p73, the MDM2-related protein MDMX, the tumor suppressor proteins pRB and PML, the ribosomal protein L5, the transcription factor E2F1, and the neuronal differentiation factor Numb, among others[73] (Fig. 12). In most cases, the effect of these interactions on cell growth has not been completed clarified. In the paragraphs below, we briefly discuss the interactions of MDM2 with these other proteins.

MDM2 Interaction with p73

P73 is a p53-related protein that has a high degree of amino acid sequence and structural similarity to p53.[74] P73 can induce apoptosis when over-expressed, and can activate transcription of p53-responsive genes.[75] These results indicate that p53 and p73 can carry out redundant functions. However, unlike p53, p73 mutations are not commonly observed in human cancers, and p73-knockout mice are not cancer prone. Instead, p73 knockout mice have neurological, pheromonal, and inflammatory defects.[76] MDM2 can bind the transactivation domain of p73 and inhibit its ability to activate transcription and induce apoptosis, but does not induce p73 degradation.[77-79] A recent report demonstrated that p73 and p63 (another p53-related protein) are required for p53-dependent apoptosis in response to DNA damage.[80] It is unclear at present whether MDM2 binding with p73 in some way inhibits the ability of p73 to cooperate with p53 in this DNA damage response.

MDM2 Interaction with pRb

The retinoblastoma protein pRb is the first characterized tumor suppressor. The Rb protein inhibits growth by sequestering the E2F family of transcription factors, thus inhibiting them from activating the expression of genes that promote G1 to S-phase progression. MDM2 binds to domains of pRb that are required for stable interaction with E2F1, suggesting that MDM2 could inhibit pRb tumor suppressor function by favoring detachment of E2F1 from pRb.[81,82] In contrast, pRb was reported to prevent MDM2-dependent degradation of p53 and inhibition of apoptosis in transient transfection studies.[83] Taken together, these findings suggest that pRb and MDM2 may antagonize each other's functions through binding one another.

MDM2 Interaction with E2F1

E2F1 is a transcription factor that functions as a dimer with DP1. Together, the E2F1/DP1 dimer can stimulate cell cycle progression by promoting the expression of genes required for the G1 to S-phase transition. In contrast, over-expression of E2F1 stimulates apoptosis, and E2F1 knockout mice are prone to develop certain cancers. These results indicate that E2F1 has both growth-promoting and growth-inhibitory capabilities. In the absence of p53, MDM2 inhibits the pro-apoptotic effect of E2F1/DP1 by favoring degradation of the heterodimer. At the same time, however, MDM2 can stimulate DNA synthesis in cooperation with E2F1/DP1. These seemingly contradictory results were reconciled by the proposal that high levels of E2F1/DP1 were reduced by MDM2 to levels appropriate for the G1-S-phase transition.[84]

MDM2 Interaction with MDMX

MDMX (also called MDM4) is an MDM2-related protein with a high degree of amino acid sequence and structural similarity with MDM2. MDMX can bind p53 but fails to promote p53 degradation.[85] Interestingly, knocking-out MDMX in the mouse resulted in early embryonic death, and this could be rescued by the concomitant knockout of p53,[86] similar to the results obtained with MDM2 knockout mice. These results suggest that the function of MDMX is to inhibit p53. MDM2 and MDMX can associate through their RING-finger domains. It is currently believed that MDMX-binding increases the ability of MDM2 to target p53 for degradation, through either stabilizing the MDM2 protein or modulating MDM2s localization.[87]

MDM2 Interaction with Ribosomal Protein L5

MDM2 forms a complex with ribosomal protein L5 and its associated 5S rRNA.[88,89] These interactions have gained interest in recent years in light of the findings that MDM2 contains a nuclear export signal and undergoes active nuclear export. Specifically, it is speculated that MDM2 nuclear export may be linked to the nuclear export of ribosome subunits following their assembly in the nucleolus, and that this may be facilitated by the interaction between MDM2 and L5.

MDM2 Interaction with Numb

Numb is a regulator of neuronal differentiation that can interact with the N terminus of MDM2. Interestingly, Numb levels are decreased in the presence of MDM2, suggesting that MDM2 may target Numb for degradation.[90] An effect of MDM2 on neuronal differentiation has not been described.

MDM2 Interaction with PML

As mentioned earlier, PML is a tumor suppressor that can activate p53 by recruiting it to multi-protein nuclear complexes termed PML-NBs.[50] MDM2 colocalized with PML in PML-NBs in cells that lacked p53 expression, and purified MDM2 protein was shown to interact with a GST-PML fusion protein generated in bacteria.[91] These results indicate that MDM2 can interact with PML in a p53-independent manner. Given that PML can activate p53, it was speculated that the MDM2:PML interaction could indirectly affect the activity of p53. However, experiments to test this possibility have not yet been carried out.

Conclusions

MDM2 is an oncoprotein that is frequently overexpressed in various human cancers. The primary function of MDM2 is to inhibit the tumor suppressor protein p53. MDM2 promotes the ubiquitination of p53 and its degradation by the proteasome, as well as the export of p53 from the nucleus to the cytoplasm. Both of these activities require the RING finger domain located in the MDM2 C terminus. DNA damage and other stresses stabilize p53 through an inhibition of MDM2-mediated degradation. MDM2 interacts with a variety of other proteins in addition to p53, and these interactions may contribute to the oncogenic potential of MDM2.

References

1. Landers JE, Cassel SL, George DL. Translational enhancement of mdm2 oncogene expression in human tumor cells containing a stabilized wild-type p53 protein. Cancer Res 1997; 57:3562-3568.
2. Cahilly-Snyder L, Yang-Feng T, Francke U et al. Molecular analysis and chromosomal mapping of amplified genes isolated from a transformed mouse 3T3 cell line. Somat Cell Molec Genet 1987; 13:235-244.
3. Momand J, Zambetti GP, Olson DC et al. The mdm-2 oncogene product forms a complex with the p53 protein and inhibits p53-mediated transactivation Cell 1992; 69:1237-1245.
4. Oliner JD, Pietenpol JA, Thiagalingam S et al. Oncoprotein MDM2 conceals the activation domain of tumour suppressor p53. Nature 1993; 362:857-860.
5. Momand J, Wu H, Dasgupta G. MDM2-master regulator of the p53 tumor suppressor protein. Gene 2000; 242:15-29.
6. Deb SP. Function and dysfunction of the human oncoprotein MDM2. Front Biosci 2002; 7:235-243.
7. Jones SN, Roe AE, Donehower LA et al. Rescue of embryonic lethality in Mdm2-deficient mice by absence of p53. Nature 1995; 378:206-208.
8. Montes de Oca Luna R, Wagner DS, Lozano G. Rescue of early embryonic lethality in mdm2-deficient mice by deletion of p53. Nature 1995; 378:203-206.
9. Perry ME, Mendrysa SM, Saucedo LJ et al. p76(MDM2) inhibits the ability of p90(MDM2) to destabilize p53. J Biol Chem 2000; 275:5733-5738.
10. Evans SC, Viswanathan M, Grier JD et al. An alternatively spliced HDM2 product increases p53 activity by inhibiting HDM2. Oncogene 2001; 20:4041-4049.
11. Deshaies RJ. SCF and cullin/RING H2-based ubiquitin ligases. Annu Rev Cell Dev Biol 1999; 15:435-467.
12. Tyers M, Jorgensen P. Proteolysis and the cell cycle: With this RING I do thee destroy. Curr Opin Genet Dev 2000; 10:54-64.
13. Jackson PK, Eldridge AG, Freed E et al. The lore of the RINGs: Substrate recognition and catalysis by ubiquitin ligases. Trends Cell Biol 2000; 10:429-439.
14. Borden KLB. RING domains: Master builders of molecular scaffolds. J Mol Biol 2000; 295:1103-1112.
15. Lorick KL, Jensen JP, Fang S et al. RING fingers mediate ubiquitin-conjugating enzyme (E2)-dependent ubiquitination. Proc Natl Acad Sci USA 1999; 96:11364-11369.
16. Joazeiro CA, Weissman AM. RING finger proteins: Mediators of ubiquitin ligase activity. Cell 2000; 102:549-552.
17. Haupt Y, Maya R, Kazaz A et al. MDM2 promotes the rapid degradation of p53. Nature 1997; 387:296-299.
18. Kubbutat MH, Jones SN, Vousden KH. Regulation of p53 stability by MDM2. Nature 1997; 387:299-303.
19. Hershko A, Ciechanover A. The ubiquitin system. Annu Rev Biochem 1998; 67:425-479.
20. Ciechanover A, Schwartz AL. Ubiquitin-mediated degradation of cellular proteins in health and disease. Hepatology 2002; 35:3-6.
21. Honda R, Tanaka H, and Yasuda H. Oncoprotein MDM2 is a ubiquitin ligase E3 for tumor suppressor p53. FEBS Lett 1997; 420:25-27.
22. Rodriguez MS, Desterro JM, Lain S et al. Multiple C-terminal lysine residues target p53 for ubiquitin-proteasome-mediated degradation. Mol Cell Biol 2000; 20:8458-8467.
23. Lai Z, Ferry KV, Diamond MA et al. Human mdm2 mediates multiple mono-ubiquitination of p53 by a mechanism requiring enzyme isomerization. J Biol Chem 2001; 27:357-367.
24. Deveraux Q, Ustrell V, Pickart C et al. A 26S protease subunit that binds ubiquitin conjugates. J Biol Chem 1994; 269:7059-7061.
25. Grossman SR, Deato ME, Brignone C et al. Polyubiquitination of p53 by a ubiquitin ligase activity of p300. Science 2003; 300:342-344.
26. Grossman SR, Perez M, Kung AL et al. p300/MDM2 complexes participate in MDM2-mediated p53 degradation. Mol Cell 1998; 2:405-415.
27. Grossman SR. p300/CBP/p53 interaction and regulation of the p53 response. Eur J Biochem 2001; 268:2773-2778.
28. Prives C, Hall PA. The p53 pathway. J Pathol 1999; 187:112-126.
29. Ryan KM, Phillips AC, Vousden KH. Regulation and function of the p53 tumor suppressor protein. Curr Opin Cell Biol 2001; 13:332-337.
30. Woods DB, Vousden KH. Regulation of p53 function. Exp Cell Res 2001; 264:56-66.
31. Lowe SW. Cancer therapy and p53. Curr Opin Oncol 1995; 7:547-553.
32. Schmitt CA, Fridman JS, Yang M et al. A senescence program controlled by p53 and p16INK4a contributes to the outcome of cancer therapy. Cell 2002; 109:335-346.
33. Maltzman W, Czyzyk L. UV irradiation stimulates levels of p53 cellular tumor antigen in nontransformed mouse cells. Mol Cell Biol 1984; 4:1689-1694.
34. Maki CG, Howley PM. Ubiquitination of p53 and p21 is differentially affected by ionizing and UV radiation. Mol Cell Biol 1997; 17:355-363.
35. Giaccia AJ, Kastan MB. The complexity of p53 modulation: Emerging patterns from divergent signals. Genes Dev 1998; 12:2973-2983.
36. Shieh SY, Ikeda M, Taya Y et al. DNA damage-induced phosphorylation of p53 alleviates inhibition by MDM2. Cell 1997; 91:325-334.
37. Chehab NH, Malikzay A, Stavridi ES et al. Phosphorylation of ser-20 mediates stabilization of human p53 in response to DNA damage. Proc Natl Acad Sci USA 1999; 96:13777-13782.
38. Unger T, Juven-Gershon T, Moallem E et al. Critical role for ser20 of p53 in the negative regulation of p53 by MDM2. EMBO J 1999; 18:1805-1814.
39. Chehab NH, Malikzay A, Appel M et al. Chk2/hCds1 functions as a DNA damage checkpoint in G(1) by stabilizing p53. Genes Dev 2000; 14:278-288.
40. Dumaz N, Meek DW. Serine 15 phosphorylation stimulates p53 transactivation but does not directly influence interaction with HDM2. EMBO J 1999; 18:7002-7010.

41. Ito A, Lai CH, Zhao X et al. p300/CBP-mediated p53 acetylation is commonly induced by p53-activating agents and inhibited by MDM2. EMBO J 2001; 20:1331-1340.
42. Khosravi R, Maya R, Gottlieb T et al. Rapid ATM-dependent phosphorylation of MDM2 precedes p53 accumulation in response to DNA damage. Proc Natl Acad Sci USA 1999; 96:14973-14977.
43. Maya R, Balass M, Kim ST et al. ATM-dependent phosphorylation of Mdm2 on serine 395: Role in p53 activation by DNA damage. Genes Dev 2001; 15:1067-1077.
44. Lowe SW. Activation of p53 by oncogenes. Endocr Relat Cancer 1999; 6:45-48.
45. Lowe SW, Sherr CJ. Tumor suppression by Ink4a-Arf: Progress and puzzles. Curr Opin Genet Dev 2003; 13:77-83.
46. Weber JD, Taylor LJ, Roussel MF et al. Nucleolar Arf sequesters Mdm2 and activates p53. Nat Cell Biol 1999; 1:20-26.
47. Zhang Y, Xiong Y. Mutations in human ARF exon 2 disrupt its nucleolar localization and impair its ability to block nuclear export of MDM2 and p53. Mol Cell 1999; 3:579-591.
48. Borden KL. Pondering the promyelocytic leukemia protein (PML) puzzle: Possible functions for PML nuclear bodies. Mol Cell Biol 2002; 22:5259-5269.
49. Salomoni P, Pandolfi PP. The role of PML in tumor suppression. Cell 2002; 108:165-170.
50. Fogal V, Gostissa M, Sandy P et al. Regulation of p53 activity in nuclear bodies by a specific PML isoform. EMBO J 2000; 19:6185-6195.
51. Doucas V, Tini M, Egan DA et al. Modulation of CREB binding protein function by the promyelocytic (PML) oncoprotein suggests a role for nuclear bodies in hormone signaling. Proc Natl Acad Sci USA 1999; 96:2627-2632.
52. Ferbeyre G, de Stanchina E, Querido E et al. PML is induced by oncogenic ras and promotes premature senescence. Genes Dev 2000; 14:2015-2027.
53. Pearson M, Carbone R, Sebastiani C et al. PML regulates p53 acetylation and premature senescence induced by oncogenic ras. Nature 2000; 406:207-210.
54. Pearson M, Pelicci PG. PML interaction with p53 and its role in apoptosis and replicative senescence. Oncogene 2001; 20:7250-7256.
55. Stommel JM, Marchenko ND, Jimenez GS et al. A leucine-rich nuclear export signal in the p53 tetramerization domain: Regulation of subcellular localization and p53 activity by NES masking. EMBO J 1999; 18:1660-1672.
56. Zhang Y, Xiong Y. A p53 amino-terminal nuclear export signal inhibited by DNA damage-induced phosphorylation. Science 2001; 292:1910-1915.
57. Freedman DA, Levine AJ. Nuclear export is required for degradation of endogenous p53 by MDM2 and human papillomavirus E6. Mol Cell Biol 1998; 18:7288-7293.
58. Roth J, Dobbelstein M, Freedman DA et al. Nucleo-cytoplasmic shuttling of the hdm2 oncoprotein regulates the levels of the p53 protein via a pathway used by the human immunodeficiency virus rev protein. EMBO J 1998; 17:554-564.
59. Tao W, Levine AJ. Nucleocytoplasmic shuttling of oncoprotein Hdm2 is required for Hdm2-mediated degradation of p53. Proc Natl Acad Sci USA 1999; 96:3077-3080.
60. Boyd SD, Tsai KY, and Jacks T. An intact HDM2 RING-finger domain is required for nuclear exclusion of p53. Nature Cell Biol 2000; 2:563-568.
61. Geyer RK, Yu ZK, Maki CG. The MDM2 RING-finger domain is required to promote p53 nuclear export. Nature Cell Biol 2000; 2:569-573.
62. Gu J, Nie L, Wiederschain D et al. Identification of p53 sequence elements that are required for MDM2-mediated nuclear export. Mol Cell Biol 2001; 21:8533-8546.
63. Lohrum MA, Woods DB, Ludwig RL et al. C-terminal ubiquitination of p53 contributes to nuclear export. Mol Cell Biol 2001; 21:8521-8532.
64. Moll UM, Riou G, Levine AJ. Two distinct mechanisms alter p53 in breast cancer: Mutation and nuclear exclusion. Proc Natl Acad Sci USA 1992; 89:7262-7266.
65. Sun XF, Carstensen JM, Zhang H et al. Prognostic significance of cytoplasmic p53 oncoprotein in colorectal adenocarcinoma. Lancet 1992; 340:1369-1373.
66. Bosari S, Viale G, Roncalli M et al. p53 gene mutations, p53 protein accumulation and compartmentalization in colorectal adenocarcinoma. Am J Pathol 1995; 147:790-798.
67. Ueda H, Ullrich SJ, Gangemi JD et al. Functional inactivation but not structural mutation of p53 causes liver cancer. Nat Genet 1995; 9:41-47.
68. Schlamp CL, Poulsen GL, Nork TM et al. Nuclear exclusion of wild-type p53 in immortalized human retinoblastoma cells. J Natl Cancer Inst 1997; 89:1530-1536.
69. Rodriguez-Lopez AM, Xenaki D, Eden TOB et al. MDM2 mediated nuclear exclusion of p53 attenuates etoposide-induced apoptosis in neuroblastoma cells. Mol Pharmocol 2000; 59:135-143.
70. Cordon-Cardo C, Latres E, Drobnjak M et al. Molecular abnormalities of mdm2 and p53 genes in adult soft tissue sarcomas. Cancer Res 1994; 54:794-799.
71. Dubs-Poterszman MC, Tocque B, Wasylyk B. MDM2 transformation in the absence of p53 and abrogation of the p107 G1 cell-cycle arrest. Oncogene 1995; 11:2445-2449.
72. Lundgren K, Montes de Oca Luna R, McNeill YB et al. Targeted expression of MDM2 uncouples S phase from mitosis and inhibits mammary gland development independent of p53. Genes Dev 1997; 11:714-725.
73. Daujat S, Neel H, Piette J. MDM2: Life without p53. Trends Genet 2001; 17:459-464.
74. Davis PK, Dowdy SF. p73. Int J Biochem Cell Biol 2001; 33:935-939.
75. Jost CA, Marin MC, Kaelin WG Jr. p73 is a simian p53-related protein that can induce apoptosis. Nature 1997; 389:122-123.
76. Yang A, Walker N, Bronson R et al. p73-deficient mice have neurological, pheromonal and inflammatory defects but lack spontaneous tumours. Nature 2000; 404:99-103.
77. Zeng X, Chen L, Jost CA et al. MDM2 suppresses p73 function without promoting p73 degradation. Mol Cell Biol 1999; 19:3257-3266.
78. Gu J, Chen D, Rosenblum J et al. Identification of a sequence element from p53 that signals for Mdm2-targeted degradation. Mol Cell Biol 2000; 20:1243-1253.
79. Dobbelstein M, Wienzek S, Konig C et al. Inactivation of the p53-homologue p73 by the mdm2-oncoprotein. Oncogene 1999; 18:2101-2106.
80. Flores ER, Tsai KY, Crowley D et al. p63 and p73 are required for p53-dependent apoptosis in response to DNA damage. Nature 2002; 416:560-564.
81. Xiao ZX, Chen J, Levine AJ et al. Interaction between the retinoblastoma protein and the oncoprotein MDM2. Nature 1995; 375:694-698.
82. Martin K, Trouche D, Hagemeier C et al. Stimulation of E2F1/ DP1 transcriptional activity by MDM2 oncoprotein. Nature 1995; 375:691-694.
83. Hsieh JK, Chan FS, O'Connor DJ et al. RB regulates the stability and the apoptotic function of p53 via MDM2. Mol Cell 1999; 3:181-193.
84. Loughran O, Thangue NB. Apoptotic and growth-promoting activity of E2F modulated by MDM2. Mol Cell Biol 2000; 20:2186-2197.
85. Jackson MW, Berberich SJ. MdmX protects p53 from Mdm2-mediated degradation. Mol Cell Biol 2000; 20:1001-1007.
86. Parant J, Chavez-Reyes A, Little NA et al. Rescue of embryonic lethality in Mdm4-null mice by loss of Trp53 suggests a nonoverlapping pathway with MDM2 to regulate p53. Nat Genet 2001; 29:92-95.
87. Gu J, Kawai H, Nie L et al. Mutual dependence of MDM2 and MDMX in their functional inactivation of p53. J Biol Chem 2002; 277:19251-19254.
88. Marechal V, Ellenbaas B, Piette J et al. The ribosomal L5 protein is associated with mdm-2 and mdm-2-p53 complexes. Mol Cell Biol 1994; 14:7414-7420.
89. Elenbaas B, Dobbelstein M, Roth J et al. The MDM2 oncoprotein binds specifically to RNA through its RING finger domain. Mol Med 1996; 2:439-451
90. Juven-Gershon T, Shifman O, Unger T et al. The Mdm2 oncoprotein interacts with the cell fate regulator Numb. Mol Cell Biol 1998; 18:3974-3982.
91. Wei X, Yu ZK, Ramalingam A et al. Physical and functional interactions between PML and MDM2. J Biol Chem March 2003 (Epub ahead of publication).

CHAPTER 35

The Zip Family of Zinc Transporters

David J. Eide

Abstract

The ZIP family of transporters plays important roles in supplying zinc to metalloproteins. These transporters are found in organisms at all phylogenetic levels including bacteria, fungi, plants, insects, and mammals. They have many conserved sequence elements and most have eight likely transmembrane domains with similar predicted topologies. Intracellular zinc homeostasis is facilitated by both transcriptional and post-transcriptional control mechanisms that regulate ZIP transporter activity. These systems work in concert to provide a constant supply of zinc in the face of changing extracellular levels and limit overaccumulation of this potentially toxic metal.

Introduction

Charged ions like Zn^{2+} do not freely diffuse across lipid bilayer membranes. Transporter proteins are required to ferry Zn^{2+} into and out of cells and into and out of intracellular compartments. One group of proteins responsible for zinc uptake in many different types of cells is the ZIP family of metal ion transporters.[1-3] ZIP transporters play an important role in supplying Zn^{2+} to metallate the various intracellular zinc binding sites that have been described throughout this book.

The "ZIP" designation stands for Zrt-, Irt-like Protein and reflects the first members of this transporter family to be identified. Zrt1 and Zrt2 are the primary zinc uptake transporters in the yeast *Saccharomyces cerevisiae* and Irt1 is the major iron uptake transporter in roots of *Arabidopsis thaliana* but has also been shown to transport zinc.[4-7] Since the initial identification of these proteins, the ZIP family has grown to over 90 members including proteins in bacteria, nematodes, insects, and mammals.[1] Among these, ZupT is a zinc uptake transporter in *Escherichia coli*[8] and several ZIPs have been implicated in zinc, iron, and/or manganese transport in plants (see below).[9-12] Database analyses have indicated that there are 14 ZIP proteins encoded by the human genome. Human genes encoding ZIP proteins are referred to by their Human Genome Organization Nomenclature Committee designation of "*solute carrier family 39*" (or *SLC39*).

Functional Characteristics

With no exceptions yet known, ZIPs transport metal ion substrates across cellular membranes into the cytoplasm. While many members are involved in the uptake of metal ions across the plasma membrane, some efflux metals from intracellular compartments. For example, the yeast Zrt3 protein is responsible for transporting stored zinc out of the lysosome-like vacuole into the cytoplasm for subsequent utilization.[13] Also in yeast, the Atx2 protein appears to transport manganese out of the Golgi.[14]

The biochemical mechanisms involved in these transport events have not been extensively investigated. In yeast, transport by Zrt1 and Zrt2 is known to be energy dependent.[4,5] In contrast, zinc uptake by the human hZip1 and hZip2 proteins was found to be energy independent.[15,16] Zinc uptake by these proteins was not dependent on K^+ or Na^+ gradients but hZip2 activity was stimulated by HCO_3^- suggesting a Zn^{2+}/HCO_3^- symport mechanism.[15] Alternatively, zinc uptake by these proteins could be driven simply by the concentration gradient of labile zinc likely to exist across the plasma membrane.[17] It is also unclear if these proteins are channels or carriers of Zn^{2+}. Their high affinity for substrate (see below) provides some support for the carrier model.

Biochemical and Structural Aspects of Zip Transporters

ZIP transporters range in size from ~300 to over 600 amino acids in length. The marked differences in lengths largely come from differences in the length of the N-terminal region before the first transmembrane domain. Most ZIP proteins have eight transmembrane domains and similar predicted topologies (Fig. 1). Aspects of this topology model have been confirmed for yeast[18,19] and some mammalian ZIP transporters.[15,16] While most loops between transmembrane domains are quite short, a longer loop region is often found between transmembrane domains III and IV. This region frequently contains a histidine-rich domain with the sequence $(HX)_n$ where n generally ranges from 3-5. Although conserved, the function of this domain is not yet clear. Mutation of this motif in the yeast Zrt1 protein resulted in slightly reduced levels of the protein accumulating on the surface of the cell but did not alter zinc uptake affinity or activity when normalized to surface protein levels.[20]

Determinants of substrate specificity have been mapped to the extracellular loop between TM II and III in Irt1.[21] For some ZIP proteins (for example the mammalian Liv-1 protein), this region is also rich in histidine residues.[22] Finally, some ZIP proteins (for example hZip4) have long N-terminal domains that are rich in potential metal-binding residues. These domains may play a role in "harvesting" the substrate from the extracellular environment prior to transport. Transmembrane domains IV and V are particularly amphipathic and contain conserved histidine residues frequently with adjacent polar or charged amino acids. Given their sequence conservation and amphipathic nature, TM IV and V are predicted to line a cavity in the transporter through which

Zinc Finger Proteins: From Atomic Contact to Cellular Function, edited by Shiro Iuchi and Natalie Kuldell.

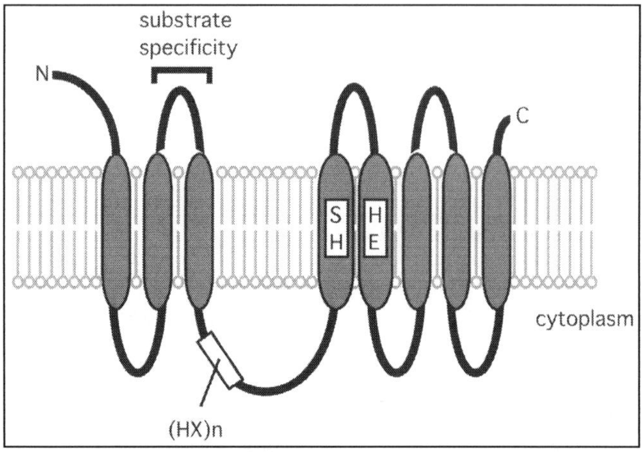

Figure 1. Topological model of ZIP proteins. Transmembrane domains are indicated in dark gray. The histidine-rich region in the cytoplasmic loop between TM III and IV, conserved histidines and charged/polar residues in TM IV and V, and the location of substrate specificity determinants mapped in Irt1 are also shown.

the substrate passes. Consistent with this hypothesis, conserved residues in these regions are essential for function of at least one ZIP transporter, Irt1.[21]

Function of Zip Proteins in Zinc Transport

As describe above, many ZIP proteins have been implicated in zinc transport. In *E. coli*, the ZupT transporter is a low affinity zinc uptake system.[8] This protein, along with the high affinity ZnuABC permease, contribute to zinc accumulation. In yeast, Zrt1 and Zrt2 are zinc uptake transporters. These proteins have high affinities for their substrate; Zrt1 has an apparent K_m of ~10 nM while Zrt2's apparent K_m is about 10-fold higher.[4,5] These two proteins provide most of the intracellular zinc used by the cell during growth. This is evident when comparing the amount of zinc required for growth of wild type cells vs. a *zrt1 zrt2* mutant. The mutant requires approximately 10^5 more zinc to grow at its optimal rate than do wild type cells. In contrast, the Zrt3 protein localizes to the yeast vacuole and is responsible for mobilization of zinc stores held in that compartment. Zinc-replete cells can store high levels of zinc in the vacuole. Under zinc-limiting conditions, Zrt3 pumps that zinc into the cytosol where it is available for use.[13] Mutants defective in Zrt3 accumulate high levels (that is mM) of zinc in their vacuoles and fail to use that zinc in deficient conditions.

In *Arabidopsis*, the Zip1, Zip2, Zip3, and Zip4 proteins have been implicated in zinc transport.[9] These ZIPs were isolated because of their ability to function in yeast as zinc transporters and complement the growth defect of a *zrt1 zrt2* mutant on low zinc media. Consistent with their role in zinc transport, *ZIP1* and *ZIP3* mRNA levels are induced in zinc deficient cells. A fourth *Arabidopsis* ZIP protein, Zip4, was not functional when expressed in yeast but was also implicated in zinc transport because its mRNA levels also rise in zinc-limited plants. In *Thlaspi caerulescens*, dysregulation of a ZIP transporter may contribute to the high levels of zinc taken up by this hyperaccumulating plant.[23]

In humans, *SLC39A1*, encoding hZip1/ZIRTL, is expressed in a wide variety of tissues and cell types.[16,24] In transfected K562 erythroleukemia cells, hZip1 is localized to the plasma membrane where it confers zinc uptake activity.[16] In other cell types (e.g., COS-7, PC3), hZip1 is localized predominantly to the endoplasmic reticulum.[25] Thus, the site of action of this transporter in different cell types is unresolved. Antisense RNA studies have indicated that hZip1 is responsible for most of the zinc uptake activity detectable in K562 cells suggesting a role in zinc uptake in other cells as well.[16] *SLC39A2*, encoding hZip2, is expressed at low levels and in only a few tissues.[15,26,27] Like hZip1, hZip2 also appears to be involved in zinc uptake. hZip2 localizes to the plasma membrane of transfected K562 cells and confers zinc uptake activity on those cells.[15]

hZip4, the product of the *SLC39A4* gene, is an important zinc uptake transporter in the human intestine and responsible for absorption of dietary zinc. This conclusion is based on the work of two groups linking *SLC39A4* to acrodermatitis enteropathica (AE), a recessive genetic disorder of zinc absorption.[28-30] Patients with AE have reduced levels of zinc absorption that results in a wide range of developmental and growth effects. Homozygosity mapping using consanguineous families with affected individuals indicated the gene responsible for AE mapped to 8q24.3.[28] Located in this region was a likely candidate gene, *SLC39A4*. Sequencing of exons from affected individuals indicated the presence of mutations that were likely to disrupt function of the protein.[29,30] These mutations were not found in unaffected samples. Moreover, the *SLC39A4* gene is expressed in tissues involved in zinc absorption, the small intestine, colon, and kidney, and the hZip4 protein was localized to the apical plasma membrane of mouse enterocytes.[29] These results strongly suggest an important role of hZip4 in dietary zinc uptake and reabsorption of zinc in the kidney. Approximately 10% of the endogenous zinc lost per day in humans is in urine and this loss is reduced under conditions of zinc deficiency suggesting homeostatic regulation of kidney zinc reabsorption. Based on EST database entries, two forms of hZip4 may be expressed that have different amino-termini. The more commonly expressed form has a 329 amino acid segment upstream of TM1 including a signal sequence. The alternative form appears to be expressed at lower levels and has an N-terminal tail of 305 amino acids with no signal sequence. This latter form may arise from transcription initiation within intron 1 of the longer transcript.[29]

Regulation of Zip Proteins to Control Zinc Acquisition and Homeostasis

Regulation of ZIP transporter activity in response to zinc status occurs at both transcriptional and post-transcriptional levels. The integration of these different levels of regulation is perhaps best understood in the yeast *S. cerevisiae*. Zinc uptake in *S. cerevisiae* is controlled at the transcriptional level in response to intracellular zinc levels. The high affinity system is induced more than 30-fold in zinc-limited cells resulting from increased transcription of the *ZRT1* gene.[5] The low affinity system is also regulated through the control of *ZRT2* transcription.[31] Regulation of these genes in response to zinc is mediated by the product of the *ZAP1* gene.[31] *ZAP1* encodes a transcriptional activator that is maximally active in zinc-deficient cells and repressed by high zinc. The molecular mechanisms of this regulation are still unclear.

A second mechanism in *S. cerevisiae* regulates zinc transporter activity at a post-translational level. In zinc-limited cells, Zrt1 is a stable plasma membrane protein. Exposure to higher levels of extracellular zinc, as occurs when these cells are reinoculated into zinc-replete media, triggers a rapid loss of Zrt1 uptake activity and protein. This inactivation does not require Zap1 and occurs through zinc-induced endocytosis of the protein and its subsequent degradation in the vacuole.[19] Mutations that inhibit the internalization step of endocytosis also inhibited zinc-induced Zrt1 inactivation and the major vacuolar proteases were required to degrade Zrt1 in response to zinc. Furthermore, immunofluorescence microscopy showed that Zrt1 is in the plasma membrane of zinc-limited cells and is transferred to the vacuole via an endosome-like compartment upon exposure to zinc. Excess zinc does not alter the stability of several other plasma membrane proteins. Therefore, zinc-induced Zrt1 inactivation is a specific regulatory mechanism to shut off zinc uptake activity in cells exposed to high extracellular zinc levels. This system thereby prevents the overaccumulation of this potentially toxic metal. Zrt1 inactivation is a relatively specific response to zinc but Cd^{2+} can also trigger the response. Zrt1 appears to be capable of transporting divalent cadmium[32] and we have recently shown that inactivation of Zrt1 by high cadmium allows zinc-limited cells to resist cadmium toxicity.[20]

It is unknown if this response is post-translational response is induced by a mechanism that senses intracellular or extracellular metal ion levels. A second unanswered question is how the zinc signal is transmitted to Zrt1. This could occur through the metal binding directly to the transporter or through an indirect signal transduction pathway. Recent studies demonstrated that Zrt1 is ubiquitinated prior to endocytosis suggesting that this modification serves as a signal for recruitment of the protein into clathrin-coated pits.[18] A similar role for ubiquitin has also been found for other *S. cerevisiae* and some mammalian plasma membrane proteins.[33] Zinc-induced ubiquitination of Zrt1 occurs on a lysine residue (K195) located in a cytosolic loop region of the protein. Therefore, the principle question currently is: how does zinc control ubiquitination of Zrt1?

There is growing evidence for regulation of ZIP proteins in other organisms. In Arabidopsis, *ZIP1*, *ZIP3*, and *ZIP4* mRNAs are induced by zinc deficiency. In human cells, *SLC39A1 (hZip1)* expression appears to be regulated by zinc status and by hormone treatment, that is mRNA levels are repressed by zinc treatment and induced by prolactin treatment of PC-3 and LNCaP prostate-derived cancer cells.[34] Similarly, *SLC39A2 (hZip2)* expression may also be regulated by zinc; treatment of the THP-1 monocytic cell line with TPEN, a cell-permeable zinc chelator, resulted in approximately 27-fold increases in *SLC39A2* mRNA levels.[35] At the time of this writing, no experiments characterizing the regulation of *SLC39A4 (hZip4)* expression have been published.

Post-translational control of ZIP protein activity may also occur in organisms other than yeast. Irt1, the Arabidopsis root iron uptake transporter may be regulated by iron in a fashion similar to that described for Zrt1. When *IRT1* mRNA is expressed from a constitutively active promoter, the protein only accumulates in iron-limited plants.[36] These results suggest translational or post-translational control of Irt1 activity occurs in response to iron. Whether similar regulation occurs for other ZIPs is the subject of ongoing studies.

Summary

The ZIP family of metal ion transporters encompasses genes from all phylogenetic levels. In numerous studies, these transporters have been implicated in zinc acquisition by microbes, plants, and mammals. The activity of these proteins is tightly regulated at transcriptional and post-transcriptional levels to maintain a constant supply of Zn^{2+} as extracellular and dietary levels of zinc change. This constant supply of Zn^{2+} provides the needed metal for function of intracellular and secreted zinc-dependent proteins.

References

1. Gaither LA, Eide DJ. Eukaryotic zinc transporters and their regulation. Biometals 2001; 14(3-4):251-270.
2. Guerinot ML. The ZIP family of metal transporters. Biochim Biophys Acta 2000; 1465:190-198.
3. Eng BH, Guerinot ML, Eide D et al. Sequence analyses and phylogenetic characterization of the ZIP family of metal ion transport proteins. J Memb Biol 1998; 166:1-7.
4. Zhao H, Eide D. The ZRT2 gene encodes the low affinity zinc transporter in Saccharomyces cerevisiae. J Biol Chem 1996; 271(38):23203-23210.
5. Zhao H, Eide D. The yeast ZRT1 gene encodes the zinc transporter of a high affinity uptake system induced by zinc limitation. Proc Natl Acad Sci USA 1996; 93:2454-2458.
6. Eide D, Broderius M, Fett J et al. A novel iron-regulated metal transporter from plants identified by functional expression in yeast. Proc Natl Acad Sci USA 1996; 93:5624-5628.
7. Vert G, Grotz N, Dedaldechamp F et al. IRT1, an Arabidopsis transporter essential for iron uptake from the soil and for plant growth. Plant Cell 2002; 14(6):1223-1233.
8. Grass G, Wong MD, Rosen BP et al. ZupT is a Zn(II) uptake system in Escherichia coli. J Bacteriol 2002; 184(3):864-866.
9. Grotz N, Fox T, Connolly E et al. Identification of a family of zinc transporter genes from Arabidopsis that respond to zinc deficiency. Proc Natl Acad Sci USA 1998; 95(12):7220-7224.
10. Vert G, Briat JF, Curie C. Arabidopsis IRT2 gene encodes a root-periphery iron transporter. Plant J 2001; 26(2):181-189.
11. Pence NS, Larsen PB, Ebbs SD et al. The molecular physiology of heavy metal transport in the Zn/Cd hyperaccumulator Thlaspi caerulescens. Proc Natl Acad Sci USA 2000; 97(9):4956-4960.
12. Lasswell J, Rogg LE, Nelson DC et al. Cloning and characterization of IAR1, a gene required for auxin conjugate sensitivity in Arabidopsis. Plant Cell 2000; 12(12):2395-2408.
13. MacDiarmid CW, Gaither LA, Eide D. Zinc transporters that regulate vacuolar zinc storage in Saccharomyces cerevisiae. EMBO J 2000; 19(12):2845-2855.
14. Lin S-J, Culotta VC. Suppression of oxidative damage by Saccharomyces cerevisiae ATX2, which encodes a manganese-trafficking protein that localizes to Golgi-like vesicles. Mol Cell Biol 1996; 16:6303-6312.
15. Gaither LA, Eide DJ. Functional expression of the human hZIP2 zinc transporter. J Biol Chem 2000; 275(8):5560-5564.
16. Gaither LA, Eide DJ. The human ZIP1 transporter mediates zinc uptake in human K562 erythroleukemia cells. J Biol Chem 2001; 276(25):22258-22264.
17. Outten CE, O'Halloran TV. Femtomolar sensitivity of metalloregulatory proteins controlling zinc homeostasis. Science 2001; 292(5526):2488-2492.
18. Gitan RS, Eide DJ. Zinc-regulated ubiquitin conjugation signals endocytosis of the yeast ZRT1 zinc transporter. Biochem J 2000; 346:329-336.
19. Gitan RS, Luo H, Rodgers J et al. Zinc-induced inactivation of the yeast ZRT1 zinc transporter occurs through endocytosis and vacuolar degradation. J Biol Chem 1998; 273:28617-28624.
20. Gitan RS, Shababi M, Kramer M et al. A cytosolic domain of the yeast Zrt1 zinc transporter is require for its post-translational inactivation in response to zinc and cadmium. J Biol Chem 2003; in press.
21. Rogers EE, Eide DJ, Guerinot ML. Altered selectivity in an Arabidopsis metal transporter. Proc Natl Acad Sci USA 2000; 97(22):12356-12360.
22. Taylor KM. LIV-1 breast cancer protein belongs to a new family of histidine-rich membrane proteins with potential to control intracellular zinc homeostasis. Life 2000; 49:249-253.
23. Lasat MM, Pence NS, Garvin DF et al. Molecular physiology of zinc transport in the Zn hyperaccumulator Thlaspi caerulescens. J Exp Bot 2000; 51(342):71-79.
24. Lioumi M, Ferguson CA, Sharpe PT et al. Isolation and characterization of human and mouse ZIRTL, a member of the IRT1 family of transporters, mapping within the epidermal differentiation complex. Genomics 1999; 62:272-280.

25. Milon B, Dhermy D, Pountney D et al. Differential subcellular localization of hZip1 in adherent and nonadherent cells. FEBS Lett 2001; 507(3):241-246.
26. Yamaguchi S. Subtraction cloning of growth arrest inducible genes in normal human epithelial cells. Kokubyo Gakkai Zasshi 1995; 62:78-93.
27. Cao J, Bobo JA, Liuzzi JP et al. Effects of intracellular zinc depletion on metallothionein and ZIP2 transporter expression and apoptosis. J Leuk Biol 2001; 70(4):559-566.
28. Wang K, Pugh EW, Griffen S et al. Homozygosity mapping places the acrodermatitis enteropathica gene on chromosomal region 8q24.3. Am J Hum Genet 2001; 68(4):1055-1060.
29. Wang K, Zhou B, Kuo YM et al. A novel member of a zinc transporter family is defective in acrodermatitis enteropathica. Am J Hum Genet 2002; 71(1):66-73.
30. Kury S, Dreno B, Bezieau S et al. Identification of SLC39A4, a gene involved in acrodermatitis enteropathica. Nature Genet 2002; 31(3):239-240.
31. Zhao H, Eide DJ. Zap1p, a metalloregulatory protein involved in zinc-responsive transcriptional regulation in Saccharomyces cerevisiae. Mol Cell Biol 1997; 17(9):5044-5052.
32. Gomes DS, Fragoso LC, Riger CJ et al. Regulation of cadmium uptake by Saccharomyces cerevisiae. Biochim Biophys Acta 2002; 1573(1):21-25.
33. Hicke L. Ubiquitin-dependent internalization and down-regulation of plasma membrane proteins. FASEB J 1997; 11:1215-1226.
34. Costello LC, Liu Y, Zou J et al. Evidence for a zinc uptake transporter in human prostate cancer cells which is regulated by prolactin and testosterone. J Biol Chem 1999; 274:17499-17504.
35. Cousins RJ, Blanchard RK, Popp MP et al. A global view of the selectivity of zinc deprivation and excess on genes expressed on genes expressed in human THP-1 mononuclear cells. Proc Natl Acad Sci USA 2003; 100:6952-6957.
36. Connolly EL, Fett JP, Guerinot ML. Expression of the IRT1 metal transporter is controlled by metals at the levels of transcript and protein accumulation. Plant Cell 2002; 14(6):1347-1357.

CHAPTER 36

Apoptosis by Zinc Deficiency

Kirsteen H. Maclean

Abstract

Zinc is an essential trace element for all forms of life. Zinc deficiency affects many systems because of the many roles it encompasses, such as in metabolism (including the activity of more than 300 enzymes), the structure of many proteins and control of genetic expression. Homeostatic regulation of this trace metal is extremely crucial, with zinc status affecting basic processes of cell division, growth, differentiation and apoptosis. The role of zinc in the regulation of apoptosis is not fully understood. The present review describes the current literature surrounding zinc deprivation and the induction of apoptotic cell death, and attempts to dissect the various molecular mechanisms involved in the process.

Introduction

Two distinct forms of cell death, necrosis and apoptosis, are extremely well characterized by their different morphological and biological features. Apoptosis is regarded as the prominent form of cell death during normal development, tissue homeostasis and in the regulation of the immune system. The phenomenon of apoptosis in cells was first described by Kerr, Wyllie and Currie over 30 years ago, when they discovered morphological changes in cells undergoing cell death that lead to membrane blebbing and digestion of the resulting vesicles by macrophages. Though it was discovered in 1972, apoptosis took almost 20 years to come into prominence in scientific research and only now can we appreciate that apoptotic cells die in a highly coordinated manner with a myriad of characteristic structural changes that include chromatin condensation, nuclear DNA fragmentation and externalization of phosphatidylserine on the surface of the cell membrane. In the last few years there has been an exponential growth in our understanding of the biochemical and molecular processes underlying apoptotic pathways, a few of which will be discussed in this chapter.

There is ample evidence that the induction and inhibition of apoptosis in many cell types are thought to involve signal transduction pathways in which the concentration of free intracellular divalent metal ions seems to be important.[1,2] The trace element zinc appears to be particularly important in this regard and unlike many other metals, zinc is virtually nontoxic.[3] The homeostatic mechanisms that regulate the entry into, distribution in, and excretion from cells and tissues are so efficient that no disorders are known to be associated with its excessive accumulation, in contrast to iron and other metals. Furthermore, its physical and chemical properties, including its general stable association with macromolecules and its coordination flexibility, make it highly adaptable to meeting the needs of proteins and enzymes that carry out diverse biological functions. These and many other chemical properties form the basis for the extensive participation of zinc in protein, nucleic acid, carbohydrate and lipid metabolism, as well as in the control of gene transcription and other fundamental biological processes.

The effects of zinc on cells is evidenced in the numerous roles this trace metal plays in basic cellular functions such as DNA replication, gene expression, cell division and cell activation. Labile pools of intracellular zinc are thought to be essential at several points in the cell cycle, since a number of enzymes involved in DNA synthesis such as thymidine kinase and DNA polymerase are affected by availability of zinc.[4] There is evidence that suggests that zinc is required for a critical process in the mid G1 phase of the cell cycle where under conditions of adequate zinc supply, cells can progress beyond this point to either replicate or differentiate depending on other environmental factors. However, if zinc availability is limiting, the cells tend to undergo apoptosis (Fig. 1). Therefore zinc may be involved in regulation of cellular homeostasis by its dual role in both proliferation and death by apoptosis. This chapter describes the current concepts of cell death and how zinc levels can affect the molecular mechanisms of apoptosis.

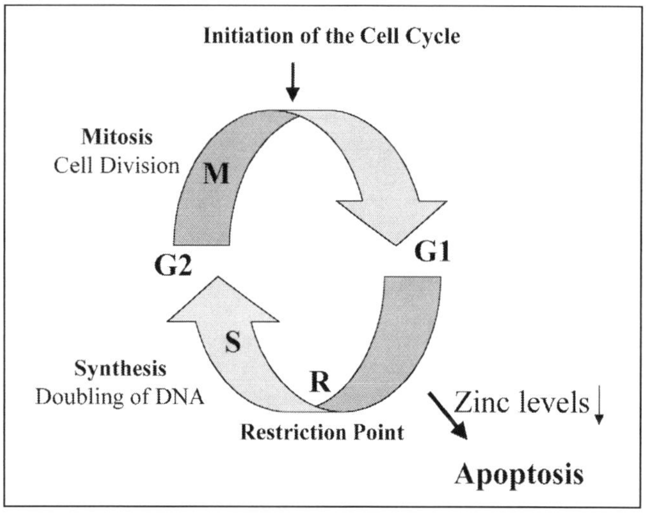

Figure 1. Relationship of zinc to the cell cycle. Requirement for zinc during the mid-G1 phase of the cell cycle and its potential to cause cell cycle exit by apoptosis.

Zinc Finger Proteins: From Atomic Contact to Cellular Function, edited by Shiro Iuchi and Natalie Kuldell.

Zinc Deficiency

The most convincing evidence that zinc plays a physiological role in apoptosis comes from zinc deprivation studies in animals, the first of which came from a study back in 1977 by Elmes.[5] She reported that zinc-deficient rats had markedly increased numbers of apoptotic cells in their small intestinal crypts, an illuminating finding considering how much scientific research was focused on the area of apoptosis at that time. Using other in vivo models, Yeiser et al showed that even a moderate decrease in zinc levels was sufficient to promote apoptosis following injury to the brain.[6] Thus far however, almost all the evidence of zinc deficiency induced apoptosis in vivo suggests that the increase in cell death is a direct consequence of a lowering of the intracellular zinc in the affected tissues. One exception is the involution of the thymus found in zinc deficient animals, which can in part be attributed to an increase in circulating glucocorticoids produced in response to the stress associated with zinc loss.[7,8] In humans, clinical outcomes of zinc deficiency include those outlined in Table 1. Given the central role of zinc in cell growth it should be of no surprise that zinc deficiency promotes an apoptotic phenotype, particularly in cells that have a rapid turnover such as those in the immune system.

There is now a plethora of evidence to support an important role for zinc in immune processes.[9-12] Adequate zinc status is essential for a number of processes including: (1) T-cell division, maturation and differentiation; (2) lymphocyte response to mitogens; (3) apoptosis of lymphoid and myeloid origins; (4) gene transcription; and (5) biomembrane function. Indeed lymphocytes are particularly sensitive in that they are one of the cells that are activated by zinc. In animals the consequences of a zinc deficient scenario in the immune system are profound and result in rapid and marked thymic atrophy, impaired cell-mediated cutaneous sensitivity and lymphopenia. Primary and secondary antibody responses are also reduced by zinc deficiency, particularly for those antigens that require T-cell help, such as those in heterologous red blood cells. In addition, antibody response and the generation of splenic cytotoxic T cells after immunization are reduced. Zinc also inhibits the production of tumor necrosis factor, which is implicated in the pathophysiology of cachexia and wasting in acquired immune deficiency syndrome.[13] An impaired immune response has been linked with low zinc levels in the plasma and several diseases that encompass noticeable zinc deficiency, particularly in the zinc malabsorption syndrome, Acrodermatitis Enteropathica. This genetic zinc deficiency syndrome is transmitted as an autosomal recessive disease. The disorder is characterized by thymic atrophy and unstable leukocyte functions associated with recurrent severe infections and decreased absorption of ingested zinc (normal adult individuals absorb 60-70% of a standard dose of ^{65}Zn taken orally and eliminate 0.7%). All of these immunological defects are reversible by means of zinc supplementation.[14] Importantly, the essentiality of zinc for differentiation, development and performance is exemplified by abnormalities that occur in zinc deficient preimplantation embryos and the teratology caused by zinc deficiency in early gestation.[15-17]

Table 1. Diseases/disorders associated with zinc deficiency

Primary	Secondary
Acrodermatitis enteropathica	Reduced appetite/ anorexia
Coeliac disease	All dermatological disorders
Common cold	Pregnancy complications
Diarrhoea	Liver cirrhosis
Infertility (male)	Sleep and behavioral disturbances
Night blindness	HIV/AIDS support
Impaired wound healing	Diabetes
Frequent and severe infections	Growth retardation

Molecular Mechanisms of Apoptosis

There are two different reasons why a cell decides to commit suicide. The first is to ensure proper development, such as the formation of the synapses between neurons in the brain require surplus cells to be eliminated by means of apoptosis. The second is to destroy cells that represent a threat to the integrity of the organism, such as occurs during zinc deficiency. There are three different mechanisms by which a cell commits suicide: the generation of internal signals, the binding of death activators to receptors at the cell surface (i.e., Fas Ligand) and the creation of reactive oxygen species. Central to each of these mechanisms is a group of highly specific proteases called Caspases (Cysteine Aspartate Specific ProteASEs), which recognize and cleave their substrate proteins at tetra-peptide sites that have characteristic sequences (Table 2), an example of which is the DXXD motif for caspase-3, where X represents any amino acid.[18] Based on the sequence of activation in apoptosis, caspases can be divided into 3 sub-types: initiator (activator) caspases; effector (executioner) caspases and cytokine processors (inflammatory). For the purposes of this chapter, we need only be concerned with initiator and effector caspases (Table 2). Initiator caspases (like caspase-8 and -9) are the upstream activators of the effector caspases (like caspase-3). As mentioned, effector caspases are the executioners in the cell and are responsible for the cleavage of proteins that actually induce apoptosis and that leads to the morphological features of apoptosis such as DNA fragmentation (Fig. 2). Indeed to illustrate just how important the caspases are to cellular homeostasis, a recent review by Fischer et al (2003) summarized the known caspase substrates that regulate key morphological changes involved in apoptosis and cell fate and comprised a list of over 280 different proteins involved in such diverse roles as apoptotic regulation, cell cycle and cell adhesion.[19] What is clear from this comprehensive review is that there is much to be learned concerning the role of caspases in apoptosis.

The initiation and efficient completion of apoptosis is generally dependent upon upstream regulators of the cell death program. Key controllers of apoptosis include the Bcl-2 family of proteins that serve to either suppress (e.g., Bcl-2, $Bcl\text{-}X_L$, and Mcl-1) or activate (Bax, Bak, Bad and PUMA) the cell death program.[20] Bcl-2 family members appear to regulate apoptosis by controlling the integrity of key organelles such as the mitochondria, therefore apoptosis triggered by internal signals represents the intrinsic or mitochondrial pathway. Most anti-apoptotic Bcl-2 family members are localized to the outer membrane of the mitochondria (and also the endoplasmic reticulum) and most proapoptotic family members disrupt the integrity of these membranes. Importantly, changes in the mitochondrial biogenesis and function are a hallmark of almost all forms of apoptosis including those induced by zinc deficiency. An initial event is a drop in

Table 2.

Class	Caspase	Regulatory Unit	Adapter Molecule	Tetrapeptide
	Caspase-2	CARD	RAIDD	DXXD
Activators	Caspase-8	DED	FADD	(I/V/L)EXD
	Caspase-9	CARD	Apaf-1	(I/V/L)EHD
	Caspase- 10	DED	FADD	(I/V/L)EXD
	Caspase-3			DEXD
Executioners	Caspase-6			(I/V/L)EXD
	Caspase-7			DEXD

CARD= Caspase recruitment domain; DED= Death effector domain

the mitochondrial membrane potential (ψAm), which is required to maintain asymmetric distribution of charges between the inner and outer mitochondrial membrane.[21,22] If disruption of the ψAm persists, the integrity of the outer membrane is compromised and results in the release of pro-apoptotic factors that include cytochrome c[23] and apoptosis inducing factor (AIF), which can induce death independent of caspase activation.[24] Release of cytochrome c from the mitochondria into the cytosol has been demonstrated to be a critical initiator of caspase-dependent death pathways. Once released, cytochrome c binds to Apaf-1 thereby triggering activation of caspase-9, which then cleaves and activates caspase-3 (Fig. 2). Activated caspase-3 then targets several proteins required for cellular integrity including Poly-ADP-ribose polymerase (PARP) and the inhibitor of calcium-activated DNAase (ICAD) which releases the endonuclease and enables it to cleave the DNA at internucleosomal sites.[25]

One of the best-understood mechanisms for initiating the proteolytic cascade of caspases involves the tumor necrosis factor (TNF) family receptors.[26] This form of apoptosis, triggered by external signals is the extrinsic or death receptor pathway. The cytosolic domains of some members of this family, such as Fas and the TNF receptor, are linked to caspases via adaptor proteins such as FADD/MORT1. Upon binding of their ligand (FasL and TNF respectively), a signal is transmitted to the cytoplasm that leads to recruitment of caspase-8 and possibly caspase-10, which then becomes activated by proteolysis. Once activated, these proximal caspases can also cleave and activate caspase-3 (Fig. 3).

Involvement of Zinc in Apoptotic Pathways

The influence of zinc on apoptosis is well documented as both in vitro and in vivo studies have shown that zinc influences apoptotic cell death.[27-32] In particular, zinc supplementation in thymocytes is known to inhibit apoptosis. In 1992, Cohen and colleagues observed that addition of zinc to dexamethasone-treated thymocytes (a potent inducer of the apoptotic program) in vitro blocked the transition from chromatin condensation to nuclear collapse. The authors attributed this finding to the specific inhibition of the ICAD, which digests the DNA leading to the classic DNA fragmentation hallmark of apoptosis.[33] However a weakness of this study and many others using alternative cell types is that the concentrations of zinc required to prevent apoptosis are extremely high, with most experimental works reporting prevention of DNA fragmentation with superfluous zinc concentrations in the range of 250mM to 5mM.[34] These are concentrations of

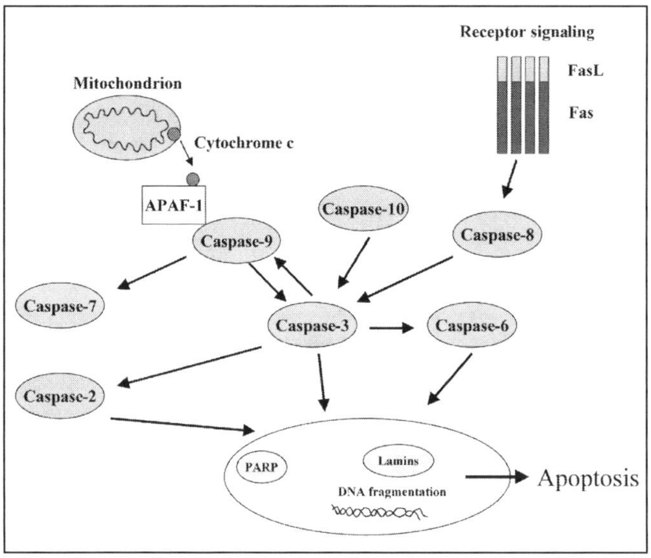

Figure 2. Involvement of caspases in apoptosis. Activation of initiator caspase such as caspase-8, -9 and-10 can activate downstream caspase members such as caspase-3, -6 and -7, which can then cleave downstream proteins associated with apoptosis.

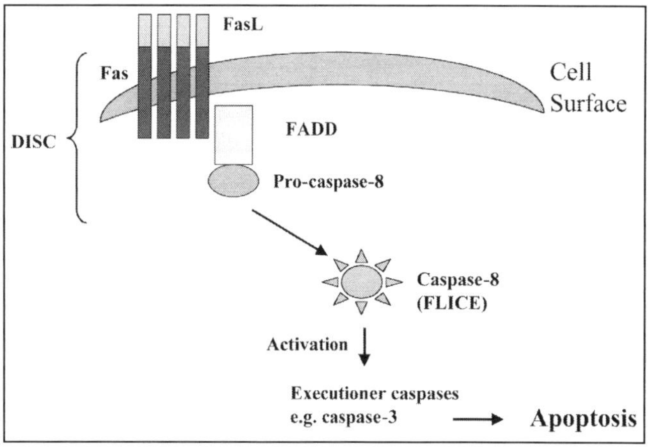

Figure 3. The Fas signal transduction pathway. When cell surface Fas binds Fas ligand (FasL), several proteins become recruited to the intracellular "death domain" of Fas to form a death-inducing signaling complex, or DISC. Of these, the Fas-associated death domain (FADD) protein acts as an adaptor, linking pro-caspase-8 to the activated Fas. This results in the proteolytic processing of pro-caspase-8 and activation of the executioner caspases.

zinc that are 10-100 fold greater than the concentration found in serum or tissues, albeit one that effectively blocks a wide array of cell death inducers for a variety of cell types. Conversely, low concentrations of intracellular zinc have been shown to accelerate apoptosis of lymphoid and myeloid cell lines[35,36] as well as human leukemia lymphocytes, rat splenocytes and rat thymocytes.[37-39] Indeed, Telford et al suggested that a zinc concentration below 300mM provides only limited protection against glucocorticoid-induced cell death. Concomitant with these studies numerous in vitro studies have now shown that depletion of intracellular zinc by culture of cells in zinc-depleted medium or by treatment with the zinc chelator N'N'N'N-tetrakis-2 pyridylmethyl-ethylenediamine (TPEN) results in apoptosis.[37-39] Several in vivo models also provide evidence for adverse effects of zinc deficiency in the whole animal.[6,40] In particular, apoptosis was shown to play an important role in the loss of precursor lymphocytes (lymphopenia) during zinc deficiency in mice.[41] Conversely, there are also instances where addition of zinc can markedly increase apoptosis. Paramanantham et al used in vitro flow cytometric studies to show that addition of zinc salts caused cell condensation and DNA fragmentation and suggested that zinc could enhance the generation of hydroxyl free radicals, results which imply that zinc supplementation could induce features resembling apoptosis.[42] Recently another group suggested that zinc induces both necrosis and apoptosis in Molt-4 cells without caspase-3 activation.[43] Collectively however, all the data currently suggest that appropriate zinc homeostasis is essential to prevent apoptosis. However, we must appreciate that although generally accepted to be a potent inhibitor of apoptosis, zinc can act as a double-edged sword, where depending on its concentration and cell context, addition or depletion can induce apoptosis

Several studies investigating zinc homeostasis imply that there is a labile or exchangeable pool of intracellular zinc that is important in the regulation of apoptosis. Many have challenged this hypothesis and suggest that this labile pool of zinc only constitutes a small part of the total cellular zinc, since most of the zinc in cells is tightly complexed to metalloenzymes and is not readily accessible.[44] Zinquin, a zinc-specific intracellular fluorophore has been used to detect labile intracellular zinc. It is easily taken up by living cells and is retained intracellularly for several hours. It is essentially nonfluorescent until it is complexed with zinc. Zalewski et al investigated the subcellular localization of zinc using fluorescence microscopy and suggested that zinc was largely restricted to the extranuclear regions.[45] However, zinquin probably only detects the less tightly complexed zinc in cells, which would include free zinc and that which is loosely associated with cellular proteins and lipids. The major pool of cellular zinc that is very tightly bound to the active sites of enzymes and in zinc finger transcription factors may not be available for interaction with zinquin. Therefore, interpretation as to which of the subcellular pools of zinc are important in the suppression of apoptosis obviously requires more experimental evidence particularly in understanding the interactions between zinc and known apoptotic regulators in these subcellular pools.

The precise step in the apoptosis signaling cascade that is sensitive to fluxes in zinc levels remains elusive, however there is a growing body of evidence that points us in the direction of some of the putative targets of zinc in apoptosis. From a historical prospective, zinc was first shown to be an effective inhibitor of the endonuclease responsible for DNA fragmentation and indeed following zinc deficiency it was this Ca/Mg-dependent enzyme that was targeted. The first evidence that zinc may affect other intracellular apoptotic mediators other than the endonuclease came from revealing studies by Lazebnik and colleagues who showed that a more sensitive target of zinc was present in the cytoplasm.[46] In these studies, they demonstrated that cytosol from cells primed to undergo apoptosis, induced chromatin condensation and DNA fragmentation in isolated healthy nuclei and that zinc suppressed these effects even when only added to the cytoplasm. Subsequent studies by others demonstrated that this mysterious component was caspase-3. Although zinc did not block the activity of caspase-3 to cleave its cellular substrates, it did block the mechanism by which caspase 3 is activated from its inactive precursor. Zinc deficiency induced apoptosis in vitro (that is, using zinc chelators) appears to be dependent on caspase-3 activation since cytosolic caspase-3 activity is increased in zinc-deficient cells, while known inhibitors of caspase-3 (that is, z-DEVD-fmk) can partially suppress apoptosis, albeit at high concentrations. Interestingly, there is also evidence in vivo where zinc deficiency induced apoptosis in rat embryos was associated with increased caspase-3 activity.[47] Many authors maintain that it is only the activation of caspase-3, which is the important determinant as to whether a cell succumbs to apoptosis following zinc deficiency. An interesting study by Chimienti et al demonstrated by western blot analysis that zinc depletion induced a strong activation of both caspase-3 and -8, but only a weak activation of caspase-9,[48] which as suggested earlier may have resulted from retroactivation by other caspases (Fig. 2). However, since the proteolytic processing and activation of effector caspases like caspase-3 is dependent on other caspases such as capase-8, -9 and caspase-6, these proteases should theoretically also be targeted during zinc deficiency. Another study established that zinc depletion from the use of chelators induced translocation of cytochrome c from the mitochondria with subsequent activation of caspases-3, -8 and –9,[49] thus it may be interesting for future experimentation if apical mediators of apoptosis are targeted by zinc depletion. One other possible target of zinc is caspase-6 which as stated earlier is also known to cleave and activate the proform of caspase-3. However, studies have demonstrated that caspase-6 activation was typically secondary to caspase-3 processing.[50] Nevertheless, what appears clear from all these experiments is that the addition of zinc salts following zinc deficiency can only effectively prevent late stages of apoptosis such as DNA fragmentation, but appear relatively inefficient at inhibiting early events such as caspase activation. Another question which warrants addressing is how important is the contribution of caspases to apoptosis potentiated by zinc deficiency? Given the limitation of most in vitro systems that are currently employed, several groups have made use of gene-targeting techniques to generate mice that are deficient in specific caspases to determine their physiological function. Caspase-3-null mice have a perinatal lethality (though this is strictly strain dependent) and exhibit altered morphological and biochemical features when undergoing apoptosis.[51] Various cell types such as mouse embryo fibroblasts (MEFs) from these mice and other examples of caspase knockout mice challenged with zinc chelators such as TPEN may give us a better understanding of the apoptotic pathways involved following zinc deficiency.

Important modulators of many apoptotic programs are the members of the Bcl-2 family of anti-apoptotic proteins and disturbances in the ratio of these proteins are indicative of whether a cell is prone to die. Fukamachi and colleagues showed that zinc supplementation increased the Bcl-2/Bax ratio in the U937 monocytic cell line, thereby increasing their resistance to cell death.[34] At this time, it has yet to be determined whether Bcl-2 levels are compromised following zinc deficiency, Ahn and colleagues have speculated that Bcl-2 is not involved in apoptosis potentiated by zinc deficiency, at least in their model of zinc-depleted induced neuronal apoptosis.[52] However, since these observations only incorporated over-expression studies of Bcl-2, it is difficult to reconcile whether levels of Bcl-2 are truly an important determinant as to the outcome of zinc-depleted apoptosis. Nevertheless, others have speculated that the comparable phenotypes of *Bcl-2* knockout mice to the zinc deficient mouse model (including, massive apoptotic involution of the thymus and spleen associated with depletion of CD4+ T cells and growth retardation) are not simply coincidental similarities.[8] Another modulator of apoptosis is p53. A recent study by Fanzo et al indicated that stress response genes, such as p53, are altered by zinc status. Here, normal human bronchial epithelial cells cultured in zinc-deficient media resulted in elevated levels of both *p53* mRNA and protein, results which suggest that reductions in zinc may trigger the DNA damage pathway to promote apoptosis.[53] p53 is activated by a diverse number of signals that stress the cell, including hyperproliferation, hypoxia, and DNA damage.[54] Normally p53 levels and function are harnessed by Mdm2,[55] which programs p53 for destruction by initiating p53 ubiquitination and shuttling it to the cytosol for degradation by the proteasome. [56] Signals that activate p53 disrupt this interaction, by inducing phosphorylation of p53[57,58] and/or Mdm2,[59] or by inducing the ARF nucleolar tumor suppressor, which binds to and inhibits Mdm2 functions.[60,61] The net result is stabilization of p53 and activation of its transcription functions. In turn, p53 either induces G_1 arrest by inducing transcription of the cyclin-dependent kinase (cdk) inhibitor $p21^{Cip1}$, allowing time for DNA repair, or eliminates cells by triggering apoptosis. p53-induced apoptosis is thought to occur through its ability to induce pro-apoptotic Bcl-2 family members, including Bax, Noxa and Puma.[54] Several studies have demonstrated that during apoptosis the cyclin-dependent kinase (cdk) inhibitors $p21^{Cip1}$ and $p27^{Kip1}$ are cleaved by caspase-3.[62] This observation was linked to zinc deficiency by Chai et al who showed that activation of caspase-3 in zinc-depleted cells was rapidly followed by initial cleavage of $p21^{Cip1}$ to a 15-kDa form, then to complete loss of $p21^{Cip1}$ protein, perhaps mediated by proteosomal degradation.[50] Moreover, one can infer that since both p53 and Mdm2 require zinc fingers for their configuration, alterations in zinc homeostasis may cause the genes to become mutated, resulting in their inactivation or suppression of their function.[63-65] Dysfunction/alterations in the p53 gene have particularly been documented in head and neck (esophageal), and gynecological (breast) cancer patients.[63,66] Interestingly, concomitant dietary zinc deficiency has also frequently been found in esophageal cancer cohorts.[67] Both correlate significantly with poorer tumor prognosis and treatment outcome. Researchers have also discovered that not only do many esophageal cancer patients harbor p53 mutations and acquire dietary zinc deficiency, but they also often have suppressed immune systems, which as indicated earlier in the chapter, is demonstrated by a disturbance in cytokine balance. There is a notable premature shift from Th-1 cytokines to Th-2 cytokines, with an accompanying poor T-cell function. The zinc-dependent hormone, thymulin, activates T-cell proliferation and function, and this same premature shift and T-cell dysfunction occurs with zinc deficiency. [68,69] Dissection of this p53 damage pathway requires much more investigation and consideration with regard to zinc deficiency.

Conclusion and Perspective

This chapter has attempted to address the current research of how apoptosis can be potentiated by zinc deficiency. While research into the various mechanisms of apoptosis is at the forefront of scientific research, recognized with the 2002 Nobel Prize for Medicine awarded for contributions to the apoptosis field, research into the cellular biology of zinc, its homeostasis and effects on apoptosis are still ambiguous. Clearly the best evidence comes from the use of animal models and primary cells. Though nonetheless important to our understanding, in vitro studies are problematic in that it has been difficult to distinguish the physiological effects from the pharmacological and toxicological effects of addition/suppression of zinc and the fact that immortalizing events associated with cell line cultures usually involve inactivation of the ARF-p53 tumor suppressor pathway, a pathway that has many ramifications on apoptosis. The utilization of cells from knock-out mouse models of apoptosis would help to define and dissect out what the important mediators are in the apoptotic response following zinc deficiency. Furthermore, the development of better methods to assess intracellular zinc deficiency, coupled with the identification of which intracellular pool(s) of zinc are important in mediating suppression of apoptosis is required.

One of the current challenges is to understand how regulators of the apoptosis pathway cooperate with metals such as zinc. Adequate zinc levels are crucial for normal human development since clinical studies have revealed that the effects of even a moderate degree of zinc deficiency may be profound. Therefore, from a therapeutic perspective the ability to target a specific apoptotic pathway may ultimately prove to be a superior measure to mere zinc supplementation for individuals harboring clinical signs of zinc deficiency (such as those individuals with Acrodermatitis Enterpathica).

References

1. Golstein P, Ojcius DM, Young JD. Cell death mechanisms and the immune system. Immunol Rev 1991; 121:29-65.
2. Orrenius S, McCabe Jr MJ, Nicotera P. Ca(2+)-dependent mechanisms of cytotoxicity and programmed cell death. Toxicol Lett 1992; 64-65(Spec No):357-364.
3. Bertholf RL. Zinc. In: Seiler HG, Sigel H, eds. Handbook on toxicity of inorganic compounds. New York: New York Dekker, 1988:788-800.
4. Chesters JK. In: Mills CF, ed. Zinc in human biology. London: Springer-Verlag, 1989:109-118.
5. Elmes ME. Apoptosis in the small intestine of zinc-deficient and fasted rats. J Pathol 1977; 123(4):219-223.
6. Yeiser EC, Vanlandingham JW, Levenson CW. Moderate zinc deficiency increases cell death after brain injury in the rat. Nutr Neurosci 2002; 5(5):345-352.
7. Fraker PJ, Telford WG. A reappraisal of the role of zinc in life and death decisions of cells. Proc Soc Exp Biol Med 1997; 215(3):229-236.
8. Truong-Tran AQ, Carter J, Ruffin RE et al. The role of zinc in caspase activation and apoptotic cell death. Biometals 2001; 14(3-4):315-330.
9. Kruse-Jarres JD. The significance of zinc for humoral and cellular immunity. J Trace Elem Electrolytes Health Dis 1989; 3(1):1-8.
10. Cunningham-Rundles S, Bockman RS, Lin A et al. Physiological and pharmacological effects of zinc on immune response. Ann N Y Acad Sci 1990; 587:113-122.
11. Wellinghausen N, Kirchner H, Rink L. The immunobiology of zinc. Immunol Today 1997; 18(11):519-521.
12. Dardenne M. Zinc and immune function. Eur J Clin Nutr 2002; 56(Suppl 3):S20-23.

13. Baum MK, Shor-Posner G, Campa A. Zinc status in human immunodeficiency virus infection. J Nutr 2000; 130(5S Suppl):1421S-1423S.
14. Neldner KH, Hambidge KM, Walravens PA. Acrodermatitis enteropathica. Int J Dermatol 1978; 17(5):380-387.
15. Hanna LA, Peters JM, Wiley LM et al. Enhancing effect of maternal zinc deficiency and 137Cs gamma-irradiation on the frequency of fetal malformations in mice. Teratog Carcinog Mutagen 1997; 17(3):127-137.
16. Peters JM, Wiley LM, Zidenberg-Cherr S et al. Influence of short-term maternal zinc deficiency on the in vitro development of preimplantation mouse embryos. Proc Soc Exp Biol Med 1991; 198(1):561-568.
17. Hurley LS, Shrader RE. Abnormal development of preimplantation rat eggs after three days of maternal dietary zinc deficiency. Nature 1975; 254(5499):427-429.
18. Thornberry NA, Lazebnik Y. Caspases: Enemies within. Science 1998; 281(5381):1312-1316.
19. Fischer U, Janicke RU, Schulze-Osthoff K. Many cuts to ruin: A comprehensive update of caspase substrates. Cell Death Differ 2003; 10(1):76-100.
20. Coultas L, Strasser A. The role of the Bcl-2 protein family in cancer. Semin Cancer Biol 2003; 13(2):115-123.
21. Vayssiere JL, Petit PX, Risler Y et al. Commitment to apoptosis is associated with changes in mitochondrial biogenesis and activity in cell lines conditionally immortalized with simian virus 40. Proc Natl Acad Sci USA 1994; 91(24):11752-11756.
22. Zamzami N, Marchetti P, Castedo M et al. Sequential reduction of mitochondrial transmembrane potential and generation of reactive oxygen species in early programmed cell death. J Exp Med 1995; 182(2):367-377.
23. Jiang X, Wang X. Cytochrome c promotes caspase-9 activation by inducing nucleotide binding to Apaf-1. J Biol Chem 2000; 275(40):31199-31203.
24. Susin SA, Lorenzo HK, Zamzami N et al. Molecular characterization of mitochondrial apoptosis-inducing factor. Nature 1999; 397(6718):441-446.
25. Janicke RU, Ng P, Sprengart ML et al. Caspase-3 is required for alpha-fodrin cleavage but dispensable for cleavage of other death substrates in apoptosis. J Biol Chem 1998; 273(25):15540-15545.
26. Peter ME, Krammer PH. The CD95(APO-1/Fas) DISC and beyond. Cell Death Differ 2003; 10(1):26-35.
27. Fraker PJ, Haas SM, Luecke RW. Effect of zinc deficiency on the immune response of the young adult A/J mouse. J Nutr 1977; 107(10):1889-1895.
28. Martin SJ, Mazdai G, Strain JJ et al. Programmed cell death (apoptosis) in lymphoid and myeloid cell lines during zinc deficiency. Clin Exp Immunol 1991; 83(2):338-343.
29. Migliorati G, Nicoletti I, Pagliacci MC et al. Interleukin-2 induces apoptosis in mouse thymocytes. Cell Immunol 1993; 146(1):52-61.
30. Kuo IC, Seitz B, LaBree L et al. Can zinc prevent apoptosis of anterior keratocytes after superficial keratectomy? Cornea 1997; 16(5):550-555.
31. Marini M, Musiani D. Micromolar zinc affects endonucleolytic activity in hydrogen peroxide-mediated apoptosis. Exp Cell Res 1998; 239(2):393-398.
32. Maclean KH, Cleveland JL, Porter JB. Cellular zinc content is a major determinant of iron chelator-induced apoptosis of thymocytes. Blood 2001; 98(13):3831-3839.
33. Cohen GM, Sun XM, Snowden RT et al. Key morphological features of apoptosis may occur in the absence of internucleosomal DNA fragmentation. Biochem J 1992; 286(Pt 2):331-334.
34. Fukamachi Y, Karasaki Y, Sugiura T et al. Zinc suppresses apoptosis of U937 cells induced by hydrogen peroxide through an increase of the Bcl-2/Bax ratio. Biochem Biophys Res Commun 1998; 246(2):364-369.
35. Trubiani O, Antonucci A, Palka G et al. Programmed cell death of peripheral myeloid precursor cells in Down patients: Effect of zinc therapy. Ultrastruct Pathol 1996; 20(5):457-462.
36. Ning ZQ, Norton JD, Li J et al. Distinct mechanisms for rescue from apoptosis in Ramos human B cells by signalling through CD40 and interleukin-4 receptor: Role for inhibition of an early response gene, Berg36. Biochem Soc Trans 1997; 25(2):306S.
37. Jiang S, Chow SC, McCabe MJ et al. Lack of Ca2+ involvement in thymocyte apoptosis induced by chelation of intracellular Zn2+. Lab Invest 1995; 73(1):111-117.
38. Telford WG, Fraker PJ. Preferential induction of apoptosis in mouse CD4+CD8+ alpha beta TCRloCD3 epsilon lo thymocytes by zinc. J Cell Physiol 1995; 164(2):259-270.
39. Mathieu J, Ferlat S, Ballester B et al. Radiation-induced apoptosis in thymocytes: Inhibition by diethyldithiocarbamate and zinc. Radiat Res 1996; 146(6):652-659.
40. Truong-Tran AQ, Ruffin RE, Foster PS et al. Altered zinc homeostasis and caspase-3 activity in murine allergic airway inflammation. Am J Respir Cell Mol Biol 2002; 27(3):286-296.
41. King LE, Fraker PJ. Zinc deficiency in mice alters myelopoiesis and hematopoiesis. J Nutr 2002; 132(11):3301-3307.
42. Paramanantham R, Sit KH, Bay BH. Adding Zn2+ induces DNA fragmentation and cell condensation in cultured human Chang liver cells. Biol Trace Elem Res 1997; 58(1-2):135-147.
43. Hamatake M, Iguchi K, Hirano K et al. Zinc induces mixed types of cell death, necrosis, and apoptosis, in molt-4 cells. J Biochem (Tokyo) 2000; 128(6):933-939.
44. Bettger WJ, O'Dell BL. A critical physiological role of zinc in the structure and function of biomembranes. Life Sci 1981; 28(13):1425-1438.
45. Zalewski PD, Forbes IJ, Betts WH. Correlation of apoptosis with change in intracellular labile Zn(II) using zinquin [(2-methyl-8-p-toluenesulphonamido-6-quinolyloxy)acetic acid], a new specific fluorescent probe for Zn(II). Biochem J 1993; 296(Pt 2):403-408.
46. Lazebnik YA, Cole S, Cooke CA et al. Nuclear events of apoptosis in vitro in cell-free mitotic extracts: A model system for analysis of the active phase of apoptosis. J Cell Biol 1993; 123(1):7-22.
47. Jankowski-Hennig MA, Clegg MS, Daston GP et al. Zinc-deficient rat embryos have increased caspase 3-like activity and apoptosis. Biochem Biophys Res Commun 2000; 271(1):250-256.
48. Chimienti F, Seve M, Richard S et al. Role of cellular zinc in programmed cell death: Temporal relationship between zinc depletion, activation of caspases, and cleavage of Sp family transcription factors. Biochem Pharmacol 2001; 62(1):51-62.
49. Kolenko VM, Uzzo RG, Dulin N et al. Mechanism of apoptosis induced by zinc deficiency in peripheral blood T lymphocytes. Apoptosis 2001; 6(6):419-429.
50. Chai F, Truong-Tran AQ, Evdokiou A et al. Intracellular zinc depletion induces caspase activation and p21 Waf1/Cip1 cleavage in human epithelial cell lines. J Infect Dis 2000; 182(Suppl 1):S85-92.
51. Zheng TS, Hunot S, Kuida K et al. Caspase knockouts: Matters of life and death. Cell Death Differ 1999; 6(11):1043-1053.
52. Ahn YH, Koh JY, Hong SH. Protein synthesis-dependent but Bcl-2-independent cytochrome C release in zinc depletion-induced neuronal apoptosis. J Neurosci Res 2000; 61(5):508-514.
53. Fanzo JC, Reaves SK, Cui L et al. Zinc status affects p53, gadd45, and c-fos expression and caspase-3 activity in human bronchial epithelial cells. Am J Physiol Cell Physiol 2001; 281(3):C751-757.
54. Vousden KH, Lu X. Live or let die: The cell's response to p53. Nat Rev Cancer 2002; 2(8):594-604.
55. Momand J, Zambetti GP, Olson DC et al. The mdm-2 oncogene product forms a complex with the p53 protein and inhibits p53-mediated transactivation. Cell 1992; 69(7):1237-1245.
56. Fuchs SY, Adler V, Buschmann T et al. JNK targets p53 ubiquitination and degradation in nonstressed cells. Genes Dev 1998; 12(17):2658-2663.
57. Banin S, Moyal L, Shieh S et al. Enhanced phosphorylation of p53 by ATM in response to DNA damage. Science 1998; 281(5383):1674-1677.
58. Canman CE, Lim DS, Cimprich KA et al. Activation of the ATM kinase by ionizing radiation and phosphorylation of p53. Science 1998; 281(5383):1677-1679.

59. Maya R, Balass M, Kim ST et al. ATM-dependent phosphorylation of Mdm2 on serine 395: Role in p53 activation by DNA damage. Genes Dev 2001; 15(9):1067-1077.
60. Pomerantz J, Schreiber-Agus N, Liegeois NJ et al. The Ink4a tumor suppressor gene product, p19Arf, interacts with MDM2 and neutralizes MDM2's inhibition of p53. Cell 1998; 92(6):713-723.
61. Weber JD, Taylor LJ, Roussel MF et al. Nucleolar Arf sequesters Mdm2 and activates p53. Nat Cell Biol 1999; 1(1):20-26.
62. Levkau B, Koyama H, Raines EW et al. Cleavage of p21Cip1/Waf1 and p27Kip1 mediates apoptosis in endothelial cells through activation of Cdk2: Role of a caspase cascade. Mol Cell 1998; 1(4):553-563.
63. Kihara C, Seki T, Furukawa Y et al. Mutations in zinc-binding domains of p53 as a prognostic marker of esophageal-cancer patients. Jpn J Cancer Res 2000; 91(2):190-198.
64. Geyer RK, Yu ZK, Maki CG. The MDM2 RING-finger domain is required to promote p53 nuclear export. Nat Cell Biol 2000; 2(9):569-573.
65. Ghosh M, Huang K, Berberich SJ. Overexpression of Mdm2 and MdmX fusion proteins alters p53 mediated transactivation, ubiquitination, and degradation. Biochemistry 2003; 42(8):2291-2299.
66. Zachos G, Spandidos DA. Transcriptional regulation of the c-H-ras1 gene by the P53 protein is implicated in the development of human endometrial and ovarian tumours. Oncogene 1998; 16(23):3013-3017.
67. Prasad AS, Beck FW, Doerr TD et al. Nutritional and zinc status of head and neck cancer patients: An interpretive review. J Am Coll Nutr 1998; 17(5):409-418.
68. Doerr TD, Marks SC, Shamsa FH et al. Effects of zinc and nutritional status on clinical outcomes in head and neck cancer. Nutrition 1998; 14(6):489-495.
69. Prasad AS. Effects of zinc deficiency on Th1 and Th2 cytokine shifts. J Infect Dis 2000; 182(Suppl 1):S62-68.

INDEX

A

B

C

D

E

F

G

H

I

J

K

L

M

N

O

P

R

S

T

U

V

W

X

Y

Z